Generalized Linear and Nonlinear Models for Correlated Data

Theory and Applications Using SAS®

Edward F. Vonesh

The correct bibliographic citation for this manual is as follows: Vonesh, Edward F. 2012. *Generalized Linear and Nonlinear Models for Correlated Data: Theory and Applications Using SAS®*. Cary, NC: SAS Institute Inc.

Generalized Linear and Nonlinear Models for Correlated Data: Theory and Applications Using SAS®

ISBN 978-1-61290-091-9 (electronic book)
ISBN 978-1-59994-647-4

SAS Institute Inc., SAS Campus Drive, Cary, North Carolina 27513-2414

1st printing, August 2012

2nd printing, June 2014

Contents

Preface

In the science that is statistics, correlation is a term used to describe the kind and degree of relationship that exists between two or more variables. As such, correlated data occurs at various levels. Correlation can occur whenever two or more dependent variables (also referred to as response variables or outcome variables) are measured on the same individual or experimental unit. Such correlation is often classified and studied under the heading of multivariate analysis. However, even when restricted to a single response variable, correlation can occur when repeated measurements of the response are taken over time on individuals (longitudinal data), or when observations from individuals are grouped into distinct homogeneous clusters (clustered data), or when the spatial location of individual observations is taken into account (spatially correlated data). Correlation can also occur between two or more independent variables (also referred to as explanatory variables, predictor variables or covariates) as well as between dependent and independent variables.

Correlation between dependent and independent variables is most often expressed in terms of regression parameters that describe how the independent or explanatory variables predict the mean response (e.g., a slope in linear regression) or the degree to which they are associated with the response (e.g., an odds ratio in logistic regression). Regression parameters, in turn, are based on formulating an appropriate statistical model which is then used to investigate the various relationships that may exist between the response variable(s) and explanatory variable(s). This book focuses on various statistical models and methods available in the SAS System which can be used to analyze correlated response data. While both theory and applications are presented, the bulk of the book is dedicated to applications using SAS. Specifically, if we exclude the appendices as well as the summary pages at the end of chapters, about 30% of the material is dedicated to the theory of estimation and inference while the remaining 70% is dedicated to applications.

The book has two primary goals: 1) to provide applied statisticians with the necessary theory, tools, and understanding to conduct "routine" and "not so routine" analyses of continuous and/or discrete correlated data under a variety of settings, particularly settings involving longitudinal data and clustered data, and 2) to illustrate these methods with a variety of "real world" applications. With an emphasis on applications requiring statistical models with a nonlinear rather than linear formulation, many of the examples in the book contain detailed SAS code so that readers can begin applying the various techniques immediately (it should be noted that the data and code including select SAS macros are all available through the author's Web page at **http://support.sas.com/publishing/authors/vonesh.html**). In the end, the ultimate goal is to bridge the gap between theory and application by demonstrating how the tools in SAS can be used to overcome what often appear to be insurmountable issues when it comes to fitting generalized linear and nonlinear models to correlated response data.

There are 7 chapters in this book. Chapter 1 starts with a description of the kinds of correlated response data covered in the book as well as an overview of the two types of models used to analyze such data, namely marginal and mixed-effects models. The remaining six chapters are divided into three parts: Part I covers linear models (chapters 2-3), Part II covers nonlinear models (chapters 4-5), and Part III covers further topics including methods for handling missing data (chapter 6) and further applications (chapter 7).

Chapters 2 and 3 cover key aspects of linear models for correlated response data. By no means are these chapters meant to be comprehensive in scope. Rather, the intent is to familiarize readers with a general modeling framework that is easily extended to generalized linear and nonlinear models. Chapter 2 focuses on marginal linear models (LM's) for correlated data which can be fit using the MIXED, GLIMMIX or GENMOD procedures. These models require one to specify a marginal covariance structure that describes variability and correlation among vectors of observations. Chapter 3 covers the linear mixed-effects (LME) model obtained by introducing random effects into the marginal LM. Because random effects are shared across observations from the same experimental unit, a type of intraclass correlation structure is introduced with LME models.

Part II of the book is devoted to nonlinear models, notably generalized linear models and generalized nonlinear models the latter of which include normal-theory nonlinear models. As was done with linear models, the material is divided into two chapters with chapter 4 focusing on marginal nonlinear models and chapter 5 with nonlinear mixed-effects models. In turn, the material in each chapter is further divided according to whether the nonlinear model belongs to a class of generalized linear models or to a class of generalized nonlinear models. The former are characterized by monotonic invertible link functions mapping a linear predictor to the mean of the response variable while the latter are characterized by more general nonlinear functions linking a set of predictors to the mean. In both cases, the response variable may be continuous or discrete and it may or may not be linked to a specific distribution such as a normal, gamma, Poisson, negative binomial or binomial distribution.

In Part III, chapter 6 covers methods for analyzing incomplete longitudinal data, i.e., data with missing values. This is a rather long chapter dealing with a host of complex issues not all of which may be of interest to everyone. For example, readers wishing to avoid some of the more thorny theoretical issues pertaining to missing data from dropout might initially skip §6.3 and focus on the material presented in §6.1 and §6.2 that describe different missing data mechanisms, and material in §6.4 and §6.5 that describe different analytical methods for handling missing data. The three case histories presented in §6.6 provide an in-depth approach to how one might analyze incomplete longitudinal data due to dropout.

Finally, chapter 7 presents further case studies that involve other nuances not already covered. This includes a discussion of how to fit mixed models in SAS when the random effects are not normally distributed. Also included in chapter 7 are two pharmacokinetic applications both of which have their own unique set of challenges, and an application that requires one to jointly model longitudinal data and survival data. Four appendices are included with the book. Appendix A provides summary matrix algebra results used throughout the book. Appendix B provides additional theoretical results pertaining to the different estimators of nonlinear mixed-effects models. Appendices C and D list, respectively, the datasets and macros used in the book. A more detailed description of the datasets listed in Appendix C and the macros listed in Appendix D is available through the author's Web page at **http://support.sas.com/publishing/authors/vonesh.html**.

Acknowledgments

I would like to express my gratitude to the editorial staff at SAS Press for all their help. I especially wish to thank George McDaniel, acquisitions editor at SAS Press, for his patience with me as I struggled to balance different priorities while writing this book. I would also like to thank Oliver Schabenberger, Jill Tao and Jim Seabolt from SAS Institute Inc. for providing their expert technical reviews of the material in the book. Their feedback and comments certainly helped improve the framework and contents of the book.

I wish to extend a special thanks to my sister Alice, her husband Gary, and my niece Stacy for their incredible support and understanding through what must have seemed an eternity in writing this book. Likewise, I would like to thank my colleagues at Northwestern University for their support as well as to my friends and former colleagues at Baxter Healthcare whom I came to rely on for feedback throughout this writing. A special thanks goes to the many colleagues I have collaborated with over the years—their insights and contributions are reflected in the work presented here. Special thanks also to Dean Follmann and Garrett Fitzmaurice for their considerable and thoughtful feedback on Chapter 6—this was undoubtedly the toughest chapter I have ever written and I am very thankful for their comments and recommendations which greatly improved the material presented. Thanks also to Eugene Demidenko for his thoughtful and critical review of Chapter 7 and Appendices A and B. Finally, I would like to thank all my friends for their support and friendship through all the difficult times while writing this book. To you the readers and practitioners, it is my hope that this book will prove to be a useful resource in your work.

Introduction

1

Correlated response data, either discrete (nominal, ordinal, counts), continuous or a combination thereof, occur in numerous disciplines and more often than not, require the use of statistical models that are nonlinear in the parameters of interest. In this chapter we briefly describe the different types of correlated response data one encounters in practice as well as the types of explanatory variables and models used to analyze such data.

1.1 Correlated response data

Within SAS, there are various models, methods and procedures that are available for analyzing any of four different types of correlated response data: (1) repeated measurements including longitudinal data, (2) clustered data, (3) spatially correlated data, and (4) multivariate data. Brief descriptions of these four types of correlated data follow.

1.1.1 Repeated measurements

Repeated measurements may be defined as data where a response variable for each experimental unit or subject is observed on multiple occasions and possibly under different experimental conditions. Within this context, the overall correlation structure among repeated measurements should be flexible enough to reflect the serial nature in which measurements are collected per subject while also accounting for possible correlation associated with subject-specific random effects. We have included longitudinal data under the heading of repeated measurements for several reasons:

Longitudinal data may be thought of as repeated measurements where the underlying metameter for the occasions at which measurements are taken is time. In this setting, interest generally centers on modeling and comparing trends over time. Consequently, longitudinal data will generally exhibit some form of serial correlation much as one would expect with time-series data. In addition, one may also model correlation induced by a random-effects structure that allows for within- and between-subject variability. *Repeated measurements* are more general than longitudinal data in that the underlying metameter may be time or it may be a set of experimental conditions (e.g., different dose levels of a drug). Within this broader context, one may be interested in modeling and comparing trends over the range of experimental conditions (e.g., dose-response curves) or simply comparing mean values across different experimental conditions (e.g., a repeated measures analysis of variance).

Some typical settings where one is likely to encounter repeated measurements and longitudinal data include:

- Pharmacokinetic studies
  Studies where the plasma concentration of a drug is measured at several time points for each subject with an objective of estimating various population pharmacokinetic parameters.
- Econometrics
  Panel data entailing both cross-sectional and time-series data together in a two-dimensional array.
- Crossover studies
  Bioavailability studies, for example, routinely employ two-period, two-treatment crossover designs (e.g., AB | BA) where each subject receives each treatment on each of two occasions.
- Growth curve studies
  - Pediatric studies examining the growth pattern of children.
  - Agricultural studies examining the growth pattern of plants.

To illustrate, we consider the orange tree growth curve data presented in Draper and Smith (1981, p. 524) and analyzed by Lindstrom and Bates (1990). The data consists of the trunk circumference (millimeters) measured over 7 different time points on each of five orange trees. As shown in Figure 1.1, growth patterns exhibit a trend of ever increasing variability over time; a pattern reflective of a random-effects structure.

While we have lumped longitudinal data together with repeated measurements, it should be noted that event-times data, i.e., data representing time to some event, may also be classified as longitudinal even though the event time may be a single outcome measure such as patient survival. We will examine the analysis of event times data within the context of joint modeling of event times with repeated measurements/longitudinal data.

1.1.2 Clustered data

Clustered dependent data occur when observations are grouped in some natural fashion into clusters resulting in within-cluster data that tend to be correlated. Correlation induced by clustering is more often than not accounted for through the specification of a random-effects model in which cluster-specific random effects are used to differentiate within- and between-cluster variability. In some instances, there may be more than one level of clustering resulting in specification of a multi-level random-effects structure. Examples of clustered data include:

- Paired data
  - Studies on twins where each pair serves as a natural cluster.
  - Ophthalmology studies where a pair of eyes serves as a cluster.

Figure 1.1 Orange tree growth data

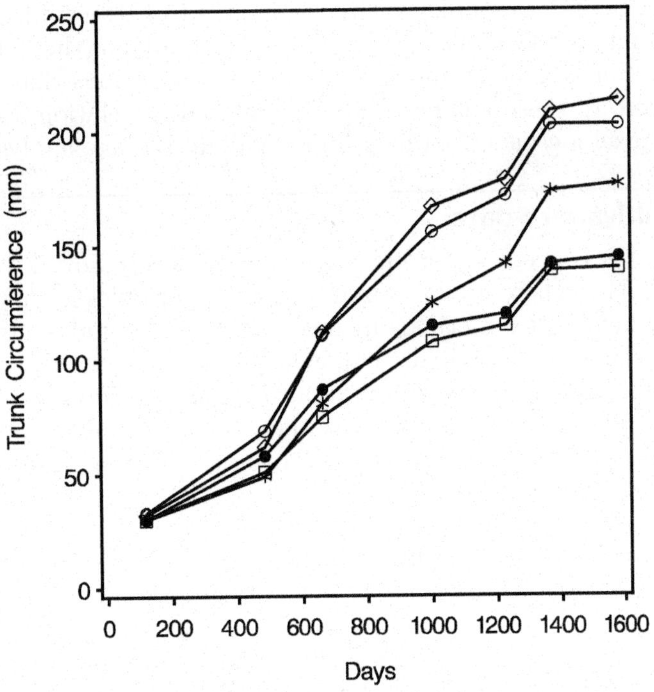

- Familial or teratologic studies
  - Studies on members of a litter of animals (e.g., toxicology studies).
  - Epidemiology studies of cancer where families serve as clusters.
- Agricultural studies
  Studies in which different plots of land serve as clusters and measurements within a plot are homogeneous.

1.1.3 Spatially correlated data

Spatially correlated data occur when observations include both a response variable and a location vector associated with that response variable. The location vector describes the position or location at which the measured response was obtained. The proximity of measured responses with one another determines the extent to which they are correlated. Lattice data, where measurements are linked to discrete regions (e.g., townships, counties, etc.) rather than some continuous coordinate system, are also considered as spatially correlated and are usually obtained from administrative data sources like census data, socio-economical data, and health data. Examples of spatially correlated data include:

- Geostatistical data
  Forestry and agronomy studies where sampling from specified (fixed) locations is used to draw inference over an entire region accounting for spatial dependencies.
  Mineral and petroleum exploration studies where the objective is more likely to be predictive in nature. Here one utilizes spatial variability patterns to help improve one's ability to predict resources in unmeasured locations.
- Epidemiological studies
  Studies aimed at describing the incidence and prevalence of a particular disease often use spatial correlation models in an attempt to smooth out region-specific counts so as to better assess potential environmental determinants and spatial patterns associated with the disease.

- Image analysis
  Image segmentation studies where the goal is to extract information about a particular region of interest from a given image. For example, in the field of medicine, image segmentation may be required to identify tissue regions that have been stained versus not stained. In these settings, modeling spatial correlation associated with a lattice array of pixel locations can help improve digital image analysis.

1.1.4 Multivariate data

Historically, the concept of correlation has been closely linked with methods for analyzing multivariate data wherein two or more response variables are measured per experimental unit or individual. Such methods include multivariate analysis of variance, cluster analysis, discriminant analysis, principal components analysis, canonical correlation analysis, etc. This book does not cover those topics but instead considers applications requiring the analysis of multivariate repeated measurements or the joint modeling of repeated measurements and one or more outcome measures that are measured only once. Examples we consider include

- Multivariate repeated measurements
  Any study where one has two or more outcome variables measured repeatedly over time.
- Joint modeling of repeated measurements and event-times data
  Studies where the primary goal is to draw inference on serial trends associated with a set of repeated measurements while accounting for possible informative censoring due to dropout.
  Studies where the primary goal is to draw joint inference on patient outcomes (e.g., patient survival) and any serial trends one might observe in a potential surrogate marker of patient outcome.

Of course one can easily imagine applications that involve two or more types of correlated data. For example, a longitudinal study may well entail both repeated measurements taken at the individual level and clustered data where groups of individuals form clusters according to some pre-determined criteria (e.g., a group randomized trial). Spatio-temporal data such as found in the mapping of disease rates over time is another example of combining two types of correlated data, namely spatially correlated data with serially correlated longitudinal data.

The majority of applications in this book deal with the analysis of repeated measurements, longitudinal data, clustered data, and to a lesser extent, spatial data. A more thorough treatment and illustration of applications involving the analysis of spatially correlated data and panel data, for example, may found in other texts including Cressie (1993), Littell et. al. (2006), Hsiao (2003), Frees (2004) and Mátyás and Sevestre (2008). We will also examine applications that require modeling multivariate repeated measurements in which two or more response variables are measured on the same experimental unit or individual on multiple occasions. As the focus of this book is on applications requiring the use of generalized linear and nonlinear models, examples will include methods for analyzing continuous, discrete and ordinal data including logistic regression for binary data, Poisson regression for counts, and nonlinear regression for normally distributed data.

1.2 Explanatory variables

When analyzing correlated data, one need first consider the kind of study from which the data were obtained. In this book, we consider data arising from two types of studies: 1) experimental studies, and 2) observational studies. Experimental studies are generally interventional in nature in that two or more treatments are applied to experimental units with a goal of comparing the mean response across different treatments. Such studies may or may not entail randomization. For example, in a parallel group randomized placebo-controlled clinical trial, the experimental study may entail randomizing individuals into two groups; those who receive the placebo control versus those who receive an active ingredient. In contrast, in a simple pre-post study, a measured response is obtained on all individuals prior to receiving a planned intervention and then, following the intervention, a second response is measured. In this case, although no randomization is performed, the study is still experimental in that it does entail a planned intervention. Whether randomization is performed or not, additional explanatory variables are usually collected so as to 1) adjust for any residual confounding that may be present with or without randomization and 2) determine if interactions exist with the intervention.

In contrast to an experimental study, an observational study entails collecting data on available individuals from some target population. Such data would include any outcome measures of interest as well as any explanatory variables thought to be associated with the outcome measures.

In both kinds of studies, the set of explanatory variables, or covariates, used to model the mean response can be broadly classified into two categories:

- *Within-unit* factors or covariates (in longitudinal studies, these are often referred to as *time-dependent* covariates or *repeated measures* factors).
  For repeated measurements and longitudinal studies, examples include time itself, different dose levels applied to the same individual in a dose-response study, different treatment levels given to the same individual in a crossover study.
  For clustered data, examples include any covariates measured on individuals within a cluster.
- *Between-unit* factors or covariates (in longitudinal studies, these are often referred to as *time-independent* covariates)
  Examples include baseline characteristics in a longitudinal study (e.g., gender, race, baseline age), different treatment levels in a randomized prospective longitudinal study, different cluster-specific characteristics, etc.

It is important to maintain the distinction between these two types of covariates for several reasons. One, it helps remind us that within-unit covariates model unit-specific trends while between-unit covariates model trends across units. Two, it may help in formulating an appropriate variance-covariance structure depending on the degree of heterogeneity between select groups of individuals or units. Finally, such distinctions are needed when designing a study. For example, sample size will be determined, in large part, on the primary goals of a study. When those goals focus on comparisons that involve within-unit covariates (either main effects or interactions), the number of experimental units needed will generally be less than when based on comparisons involving strictly between-unit comparisons.

1.3 Types of models

While there are several approaches to modeling correlated response data, we will confine ourselves to two basic approaches, namely the use of 1) marginal models and 2) mixed-effects models. With marginal models, the emphasis is on *population-averaged* (PA) inference where one focuses on the *marginal expectation of the responses.* Correlation is accounted for solely through specification of a *marginal* variance-covariance structure. The regression parameters of marginal models describe the population mean response and are most applicable in settings where the data are used to derive public policies. In contrast,

Figure 1.2 Individual and marginal mean responses under a simple negative exponential decay model with random decay rates

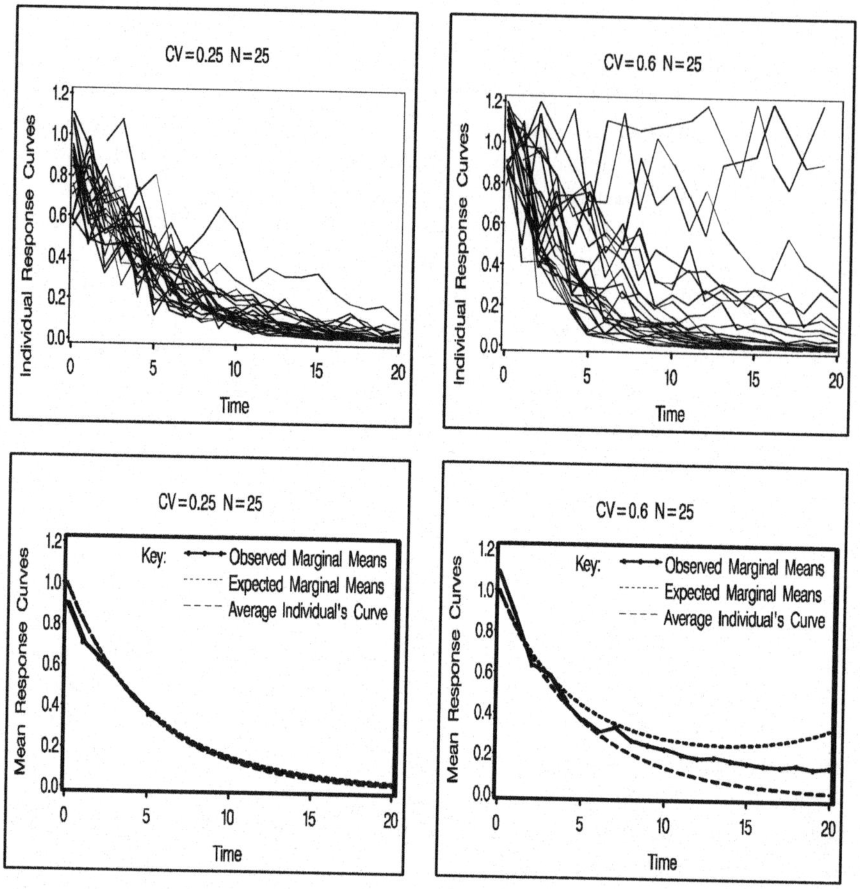

with a mixed-effects model, inference is more likely to be *subject-specific* (SS) or *cluster-specific* in scope with the focus centering on the *individual's* mean response. Here correlation is accounted for through specification of subject-specific random effects and possibly on an intra-subject covariance structure. Unlike marginal models, the fixed-effects regression parameters of mixed-effects models describe the average individual's response and are more informative when advising individuals of their expected outcomes. When a mixed-effects model is strictly linear in the random effects, the regression parameters will have both a population-averaged and subject-specific interpretation.

1.3.1 Marginal versus mixed-effects models

In choosing between marginal and mixed-effects models, one need carefully assess the type of inference needed for a particular application and weigh this against the complexities associated with running each type of model. For example, while a mixed-effects model may make perfect sense as far as its ability to describe heterogeneity and correlation, it may be extremely difficult to draw any population-based inference unless the model is strictly linear in the random effects. Moreover, population-based inference on, say, the marginal means may make little sense in applications where a mixed-effects model is assumed at the start. We illustrate this with the following example.

Shown in Figure 1.2 are the individual and marginal mean responses from a randomly generated sample of 25 subjects assuming a simple negative exponential decay model with random decay rates. The model used to generate the data is given by:

$$y_{ij}|b_i = \exp\{-\beta_i t_{ij}\}[1+\epsilon_{ij}] \tag{1.1}$$

$$\beta_i = \beta + b_i \quad \text{with } \beta = 0.20,\ b_i \sim N(0,\sigma_b^2),\ \epsilon_{ij} \sim \text{iid}\, N(0,\sigma_\epsilon^2)$$

where y_{ij} is the response for the ith subject ($i = 1, \ldots, n$) on the jth occasion, t_{ij} ($j = 1, \ldots, p$), β_i is the ith subject's decay rate, β is a population parameter decay rate, and b_i is a subject-specific random effect describing how far the ith individual's decay rate deviates from the population parameter decay rate. Under this model, subject-specific (SS) inference targets the average individual's response curve while population-averaged (PA) inference targets the average response in the population. Specifically we have:

SS Inference - Average Subject's Response Curve:

$$E_{y|b}[y_{ij}|b_i = 0] = E_{y|b}[y_{ij}|b_i = E_b(b_i)] = \exp\{-\beta t_{ij}\},$$

PA Inference - Population-Averaged Response Curve:

$$E_y[y_{ij}] = E_b[E_{y|b}(y_{ij}|b_i)] = \exp\{-\beta t_{ij} + \tfrac{1}{2}\sigma_b^2 t_{ij}^2\}.$$

Notice that for the average subject's response curve, expectation with respect to random effects is applied within the conditional argument (i.e., we are evaluating a conditional mean at the average random effect) while for the population-averaged response curve, expectation with respect to random effects is applied to the conditional means (i.e., we are evaluating the average response across subjects). Hence under model (1.1), the population-averaged (i.e., marginal) mean response depends on both the first and second moments of the subject-specific parameter, $\beta_i \sim N(\beta, \sigma_b^2)$.

To contrast the SS response curves and PA response curves, we simulated data assuming 1) $\sigma_b/\beta = 0.25$ (i.e., a coefficient of variation, CV, of 25% with respect to β_i), and 2) $\sigma_b/\beta = 0.60$ (CV = 60%). As indicated in Figure 1.2, when the coefficient of variation of β_i increases, there is a clear divergence between the average subject's response curve and the mean response in the population. This reflects the fact that the marginal mean is a logarithmically convex function in t for all $t > \beta/\sigma_b^2$. It is also of interest to note that as the CV increases, the sample size required for the observed means to approximate the expected population means also increases (see Figure 1.3). One may question the validity of using simulated data where the response for some individuals actually increases over time despite the fact that the population parameter β depicts an overall decline over time (here, this will occur whenever the random effect $b_i < -0.20$ as $-\beta_i$ will then be positive). However, such phenomena do occur. For example, in section 1.4, we consider a study among patients with end-stage renal disease where their remaining kidney function, as measured by the glomerular filtration rate (GFR), tends to decline exponentially over time but for a few patients, their GFR actually increases as they regain their renal function.

1.3.2 Models in SAS

There are a variety of SAS procedures and macros available to users seeking to analyze correlated data. Depending on the type of data (discrete, ordinal, continuous, or a combination thereof), one can choose from one of four basic categories of models available in SAS: linear models, generalized linear models, nonlinear models and generalized nonlinear models. Within each basic category one can also choose to run a marginal model or a mixed-effects model resulting in the following eight classes of models available in SAS: 1) Linear Models (LM); 2) Linear Mixed-Effects Models (LME models); 3) Generalized Linear Models (GLIM); 4) Generalized Linear Mixed-Effects Models (GLME models); 5) Nonlinear Models (NLM); 6) Nonlinear Mixed-Effects Models (NLME models); 7) Generalized Nonlinear Models (GNLM); and 8) Generalized Nonlinear Mixed-Effects

Figure 1.3 Observed and expected marginal means for different sample sizes

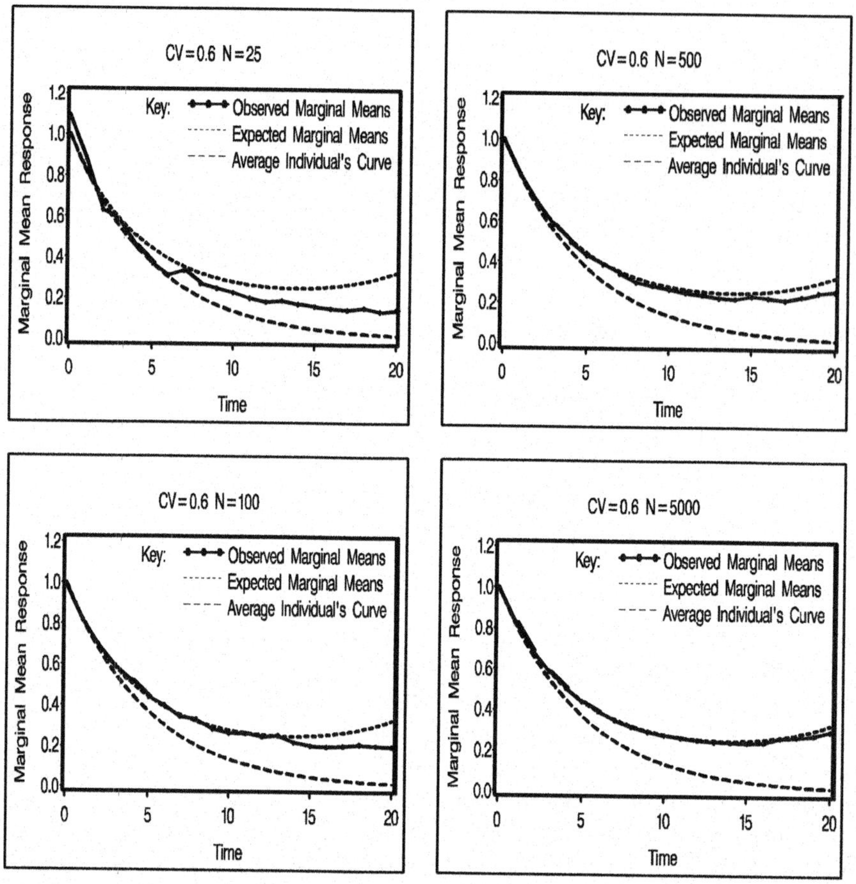

Models (GNLME models). The models within any one class are determined through specification of the moments and possibly the distribution functions under which the data are generated. Moment-based specifications usually entail specifying unconditional (marginal) or conditional (mixed-effects) means and variances in terms of their dependence on covariates and/or random effects. Using the term "subject" to refer to an individual, subject, cluster or experimental unit, we adopt the following general notation which we use to describe marginal or conditional moments and/or likelihood functions.

Notation

$\boldsymbol{y}$ is a response vector for a given subject. Within a given context or unless otherwise noted, lower case lettering $\boldsymbol{y}$ (or y) will be used to denote either the underlying random vector (or variable) or its realization.
$\boldsymbol{\beta}$ is a vector of fixed-effects parameters associated with first-order moments (i.e., marginal or conditional means).
$\boldsymbol{b}$ is a vector of random-effects (possibly multi-level) which, unless otherwise indicated, is assumed to have a multivariate normal distribution, $\boldsymbol{b} \sim N(\boldsymbol{0}, \boldsymbol{\Psi})$, with variance-covariance matrix $\boldsymbol{\Psi} = \boldsymbol{\Psi}(\boldsymbol{\theta}_b)$ that depends on a vector of between-subject variance-covariance parameters, say $\boldsymbol{\theta}_b$.
$\boldsymbol{X}$ is a matrix of within- and between-subject covariates linking the fixed-effects parameter vector $\boldsymbol{\beta}$ to the marginal or conditional mean.
$\boldsymbol{Z}$ is a matrix of within-subject covariates contained in $\boldsymbol{X}$ that directly link the random effects to the conditional mean.
$\boldsymbol{\mu}(\boldsymbol{X}, \boldsymbol{\beta}, \boldsymbol{Z}, \boldsymbol{b}) = \boldsymbol{\mu}(\boldsymbol{\beta}, \boldsymbol{b}) = E(\boldsymbol{y}|\boldsymbol{b})$ is the conditional mean of $\boldsymbol{y}$ given random effects $\boldsymbol{b}$.

Table 1.1 Hierarchy of models in SAS according to mean structure and cumulative distribution function (CDF)

Marginal Models			Mixed-Effects Models		
Model	$\mu(\boldsymbol{X},\boldsymbol{\beta})$	CDF	Model	$\mu(\boldsymbol{X},\boldsymbol{\beta},\boldsymbol{Z},\boldsymbol{b})$	CDF
LM	$\boldsymbol{X\beta}$	Normal	LME	$\boldsymbol{X\beta}+\boldsymbol{Zb}$	Normal
GLIM	$g^{-1}(\boldsymbol{X\beta})$	General	GLME	$g^{-1}(\boldsymbol{X\beta}+\boldsymbol{Zb})$	General
NLM	$f(\boldsymbol{X},\boldsymbol{\beta})$	Normal	NLME	$f(\boldsymbol{X},\boldsymbol{\beta},\boldsymbol{b})$	Normal
GNLM	$f(\boldsymbol{X},\boldsymbol{\beta})$	General	GNLME	$f(\boldsymbol{X},\boldsymbol{\beta},\boldsymbol{b})$	General

$\boldsymbol{\Lambda}(\boldsymbol{X},\boldsymbol{\beta},\boldsymbol{Z},\boldsymbol{b},\boldsymbol{\theta}_w)=\boldsymbol{\Lambda}(\boldsymbol{\beta},\boldsymbol{b},\boldsymbol{\theta}_w)=Var(\boldsymbol{y}|\boldsymbol{b})$ is the conditional covariance matrix of $\boldsymbol{y}$ given random effects $\boldsymbol{b}$. This matrix may depend on an additional vector, $\boldsymbol{\theta}_w$, of within-subject variance-covariance parameters.

$\boldsymbol{\mu}(\boldsymbol{X},\boldsymbol{\beta})=\boldsymbol{\mu}(\boldsymbol{\beta})=E(\boldsymbol{y})$ is the marginal mean of $\boldsymbol{y}$ except for mixed models where the marginal mean $\boldsymbol{\mu}(\boldsymbol{X},\boldsymbol{\beta},\boldsymbol{Z},\boldsymbol{\theta}_b)=E(\boldsymbol{y})$ may depend on $\boldsymbol{\theta}_b$ as well as $\boldsymbol{\beta}$ (e.g., see the PA mean response for model (1.1)).

$\boldsymbol{\Sigma}(\boldsymbol{\beta},\boldsymbol{\theta})=Var(\boldsymbol{y})$ is the marginal variance-covariance of $\boldsymbol{y}$ that depends on between- and/or within-subject covariance parameters, $\boldsymbol{\theta}=(\boldsymbol{\theta}_b,\boldsymbol{\theta}_w)$ and possibly on the fixed-effects regression parameters $\boldsymbol{\beta}$.

There is an inherent hierarchy to these models as suggested in Table 1.1. Specifically, linear models can be considered a special case of generalized linear models in that the latter allow for a broader class of distributions and a more general mean structure given by an inverse link function, $g^{-1}(\boldsymbol{X\beta})$, which is a monotonic invertible function linking the mean, $E(\boldsymbol{y})$, to a linear predictor, $\boldsymbol{X\beta}$, via the relationship $g(E(\boldsymbol{y}))=\boldsymbol{X\beta}$. Likewise Gaussian-based linear models are a special case of Gaussian-based nonlinear models in that the latter allow for a more general nonlinear mean structure, $f(\boldsymbol{X},\boldsymbol{\beta})$, rather than one that is strictly linear in the parameters of interest. Finally, generalized linear models are a special case of generalized nonlinear models in that the latter, in addition to allowing for a broader class of distributions, also allow for more general nonlinear mean structures of the form $f(\boldsymbol{X},\boldsymbol{\beta})$. That is, generalized nonlinear models do not require a mean structure that is an invertible monotonic function of a linear predictor as is the case for generalized linear models. In like fashion, within any row of Table 1.1, one can consider the marginal model as a special case of the mixed-effect model in that the former can be obtained by merely setting the random effects of the mixed-effects model to 0. We also note that since nonlinear mixed models do not require specification of a conditional linear predictor, $\boldsymbol{X\beta}+\boldsymbol{Zb}$, the conditional mean, $f(\boldsymbol{X},\boldsymbol{\beta},\boldsymbol{b})$, may be specified without reference to $\boldsymbol{Z}$. This is because the fixed and random effects parameters, $\boldsymbol{\beta}$ and $\boldsymbol{b}$, will be linked to the appropriate covariates through specification of the function f and its relationship to a design matrix $\boldsymbol{X}$ that encompasses $\boldsymbol{Z}$.

The generalized nonlinear mixed-effects (GNLME) model is the most general model considered in that it combines the flexibility of a generalized linear model in terms of its ability to specify non-Gaussian distributions, and the flexibility of a nonlinear model in terms of its ability to specify more general mean structures. One might wonder, then, why SAS does not develop a single procedure based on the GNLME model rather than the various procedures currently available in SAS. The answer is simple. The various procedures specific to linear (MIXED, GENMOD and GLIMMIX) and generalized linear models (GENMOD, GLIMMIX) offer specific options and computational features that take full advantage of the inherent structure of the underlying model (e.g., the linearity of all parameters, both fixed and random, in the LME model, or the monotonic transformation that links the mean function to a linear predictor in a generalized linear model). Such flexibility is next to impossible to incorporate under NLME and GNLME models, both of which can be fit to data using either the SAS procedure NLMIXED or the SAS macro %NLINMIX. A "road map" linking the SAS procedures and their key statements/options to these various models is presented in Table 1.2 of the summary section at the end of this chapter.

Finally, we shall assume throughout that the design matrices, $\boldsymbol{X}$ and $\boldsymbol{Z}$, are of full rank such that, where indicated, all matrices are invertible. In those cases where we are dealing with less than full rank design matrices and matrices of the form $\boldsymbol{X}'\boldsymbol{A}\boldsymbol{X}$, for example, are not of full rank, the expression $(\boldsymbol{X}'\boldsymbol{A}\boldsymbol{X})^{-1}$ will be understood to represent a generalized inverse of $\boldsymbol{X}'\boldsymbol{A}\boldsymbol{X}$.

1.3.3 Alternative approaches

Alternative approaches to modeling correlated response data include the use of conditional and/or transition models as well as hierarchical Bayesian models. Texts by Diggle, Liang and Zeger (1994) and Molenberghs and Verbeke (2005), for example, provide an excellent source of information on conditional and transition models for longitudinal data. Also, with the advent of Markov Chain Monte Carlo (MCMC) and other techniques for generating samples from Bayesian posterior distributions, interested practitioners can opt for a full Bayesian approach to modeling correlated data as exemplified in the texts by Carlin and Louis (2000) and Gelman et. al. (2004). A number of Bayesian capabilities were made available with the release of SAS 9.2 including the MCMC procedure. With PROC MCMC, users can fit a variety of Bayesian models using a general purpose MCMC simulation procedure.

1.4 Some examples

In this section, we present just a few examples illustrating the different types of response data, covariates, and models that will be discussed in more detail in later chapters.

Soybean Growth Data

Davidian and Giltinan (1993, 1995) describe an experimental study in which the growth patterns of two genotypes of soybeans were to be compared. The essential features of the study are as follows:

- The experimental unit or cluster is a plot of land
- Plots were sampled 8-10 occasions (times) within a calendar year
- Six plants were randomly selected at each occasion and the average leaf weight per plant was calculated for a plot
- Response variable:
  - y_{ij} = average leaf weight (g) per plant for i^{th} plot on the jth occasion (time) within a calendar year ($i = 1, \ldots, n = 48$ plots; 16 plots per each of the calendar years 1988, 1989 and 1990; $j = 1, \ldots, p_i$ with $p_i = 8$ to 10 measurements per calendar year)
- One within-unit covariate:
  - t_{ij} =days after planting for i^{th} plot on the jth occasion
- Two between-unit covariates:
  - Genotype of Soybean (F=commercial, P=experimental) denoted by

$$a_{1i} = \begin{cases} 0, & \text{if commercial (F)} \\ 1, & \text{if experimental (P)} \end{cases}$$

Figure 1.4 Soybean growth data

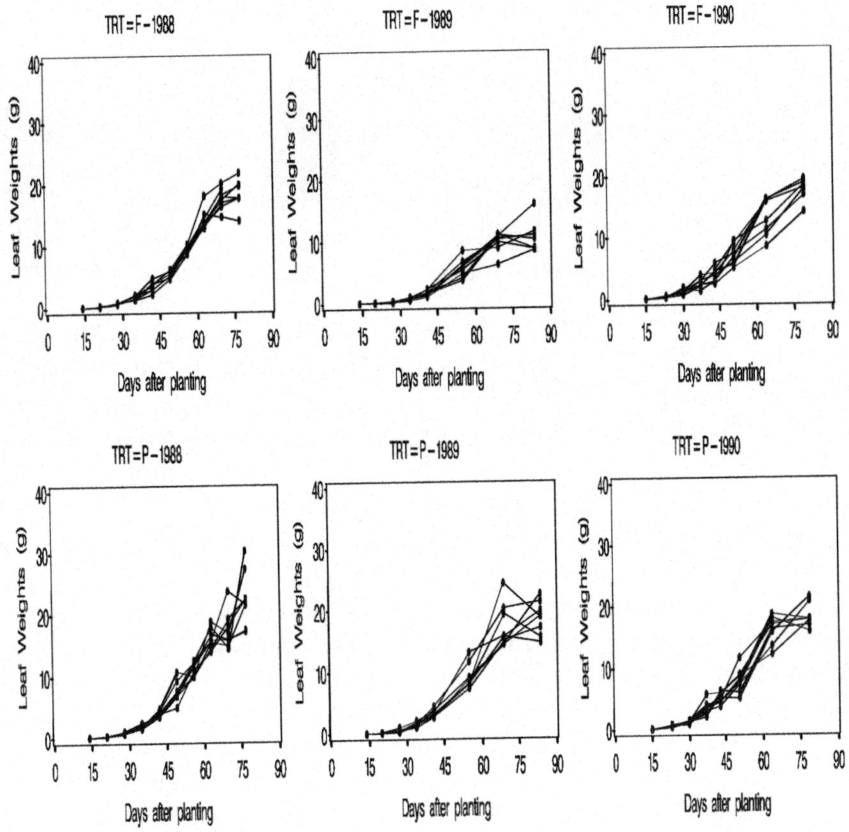

– Calendar Year (1988, 1989, 1990) denoted by two indicator variables,

$$a_{2i} = \begin{cases} 0, & \text{if year is 1988 or 1990} \\ 1, & \text{if year is 1989} \end{cases}$$

$$a_{3i} = \begin{cases} 0, & \text{if year is 1988 or 1989} \\ 1, & \text{if year is 1990} \end{cases}$$

- Goal: compare the growth patterns of the two genotypes of soybean over the three growing seasons represented by calendar years 1988-1990.

This is an example of an experimental study involving clustered longitudinal data in which the response variable, y =average leaf weight per plant (g), is measured over time within plots (clusters) of land. In each of the three years, 1988, 1989 and 1990, 8 different plots of land were seeded with the genotype Forrest (F) representing a commercial strain of seeds and 8 different plots were seeded with genotype Plant Introduction #416937 (P), an experimental strain of seeds. During the growing season of each calendar year, six plants were randomly sampled from each plot on a weekly basis (starting roughly two weeks after planting) and the leaves from these plants were collected and weighed yielding an average leaf weight per plant, y, per plot. A plot of the individual profiles, shown in Figure 1.4, suggest a nonlinear growth pattern which Davidian and Giltinan modeled using a nonlinear mixed-effects logistic growth curve model. One plausible form of this model might be

$$\begin{aligned} y_{ij} &= f(\boldsymbol{x}'_{ij}, \boldsymbol{\beta}, \boldsymbol{b}_i) + \epsilon_{ij} \\ &= f(\boldsymbol{x}'_{ij}, \boldsymbol{\beta}_i) + \epsilon_{ij} \\ &= \frac{\beta_{i1}}{1 + \exp\{\beta_{i3}(t_{ij} - \beta_{i2})\}} + \epsilon_{ij} \end{aligned} \tag{1.2}$$

$$\boldsymbol{\beta}_i = \begin{pmatrix} \beta_{i1} \\ \beta_{i2} \\ \beta_{i3} \end{pmatrix} = \begin{pmatrix} \beta_{01} + \beta_{11}a_{1i} + \beta_{21}a_{2i} + \beta_{31}a_{3i} \\ \beta_{02} + \beta_{12}a_{1i} + \beta_{22}a_{2i} + \beta_{32}a_{3i} \\ \beta_{03} + \beta_{13}a_{1i} + \beta_{23}a_{2i} + \beta_{33}a_{3i} \end{pmatrix} + \begin{pmatrix} b_{i1} \\ b_{i2} \\ b_{i3} \end{pmatrix}$$

where y_{ij} is the average leaf weight per plant for the ith plot on the jth occasion, $f(\boldsymbol{x}'_{ij}, \boldsymbol{\beta}, \boldsymbol{b}_i) = f(\boldsymbol{x}'_{ij}, \boldsymbol{\beta}_i) = \beta_{i1}/\left[1 + \exp\{\beta_{i3}(t_{ij} - \beta_{i2})\}\right]$ is the conditional mean response for the ith plot on the jth occasion, $\boldsymbol{x}'_{ij} = (1 \quad t_{ij} \quad a_{1i} \quad a_{2i} \quad a_{3i})$ is the vector of within- and between-cluster covariates associated with the population parameter vector, $\boldsymbol{\beta}' = (\beta_{01} \quad \beta_{11} \quad \beta_{21} \quad \beta_{31} \quad \beta_{02} \quad \ldots \quad \beta_{23} \quad \beta_{33})$ on the jth occasion. The first two columns of $\boldsymbol{x}'_{ij}$, say $\boldsymbol{z}'_{ij} = (1 \quad t_{ij})$, is the vector of within-cluster covariates associated with the cluster-specific random effects, $\boldsymbol{b}'_i = (b_{i1} \quad b_{i2} \quad b_{i3})$ on the jth occasion, and ϵ_{ij} is the intra-cluster (within-plot or within-unit) error on the jth occasion. The vector of random-effects are assumed to be iid $N(\boldsymbol{0}, \boldsymbol{\Psi})$ where $\boldsymbol{\Psi}$ is a 3×3 arbitrary positive definite covariance matrix. One should note that under this particular model, we are investigating the main effects of genotype and calendar year on the mean response over time.

The vector $\boldsymbol{\beta}'_i = (\beta_{i1} \quad \beta_{i2} \quad \beta_{i3})$ may be regarded as a cluster-specific parameter vector that uniquely describes the mean response for the ith plot with $\beta_{i1} > 0$ representing the limiting growth value (asymptote), $\beta_{i2} > 0$ representing soybean "half-life" (i.e., the time at which the soybean reaches half its limiting growth value), and $\beta_{i3} < 0$ representing the growth rate. It may be written in terms of the linear random-effects model

$$\boldsymbol{\beta}_i = \boldsymbol{A}_i\boldsymbol{\beta} + \boldsymbol{B}_i\boldsymbol{b}_i = \boldsymbol{A}_i\boldsymbol{\beta} + \boldsymbol{b}_i$$

where

$$\boldsymbol{A}_i = \begin{pmatrix} 1 & a_{1i} & a_{2i} & a_{3i} & 0 & 0 & 0 & 0 & 0 & 0 & 0 & 0 \\ 0 & 0 & 0 & 0 & 1 & a_{1i} & a_{2i} & a_{3i} & 0 & 0 & 0 & 0 \\ 0 & 0 & 0 & 0 & 0 & 0 & 0 & 0 & 1 & a_{1i} & a_{2i} & a_{3i} \end{pmatrix}$$

is a between-cluster design matrix linking the between-cluster covariates, genotype and calendar year, to the population parameters $\boldsymbol{\beta}$ while $\boldsymbol{B}_i$ is an incidence matrix of 0's and 1's indicating which components of $\boldsymbol{\beta}_i$ are random and which are strictly functions of the fixed-effect covariates. In our current example, all three components of $\boldsymbol{\beta}_i$ are assumed random and hence $\boldsymbol{B}_i = \boldsymbol{I}_3$ is simply the identity matrix. Suppose, however, that the half-life parameter, β_{i2}, was assumed to be a function solely of the fixed-effects covariates. Then $\boldsymbol{B}_i$ would be given by

$$\boldsymbol{B}_i = \begin{pmatrix} 1 & 0 \\ 0 & 0 \\ 0 & 1 \end{pmatrix}$$

with $\boldsymbol{b}'_i = (b_{i1} \quad b_{i3})$. Note, too, that we can express the between-unit design matrix more conveniently as $\boldsymbol{A}_i = \boldsymbol{I}_3 \otimes (1 \quad a_{1i} \quad a_{2i} \quad a_{3i})$ where $\otimes$ is the direct product operator (or Kronecker product) linking the dimension of $\boldsymbol{\beta}_i$ via the identity matrix $\boldsymbol{I}_3$ to the between-unit covariate vector, $\boldsymbol{a}'_i = (1 \quad a_{1i} \quad a_{2i} \quad a_{3i})$. The $\otimes$ operator is a useful tool which we will have recourse to use throughout the book and which is described more fully in Appendix A.

Finally, by assuming the intra-cluster errors ϵ_{ij} are iid $N(0,\sigma_w^2)$, model (1.2) may be classified as a nonlinear mixed-effects model (NLME) having conditional means expressed in terms of a nonlinear function, i.e., $E(y_{ij}|\boldsymbol{b}_i) = \mu(\boldsymbol{x}'_{ij},\boldsymbol{\beta},\boldsymbol{z}'_{ij},\boldsymbol{b}_i) = f(\boldsymbol{x}'_{ij},\boldsymbol{\beta},\boldsymbol{b}_i)$, as represented using the general notation of Table 1.1. In this case, inference with respect to the population parameters $\boldsymbol{\beta}$ and $\boldsymbol{\theta} = (\boldsymbol{\theta}_b,\boldsymbol{\theta}_w) = (vech(\boldsymbol{\Psi}),\sigma_w^2)$ will be cluster-specific in scope in that the function f, when evaluated at $\boldsymbol{b}_i = \mathbf{0}$, describes what the average soybean growth pattern is for a "typical plot."

Respiratory Disorder Data

Koch et. al. (1990) and Stokes et. al. (2000) analyzed data from a multicenter randomized controlled trial for patients with a respiratory disorder. The trial was conducted in two centers in which patients were randomly assigned to one of two treatment groups: an active treatment group or a placebo control group (0 = placebo, 1 = active). The initial outcome variable of interest was patient status which is defined in terms of the ordinal categorical responses: 0 = terrible, 1 = poor, 2 = fair, 3 = good, 4 = excellent. This categorical response was obtained both at baseline and at each of four visits (visit 1, visit 2, visit 3, visit 4) during the course of treatment. Here we consider an analysis obtained by collapsing the data into a discrete binary outcome with the primary focus being a comparison of the average response of patients. The essential components of the study are listed below (the SAS variables are listed in parentheses).

- The experimental unit is a patient (identified by SAS variable ID)
- The initial outcome variable was patient status defined by the ordinal response: 0 = terrible, 1 = poor, 2 = fair, 3 = good, 4 = excellent
- Response variable (y):
  - The data were collapsed into a simple binary response as

$$y_{ij} = \begin{cases} 0 = \text{negative response if (terrible, poor, or fair)} \\ 1 = \text{positive response if (good, excellent)} \end{cases}$$

  which is the ith patient's response obtained on the jth visit
  - One within-unit covariate:
  - Visit (1, 2, 3, or 4) defined here by the indictor variables (Visit)

$$v_{ij} = \begin{cases} 1, & \text{if visit } j \\ 0, & \text{otherwise} \end{cases}$$

- Five between-unit covariates:
  - Treatment group (Treatment) defined as 'P' for placebo or 'A' for active.

$$a_{1i} = \begin{cases} 0, & \text{if placebo} \\ 1, & \text{if active} \end{cases} \quad \text{(this indicator is labeled Drug0)}$$

  - Center (Center)

$$a_{2i} = \begin{cases} 0, & \text{if center 1} \\ 1, & \text{if center 2} \end{cases} \quad \text{(this indicator is labeled Center0)}$$

  - Gender (Gender) defined as 0 = male, 1 = female

$$a_{3i} = \begin{cases} 0, & \text{if male} \\ 1, & \text{if female} \end{cases} \quad \text{(this indicator is labeled Sex)}$$

  - Age at baseline (Age)

    a_{4i} =patient age

Figure 1.5 A plot of the proportion of positive responses by visit and drug for the respiratory disorder data

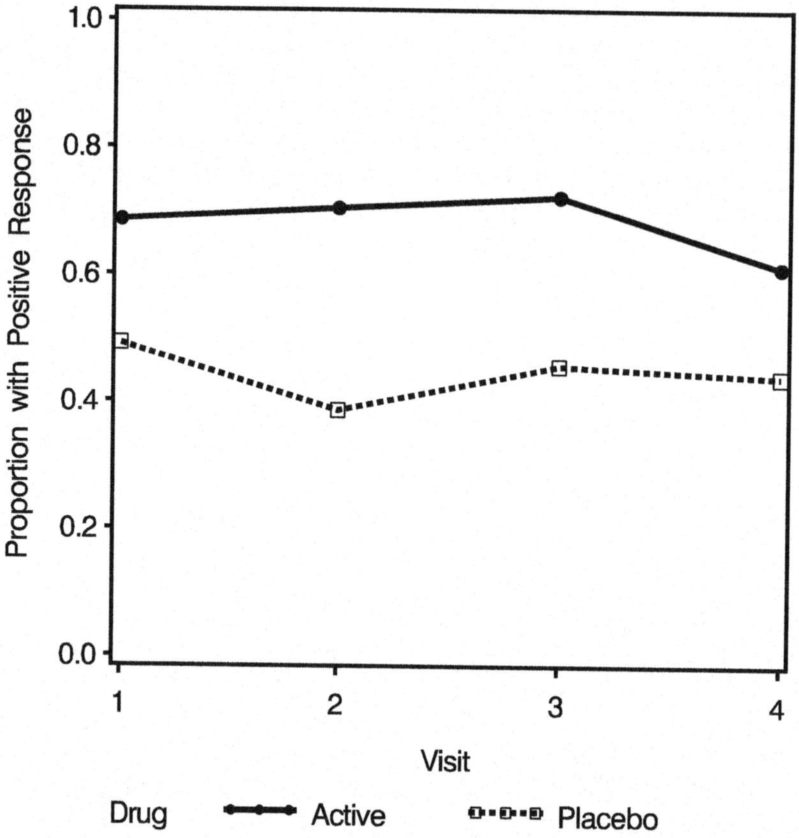

– Response at baseline (Baseline),

$$a_{5i} = \begin{cases} 0 = \text{negative}, \\ 1 = \text{positive} \end{cases} \quad \text{(this indicator is labeled y0)}$$

- Goal: determine if there is a treatment effect after adjusting for center, gender, age and baseline differences.

A plot of the proportion of positive responses at each visit according to treatment group is shown in Figure 1.5. In this example, previous authors (Koch et. al., 1977; Koch et. al., 1990) fit the data using several different marginal models resulting in a population-averaged approach to inference. One family of such models is the family of marginal generalized linear models (GLIM's) with working correlation structure. For example, under the assumption of a working independence structure across visits and assuming there is no visit effect, one might fit this data to a binary logistic regression model of the form

$$\begin{aligned} E(y_{ij}) &= \mu_{ij}(\boldsymbol{x}'_{ij}, \boldsymbol{\beta}) = g^{-1}(\boldsymbol{x}'_{ij}\boldsymbol{\beta}) \\ &= \frac{\exp\{\boldsymbol{x}'_{ij}\boldsymbol{\beta}\}}{1+\exp\{\boldsymbol{x}'_{ij}\boldsymbol{\beta}\}} \end{aligned} \tag{1.3}$$

$$\boldsymbol{x}'_{ij}\boldsymbol{\beta} = \beta_0 v_{ij} + \beta_1 a_{1i} + \beta_2 a_{2i} + \beta_3 a_{3i} + \beta_4 a_{4i} + \beta_5 a_{5i}$$

$$Var(y_{ij}) = \mu_{ij}(\boldsymbol{x}'_{ij}, \boldsymbol{\beta})[1 - \mu_{ij}(\boldsymbol{x}'_{ij}, \boldsymbol{\beta})] = \mu_{ij}(1-\mu_{ij})$$

where $\mu_{ij} = \mu_{ij}(\boldsymbol{x}'_{ij}, \boldsymbol{\beta}) = \Pr(y_{ij} = 1|\boldsymbol{x}_{ij})$ is the probability of a positive response on the jth visit, $\boldsymbol{x}'_{ij} = \begin{pmatrix} v_{ij} & a_{1i} & a_{2i} & a_{3i} & a_{4i} & a_{5i} \end{pmatrix}$ is the design vector of within- and between-unit covariates on the jth visit and $\boldsymbol{\beta}' = (\beta_0 \quad \beta_1 \quad \beta_2 \quad \beta_3 \quad \beta_4 \quad \beta_5)$ is the parameter vector associated with the mean response. Here, we do not model a visit effect but rather assume a common visit effect that is reflected in the overall intercept parameter, β_0 (i.e., since we always have $v_{ij} = 1$ on the jth visit, we can simply replace $\beta_0 v_{ij}$ with β_0). In matrix notation, model (1.3) may be written as

$$E(\boldsymbol{y}_i) = \mu_i(\boldsymbol{X}_i, \boldsymbol{\beta}) = g^{-1}(\boldsymbol{X}_i\boldsymbol{\beta})$$
$$Var(\boldsymbol{y}_i) = \boldsymbol{\Sigma}_i(\boldsymbol{\beta}) = \boldsymbol{H}_i(\boldsymbol{\mu}_i)^{1/2}\boldsymbol{I}_4\boldsymbol{H}_i(\boldsymbol{\mu}_i)^{1/2}$$

where

$$\boldsymbol{y}_i = \begin{pmatrix} y_{i1} \\ y_{i2} \\ y_{i3} \\ y_{i4} \end{pmatrix}, \ \boldsymbol{X}_i = \begin{pmatrix} \boldsymbol{x}'_{i1} \\ \boldsymbol{x}'_{i2} \\ \boldsymbol{x}'_{i3} \\ \boldsymbol{x}'_{i4} \end{pmatrix},$$

and $\boldsymbol{H}_i(\boldsymbol{\mu}_i)$ is the 4×4 diagonal variance matrix with $Var(y_{ij}) = \mu_{ij}(1 - \mu_{ij})$ as the jth diagonal element and $\boldsymbol{H}_i(\boldsymbol{\mu}_i)^{1/2}$ is the square root of $\boldsymbol{H}_i(\boldsymbol{\mu}_i)$ which, for arbitrary positive definite matrices, may be obtained via the Cholesky decomposition. To accommodate possible correlation among binary responses taken on the same subject across visits, we can use a generalized estimating equation (GEE) approach in which robust standard errors are computed using the so-called empirical sandwich estimator (e.g., see §4.2 of Chapter 4). An alternative GEE approach might assume an overdispersion parameter ϕ and "working" correlation matrix $\boldsymbol{R}_i(\boldsymbol{\alpha})$ such that

$$Var(\boldsymbol{y}_i) = \boldsymbol{\Sigma}_i(\boldsymbol{\beta}, \boldsymbol{\theta}) = \phi\boldsymbol{H}_i(\boldsymbol{\mu}_i)^{1/2}\boldsymbol{R}_i(\boldsymbol{\alpha})\boldsymbol{H}_i(\boldsymbol{\mu}_i)^{1/2}$$

where $\boldsymbol{\theta} = (\boldsymbol{\alpha}, \phi)$. Commonly used working correlation structures include working independence [i.e., $\boldsymbol{R}_i(\boldsymbol{\alpha}) = \boldsymbol{I}_4$ as above], compound symmetry and first-order autoregressive. Alternatively, one can extend the GLIM in (1.3) to a generalized linear mixed-effects model (GLME) by simply adding a random intercept term, b_i, to the linear predictor, i.e.,

$$\boldsymbol{x}'_{ij}\boldsymbol{\beta} + b_i = \beta_0 v_{ij} + \beta_1 a_{1i} + \beta_2 a_{2i} + \beta_3 a_{3i} + \beta_4 a_{4i} + \beta_5 a_{5i} + b_i.$$

This will induce a correlation structure among the repeated binary responses via the random intercepts shared by patients across their visits.

In Chapters 4-5, we shall consider analyses based on both marginal and mixed-effects logistic regression thereby allowing one to contrast PA versus SS inference. With respect to marginal logistic regression, we will present results from a first-order and second-order generalized estimating equation approach.

Epileptic Seizure Data

Thall and Vail (1990) and Breslow and Clayton (1993) used Poisson regression to analyze epileptic seizure data from a randomized controlled trial designed to compare the effectiveness of progabide versus placebo to reduce the number of partial seizures occurring over time. The key attributes of the study are summarized below (SAS variables denoted in parentheses).

- The experimental unit is a patient (ID)
- The data consists of the number of partial seizures occurring over a two week period on each of four successive visits made by patients receiving one of two treatments (progabide, placebo).

- Response variable (y):
  – y_{ij} =number of partial seizures in a two-week interval for the ith patient as recorded on the jth visit
- One within-unit covariate:
  – Visit (1, 2, 3, or 4) defined here by the indictor variables (Visit)
  $$v_{ij} = \begin{cases} 1, & \text{if visit } j \\ 0, & \text{otherwise} \end{cases}$$
- Three between-unit covariates:
  – Treatment group (Trt)
  $$a_{1i} = \begin{cases} 0, & \text{if placebo} \\ 1, & \text{if progabide} \end{cases}$$
  – Age at baseline (Age)
  a_{2i} =patient age
  – Baseline seizure counts (bline) normalized to a two week period
  $a_{3i} = \frac{1}{4}\times$baseline seizure counts over an 8 week period (y0)
- Goal: determine if progabide is effective in reducing the number of seizures after adjustment for relevant baseline covariates

In Chapters 4-5, we will consider several different models for analyzing this data using both a marginal and mixed-effects approach. For example, we consider several mixed-effects models in which the count data (y_{ij} =number of seizures per 2-week interval) were fitted to a mixed-effects log-linear model with conditional means of the form

$$\begin{aligned} E(y_{ij}|b_i) =& \mu_{ij}(\boldsymbol{x}'_{ij}, \boldsymbol{\beta}, z_{ij}, b_i) = g^{-1}(\boldsymbol{x}'_{ij}\boldsymbol{\beta} + z_{ij}b_i) \\ =& \exp\{\boldsymbol{x}'_{ij}\boldsymbol{\beta} + z_{ij}b_i\} \\ \boldsymbol{x}'_{ij}\boldsymbol{\beta} + z_{ij}b_i =& \beta_0 + \beta_1 a_{1i} + \beta_2\log(a_{2i}) + \beta_3\log(a_{3i}) + \\ & \beta_4 a_{1i}\log(a_{3i}) + \beta_5 v_{i4} + \log(2) + b_i \end{aligned} \tag{1.4}$$

where $\mu_{ij} = \mu_{ij}(\boldsymbol{x}'_{ij}, \boldsymbol{\beta}, z_{ij}, b_i)$ is the average number of seizures per two week period on the jth visit, $\boldsymbol{x}'_{ij} = (1 \quad a_{1i} \quad \log(a_{2i}) \quad \log(a_{3i}) \quad a_{1i}\log(a_{3i}) \quad v_{i4})$ is the design vector of within- and between-unit covariates on the jth visit, $z_{ij} \equiv 1$ for each j, b_i is a subject-specifc random intercept, and $\boldsymbol{\beta}' = (\beta_0 \quad \beta_1 \quad \beta_2 \quad \beta_3 \quad \beta_4 \quad \beta_5)$ is the parameter vector associated with the mean response. Following Thall and Vail (1990) and Breslow and Clayton (1993), this model assumes that only visit 4 has an effect on seizure counts. The term, log(2), is an offset that is included to reflect that the mean count is over a two-week period. In matrix notation, the conditional means across all four visits for a given subject may be written as

$$E(\boldsymbol{y}_i|b_i) = \boldsymbol{\mu}_i(\boldsymbol{X}_i, \boldsymbol{\beta}, \boldsymbol{Z}_i, b_i) = \boldsymbol{\mu}_i(\boldsymbol{\beta}, \boldsymbol{b}_i) = g^{-1}(\boldsymbol{X}_i\boldsymbol{\beta} + \boldsymbol{Z}_i b_i)$$

where

$$\boldsymbol{y}_i = \begin{pmatrix} y_{i1} \\ y_{i2} \\ y_{i3} \\ y_{i4} \end{pmatrix}, \ \boldsymbol{X}_i = \begin{pmatrix} \boldsymbol{x}'_{i1} \\ \boldsymbol{x}'_{i2} \\ \boldsymbol{x}'_{i3} \\ \boldsymbol{x}'_{i4} \end{pmatrix}, \ \boldsymbol{Z}_i = \begin{pmatrix} 1 \\ 1 \\ 1 \\ 1 \end{pmatrix}.$$

One could consider fitting this model assuming any one of three conditional variance structures, $Var(y_{ij}|b_i) = \boldsymbol{\Lambda}_i(\boldsymbol{x}_{ij}, \boldsymbol{\beta}, \boldsymbol{z}_{ij}, b_i, \boldsymbol{\theta}_w) = \boldsymbol{\Lambda}_i(\boldsymbol{\beta}, \boldsymbol{b}_i, \boldsymbol{\theta}_w)$,

Case 1: $Var(y_{ij}|b_i) = \mu_{ij}$ (standard Poisson variation)
Case 2: $Var(y_{ij}|b_i) = \phi\mu_{ij}$ (extra-Poisson variation)
Case 3: $Var(y_{ij}|b_i) = \mu_{ij}(1+\alpha\mu_{ij})$ (negative binomial variation).

In the first case, $\mathbf{\Lambda}_i(\boldsymbol{\beta}, b_i, \boldsymbol{\theta}_w) = \mu_{ij}(\boldsymbol{\beta}, b_i)$ and there is no $\boldsymbol{\theta}_w$ parameter. In the second case, we allow for conditional overdispersion in the form of $\mathbf{\Lambda}_i(\boldsymbol{\beta}, b_i, \boldsymbol{\theta}_w) = \phi\mu_{ij}(\boldsymbol{\beta}, b_i)$ where $\boldsymbol{\theta}_w = \phi$ while in the third case, over-dispersion in the form of a negative binomial model with $\mathbf{\Lambda}_i(\boldsymbol{\beta}, b_i, \boldsymbol{\theta}_w) = \mu_{ij}(\boldsymbol{\beta}, b_i)(1 + \alpha\mu_{ij}(\boldsymbol{\beta}, b_i))$ is considered with $\boldsymbol{\theta}_w = \alpha$. The conditional negative binomial model coincides with a conditional gamma-Poisson model which is obtained by assuming the conditional rates within each two-week interval are further distributed conditionally as a gamma random variable. This allows for a specific form of conditional overdispersion which may or may not make much sense in this setting. Assuming $b_i \sim$iid $N(0, \sigma_b^2)$, all three cases result in models that belong to the class of GLME models. However, the models in the first and third cases are based on a well-defined conditional distribution (Poisson in case 1 and negative binomial in case 3) which allows one to estimate the model parameters using maximum likelihood estimation. Under this same mixed-effects setting, other covariance structures could also be considered some of which when combined with the assumption of a conditional Poisson distribution yield a GNLME model generally not considered in most texts.

ADEMEX Data

Paniagua et. al. (2002) summarized results of the ADEMEX trial, a randomized multi-center trial of 965 Mexican patients designed to compare patient outcomes (e.g., survival, hospitalization, quality of life, etc.) among end-stage renal disease patients randomized to one of two dose levels of continuous ambulatory peritoneal dialysis (CAPD). While patient survival was the primary endpoint, there were a number of secondary and exploratory endpoints investigated as well. One such endpoint was the estimation and comparison of the decline in the glomerular filtration rate (GFR) among incident patients randomized to the standard versus high dose of dialysis. There were several challenges with this objective as described below. The essential features of this example are as follows:

- The experimental unit is an incident patient
- Response variables:
  - y_{ij} =glomerular filtration rate (GFR) of the kidney (ml/min) for i^{th} subject on the jth occasion
  - T_i =patient survival time in months
- One within-unit covariate:
  - t_{ij} =months after randomization
- Six between-unit covariates:
  - Treatment group (standard dose, high dose)

    $$a_{1i} = \begin{cases} 0, & \text{if control = standard dose} \\ 1, & \text{if intervention = high dose} \end{cases}$$
  - Gender

    $$a_{2i} = \begin{cases} 0, & \text{if male} \\ 1, & \text{if female} \end{cases}$$
  - Age at baseline

    a_{3i} =patient age
  - Presence or absence of diabetes at baseline

    $$a_{4i} = \begin{cases} 0, & \text{if non-diabetic} \\ 1, & \text{if diabetic} \end{cases}$$
  - Baseline value of albumin

    a_{5i} =serum albumin (g/dL)
  - Baseline value of normalized protein nitrogen appearance (nPNA)

    a_{6i} =nPNA (g/kg/day)

Figure 1.6 SS and PA mean GFR profiles among control patients randomized to the standard dose of dialysis.

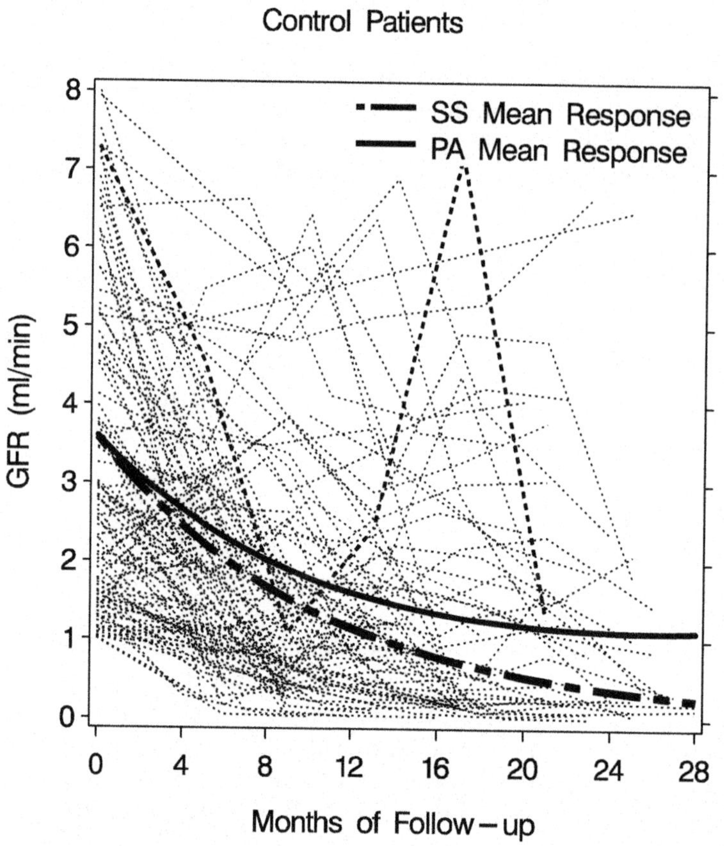

- Goal: Estimate the rate of decline in GFR and assess whether this rate differentially affects patient survival according to dose of dialysis
- Issues:
  1. Past studies have linked low GFR with increased mortality
  2. The analysis requires one to jointly model GFR and patient survival in order to determine if a) the rate of decline in GFR is associated with survival and b) if the rate of decline is affected by the dose of dialysis

As shown in Figure 1.6, the decline in GFR appears to occur more rapidly early on and then gradually reaches a value of 0 provided the patient lives long enough (i.e., the patient becomes completely anuric). Such data might be reasonably fit assuming a nonlinear exponential decay model with a random intercept and random decay rate. One such model is given by

$$\begin{aligned} y_{ij} &= f(\boldsymbol{x}'_{ij}, \boldsymbol{\beta}, \boldsymbol{b}_i) + \epsilon_{ij} \\ &= f(\boldsymbol{x}'_{ij}, \boldsymbol{\beta}_i) + \epsilon_{ij} \\ &= \beta_{i1}\exp(-\beta_{i2}t_{ij}) + \epsilon_{ij} \end{aligned} \tag{1.5}$$

$$\boldsymbol{\beta}_i = \begin{pmatrix} \beta_{i1} \\ \beta_{i2} \end{pmatrix} = \begin{pmatrix} \boldsymbol{a}'_i\boldsymbol{\beta}_1 \\ \boldsymbol{a}'_i\boldsymbol{\beta}_2 \end{pmatrix} + \begin{pmatrix} b_{i1} \\ b_{i2} \end{pmatrix} = \boldsymbol{A}_i\boldsymbol{\beta} + \boldsymbol{b}_i$$

where $\boldsymbol{a}_i' = (1 \quad a_{1i} \quad a_{2i} \quad a_{3i} \quad a_{4i} \quad a_{5i} \quad a_{6i})$ is the between-subject design vector, $\boldsymbol{x}_{ij}' = (t_{ij}, \boldsymbol{a}_i')$ is the vector of within- and between-subject covariates on the jth occasion,

$$\begin{aligned} \boldsymbol{a}_i'\boldsymbol{\beta}_1 + b_{i1} &= \beta_{01} + \beta_{11}a_{1i} + \beta_{21}a_{2i} + \beta_{31}a_{3i} + \beta_{41}a_{4i} + \beta_{51}a_{5i} + \beta_{61}a_{6i} + b_{i1} \\ \boldsymbol{a}_i'\boldsymbol{\beta}_2 + b_{i2} &= \beta_{02} + \beta_{12}a_{1i} + \beta_{22}a_{2i} + \beta_{32}a_{3i} + \beta_{42}a_{4i} + \beta_{52}a_{5i} + \beta_{62}a_{6i} + b_{i2} \end{aligned}$$

are linear predictors for the intercept (β_{i1}) and decay rate (β_{i2}) parameters, respectively, $\boldsymbol{b}_i' = (b_{i1} \quad b_{i2})$ are the random intercept and decay rate effects, and $\boldsymbol{\beta}' = (\boldsymbol{\beta}_1', \boldsymbol{\beta}_2')$ is the corresponding population parameter vector with $\boldsymbol{\beta}_1' = (\beta_{01} \quad \beta_{11} \quad \beta_{21} \quad \beta_{31} \quad \beta_{41} \quad \beta_{51} \quad \beta_{61})$ denoting the population intercept parameters and $\boldsymbol{\beta}_2' = (\beta_{02} \quad \beta_{12} \quad \beta_{22} \quad \beta_{32} \quad \beta_{42} \quad \beta_{52} \quad \beta_{62})$ the population decay rate parameters. Here $\boldsymbol{A}_i = \boldsymbol{I}_2 \otimes \boldsymbol{a}_i'$ is the between-subject design matrix linking the covariates $\boldsymbol{a}_i$ to $\boldsymbol{\beta}$ while $\epsilon_{ij} \sim$iid $N(0, \sigma_w^2)$ independent of $\boldsymbol{b}_i$.

Figure 1.6 reflects a reduced version of this model in which the subject-specific intercept and decay rate parameters are modeled as a simple linear function of the treatment group only, i.e.,

$$\begin{aligned} \beta_{i1} &= \beta_{01} + \beta_{11}a_{1i} + b_{i1} \\ \beta_{i2} &= \beta_{02} + \beta_{12}a_{1i} + b_{i2}. \end{aligned}$$

What is shown in Figure 1.6 is the estimated marginal mean profile (i.e., PA mean response) for the patients randomized to the control group. This PA mean response is obtained by averaging over the individual predicted response curves while the average patient's mean profile (i.e., the SS mean response) is obtained by simply plotting the predicted response curve achieved when one sets the random effects b_{i1} and b_{i2} equal to 0. As indicated previously (page 7), this example illustrates how one can have a random-effects exponential decay model that can predict certain individuals as having an increasing response over time. In this study, for example, two patients (one from each group) had a return of renal function while others showed either a modest rise or no change in renal function. The primary challenge here is to jointly model the decline in GFR and patient survival using a generalized nonlinear mixed-effects model that allows one to account for correlation in GFR values over time as well as determine if there is any association between the rate of decline in GFR over time and patient survival.

1.5 Summary features

We summarize here a number of features associated with the analysis of correlated data. First, since the response variables exhibit some degree of dependency as measured by correlation among the responses, most analyses may be classified as being essentially multivariate in nature. With repeated measurements and clustered data, for example, the analysis requires combining cross-sectional (between-cluster, between-subject, inter-subject) methods with time-series (within-cluster, within-subject, intra-subject) methods.

Second, the type of model one uses, marginal versus mixed, determines and/or limits the type of correlation structure one can model. In marginal models, correlation is accounted for by directly specifying an intra-subject or intra-cluster covariance structure. In mixed-effects models, a type of intraclass correlation structure is introduced through specification of subject-specific random effects. Specifically, intraclass correlation occurs as a result of having random effect variance components that are shared across measurements within subjects. Along with specifying what type of model, marginal or mixed, is needed for inferential purposes, one must also select what class of models is most appropriate based on the type of response variable being measured (e.g., continuous, ordinal, count or nominal) and its underlying mean structure. By specifying both the class of models and

Table 1.2 A summary of the different classes of models and the type of model within a class that are available for the analysis of correlated response data. [1.] The SAS macro %NLINMIX iteratively calls the MIXED procedure when fitting nonlinear models. [2.] NLMIXED can be adpated to analyze marginal correlation structures (see §4.4, §4.5.2, §4.5.3, §4.5.4, §5.3.3).

Class of Models	Types of Data	SAS PROC	Type of Model: Marginal[2.]	Type of Model: Mixed
Linear	Continuous	MIXED	REPEATED	RANDOM
		GLIMMIX	RANDOM/RSIDE	RANDOM
Generalized Linear	Continuous	GENMOD	REPEATED	–
	Ordinal	GLIMMIX	RANDOM/RSIDE	RANDOM
	Count	NLMIXED	–	RANDOM
	Binary			
Generalized Nonlinear	Continuous	NLMIXED	–	RANDOM
	Ordinal	%NLINMIX[1.]	REPEATED	RANDOM
	Count			
	Binary			

type of model within a class, an appropriate SAS procedure can then be used to analyze the data. Summarized in Table 1.2 are the three major classes of models used to analyze correlated response data and the SAS procedure(s) and corresponding procedural statements/options for conducting such an analysis.

Another feature worth noting is that studies involving clustered data or repeated measurements generally lead to efficient within-unit comparisons but inefficient between-unit comparisons. This is evident with split-plot designs where we have efficient split-plot comparisons but inefficient whole plot comparisons. In summary, some of the features we have discussed in this chapter include:

- Repeated measurements, clustered data and spatial data are essentially multivariate in nature due to the correlated outcome measures
- Analysis of repeated measurements and longitudinal data often requires combining cross-sectional methods with time-series methods
- The analysis of correlated data, especially that of repeated measurements, clustered data and spatial data can be based either on marginal or mixed-effects models. Parameters from these two types of models generally differ in that
  1) Marginal models target population-averaged (PA) inference
  2) Mixed-effects models target subject-specific (SS) inference
- Marginal models can accommodate correlation via
  1) Direct specification of an intra-subject correlation structure
- Mixed-effects models can accommodate correlation via
  1) Direct specification of an intra-subject correlation structure
  2) Intraclass correlation resulting from random-effects variance components that are shared across observations within subjects
- The family of models available within SAS include linear, generalized linear, nonlinear, and generalized nonlinear models
- Repeated measurements and clustered data lead to efficient within-unit comparisons (including interactions of within-unit and between-unit covariates) but inefficient between-unit comparisons.

Part I

Linear Models

Marginal Linear Models—Normal Theory

2

In this chapter and the next, we examine various types of linear models used for the analysis of correlated response data assuming the data are normally distributed or at least approximately so. These two chapters are by no means intended to provide a comprehensive treatment of linear models for correlated response data. Indeed such texts as those by Verbeke and Molenberghs (1997), Davis (2002), Demidenko (2004), and Littell et. al. (2006) provide a far more in-depth treatment of such models, especially of linear mixed-effects models. Rather, our goals are to 1) provide a basic introduction to linear models; 2) provide several examples illustrating both the basic and more advanced capabilities within SAS for analyzing linear models, and 3) provide a general modeling framework that can be extended to the generalized linear and nonlinear model settings.

2.1 The marginal linear model (LM)

As noted in Chapter 1, marginal or population-averaged (PA) models are defined by parameters that describe what the average response in the population is and by a variance-covariance structure that is directly specified in terms of the marginal second-order moments. This is true whether the data are correlated or not. To set the stage for linear models for correlated response data, we first consider the univariate general linear model.

Linear models – univariate data

For univariate data consisting of a response variable y measured across n independent experimental units and related to a set of s explanatory variables, $x_1, x_2, \ldots, x_s$, the usual general linear model may be written as

$$y_i = \beta_0 + \beta_1 x_{i1} + \ldots + \beta_s x_{is} + \epsilon_i,\ i = 1, \ldots, n$$

with $\epsilon_i \sim$iid $N(0, \sigma^2)$. One can write this more compactly using matrix notation (see Appendix A) as

$$\boldsymbol{y} = \boldsymbol{X}\boldsymbol{\beta} + \boldsymbol{\epsilon},\ \boldsymbol{\epsilon} \sim N(\boldsymbol{0}, \sigma^2 \boldsymbol{I}_n)$$

where

$$\boldsymbol{y} = \begin{pmatrix} y_1 \\ y_2 \\ \vdots \\ y_n \end{pmatrix}, \boldsymbol{X} = \begin{pmatrix} 1 & x_{11} & \dots & x_{1s} \\ 1 & x_{21} & \dots & x_{2s} \\ \vdots & \vdots & \vdots & \vdots \\ 1 & x_{n1} & \dots & x_{ns} \end{pmatrix}, \boldsymbol{\beta} = \begin{pmatrix} \beta_0 \\ \beta_1 \\ \vdots \\ \beta_s \end{pmatrix}, \boldsymbol{\epsilon} = \begin{pmatrix} \epsilon_1 \\ \epsilon_2 \\ \vdots \\ \epsilon_n \end{pmatrix}$$

and $\boldsymbol{I}_n$ is the $n \times n$ identity matrix. Estimation under such models is usually carried out using ordinary least squares (OLS) which may be implemented in SAS using the ANOVA, GLM, or REG procedures.

Linear models – correlated response data

Following the same basic scheme, the marginal linear model (LM) for correlated response data may be written as

$$\boldsymbol{y}_i = \boldsymbol{X}_i\boldsymbol{\beta} + \boldsymbol{\epsilon}_i, \quad i = 1, \dots, n \text{ units} \tag{2.1}$$

where

$\boldsymbol{y}_i$ is a $p_i \times 1$ response vector, $\boldsymbol{y}_i = (y_{i1} \dots y_{ip_i})'$, on the ith unit
$\boldsymbol{X}_i$ is a $p_i \times s$ design matrix of within- and between-unit covariates,
$\boldsymbol{\beta}$ is an $s \times 1$ unknown parameter vector,
$\boldsymbol{\epsilon}_i$ a $p_i \times 1$ random error vector with $\boldsymbol{\epsilon}_i \sim \text{ind } N_{p_i}(\mathbf{0}, \boldsymbol{\Sigma}_i(\boldsymbol{\theta}))$, $i = 1, \dots, n$

and $\boldsymbol{\theta}$ is a $d \times 1$ vector of variance-covariance parameters that defines the variability and correlation across and within individual units. Here we differentiate experimental units, subjects or clusters by indexing them with the subscript i. Essentially, model (2.1) mimics the structure of the univariate general linear model except that we now think of a response vector, $\boldsymbol{y}_i$, a matrix of explanatory variables, $\boldsymbol{X}_i$, and a vector of within-unit errors, $\boldsymbol{\epsilon}_i$ for each of n independent experimental units. The key difference is that the residual vectors $\boldsymbol{\epsilon}_i$ are assumed to be independent across units but within-unit residuals may be correlated, i.e., $Cov(\epsilon_{ij}, \epsilon_{ij'}) \neq 0$ for $j \neq j'$. If we set

$$\boldsymbol{y} = \begin{pmatrix} \boldsymbol{y}_1 \\ \boldsymbol{y}_2 \\ \vdots \\ \boldsymbol{y}_n \end{pmatrix}, \boldsymbol{X} = \begin{pmatrix} \boldsymbol{X}_1 \\ \boldsymbol{X}_2 \\ \vdots \\ \boldsymbol{X}_n \end{pmatrix}, \boldsymbol{\epsilon} = \begin{pmatrix} \boldsymbol{\epsilon}_1 \\ \boldsymbol{\epsilon}_2 \\ \vdots \\ \boldsymbol{\epsilon}_n \end{pmatrix}, \boldsymbol{\Sigma}(\boldsymbol{\theta}) = \begin{pmatrix} \boldsymbol{\Sigma}_1(\boldsymbol{\theta}) & \mathbf{0} & \dots & \mathbf{0} \\ \mathbf{0} & \boldsymbol{\Sigma}_2(\boldsymbol{\theta}) & \dots & \mathbf{0} \\ \vdots & \vdots & \vdots & \vdots \\ \mathbf{0} & \mathbf{0} & \dots & \boldsymbol{\Sigma}_n(\boldsymbol{\theta}) \end{pmatrix}$$

then we can write model (2.1) more succinctly as

$$\boldsymbol{y} = \boldsymbol{X}\boldsymbol{\beta} + \boldsymbol{\epsilon}, \ \boldsymbol{\epsilon} \sim N_N(\mathbf{0}, \boldsymbol{\Sigma}(\boldsymbol{\theta}))$$

where $\boldsymbol{y}$ and $\boldsymbol{\epsilon}$ are $N \times 1$ vectors; $\boldsymbol{\Sigma}(\boldsymbol{\theta})$ is an $N \times N$ positive definite matrix, and $N = \sum_{i=1}^{n} p_i$ represents the total number of observations.

Model (2.1) is fairly general and offers users a wide variety of options for analyzing correlated data including such familiar techniques as

- Generalized multivariate analysis of variance (GMANOVA)
- Repeated measures analysis of variance (RM-ANOVA)
- Repeated measures analysis of covariance (RM-ANCOVA)
- Growth curve analysis
- Response surface regression

Table 2.1 Some marginal variance-covariance structures available in SAS

Type	Structure: $(k,l)^{\text{th}}$ Element	$\boldsymbol{\theta}$
Unstructured ($p_i = p$)	$\boldsymbol{\Sigma}_i(\boldsymbol{\theta}) = \boldsymbol{\Sigma} = (\sigma_{kl});\ k,l = 1,\ldots p$	σ_{kl}
Compound Symmetry	$\boldsymbol{\Sigma}_i(\boldsymbol{\theta}) = \sigma^2(1-\rho)\boldsymbol{I}_{p_i} + \sigma^2\rho\boldsymbol{J}_{p_i}$	σ^2, ρ
Discrete time AR(1)	$\boldsymbol{\Sigma}_i(\boldsymbol{\theta}) = \sigma^2(\rho_{kl}),\ \rho_{kl} = \rho^{\lvert k-l\rvert}$	σ^2, ρ
Spatial (d_{kl})	$\boldsymbol{\Sigma}_i(\boldsymbol{\theta}) = \sigma^2(\rho_{kl}),\ \rho_{kl} = \exp\{-\rho d_{kl}\}$ where d_{kl} =Euclidian distance	σ^2, ρ
Banded Toeplitz (h+1)	$\boldsymbol{\Sigma}_i(\boldsymbol{\theta}) = \sigma^2(\rho_{kl}),\ \rho_{kl} = \begin{cases}\rho_m & m \le h \\ 0 & m > h\end{cases}$ $m = \lvert k-l\rvert,\ \rho_0 = 1,\ \boldsymbol{\rho} = (\rho_1 \ldots \rho_h)'$	$\sigma^2, \boldsymbol{\rho}$
Linear Covariances (dimension= h)	$\boldsymbol{\Sigma}_i(\boldsymbol{\theta}) = \theta_1\boldsymbol{H}_{i1} + \ldots + \theta_h\boldsymbol{H}_{ih}$ for known matrices, $\boldsymbol{H}_{i1},\ldots,\boldsymbol{H}_{ih}$ and unknown parameter vector $\boldsymbol{\theta} = (\theta_1 \ldots \theta_h)'$	$\boldsymbol{\theta}$

associated with data from repeated measures designs, crossover designs, longitudinal studies, group randomized trials, etc. Depending on the approach one takes and the covariance structure one assumes, the marginal LM (2.1) can be fit using a least squares approach (the GLM procedure with a REPEATED statement), a generalized estimating equations (GEE) approach (the GENMOD procedure) or a likelihood approach (the MIXED procedure).

There are a number of possible variance-covariance structures one can specify under (2.1) including some of the more common ones shown in Table 2.1. The MIXED procedure offers users the greatest flexibility in terms of specifying a suitable variance-covariance structure. For instance, one can choose from 23 different specifications suitable for applications involving repeated measurements, longitudinal data and clustered data while 13 different spatial covariance structures are available for applications involving spatially correlated data. A complete listing of the available covariance structures in SAS 9.2 are available in the MIXED documentation. Included are three specifications (UN@AR(1), UN@CS, UN@UN) ideally suited for applications involving multivariate repeated measurements.

2.1.1 Estimation

Estimation under the marginal LM (2.1) can be carried out using a generalized least squares (GLS) approach, a generalized estimating equations (GEE) approach or a likelihood approach, which can be either maximum likelihood (ML) or restricted maximum likelihood (REML). Under normality assumptions, all three approaches are closely related. In this and the following section, we briefly summarize methods of estimation and inference under the marginal linear model (2.1). A more comprehensive treatment of the theory of estimation and inference for marginal linear models can be found in such classical texts as that of Searle (1971, 1987), Rao (1973) and Graybill (1976) as well as recent texts by Vonesh and Chinchilli (1997), Timm (2002), Muller and Stewart (2006) and Kim and Timm (2007).

Generalized Least Squares (GLS):

Generalized least squares is an extension of least squares in that it seeks to minimize, with respect to regression parameters $\boldsymbol{\beta}$, an objective function representing a weighted sum of squared distances between observations and their mean. Consequently, the resulting GLS estimator of $\boldsymbol{\beta}$ may be classified as a least mean distance estimator. Unlike ordinary least squares (OLS), GLS accommodates heterogeneity and correlation via weighted least squares where the weights correspond to the inverse of the variance-covariance matrix.

Estimation must be considered under two cases: 1) when the vector of variance-covariance parameters, $\boldsymbol{\theta}$, is known and 2) when $\boldsymbol{\theta}$ is unknown.

Case 1 $\boldsymbol{\theta}$ *known:*

Minimize the generalized least squares (GLS) objective function:

$$Q_{\text{GLS}}(\boldsymbol{\beta}|\boldsymbol{\theta}) = \sum_{i=1}^{n} (\boldsymbol{y}_i - \boldsymbol{X}_i\boldsymbol{\beta})'\boldsymbol{\Sigma}_i(\boldsymbol{\theta})^{-1}(\boldsymbol{y}_i - \boldsymbol{X}_i\boldsymbol{\beta}) \tag{2.2}$$

where $\boldsymbol{\Sigma}_i(\boldsymbol{\theta})= Var(\boldsymbol{y}_i)$. The GLS estimator $\widehat{\boldsymbol{\beta}}(\boldsymbol{\theta})$ is the solution to the "normal" estimating equations:

$$U_{\text{GLS}}(\boldsymbol{\beta}|\boldsymbol{\theta}) = \sum_{i=1}^{n} \boldsymbol{X}_i'\boldsymbol{\Sigma}_i(\boldsymbol{\theta})^{-1}(\boldsymbol{y}_i - \boldsymbol{X}_i\boldsymbol{\beta}) = \boldsymbol{0} \tag{2.3}$$

obtained by differentiating (2.2) with respect to $\boldsymbol{\beta}$, setting the result to $\boldsymbol{0}$, and solving for $\boldsymbol{\beta}$. The resulting solution to $U_{\text{GLS}}(\boldsymbol{\beta}|\boldsymbol{\theta}) = \boldsymbol{0}$ is

$$\widehat{\boldsymbol{\beta}}_{\text{GLS}} = \widehat{\boldsymbol{\beta}}(\boldsymbol{\theta}) = \Big(\sum_{i=1}^{n} \boldsymbol{X}_i'\boldsymbol{\Sigma}_i(\boldsymbol{\theta})^{-1}\boldsymbol{X}_i\Big)^{-1} \sum_{i=1}^{n} \boldsymbol{X}_i'\boldsymbol{\Sigma}_i(\boldsymbol{\theta})^{-1}\boldsymbol{y}_i \tag{2.4}$$

and the model-based variance-covariance matrix of $\widehat{\boldsymbol{\beta}}(\boldsymbol{\theta})$ is

$$Var(\widehat{\boldsymbol{\beta}}_{\text{GLS}}) = \boldsymbol{\Omega}(\widehat{\boldsymbol{\beta}}_{\text{GLS}}) = \boldsymbol{\Omega}(\widehat{\boldsymbol{\beta}}(\boldsymbol{\theta})) = \Big(\sum_{i=1}^{n} \boldsymbol{X}_i'\boldsymbol{\Sigma}_i(\boldsymbol{\theta})^{-1}\boldsymbol{X}_i\Big)^{-1}. \tag{2.5}$$

One can obtain the GLS estimator (2.4) and its corresponding covariance matrix (2.5) within the MIXED procedure of SAS by specifying the variance-covariance parameters directly using the PARMS statement along with the HOLD= or EQCONS= option and/or the NOITER option (see SAS documentation for MIXED). Rather than basing inference on the model-based variance-covariance matrix (2.5), one can estimate $Var(\widehat{\boldsymbol{\beta}}(\boldsymbol{\theta}))$ using the robust "sandwich" estimator (also called the empirical "sandwich" estimator)

$$\begin{aligned} Var(\widehat{\boldsymbol{\beta}}_{\text{GLS}}) &= \boldsymbol{\Omega}_R(\widehat{\boldsymbol{\beta}}_{\text{GLS}}) \\ &= \boldsymbol{\Omega}_R(\widehat{\boldsymbol{\beta}}(\boldsymbol{\theta})) \\ &= \boldsymbol{\Omega}(\widehat{\boldsymbol{\beta}}_{\text{GLS}})\Big(\sum_{i=1}^{n} \boldsymbol{X}_i'\boldsymbol{\Sigma}_i(\boldsymbol{\theta})^{-1}\widehat{\boldsymbol{e}}_i\widehat{\boldsymbol{e}}_i'\boldsymbol{\Sigma}_i(\boldsymbol{\theta})^{-1}\boldsymbol{X}_i\Big)\boldsymbol{\Omega}(\widehat{\boldsymbol{\beta}}_{\text{GLS}}) \end{aligned} \tag{2.6}$$

where $\widehat{\boldsymbol{e}}_i = \boldsymbol{y}_i - \boldsymbol{X}_i\widehat{\boldsymbol{\beta}}(\boldsymbol{\theta})$ is the residual vector reflecting error between the observed value $\boldsymbol{y}_i$ and its predicted value, $\widehat{\boldsymbol{y}}_i = \boldsymbol{X}_i\widehat{\boldsymbol{\beta}}(\boldsymbol{\theta})$. Use of this robust estimator provides some protection against model misspecification with respect to the assumed variance-covariance structure of $\boldsymbol{y}_i$. The empirical sandwich estimator (2.6) is implemented whenever one specifies the EMPIRICAL option of the PROC MIXED statement.

Case 2 $\boldsymbol{\theta}$ *unknown:*

For some consistent estimate, $\widehat{\boldsymbol{\theta}}$ of $\boldsymbol{\theta}$, minimize the estimated generalized least squares (EGLS) objective function:

$$Q_{\text{EGLS}}(\boldsymbol{\beta}|\widehat{\boldsymbol{\theta}}) = \sum_{i=1}^{n} (\boldsymbol{y}_i - \boldsymbol{X}_i\boldsymbol{\beta})'\boldsymbol{\Sigma}_i(\widehat{\boldsymbol{\theta}})^{-1}(\boldsymbol{y}_i - \boldsymbol{X}_i\boldsymbol{\beta}) \tag{2.7}$$

where $\Sigma_i(\widehat{\theta}) = Var(y_i)$ is evaluated at the initial estimate $\widehat{\theta}$. Minimizing (2.7) with respect to β corresponds to solving the EGLS estimating equations:

$$U_{\text{EGLS}}(\beta|\widehat{\theta}) = \sum_{i=1}^{n} X_i'\Sigma_i(\widehat{\theta})^{-1}(y_i - X_i\beta) = 0.$$

The solution to $U_{\text{EGLS}}(\beta|\widehat{\theta}) = 0$ is the EGLS estimator

$$\widehat{\beta}_{\text{EGLS}} = \widehat{\beta}(\widehat{\theta}) = \Big(\sum_{i=1}^{n} X_i'\Sigma_i(\widehat{\theta})^{-1}X_i\Big)^{-1}\sum_{i=1}^{n} X_i'\Sigma_i(\widehat{\theta})^{-1}y_i \tag{2.8}$$

where we write $\widehat{\beta}_{\text{EGLS}} = \widehat{\beta}(\widehat{\theta})$ to highlight its dependence on $\widehat{\theta}$. The model-based variance-covariance matrix of $\widehat{\beta}_{\text{EGLS}}$ may be estimated as

$$Var(\widehat{\beta}_{\text{EGLS}}) \simeq \widehat{\Omega}(\widehat{\beta}_{\text{EGLS}}) = \widehat{\Omega}(\widehat{\beta}(\widehat{\theta})) = \Big(\sum_{i=1}^{n} X_i'\Sigma_i(\widehat{\theta})^{-1}X_i\Big)^{-1}. \tag{2.9}$$

As with the GLS estimator, one can also compute a robust sandwich estimate $\widehat{\Omega}_R(\widehat{\beta}_{\text{EGLS}})$ of the variance-covariance of $\widehat{\beta}_{\text{EGLS}}$ using (2.6) with $\Omega(\widehat{\beta}_{\text{GLS}}), \Sigma(\theta)$ and $\widehat{\beta}(\theta)$ replaced by $\widehat{\Omega}(\widehat{\beta}_{\text{EGLS}}), \Sigma_i(\widehat{\theta})$ and $\widehat{\beta}(\widehat{\theta})$, respectively.

Typically $\widehat{\theta} = \widehat{\theta}(\widehat{\beta}_0)$, where $\widehat{\beta}_0$ is an initial unbiased estimate of β (e.g., the OLS estimate $\widehat{\beta}_0 = \widehat{\beta}_{\text{OLS}}$) and $\widehat{\theta}(\widehat{\beta}_0)$ is a noniterative method of moments (MM) type estimator that is consistent for θ (i.e., $\widehat{\theta}(\widehat{\beta}_0)$ converges in probability to θ as $n \to \infty$). One can specify the components of $\widehat{\theta}$ within the PARMS statement of the MIXED procedure and use the HOLD= or EQCONS= option and/or the NOITER option to instruct SAS to obtain an EGLS estimator of θ based on the estimated value $\widehat{\theta}$. Alternatively, if one specifies the option METHOD=MIVQUE0 under PROC MIXED, SAS will compute the minimum variance quadratic unbiased estimator of the covariance parameters which is a noniterative method of moments type estimator of θ (e.g., Rao, 1972). SAS will then produce an EGLS estimate of β using this MINQUE0 estimate of θ.

Generalized Estimating Equations (GEE):

To remove any inefficiencies associated with using weights $\Sigma_i(\widehat{\theta})^{-1}$ that are based on an initial and perhaps imprecise estimate $\widehat{\theta} = \widehat{\theta}(\widehat{\beta}_0)$, an alternative approach to EGLS is to perform some sort of iteratively reweighted generalized least squares (IRGLS) where one updates $\widehat{\theta} = \widehat{\theta}(\widehat{\beta})$ based on a current estimate $\widehat{\beta}$ of β. Specifically, if for fixed β, $\widehat{\theta}(\beta)$ is any consistent estimator of θ, then one may improve on the EGLS procedure by simply updating $\Sigma_i(\widehat{\theta})$ using $\widehat{\theta} = \widehat{\theta}(\widehat{\beta})$ prior to estimating β using (2.8). This leads to the following iteratively reweighted generalized least squares (IRGLS) algorithm:

IRGLS Algorithm:

Step 1: Given $\widehat{\beta}^{(k)}$ apply method of moments (MM) or some other technique to obtain an updated estimate $\widehat{\theta}(\widehat{\beta}^{(k)})$ of θ

Step 2: Given $\widehat{\theta}(\widehat{\beta}^{(k)})$, apply the EGLS estimate (2.8) to update $\widehat{\beta}^{(k+1)}$.

The IRGLS procedure corresponds to solving the set of *generalized estimating equations* (GEE):

$$U_{\text{GEE}}(\beta,\widehat{\theta}(\beta)) = \sum_{i=1}^{n} X_i'\Sigma_i(\widehat{\theta}(\beta))^{-1}(y_i - X_i\beta) = 0 \tag{2.10}$$

where $\widehat{\theta}(\beta)$ is any consistent estimator of θ given β. Let $\widehat{\theta}_{\text{GEE}} = \widehat{\theta}(\beta)$ denote the estimator of θ one chooses when updating $\Sigma_i(\widehat{\theta}(\beta))$ within the GEE (2.10). At the point of convergence, the GEE estimator $\widehat{\beta}_{\text{GEE}} = \widehat{\beta}(\widehat{\theta}_{\text{GEE}})$ will equal the EGLS estimator (2.8) evaluated at the final value of $\widehat{\theta}_{\text{GEE}}$. The variance-covariance matrix of $\widehat{\beta}_{\text{GEE}}$ may be estimated as $\widehat{\Omega}(\widehat{\beta}_{\text{GEE}}) = \widehat{\Omega}(\widehat{\beta}(\widehat{\theta}_{\text{GEE}}))$ where $\widehat{\Omega}(\widehat{\beta}(\widehat{\theta}_{\text{GEE}}))$ has the same form as that of the GLS estimator, namely (2.5), but evaluated at $\theta = \widehat{\theta}_{\text{GEE}}$. The robust or empirical "sandwich" estimate $\widehat{\Omega}_R(\widehat{\beta}_{\text{GEE}})$ would likewise be given by (2.6) but with $\Omega(\widehat{\beta}_{\text{GLS}}), \Sigma(\theta)$ and $\widehat{\beta}(\theta)$ now replaced by $\widehat{\Omega}(\widehat{\beta}_{\text{GEE}}), \Sigma_i(\widehat{\theta}_{\text{GEE}})$ and $\widehat{\beta}(\widehat{\theta}_{\text{GEE}})$, respectively. For a number of different marginal variance-covariance structures, one can obtain GEE-based estimates using the REPEATED statement in GENMOD or by using the RSIDE option of the RANDOM statement in the GLIMMIX procedure. In both procedures, inference can be made using either a model-based estimate of the variance-covariance of $\widehat{\beta}_{\text{GEE}}$ or a robust/empirical "sandwich" estimate (see Chapter 4, §4.2 for more details).

Maximum Likelihood (ML):

Under normality assumptions, the joint maximum likelihood estimate (MLE) of β and θ is obtained by maximizing the log-likelihood function

$$\begin{aligned} L(\beta,\theta;y) &= -\frac{1}{2}N\,\log(2\pi) - \frac{1}{2}\sum_{i=1}^{n}\left\{(y_i - X_i\beta)'\Sigma_i(\theta)^{-1}(y_i - X_i\beta)\right\} \qquad (2.11)\\ &\quad -\frac{1}{2}\sum_{i=1}^{n}\left\{\log|\Sigma_i(\theta)|\right\}\\ &= -\frac{1}{2}N\,\log(2\pi) - \frac{1}{2}\sum_{i=1}^{n}Q_{i,\text{GLS}}(\beta|\theta) - \frac{1}{2}\sum_{i=1}^{n}\left\{\log|\Sigma_i(\theta)|\right\}\\ &= -\frac{1}{2}N\,\log(2\pi) - \frac{1}{2}Q_{\text{GLS}}(\beta|\theta) - \frac{1}{2}\sum_{i=1}^{n}\left\{\log|\Sigma_i(\theta)|\right\} \end{aligned}$$

where $N = \sum_{i=1}^{n} p_i$, $Q_{i,\text{GLS}}(\beta|\theta) = (y_i - X_i\beta)'\Sigma_i(\theta)^{-1}(y_i - X_i\beta)$. It is apparent from (2.11) that for known θ, the MLE of β is equivalent to the GLS estimator (2.4) since minimizing (2.2) is equivalent to maximizing (2.11) when θ is fixed and known. Similarly, when θ is unknown and must be estimated jointly with β, it can be shown that the EGLS estimate (2.8) of β will be equivalent to the MLE of β provided $\widehat{\theta}$ in (2.8) is the MLE of θ. One can use this feature to profile β out of the log-likelihood function and maximize the profile log-likelihood,

$$L(\theta;y) = -\frac{1}{2}\left\{(N-s)\log(2\pi) + \sum_{i=1}^{n}\left(r_i'\Sigma_i(\theta)^{-1}r_i + \log\left|\Sigma_i(\theta)\right|\right)\right\} \qquad (2.12)$$

where $r_i = y_i - X_i\left(\sum_{k=1}^{n} X_k'\Sigma_k(\theta)^{-1}X_k\right)^{-1}\sum_{k=1}^{n} X_k'\Sigma_k(\theta)^{-1}y_k = y_i - X_i\widehat{\beta}(\theta)$. Estimation may be carried out using an EM algorithm, a Newton-Raphson algorithm or a Fisher scoring algorithm (e.g., Jennrich and Schluchter, 1986; Laird, Lange and Stram, 1987; Lindstrom and Bates, 1988). Within the SAS procedure MIXED, the Newton-Raphson algorithm is the default algorithm but one can also request the use of Fisher's scoring algorithm by specifying the SCORING= option under PROC MIXED. If we let $\widehat{\theta}_{\text{MLE}}$ denote the MLE of θ obtained by maximizing the profile log-likelihood,

$L(\boldsymbol{\theta};\boldsymbol{y})$, then the estimated variance-covariance matrix of the MLE of $\boldsymbol{\beta}$ is simply

$$Var(\widehat{\boldsymbol{\beta}}(\widehat{\boldsymbol{\theta}}_{\text{MLE}})) \simeq \widehat{\boldsymbol{\Omega}}(\widehat{\boldsymbol{\beta}}_{\text{MLE}}) = \widehat{\boldsymbol{\Omega}}(\widehat{\boldsymbol{\beta}}(\widehat{\boldsymbol{\theta}}_{\text{MLE}})) = \Big(\sum_{i=1}^{n} \boldsymbol{X}_i'\boldsymbol{\Sigma}_i(\widehat{\boldsymbol{\theta}}_{\text{MLE}})^{-1}\boldsymbol{X}_i\Big)^{-1}$$

where we denote the MLE of $\boldsymbol{\beta}$ by $\widehat{\boldsymbol{\beta}}_{\text{MLE}} = \widehat{\boldsymbol{\beta}}(\widehat{\boldsymbol{\theta}}_{\text{MLE}})$. Note that $\widehat{\boldsymbol{\beta}}(\widehat{\boldsymbol{\theta}}_{\text{MLE}})$ is simply the EGLS estimator (2.8) with $\widehat{\boldsymbol{\theta}}$ replaced by $\widehat{\boldsymbol{\theta}}_{\text{MLE}}$. Similarly, a robust sandwich estimator, $\widehat{\boldsymbol{\Omega}}_R(\widehat{\boldsymbol{\beta}}_{\text{MLE}})$, of $Var(\widehat{\boldsymbol{\beta}}(\widehat{\boldsymbol{\theta}}_{\text{MLE}}))$ would be given by (2.6) with $\boldsymbol{\theta}$ evaluated at $\widehat{\boldsymbol{\theta}}_{\text{MLE}}$.

Restricted Maximum Likelihood (REML):

The default method of estimation in the MIXED procedure of SAS is the restricted maximum likelihood (REML) method. In small samples, ML estimation generally leads to small-sample bias in the estimated variance components. Consider, for example, a simple random sample of observations, $y_1,\ldots,y_n$, which are independent and identically distributed as $N(\mu,\sigma^2)$. The MLE of μ is the sample mean $\bar{y} = \frac{1}{n}\sum_{i=1}^{n} y_i$ which is known to be an unbiased estimate of μ. However the MLE of σ^2 is $\frac{1}{n}\sum_{i=1}^{n}(y_i - \bar{y})^2$ which has expectation $\sigma^2(n-1)/n$ and hence underestimates the variance by a factor of $-\sigma^2/n$ in small samples (see, for example, Littell et. al., pp. 747-750, 2006). To minimize such bias, the method of restricted maximum likelihood was pioneered by Patterson and Thompson (1971). Briefly, their approach entails maximizing a reduced log-likelihood function obtained by transforming $\boldsymbol{y}$ to $\boldsymbol{y}^* = \boldsymbol{T}\boldsymbol{y}$ where the distribution of $\boldsymbol{y}^*$ is independent of $\boldsymbol{\beta}$. This approach adjusts for the loss in degrees of freedom due to estimating $\boldsymbol{\beta}$ and reduces the bias associated with ML estimation of $\boldsymbol{\theta}$. When we take the transformation $\boldsymbol{T}\boldsymbol{y} = (\boldsymbol{I} - \boldsymbol{X}(\boldsymbol{X}'\boldsymbol{X})^{-1}\boldsymbol{X}')\boldsymbol{y}$, it can be shown that the REML estimate of $\boldsymbol{\theta}$ is obtained by maximizing the restricted profile log-likelihood, $L_R(\boldsymbol{\theta};\boldsymbol{y})$

$$L_R(\boldsymbol{\theta};\boldsymbol{y}) = -\frac{1}{2}\Bigg\{(N-s)\log(2\pi) + \sum_{i=1}^{n}\Big(\boldsymbol{r}_i'\boldsymbol{\Sigma}_i(\boldsymbol{\theta})^{-1}\boldsymbol{r}_i + \log\big|\boldsymbol{\Sigma}_i(\boldsymbol{\theta})\big|\Big) \qquad (2.13)$$
$$+ \log\Big|\sum_{i=1}^{n}\boldsymbol{X}_i'\boldsymbol{\Sigma}_i(\boldsymbol{\theta})^{-1}\boldsymbol{X}_i\Big|\Bigg\}.$$

A REML-based estimate of $\boldsymbol{\beta}$ is then given by $\widehat{\boldsymbol{\beta}}_{\text{REML}} = \widehat{\boldsymbol{\beta}}(\widehat{\boldsymbol{\theta}}_{\text{REML}})$ where $\widehat{\boldsymbol{\beta}}(\widehat{\boldsymbol{\theta}}_{\text{REML}})$ is the EGLS estimator (2.8) evaluated at the REML estimate $\widehat{\boldsymbol{\theta}}_{\text{REML}}$. Continuing with our process of substitution, the model-based and robust sandwich estimators of the variance-covariance of $\widehat{\boldsymbol{\beta}}_{\text{REML}}$ may be written as $\widehat{\boldsymbol{\Omega}}(\widehat{\boldsymbol{\beta}}_{\text{REML}})$ and $\widehat{\boldsymbol{\Omega}}_R(\widehat{\boldsymbol{\beta}}_{\text{REML}})$,respectively, where $\widehat{\boldsymbol{\Omega}}(\widehat{\boldsymbol{\beta}}_{\text{REML}})$ is given by (2.5) and $\widehat{\boldsymbol{\Omega}}_R(\widehat{\boldsymbol{\beta}}_{\text{REML}})$ by (2.6) but with each evaluated at $\boldsymbol{\theta} = \widehat{\boldsymbol{\theta}}_{\text{REML}}$.

2.1.2 Inference and test statistics

Inference under the marginal LM (2.1) is usually based on large-sample theory since the exact small-sample distribution of the estimates of $\boldsymbol{\beta}$ and $\boldsymbol{\theta}$ are generally unknown although in certain cases (e.g., certain types of balanced growth curve data), exact distributional results are available.

Inference about Fixed Effects Parameters, $\boldsymbol{\beta}$

Under certain regularity conditions [most notably the assumption that the information matrix $\sum_{i=1}^{n}(\boldsymbol{X}_i'\boldsymbol{\Sigma}_i(\boldsymbol{\theta})^{-1}\boldsymbol{X}_i)$ converges to a nonsingular matrix and that $\boldsymbol{\Sigma}_i(\boldsymbol{\theta})$ is the correct covariance matrix], each of the estimation techniques presented in the preceding section will yield a consistent estimator $\widehat{\boldsymbol{\beta}}$ that is asymptotically normally distributed as

$$\sqrt{n}(\widehat{\boldsymbol{\beta}} - \boldsymbol{\beta}) \xrightarrow{d} N_s(\boldsymbol{0}, \boldsymbol{\Omega}_0(\boldsymbol{\beta}))$$

where $\xrightarrow{d}$ denotes convergence in distribution, $\boldsymbol{\Omega}_0(\boldsymbol{\beta}) = \lim_{n\to\infty}\{n\boldsymbol{I}_0(\boldsymbol{\beta})^{-1}\}$ and $\boldsymbol{I}_0(\boldsymbol{\beta}) = \sum_{i=1}^{n}(\boldsymbol{X}_i'\boldsymbol{\Sigma}_i(\boldsymbol{\theta})^{-1}\boldsymbol{X}_i)$ denotes Fisher's information matrix for $\boldsymbol{\beta}$ (e.g., Vonesh and Chinchilli, pp. 244-249; 1997). Thus valid large-sample tests of the general linear hypothesis

$$H_0 : \boldsymbol{L\beta} = \boldsymbol{0} \quad vs \qquad (2.14)$$
$$H_1 : \boldsymbol{L\beta} \neq \boldsymbol{0}$$

can be carried out using the Wald chi-square test

$$T^2(\widehat{\boldsymbol{\beta}}) = (\boldsymbol{L}\widehat{\boldsymbol{\beta}})'(\boldsymbol{L}\widehat{\boldsymbol{\Omega}}(\widehat{\boldsymbol{\beta}})\boldsymbol{L}')^{-1}(\boldsymbol{L}\widehat{\boldsymbol{\beta}}) \qquad (2.15)$$

where $\boldsymbol{L} = \boldsymbol{L}_{r\times s} = \begin{pmatrix} \boldsymbol{l}_1' \\ \boldsymbol{l}_2' \\ \vdots \\ \boldsymbol{l}_r' \end{pmatrix}$ is a $r \times s$ hypothesis or contrast matrix of rank $r \leq s$ and $\widehat{\boldsymbol{\Omega}}(\widehat{\boldsymbol{\beta}})$ is the model-based estimated variance-covariance matrix of $\widehat{\boldsymbol{\beta}}$. For $\boldsymbol{L\beta}$ to be testable, the row vectors of $\boldsymbol{L}$ must each satisfy the estimability requirement that $\boldsymbol{l}_k'\boldsymbol{\beta}$ be an estimable linear function ($k = 1,\ldots r$). A linear combination, $\boldsymbol{l}'\boldsymbol{\beta}$, is said to be an estimable linear function if $\boldsymbol{l}'\boldsymbol{\beta} = \boldsymbol{k}'E(\boldsymbol{y}) = \boldsymbol{k}'\boldsymbol{X\beta}$ for some vector of constants, $\boldsymbol{k}$ (Searle, 1971). Under these conditions, it can be shown that $T^2(\widehat{\boldsymbol{\beta}})\xrightarrow{d}\chi^2(r)$ as $n \to \infty$ implying one can conduct valid large-sample chi-square tests of hypotheses. When the rank of $\boldsymbol{L}$ is 1 such that $\boldsymbol{L\beta} = \boldsymbol{l}'\boldsymbol{\beta}$ is a single estimable linear function of $\boldsymbol{\beta}$, an asymptotically valid α-level confidence interval can be constructed as

$$\boldsymbol{l}'\widehat{\boldsymbol{\beta}} \pm z_{\alpha/2}\sqrt{\boldsymbol{l}'\widehat{\boldsymbol{\Omega}}(\widehat{\boldsymbol{\beta}})\boldsymbol{l}} \qquad (2.16)$$

where $z_{\alpha/2}$ is the $100(1-\alpha/2)$ percentile of the standard normal distribution.

A more robust approach to inference that is less sensitive to the assumed covariance structure of $\boldsymbol{y}_i$ can be achieved by simply replacing $\widehat{\boldsymbol{\Omega}}(\widehat{\boldsymbol{\beta}})$ in (2.15)-(2.16) with a robust sandwich estimator, $\widehat{\boldsymbol{\Omega}}_R(\widehat{\boldsymbol{\beta}})$. Specifically, it can be shown that, regardless of the assumed covariance structure of $\boldsymbol{y}_i$,

$$\sqrt{n}(\widehat{\boldsymbol{\beta}} - \boldsymbol{\beta})\xrightarrow{d}N_s(\boldsymbol{0}, \boldsymbol{\Omega}_1(\boldsymbol{\beta}))$$

where

$$\boldsymbol{\Omega}_1(\boldsymbol{\beta}) = \lim_{n\to\infty}\left\{n\boldsymbol{I}_0(\boldsymbol{\beta})^{-1}\Big(\sum_{i=1}^{n}(\boldsymbol{X}_i'\boldsymbol{\Sigma}_i(\boldsymbol{\theta})^{-1}Var(\boldsymbol{y}_i)\boldsymbol{\Sigma}_i(\boldsymbol{\theta})^{-1}\boldsymbol{X}_i\Big)\boldsymbol{I}_0(\boldsymbol{\beta})^{-1}\right\}$$

is the asymptotic variance-covariance of $\widehat{\boldsymbol{\beta}}(\widehat{\boldsymbol{\theta}})$ provided the limit exists. Since under mild regularity conditions $n\widehat{\boldsymbol{\Omega}}_R(\widehat{\boldsymbol{\beta}})$ converges in probability to $\boldsymbol{\Omega}_1(\boldsymbol{\beta})$ [denoted $n\widehat{\boldsymbol{\Omega}}_R(\widehat{\boldsymbol{\beta}})\xrightarrow{p}\boldsymbol{\Omega}_1(\boldsymbol{\beta})$], the Wald chi-square test based on a robust sandwich estimator will always lead to asymptotically valid inference (i.e., reasonable p-values and confidence limits) in large samples provided one has correctly specified the marginal means. However, there may be some loss in efficiency when using a robust sandwich estimator compared to the model-based estimator. Since $\boldsymbol{\Omega}_1(\boldsymbol{\beta}) = \boldsymbol{\Omega}_0(\boldsymbol{\beta})$ whenever $Var(\boldsymbol{y}_i) = \boldsymbol{\Sigma}_i(\boldsymbol{\theta})$, it follows that by measuring how "close" $\widehat{\boldsymbol{\Omega}}_R(\widehat{\boldsymbol{\beta}})$ is to $\widehat{\boldsymbol{\Omega}}(\widehat{\boldsymbol{\beta}})$, one may gain valuable insight into how appropriate an assumed covariance structure for $\boldsymbol{y}_i$ is. If these two matrices are reasonably "close," one may achieve greater efficiency using the model-based estimator. Vonesh et. al. (1996), and Vonesh and Chinchilli (1997) offer some practical tools for assessing how close

$\widehat{\boldsymbol{\Omega}}_R(\widehat{\boldsymbol{\beta}})$ is to $\widehat{\boldsymbol{\Omega}}(\widehat{\boldsymbol{\beta}})$ and an adaptation of these techniques is offered in the COVB(DETAILS) option of the MODEL statement in the GLIMMIX procedure.

With the SAS procedures MIXED and GLIMMIX, one can conduct inference in the form of statistical hypothesis testing using the large-sample Wald chi-square test (2.15) by specifying the CHISQ option of the MODEL statement. However, by default, both procedures use an approximate F-test based on the scaled test statistic

$$F(\widehat{\boldsymbol{\beta}}) = T^2(\widehat{\boldsymbol{\beta}})/r \tag{2.17}$$

to test the general linear hypothesis (2.14). Depending on the type of data (e.g., balanced and complete), and the assumed mean and covariance structure, the F-statistic (2.17) may have an exact F-distribution with r numerator degrees of freedom and a fixed known denominator degrees of freedom, say v_e. In most cases, however, the F-statistic (2.17) will only have an approximate F distribution with r numerator degrees of freedom and an approximate denominator degrees of freedom, $\widehat{v}_e$, that must either be specified or estimated using either the DDF or DDFM option of the MODEL statement. The DDF option allows users to specify their own denominator degrees of freedom specific to each of the fixed effects in the model while the DDFM option instructs what method SAS should use to estimate $\widehat{v}_e$. Choices under the DDFM option are

- DDFM=BETWITHIN
- DDFM=CONTAIN
- DDFM=KENWARDROGER<(FIRSTORDER)>
- DDFM=RESIDUAL
- DDFM=SATTERTHWAITE

For the marginal LM (2.1) in which the covariance structure is specified via the REPEATED statement in MIXED or the RSIDE option of the RANDOM statement in GLIMMIX, the default approach is to estimate $\widehat{v}_e$ using DDFM=BETWITHIN. With this option, the residual degrees of freedom, $N-\text{rank}(\boldsymbol{X})$, is divided into between-subject and within-subject degrees of freedom corresponding to whether the fixed-effect covariate is a between-subject or within-subject covariate. When the data are balanced and complete with $p_i = p$ and the subject-specific design matrices are all of full rank s (i.e., $\text{rank}(\boldsymbol{X}_i) = \text{rank}(\boldsymbol{X}) = s$), the residual degrees of freedom would be $\widehat{v}_e = np - s$. Suppose, in this case, one had one between-subject class variable X_1 at two levels and one within-subject class variable X_2 at four levels and the MIXED statements looked like:

Program 2.1

```
proc mixed;
  class subject X1 X2;
  model y = X1 X2 X1*X2 / ddfm=betwithin;
  repeated X2 / sub=subject type=cs;
run;
```

In this setting, $p = 4$, $s = 8$ (including the intercept term) and the denominator degrees of freedom for testing X_1, X_2 and $X_1 \times X_2$ would be $n-2$, $n4-8-(n-2)$, and $n4-8-(n-2)$, respectively, while the numerator degrees of freedom would be $1, 3$ and 3, respectively. Note, however, that if one specifies an unstructured covariance (TYPE=UN) matrix using the REPEATED statement in MIXED or the RSIDE option of the RANDOM statement in GLIMMIX, SAS will assign the between-subject degrees of freedom for all fixed-effects. Hence if we replaced `type=cs` with `type=un` above, the denominator degrees of freedom would be $n-2$ for all three tests. There are other options as well including the ANOVAF option of the MIXED procedure which we will discuss in the examples.

Large-sample confidence intervals of the form (2.16) are available in GLIMMIX but not in MIXED. By default, both procedures compute t-tests and corresponding confidence intervals for simple linear combinations. The form of the t-test is

$$t = \boldsymbol{l}'\widehat{\boldsymbol{\beta}}/\sqrt{\boldsymbol{l}'\widehat{\boldsymbol{\Omega}}(\widehat{\boldsymbol{\beta}})\boldsymbol{l}} \tag{2.18}$$

which is distributed approximately as a Student-t with degrees of freedom, $\widehat{v}_e$, estimated using the same DDFM calculations as used for the F-test approximations. A corresponding confidence interval is given by (2.16) but with $z_{\alpha/2}$ replaced by $t_{\alpha/2}(\widehat{v}_e)$, the $100(1-\alpha/2)$ percentile of the student-t distribution.

When comparing nested models with respect to the mean response, one can also choose to run the MIXED and GLIMMIX procedures twice in order to compute a likelihood ratio test (LRT). The LRT is achieved by running ML estimation under each of two models, a full model and a reduced model. Specifically, if one specifies full and reduced models as

$$\text{Model 1: } \boldsymbol{y}_i = \boldsymbol{X}_{1i}\boldsymbol{\beta}_1 + \boldsymbol{X}_{2i}\boldsymbol{\beta}_2 + \boldsymbol{\epsilon}_i$$
$$\text{Model 2: } \boldsymbol{y}_i = \boldsymbol{X}_{1i}\boldsymbol{\beta}_1 + \boldsymbol{\epsilon}_i$$

where $\boldsymbol{\epsilon}_i \sim$ind $N(\boldsymbol{0}, \boldsymbol{\Sigma}_i(\boldsymbol{\theta}))$, $\boldsymbol{\beta}_1$ and $\boldsymbol{\beta}_2$ are $s_1 \times 1$ and $s_2 \times 1$ parameter vectors associated with covariate matrices $\boldsymbol{X}_{1i}$ and $\boldsymbol{X}_{2i}$, then the usual likelihood ratio test for testing $\boldsymbol{\beta}_2 = \boldsymbol{0}$ is given by

$$2L_1(\widehat{\boldsymbol{\beta}}_1, \widehat{\boldsymbol{\beta}}_2, \widehat{\boldsymbol{\theta}}) - 2L_2(\widehat{\boldsymbol{\beta}}_1, \widehat{\boldsymbol{\theta}}) \tag{2.19}$$

which will be approximately distributed as chi-square with s_2 degrees of freedom. Here L_1 and L_2 are the log-likelihood functions given by (2.11) for the full and reduced models, respectively. When conducting a likelihood ratio test, one should avoid using REML when estimating the variance components, $\boldsymbol{\theta}$, since the REML estimates of $\boldsymbol{\theta}$ can be viewed as MLE's of $\boldsymbol{\theta}$ under two different transformations of the data (e.g., Littell et. al., 2006).

One can also use GENMOD to analyze the marginal LM (2.1) with certain limitations. First, GENMOD was developed to handle data that are not necessarily normally distributed nor linear in the parameters of interest. Specifically, it allows users to fit continuous, discrete or ordinal data from univariate distributions within the exponential family using a generalized linear model (GLIM) approach (see Chapter 4). When observations are independently distributed according to one of the univariate distributions available within GENMOD, the procedure employs an IRGLS procedure equivalent to Fisher's scoring algorithm for solving an appropriate set of log-likelihood estimating equations. Hence, in the univariate setting, GENMOD extends ML estimation to a broader class of distributions that includes, as a special case, the normal-theory univariate linear model. However, when observations are thought to be correlated such as with repeated measurements, GENMOD utilizes a GEE approach in which a "working" correlation matrix is introduced as a means of describing possible correlation between observations from the same subject or unit. For the marginal LM (2.1), this implies a variance-covariance structure of the form

$$\boldsymbol{\Sigma}_i(\boldsymbol{\theta}) = \sigma^2 \boldsymbol{R}(\boldsymbol{\alpha})$$

where σ^2 is the assumed common scale parameter from the exponential family assuming normality, $\boldsymbol{R}(\boldsymbol{\alpha})$ is a "working" correlation matrix with parameter vector $\boldsymbol{\alpha}$, and $\boldsymbol{\theta} = (\sigma^2, \boldsymbol{\alpha})$. Given this restriction, GENMOD allows users to specify one of five working correlation structures, first-order autoregressive (TYPE=AR(1)), compound-symmetry or exchangeable (TYPE=CS), m-dependent (TYPE=MDEP(number)), independent

(TYPE=IND), and a user-specified correlation (TYPE=USER(matrix)). Method of moments is used to estimate the variance parameter, σ^2, and the vector of correlation parameters, $\boldsymbol{\alpha}$. GEE is then used to estimate the fixed-effects regression parameters, $\boldsymbol{\beta}$. Large-sample inference is carried out using the Wald chi-square test (2.15) and z-based confidence intervals (2.16).

Inference about Variance-Covariance Parameters, θ

With respect to the vector of variance-covariance parameters $\boldsymbol{\theta}$, inference in the form of statistical hypothesis testing and confidence intervals is based almost exclusively on large-sample theory in conjunction with normal theory likelihood estimation. In particular, under the marginal LM (2.1), the ML or REML estimate of $\boldsymbol{\theta}$ will satisfy

$$\sqrt{n}(\widehat{\boldsymbol{\theta}} - \boldsymbol{\theta}) \xrightarrow{d} N_d\Big(\mathbf{0}, \boldsymbol{\Omega}_0(\boldsymbol{\theta})\Big)$$

where $\boldsymbol{\Omega}_0(\boldsymbol{\theta}) = \lim_{n\to\infty}\Big\{n\boldsymbol{I}_0(\boldsymbol{\theta})^{-1}\Big\}$, $\boldsymbol{I}_0(\boldsymbol{\theta})^{-1}$ is the inverse of Fisher's information matrix for $\boldsymbol{\theta}$, and d is the dimension of $\boldsymbol{\theta}$. For ML estimation, the explicit form of $\boldsymbol{I}_0(\boldsymbol{\theta})^{-1}$ is given by

$$\boldsymbol{I}_0(\boldsymbol{\theta})^{-1} = \Big(\sum_{i=1}^{n} \boldsymbol{E}_i(\boldsymbol{\theta})'\boldsymbol{V}_i(\boldsymbol{\theta})^{-1}\boldsymbol{E}_i(\boldsymbol{\theta})\Big)^{-1}$$

where $\boldsymbol{E}_i(\boldsymbol{\theta}) = \partial Vec(\boldsymbol{\Sigma}_i(\boldsymbol{\theta}))/\partial\boldsymbol{\theta}'$ and $\boldsymbol{V}_i(\boldsymbol{\theta}) = 2\boldsymbol{\Sigma}_i(\boldsymbol{\theta}) \otimes \boldsymbol{\Sigma}_i(\boldsymbol{\theta})$ ($\otimes$ is the direct or Kronecker product—see Appendix A).

Estimates of the standard errors of the elements of $\widehat{\boldsymbol{\theta}}$ may be obtained by simply evaluating the inverse of Fisher's information matrix at the final estimates, i.e., $\widehat{\boldsymbol{\Omega}}(\widehat{\boldsymbol{\theta}}) = \boldsymbol{I}_0(\widehat{\boldsymbol{\theta}})^{-1}$. Alternatively, one can use standard errors based on the inverse of the observed information matrix evaluated at $\widehat{\boldsymbol{\theta}}$; namely $\widehat{\boldsymbol{\Omega}}(\widehat{\boldsymbol{\theta}}) = -\boldsymbol{H}(\widehat{\boldsymbol{\theta}})^{-1}$ where $\boldsymbol{H}(\widehat{\boldsymbol{\theta}}) = \frac{\partial^2}{\partial\boldsymbol{\theta}\partial\boldsymbol{\theta}'} ll(\boldsymbol{\theta})\Big|_{\boldsymbol{\theta}=\widehat{\boldsymbol{\theta}}}$ is the Hessian matrix of the appropriate log-likelihood $ll(\boldsymbol{\theta})$. For ML estimation, $ll(\boldsymbol{\theta})$ is defined by the profile log-likelihood $L(\boldsymbol{\theta};\boldsymbol{y})$ while for REML estimation $ll(\boldsymbol{\theta})$ is defined by the restricted profile log-likelihood $L_R(\boldsymbol{\theta};\boldsymbol{y})$ (e.g., Jennrich and Schluchter, 1986; Wolfinger, Tobias, and Sall, 1994). By default, the MIXED and GLIMMIX procedures use different algorithms (depending on the model in GLIMMIX) but both procedures use the inverse of the observed information matrix, $-\boldsymbol{H}(\boldsymbol{\theta})^{-1}$, to conduct inference with respect to $\boldsymbol{\theta}$. If, however, one specifies the SCORING<=number> option of MIXED or GLIMMIX and convergence is reached while still in the scoring mode, then standard errors will be based on the inverse of Fisher's information matrix. It should be noted that in the SAS documentation for MIXED, the observed inverse information matrix is defined as being twice the inverse Hessian. This merely reflects the fact that the MIXED procedure minimizes the objective function $-2ll(\boldsymbol{\theta})$ rather than maximize the log-likelihood $ll(\boldsymbol{\theta})$ and hence the Hessian matrix referred to in the SAS documentation is $\boldsymbol{H}^*(\boldsymbol{\theta}) = \frac{\partial^2}{\partial\boldsymbol{\theta}\partial\boldsymbol{\theta}'}\{-2ll(\boldsymbol{\theta})\}$ which implies $-\boldsymbol{H}(\boldsymbol{\theta})^{-1} = 2\boldsymbol{H}^*(\boldsymbol{\theta})^{-1}$.

For ML or REML estimation, large sample Wald chi-square tests and z-based confidence intervals are available with the COVTEST option of PROC MIXED and the WALD option of the COVTEST statement in GLIMMIX. These tests and confidence intervals are analogous to those defined by (2.15)-(2.16) but with $\widehat{\boldsymbol{\theta}}$ and $\widehat{\boldsymbol{\Omega}}(\widehat{\boldsymbol{\theta}})$ replacing $\widehat{\boldsymbol{\beta}}$ and $\widehat{\boldsymbol{\Omega}}(\widehat{\boldsymbol{\beta}})$, respectively. Alternatively, one can implement a likelihood ratio test (LRT) provided the mean structure remains the same and the covariance structure one wishes to test is a special case of a more general covariance structure. One can perform this with the MIXED procedure by running it twice, once with the full covariance structure and once with the reduced covariance structure. Output from the two procedures can then be used to construct the appropriate LRT test. There are some restrictions one should be aware of such as when the hypothesized covariance structure of the reduced model contains parameters that are on the boundary space of the more general covariance structure (see SAS documentation for details). The COVTEST statement of GLIMMIX offers users greater flexibility in terms of providing likelihood-based inference with respect to the covariance parameters including safeguarding p-values from the kinds of boundary space problems one may encounter (see the GLIMMIX documentation and references therein). These issues are more likely to occur with the linear mixed-effects model (e.g., Chapter 3).

Model selection, goodness-of-fit and model diagnostics

Assuming normality, model selection and goodness-of-fit can be carried out using the likelihood ratio test (2.17) for nested models, and Akaike's information criterion (AIC) or Schwarz's Bayesian criterion (SBC) for non-nested models. In addition, one can use the SAS macro %GOF (see Appendix D) with select output from MIXED and GLIMMIX to obtain R-square and concordance correlation goodness-of-fit measures described by Vonesh and Chinchilli (1997) for assessing the model fit. As noted previously, the COVB(DETAILS) option of the MODEL statement of GLIMMIX also offers users tools for assessing how well an assumed covariance structure fits the data. These tools are based on measuring how close the empirical sandwich estimator $\widehat{\boldsymbol{\Omega}}_R(\widehat{\boldsymbol{\beta}})$ of the variance-covariance of $\widehat{\boldsymbol{\beta}}$ is to the model-based estimator, $\widehat{\boldsymbol{\Omega}}(\widehat{\boldsymbol{\beta}})$ [e.g., Vonesh et. al. (1996), and Vonesh and Chinchilli (1997)].

Both the MIXED and GLIMMIX procedures provide users with a number of options for assessing how influential observations are in terms of their impact on various parameter estimates. Through specification of the PLOTS option in both MIXED and GLIMMIX, one can examine various diagnostic plots based on marginal residuals which one can use to check normality assumptions and/or detect possible outliers. In addition, the INFLUENCE option of the MODEL statement in MIXED provides users with a number of influence and case deletion diagnostics. While time and space do not permit an exhaustive treatment of these options, they are described at length in the SAS documentation accompanying both procedures.

2.2 Examples

In this section we present several examples illustrating the various tools available in SAS for analyzing marginal LM's. We begin by considering the classic dental growth curve data taken from Potthoff and Roy (1964). We use this data to illustrate how one can conduct a simple repeated measures profile analysis for balanced repeated measurements and how one can extend that analysis to the generalized multivariate analysis of variance (GMANOVA) setting. We follow this with several more complex examples involving unbalanced longitudinal data in which different objectives require different modeling strategies. We conclude with a novel example of spatial data wherein the primary goal is to evaluate the overall reproducibility of a biomarker for prostate cancer by means of a concordance correlation coefficient.

2.2.1 Dental growth data

One of the earliest marginal models used to analyze correlated data, particularly balanced repeated measurements and longitudinal data, was the generalized multivariate analysis of variance (GMANOVA) model (Potthoff and Roy, 1964; Grizzle and Allen, 1969). The GMANOVA model is usually written in matrix notation as

$$\boldsymbol{Y} = \boldsymbol{A}\boldsymbol{\Gamma}\boldsymbol{X} + \boldsymbol{\epsilon} \qquad (2.20)$$

where

$\boldsymbol{Y}$ is an $n \times p$ response matrix, $[\boldsymbol{y}_1 \dots \boldsymbol{y}_n]'$
$\boldsymbol{A} = [\boldsymbol{a}_1 \dots \boldsymbol{a}_n]'$ is an $n \times q$ full rank between-subject design matrix,
$\boldsymbol{\Gamma}$ is a $q \times t$ unknown parameter matrix,
$\boldsymbol{X}$ is a $t \times p$ within-subject design matrix of full rank t $(\leq p)$,
$\boldsymbol{\epsilon} = [\boldsymbol{\epsilon}_1 \dots \boldsymbol{\epsilon}_n]'$ is an $n \times p$ random error matrix with $\boldsymbol{\epsilon}_i \sim iid\, N_p(0, \boldsymbol{\Sigma})$ where $\boldsymbol{\Sigma}$ is an unstructured $p \times p$ covariance matrix.

The GMANOVA model can also be written in terms of model (2.1) as:

$$\begin{aligned} \boldsymbol{y}_i &= \boldsymbol{X}_i\boldsymbol{\beta} + \boldsymbol{\epsilon}_i \\ &= \boldsymbol{X}'(\boldsymbol{I}_t \otimes \boldsymbol{a}_i')Vec(\boldsymbol{\Gamma}) + \boldsymbol{\epsilon}_{i,} \\ &= \boldsymbol{X}'(\boldsymbol{a}_i' \otimes \boldsymbol{I}_t)Vec(\boldsymbol{\Gamma}') + \boldsymbol{\epsilon}_i \end{aligned} \tag{2.21}$$

where either $\boldsymbol{X}_i = \boldsymbol{X}'(\boldsymbol{I}_t \otimes \boldsymbol{a}_i')$ with $\boldsymbol{\beta} = Vec(\boldsymbol{\Gamma})$ or, equivalently, $\boldsymbol{X}_i = \boldsymbol{X}'(\boldsymbol{a}_i' \otimes \boldsymbol{I}_t)$ with $\boldsymbol{\beta} = Vec(\boldsymbol{\Gamma}')$. In either case, $\boldsymbol{\beta}$ is an $s \times 1$ parameter vector $(s = qt)$ based on $\boldsymbol{\Gamma}$ and the $\boldsymbol{a}_i$ are the $q \times 1$ vectors of between-subject covariates. Here we see how $\boldsymbol{X}_i$ of model (2.1) combines both between-unit and within-unit covariates from the GMANOVA model (2.20).

To illustrate an analysis using the GMANOVA model, we consider the dental growth data of Potthoff and Roy (1964) in which orthodontic measurements of 16 boys and 11 girls were measured successively at ages 8, 10, 12 and 14 years. The data shown in Output 2.1 is from investigators at the University of North Carolina Dental School and consists of the distance (mm) from the center of the pituitary to the pteryomaxillary fissure.

Output 2.1: Dental Growth Data

		age			
sex	person	8	10	12	14
boys	1	26.0	25.0	29.0	31.0
	2	21.5	22.5	23.0	26.5
	3	23.0	22.5	24.0	27.5
	4	25.5	27.5	26.5	27.0
	5	20.0	23.5	22.5	26.0
	6	24.5	25.5	27.0	28.5
	7	22.0	22.0	24.5	26.5
	8	24.0	21.5	24.5	25.5
	9	23.0	20.5	31.0	26.0
	10	27.5	28.0	31.0	31.5
	11	23.0	23.0	23.5	25.0
	12	21.5	23.5	24.0	28.0
	13	17.0	24.5	26.0	29.5
	14	22.5	25.5	25.5	26.0
	15	23.0	24.5	26.0	30.0
	16	22.0	21.5	23.5	25.0
girls	1	21.0	20.0	21.5	23.0
	2	21.0	21.5	24.0	25.5
	3	20.5	24.0	24.5	26.0
	4	23.5	24.5	25.0	26.5
	5	21.5	23.0	22.5	23.5
	6	20.0	21.0	21.0	22.5

7	21.5	22.5	23.0	25.0
8	23.0	23.0	23.5	24.0
9	20.0	21.0	22.0	21.5
10	16.5	19.0	19.0	19.5
11	24.5	25.0	28.0	28.0

One might consider performing a simple repeated measures ANOVA assuming either a structured or unstructured covariance matrix. Since the GMANOVA model (2.20) includes, as a special case, the repeated measures MANOVA model with unstructured covariance, we first test whether there is an effect due to age, gender or age by gender interaction using a RM-MANOVA. This can be done using either the REPEATED statement in the GLM procedure or the REPEATED statements in MIXED or GLIMMIX.

A partial listing of the data (Appendix C, Dataset C.1) reveals a data structure compatible with the MIXED, GLIMMIX and GENMOD procedures, namely one in which the response variable is arranged vertically according to each child. To apply a MANOVA using GLM, we first transpose the data such that the four measurements are arranged horizontally for each child. The program statements that transpose the data as shown in Output 2.1 and perform the necessary MANOVA are shown below.

Program 2.2

```
data dental_data;
 input gender person age y;
 if gender=1 then sex='boys ';
             else sex='girls';
 _age_=age;
cards;
1 1   8 26
1 1  10 25
1 1  12 29
1 1  14 31
1 2   8 21.5
1 2  10 22.5
1 2  12 23
1 2  14 26.5
     :
proc sort data=dental_data out=example2_2_1;
 by sex person;
run;
proc transpose data=example2_2_1 out=dental prefix=y;
 by sex person;
 var y;
run;
proc report data=dental split = '|' nowindows spacing=1;
 column sex person ('age' y1 y2 y3 y4);
 define sex /group 'sex';
 define person /display 'person';
 define y1 /display '8';
 define y2 /display '10' ;
 define y3 /display '12';
 define y4 /display '14';
 format y1--y4 4.1;
run;quit;
```

```
proc glm data=dental;
 class sex;
 model y1 y2 y3 y4=sex/nouni;
 repeated age 4 (8 10 12 14);
 manova;
run;
```

The GLM REPEATED statement shown above defines a GMANOVA model in which the within-subject design matrix, $\boldsymbol{X}$, is simply the identity matrix $\boldsymbol{I}_4$ one would get by assigning a different indicator variable for each age at which measurements are taken. The corresponding output shown below is simply a traditional repeated measures MANOVA corresponding to the usual hypothesis tests of interest, namely that of no `sex`, `age` or `age*sex` effects. The `Class Level Information` shown in Output 2.2 describes the between-subject covariates while the `Repeated Measures Level Information` describes the within-subject covariates as defined by the SAS statement: `repeated age 4 (8 10 12 14);` shown above.

Output 2.2: Repeated Measures MANOVA of Dental Growth Data

```
   Class Level Information
 Class    Levels    Values
 sex           2    boys girls
 Number of Observations Read    27
 Number of Observations Used    27
  Repeated Measures Level Information
 Dependent Variable    y1    y2    y3    y4
       Level of age     8    10    12    14
   MANOVA Test Criteria and Exact F Statistics for the Hypothesis of no age Effect
                  H = Type III SSCP Matrix for age
                      E = Error SSCP Matrix

                        S=1 M=0.5 N=10.5

 Statistic                       Value    F Value    Num DF    Den DF            Pr > F
 Wilks' Lambda              0.19479424      31.69         3        23            <.0001
 Pillai's Trace             0.80520576      31.69         3        23            <.0001
 Hotelling-Lawley Trace     4.13362211      31.69         3        23            <.0001
 Roy's Greatest Root        4.13362211      31.69         3        23            <.0001
  MANOVA Test Criteria and Exact F Statistics for the Hypothesis of no age*sex Effect
                  H = Type III SSCP Matrix for age*sex
                      E = Error SSCP Matrix

                        S=1 M=0.5 N=10.5

 Statistic                       Value    F Value    Num DF    Den DF            Pr > F
 Wilks' Lambda              0.73988739       2.70         3        23            0.0696
 Pillai's Trace             0.26011261       2.70         3        23            0.0696
 Hotelling-Lawley Trace     0.35155702       2.70         3        23            0.0696
 Roy's Greatest Root        0.35155702       2.70         3        23            0.0696
 Source    DF    Type III SS    Mean Square    F Value    Pr > F
 sex        1    140.4648569    140.4648569       9.29    0.0054
 Error     25    377.9147727     15.1165909
 Source          DF    Type III SS    Mean Square    F Value    Pr > F       Adj Pr > F
                                                                          G - G     H - F

 age              3    209.4369739     69.8123246      35.35    <.0001    <.0001    <.0001
 age*sex          3     13.9925295      4.6641765       2.36    0.0781    0.0878    0.0781
 Error(age)      75    148.1278409      1.9750379
```

Following the class level information, the MANOVA statement specifies that three tests of hypotheses are to be carried out. The first two tests are MANOVA tests for comparing the effects of `age` and `age*sex` while the third test is an ANOVA test comparing the effects of the variable, `sex`. All three tests of hypotheses correspond to the general linear hypothesis

$$\begin{aligned} H_0 &: \boldsymbol{C\Gamma U} = \mathbf{0} \quad vs \\ H_1 &: \boldsymbol{C\Gamma U} \neq \mathbf{0} \end{aligned} \tag{2.22}$$

where $\boldsymbol{C}$ is a $c \times q$ matrix [rank($\boldsymbol{C}$) $= c \leq q$] that forms linear combinations of $\boldsymbol{\Gamma}$ associated with the between-subject fixed effects in $\boldsymbol{A}$ (in this example, sex) and $\boldsymbol{U}$ is a $p \times u$ matrix [rank($\boldsymbol{U}$) $= u \leq p$] that forms linear combinations of $\boldsymbol{\Gamma}$ associated with the within-subject fixed effects (in this example, age) in $\boldsymbol{X}$. Whenever the rank of the matrix $\boldsymbol{U}$ is one, the resulting test reduces to a simple analysis of variance test as in the case of testing the effect of sex whereas whenever the rank of $\boldsymbol{U}$ exceeds one, a MANOVA test is required as in the case for testing the effects of age and age*sex.

To illustrate, the full rank design matrices $\boldsymbol{A}$ and $\boldsymbol{X}$ of the GMANOVA model for this example could be

$$\boldsymbol{A}_{27\times 2} = \begin{pmatrix} 1 & 0 \\ \vdots & \vdots \\ 1 & 0 \\ 0 & 1 \\ \vdots & \vdots \\ 0 & 1 \end{pmatrix} = \begin{pmatrix} \mathbf{1}_{16\times 1} & \mathbf{0}_{16\times 1} \\ \mathbf{0}_{11\times 1} & \mathbf{1}_{11\times 1} \end{pmatrix} = \begin{pmatrix} \boldsymbol{a}_1' \\ \boldsymbol{a}_2' \\ \vdots \\ \boldsymbol{a}_{27}' \end{pmatrix}, \boldsymbol{X} = \boldsymbol{I}_4$$

where $\mathbf{1}_{n\times 1}$ is an $n \times 1$ vector of 1's, $\mathbf{0}_{n\times 1}$ is an $n \times 1$ vector of 0's, and $\boldsymbol{a}_i' = \begin{pmatrix} a_i & 1 - a_i \end{pmatrix}$ with $a_i = 1$ if male and $a_i = 0$ if female. The design matrix $\boldsymbol{X}$ is simply the 4×4 identity matrix accounting for all 4 age levels. The parameter matrix, $\boldsymbol{\Gamma}$, is

$$\boldsymbol{\Gamma}_{2\times 4} = \begin{pmatrix} \beta_{11} & \beta_{12} & \beta_{13} & \beta_{14} \\ \beta_{21} & \beta_{22} & \beta_{23} & \beta_{24} \end{pmatrix}$$

with

$$\boldsymbol{\mu}_1' = \begin{pmatrix} \beta_{11} & \beta_{12} & \beta_{13} & \beta_{14} \end{pmatrix}$$

representing the mean response vector among boys at ages 8, 10, 12 and 14 and

$$\boldsymbol{\mu}_2' = \begin{pmatrix} \beta_{21} & \beta_{22} & \beta_{23} & \beta_{24} \end{pmatrix}$$

representing the mean response vector among girls at the same ages.

Given this parameterization, the test for no gender effect would require $\boldsymbol{C} = \begin{pmatrix} 1 & -1 \end{pmatrix}$, and $\boldsymbol{U} = \mathbf{1}_{4\times 1}' = \begin{pmatrix} 1 & 1 & 1 & 1 \end{pmatrix}$. In this case and, more generally, for any test involving only between-subject fixed effects, the post-hypothesis matrix will have the form $\boldsymbol{U} = \mathbf{1}_{p\times 1}'$ (or more generally, $\boldsymbol{U} = \boldsymbol{u}_{p\times 1}'$) with rank=1 and the resulting MANOVA test would simply correspond to a standard univariate ANOVA F-test with numerator degrees of freedom, c =rank($\boldsymbol{C}$), and denominator degrees of freedom, $v_e = n-$rank($\boldsymbol{A}$). For testing no overall age effect, $\boldsymbol{C} = \mathbf{1}_{2\times 1}' = \begin{pmatrix} 1 & 1 \end{pmatrix}$, and $\boldsymbol{U}$ is the contrast matrix

$$\boldsymbol{U}_{3\times 4} = \begin{pmatrix} 1 & 0 & 0 & -1 \\ 0 & 1 & 0 & -1 \\ 0 & 0 & 1 & -1 \end{pmatrix}.$$

In general, for tests involving only within-subject effects, $\boldsymbol{C}$ will usually assume the form $\boldsymbol{C} = \mathbf{1}_{q\times 1}'$ while $\boldsymbol{U}$ would be an appropriately chosen full rank contrast matrix. Finally, for testing whether there is any age by sex interaction, one would use $\boldsymbol{C}$ defined by the test for no gender effect, and $\boldsymbol{U}$ defined by the test for no age effect.

For those tests that involve contrasts among any within-subject fixed effects (in our example, age and age*sex), the REPEATED statement of the GLM procedure produces four different multivariate test statistics (as shown in Output 2.2) associated with a MANOVA. These tests are the Wilk's Lambda, Pillai's Trace, Hotelling-Lawley Trace and Roy's Greatest Root tests. In general, none of these tests is uniformly most powerful and p-values will be based on an F-approximation to each test (e.g., Anderson, Chapter 8, 1984; Vonesh and Chinchilli, Chapter 2, 1997; Timm, Chapter 4, 2002; see also Muller and Stewart, 2006 and Kim and Timm, 2007). However, whenever $c = \text{rank}(\boldsymbol{C}) = 1$ or $u = \text{rank}(\boldsymbol{U}) = 1$ (as in this example), then these four test statistics will all be equivalent to an exact F-test which is uniformly most powerful and invariant under normality assumptions.

In addition to the four standard MANOVA tests comparing the within-subject fixed effects defined by `age` and `age*sex`, the REPEATED statement of the GLM procedure also instructs SAS to provide approximate univariate tests of these effects. These tests are based on the approximate F-tests of Greenhouse and Geisser (1959) and Huynh and Feldt (1970, 1976) and are shown at the bottom of Output 2.2.

As an alternative to the GLM procedure, one can use PROC MIXED to obtain p-values based on an F-test approximation to the multivariate Hotelling-Lawley Trace test statistic as well as p-values based on the univariate F test approximation of Greenhouse-Geisser. To accomplish this, one will need to use the original vertically arrayed dataset, `data=example2_2_1`. One can then request the MIXED procedure to compute one of two F-approximations to the multivariate Hotelling-Lawley Trace test statistic using the HLM and HLPS options of the REPEATED statement. Likewise, by specifying the ANOVAF option in MIXED, one will get p-values corresponding to the Greenhouse-Geisser univariate F test. The SAS program is as follows.

Program 2.3

```
proc sort data=example2_2_1;
 by sex person;
run;
proc mixed data=example2_2_1 method=reml ANOVAF scoring=200;
 class person sex age;
 model y = sex age sex*age ;
 repeated age / type=un subject=person(sex) hlm hlps r;
run;
```

The above call to MIXED fits the marginal LM (2.1) to the data assuming $\boldsymbol{X}_i\boldsymbol{\beta} = (\boldsymbol{I}_4 \otimes \boldsymbol{a}_i')\boldsymbol{\beta}$ where $\boldsymbol{a}_i' = \begin{pmatrix}1 & 1-a_i\end{pmatrix}$ and $\boldsymbol{\beta} = Vec(\boldsymbol{\Gamma})$ is the parameter vector obtained by stacking the four column vectors of $\boldsymbol{\Gamma}$ below one another. The general linear hypothesis corresponding to (2.22) is simply $H_0 : \boldsymbol{L\beta} = \boldsymbol{0}$ where $\boldsymbol{L} = \boldsymbol{U}' \otimes \boldsymbol{C}$ for each choice of $\boldsymbol{C}$ and $\boldsymbol{U}$ described above. The MANOVA results corresponding to the options HLM and HLPS are shown in Output 2.3.

Output 2.3: Selected output from PROC MIXED corresponding to GLM MANOVA output.

Estimated R Matrix for person(sex) 1 boys

Row	Col1	Col2	Col3	Col4
1	5.4155	2.7168	3.9102	2.7102
2	2.7168	4.1848	2.9272	3.3172
3	3.9102	2.9272	6.4557	4.1307
4	2.7102	3.3172	4.1307	4.9857

Type 3 Hotelling-Lawley-McKeon Statistics

Effect	Num DF	Den DF	F Value	Pr > F
age	3	23	31.69	<.0001
sex*age	3	23	2.70	0.0696

Type 3 Hotelling-Lawley-Pillai-Samson Statistics

Effect	Num DF	Den DF	F Value	Pr > F
age	3	23	31.69	<.0001
sex*age	3	23	2.70	0.0696

Included with the MANOVA test results shown above is the estimated variance-covariance matrix, $\Sigma(\widehat{\theta})$, corresponding to subject 1. SAS labels this as the R matrix consistent with its identification of the R-side covariance parameters coinciding with a specified intra-subject covariance structure (in this case, TYPE=UN yields an unstructured positive-definite covariance matrix). The MIXED procedure also computes approximate F-tests based on the default HTYPE=3 type of hypothesis test to perform on the fixed effects. In addition, when one specifies the option ANOVAF, approximate F-tests are presented which, for an unstructured covariance matrix in the repeated measures setting, are identical to a multivariate MANOVA where degrees of freedom are corrected with the Greenhouse-Geisser adjustment (Greenhouse and Geisser, 1959). The results of both the standard ANOVA F-tests and approximate Greenhouse-Geisser F-tests are shown in Output 2.4.

Output 2.4: Selected output from PROC MIXED corresponding to GLM repeated measures ANOVA F-tests.

Effect	Num DF	Den DF	F Value	Pr > F
sex	1	25	9.29	0.0054
age	3	25	34.45	<.0001
sex*age	3	25	2.93	0.0532

			ANOVA F		
Effect	Num DF	Den DF	F Value	Pr > F(DDF)	Pr > F(infty)
sex	1	25	9.29	0.0054	0.0023
age	2.6	65	35.35	<.0001	<.0001
sex*age	2.6	65	2.36	0.0878	0.0783

The results shown in Output 2.3 and Output 2.4 are identical with those of the GLM-based results shown in Output 2.2. The default type 3 tests of the fixed effects `sex`, `age` and `sex*age` are F-approximations using the default DFFM=BETWITHIN. If one were to replace the model statement with

```
model y = sex age sex*age / ddfm=kenwardroger;
```

then because the data are complete with only two groups, males and females, the default type 3 tests of the within-subject fixed effects (i.e. `age` and `sex*age`) would be equivalent to the exact Hotelling-Lawley based F-tests shown in Output 2.2 and 2.3. In fact, these tests can also be obtained using GLIMMIX as indicated by the program and output shown below.

Figure 2.1: Dental Growth Data

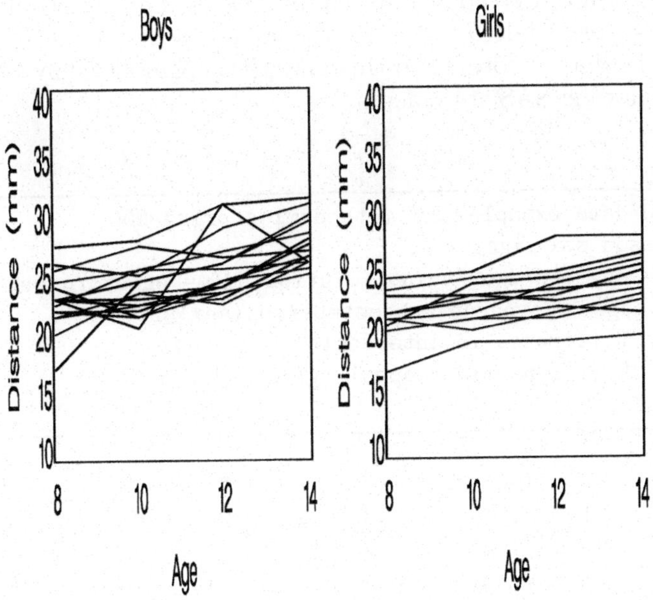

Program 2.4

```
proc glimmix data=example2_2_1 method=rmpl scoring=200;
 class person sex age;
 model y = sex age sex*age / ddfm=kenwardroger;
 random age / type=un subject=person(sex) rside;
run;
```

Output 2.5: GLIMMIX default output using DDFM=kenwardroger

Type III Tests of Fixed Effects

Effect	Num DF	Den DF	F Value	Pr > F
sex	1	25	9.29	0.0054
age	3	23	31.69	<.0001
sex*age	3	23	2.70	0.0696

A major advantage of using the marginal LM (2.1) in combination with the MIXED or GLIMMIX procedure is the flexibility with which one can extend the traditional RM-MANOVA to the more general RM-GMANOVA where one may be interested in testing for trends among repeated measurements. As shown in Figure 2.1, there appears to be a linear trend with increasing age. Suppose, then, we are interested in estimating a linear trend with age that is allowed to differ between boys and girls. We can accomplish this with the GMANOVA model by simply defining

$$\boldsymbol{A} = \begin{pmatrix} \boldsymbol{a}_1' \\ \boldsymbol{a}_2' \\ \vdots \\ \boldsymbol{a}_n' \end{pmatrix}, \boldsymbol{X}_{2\times 4} = \begin{pmatrix} 1 & 1 & 1 & 1 \\ 8 & 10 & 12 & 14 \end{pmatrix} \text{ and } \boldsymbol{\Gamma} = \begin{pmatrix} \beta_{10} & \beta_{11} \\ \beta_{20} & \beta_{21} \end{pmatrix}.$$

where we redefine $\boldsymbol{a}_i' = \begin{pmatrix} a_i & 1-a_i \end{pmatrix}$ so that $\boldsymbol{A}$ defines a cell means design matrix (i.e., one without an overall intercept). The parameters β_{10} and β_{11} would represent the intercept and slope for boys while β_{20} and β_{21} would represent the intercept and slope for girls. Analyzing the data according to this GMANOVA model may be accomplished in MIXED using the following SAS statements.

Program 2.5

```
proc mixed data=example2_2_1 method=ml scoring=200;
 class person sex _age_;
 model y = sex sex*age /noint solution ddfm=kenwardroger;
 repeated _age_ / type=un subject=person(sex) r;
 estimate 'Difference in intercepts' sex 1 -1;
 estimate 'Difference in slopes' age*sex 1 -1;
run;
```

Before we look at the corresponding output, there are several important programming aspects we need to keep in mind when running marginal linear models with PROC MIXED or PROC GLIMMIX. First, one should always sort by unique subjects. In this example, we have a subject variable, person, that is only unique within gender. Hence we should sort the data by sex and person using the statements:

```
proc sort data=example2_2_1;
 by sex person;
run;
```

before running PROC MIXED. Second, one must always define who the unique subjects are with the SUBJECT= option of the REPEATED statement. In the above SAS statements, we have used the specification '`subject = person(sex)`' so as to identify who the unique subjects are. If we had specified '`subject = person`' then the MIXED procedure would have treated person 1 as a unique subject regardless of gender. For example, if we replaced the above REPEATED statement with

```
repeated _age_ / type=un subject=person r;
```

we would have received the following warning in the SAS log:

```
NOTE: An infinite likelihood is assumed in iteration 0 because of a
nonpositive definite estimated R matrix for person 1.
NOTE: PROCEDURE MIXED used (Total process time):
      real time            0.01 seconds
      cpu time             0.01 seconds
```

Finally, while not absolutely necessary, it is advisable to always include a repeated measures or within-subject factor in both the CLASS and REPEATED statements that can be used to identify the proper location of non-missing repeated measurements (see SAS documentation). In our current example, we created a duplicate variable, `_age_=age`, and specified `_age_` within the CLASS and REPEATED statements so that repeat measurements would align correctly with the specifications of an unstructured covariance matrix.

Returning to our example, the GMANOVA model implemented with the above programming statements yields ML estimates of $\boldsymbol{\beta}$ and $\boldsymbol{\Sigma}$ as well as tests of hypotheses on the components of $\boldsymbol{\beta}$ (see Output 2.6). The MLE $\widehat{\boldsymbol{\Sigma}} = \boldsymbol{\Sigma}(\widehat{\boldsymbol{\theta}})$, shown in matrix form as the R matrix in Output 2.6 is the same as that obtained by Jennrich and Schluchter (1986). This same information is displayed in vector form under the output heading '`Covariance Parameter Estimates`' and is simply a rearrangement of the vector estimate, $\widehat{\boldsymbol{\theta}}$, which is the vector of unique components of $\widehat{\boldsymbol{\Sigma}}$, namely $\widehat{\boldsymbol{\theta}} = Vech(\widehat{\boldsymbol{\Sigma}})$. See Appendix A for a description of the $Vech$ and Vec matrix operators.

Output 2.6: Selected output from MIXED corresponding to a GMANOVA model incorporating linear trends with age

```
              Model Information
Data Set                         WORK.EXAMPLE2_2_1
Dependent Variable               y
Covariance Structure             Unstructured
Subject Effect                   person(sex)
Estimation Method                ML
Residual Variance Method         None
Fixed Effects SE Method          Kenward-Roger
Degrees of Freedom Method        Kenward-Roger
                 Class Level Information
Class     Levels   Values
person        16   1 2 3 4 5 6 7 8 9 10 11 12 13 14 15 16
sex            2   boys girls
_age_          4   8 10 12 14
Estimated R Matrix for person(sex) 1 boys
Row      Col1      Col2      Col3       Col4
  1    5.1191    2.4409    3.6105     2.5223
  2    2.4409    3.9280    2.7175     3.0623
  3    3.6105    2.7175    5.9798     3.8235
  4    2.5223    3.0623    3.8235     4.6180
  Covariance Parameter Estimates
Cov Parm   Subject          Estimate
UN(1,1)    person(sex)        5.1191
UN(2,1)    person(sex)        2.4409
UN(2,2)    person(sex)        3.9280
UN(3,1)    person(sex)        3.6105
UN(3,2)    person(sex)        2.7175
UN(3,3)    person(sex)        5.9798
UN(4,1)    person(sex)        2.5223
UN(4,2)    person(sex)        3.0623
UN(4,3)    person(sex)        3.8235
UN(4,4)    person(sex)        4.6180
         Fit Statistics
-2 Log Likelihood              419.5
AIC (smaller is better)        447.5
AICC (smaller is better)       452.0
BIC (smaller is better)        465.6
                    Solution for Fixed Effects
Effect     sex      Estimate   Standard Error   DF   t Value   Pr > |t|
sex        boys      15.8423           1.0025   27     15.80     <.0001
sex        girls     17.4254           1.2091   27     14.41     <.0001
age*sex    boys       0.8268          0.08477   27      9.75     <.0001
age*sex    girls      0.4764           0.1022   27      4.66     <.0001
      Type 3 Tests of Fixed Effects
Effect     Num DF   Den DF   F Value   Pr > F
sex             2       27    228.72   <.0001
age*sex         2       27     58.42   <.0001
                                 Estimates
Label                       Estimate   Standard Error   DF   t Value   Pr > |t|
Difference in intercepts     -1.5830           1.5706   27     -1.01     0.3225
Difference in slopes          0.3504           0.1328   27      2.64     0.0137
```

Figure 2.2: Individual Profiles of TBBMD Among 55 Women on Calcium Supplement Versus 57 on Placebo

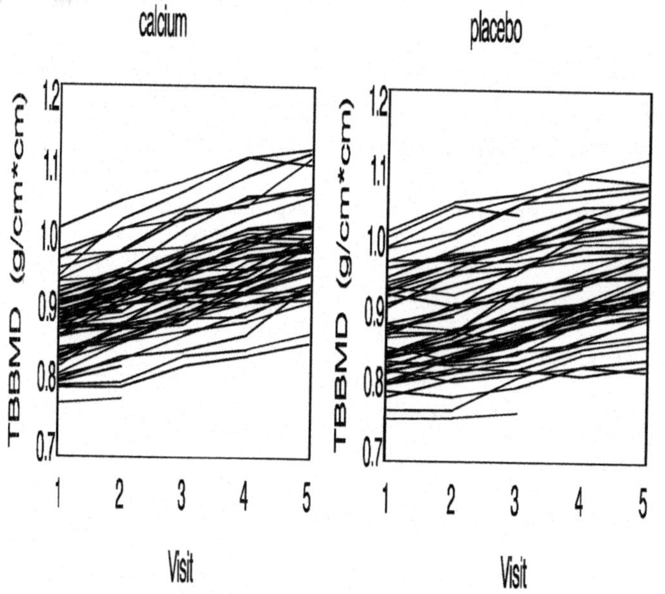

The NOINT option combined with the fixed-effects specifications SEX SEX*AGE instructs MIXED to use the cell means design matrix $\boldsymbol{A}$ resulting in separate intercept and slope estimates for boys and girls. Consequently, the default Type 3 tests of the fixed effects correspond to testing whether the intercepts are jointly equal to 0 and likewise whether the slopes are jointly equal to 0. The more interesting comparisons, of course, are whether the intercepts and slopes differ between boys and girls. This is accomplished using the two ESTIMATE statements. The resulting F-tests for comparing the intercepts and slopes between boys and girls are shown at the bottom of Output 2.6.

The MLE's of $\boldsymbol{\beta}$ and $\boldsymbol{\Sigma}$ calculated using PROC MIXED will be equivalent to the closed form MLE's derived by Khatri (1966) and Grizzle and Allen (1969) provided the data are balanced and complete. Indeed, the MLE's of $\boldsymbol{\beta}$ and $\boldsymbol{\Sigma}$ may be expressed in terms of fully weighted least squares estimates (e.g., Vonesh and Chinchilli, pp. 204-207, 1997). However, such estimates are only available with balanced and complete data. In the following example, we consider the GMANOVA model when the data are balanced but incomplete.

2.2.2 Bone mineral density data

Lloyd et. al. (1993) presented results of a randomized controlled trial in which 112 healthy adolescent women were randomized to receive a daily calcium supplement (500 mg calcium citrate malate) or placebo over the course of two years. Data were obtained in roughly 6-month intervals over two years including baseline (visit 1). The primary objective was to determine if administration of a daily calcium supplement would improve bone density growth as measured by total body bone mineral density or TBBMD (g/cm^2).

The data taken from Vonesh and Chinchilli (1997) and depicted in Figure 2.2 suggest there is an approximate linear increase in total body bone mineral density over the first two years of follow-up. Although the data are unbalanced in that measurements were not exactly six months apart, we follow Vonesh and Chinchilli (1997) and consider an analysis based on the GMANOVA model (2.21) treating the data as though it were balanced with respect to time. Specifically, the data were analyzed based on the following features and assumptions.

- Two-year randomized controlled trial designed to evaluate the effects of daily calcium supplement in healthy adolescent women.
- The experimental unit is a subject
- Response variable:
  - y_{ij} =total body bone mineral density, TBBMD (g/cm^2), for the ith subject on the jth occasion
- One within-unit covariate:
  - t_{ij} =time 0 (baseline), 0.5, 1.0, 1.5, 2.0 years (assuming balanced data)
- One between-unit covariate:
  - Treatment Group (C=calcium, P=placebo)

$$a_{1i} = \begin{cases} 0, & \text{if placebo (P)} \\ 1, & \text{if calcium (C)} \end{cases}$$

- Structural Model: $TBBMD = \beta_1 + \beta_2 \times t$
- Objective: Determine if daily calcium supplementation improves the rate (β_2) of bone gain during early adolescence

We first consider an analysis based on balanced and complete data as analyzed by Vonesh and Chinchilli (Chapter 5, pp. 228-233, 1997). The dataset as given in Appendix C (Dataset C.2) is arranged horizontally with respect to the dates and repeated measurements. To carry out a GMANOVA analysis using MIXED, we must first convert the data to the vertical format as required by the procedure. The SAS statements required to do this and perform a GMANOVA on the balanced data are as follows.

Program 2.6

```
data TBBMD;
 input (subject group date1 tbbmd1 date2 tbbmd2 date3 tbbmd3
        date4 tbbmd4 date5 tbbmd5)
       (3. @5 $char1. @7 date7. @15 5.3 @21 date7.
           @29 5.3 @35 date7. @43 5.3 @49 date7. @57 5.3
           @63 date7. @71 5.3);
 if tbbmd1^=. & tbbmd2^=. & tbbmd3^=. &
    tbbmd4^=. & tbbmd5^=. then data='Complete  ';
 else data='Incomplete';
 format date1 date2 date3 date4 date5 date7.;
cards;
101 C 01MAY90 0.815 05NOV90 0.875 24APR91 0.911 30OCT91 0.952 29APR92 0.970
102 P 01MAY90 0.813 05NOV90 0.833 15APR91 0.855 21OCT91 0.881 13APR92 0.901
103 P 02MAY90 0.812 05NOV90 0.812 17APR91 0.843 23OCT91 0.855 15APR92 0.895
104 C 09MAY90 0.804 12NOV90 0.847 29APR91 0.885 11NOV91 0.920 19JUN92 0.948
105 C 14MAY90 0.904 10DEC90 0.927 06MAY91 0.952 04NOV91 0.955 27APR92 1.002
106 P 15MAY90 0.831 26NOV90 0.855 20MAY91 0.890 17DEC91 0.908 15JUL92 0.933
107 P 23MAY90 0.777 05DEC90 0.803 15MAY91 0.817 27NOV91 0.809 04MAY92 0.823
108 C 23MAY90 0.792 03DEC90 0.814
109 C 23MAY90 0.821 03DEC90 0.850 06MAY91 0.865 05NOV91 0.879 29APR92 0.908
110 P 30MAY90 0.823 10DEC90 0.827 21MAY91 0.839 25NOV91 0.885 15MAY92 0.923
111 C 30MAY90 0.828 05DEC90 0.873 07AUG91 0.935 12FEB92 0.952

112 P 04JUN90 0.797 17DEC90 0.818 13MAY91 0.817 16DEC91 0.847 18MAY92 0.862
 :
 :
data example2_2_2;
 set TBBMD;
 array d date1-date5;
 array t tbbmd1-tbbmd5;
```

```
  do visit=1 to 5 by 1;
     date=d[visit];
     tbbmd=t[visit];
     years=0.5*(visit-1);
     format date date7.;
     keep subject group data visit date years tbbmd;
    output;
  end;
run;
proc sort data=example2_2_2;
 by subject visit;
run;
proc mixed data=example2_2_2 method=ml;
 where data='Complete';
 class subject group visit;
 model tbbmd=group group*years/noint solution;
 repeated visit /subject=subject type=unr;
 estimate 'slope diff' group*years 1 -1;
run;
```

The above MIXED statements are similar to that used for the GMANOVA analysis of the dental growth data (see page 42) in that we specify a cell means representation for the intercepts and slopes using the NOINT option of the MODEL statement. However, we use the default DDFM=BETWITHIN for estimating the denominator degrees of freedom for fixed-effects comparisons and we specify TYPE=UNR within the REPEATED statement so as display the unstructured covariance matrix parameter estimates in terms of the variances and correlation coefficients rather than variances and covariances. The results of this GMANOVA based on the balanced and compete data of 91 women are shown in Output 2.7. Subject to rounding errors, these results are similar to the results obtained by Vonesh and Chinchilli (1997; Table 5.4.2, page 232) using calculations in PROC IML.

Output 2.7: Selected output from MIXED showing results of a GMANOVA for 91 women with balanced bone mineral density data

```
                              Model Information
     Data Set                       WORK.EXAMPLE2_2_2
     Dependent Variable             tbbmd
     Covariance Structure           Unstructured using Correlations
     Subject Effect                 subject
     Estimation Method              ML
     Residual Variance Method       None
     Fixed Effects SE Method        Model-Based
     Degrees of Freedom Method      Between-Within
               Dimensions
     Covariance Parameters     15
     Columns in X               4
     Columns in Z               0
     Subjects                  91
     Max Obs Per Subject        5
     Covariance Parameter Estimates
     Cov Parm     Subject    Estimate
     Var(1)       subject    0.003694
     Var(2)       subject    0.004369
     Var(3)       subject    0.004729
     Var(4)       subject    0.005108
     Var(5)       subject    0.004631
     Corr(2,1)    subject      0.9671
     Corr(3,1)    subject      0.9404
```

Corr(3,2)	subject	0.9724
Corr(4,1)	subject	0.9217
Corr(4,2)	subject	0.9558
Corr(4,3)	subject	0.9794
Corr(5,1)	subject	0.8946
Corr(5,2)	subject	0.9384
Corr(5,3)	subject	0.9568
Corr(5,4)	subject	0.9726

Fit Statistics

-2 Log Likelihood	−2245.7
AIC (smaller is better)	−2207.7
AICC (smaller is better)	−2206.0
BIC (smaller is better)	−2160.0

Solution for Fixed Effects

Effect	group	Estimate	Standard Error	DF	t Value	Pr > |t|
group	C	0.8765	0.008568	89	102.30	<.0001
group	P	0.8650	0.008290	89	104.34	<.0001
years*group	C	0.05319	0.002221	89	23.95	<.0001
years*group	P	0.04459	0.002149	89	20.75	<.0001

Estimates

Label	Estimate	Standard Error	DF	t Value	Pr > |t|
slope diff	0.008597	0.003090	89	2.78	0.0066

In order to avoid potential selection bias that may occur when one excludes adolescent women who fail to complete all 5 visits and at the same time increase statistical efficiency and power, we re-ran the GMANOVA analysis above for all 112 women by simply dropping the above WHERE clause following the call to MIXED. The results of this full analysis are shown in Output 2.8 below. These results differ slightly from those reported by Vonesh and Chinchilli (1997). This can be explained by the fact that we elected to use ML estimation (versus REML estimation) and we chose not to use Fisher's scoring algorithm which one can implement using the MIXED option, SCORING=5 (as done by Vonesh and Chinchilli). Comparing the results in Output 2.7 with that in Output 2.8, we find that the results for all 112 women are qualitatively similar to results from the 91 women having complete data. In particular, based on all 112 women, the estimated rate of change in total body bone mineral density, or TBBMD (g/cm^2), was significantly greater in women receiving the daily calcium supplement (slope = 0.05287 g/cm^2/year) versus those receiving the placebo control (slope = 0.04416 g/cm^2/year) with a mean rate increase of 0.008704 $\pm$ 0.003012 g/cm^2/year for women randomized to the calcium supplement (p=0.0047).

Output 2.8: Selected output from MIXED showing GMANOVA results based on all 112 women

Estimated R Matrix for subject 101

Row	Col1	Col2	Col3	Col4	Col5
1	0.003892	0.004128	0.004114	0.004194	0.003892
2	0.004128	0.004656	0.004650	0.004749	0.004453
3	0.004114	0.004650	0.004913	0.005002	0.004666
4	0.004194	0.004749	0.005002	0.005299	0.004922
5	0.003892	0.004453	0.004666	0.004922	0.004819

Covariance Parameter Estimates

Cov Parm	Subject	Estimate
Var(1)	subject	0.003892
Var(2)	subject	0.004656
Var(3)	subject	0.004913
Var(4)	subject	0.005299
Var(5)	subject	0.004819
Corr(2,1)	subject	0.9697
Corr(3,1)	subject	0.9408
Corr(3,2)	subject	0.9723

```
Corr(4,1)   subject   0.9235
Corr(4,2)   subject   0.9561
Corr(4,3)   subject   0.9803
Corr(5,1)   subject   0.8987
Corr(5,2)   subject   0.9400
Corr(5,3)   subject   0.9590
Corr(5,4)   subject   0.9738
           Fit Statistics
-2 Log Likelihood            -2431.3
AIC (smaller is better)      -2393.3
AICC (smaller is better)     -2391.8
BIC (smaller is better)      -2341.7
                      Solution for Fixed Effects
Effect        group  Estimate  Standard Error    DF   t Value   Pr > |t|
group         C        0.8774        0.007889   110    111.21     <.0001
group         P        0.8665        0.007745   110    111.87     <.0001
years*group   C       0.05287        0.002156   110     24.52     <.0001
years*group   P       0.04416        0.002104   110     20.99     <.0001
                          Estimates
Label        Estimate  Standard Error   DF   t Value   Pr > |t|
slope diff   0.008704        0.003012  110      2.89     0.0047
```

To further increase statistical efficiency, one might consider running alternative models using more parsimonious covariance structures. For example, based on the unstructured variance-covariance matrix (the R matrix for subject 101) and correlation coefficients shown in Output 2.8, one might consider running a marginal linear model with a compound symmetric structure, a heterogeneous compound-symmetric structure or a first-order autoregressive structure. Using the COVTEST statement in conjunction with the GLIMMIX procedure, one can readily test whether the first two structures are reasonable by fitting an unstructured covariance matrix and testing whether one of these reduced structures holds using a likelihood ratio chi-square test. This can be accomplished in GLIMMIX using the following statements:

Program 2.7

```
proc glimmix data=example2_2_2 method=mmpl scoring=100;
 class subject group visit;
 model tbbmd=group group*years /noint solution
                               covb(details);
 random visit /subject=subject type=unr rside;
 nloptions technique=newrap;
 covtest 'Compound symmetry'
 general  1 -1                                              ,
          1  0 -1                                           ,
          1  0  0 -1                                        ,
          1  0  0  0 -1                                     ,
          0  0  0  0  0 1 -1                                ,
          0  0  0  0  0 1  0 -1                             ,
          0  0  0  0  0 1  0  0 -1                          ,
          0  0  0  0  0 1  0  0  0 -1                       ,
          0  0  0  0  0 1  0  0  0  0 -1                    ,
          0  0  0  0  0 1  0  0  0  0  0 -1                 ,
          0  0  0  0  0 1  0  0  0  0  0  0 -1              ,
          0  0  0  0  0 1  0  0  0  0  0  0  0 -1           ,
          0  0  0  0  0 1  0  0  0  0  0  0  0  0 -1 /
 estimates;
```

```
 covtest 'Heterogeneous compound symmetry'
 general   0 0 0 0  0 1 -1                        ,
           0 0 0 0  0 1  0 -1                     ,
           0 0 0 0  0 1  0  0 -1                  ,
           0 0 0 0  0 1  0  0  0 -1               ,
           0 0 0 0  0 1  0  0  0  0 -1            ,
           0 0 0 0  0 1  0  0  0  0  0 -1         ,
           0 0 0 0  0 1  0  0  0  0  0  0 -1      ,
           0 0 0 0  0 1  0  0  0  0  0  0  0 -1   ,
           0 0 0 0  0 1  0  0  0  0  0  0  0  0 -1 /
 estimates;
 estimate 'slope diff' group*years 1 -1;
run;
```

In general, the COVTEST statement of the GLIMMIX procedure allows users to test select hypotheses with respect to the variance-covariance parameters from a GLME model including the marginal linear model of this chapter. In addition to a number of default tests of hypotheses available in GLIMMIX, the COVTEST statement also allows users the option to test a specific set of contrasts of interest. Specifically, if $\boldsymbol{\theta}$ is the vector of unique variance-covariance parameters from the marginal LM (2.1), the COVTEST statement can be used to test the hypothesis

$$H_0 : \boldsymbol{C\theta} = \mathbf{0}$$

where $\boldsymbol{C}$ is a contrast matrix that tests whether select linear combinations of the variance-covariance parameters are zero. In our current example, the first COVTEST statement above defines a test of the hypothesis

$$H_0 : \boldsymbol{\Sigma}(\boldsymbol{\theta}) = \boldsymbol{\Sigma}(\boldsymbol{\theta}_0) = \sigma^2(1-\rho)\boldsymbol{I}_5 + \sigma^2\rho\boldsymbol{J}_5$$

where $\boldsymbol{\Sigma}(\boldsymbol{\theta}) = Diag(\boldsymbol{\sigma}^*)\boldsymbol{R}(\boldsymbol{\rho}^*)Diag(\boldsymbol{\sigma}^*)$ is a 5×5 unstructured variance-covariance matrix corresponding to the 5 possible time points and expressed in Output 2.8 in terms of a variance components vector $\boldsymbol{\sigma}^*$ and a correlation components vector $\boldsymbol{\rho}^*$. If we let $\boldsymbol{\theta} = Vech(\boldsymbol{\Sigma}(\boldsymbol{\theta}))$ represent the 15×1 vector of unique variance-covariance parameters with $\sigma_k^2 = \sigma_{kk} = Var(y_{ik})$ and $\sigma_{kl} = Cov(y_{ik}, y_{il}) = \sqrt{\sigma_{kk}}\sqrt{\sigma_{ll}}\rho_{kl}$, and if we arrange the elements of the correlation vector $\boldsymbol{\rho}^*$ in a manner consistent with how GLIMMIX prints the results (see Output 2.8), then we can write $\boldsymbol{\Sigma}(\boldsymbol{\theta})$ and $\boldsymbol{\theta}$ as

$$\boldsymbol{\Sigma}(\boldsymbol{\theta}) = \begin{pmatrix} \sigma_{11} & \sigma_{12} & \sigma_{13} & \sigma_{14} & \sigma_{15} \\ \sigma_{21} & \sigma_{22} & \sigma_{23} & \sigma_{24} & \sigma_{25} \\ \sigma_{31} & \sigma_{32} & \sigma_{33} & \sigma_{34} & \sigma_{35} \\ \sigma_{41} & \sigma_{42} & \sigma_{43} & \sigma_{44} & \sigma_{45} \\ \sigma_{51} & \sigma_{52} & \sigma_{53} & \sigma_{54} & \sigma_{55} \end{pmatrix}$$

$$\boldsymbol{\theta}' = \begin{pmatrix} \sigma_{11} & \sigma_{21} & \sigma_{31} & \sigma_{41} & \sigma_{51} & \sigma_{22} & \sigma_{32} & \sigma_{42} & \sigma_{52} & \sigma_{33} & \sigma_{43} & \sigma_{53} & \sigma_{44} & \sigma_{54} & \sigma_{55} \end{pmatrix}$$

while $\boldsymbol{\sigma}^*$ and $\boldsymbol{\rho}^*$ are given by

$$\boldsymbol{\sigma}^{*'} = \begin{pmatrix} \sigma_{11} & \sigma_{22} & \sigma_{33} & \sigma_{44} & \sigma_{55} \end{pmatrix}$$

$$\boldsymbol{\rho}^{*'} = \begin{pmatrix} \rho_{21} & \rho_{31} & \rho_{32} & \rho_{41} & \rho_{42} & \rho_{43} & \rho_{51} & \rho_{52} & \rho_{53} & \rho_{54} \end{pmatrix}.$$

A test of $\boldsymbol{\Sigma}(\boldsymbol{\theta}) = \boldsymbol{\Sigma}(\boldsymbol{\theta}_0)$ where $\boldsymbol{\theta}_0' = \begin{pmatrix} \sigma^2 & \sigma^2\rho \end{pmatrix}$ is the vector of reduced variance-covariance parameters assuming compound symmetry may be obtained by jointly testing

Table 2.2 Example of the first 4 contrast vectors of the COVTEST statement used to test for compound symmetry

Variance-Covariance Estimates			Parameters	Contrasts			
Cov Parm	Subject	Estimate	$\boldsymbol{\theta}^*$	C_1	C_2	C_3	C_4
Var(1)	subject	0.003892	σ_{11}	1	1	1	1
Var(2)	subject	0.004656	σ_{22}	-1	0	0	0
Var(3)	subject	0.004913	σ_{33}	0	-1	0	0
Var(4)	subject	0.005299	σ_{44}	0	0	-1	0
Var(5)	subject	0.004819	σ_{55}	0	0	0	-1
Corr(2,1)	subject	0.9697	ρ_{21}	0	0	0	0
Corr(3,1)	subject	0.9408	ρ_{31}	0	0	0	0
Corr(3,2)	subject	0.9723	ρ_{32}	0	0	0	0
Corr(4,1)	subject	0.9235	ρ_{41}	0	0	0	0
Corr(4,2)	subject	0.9561	ρ_{42}	0	0	0	0
Corr(4,3)	subject	0.9803	ρ_{43}	0	0	0	0
Corr(5,1)	subject	0.8987	ρ_{51}	0	0	0	0
Corr(5,2)	subject	0.9400	ρ_{52}	0	0	0	0
Corr(5,3)	subject	0.9590	ρ_{53}	0	0	0	0
Corr(5,4)	subject	0.9738	ρ_{54}	0	0	0	0

$$\boldsymbol{C}_1\boldsymbol{\sigma}^* = \begin{pmatrix} 1 & -1 & 0 & 0 & 0 \\ 1 & 0 & -1 & 0 & 0 \\ 1 & 0 & 0 & -1 & 0 \\ 1 & 0 & 0 & 0 & -1 \end{pmatrix} \begin{pmatrix} \sigma_{11} \\ \sigma_{22} \\ \sigma_{33} \\ \sigma_{44} \\ \sigma_{55} \end{pmatrix} = \begin{pmatrix} 0 \\ 0 \\ 0 \\ 0 \end{pmatrix}$$

and

$$\boldsymbol{C}_2\boldsymbol{\rho}^* = \begin{pmatrix} 1 & -1 & 0 & 0 & 0 & 0 & 0 & 0 & 0 & 0 \\ 1 & 0 & -1 & 0 & 0 & 0 & 0 & 0 & 0 & 0 \\ 1 & 0 & 0 & -1 & 0 & 0 & 0 & 0 & 0 & 0 \\ 1 & 0 & 0 & 0 & -1 & 0 & 0 & 0 & 0 & 0 \\ 1 & 0 & 0 & 0 & 0 & -1 & 0 & 0 & 0 & 0 \\ 1 & 0 & 0 & 0 & 0 & 0 & -1 & 0 & 0 & 0 \\ 1 & 0 & 0 & 0 & 0 & 0 & 0 & -1 & 0 & 0 \\ 1 & 0 & 0 & 0 & 0 & 0 & 0 & 0 & -1 & 0 \\ 1 & 0 & 0 & 0 & 0 & 0 & 0 & 0 & 0 & -1 \end{pmatrix} \begin{pmatrix} \rho_{21} \\ \rho_{31} \\ \rho_{32} \\ \rho_{41} \\ \rho_{42} \\ \rho_{43} \\ \rho_{51} \\ \rho_{52} \\ \rho_{53} \\ \rho_{54} \end{pmatrix} = \begin{pmatrix} 0 \\ 0 \\ 0 \\ 0 \\ 0 \\ 0 \\ 0 \\ 0 \\ 0 \end{pmatrix}.$$

In this case, if we express $\boldsymbol{\theta}$ in terms of $\boldsymbol{\sigma}^*$ and $\boldsymbol{\rho}^*$ as $\boldsymbol{\theta}^* = \begin{pmatrix} \boldsymbol{\sigma}^* \\ \boldsymbol{\rho}^* \end{pmatrix}$, then testing the hypothesis $H_0 : \boldsymbol{C\theta} = \boldsymbol{0}$ is equivalent to testing $H_0 : \boldsymbol{C}^*\boldsymbol{\theta}^* = \boldsymbol{0}$ where $\boldsymbol{C}^* = \begin{pmatrix} \boldsymbol{C}_1 & \boldsymbol{0}_{4\times 10} \\ \boldsymbol{0}_{9\times 5} & \boldsymbol{C}_2 \end{pmatrix}$.
For example, shown in Table 2.2 are the four row contrast vectors associated with $\boldsymbol{C}^*$ [i.e. the rows of $(\boldsymbol{C}_1 \;\; \boldsymbol{0}_{4\times 10})$] used to test for a common variance.

A test for heterogenous compound symmetry, a structure that assumes unequal variances but a common correlation coefficient, corresponds to the hypothesis $H_0 : \boldsymbol{C}^{**}\boldsymbol{\theta}^* = \boldsymbol{0}$ where $\boldsymbol{C}^{**} = (\boldsymbol{0} \;\; \boldsymbol{C}_2)$ is obtained by simply removing the first four rows of $\boldsymbol{C}^*$ (i.e., the set of contrasts shown in the second COVTEST statement above). Note that when writing SAS code, leaving blank values within a contrast statement (such as shown above) is equivalent to assigning values of 0.

The results from the above call to GLIMMIX are shown in Output 2.9. As indicated by the likelihood ratio tests, both forms of compound symmetry are rejected indicating the unstructured covariance matrix is preferred to these more parsimonious covariance structures. It should be noted that to replicate the same results obtained from the MIXED procedure shown in Output 2.8, one must specify the TECHNIQUE=NEWRAP option of the NLOPTIONS statement so as to override the default quasi-Newton optimization technique that GLIMMIX uses whenever there is a RANDOM statement.

Output 2.9: Selected output from GLIMMIX for testing whether a reduced covariance structure, in the form of compound symmetry, is feasible.

```
                 Model Information
Data Set                        WORK.EXAMPLE2_2_2
Response Variable               tbbmd
Response Distribution           Gaussian
Link Function                   Identity
Variance Function               Default
Variance Matrix Blocked By      subject
Estimation Technique            Maximum Likelihood
Degrees of Freedom Method       Between-Within
             Optimization Information
Optimization Technique          Newton-Raphson
Fisher Scoring                  to iteration 100
Parameters in Optimization      15
Lower Boundaries                15
Upper Boundaries                10
Fixed Effects                   Profiled
Starting From                   Data
              Fit Statistics
-2 Log Likelihood             -2431.34
AIC (smaller is better)       -2393.34
AICC (smaller is better)      -2391.76
BIC (smaller is better)       -2341.69
CAIC (smaller is better)      -2322.69
HQIC (smaller is better)      -2372.38
Generalized Chi-Square          501.00
Gener. Chi-Square / DF            1.00
            Covariance Parameter Estimates
Cov Parm     Subject    Estimate    Standard Error
Var(1)       subject    0.003892          0.000520
Var(2)       subject    0.004656          0.000625
Var(3)       subject    0.004912          0.000663
Var(4)       subject    0.005299          0.000720
Var(5)       subject    0.004819          0.000660
Corr(2,1)    subject      0.9697          0.005745
Corr(3,1)    subject      0.9408           0.01118
Corr(3,2)    subject      0.9723          0.005357
Corr(4,1)    subject      0.9235           0.01446
Corr(4,2)    subject      0.9561          0.008489
Corr(4,3)    subject      0.9803          0.003873
Corr(5,1)    subject      0.8987           0.01905
Corr(5,2)    subject      0.9400           0.01164
Corr(5,3)    subject      0.9590          0.008066
Corr(5,4)    subject      0.9738          0.005198
                           Solutions for Fixed Effects
Effect        group   Estimate   Standard Error    DF   t Value   Pr > |t|
group         C         0.8774         0.007889   110    111.21     <.0001
group         P         0.8665         0.007745   110    111.87     <.0001
years*group   C        0.05287         0.002156   110     24.52     <.0001
years*group   P        0.04416         0.002104   110     20.99     <.0001
```

```
                         Estimates
Label        Estimate   Standard Error    DF    t Value    Pr > |t|
slope diff   0.008704         0.003012   110       2.89      0.0047
        Tests of Covariance Parameters          Based on the Likelihood
Label                                 DF   -2 Log Like    ChiSq   Pr > ChiSq   Note
Compound symmetry                     13     -2275.57    155.77       <.0001   DF
Heterogeneous compound symmetry        9     -2295.22    136.12       <.0001   DF
```

Finally, given the pattern of decreasing correlation with increasing time between measurements, we consider running the marginal LM assuming a first-order autoregressive (i.e., AR(1)) structure. The syntax for running this model along with an option for performing robust inference using empirical "sandwich" estimates of the standard error of $\widehat{\beta}$ are shown below.

Program 2.8

```
proc glimmix data=example2_2_2 method=mmpl scoring=100
             empirical;
 class subject group visit;
 model tbbmd=group group*years /noint solution
                                  covb(details);
 random visit /subject=subject type=ar(1) rside;
 nloptions technique=newrap;
 estimate 'slope diff' group*years 1 -1;
 output out=pred /allstats;
 ods output CovBDetails=gof;
 ods output ParameterEstimates=pe;
 ods output CovParms=cov;
 ods output dimensions=n;
run;
```

The MODEL statement option, COVB(DETAILS), provides additional criteria with which to judge goodness-of-fit of the assumed covariance structure based on the estimated variance-covariance of the estimated regression coefficients, $\widehat{\beta}(\widehat{\theta})$ (see page 226) while the GLIMMIX option, EMPIRICAL, instructs SAS to use the classical empirical sandwich estimator of the variance-covariance matrix of $\widehat{\beta}(\widehat{\theta})$ to conduct robust inference (see page 29). Results from this analysis are shown in Output 2.10.

Output 2.10: Selected output from GLIMMIX for assessing the goodness-of-fit of the AR(1) covariance structure.

```
                 Model Information
Data Set                        WORK.EXAMPLE2_2_2
Response Variable               tbbmd
Response Distribution           Gaussian
Link Function                   Identity
Variance Function               Default
Variance Matrix Blocked By      subject
Estimation Technique            Maximum Likelihood
Degrees of Freedom Method       Between-Within
Fixed Effects SE Adjustment     Sandwich - Classical
             Optimization Information
Optimization Technique         Newton-Raphson
Fisher Scoring                 to iteration 100
Parameters in Optimization     1
Lower Boundaries               1
Upper Boundaries               1
Fixed Effects                  Profiled
Residual Variance              Profiled
Starting From                  Data
```

Model Based Covariance Matrix for Fixed Effects (Unadjusted)

Effect	group	Row	Col1	Col2	Col3	Col4
group	C	1	0.000079		−4.56E−6	
group	P	2		0.000076		−4.39E−6
years*group	C	3	−4.56E−6		5.245E−6	
years*group	P	4		−4.39E−6		5.005E−6

Empirical Covariance Matrix for Fixed Effects

Effect	group	Row	Col1	Col2	Col3	Col4
group	C	1	0.000064		−8.48E−7	
group	P	2		0.000075		8.416E−7
years*group	C	3	−8.48E−7		5.387E−6	
years*group	P	4		8.416E−7		4.327E−6

Diagnostics for Covariance Matrices of Fixed Effects

		Model-Based	Adjusted
Dimensions	Rows	4	4
	Non-zero entries	8	8
Summaries	Trace	0.0002	0.0001
	Log determinant	−43.4	−43.64
Eigenvalues	> 0	4	4
	max abs	0.0001	0.0001
	min abs non-zero	473E−8	432E−8
	Condition number	16.7	17.376
Norms	Frobenius	0.0001	0.0001
	Infinity	0.0001	0.0001
Comparisons	Concordance correlation		0.9824
	Discrepancy function		0.1289
	Frobenius norm of difference		175E−7
	Trace(Adjusted Inv(MBased))		3.8851

Fit Statistics

-2 Log Likelihood	−2402.14
AIC (smaller is better)	−2390.14
AICC (smaller is better)	−2389.97
BIC (smaller is better)	−2373.83
CAIC (smaller is better)	−2367.83
HQIC (smaller is better)	−2383.52
Generalized Chi-Square	2.17
Gener. Chi-Square / DF	0.00

Covariance Parameter Estimates

Cov Parm	Subject	Estimate	Standard Error
AR(1)	subject	0.9698	0.004353
Residual		0.004333	0.000551

Solutions for Fixed Effects

Effect	group	Estimate	Standard Error	DF	t Value	Pr > |t|
group	C	0.8805	0.007988	110	110.23	<.0001
group	P	0.8701	0.008660	110	100.47	<.0001
years*group	C	0.05307	0.002321	387	22.87	<.0001
years*group	P	0.04352	0.002080	387	20.92	<.0001

Estimates

Label	Estimate	Standard Error	DF	t Value	Pr > |t|
slope diff	0.009555	0.003117	387	3.07	0.0023

Based on the results shown in Output 2.9 and 2.10, a likelihood ratio chi-square test comparing the full model with unstructured covariance with that of the reduced model assuming an AR(1) structure is
$-2\log(\widehat{\lambda}) = 2ll(\widehat{\boldsymbol{\theta}}_{\mathrm{UN}}) - 2ll(\widehat{\boldsymbol{\theta}}_{\mathrm{AR(1)}}) = 2431.34 - 2402.14 = 29.2$ on 13 degrees of freedom yielding a p-value of 0.00613. An alternative test based on the discrepancy function (=0.1289) shown in Output 2.10 is the pseudo-likelihood ratio test described by Vonesh et. al. (1996). This test compares the model-based covariance matrix of the fixed-effects to the

empirical covariance matrix and does not require one to fit the full model as shown in Output 2.9. This test may be implemented using either the SAS macro %GLIMMIX_GOF or %GOF in combination with the OUTPUT OUT= and ODS OUTPUT statements shown above (see Appendix D for a complete description of %GLIMMIX_GOF and %GOF). Below is the call to the GLIMMIX_GOF macro.

Program 2.9

```
%GLIMMIX_GOF(dimension=n,
             parms=pe,
             covb_gof=gof,
             output=pred,
             response=tbbmd,
             pred_ind=PredMu,
             pred_avg=PredMuPA);
proc print data=_fitting noobs;
run;
```

The results, shown in Output 2.11, include two versions of the pseudo-likelihood ratio test. The first test is that initially proposed by Vonesh et. al. (1996) and is based on degrees of freedom $= s(s+1)/2$ where $s = rank(\widehat{\boldsymbol{\Omega}}_R) = 4$ and $\widehat{\boldsymbol{\Omega}}_R$ is the robust variance-covariance matrix of the fixed-effects parameter estimates. The second test modifies the degrees of freedom to be $s + s_1 = 6$ where s_1 is the number of unique non-zero off-diagonal elements of $\widehat{\boldsymbol{\Omega}}_R$ as shown in Output 2.10.

Output 2.11: Additional goodness-of-fit statistics for the assumed variance-covariance matrix based on Vonesh et. al. (1996)

DESCRIPTION	Value
Total Observations	501.000
N (number of subjects)	112.000
Number of Fixed-Effects Parameters	4.000
Average Model R-Square:	0.234
Average Model Adjusted R-Square:	0.228
Average Model Concordance Correlation:	0.370
Average Model Adjusted Concordance Correlation:	0.365
Conditional Model R-Square:	0.234
Conditional Model Adjusted R-Square:	0.228
Conditional Model Concordance Correlation:	0.370
Conditional Model Adjusted Concordance Correlation:	0.365
Variance-Covariance Concordance Correlation:	0.982
Discrepancy Function	0.129
s = Rank of robust sandwich estimator, OmegaR	4.000
s1 = Number of unique non-zero off-diagonal elements of OmegaR	2.000
Approx. Chi-Square for H0: Covariance Structure is Correct	14.434
DF1 = s(s+1)/2, per Vonesh et al (Biometrics 52:572-587, 1996)	10.000
Pr > Chi Square based on degrees of freedom, DF1	0.154
DF2 = s+s1, a modified degress of freedom	6.000
Pr > Chi Square based on modified degrees of freedom, DF2	0.025

As judged by both the above likelihood ratio test and the pseudo-likelihood ratio test (with modified degrees of freedom) for evaluating the AR(1) structure as well as the two likelihood ratio tests in Output 2.10 for evaluating compound symmetry, an unstructured variance-covariance seems to provide the best fit to the data for this particular application. Note that despite the evidence suggesting the AR(1) covariance structure is misspecified, the standard errors used to draw inference about the parameters $\boldsymbol{\beta}$ are reasonably close to those in Output 2.9. This is not surprising in consideration of the fact that the standard errors are calculated using the robust sandwich estimator as defined by the EMPIRICAL option of PROC GLIMMIX. This option provides users some protection against model misspecification with respect to the modeled variance-covariance structure, $\boldsymbol{\Sigma}(\boldsymbol{\theta})$.

2.2.3 ADEMEX adequacy data

In the previous two examples, we considered applications where a GMANOVA model can be applied to balanced but possibly incomplete correlated data using the GLM, MIXED or GLIMMIX procedures. In the context of repeated measurements and clustered data, we define correlated data as being balanced whenever the within-subject or within-cluster design matrix is the same for each individual. In such cases the data may be incomplete in that observations may be missing at planned visits but nonetheless the within-subject design remains the same across individuals. In this example, we consider a case where one has unbalanced data in that measurements are obtained at irregularly spaced intervals across individuals.

The ADEMEX (ADEquacy of PD in MEXico) study, as described in detail by Paniagua et. al. (2002) (see also section 1.4, page 17), is a randomized controlled trial whose primary objective is determining whether an increased dose of continuous ambulatory peritoneal dialysis (CAPD) will improve survival among patients with end-stage renal disease. Briefly, peritoneal dialysis is an artificial replacement therapy for lost kidney function due to renal failure. It consists of performing daily exchanges in which a dialysis solution is infused into a patient's peritoneal cavity, allowed to dwell for a period of time and then is drained out of the peritoneum. During each exchange, an osmotically active agent in the dialysis solution serves to filter a patient's blood of toxic solute waste and excess plasma water via diffusion and osmosis, respectively. In the ADEMEX trial, a total of 965 patients were randomized to either a control group (n=484) or test group (n=481). Patients in the control group received a conventional CAPD treatment consisting of four daily exchanges of a 2-liter dialysis solution. Patients in the intervention or test group received an increased dose of dialysis in which either the volume of the infused solution was increased from 2 liters to 2.5 liters or 3.0 liters, or the number of exchanges per day was increased from four to five or six exchanges. In some cases, patients received an increase in both the fill volume and number of exchanges. The goal was to increase the dose of dialysis in the test group to a level whereby the patient's measured peritoneal creatinine clearance (pCCr) would equal or exceed 60 liters per week per 1.73 m^2 body surface area—a level that was generally believed to provide adequate therapy in terms of improved patient outcomes.

To determine if patient survival improves with an increased dose of dialysis, a comparison was first made to determine if the planned intervention succeeded in increasing pCCR to a level beyond that achieved with conventional peritoneal dialysis. To that end, follow-up visits were scheduled at 2-month intervals ($\pm$ 2 week window) in order to measure pCCr as well as other indicators of dialysis adequacy. These planned visits began immediately after randomization for the control group but only after stabilization with a final prescription for the intervention group. This is because, in some cases, patients in the test group required several prescription changes in an attempt to reach a target pCCr of 60L/week/1.73m^2. Consequently, the stabilization period varied among test patients resulting in an unbalanced timing of visits. Also, some patients arrived before or after the allowable 2-week window per scheduled visit resulting in some visits being 1 or 3 months apart rather than 2 months as scheduled. The study was designed to have a minimum two-year follow-up following the end of a 4-month accrual period and measurements of pCCr were analyzed through 28 months. The essential components of this study with respect to comparing pCCr are summarized below. A description of the data is presented in Appendix C (Dataset C.3).

Figure 2.3 Least squares means comparing pCCr (L/week/1.73 m. sq.) for control (n=484) and test (n=481) patients

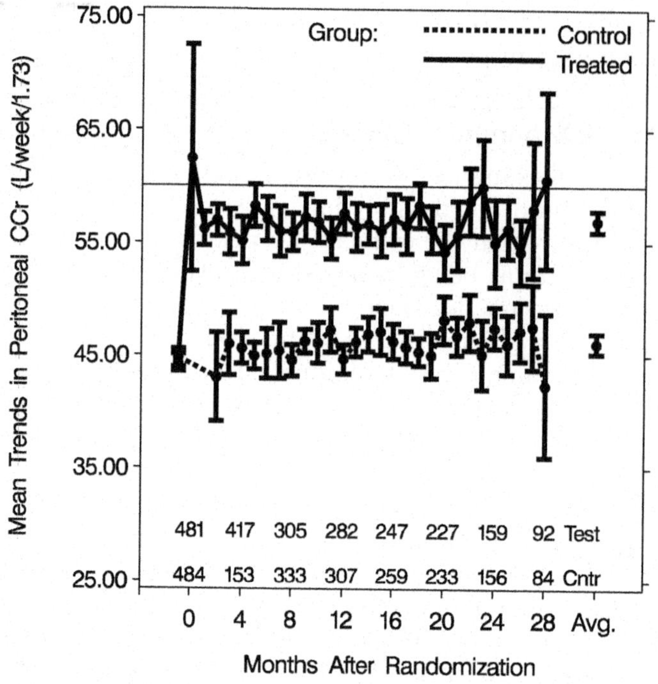

- A two-year randomized controlled trial designed to evaluate the effects of increased dose of peritoneal clearance on patient survival among patients with end-stage renal disease (ESRD).
- The experimental unit is a patient with end-stage renal disease
- Response variable:
  - y_{ij} =peritoneal creatinine clearance or pCCr (L/week/1.73m^2) for the ith subject on the jth occasion
- One within-unit covariate:
  - t_{ij} =months after randomization (targeting 2-month intervals)
- One between-unit covariate:
  - Treatment Group (Control, Treated or Intervention)

$$a_{1i} = \begin{cases} 0, & \text{if control (standard } 4 \times 2\text{L CAPD)} \\ 1, & \text{if intervention (target pCCr} = 60 \text{ L/week/1.73m}^2) \end{cases}$$

• Objective: The goal of this analysis was to compare mean profiles of pCCr over time between control and treated patients

To compare the mean profile of pCCr between control and treated patients, a repeated measures analysis of variance (RM-ANOVA) was performed. A basic RM-ANOVA model for this example would be

$$y_{ijk} = \mu + \tau_j + \gamma_k + (\tau\gamma)_{jk} + \epsilon_{ijk},\ (i = 1, \ldots, 965;\ j = 1, 2;\ k = 1, \ldots p_i)$$

where μ is an overall mean, τ_j is the added effect associated with the jth treatment group (control vs. treated), γ_k is the added effect at the kth possible time point (month) after randomization, $(\tau\gamma)_{jk}$ is the added effect of a treatment by time interaction, and ϵ_{ijk} is within-subject experimental error associated with the pCCr response, y_{ijk}, from subject i within treatment j on the kth month following randomization. Under the assumption of compound symmetry such that the $\epsilon_{ijk} \sim N(0, \sigma_w^2)$ with $Cov(\epsilon_{ijk}, \epsilon_{ijk'}) = \sigma_w^2 \rho$ for $k \neq k'$, we could write this in terms of an unbalanced cell-means GMANOVA model and subsequently in terms of our marginal LM (2.1) as

$$\begin{aligned} \boldsymbol{y}_i &= \boldsymbol{D}_i' \boldsymbol{X}' (\boldsymbol{I}_p \otimes \boldsymbol{a}_i') \boldsymbol{\beta} + \boldsymbol{\epsilon}_i \\ &= \boldsymbol{X}_i \boldsymbol{\beta} + \boldsymbol{\epsilon}_i \end{aligned}$$

where $\boldsymbol{X} = \boldsymbol{I}_p$ is the $p \times p$ identity matrix associated with a balanced RM-ANOVA within-subject design matrix for modeling the additive effects of the maximum possible time points (i.e., months after randomization) that pCCr measurements could be made, $\boldsymbol{a}_i' = \begin{pmatrix} 1 & 1 - a_i \end{pmatrix}$ is the between-subject design vector for the ith subject, and $\boldsymbol{D}_i'$ is formed from those rows of $\boldsymbol{X}$ corresponding to the time points at which pCCr measurements were obtained for the ith subject. Here, we have set up a cell-means parameterization by setting $\mu_{jk} = \mu + \tau_j + \gamma_k + (\tau\gamma)_{jk}$ to be the mean pCCr for the jth treatment group at the kth possible time point studied within the jth group.

Because of the large sample size and the desire to compare mean profiles in a fairly robust manner, a GEE approach was used to assess the effects of time, treatment and their interaction on pCCr. Using the GENMOD procedure (see code below), two different models were run, a main-effects model in which the overall treatment effect post-baseline (SAS where clause: `where Month>-1;`) is evaluated across all visits (33 visits in all) and a model that includes a treatment by time interaction up through month 28 (the test for a treatment by time interaction was restricted to the first 28 months due to very sparse data after month 28). In both models, the variance-covariance structure was assumed to be compound symmetric (TYPE=CS) although empirical sandwich estimators of the standard errors were used for all subsequent inference. Least squares means were computed under both models and the results sent to output files using the output delivery system statements (i.e., ODS OUTPUT 'name') shown in the SAS code below. The appropriate least squares means were then pooled from the two models and the results plotted in Figure 2.3.

Program 2.10

```
proc sort data=SASdata.ADEMEX_Adequacy_Data out=example2_2_3;
 by Trt ptid Month Visit;
run;
ods select Type3 LSMeans;
ods output LSMeans=lsAvgout LSMeanDiffs=lsAvgdiff;
proc genmod data=example2_2_3;
 where Month>-1;
 class Trt Month ptid;
 model pCCR = Trt Month / type3;
 repeated subject=ptid / type=cs;
 lsmeans Trt /diff cl;
run;
ods select Type3;
ods output LSMeans=lsout LSMeanDiffs=lsdiff;
proc genmod data=example2_2_3; where Month<=28;
 class Trt Month ptid;
 model pCCR = Trt Month Trt*Month / type3;
 repeated subject=ptid / type=cs;
 lsmeans Trt*Month /diff cl;
run;
```

The type III chi-square tests shown in Output 2.12 suggest there is a significant treatment by time interaction but this is due to the inclusion of the baseline period (month= -1) where we find no difference in pCCr between control and test subjects. Indeed, when we restrict the analysis to the post-baseline phase using the WHERE statement

```
where Month>-1 and Month<=28;
```

immediately following PROC GENMOD, we find no significant treatment by time interaction (p=0.9045, results not shown). Indeed, as suggested in Figure 2.3, the mean profiles for the two groups remain fairly stable and constant following month 0 indicating there is no treatment by time interaction during the treatment phase of the trial. One might notice an apparent discrepancy in degrees of freedom associated with the second set of Type 3 score tests in Output 2.12 (corresponding to the second call to GENMOD). Observe that the main-effects degrees of freedom for the MONTH effect is 29 while the interaction degrees of freedom for the TRT*MONTH interaction effect is 27. This is because mean pCCr values were available for at least one of the two treatment groups at all 30 months (months = -1 to 28) but there were only 28 months where both treatment groups had mean pCCr values evaluated (see Figure 2.3).

Output 2.12: Select output from GENMOD (using above ODS SELECT statements) for the analysis of ADEMEX adequacy data

```
Score Statistics For Type 3 GEE Analysis
Source   DF   Chi-Square       Pr > ChiSq
Trt       1       247.78           <.0001
Month    32        23.51           0.8616
                                 Least Squares Means
Effect    Trt        Estimate          Standard Error    DF    Chi-Square    Pr > ChiSq
                  Mean      L'Beta
Trt        0    46.0840   46.0840               0.4513    1         10429        <.0001
Trt        1    56.9399   56.9399               0.4843    1         13823        <.0001
Score Statistics For Type 3 GEE Analysis
Source        DF   Chi-Square     Pr > ChiSq
Trt            1       204.67         <.0001
Month         29       328.28         <.0001
Trt*Month     27       271.36         <.0001
```

The results of this analysis demonstrate that the mean pCCr profile among test patients, although slightly below the target of 60L/week/1.73m^2, was significantly elevated and sustained for the duration of the ADEMEX trial when compared to the control group.

2.2.4 MCM2 biomarker data

Each of the three previous examples were concerned primarily with estimating and comparing mean profiles in a repeated measurements setting. We now consider an example where emphasis is on estimating the parameters of an appropriately selected variance-covariance structure for the purpose of evaluating the reproducibility of a prostate cancer biomarker coming from the prostate glands of subjects having prostate biopsies.

Helenowski et. al. (2011) considered the problem of estimating the overall reproducibility of the biomarker, minichromosome maintenance protein 2 (MCM2), obtained from the prostate glands of 7 subjects having prostate biopsies between 2002-2003. By estimating the reproducibility of a potential biomarker of a given disease, in this case prostate cancer, one can determine whether such a marker can provide an accurate indication of the underlying disease progression. When biomarker measurements involve obtaining samples of tissue at random from the organ of interest, sampling variability based on a continuum of time and spatiality can affect this reproducibility. This situation arises with biomarkers of prostate cancer in that the prostate gland is evaluated by multiple random needle biopsies. A key question for investigators is determining whether concentrations of a potential biomarker like MCM2 are similar (i.e., reproducible) throughout the prostate gland. If so, then investigators could examine fewer samples from anywhere in the gland. This, in turn, would greatly reduce the mental and physical stress associated with having multiple needle biopsies. Following the work of Helenowski et. al. (2011), we present a general approach toward estimating reproducibility in the presence of different variance-covariance structures needed to account for possible spatial and/or temporal variation and correlation.

Biomarker reproducibility can best be estimated using the concordance correlation coefficient (CCC) as originally described by Lin (1989, 2000). The CCC for bivariate measurements, Y_1 and Y_2, is defined by Lin (1989) to be

$$\begin{aligned}\rho_c &= 1 - \frac{E(Y_1 - Y_2)^2}{\sigma_1^2 + \sigma_2^2 + (\mu_1 - \mu_2)^2} \\ &= \frac{2\sigma_{12}}{\sigma_1^2 + \sigma_2^2 + (\mu_1 - \mu_2)^2} \\ &= \frac{2\rho\sigma_1\sigma_2}{\sigma_1^2 + \sigma_2^2 + (\mu_1 - \mu_2)^2} \\ &= \rho C_b\end{aligned} \tag{2.23}$$

where $C_b = \frac{2\sigma_1\sigma_2}{\sigma_1^2+\sigma_2^2+(\mu_1-\mu_2)^2}$ is a measure of accuracy representing deviation from the line of identity, and ρ =Pearson's correlation which is a measure of precision representing deviation from the best fit line. Conceptually, a concordance correlation coefficient directly measures the level of agreement between two sets of measurements based on their scatter about a 45 $^\circ$ line (i.e., the line of identity). Estimation can be carried out by replacing the population means, μ_1 and μ_2, variances, σ_1^2 and σ_2^2, and correlation ρ using method of moments (MM) as was done initially by Lin (1989) or by replacing them with ML or REML estimates assuming normality. In the case of balanced complete data, substitution of MM and REML estimates will be equivalent under normality. If we let $\widehat{\rho}_c$ be the estimate of ρ_c, inference in terms of confidence intervals can be based on using the inverse hyperbolic tangent transformation $\widehat{Z} = \tanh(\widehat{\rho}_c)$ (Lin, 1989).

Lin (1989, 2000) and Barnhart et. al. (2002) also presented an overall concordance correlation coefficient (OCCC) which one can use to assess the overall agreement between p sets of measurements ($p \geq 2$). The overall concordance correlation coefficient (OCCC) is defined as

$$\rho_{c,overall} = \frac{2\sum_{i=1}^{p-1}\sum_{j=i+1}^{p}\sigma_{ij}}{(p-1)\sum_{i=1}^{p}\sigma_i^2 + 2\sum_{i=1}^{p-1}\sum_{j=i+1}^{p}(\mu_i - \mu_j)^2}. \tag{2.24}$$

As in the bivariate case, if we let $\boldsymbol{y}_i \sim N_p(\boldsymbol{\mu}, \boldsymbol{\Sigma})$ for $i = 1, \ldots, n$ where $\boldsymbol{y}_i$ is a set of p measurements on the ith individual (e.g., ratings from p different judges, or measurements from p different locations of a prostate gland) with mean $\boldsymbol{\mu}$ and unstructured variance-covariance $\boldsymbol{\Sigma}$, we can estimate a CCC matrix with elements representing pairwise CCC's between the components of the $\boldsymbol{y}_i$'s by simply substituting MM, ML or REML estimates of the corresponding components of $\boldsymbol{\mu}$ and $\boldsymbol{\Sigma}$ into the CCC matrix using (2.23). Likewise, the overall CCC (2.24) may be estimated directly on the basis of MM, ML or REML estimates of $\boldsymbol{\mu}$ and $\boldsymbol{\Sigma}$. Extensions of these basic measures to account for covariate adjustment were made by Barnhart and Williamson (2001) and Barnhart et. al. (2002)

Figure 2.4 Orientation of prostate gland sections from which MCM2 data are taken.

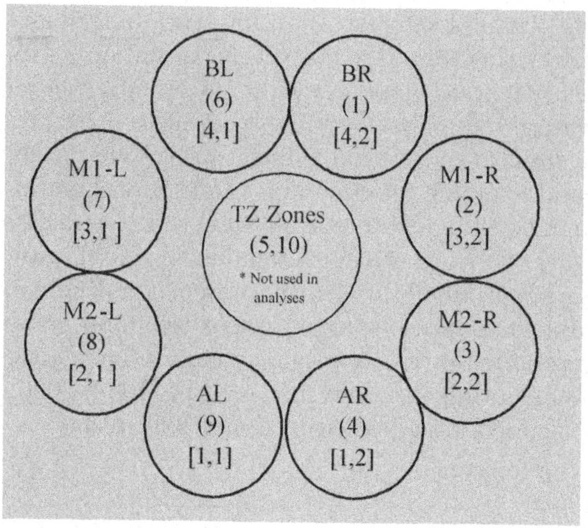

using GEE. Carrasco and Jover (2003) showed how one can express pairwise estimates of the CCC as well as the overall CCC in terms of an intra-class correlation coefficient using random-effects models that likewise can include adjustments for covariates.

Similar to Barnhart et. al. (2002) and Carrasco and Jover (2003), Helenowski et. al. (2011) considered estimation of a CCC matrix and the OCCC based on the marginal linear model

$$y_{ij} = \mu + \tau_j + \epsilon_{ij} \tag{2.25}$$

where y_{ij} is a measurement from the ith individual ($i = 1, \ldots, n$), j^{th} source ($j = 1, \ldots, p$), μ is an overall mean, τ_j is the effect due to the jth source (e.g., different raters or instruments or, in our example, different prostate gland sections), and the ϵ_{ij} are random error terms with $\boldsymbol{\epsilon}_i = (\epsilon_{i1}, \ldots \epsilon_{ip})'$ assumed to be normally distributed as $N_p(\mathbf{0}, \boldsymbol{\Sigma}(\boldsymbol{\theta}))$ for some specified variance-covariance structure $\boldsymbol{\Sigma}(\boldsymbol{\theta})$. One can include covariate adjustments by simply extending model (2.25) as

$$y_{ij} = \mu + \tau_j + \sum_{k=1}^{s} \beta_k x_{ik} + \epsilon_{ij} \tag{2.26}$$

where $\{x_{ik}; k = 1, \ldots, s\}$ are a set of select covariates one wishes to adjust for. Notice that (2.26) is simply a special case of the marginal LM (2.1). One can estimate the unadjusted (model 2.25) or adjusted (model 2.26) means, $\mu_j = \mu + \tau_j$, the variances, σ_j^2, and correlations, $\rho_{jj'}$, between any two sources j and j' using PROC MIXED. Using ODS OUTPUT statements to output key parameter estimates, one can then estimate a pairwise CCC matrix and OCCC under an assumed covariance structure using the SAS macro %CCC (see Appendix D) as illustrated below with the MCM2 biomarker data.

As described by Helenowski et. al. (2011), MCM2 biomarker data were obtained from seven subjects having prostate biopsies performed between 2002 and 2003. For the biopsies, needle core samples were taken from each of ten sections of the prostate. Figure 2.4 shows the orientation of each section. We restrict our analysis to the eight peripheral sections of the prostate gland. Four of these peripheral sections are on the left side of the prostate gland and four sections are on the right side. The two central (transitional) sections from the normal compartment and sections that have evidence of pre-neoplastic or cancerous features were not used in the analysis. Referring to Figure 2.4, the numbers in the parentheses indicate the number of the biopsy needle core designated to take a tissue sample from that particular prostate gland section of a subject. The pair of numbers in brackets represent the spatial coordinates when the prostate is viewed as a 4 x 2 grid consisting of 4 rows and 2 columns. The columns represent the left and right side of the gland. We arbitrarily chose core 9 to represent row 1, column 1 (i.e., coordinate [1,1]). Based on this orientation, core 4 represents row 1, column 2 (i.e., coordinate [1,2]); core 8 represents row 2, column 1 (i.e., coordinate [2,1]), etc. A partial listing of the raw data (subject 16 and 17 only) consisting of an MCM2 index and spatial coordinates used to represent the section of the gland corresponding to the MCM2 index (defined as the percentage of cells taken from a section positively staining for MCM2) are shown in Output 2.13 (see also Appendix C, Dataset C.4).

The primary goal of the analysis is to evaluate the reproducibility of the MCM2 index between any two sections of the gland using pairwise estimates of the concordance correlation coefficient (2.23). Likewise, it was of interest to evaluate the overall CCC (2.24) across all eight sections of the gland.

Output 2.13: MCM2 index data from 2 representative individuals across 8 core samples. Included are the spatial coordinates of the data as pictured in Figure 2.4.

Subject	Core	Row	Col	MCM2Index
16	1	4	2	23.8908
16	2	3	2	26.1251
16	3	2	2	31.0541
16	4	1	2	32.0144
16	6	4	1	.
16	7	3	1	31.9534
16	8	2	1	26.7631
16	9	1	1	32.5031
17	1	4	2	52.9183
17	2	3	2	36.4055
17	3	2	2	52.8875
17	4	1	2	59.8259
17	6	4	1	45.0262
17	7	3	1	41.9355
17	8	2	1	46.4435
17	9	1	1	40.3614

One can evaluate pairwise and overall reproducibility assuming a completely unstructured variance-covariance matrix or by assuming some structured covariance matrix. In this example, the goal was to evaluate the OCCC assuming an unstructured covariance matrix, a compound symmetric covariance matrix and a spatially linear covariance structure (so as to account for possible spatially correlated data). Because there are up to 8 measurements per subject but only 7 subjects, one can not fit the data assuming an unstructured variance-covariance matrix as the observation matrix, $\boldsymbol{Y}_{n \times p}$, will have rank $n < p$ ($n = 7$, $p = 8$). However, one can fit pairwise CCC's using a SAS macro %CONCORR (see Appendix D for a description). To do so requires transposing the original data so that there is one record per subject (i.e., the 8 MCM2 values are arranged horizontally rather than vertically). The SAS code required to run the macro %CONCORR is shown below followed by the estimated pairwise CCC's shown in Output 2.14.

Program 2.11

```
proc sort data=MCM2_data out=Example2_2_4;
 by Subject;
run;
proc transpose data=Example2_2_4 out=pairwise prefix=Core_;
 var MCM2Index;
 by Subject;
 id Core;
run;
title 'Pairwise CCC on original data';
%CONCORR(DATA=Pairwise,
   VAR=Core_1 Core_2 Core_1 Core_3 Core_1 Core_4
       Core_1 Core_6 Core_1 Core_7 Core_1 Core_8
       Core_1 Core_9
       Core_2 Core_3 Core_2 Core_4 Core_2 Core_6
       Core_2 Core_7 Core_2 Core_8 Core_2 Core_9
       Core_3 Core_4 Core_3 Core_6 Core_3 Core_7
       Core_3 Core_8 Core_3 Core_9
       Core_4 Core_6 Core_4 Core_7 Core_4 Core_8
       Core_4 Core_9
       Core_6 Core_7 Core_6 Core_8 Core_6 Core_9
       Core_7 Core_8 Core_7 Core_9
       Core_8 Core_9,
       FORMAT=5.2);
```

Output 2.14: Pairwise CCC's of MCM2 data using %CONCORR

Summary Statistics (N, Mean, SD), Correlation (R) and Concordance Correlation (Rc)

Core	N	MEAN	STD	With	N	MEAN	STD	R	Rc
Core_1	7.00	35.74	13.32	Core_2	7.00	33.99	6.62	0.09	0.07
Core_1	7.00	35.74	13.32	Core_3	7.00	36.36	7.77	0.61	0.53
Core_1	6.00	33.90	13.58	Core_4	6.00	36.90	12.68	0.92	0.89
Core_1	6.00	37.72	13.42	Core_6	6.00	34.43	6.61	0.67	0.50
Core_1	7.00	35.74	13.32	Core_7	7.00	35.25	7.22	0.50	0.42
Core_1	7.00	35.74	13.32	Core_8	7.00	34.83	7.34	0.63	0.53
Core_1	7.00	35.74	13.32	Core_9	7.00	34.39	4.06	0.38	0.21
Core_2	7.00	33.99	6.62	Core_3	7.00	36.36	7.77	0.31	0.28
Core_2	6.00	34.10	7.24	Core_4	6.00	36.90	12.68	0.07	0.05
Core_2	6.00	35.30	6.17	Core_6	6.00	34.43	6.61	−0.03	−0.03
Core_2	7.00	33.99	6.62	Core_7	7.00	35.25	7.22	0.16	0.15
Core_2	7.00	33.99	6.62	Core_8	7.00	34.83	7.34	0.66	0.65
Core_2	7.00	33.99	6.62	Core_9	7.00	34.39	4.06	0.66	0.58
Core_3	6.00	36.00	8.45	Core_4	6.00	36.90	12.68	0.82	0.76
Core_3	6.00	37.24	8.12	Core_6	6.00	34.43	6.61	0.62	0.56
Core_3	7.00	36.36	7.77	Core_7	7.00	35.25	7.22	0.33	0.33
Core_3	7.00	36.36	7.77	Core_8	7.00	34.83	7.34	0.68	0.66
Core_3	7.00	36.36	7.77	Core_9	7.00	34.39	4.06	0.73	0.56
Core_4	5.00	37.88	13.92	Core_6	5.00	35.22	7.07	0.84	0.66
Core_4	6.00	36.90	12.68	Core_7	6.00	35.23	7.91	0.43	0.38
Core_4	6.00	36.90	12.68	Core_8	6.00	35.50	7.81	0.77	0.68
Core_4	6.00	36.90	12.68	Core_9	6.00	34.26	4.43	0.49	0.29
Core_6	6.00	34.43	6.61	Core_7	6.00	35.80	7.75	0.80	0.77
Core_6	6.00	34.43	6.61	Core_8	6.00	36.17	7.04	0.69	0.66
Core_6	6.00	34.43	6.61	Core_9	6.00	34.70	4.35	0.56	0.51
Core_7	7.00	35.25	7.22	Core_8	7.00	34.83	7.34	0.40	0.40
Core_7	7.00	35.25	7.22	Core_9	7.00	34.39	4.06	0.62	0.53
Core_8	7.00	34.83	7.34	Core_9	7.00	34.39	4.06	0.66	0.56

Moreover, although one can not estimate an OCCC under model (2.25) with an unstructured covariance matrix, Helenowski et. al. (2011) showed that for balanced complete data, the OCCC computed under an unstructured covariance matrix will be equivalent to the OCCC computed under compound symmetry. This follows by noting that for balanced complete data, the variance and covariance parameters under compound symmetry can be treated as the average of the variances and covariances of an unstructured covariance matrix. Hence using the SAS macro %CCC in combination with output from PROC MIXED, one can fit the MCM2 data to model (2.25) assuming compound symmetry and obtain 1) an estimated CCC matrix assuming compound symmetry and 2) obtain a reasonable estimate of the overall CCC corresponding to the pairwise estimates shown in Output (2.13). In similar fashion, we can investigate what the CCC matrix and overall CCC would be assuming some form of spatial correlation. Presented below are the SAS statements needed to fit the MCM2 data to model (2.25) assuming either compound symmetry or spatial linearity and compute the corresponding CCC matrix and OCCC.

Program 2.12

```
ods output solutionF=pe1;
ods output covparms=var1;
ods output R=rvar1;
ods output infocrit=IC1;
ods output Dimensions=no_subjects1;
proc mixed data=Example2_2_4 ic;
  class Subject Core ;
  model MCM2Index = Core /solution;
  repeated Core / subject=Subject r=2 rcorr=2 type=cs;
run;

ods output solutionF=pe2;
ods output covparms=var2;
ods output R=rvar2;
ods output infocrit=IC2;
ods output Dimensions=no_subjects2;
proc mixed data=Example2_2_4 ic;
  class Subject Core ;
  model MCM2Index = Core /solution;
  repeated /subject=Subject r=2 rcorr=2 type=sp(lin) (row col);
run;

data IC1;
 set IC1;
 Model=1;
 Covariance='Compound Symmetry';
run;
data IC2;
 set IC2;
 Model=2;
 Covariance='Spatial Linearity';
run;
```

```
 title 'CCC matrix and OCCC assuming compound symmetry';
%CCC(structure=compound symmetric,
     effect=Core,
     cov=Rvar1,
     mean=pe1,
     n=no_subjects1,
     AIC=ic1);
run;
title 'CCC matrix and OCCC assuming spatial linearity';
%CCC(structure=spatially linear,
     effect=Core,
     cov=Rvar2,
     mean=pe2,
     n=no_subjects2,
     AIC=ic2);
run;
data AIC;
 set ic1 ic2;
run;
proc print data=AIC;
  id Model Covariance;
  var Parms Neg2LogLike AIC AICC HQIC BIC;
run;
```

The results of this analysis are shown in Output 2.15. At first glance, it appears as though there is better overall agreement assuming spatial linearity. However, when one compares the Akaike information criteria (AIC) and other likelihood-based information criteria, we find a better fit to the data assuming compound symmetry versus spatial linearity. On this basis and the fact that for balanced data, the OCCC corresponding to compound symmetry is equivalent to an OCCC for a completely unstructured covariance matrix, it is preferable to report the estimated OCCC assuming compound symmetry. In this example, we have a very limited number of subjects and hence we can not rule out the possibility that there may exist some degree of spatial correlation. Moreover, with so few subjects, we likely do not have enough spread in the data to better reflect the degree of variation and correlation in the MCM2 index across different sections of the prostate gland. Nonetheless, despite these limitations, there is some evidence to suggest the MCM2 biomarker is fairly reproducible across the different sections of the prostate gland—a result that needs further evaluation in larger numbers of individuals. Finally, we point out that the results presented in Output 2.15 differ from that reported by Helenowski et. al. (2011) in that we based results on the untransformed MCM2 data whereas Helenowski et. al. based their results on transformed rank-ordered data.

Output 2.15: CCC Matrices and OCCC under compound symmetry and spatial linearity. Included are likelihood-based information criteria for comparing how well each model fits the data.

```
CCC matrix and OCCC assuming compound symmetry
Summary statistics (N, Mean, SD), Correlation (R) matrix,
and Concordance Correlation (Rc) matrix.
CCC's and OCCC assuming a compound symmetric covariance matrix.
```

Core	N	Mean	SD
1	7.00	35.74	8.62
2	7.00	33.99	8.62
3	7.00	36.36	8.62
4	7.00	37.04	8.62
6	7.00	33.54	8.62
7	7.00	35.25	8.62
8	7.00	34.83	8.62
9	7.00	34.39	8.62

Core	R							
1	1.0000	0.4800	0.4800	0.4800	0.4800	0.4800	0.4800	0.4800
2	0.4800	1.0000	0.4800	0.4800	0.4800	0.4800	0.4800	0.4800
3	0.4800	0.4800	1.0000	0.4800	0.4800	0.4800	0.4800	0.4800
4	0.4800	0.4800	0.4800	1.0000	0.4800	0.4800	0.4800	0.4800
6	0.4800	0.4800	0.4800	0.4800	1.0000	0.4800	0.4800	0.4800
7	0.4800	0.4800	0.4800	0.4800	0.4800	1.0000	0.4800	0.4800
8	0.4800	0.4800	0.4800	0.4800	0.4800	0.4800	1.0000	0.4800
9	0.4800	0.4800	0.4800	0.4800	0.4800	0.4800	0.4800	1.0000

Core	Rc							
1	1.0000	0.4703	0.4788	0.4747	0.4648	0.4793	0.4773	0.4741
2	0.4703	1.0000	0.4626	0.4518	0.4794	0.4749	0.4778	0.4795
3	0.4788	0.4626	1.0000	0.4785	0.4557	0.4761	0.4726	0.4678
4	0.4747	0.4518	0.4785	1.0000	0.4435	0.4700	0.4648	0.4584
6	0.4648	0.4794	0.4557	0.4435	1.0000	0.4707	0.4747	0.4777
7	0.4793	0.4749	0.4761	0.4700	0.4707	1.0000	0.4794	0.4776
8	0.4773	0.4778	0.4726	0.4648	0.4747	0.4794	1.0000	0.4794
9	0.4741	0.4795	0.4678	0.4584	0.4777	0.4776	0.4794	1.0000

	Overall Rc (OCCC)
Overall Measure of Agreement	0.5166

Summary of Goodness-of-Fit for an assumed compound symmetric covariance matrix

Information Criteria

	Neg2LogLike	Parms	AIC	AICC	BIC	Model
Criteria	326.42781	2	330.42781	330.70688	330.31963	1

CCC matrix and OCCC assuming spatial linearity
Summary statistics (N, Mean, SD), Correlation (R) matrix,
and Concordance Correlation (Rc) matrix.
CCC's and OCCC assuming a spatially linear covariance matrix.

Core	N	Mean	SD
1	7.00	35.74	12.44
2	7.00	33.99	12.44
3	7.00	36.36	12.44
4	7.00	37.16	12.44
6	7.00	33.38	12.44
7	7.00	35.25	12.44
8	7.00	34.83	12.44
9	7.00	34.39	12.44

Core	R							
1	1.0000	0.7209	0.4418	0.1627	0.7209	0.6053	0.3759	0.1175
2	0.7209	1.0000	0.7209	0.4418	0.6053	0.7209	0.6053	0.3759
3	0.4418	0.7209	1.0000	0.7209	0.3759	0.6053	0.7209	0.6053
4	0.1627	0.4418	0.7209	1.0000	0.1175	0.3759	0.6053	0.7209
6	0.7209	0.6053	0.3759	0.1175	1.0000	0.7209	0.4418	0.1627
7	0.6053	0.7209	0.6053	0.3759	0.7209	1.0000	0.7209	0.4418
8	0.3759	0.6053	0.7209	0.6053	0.4418	0.7209	1.0000	0.7209
9	0.1175	0.3759	0.6053	0.7209	0.1627	0.4418	0.7209	1.0000

Core	Rc							
1	1.0000	0.7138	0.4413	0.1617	0.7082	0.6048	0.3749	0.1168
2	0.7138	1.0000	0.7081	0.4280	0.6046	0.7172	0.6039	0.3758
3	0.4413	0.7081	1.0000	0.7194	0.3655	0.6029	0.7155	0.5978
4	0.1617	0.4280	0.7194	1.0000	0.1123	0.3716	0.5949	0.7035
6	0.7082	0.6046	0.3655	0.1123	1.0000	0.7129	0.4389	0.1622
7	0.6048	0.7172	0.6029	0.3716	0.7129	1.0000	0.7205	0.4408
8	0.3749	0.6039	0.7155	0.5949	0.4389	0.7205	1.0000	0.7205
9	0.1168	0.3758	0.5978	0.7035	0.1622	0.4408	0.7205	1.0000

```
                                Overall Rc (OCCC)
Overall Measure of Agreement              0.5612
Summary of Goodness-of-Fit for an assumed spatially linear covariance matrix
                             Information Criteria
           Neg2LogLike   Parms        AIC        AICC         BIC   Model
Criteria     341.23659       2  345.23659   345.51566   345.12841       2
Model   Covariance           Parms   Neg2LogLike     AIC     AICC     BIC
    1   Compound Symmetry      2          326.4    330.4    330.7   330.3
    2   Spatial Linearity      2          341.2    345.2    345.5   345.1
```

2.3 Summary

In this chapter, we present a class of marginal linear models for applications involving correlated data. The marginal linear model (2.1) mimics the usual univariate linear model except that we now consider a vector of observations per individual rather than a single observation. Consequently, while observations from different experimental units or individuals are assumed to be independent, observations within individuals are dependent and correlated. The marginal LM allows one to directly specify a variance-covariance structure describing the variability and correlation among the vector of observations per unit. Under normality assumptions, one can estimate the population parameters and conduct inference using estimated generalized least squares (EGLS), maximum or restricted maximum likelihood (ML, REML), or generalized estimating equations (GEE). The EGLS, ML and REML techniques may be implemented for balanced or unbalanced data using the SAS procedures MIXED or GLIMMIX, while GEE is the method of choice with the GENMOD procedure. In each of these procedures, inference in the form of hypothesis testing and confidence intervals can be carried out under the assumed covariance structure or one can perform robust inference using a so-called empirical "sandwich" estimator of the variance-covariance matrix of the estimated regression parameters.

The methods of estimation and inference presented in this chapter ignore any weighting of the observations one may wish to include in the analysis. This is easily accomplished by simply replacing $\boldsymbol{\Sigma}_i(\boldsymbol{\theta})$ with $\boldsymbol{\Sigma}_i^*(\boldsymbol{\theta}) = \boldsymbol{W}_i^{-1/2}\boldsymbol{\Sigma}_i(\boldsymbol{\theta})\,\boldsymbol{W}_i^{-1/2}$ where $\boldsymbol{W}_i$ is a diagonal matrix of weights associated with $\boldsymbol{y}_i$, the vector of observations on the ith subject/cluster.

As we have illustrated in three of the four examples presented in this chapter, the marginal LM (2.1) may be used to perform a multivariate analysis of variance (MANOVA) or, more generally, a generalized multivariate analysis of variance (GMANOVA) for applications involving repeated measurements and longitudinal data. While the repeated measures MANOVA and GMANOVA models were initially developed for balanced and complete data, the MIXED, GLIMMIX and GENMOD procedures can all be applied to unbalanced data. The fourth example (i.e., the MCM2 data) illustrates an application in which a single parameter, the concordance correlation coefficient, is a function of the marginal means, variances and covariances. This example illustrates the flexibility with which one can model an appropriate covariance structure suitable for estimating the overall reproducibility of a biomarker. While one can certainly think of numerous other applications, it is hoped that these examples will demonstrate just how powerful the various options in MIXED, GLIMMIX and GENMOD are for handling marginal linear models for correlated data.

3

Linear Mixed-Effects Models—Normal Theory

Within linear models, there are two distinct stochastic mechanisms for determining the overall correlation structure associated with repeated measurements and/or clustered data. In the marginal linear model of the preceding chapter, we account for correlation among observations from the same individual or unit by specifying a within-unit random error vector and its covariance structure. This approach assumes that random error vectors between units are independently distributed according to a multivariate normal distribution that varies between units only its dimension. In the repeated measurements setting, for example, we might regard the vector of observations per subject as representing a single time series in which each subject's random error vector is assumed to be normally distributed and serial correlation can be described by a first-order autoregressive structure. Although the resulting marginal covariance matrix may vary in dimension between individuals, it will still share a common set of covariance parameters across individuals. Likewise for clustered data, one may think of within-cluster observations as having a common variance and correlation while observations from different clusters are assumed to be independent. This results in a marginal LM having a compound symmetric covariance structure that varies from cluster to cluster only in dimension. Both these examples illustrate one stochastic mechanism for describing correlated data, namely specifying the marginal distribution of a random error vector associated with each unit or subject.

A second mechanism would be to include a set of subject-specific or cluster-specific random effects into the linear model framework. For example, the inclusion of a single subject-specific random effect into a marginal LM would result in 1) an additional source of variation between subjects and 2) correlation between those measurements that share a common random effect from the same subject. In this chapter, we focus on the linear-mixed-effects (LME) model obtained by including one or more subject-specific random effects into the marginal LM (2.1) of the previous chapter. As with the marginal LM, we will consider various aspects of population-averaged (PA) inference as it pertains to population-based estimation and hypothesis testing under the LME model. In addition, we also consider subject-specific (SS) inference as it pertains to individual predictions based on random-effect estimation. Several examples illustrating both PA and SS inference will be presented. Note that here and throughout the remainder of this book, we shall use the generic term "subject" to mean any distinct experimental unit, individual, or cluster over which two or more correlated observations are taken. Likewise, the term "subject-specific" will be used when referring to parameters and/or covariates that are specific to a given "subject." Thus, depending on the context, "subject-specific" may be used interchangeably with "unit-specific" and "cluster-specific."

3.1 The linear mixed-effects (LME) model

As noted above, a LME model may be obtained as an extension of a marginal LM by including one or more random effects that are shared across observations from the same subject (cluster, block, etc.). These unobserved random effects are not parameters in that they are NOT fixed unknown constants that describe a mean response across all subjects coming from the same population such as the set of fixed-effects regression parameters from a marginal LM. Rather, these are random values associated with randomly selected individuals and each individual's set of random effects uniquely determines that individual's average response (i.e., subject-specific mean response). Inclusion of such random effects may be dictated by study design such as with randomized block designs where random blocking factors are used to form more homogeneous responses within blocks. Alternatively, random effects may be required in either experimental or observational settings as a means for explaining heterogeneity across observations. Regardless of the underlying reason for including random effects, it is clear from the following general relation,

$$Cov(y_1 + b,\, y_2 + b) = Cov(y_1, y_2) + Cov(y_1, b) + Cov(b, y_2) + Var(b),$$

that observations that share a common random effect will exhibit some level of intra-class correlation even when the random components of those observations are mutually independent (such as determined by the non-zero variance component, $Var(b)$, above when y_1, y_2 and b are all independent).

Linear mixed-effects (LME) models—correlated data

Linear mixed-effects models have been around in one form or another for a number of years principally in the areas of random effects and variance components models (e.g., Henderson, 1953; Harville, 1977), and random coefficient growth curve models (e.g., Rao, 1965; Swamy, 1970; Fearn, 1975). An excellent overview of both the methodology and history of random effects models and variance components models may be found in Searle, Casella, and McCulloch (1992). However, it was not until the landmark papers of Harville (1976, 1977) and Laird and Ware (1982) and the subsequent advances in computational software that linear mixed-effects models for correlated data, particularly repeated measurements and longitudinal data, came into widespread use. Following Laird and Ware (1982), the LME model may be written as:

$$\boldsymbol{y}_i = \boldsymbol{X}_i\boldsymbol{\beta} + \boldsymbol{Z}_i\boldsymbol{b}_i + \boldsymbol{\epsilon}_i, \;\; i = 1, \ldots, n \text{ units (or subjects)} \tag{3.1}$$

where

$\boldsymbol{y}_i$ is a $p_i \times 1$ response vector, $\boldsymbol{y}_i = (y_1 \ldots y_{p_i})'$, on the ith subject,
$\boldsymbol{\beta}$ is an $s \times 1$ unknown parameter vector,
$\boldsymbol{b}_i$ is a $v \times 1$ vector of subject-specific random-effects (possibly multi-level). The $\boldsymbol{b}_i$'s are assumed to be independent and identically distributed as $\boldsymbol{b}_i \sim$iid $N_v(\boldsymbol{0}, \boldsymbol{\Psi}(\boldsymbol{\theta}_b))$ with covariance matrix $\boldsymbol{\Psi}(\boldsymbol{\theta}_b)$ that depends on a vector of between-subject covariance parameters, $\boldsymbol{\theta}_b$,

$\boldsymbol{X}_i$ is a $p_i \times s$ design matrix of within- and between-subject covariates linking the fixed-effects parameter vector $\boldsymbol{\beta}$ to the marginal (i.e., $E(\boldsymbol{y}_i) = \boldsymbol{X}_i\boldsymbol{\beta}$) and conditional (i.e., $E(\boldsymbol{y}_i|\boldsymbol{b}_i) = \boldsymbol{X}_i\boldsymbol{\beta} + \boldsymbol{Z}_i\boldsymbol{b}_i$) means,
$\boldsymbol{Z}_i$ is a $p_i \times v$ matrix of constants containing within-subject covariates linking the random effects to the conditional means,
$\boldsymbol{\epsilon}_i$ is a $p_i \times 1$ within-subject random error vector. The $\boldsymbol{\epsilon}_i$'s are assumed to be distributed as $\boldsymbol{\epsilon}_i \sim$ind $N_{p_i}(\boldsymbol{0}, \boldsymbol{\Lambda}_i(\boldsymbol{\theta}_w))$ independently of the $\boldsymbol{b}_i$'s and of each other. The variance-covariance matrices, $\boldsymbol{\Lambda}_i(\boldsymbol{\theta}_w)$, depend on a common vector of within-subject covariance parameters, $\boldsymbol{\theta}_w$.

As with the marginal LM, we differentiate experimental units, subjects or clusters by indexing them with the subscript i. Based on this formulation, the conditional distribution of $\boldsymbol{y}_i$ given $\boldsymbol{b}_i$ is

$$\boldsymbol{y}_i|\boldsymbol{b}_i \sim N_{p_i}(\boldsymbol{X}_i\boldsymbol{\beta} + \boldsymbol{Z}_i\boldsymbol{b}_i,\ \boldsymbol{\Lambda}_i(\boldsymbol{\theta}_w)) \tag{3.2}$$

independently for each subject. By taking expectations with respect to the random effects, it is straightforward to show that under the assumption of independence between the random effects $\boldsymbol{b}_i$ and within-subject errors $\boldsymbol{\epsilon}_i$, the marginal distribution of $\boldsymbol{y}_i$ is

$$\boldsymbol{y}_i \sim N_{p_i}(\boldsymbol{X}_i\boldsymbol{\beta}, \boldsymbol{\Sigma}_i(\boldsymbol{\theta})) \tag{3.3}$$

where

$$\boldsymbol{\Sigma}_i(\boldsymbol{\theta}) = \boldsymbol{Z}_i\boldsymbol{\Psi}(\boldsymbol{\theta}_b)\boldsymbol{Z}_i' + \boldsymbol{\Lambda}_i(\boldsymbol{\theta}_w) \tag{3.4}$$

is the marginal variance-covariance matrix of $\boldsymbol{y}_i$. Here $\boldsymbol{\Sigma}_i(\boldsymbol{\theta})$ depends on both $\boldsymbol{\theta} = (\boldsymbol{\theta}_b, \boldsymbol{\theta}_w)$ and the matrix $\boldsymbol{Z}_i$ of within-subject covariates. While some degree of correlation may be accounted for via direct specification of the intra-subject covariance matrix, $\boldsymbol{\Lambda}_i(\boldsymbol{\theta}_w)$, correlation in the form of the singular matrix, $\boldsymbol{Z}_i\boldsymbol{\Psi}(\boldsymbol{\theta}_b)\boldsymbol{Z}_i'$, will occur as a result of having shared random effects across observations from the same subject. As a matter of notation, readers should be aware that within the SAS documentation for MIXED and GLIMMIX, the random-effects (G-side) covariance matrix $\boldsymbol{\Psi}(\boldsymbol{\theta}_b)$ is denoted by $\boldsymbol{G}$ while the within-subject (R-side) covariance matrix $\boldsymbol{\Lambda}_i(\boldsymbol{\theta}_w)$ is denoted by $\boldsymbol{R}$.

3.1.1 Features of the LME model

Before we discuss estimation and inference under the LME model, we first consider some of its unique features. We start by considering some of the more common LME models that researchers frequently encounter in practice. We examine how these models compare with similarly structured marginal LM's in terms of overall interpretability and ease of use. We also consider the variety of variance-covariance structures available within the family of LME models paying particular attention to the difficulties associated with determining an appropriate structure.

Repeated measures ANOVA models

As discussed in the previous chapter, one of the most commonly used tools for analyzing repeated measurements data is the repeated measures analysis of variance (RM-ANOVA) model. Such models occur frequently in both regulatory and non-regulatory clinical trials designed to compare different treatments administered both within and between subjects. Consider, for example, a single-sample repeated measures design whereby n subjects each receive p different treatments according to some subject-specific randomization scheme. Assuming sufficient washout between different treatments, this crossover study can be viewed as having risen from a randomized complete block design where there are n blocks (i.e., subjects) and within each block, p treatments are applied in random order over time. Assuming the subjects are chosen at random from a population of subjects, the linear model for this randomized block design would be

$$y_{ij} = \mu + b_i + \tau_j + \epsilon_{ij}, \; (i = 1, \ldots, n; \; j = 1, \ldots p)$$

where μ is an overall mean, b_i is a random effect associated with the ith block or subject, τ_j is a fixed effect associated with the jth treatment, $\mu_j = \mu + \tau_j$ is the jth treatment mean, and ϵ_{ij} is within-subject experimental error associated with response y_{ij} from subject i following treatment j. Under the assumption that the $b_i \sim$iid $N(0, \sigma_b^2)$ and that the $\epsilon_{ij} \sim$iid $N(0, \sigma_w^2)$ independent of b_i, one can easily verify that
$Var(y_{ij}) = \sigma_b^2 + \sigma_w^2$, $Cov(y_{ij}, y_{ij'}) = \sigma_b^2$ for $j \neq j'$ and $Cov(y_{ij}, y_{i'j'}) = 0$ for all $i \neq i'$.

This is an example of a linear mixed-effects (LME) model in that, unlike a marginal LM, it entails both random- and fixed-effects parameters. Following the matrix notation of the LME model (3.1), it may be written in terms of the cell means model

$$\boldsymbol{y}_i = \boldsymbol{\beta} + \mathbf{1}_p b_i + \boldsymbol{\epsilon}_i, \;\; b_i \sim N(0, \sigma_b^2), \; \boldsymbol{\epsilon}_i \sim N_p(\mathbf{0}, \sigma_w^2 \boldsymbol{I}_p) \tag{3.5}$$

where $\mathbf{1}_p$ is a $p \times 1$ within-block design vector of all one's (i.e., $\mathbf{1}_p = \boldsymbol{Z}_i$), $\boldsymbol{\beta}' = (\mu + \tau_1, \ldots, \mu + \tau_p) = (\mu_1, \ldots, \mu_p)$ is the vector of treatment means and $\boldsymbol{\epsilon}_i$ is the $p \times 1$ within-subject vector of experimental errors associated with the ith block or subject. The marginal variance-covariance matrix of $\boldsymbol{y}_i$ under (3.5) is

$$\boldsymbol{\Sigma}(\boldsymbol{\theta}) = \begin{pmatrix} \sigma_b^2 + \sigma_w^2 & \sigma_b^2 & \ldots & \sigma_b^2 \\ \sigma_b^2 & \sigma_b^2 + \sigma_w^2 & \ldots & \sigma_b^2 \\ \vdots & \vdots & \vdots & \vdots \\ \sigma_b^2 & \sigma_b^2 & \ldots & \sigma_b^2 + \sigma_w^2 \end{pmatrix}. \tag{3.6}$$

where $\boldsymbol{\theta} = (\sigma_b^2, \sigma_w^2)$ is the vector of between- and within-subject variance components.

This LME model and its covariance structure resembles that of a marginal LM with compound symmetry in that one has a common variance $\sigma^2 = \sigma_b^2 + \sigma_w^2$ and common intraclass correlation $\rho = \sigma_b^2/(\sigma_b^2 + \sigma_w^2)$ across each subject's vector of observations [i.e., $\boldsymbol{y}_i' = (y_{i1}, \ldots, y_{ip})$]. In fact, from Chapter 2, we can represent this single-sample repeated measures design directly in terms of the marginal LM:

$$\boldsymbol{y}_i = \boldsymbol{\beta} + \boldsymbol{\epsilon}_i, \;\; \boldsymbol{\epsilon}_i \sim N_p(\mathbf{0}, \boldsymbol{\Sigma}(\boldsymbol{\theta}^*)) \tag{3.7}$$

where $\boldsymbol{\Sigma}(\boldsymbol{\theta}^*)$ is the compound symmetric covariance matrix

$$\boldsymbol{\Sigma}(\boldsymbol{\theta}^*) = \sigma^2(1 - \rho)\boldsymbol{I}_p + \sigma^2 \rho \boldsymbol{J}_p,$$

$\boldsymbol{J}_p$ is a $p \times p$ matrix of one's, and $\boldsymbol{\theta}^* = (\sigma^2, \rho)$ is the vector of marginal variance-covariance parameters.

In comparing these two models we notice one subtle but important difference. Under the LME model (3.5), the covariance parameter, σ_b^2, is a variance component associated with the random block effects (i.e., subjects) and hence must be non-negative. Consequently, the intraclass correlation coefficient, $\rho = \sigma_b^2/(\sigma_b^2 + \sigma_w^2)$, must also be non-negative. In contrast, under the marginal LM (3.7), the intraclass correlation coefficient ρ is interpreted directly as a marginal correlation between any two measurements from the same subject and, as such, ρ may range from -1 to 1. If, for a given application, the intraclass correlation is truly negative, then it is entirely possible under a LME model to obtain a negative variance component estimate. When this occurs, the MIXED procedure will, by default, set the variance component to 0. However, there are a number of issues pertaining to this including the possibility of having inflated type I or type II errors depending on how one handles the presence of negative variance components (see, for example, Chapter 4, section 4.7 of Littell et. al., 2006). In addressing these issues, there are options in PROC MIXED such as the NOBOUND option and the METHOD=TYPE3 option that enable one to conduct inference under a LME model using negative variance component estimates. One scenario where we should allow for negative variance components is illustrated below for a repeated measures split-plot ANOVA.

Consider a two-group repeated measures design where n subjects from each of two groups ($2n$ subjects in total) have measurements taken on each of p different occasions corresponding, possibly, to p different experimental conditions. We may view this two-group repeated measures design as representing a "split-plot" design with subjects serving as random whole plots, the two groups of subjects serving as whole-plot treatment groups and the p occasions or conditions within each subject serving as "split-plot" treatments. The "split-plot" repeated measures ANOVA would then look like

$$y_{ijk} = \mu + \gamma_j + b_{i(j)} + \tau_k + (\gamma\tau)_{jk} + \epsilon_{ijk},\ (i = 1, \ldots, n; j = 1, 2; k = 1, \ldots, p)$$

where μ is the overall mean, γ_j is the jth "whole-plot" treatment group effect ($j = 1, 2$), $b_{i(j)}$ is the subject-specific random effect ("whole-plot error") associated with the ith subject within the jth group, τ_k is the kth "split-plot" treatment effect applied to each subject, $(\gamma\tau)_{jk}$ is the interaction term and ϵ_{ijk} is within-subject experimental error ("split-plot error"). Expressing the mean response of y_{ijk} in terms of the "cell means," $\mu_{jk} = \mu + \gamma_j + \tau_k + (\gamma\tau)_{jk}$, we can write this split-plot ANOVA model in terms of the following LME model

$$\begin{aligned} \boldsymbol{y}_i &= a_i\boldsymbol{\mu}_1 + (1 - a_i)\boldsymbol{\mu}_2 + \mathbf{1}_p b_i + \boldsymbol{\epsilon}_i,\ (i = 1, \ldots, 2n) \\ &= \boldsymbol{X}_i\boldsymbol{\beta} + \boldsymbol{Z}_i b_i + \boldsymbol{\epsilon}_i \end{aligned} \tag{3.8}$$

where

$\boldsymbol{y}_i$ is the $p \times 1$ vector of observations for the ith subject ($i = 1, \ldots, 2n$),
$\boldsymbol{\mu}_j = (\mu_{j1}, \ldots, \mu_{jp})'$ is the "cell" mean vector for the jth group across p treatment conditions ($j = 1, 2$),
$a_i = \begin{cases} 1 & \text{if subject } i \text{ is in group 1} \\ 0 & \text{if subject } i \text{ is in group 2} \end{cases}$,
$\boldsymbol{X}_i = (\boldsymbol{a}_i' \otimes \boldsymbol{I}_p)$ is the fixed-effects design matrix with $\boldsymbol{a}_i' = (a_i \quad 1 - a_i)$
$\boldsymbol{\beta} = \begin{pmatrix} \boldsymbol{\mu}_1 \\ \boldsymbol{\mu}_2 \end{pmatrix}$ is the $2p \times 1$ vector of fixed-effects "cell mean" parameters,
$\boldsymbol{Z}_i = \mathbf{1}_p$ is the $p \times 1$ within-subject design vector of all one's
b_i are subject-specific random effects satisfying $b_i \sim$iid $N(0, \sigma_b^2)$,
$\boldsymbol{\epsilon}_i \sim$iid $N_p(\mathbf{0}, \sigma_w^2\boldsymbol{I}_p)$ are within-subject error vectors assumed to be independently distributed from b_i.

Under the above assumptions, the marginal variance-covariance matrix of $\boldsymbol{y}_i$ will be the same as that of the single-sample repeated measures design as shown in (3.6).

If we now simply drop off the random effects, b_i, from model (3.8) and replace the intra-subject covariance matrix $\boldsymbol{\Lambda}_i(\boldsymbol{\theta}_w) = \sigma_w^2\boldsymbol{I}_p$ with the marginal variance-covariance matrix $\boldsymbol{\Sigma}(\boldsymbol{\theta}^*) = \sigma^2(1-\rho)\boldsymbol{I}_p + \sigma^2\rho\boldsymbol{J}_p$, we end up with a marginal LM analogous to the GMANOVA model (2.20) with a structured variance-covariance matrix, namely

$$\boldsymbol{Y} = \boldsymbol{A\Gamma X} + \boldsymbol{\epsilon} \tag{3.9}$$

with $\boldsymbol{Y} = [\boldsymbol{y}_1 \ldots \boldsymbol{y}_{2n}]'$, $\boldsymbol{A} = [\boldsymbol{a}_1 \ldots \boldsymbol{a}_{2n}]'$, $\boldsymbol{\Gamma} = \begin{pmatrix} \boldsymbol{\mu}_1' \\ \boldsymbol{\mu}_2' \end{pmatrix}$, $\boldsymbol{X} = \boldsymbol{I}_p$ and $Vec(\boldsymbol{\epsilon}) \sim N_{2np}(\mathbf{0}, \boldsymbol{\Sigma}(\boldsymbol{\theta}^*) \otimes \boldsymbol{I}_{2n})$. As shown in (2.21), this may be rewritten in terms of the marginal LM as

$$\boldsymbol{y}_i = \boldsymbol{X}_i\boldsymbol{\beta} + \boldsymbol{\epsilon}_i, \ (i = 1, \ldots, 2n) \tag{3.10}$$

where $\boldsymbol{X}_i = (\boldsymbol{a}_i' \otimes \boldsymbol{I}_p)$, $\boldsymbol{\beta} = Vec(\boldsymbol{\Gamma}') = \begin{pmatrix}\boldsymbol{\mu}_1\\ \boldsymbol{\mu}_2\end{pmatrix}$ and $\boldsymbol{\epsilon}_i \sim N_p(\boldsymbol{0}, \boldsymbol{\Sigma}(\boldsymbol{\theta}^*))$. Based on results from the previous chapter, it can be shown that under the GMANOVA model (3.9) or, equivalently, the marginal LM (3.10), the null hypothesis of no group main effect, $H_0: \boldsymbol{C\Gamma U} = \boldsymbol{U}'\boldsymbol{\Gamma}'\boldsymbol{C}' = \boldsymbol{L\beta} = 0$ [where $\boldsymbol{C} = (1, -1)$, $\boldsymbol{U} = \boldsymbol{1}_p$ and $\boldsymbol{L\beta} = [\boldsymbol{C} \otimes \boldsymbol{U}']\boldsymbol{\beta} = \boldsymbol{1}_p'(\boldsymbol{\mu}_1 - \boldsymbol{\mu}_2)$], can be tested via a two-sample t-test as

$$t = \sqrt{\frac{n}{2}}\boldsymbol{1}_p'(\widehat{\boldsymbol{\mu}}_1 - \widehat{\boldsymbol{\mu}}_2)/\sqrt{\boldsymbol{1}_p'\boldsymbol{\Sigma}(\widehat{\boldsymbol{\theta}}^*)\boldsymbol{1}_p}$$

where $\widehat{\boldsymbol{\beta}} = \begin{pmatrix}\widehat{\boldsymbol{\mu}}_1\\ \widehat{\boldsymbol{\mu}}_2\end{pmatrix}$ and $\boldsymbol{\Sigma}(\widehat{\boldsymbol{\theta}}^*) = \widehat{\sigma}^2(1-\widehat{\rho})\boldsymbol{I}_p + \widehat{\sigma}^2\widehat{\rho}\boldsymbol{J}_p$ are the ML or REML estimates of $\boldsymbol{\beta}$ and $\boldsymbol{\Sigma}(\boldsymbol{\theta}^*)$, respectively. The same test will hold for the LME model (3.9) based on $\boldsymbol{\Sigma}(\widehat{\boldsymbol{\theta}}) = \widehat{\sigma}_w^2\boldsymbol{I}_p + \widehat{\sigma}_b^2\boldsymbol{J}_p$. Moreover, for this balanced design case, we have $\boldsymbol{\Sigma}(\widehat{\boldsymbol{\theta}}^*) = \boldsymbol{\Sigma}(\widehat{\boldsymbol{\theta}})$ with $\widehat{\sigma}^2(1-\widehat{\rho}) = \widehat{\sigma}_w^2$ and $\widehat{\sigma}^2\widehat{\rho} = \widehat{\sigma}_b^2$.

At this point we note that if the estimated correlation coefficient $\widehat{\rho}$ is negative under the marginal LM (3.10), then the estimated covariance $\widehat{\sigma}^2\widehat{\rho}$ will also be negative. Similarly, if one uses the NOBOUND option in PROC MIXED so as to not constrain the variance component estimates, the estimated covariance parameter $\widehat{\sigma}_b^2$ of the LME model will likewise be negative despite the fact that σ_b^2 is the variance component associated with the random effect parameter, b_i. For either model, the t statistic for testing no group main effect involves the estimated standard error of $\boldsymbol{1}_p'(\widehat{\boldsymbol{\mu}}_1 - \widehat{\boldsymbol{\mu}}_2)$, namely

$$\begin{aligned}\sqrt{2(\boldsymbol{1}_p'\boldsymbol{\Sigma}(\widehat{\boldsymbol{\theta}}^*)\boldsymbol{1}_p)/n} &= \sqrt{2p\widehat{\sigma}^2[1+(p-1)\widehat{\rho}]/n}\\ &= \sqrt{2(p\widehat{\sigma}_w^2 + p^2\widehat{\sigma}_b^2)/n}\\ &= \sqrt{2(\boldsymbol{1}_p'\boldsymbol{\Sigma}(\widehat{\boldsymbol{\theta}})\boldsymbol{1}_p)/n}.\end{aligned}$$

It therefore follows that whenever $\widehat{\rho} < 0$ or, equivalently, $\widehat{\sigma}_b^2 < 0$, and one uses the default value of $\widehat{\sigma}_b^2 = 0$ in SAS when running the LME model (3.9), then one will have effectively inflated the estimated standard error and type II error for testing the null hypothesis of no group main effect. This may lead to incorrect inference depending on whether the intraclass correlation truly is negative as reflected in the corresponding noncentrality parameter

$$\delta(\boldsymbol{\Sigma}(\boldsymbol{\theta})) = \sqrt{\frac{n}{2}}\boldsymbol{1}_p'(\boldsymbol{\mu}_1 - \boldsymbol{\mu}_2)/\sqrt{\boldsymbol{1}_p'\boldsymbol{\Sigma}(\boldsymbol{\theta})\boldsymbol{1}_p} = \sqrt{\frac{n}{2}}\Delta/\sqrt{p\sigma_w^2 + p^2\sigma_b^2}$$

where $\Delta = \boldsymbol{1}_p'(\boldsymbol{\mu}_1 - \boldsymbol{\mu}_2)$. Since the noncentrality parameter increases with decreasing values of σ_b^2, the power of the test increases. Hence setting $\widehat{\sigma}_b^2 = 0$ when $\widehat{\sigma}_b^2$ is negative can yield very misleading results. Similar arguments can be used to show that setting a negative value of $\widehat{\sigma}_b^2$ to 0 will decrease the standard errors and artificially increase the power associated with tests of hypotheses on within-subject treatment comparisons.

As noted above, one can circumvent these problems by simply specifying the NOBOUND option within the MIXED or GLIMMIX procedures when analyzing RM-ANOVA models using a LME model formulation. Alternatively, one may simply choose to use a marginal LM specification for a RM-ANOVA as shown above for both the single-sample and two-sample repeated measures designs. A marginal LM will yield estimates, tests of hypotheses and confidence intervals equivalent to that of a LME model with negative variance components but with the advantage of having greater interpretability (i.e., not having to explain a negative variance component). Of course, in most repeated measures settings and, more generally, in the split-plot design setting, one would expect a positive correlation which explains why split-plot designs are generally thought to be inefficient with respect to whole-plot (between-subject) treatment comparisons and efficient with respect to split-plot (within-subject) treatment comparisons.

Random coefficient growth curve models

The random coefficient growth curve model, as developed by Rao (1965), Swamy (1970), Fearn (1975) and others, is another commonly encountered model within the LME family of models. Developed around the same time as the GMANOVA model, random coefficient regression models were seen by many authors as representing either a Bayesian or empirical Bayesian alternative to the GMANOVA model (e.g., Geisser, 1970; Lindley and Smith, 1972; Rosenberg, 1973; Fearn, 1975). Swamy (1970) considered random coefficient regression models for econometric panel data viewing such data as time series of cross sections. A major advantage with these models was the relaxation of the balanced data structure of the GMANOVA model.

While there are various formulations of the random coefficient growth curve model, we adopt the general two-stage approach described by Vonesh and Chinchilli (1997). Specifically, by considering a separate linear regression model for each subject or cluster in the first stage and then modeling the first-stage subject-specific regression parameters using multivariate regression in the second stage, we end up with the following description of a random coefficient growth curve model:

$$\begin{aligned} &\text{Stage 1: } \boldsymbol{y}_i|\boldsymbol{\beta}_i = \boldsymbol{Z}_i\boldsymbol{\beta}_i + \boldsymbol{\epsilon}_i,\ (i = 1, \ldots, n) \\ &\text{Stage 2: } \boldsymbol{\beta}_i = \boldsymbol{A}_i\boldsymbol{\beta} + \boldsymbol{b}_i \end{aligned} \tag{3.11}$$

where $\boldsymbol{y}_i$ is our usual $p_i \times 1$ vector of observations on the ith subject or cluster, $\boldsymbol{Z}_i$ is a $p_i \times v$ full-rank design matrix of within-subject regressors or time-dependent covariates with $\text{rank}(\boldsymbol{Z}_i) = v \le p_i$, $\boldsymbol{\beta}_i$ is a $v \times 1$ set of subject-specific regression coefficients for the ith subject, $\boldsymbol{A}_i$ is a $v \times s$ between-subject design matrix of time-independent covariates, $\boldsymbol{\beta}$ is an $s \times 1$ vector of fixed-effects regression parameters and $\boldsymbol{b}_i$ is a $v \times 1$ vector of subject-specific random effects. The within-subject error vectors, $\boldsymbol{\epsilon}_i$, are assumed to be conditionally iid $N_{p_i}(\boldsymbol{0}, \sigma_w^2\boldsymbol{I}_{p_i})$ independent of the $\boldsymbol{b}_i$ which are assumed to be iid $N_v(\boldsymbol{0}, \boldsymbol{\Psi}(\boldsymbol{\theta}_b))$.

Rao (1965) was the first to recognize this two-stage model as representing a special case of the LME model. Specifically, if we simply replace $\boldsymbol{\beta}_i$ in the first stage with its fixed and stochastic components, $\boldsymbol{A}_i\boldsymbol{\beta} + \boldsymbol{b}_i$, from the second stage, we can write the random coefficient growth curve model in terms of our LME model as

$$\begin{aligned} \boldsymbol{y}_i &= \boldsymbol{Z}_i\boldsymbol{A}_i\boldsymbol{\beta} + \boldsymbol{Z}_i\boldsymbol{b}_i + \boldsymbol{\epsilon}_i,\ (i = 1, \ldots, n) \\ &= \boldsymbol{X}_i\boldsymbol{\beta} + \boldsymbol{Z}_i\boldsymbol{b}_i + \boldsymbol{\epsilon}_i \end{aligned} \tag{3.12}$$

where $\boldsymbol{X}_i = \boldsymbol{Z}_i\boldsymbol{A}_i$ is the fixed-effects design matrix of both within-subject ($\boldsymbol{Z}_i$) and between-subject ($\boldsymbol{A}_i$) covariates. As with the general LME model, taking expectations with respect to the random effects, $\boldsymbol{b}_i$, yields the marginal distribution $\boldsymbol{y}_i \sim N_{p_i}(\boldsymbol{X}_i\boldsymbol{\beta}, \boldsymbol{\Sigma}_i(\boldsymbol{\theta}))$ with $\boldsymbol{\Sigma}_i(\boldsymbol{\theta}) = \boldsymbol{Z}_i\boldsymbol{\Psi}(\boldsymbol{\theta}_b)\boldsymbol{Z}_i' + \sigma_w^2\boldsymbol{I}_{p_i}$ where $\boldsymbol{\theta} = (\boldsymbol{\theta}_b, \sigma_w^2)$.

Several aspects of the random coefficient growth curve model are worth mentioning. First, the inherent structure of the variance-covariance matrix $\boldsymbol{\Sigma}_i(\boldsymbol{\theta})$ suggests that variation will usually increase with increasing values of the covariates contained within $\boldsymbol{Z}_i$. To illustrate, consider the simple random coefficient linear regression model

$$y_{ij} = \beta_{i0} + \beta_{i1}t_{ij} + \epsilon_{ij},\ (i = 1, \ldots, n; j = 1, \ldots, p_i) \tag{3.13}$$

where $\beta_{i0} = \beta_0 + b_{i0}$ is a random intercept and $\beta_{i1} = \beta_1 + b_{i1}$ is a random slope for the ith subject. Assuming $\boldsymbol{b}_i = \begin{pmatrix} b_{i0} \\ b_{i1} \end{pmatrix} \sim N_2(\mathbf{0}, \boldsymbol{\Psi}(\boldsymbol{\theta}_b))$ independent of $\epsilon_{ij} \sim N(0, \sigma_w^2)$ where $\boldsymbol{\Psi}(\boldsymbol{\theta}_b) = \begin{pmatrix} \sigma_{b_0}^2 & \sigma_{b_0 b_1} \\ \sigma_{b_0 b_1} & \sigma_{b_1}^2 \end{pmatrix}$, then for subject i having two observations at t_{i1} and t_{i2}, we have

$$\begin{aligned}
\boldsymbol{\Sigma}_i(\boldsymbol{\theta}) &= \boldsymbol{Z}_i \boldsymbol{\Psi}(\boldsymbol{\theta}_b) \boldsymbol{Z}_i' + \sigma_w^2 \boldsymbol{I}_2 \\
&= \begin{pmatrix} 1 & t_{i1} \\ 1 & t_{i2} \end{pmatrix} \begin{pmatrix} \sigma_{b_0}^2 & \sigma_{b_0 b_1} \\ \sigma_{b_0 b_1} & \sigma_{b_1}^2 \end{pmatrix} \begin{pmatrix} 1 & 1 \\ t_{i1} & t_{i2} \end{pmatrix} + \sigma_w^2 \begin{pmatrix} 1 & 0 \\ 0 & 1 \end{pmatrix} \\
&= \begin{pmatrix} \sigma_{b_0}^2 + 2\sigma_{b_0 b_1} t_{i1} + \sigma_{b_1}^2 t_{i1}^2 + \sigma_w^2 & \sigma_{b_0}^2 + \sigma_{b_0 b_1}(t_{i1} + t_{i2}) + \sigma_{b_1}^2 t_{i1} t_{i2} \\ \sigma_{b_0}^2 + \sigma_{b_0 b_1}(t_{i1} + t_{i2}) + \sigma_{b_1}^2 t_{i1} t_{i2} & \sigma_{b_0}^2 + 2\sigma_{b_0 b_1} t_{i2} + \sigma_{b_1}^2 t_{i2}^2 + \sigma_w^2 \end{pmatrix}.
\end{aligned}$$

Notice that in this example, if the two random effects, b_{i0} and b_{i1}, are either uncorrelated or positively correlated, then the marginal variances of the y_{ij} will always increase with increasing time t_{ij}, a phenomenon that occurs quite often with random coefficient growth curve models (see, for example, Figure 1.1 of the orange tree growth data).

A second feature of the random coefficient growth curve model is that, unlike the original GMANOVA model of Potthoff and Roy (1964), it readily accommodates unbalanced as well as incomplete data. It is easy to verify that whenever the data are balanced and complete with $p_i = p$ and $\boldsymbol{Z}_i = \boldsymbol{Z}$ for all i and whenever the between-subject design matrix has the form $\boldsymbol{A}_i = (\boldsymbol{I}_p \otimes \boldsymbol{a}_i')$ for a $q \times 1$ vector of unique between-subject covariates, then the random coefficient growth curve model written in terms of the LME model (3.12) will be equivalent to the GMANOVA model (2.20) with $\boldsymbol{X} = \boldsymbol{Z}'$ and $\boldsymbol{\Sigma} = \boldsymbol{Z}\boldsymbol{\Psi}\boldsymbol{Z}' + \sigma_w^2 \boldsymbol{I}_p$. Moreover, the random coefficient growth curve model is easily interpretable as a model where individuals each have their own regression curve and the overall (i.e., marginal) model depicts what the average regression curve will look like for individuals having the same set of covariates, $\boldsymbol{a}_i$. Thus, in the simple random coefficient linear regression model above, the parameters β_0 and β_1 describe what the average regression line looks like.

Finally, while the assumption is often made that the within-subject errors ϵ_{ij} are conditionally independent given $\boldsymbol{b}_i$, this assumption may be relaxed by allowing $Var(\boldsymbol{y}_i|\boldsymbol{b}_i) = \boldsymbol{\Lambda}_i(\boldsymbol{\theta}_w)$ where $\boldsymbol{\Lambda}_i(\boldsymbol{\theta}_w)$ follows some specified variance-covariance structure (e.g., AR(1), Banded Toeplitz, etc.). As discussed in the following section, the use of some sort of within-subject serial correlation may be more appropriate within certain random coefficient growth curve settings.

Variance-covariance structures

The LME model (3.1) offers a wide range of possible variance-covariance structures from which to choose. By specifying random effects in combination with an intra-subject covariance matrix, a more general covariance structure is possible including structures where the variances and covariances depend on within-subject covariates as we have seen for the random coefficient growth curve model. A number of authors have advocated more complex within-subject covariance structures for longitudinal studies including Diggle (1988), Chi and Reinsel (1989), Jones and Boadi-Boteng (1991), and Diggle, Liang and Zeger (1994) For instance, using the REPEATED statement of MIXED, one can specify an autoregressive structure (e.g., AR(1)) to reflect serial correlation over time within subjects and a RANDOM statement to specify subject-specific random intercepts and slopes to reflect increasing variation over time. With this approach, it may be difficult in some instances to identify or differentiate between the various sources of variation and correlation (e.g., Jones, 1990).

With the SAS procedure MIXED, one can specify any of the covariance structures available with the TYPE= option of the REPEATED or RANDOM statements. However,

Table 3.1 Some intra-subject variance-covariance structures available in SAS

Type	Structure: $(k,l)^{\text{th}}$ Element	$\boldsymbol{\theta}_w$
Independence	$\boldsymbol{\Lambda}_i(\boldsymbol{\theta}_w) = \sigma^2 \boldsymbol{I}_{p_i}$	σ^2
Compound Symmetry	$\boldsymbol{\Lambda}_i(\boldsymbol{\theta}_w) = \sigma^2(1-\rho)\boldsymbol{I}_{p_i} + \sigma^2\rho\boldsymbol{J}_{p_i}$	σ^2, ρ
Discrete time AR(1)	$\boldsymbol{\Lambda}_i(\boldsymbol{\theta}_w) = \sigma^2(\rho_{kl}),\ \rho_{kl} = \rho^{\lvert k-l \rvert}$	σ^2, ρ
Spatial (d_{kl})	$\boldsymbol{\Lambda}_i(\boldsymbol{\theta}_w) = \sigma^2(\rho_{kl}),\ \rho_{kl} = \exp\{-\rho d_{kl}\}$ where d_{kl} =Euclidian distance	σ^2, ρ
Banded Toeplitz (h+1)	$\boldsymbol{\Lambda}_i(\boldsymbol{\theta}_w) = \sigma^2(\rho_{kl}),\ \rho_{kl} = \begin{cases} \rho_m & m \le h \\ 0 & m > h \end{cases}$ $m = \lvert k-l \rvert,\ \rho_0 = 1,\ \boldsymbol{\rho} = (\rho_1 \ldots \rho_h)'$	$\sigma^2, \boldsymbol{\rho}$
Linear Covariances (dimension= h)	$\boldsymbol{\Lambda}_i(\boldsymbol{\theta}_w) = \theta_1 \boldsymbol{H}_{i1} + \ldots + \theta_h \boldsymbol{H}_{ih}$ for known matrices, $\boldsymbol{H}_{i1}, \ldots, \boldsymbol{H}_{ih}$ and unknown parameter vector $\boldsymbol{\theta}_w = (\theta_1 \ldots \theta_h)'$	$\boldsymbol{\theta}_w$

extreme care should be taken when specifying these structures in combination with each other as many of the available options will make little or no sense for the application in hand. Under the LME model (3.1), the most common covariance structures for the intra-subject covariance matrix, $\boldsymbol{\Lambda}_i(\boldsymbol{\theta}_w)$, are those shown in Table 3.1. With respect to the random effects variance-covariance matrix, $\boldsymbol{\Psi}(\boldsymbol{\theta}_b)$, the most common choices are the variance components structure (TYPE=VC) and the unstructured covariance (TYPE=UN).

A particularly useful feature with the MIXED procedure is the ability to model heterogeneity with respect to $\boldsymbol{\Lambda}_i(\boldsymbol{\theta}_w)$ and/or $\boldsymbol{\Lambda}_i(\boldsymbol{\theta}_b)$ between select groups of subjects. This can be accomplished using the GROUP= option of either the REPEATED or RANDOM statements. Other SAS features for specifying the variance-covariance structure of a LME model (or of a marginal LM) include the LOCAL=EXP(<effects>) option and the LOCAL=POM option of the REPEATED statement. The former allows users to model heterogeneity that occurs whenever the intra-subject variance varies with one or more covariates. In this case, the intra-subject variance structure has the form

$$\boldsymbol{\Lambda}_i(\boldsymbol{\theta}_w) = \sigma_w^2 \boldsymbol{H}_i(\boldsymbol{u}_i'\boldsymbol{\theta}_{w^*})$$

where $\boldsymbol{H}_i(\boldsymbol{u}_i'\boldsymbol{\theta}_{w^*}) = Diag[\exp\{\boldsymbol{u}_i'\boldsymbol{\theta}_{w^*}]$ is a diagonal matrix with $\boldsymbol{u}_i$ representing the ith row of some known design matrix of covariates, and $\boldsymbol{\theta}_{w^*}$ is an additional vector of unknown dispersion effects parameters. For example, rather than consider a simple random coefficient linear regression model as in (3.13) above, one might consider the alternative model

$$y_{ij} = \beta_{0i} + \beta_1 t_{ij} + \epsilon_{ij},\ (i = 1, \ldots, n; j = 1, \ldots, p_i)$$

where $\beta_{0i} = \beta_0 + b_i$ is a random intercept and β_1 is a fixed population slope. Assuming $b_i \sim N(0, \sigma_b^2)$ independent of ϵ_{ij} ~ind $N(0, \sigma_w^2 \exp(\theta_w \log(t_{ij}^2))$, then for subject i having two observations at t_{i1} and t_{i2}, we have

$$\begin{aligned}
\boldsymbol{\Sigma}_i(\boldsymbol{\theta}) &= \boldsymbol{Z}_i \boldsymbol{\Psi}(\boldsymbol{\theta}_b)\boldsymbol{Z}_i' + \sigma_w^2 \boldsymbol{I}_2 \\
&= \begin{pmatrix} 1 \\ 1 \end{pmatrix} \sigma_b^2 \begin{pmatrix} 1 & 1 \end{pmatrix} + \sigma_w^2 \begin{pmatrix} \exp(\theta_w)t_{i1}^2 & 0 \\ 0 & \exp(\theta_w)t_{i2}^2 \end{pmatrix} \\
&= \begin{pmatrix} \sigma_b^2 + \sigma_w^2 \exp(\theta_w)t_{i1}^2 & \sigma_b^2 \\ \sigma_b^2 & \sigma_b^2 + \sigma_w^2 \exp(\theta_w)t_{i2}^2 \end{pmatrix}.
\end{aligned}$$

Similar to the random coefficient linear regression model (3.13), here we have a model in which the variance also increases with time. Depending on the covariance parameter $\sigma_{b_0 b_1}$

of model (3.13), it may difficult to choose between these two covariance structures especially with regards to the marginal variances.

The option, LOCAL=POM, enables users to specify a power of the mean structure of the form

$$\boldsymbol{\Lambda}_i(\boldsymbol{\theta}_w) = \sigma_w^2 \boldsymbol{H}_i(\boldsymbol{\beta}^*, \theta)$$

where $\boldsymbol{H}_i(\boldsymbol{\beta}^*, \theta) = Diag\{|\boldsymbol{x}_i'\boldsymbol{\beta}^*|^\theta\}$ is a diagonal matrix containing the marginal means, in absolute value, raised to some power θ. With this option, the user must specify a value $\boldsymbol{\beta}^*$ of the fixed-effects parameters $\boldsymbol{\beta}$ as otherwise the model would be nonlinear in the parameter $\boldsymbol{\beta}$ [see Chapters 4-6 for more general approaches; see also Littell et. al., (2006, pp. 359-365)].

3.1.2 Estimation

Since the LME model (3.1) entails both fixed-effects parameters $\boldsymbol{\beta}$ and random-effects $\boldsymbol{b}_i$, we need to consider the problem of jointly estimating $\boldsymbol{\beta}$, $\boldsymbol{\theta} = (\boldsymbol{\theta}_b, \boldsymbol{\theta}_w)$ and $\boldsymbol{b}' = (\boldsymbol{b}_1', \ldots \boldsymbol{b}_n')$. This is because unlike the marginal LM, inference under a LME model need not be restricted to the population parameters $\boldsymbol{\beta}$ and $\boldsymbol{\theta}$. Indeed, under the LME we may wish to focus our inference on estimable functions of the fixed-effects parameters $\boldsymbol{\beta}$ that describe the population-averaged mean response

$$E(\boldsymbol{y}_i) = \boldsymbol{X}_i\boldsymbol{\beta}$$

or we may wish to focus inference on functions of both $\boldsymbol{\beta}$ and $\boldsymbol{b}_i$ that describe the individual's conditional mean response

$$E(\boldsymbol{y}_i|\boldsymbol{b}_i) = \boldsymbol{X}_i\boldsymbol{\beta} + \boldsymbol{Z}_i\boldsymbol{b}_i.$$

The latter form of inference is *subject-specific* in scope whereas the former is *population-wide* in scope. Consequently, we need consider estimating both the fixed-effects parameters $\boldsymbol{\beta}$ and the random effects $\boldsymbol{b}_i$ $(i = 1, \ldots, n)$. It should be noted that some authors prefer to describe the problem as one of estimating the fixed-effects parameters and "predicting" (rather than "estimating") the random effects since the random effects are not parameters per se. Here, we shall use both terms interchangeably since, as has been pointed out by Harville (1977) and others, one can think of the random effects as subject-specific "parameters."

There are three basic estimation strategies that have been adopted for the LME model. These strategies are:

1. Generalized Least Squares (GLS): A two-step approach in which an extended Gauss-Markov set-up is invoked resulting in a best linear unbiased estimator (BLUE) of $\boldsymbol{\beta}$ and a best linear unbiased predictor (BLUP) of $\boldsymbol{b}$ with estimation of $\boldsymbol{\theta}$ based on either method of moments or maximum likelihood (Henderson, 1963; Harville, 1976, 1977).
2. Likelihood-based estimation (ML, REML): A two-step approach in which $\boldsymbol{\beta}$ and $\boldsymbol{\theta}$ are jointly estimated by solving a set of maximum or restricted maximum likelihood estimating equations and the $\boldsymbol{b}_i$ are "estimated" via empirical Bayes estimation (Laird and Ware, 1982).
3. ML and REML estimation under a 'Bayesian' framework (Laird and Ware, 1982).

Here we consider the first strategy as this is the strategy implemented in the MIXED procedure of SAS. For the moment, let us fix the variance covariance parameters $\boldsymbol{\theta}$. Under the extended Gauss-Markov set-up of Harville (1976; 1977), joint estimates of the fixed-effects parameters $\boldsymbol{\beta}$ and random effects $\boldsymbol{b}$ are obtained by solving the so-called mixed model equations:

$$\begin{pmatrix} \boldsymbol{X}'\boldsymbol{\Lambda}_D^{-1}\boldsymbol{X} & \boldsymbol{X}'\boldsymbol{\Lambda}_D^{-1}\boldsymbol{Z} \\ \boldsymbol{Z}'\boldsymbol{\Lambda}_D^{-1}\boldsymbol{X} & \boldsymbol{Z}'\boldsymbol{\Lambda}_D^{-1}\boldsymbol{Z}+\boldsymbol{\Psi}_D^{-1} \end{pmatrix} \begin{pmatrix} \widehat{\boldsymbol{\beta}}(\boldsymbol{\theta}) \\ \widehat{\boldsymbol{b}}(\boldsymbol{\theta}) \end{pmatrix} = \begin{pmatrix} \boldsymbol{X}'\boldsymbol{\Lambda}_D^{-1}\boldsymbol{y} \\ \boldsymbol{Z}'\boldsymbol{\Lambda}_D^{-1}\boldsymbol{y} \end{pmatrix} \tag{3.14}$$

where,

$\widehat{\boldsymbol{\beta}}(\boldsymbol{\theta})$ is the MLE (BLUE) of the fixed-effects parameters,
$\widehat{\boldsymbol{b}}(\boldsymbol{\theta})' = \left[\widehat{\boldsymbol{b}}_1(\boldsymbol{\theta})' \cdots \widehat{\boldsymbol{b}}_n(\boldsymbol{\theta})'\right]$ are the best linear unbiased predictors (BLUP) of the random-effects,
$\boldsymbol{X}' = \left[\boldsymbol{X}_1' \cdots \boldsymbol{X}_n'\right]$, $\boldsymbol{Z}' = Diag\{\boldsymbol{Z}_1', \ldots, \boldsymbol{Z}_n'\}$, $\boldsymbol{y}' = \left[\boldsymbol{y}_1' \cdots \boldsymbol{y}_n'\right]$,
$\boldsymbol{\Lambda}_D = Diag\{\boldsymbol{\Lambda}_1(\boldsymbol{\theta}_w), \ldots, \boldsymbol{\Lambda}_n(\boldsymbol{\theta}_w)\}$ and $\boldsymbol{\Psi}_D = Diag\{\boldsymbol{\Psi}(\boldsymbol{\theta}_b), \ldots, \boldsymbol{\Psi}(\boldsymbol{\theta}_b)\}$.

Assuming a full-rank model, solutions to the mixed model equations (3.14) are

$$\begin{aligned} \widehat{\boldsymbol{\beta}}(\boldsymbol{\theta}) &= \Big(\sum_{i=1}^n \boldsymbol{X}_i'\boldsymbol{\Sigma}_i(\boldsymbol{\theta})^{-1}\boldsymbol{X}_i\Big)^{-1} \sum_{i=1}^n \boldsymbol{X}_i'\boldsymbol{\Sigma}_i(\boldsymbol{\theta})^{-1}\boldsymbol{y}_i \\ &= (\boldsymbol{X}'\boldsymbol{\Sigma}_D^{-1}\boldsymbol{X})^{-1}\boldsymbol{X}'\boldsymbol{\Sigma}_D^{-1}\boldsymbol{y} \\ \widehat{\boldsymbol{b}}(\boldsymbol{\theta}) &= \boldsymbol{\Psi}_D\boldsymbol{Z}'\boldsymbol{\Sigma}_D^{-1}(\boldsymbol{y} - \boldsymbol{X}\widehat{\boldsymbol{\beta}}(\boldsymbol{\theta})) \end{aligned} \tag{3.15}$$

where $\boldsymbol{\Sigma}_D = \boldsymbol{Z}\boldsymbol{\Psi}_D\boldsymbol{Z}' + \boldsymbol{\Lambda}_D = Diag\{\boldsymbol{\Sigma}_1(\boldsymbol{\theta}), \ldots, \boldsymbol{\Sigma}_n(\boldsymbol{\theta})\}$. The individual random effect estimates are $\widehat{\boldsymbol{b}}_i(\boldsymbol{\theta}) = \boldsymbol{\Psi}\boldsymbol{Z}_i'\boldsymbol{\Sigma}_i(\boldsymbol{\theta})^{-1}(\boldsymbol{y}_i - \boldsymbol{X}_i\widehat{\boldsymbol{\beta}}(\boldsymbol{\theta}))$. For fixed $\boldsymbol{\theta}$, the variance-covariance matrix of $\begin{pmatrix} \widehat{\boldsymbol{\beta}}(\boldsymbol{\theta}) \\ \widehat{\boldsymbol{b}}(\boldsymbol{\theta}) \end{pmatrix}$ is

$$\boldsymbol{\Upsilon}(\widehat{\boldsymbol{\beta}}, \widehat{\boldsymbol{b}}) = \begin{pmatrix} \boldsymbol{\Omega} & -\boldsymbol{\Omega}\boldsymbol{X}'\boldsymbol{\Sigma}_D^{-1}\boldsymbol{Z}\boldsymbol{\Psi}_D \\ -\boldsymbol{\Psi}_D\boldsymbol{Z}'\boldsymbol{\Sigma}_D^{-1}\boldsymbol{X}\boldsymbol{\Omega} & \boldsymbol{Q} + \boldsymbol{\Psi}_D\boldsymbol{Z}'\boldsymbol{\Sigma}_D^{-1}\boldsymbol{X}\boldsymbol{\Omega}\boldsymbol{X}'\boldsymbol{\Sigma}_D^{-1}\boldsymbol{Z}\boldsymbol{\Psi}_D \end{pmatrix} \tag{3.16}$$

where $\boldsymbol{\Omega} = \boldsymbol{\Omega}(\widehat{\boldsymbol{\beta}}(\boldsymbol{\theta})) = (\boldsymbol{X}'\boldsymbol{\Sigma}_D^{-1}\boldsymbol{X})^{-1}$ is the variance-covariance matrix of $\widehat{\boldsymbol{\beta}}(\boldsymbol{\theta})$ and $\boldsymbol{Q} = \boldsymbol{Q}(\boldsymbol{\theta}) = (\boldsymbol{Z}'\boldsymbol{\Lambda}_D^{-1}\boldsymbol{Z} + \boldsymbol{\Psi}_D^{-1})^{-1}$. Since (3.15) provides closed form estimates of $\boldsymbol{\beta}$ and $\boldsymbol{b}$ for a given $\boldsymbol{\theta}$, final estimates may be obtained by simply replacing $\boldsymbol{\theta}$ with any $\sqrt{n}$-consistent estimator, say $\widehat{\boldsymbol{\theta}}$. As with the marginal linear model, one can adopt either a method of moments (MM) approach or a likelihood-based approach to estimating the components of $\boldsymbol{\theta}$. Estimating $\boldsymbol{\theta}$ by a MM approach (e.g., using the option METHOD=MIVQUE0 in PROC MIXED) may be interpreted as estimating $\boldsymbol{\beta}$ via estimated generalized least squares (EGLS) or iteratively reweighted generalized least squares (IRGLS) with the latter being equivalent to solving a set of GEE's. Alternatively, estimating $\boldsymbol{\theta}$ by either ML or REML (i.e., by specifying either METHOD=ML or METHOD=REML) leads to a MLE for $\boldsymbol{\beta}$. In either case, we refer to $\widehat{\boldsymbol{\beta}}(\widehat{\boldsymbol{\theta}})$ as the empirical (or estimated) best linear unbiased estimator (EBLUE) of $\boldsymbol{\beta}$. Likewise we refer to the resulting estimated predictors, $\widehat{\boldsymbol{b}}_i(\widehat{\boldsymbol{\theta}})$, of the random effects as empirical (or estimated) best linear unbiased predictors or EBLUP's.

With respect to estimation of $\boldsymbol{\theta}$, since the marginal distribution of $\boldsymbol{y}_i$ under the LME model (3.1) has exactly the same general form as that of the marginal LM (2.1), ML and REML estimation of the variance-covariance parameters $\boldsymbol{\theta} = (\boldsymbol{\theta}_b, \boldsymbol{\theta}_w)$ can be carried out in the same manner as described in section (2.1.1) of Chapter 2.

3.1.3 Inference and test statistics

As with the marginal linear model, inference under the LME model (3.1) is almost always based on large-sample theory since exact small-sample distributional properties of the estimates are generally unknown. Of course there are exceptions and one must consider each application before deciding on what hypothesis testing option(s) one should use with PROC MIXED or PROC GLIMMIX. However, as noted in the previous section, inference under the LME is not necessarily restricted to estimable functions of the fixed-effects parameters, $\boldsymbol{\beta}$. Indeed, one can conduct tests of hypotheses and construct confidence intervals on linear combinations involving both the fixed-effects parameters and random-effects. We consider each case below.

Inference about fixed-effects parameters, $\boldsymbol{\beta}$

Tests of hypotheses and confidence intervals for select estimable functions of the fixed-effects parameters, β, are presented for both the MIXED and GLIMMIX procedures in the preceding chapter (see page 29). Briefly, one can specify tests of hypotheses based on the large-sample Wald chi-square test (2.15) by specifying the CHISQ option of the MODEL statement in both procedures. The default test of hypothesis for both procedures will be based on an approximate F-test using the scaled test statistic

$$F(\widehat{\boldsymbol{\beta}}) = T^2(\widehat{\boldsymbol{\beta}})/r \tag{3.17}$$

to test the general linear hypothesis (2.14). For certain types of balanced and complete data, the F-statistic (3.17) may have an exact F-distribution with r numerator degrees of freedom and a fixed known denominator degrees of freedom, say v_e. In most cases, however, the F-statistic (3.17) will only have an approximate F distribution with r numerator degrees of freedom and an approximate denominator degrees of freedom, $\widehat{v}_e$, that must either be specified or estimated using either the DDF or DDFM option of the MODEL statement. The DDF option allows users to specify their own denominator degrees of freedom specific to each of the fixed effects in the model while the DDFM option instructs what method SAS should use to estimate $\widehat{v}_e$. For the LME model with random effects specified (i.e., when a RANDOM statement is specified), the DDFM=CONTAIN is the default method by which the MIXED and GLIMMIX procedures estimate $\widehat{v}_e$. With this option, the denominator degrees of freedom is estimated using a containment method explained in the on-line SAS Help and documentation. As noted in the documentation, this default approach provides denominator degrees of freedom that match the F-tests one would perform for a balanced split-plot design and should be adequate for moderately unbalanced designs.

For any estimable linear function $\boldsymbol{l}'\boldsymbol{\beta}$ such that $\boldsymbol{l}'\boldsymbol{\beta} = \boldsymbol{k}'E(\boldsymbol{y}) = \boldsymbol{k}'\boldsymbol{X}\boldsymbol{\beta}$ for some vector of constants, $\boldsymbol{k}$, an asymptotically valid α-level confidence interval can be constructed as

$$\boldsymbol{l}'\widehat{\boldsymbol{\beta}} \pm z_{\alpha/2}\sqrt{\boldsymbol{l}'\widehat{\boldsymbol{\Omega}}^*(\widehat{\boldsymbol{\beta}})\boldsymbol{l}} \tag{3.18}$$

where $z_{\alpha/2}$ is the $100(1-\alpha/2)$ percentile of the standard normal distribution and $\widehat{\boldsymbol{\Omega}}^*(\widehat{\boldsymbol{\beta}})$ is either the model-based estimator, $\widehat{\boldsymbol{\Omega}}(\widehat{\boldsymbol{\beta}})$, or the robust sandwich estimator, $\widehat{\boldsymbol{\Omega}}_R(\widehat{\boldsymbol{\beta}})$, of the variance-covariance matrix of $\widehat{\boldsymbol{\beta}}$ (see, for example, equations 2.5 and 2.6, respectively but with $\widehat{\boldsymbol{\beta}}(\boldsymbol{\theta})$ replaced by $\widehat{\boldsymbol{\beta}}(\widehat{\boldsymbol{\theta}})$ where $\widehat{\boldsymbol{\theta}}$ is the ML or REML estimator of $\boldsymbol{\theta}$). To better reflect small sample behavior, the default approach in SAS is to replace $z_{\alpha/2}$ with a Student-t value, $t_{\alpha/2}(\widehat{v}_e)$. When running GLIMMIX, one can specifically request z-based confidence intervals using the DDFM=NONE option (no such option is available with MIXED).

Inference about fixed-effects parameters β and random-effects b

Under the LME model, one may be interested in a more narrow *subject-specific* base of inference involving linear combinations of both fixed and random effects. In this case, tests of hypotheses and confidence intervals are restricted to predictable functions involving linear combinations of fixed and random effects of the form

$$\boldsymbol{l}'\boldsymbol{\beta} + \boldsymbol{m}'\boldsymbol{b} \tag{3.19}$$

where $\boldsymbol{l}'\boldsymbol{\beta}$ is an estimable function of $\boldsymbol{\beta}$ (i.e., $\boldsymbol{l}'\boldsymbol{\beta} = \boldsymbol{k}'E(\boldsymbol{y}) = \boldsymbol{k}'\boldsymbol{X}\boldsymbol{\beta}$ for some vector of constants, $\boldsymbol{k}$).

Under the LME model, tests of hypotheses involving predictable functions of the form

$$H_0 : \boldsymbol{L}\begin{pmatrix}\boldsymbol{\beta}\\ \boldsymbol{b}\end{pmatrix} = \boldsymbol{0} \tag{3.20}$$

may be carried out using the Wald chi-square test

$$T^2(\widehat{\boldsymbol{\beta}}, \widehat{\boldsymbol{b}}) = \begin{pmatrix}\widehat{\boldsymbol{\beta}}\\ \widehat{\boldsymbol{b}}\end{pmatrix}' \boldsymbol{L}'(\boldsymbol{L}\widehat{\boldsymbol{\Upsilon}}(\widehat{\boldsymbol{\beta}}, \widehat{\boldsymbol{b}})\boldsymbol{L}')^{-1}\boldsymbol{L}\begin{pmatrix}\widehat{\boldsymbol{\beta}}\\ \widehat{\boldsymbol{b}}\end{pmatrix} \tag{3.21}$$

where $\boldsymbol{L} = \boldsymbol{L}_{r\times snv}$ is a $r \times snv$ hypothesis or contrast matrix of rank $r \leq snv$ of predictable functions and $\widehat{\boldsymbol{\Upsilon}}(\widehat{\boldsymbol{\beta}}, \widehat{\boldsymbol{b}})$ is the estimated variance-covariance matrix of $(\widehat{\boldsymbol{\beta}}(\widehat{\boldsymbol{\theta}})' \;\; \widehat{\boldsymbol{b}}(\widehat{\boldsymbol{\theta}})')$ obtained from equation (3.16) by evaluating $\boldsymbol{\Psi}_D$ and $\boldsymbol{\Lambda}_D$ at $\widehat{\boldsymbol{\theta}}$. Note, however, that when one specifies the EMPIRICAL option, $\widehat{\boldsymbol{\Upsilon}}(\widehat{\boldsymbol{\beta}}, \widehat{\boldsymbol{b}})$ is evaluated using the robust covariance matrix, $\widehat{\boldsymbol{\Omega}}_R(\widehat{\boldsymbol{\beta}})$, rather than $\widehat{\boldsymbol{\Omega}}(\widehat{\boldsymbol{\beta}})$. Both MIXED and GLIMMIX test the general linear hypothesis using the default F-test, $F(\widehat{\boldsymbol{\beta}}, \widehat{\boldsymbol{b}}) = T^2(\widehat{\boldsymbol{\beta}}, \widehat{\boldsymbol{b}})/r$ with denominator degrees of freedom determined by the DDF or DDFM option. The option DDFM=KR is suggested for subject-specific inference as the random effect components of predictable functions may involve a linear combination of variance components. An excellent discussion of subject-specific inference under the LME model along with several examples illustrating tests of hypotheses of the form (3.20) are given in Chapter 6 of Littell, et. al. (2006; pp. 206-241).

Inference with respect to the variance-covariance parameters, $\boldsymbol{\theta} = (\boldsymbol{\theta}_b, \boldsymbol{\theta}_w)$, may be carried out using methods described for the marginal LM of Chapter 2 (see page 33). Likewise, goodness-of-fit criteria and other techniques for assessing model adequacy appropriate for the LME model may be found on page 34 of Chapter 2 as well as model diagnostics features as documented in the MIXED and GLIMMIX procedures.

3.2 Examples

As the focus of this book is on generalized linear and nonlinear models for correlated data, we restrict ourselves to just three applications requiring the use of a LME model. In the first two examples we revisit the dental growth data and bone mineral density data examples of Chapter 2 within the context of the random coefficient growth curve setting. In addition, we consider a third application involving the use of nested random effects from a study examining the relationship between nipple aspirate fluid (NAF) hormone levels to those of serum and saliva hormone levels in healthy premenopausal women. For numerous other examples illustrating the variety of LME models at one's disposal, the reader is encouraged to look over the book by Littell et. al. (2006).

3.2.1 Dental growth data—continued

In §2.2.1, we applied several different marginal LM's to the dental growth curve data of Potthoff and Roy (1964) to determine the magnitude of the effect that age and gender have on dental growth as measured by the distance (mm) from the center of the pituitary to the pteryomaxillary fissure. Based on the subject-specific profiles shown in Figure 2.1, we fit a GMANOVA model assuming a linear trend with age. The results in Output 2.6 show that under an unstructured variance-covariance matrix, there is a significant relationship between the response variable, fissure distance, and a child's age and that this relationship is different for boys and girls.

Here we consider the same linear relationship between fissure distance and age but we do so within the random coefficient growth curve setting. Specifically, rather than assume an unstructured marginal variance-covariance matrix for the response y, we fit a more parsimonious random coefficient growth curve model in which each child's response can be described by a regression line specific to that child. Each child's regression line is determined by the child's random intercept and random slope both of which represent deviations from a population-wide intercept and slope. The basic structure follows that of model (3.13) except that a separate population intercept and slope are fit to boys and girls, respectively. The SAS code required to fit this LME model is given as follows.

Program 3.1

```
proc mixed data=example2_2_1 method=ml scoring=200;
 class person sex _age_;
 model y = sex sex*age /noint solution ddfm=kenwardroger;
 random intercept age / type=un subject=person(sex);
 estimate 'Difference in intercepts' sex 1 -1;
 estimate 'Difference in slopes' age*sex 1 -1;
run;
```

The above RANDOM statement instructs the MIXED procedure to treat the intercept and slope (AGE) as random effects having an unstructured variance-covariance matrix (TYPE=UN). We fit this model using maximum likelihood estimation (METHOD=ML) so as to compare results against the marginal LM with an unstructured variance-covariance matrix. The results from this random coefficient growth curve model, shown in Output 3.1 below, are qualitatively similar to that of the marginal LM (Output 2.6).

Output 3.1: Random coefficient growth curve analysis of dental growth data

```
                Model Information
  Data Set                    WORK.EXAMPLE2_2_1
  Dependent Variable          y
  Covariance Structure        Unstructured
  Subject Effect              person(sex)
  Estimation Method           ML
  Residual Variance Method    Profile
  Fixed Effects SE Method     Kenward-Roger
  Degrees of Freedom Method   Kenward-Roger
                  Class Level Information
  Class     Levels   Values
  person        16   1 2 3 4 5 6 7 8 9 10 11 12 13 14 15 16
  sex            2   boys girls
  _age_          4   8 10 12 14
     Covariance Parameter Estimates
  Cov Parm   Subject        Estimate
  UN(1,1)    person(sex)      4.5569
  UN(2,1)    person(sex)     -0.1983
  UN(2,2)    person(sex)     0.02376
  Residual                    1.7162
             Fit Statistics
  -2 Log Likelihood           427.8
  AIC (smaller is better)     443.8
  AICC (smaller is better)    445.3
  BIC (smaller is better)     454.2
```

Solution for Fixed Effects

Effect	sex	Estimate	Standard Error	DF	t Value	Pr > |t|
sex	boys	16.3406	0.9801	27	16.67	<.0001
sex	girls	17.3727	1.1820	27	14.70	<.0001
age*sex	boys	0.7844	0.08275	27	9.48	<.0001
age*sex	girls	0.4795	0.09980	27	4.80	<.0001

Type 3 Tests of Fixed Effects

Effect	Num DF	Den DF	F Value	Pr > F
sex	2	27	247.00	<.0001
age*sex	2	27	56.46	<.0001

Estimates

Label	Estimate	Standard Error	DF	t Value	Pr > |t|
Difference in intercepts	−1.0321	1.5355	27	−0.67	0.5072
Difference in slopes	0.3048	0.1296	27	2.35	0.0263

A formal likelihood ratio chi-square test comparing the full model with unstructured covariance with that of the reduced model assuming random intercepts and slopes, i.e., the random coefficient growth curve model (RCGCM) is $-2\log(\widehat{\lambda}) = 2ll(\widehat{\boldsymbol{\theta}}_{\mathrm{UN}}) - 2ll(\widehat{\boldsymbol{\theta}}_{\mathrm{RCGCM}}) = -419.5 - (-427.8) = 8.3$ on $14{-}8 = 6$ degrees of freedom yielding a p-value of 0.21694. This result along with a comparison of Akaike's information criteria ($\mathrm{AIC}_{\mathrm{UN}} = 447.5$ versus $\mathrm{AIC}_{\mathrm{RCGCM}} = 443.8$) suggests the linear trend between fissure distance and age is better explained on the basis of a random coefficient growth curve model than the marginal GMANOVA model.

3.2.2 Bone mineral density data—continued

For the balanced but incomplete bone mineral density data of Lloyd et. al., we found that a marginal linear GMANOVA model with unstructured variance-covariance provided the best fit to the data compared to several candidate models having reduced marginal covariance structures (see §2.2.2 and results therein). Here, we test whether a random coefficient growth curve model (RCGCM) might suffice in better describing variability associated with linear trends in total body bone mineral density (TBBMD) versus time. The SAS code and select output are shown below.

Program 3.2

```
proc mixed data=example2_2_2 method=ml;
 class subject group visit;
 model tbbmd=group group*years /noint solution;
 random intercept years / type=un subject=subject;
 estimate 'slope diff' group*years 1 -1;
run;
quit;
```

Output 3.2: Random coefficient growth curve analysis of bone mineral density data

Covariance Parameter Estimates

Cov Parm	Subject	Estimate
UN(1,1)	subject	0.004084
UN(2,1)	subject	0.000102
UN(2,2)	subject	0.000186
Residual		0.000125

Fit Statistics

-2 Log Likelihood	−2380.5
AIC (smaller is better)	−2364.5
AICC (smaller is better)	−2364.2
BIC (smaller is better)	−2342.7

Solution for Fixed Effects

Effect	group	Estimate	Standard Error	DF	t Value	Pr > |t|
group	C	0.8811	0.008701	284	101.26	<.0001
group	P	0.8695	0.008547	284	101.73	<.0001
years*group	C	0.05383	0.002225	284	24.19	<.0001
years*group	P	0.04484	0.002173	284	20.63	<.0001

Estimates

Label	Estimate	Standard Error	DF	t Value	Pr > |t|
slope diff	0.008990	0.003110	284	2.89	0.0041

The SAS code is similar to that used for the dental growth data in that the RANDOM statement specifies that the intercept and slope (YEARS) are to be treated as random effects with an unstructured variance-covariance matrix. The likelihood ratio chi-square test comparing the full GMANOVA model of §2.2.2 with this LME model specification assuming random intercepts and slopes is
$-2\log(\widehat{\lambda}) = 2ll(\widehat{\boldsymbol{\theta}}_{\text{UN}}) - 2ll(\widehat{\boldsymbol{\theta}}_{\text{RCGCM}}) = 2431.34 - 2380.5 = 50.8$ on $19-8 = 11$ degrees of freedom (p-value< 0.0001). It is clear that in this case, this particular form of the RCGCM does not adequately explain variation within and between subjects. As an alternative, we might consider fitting a RCGCM in which we replace the assumption of conditional independence that is implicitly assumed in the absence of a REPEATED statement with a model that assumes a within-subject first-order autoregressive structure. We can fit this alternative RCGCM by simply including the following REPEATED statement immediately after the RANDOM statement shown above.

```
repeated visit /subject=subject type=ar(1);
```

As indicated by both the note that appears in the SAS log

```
NOTE: Estimated G matrix is not positive definite
```

and the output shown below, assuming an AR(1) within-subject covariance structure appears to overwhelm any variation that might be accounted for by also assuming a random intercept and random slope. A similar result is obtained if we simply assume random intercepts and a fixed slope.

Output 3.3: Random coefficient growth curve analysis of bone mineral density data assuming an AR(1) within-subject covariance structure.

Covariance Parameter Estimates

Cov Parm	Subject	Estimate
UN(1,1)	subject	0
UN(2,1)	subject	0.000264
UN(2,2)	subject	0
AR(1)	subject	0.9665
Residual		0.003908

Fit Statistics

-2 Log Likelihood	−2408.2
AIC (smaller is better)	−2394.2
AICC (smaller is better)	−2394.0
BIC (smaller is better)	−2375.2

Solution for Fixed Effects

Effect	group	Estimate	Standard Error	DF	t Value	Pr > |t|
group	C	0.8805	0.008429	284	104.46	<.0001
group	P	0.8701	0.008280	284	105.09	<.0001
years*group	C	0.05310	0.002291	284	23.18	<.0001
years*group	P	0.04351	0.002238	284	19.44	<.0001

Estimates

Label	Estimate	Standard Error	DF	t Value	Pr > |t|
slope diff	0.009595	0.003202	284	3.00	0.0030

Of course one can try fitting numerous other covariance structures in an attempt to find a more parsimonious model but it would appear based on the preceding set of analyses that a random coefficient growth curve model is not well suited to this particular application. Indeed, we leave it as an exercise to show that a marginal LM with a heterogeneous AR(1) structure (TYPE=ARH(1)) does, in fact, provide a more parsimonious model compared to the unstructured GMANOVA model of §2.2.2.

3.2.3 Estrogen levels in healthy premenopausal women

Gann et. al. (2006) evaluated the relationship of nipple aspirate fluid (NAF) estrogen and progesterone levels to those in serum and saliva from a group of 47 healthy premenopausal women. In this example, we fit NAF estradiol levels to a basic repeated measures ANOVA model in order to account for correlation among paired measurements obtained from the left and right breast of women measured repeatedly over time. We then extend this model to a repeated measures analysis of covariance (ANCOVA) model by including serum and/or saliva estradiol levels as covariates to determine to what extent these values may be used in tracking breast cancer risk. Presented below is a brief description of the dataset used and other features that make this example somewhat unique.

The data, as described by Gann et. al. (2006), are from the Diet and Hormone Study (designated as the DHSI study). The DHSI study was a randomized controlled trial (RCT) that entailed randomizing women to one of four (4) treatment groups with planned longitudinal follow-up at four visits over approximately a 15-month period. Visit 1 corresponded to a baseline visit, while visit 2 corresponded to a woman's 4th menstrual cycle, visits 3 corresponded to the 12th menstrual cycle and visit 4 corresponded to the 15th menstrual cycle. The four treatment groups are defined according to two sequential phases with the first phase (visits 1-4) corresponding to dietary intervention (usual diet versus low-fat diet) and the second phase (visits 3-4) corresponding to an additional soy protein intervention (full soy protein versus isoflavone-free soy protein). The dataset presented for analysis designated the phase 1 dietary groups as control (defined here as normal diet) and intervention (defined here as low-fat diet). Likewise, the dataset designated the phase 2 soy protein assignment as control (full soy protein) and intervention (isoflavone-free soy protein). The four groups are therefore defined as follows:

Group 1 = Normal Diet + Soy(Control) Supplement
Group 2 = Normal Diet + Soy(Intervention) Supplement
Group 3 = Low Fat Diet + Soy(Control) Supplement
Group 4 = Low Fat Diet + Soy(Intervention) Supplement.

Details of the study and the key variables analyzed in this example are as follows.

- A 15-month randomized controlled trial designed to evaluate the effects of a normal versus low-fat diet in combination with use of a control soy supplement and an intervention soy supplement.
- The experimental unit is a woman ($n = 47$ women were studied)
- Response variable:
  - y_{ijkl} =log NAF estradiol levels for the ith woman ($i = 1, \ldots, 47$) from the jth treatment group ($j = 1, 2, 3, 4$) at the kth visit ($k = 1, 2, 3, 4$) corresponding to the woman's l^{th} breast ($l = 1$ for left breast, 2 for right breast).
- Four within-unit covariates:
  - Visit (1, 2, 3, or 4) with visit 1=baseline, visit 2=4th menstrual cycle, visit 3=12th menstrual cycle, and visit 4=15th menstrual cycle. We may define each visit by the indictor variables

$$v_{ik} = \begin{cases} 1, & \text{if visit } k \ (k = 1, \ldots, 4) \\ 0, & \text{otherwise} \end{cases}$$

- Breast (Left, Right) defined by the indicator variable

$$B_{il} = \begin{cases} 1, & \text{if left breast } (l = 1) \\ 0, & \text{if right breast } (l = 2) \end{cases}$$

- x_{1ijk} =log serum estradiol level for the ith woman from the jth treatment group ($j = 1, 2, 3, 4$) at the kth visit ($k = 1, 2, 3, 4$)
- x_{2ijk} =log saliva estradiol level for the ith woman from the jth treatment group ($j = 1, 2, 3, 4$) at the kth visit ($k = 1, 2, 3, 4$)

- One between-unit covariate:
  - Treatment Groups as uniquely defined by the two indicator variables

$$a_{1i} = \begin{cases} 0, & \text{if control diet (normal diet)} \\ 1, & \text{if intervention diet (low-fat diet)} \end{cases}$$

$$a_{2i} = \begin{cases} 0, & \text{if control soy (full soy protein)} \\ 1, & \text{if intervention soy (isoflavone-free soy protein)} \end{cases}$$

  With these two indicator variables, a cell means representation of the four treatment groups as defined above is possible.
- All estradiol values, measured in pg/mL, were log-transformed following preliminary univariate analyses that suggest the log transformation yields approximately normally distributed data.
- Objectives: Determine the extent to which serum and saliva estradiol values are correlated with each other and whether or not these values are predictive of NAF estradiol levels.

A partial listing of the data is shown below. A description of the variables is available in Appendix C of the author's Web page.

Output 3.4: Partial listing (first 3 subjects) of the estrogen hormone data. The variable GROUP is coded according to the four treatment groups defined above.

Subject_ID	Group	Visit	Breast	NAF_Estradiol	Saliva_Estradiol	Serum_Estradiol
1	2	1	Left	5.12	.	4.43
1	2	1	Right	5.09	.	4.43
1	2	2	Left	5.89	.	4.50
1	2	2	Right	5.49	.	4.50
1	2	3	Left	5.29	.	4.43
1	2	3	Right	5.40	.	4.43
1	2	4	Left	4.82	.	4.31
1	2	4	Right	5.78	.	4.31
2	4	1	Left	5.01	1.06	4.53
2	4	1	Right	5.53	1.06	4.53
2	4	2	Left	5.15	.	4.85
2	4	2	Right	5.11	.	4.85
2	4	3	Left	5.21	1.29	4.59
2	4	3	Right	.	1.29	4.59
2	4	4	Left	.	0.78	4.62
2	4	4	Right	.	0.78	4.62
3	3	1	Left	.	2.05	4.96
3	3	1	Right	6.06	2.05	4.96
3	3	2	Left	.	2.25	4.73
3	3	2	Right	6.02	2.25	4.73
3	3	3	Left	.	2.88	4.83
3	3	3	Right	.	2.88	4.83
3	3	4	Left	6.28	2.26	4.56
3	3	4	Right	.	2.26	4.56

The first thing to note is the fact that salivary and serum samples are common to both the left and right breast NAF samples. Consequently, there is only one set of values per visit of serum and salivary estradiol levels (hence the repeating values of these two variables by breast). The second thing to note is the highly unbalanced nature of the data. Subject 1, for example, has complete data on NAF estradiol levels for both breasts at all four visits as well as complete serum estradiol measurements at all four visits but has no salivary estradiol measurements. Subject 3, on the other hand, has incomplete NAF estradiol measurements but complete serum and salivary estradiol measurements at all four visits. In fact, based on the following two calls to TABULATE, we see just how incomplete the data are (Output 3.5).

Program 3.3

```
proc tabulate data=example3_2_3 ;
 class Group Visit Breast;
 var NAF_Estradiol;
 table Group*Visit,
       Breast*(NAF_Estradiol)*(N*F=2.0 Mean*F=5.2 Std*F=5.2)
      /rts=16;
run;
proc tabulate data=example3_2_3 ;
 where Breast='Left';
 class Group Visit;
 var Saliva_Estradiol Serum_Estradiol;
 table Group*Visit,
       (Serum_Estradiol Saliva_Estradiol)*(N*F=2.0 Mean*F=6.2 Std*F=6.2)
      /rts=16;
run;
```

Output 3.5: Summary statistics on log estradiol levels according to group, visit and breast (for NAF estradiol only).

		Breast					
		Left			Right		
		NAF_Estradiol			NAF_Estradiol		
		N	Mean	Std	N	Mean	Std
Group	Visit						
1	1	5	5.22	0.57	7	4.62	0.62
	2	5	5.22	0.45	7	4.28	1.98
	3	5	5.37	0.74	6	4.61	1.08
	4	5	4.49	1.39	5	4.73	0.98
2	1	9	4.73	0.79	9	4.76	0.97
	2	10	4.78	0.85	9	4.38	1.53
	3	8	4.37	1.15	7	4.36	0.78
	4	9	4.83	0.63	4	5.34	0.66
3	1	2	4.30	0.41	7	4.10	1.17
	2	3	4.27	0.91	6	4.32	1.29
	3	4	4.74	1.35	7	4.00	0.91
	4	3	4.57	1.58	6	3.78	1.25
4	1	8	3.73	1.57	8	3.90	1.54
	2	6	4.34	1.38	7	4.54	0.81
	3	7	4.43	0.64	6	3.85	1.07
	4	5	4.03	0.97	5	4.05	1.08

		Serum_Estradiol			Saliva_Estradiol		
		N	Mean	Std	N	Mean	Std
Group	Visit						
1	1	10	4.72	0.51	6	0.84	1.13
	2	10	4.65	0.49	6	1.27	0.65
	3	10	4.59	0.42	6	1.25	0.34
	4	9	4.47	0.43	5	1.22	0.50
2	1	16	4.38	0.43	12	1.62	0.79
	2	16	4.46	0.25	11	1.57	0.75
	3	15	4.31	0.48	11	1.61	0.64
	4	13	4.34	0.34	7	1.69	0.92
3	1	8	4.72	0.30	4	1.91	0.20
	2	9	4.78	0.49	5	1.68	0.41
	3	9	4.66	0.35	5	1.73	0.65
	4	8	4.67	0.46	4	1.81	0.58
4	1	12	4.49	0.46	11	1.37	0.66
	2	12	4.47	0.46	10	1.48	0.88
	3	12	4.60	0.41	11	1.16	0.84
	4	12	4.46	0.36	10	1.20	0.79

This level of incompleteness poses serious challenges to model validation and model building. For example, it may be extremely difficult to fit models with a complex covariance structure given the degree of imbalance in observations across visits and between the left and right breast. Also, owing to the sparseness of salivary levels among women, results from models of NAF estradiol which include both serum and salivary estradiol levels as covariates will be based on far fewer subjects than would models that include only serum estradiol as a covariate. These features will become more clear as we attempt to investigate the relationship between NAF estradiol levels versus serum and salivary estradiol levels.

Preliminary analyses

Prior to investigating how well serum and/or saliva estradiol levels predict NAF estradiol levels, we first carry out an investigation to determine just how correlated serum and saliva estradiol levels are to each other over time. This is done so as to avoid issues with possible multicollinearity associated with the inclusion of these two variables as covariates in a repeated measures ANCOVA. To accomplish this, we first fit the following marginal LM model to just the serum and saliva estradiol values:

$$x_{1ijk} = \mu + \tau_j + \nu_k + (\tau\nu)_{jk} + \beta x_{2ijk} + \epsilon_{ijk}. \tag{3.22}$$

Under this repeated measures ANCOVA model, x_{1ijk}, which is the log serum estradiol for the ith woman in the jth treatment group at the kth visit, serves as our dependent response variable; μ is an overall mean intercept parameter; τ_j is the jth cell means treatment group effect ($j = 1, 2, 3, 4$); ν_k is the kth visit effect, $(\tau\nu)_{jk}$ is a treatment group by visit interaction effect, x_{2ijk} is the log salivary estradiol concentration which serves as a time-dependent covariate, and ϵ_{ijk} is an error term assumed to be normally distributed. The errors may be grouped into vector form according to the possibly four repeated measurements per woman as $\boldsymbol{\epsilon}_{ij} = \begin{pmatrix}\epsilon_{ij1} & \epsilon_{ij2} & \epsilon_{ij3} & \epsilon_{ij4}\end{pmatrix}'$ which is assumed to have a multivariate normal distribution with some specified variance-covariance structure.

Using this basic marginal LM model, three assumed variance-covariance structures were fit to the data. The first is an unstructured (UN) variance-covariance matrix in which the variances and correlations of log serum estradiol concentrations were allowed to vary across visits. This is the most general covariance structure and it entails 10 parameters, 4 variances and 6 correlations. The second variance-covariance structure is a first-order autoregressive structure (AR) which assumes there is a common variance across visits and a first-order autocorrelation parameter describing autocorrelation over time (this results in 2 parameters). The third covariance structure is that of compound symmetry (CS) which assumes a common variance and a common correlation across visits (also resulting in 2 parameters). The code for running this preliminary analysis is shown below.

Program 3.4

```
%macro AIC(type=un);
  ods listing close;
  ods output InfoCrit=AIC_i;
  proc mixed data=example3_2_3 scoring=100 covtest ic;
   where Breast='Left';
   class Subject_ID Group Visit;
   model Serum_Estradiol = Group Visit Group*Visit
                           Saliva_Estradiol /solution;
   repeated Visit / subject=Subject_ID type=&type;
  run;
  data AIC_i;
   set AIC_i;
   Type="&type";
  run;
  data AIC;
   set AIC AIC_i;
   if Type=' ' then delete;
  run;
  ods listing;
%mend;
data AIC;
 %AIC(type=UN);
 %AIC(type=AR(1));
 %AIC(type=CS);
proc print data=AIC split='|';
 id Type;
run;
```

Notice that in the above code, we subset the data using the WHERE statement (`where Breast='Left';`). This is because, as noted previously, the salivary and serum samples are common to both the left and right breast NAF samples.

Shown in Output 3.6 is a comparison of how well these three covariance structures fit the data. Using information criteria such as the AIC (Akaike's Information Criterion), one can compare how well the three models fit by choosing the one having the lowest information criterion. Based on any one of the five information criteria shown in Output 3.6, the compound symmetric (CS) structure is seen to provide the most parsimonious and best fit to the data.

Output 3.6: Information criteria for assessing variance-covariance goodness-of-fit

Type	Neg2LogLike	Parms	AIC	AICC	HQIC	BIC	CAIC
UN	103.6	10	123.6	125.9	128.8	138.8	148.8
AR	116.6	2	120.6	120.7	121.6	123.6	125.6
CS	110.6	2	114.6	114.7	115.6	117.6	119.6

Results from fitting model (3.22) assuming compound symmetry show that the estimated regression coefficient (± standard error) describing the relation between log serum estradiol and log salivary estradiol is $\widehat{\beta} = 0.07793$ (± 0.06191) which is not statistically significant (p=0.2119). This suggests there is very low correlation between serum and salivary estradiol levels.

In order to more directly estimate the correlation between log serum and log salivary estradiol concentrations over time, a multivariate repeated measures ANOVA model was fit to the data using the same set of fixed effects as defined by model (3.22) except that log serum and log salivary estradiol values were modeled as bivariate dependent variables directly rather than having log salivary estradiol levels serve as a time-dependent covariate. Based on the preceding analysis, it appears reasonable to assume that log serum estradiol levels will share a common correlation across time. Consequently, we fit a multivariate repeated measures ANOVA by assuming a 2×2 unstructured covariance matrix for the bivariate values of log serum and log salivary estradiol levels at any visit and by assuming a common correlation coefficient between any two log serum estradiol levels over time. This same correlation coefficient is also assumed to be common between any two log salivary estradiol levels over time. The overall variance-covariance structure is therefore a direct product covariance structure which can be expressed via the Kronecker product (i.e., direct product—see Appendix A) as

$$\boldsymbol{\Sigma}(\boldsymbol{\theta}) = \boldsymbol{\Sigma}_{\mathrm{UN}} \otimes \boldsymbol{R}(\rho) = \begin{pmatrix} \sigma_1^2 & \sigma_{12} \\ \sigma_{12} & \sigma_2^2 \end{pmatrix} \otimes \begin{pmatrix} 1 & \rho & \rho & \rho \\ \rho & 1 & \rho & \rho \\ \rho & \rho & 1 & \rho \\ \rho & \rho & \rho & 1 \end{pmatrix} \tag{3.23}$$

$$= \begin{pmatrix} \sigma_1^2 & \sigma_1^2\rho & \sigma_1^2\rho & \sigma_1^2\rho & \sigma_{12} & \sigma_{12}\rho & \sigma_{12}\rho & \sigma_{12}\rho \\ \sigma_1^2\rho & \sigma_1^2 & \sigma_1^2\rho & \sigma_1^2\rho & \sigma_{12}\rho & \sigma_{12} & \sigma_{12}\rho & \sigma_{12}\rho \\ \sigma_1^2\rho & \sigma_1^2\rho & \sigma_1^2 & \sigma_1^2\rho & \sigma_{12}\rho & \sigma_{12}\rho & \sigma_{12} & \sigma_{12}\rho \\ \sigma_1^2\rho & \sigma_1^2\rho & \sigma_1^2\rho & \sigma_1^2 & \sigma_{12}\rho & \sigma_{12}\rho & \sigma_{12}\rho & \sigma_{12} \\ \sigma_{12} & \sigma_{12}\rho & \sigma_{12}\rho & \sigma_{12}\rho & \sigma_2^2 & \sigma_2^2\rho & \sigma_2^2\rho & \sigma_2^2\rho \\ \sigma_{12}\rho & \sigma_{12} & \sigma_{12}\rho & \sigma_{12}\rho & \sigma_2^2\rho & \sigma_2^2 & \sigma_2^2\rho & \sigma_2^2\rho \\ \sigma_{12}\rho & \sigma_{12}\rho & \sigma_{12} & \sigma_{12}\rho & \sigma_2^2\rho & \sigma_2^2\rho & \sigma_2^2 & \sigma_2^2\rho \\ \sigma_{12}\rho & \sigma_{12}\rho & \sigma_{12}\rho & \sigma_{12} & \sigma_2^2\rho & \sigma_2^2\rho & \sigma_2^2\rho & \sigma_2^2 \end{pmatrix}.$$

Here $\boldsymbol{\Sigma}_{\mathrm{UN}}$ is the unstructured covariance matrix for bivariate values of log salivary and serum estradiol levels on any given visit and ρ is the common correlation for either variable across visits. This structure may be obtained by specifying TYPE=UN@CS in the REPEATED statement of MIXED. Presented below is the SAS code used to fit this model.

Program 3.5

```
data example3_2_3_new;
 set example3_2_3;
 if Breast='Left';
 Source='Saliva'; Estradiol=Saliva_Estradiol; output;
 Source='Serum '; Estradiol=Serum_Estradiol;  output;
run;
proc sort data=example3_2_3_new;
 by Subject_ID Group Visit Source;
run;
ods output R=R;
ods output Rcorr=Rcorr;
ods output CovParms=CovParms;
proc mixed data=example3_2_3_new scoring=100 covtest;
 class Subject_ID Group Visit Source;
 model Estradiol = Source Group Source*Group Visit
    Source*Visit Group*Visit Source*Group*Visit / solution;
 repeated Source Visit / subject=Subject_ID type=un@cs r=3
 rcorr=3;
run;
```

```
data R1;
 set R;
 Type='UN@CS';
 Matrix='1 Cov ';
run;
data Rcorr1;
 set Rcorr;
 Type='UN@CS';
 Matrix='2 Corr';
run;
data Covparm;
 set Covparms;
 Type='UN@CS';
run;
data R_all;
 set R1 Rcorr1;
 drop index;
run;
data Covparm;
 set Covparm;
 drop subject;
run;
proc print data=covparm noobs;
run;
proc print data=R_all noobs;
 by Matrix;
 id Matrix;
 var Row Col1-Col8;
 format Col1-Col8 5.3;
run;
```

There are several things one should note regarding the above SAS code. First, in order to specify the direct product covariance structure shown in (3.23), we must first create a new dataset in which the two SAS variables, SALIVA_ESTRADIOL and SERUM_ESTRADIOL, are "converted" to a single "univariate" variable, ESTRADIOL. Second, we must create a second variable, SOURCE, that uniquely identifies which of these bivariate observations the value of ESTRADIOL corresponds to. With this newly created "univariate" dataset, we can specify the covariance structure (3.23) using PROC MIXED by specifying two distinct REPEATED effects, both of which must be included in the CLASS statement. The first effect, SOURCE, indicates the bivariate observations, and the second effect, `VISIT`, identifies the levels of our repeated measures factor. Third, we include the variable SOURCE and its associated interactions with the other fixed effects within the model so as to compute separate parameter estimates for the bivariate observations. Finally, with the above ODS OUTPUT statements, we create three new datasets, COVPARMS, R and RCORR, containing, respectively, estimates of the variance-covariance parameters $\boldsymbol{\theta} = (\sigma_1^2, \sigma_{12}, \sigma_2^2, \rho)$, an estimate of the overall marginal covariance matrix, and an estimate of the corresponding overall correlation matrix. To obtain an output dataset containing the complete 8×8 covariance and correlation matrices, it is essential to identify a subject having complete data for both variables at all four visits. In our example, the REPEATED statement

```
repeated Source Visit / subject=Subject_ID type=un@cs r=3 rcorr=3;
```

instructs SAS to print the estimated variance-covariance matrix and correlation matrix for the third subject as identified by the SUBJECT = SUBJECT_ID option in combination with the R=3 and RCORR=3 options. In this case, we specified the third subject as this was the first subject to have complete data on both log salivary and log serum estradiol concentrations at all four visits.

Select output from this analysis is presented in Output 3.7. The variance of log salivary estradiol corresponding to CovParm UN(1,1) was estimated to be $\widehat{\sigma}_1^2 = 0.4514$, while the variance of log serum estradiol corresponding to CovParm UN(2,2) was estimated to be $\widehat{\sigma}_2^2 = 0.2112$. The covariance between these two variables was estimated to be $\widehat{\sigma}_{12} = 0.04304$ implying a correlation of 0.1394 between log salivary and serum estradiol levels at any given visit. The common correlation between any two log salivary estradiol levels at different visits or between any two log serum estradiol levels at different visits was estimated to be $\widehat{\rho} = 0.6603$. This implies that the correlation between a log salivary estradiol level at one visit and the log serum estradiol level at another visit is just 0.092 (obtained as the product of correlations subject to rounding errors).

Output 3.7: Parameter estimates from the direct product covariance structure as specified in (3.23).

CovParm	Estimate	StdErr	ZValue	ProbZ	Type
Source UN(1,1)	0.4514	0.07419	6.08	<.0001	UN@CS
UN(2,1)	0.04304	0.03008	1.43	0.1524	UN@CS
UN(2,2)	0.2112	0.03071	6.88	<.0001	UN@CS
Visit Corr	0.6603	0.04945	13.35	<.0001	UN@CS

Matrix	Row	Col1	Col2	Col3	Col4	Col5	Col6	Col7	Col8
1 Cov	1	0.451	0.043	0.298	0.028	0.298	0.028	0.298	0.028
	2	0.043	0.211	0.028	0.139	0.028	0.139	0.028	0.139
	3	0.298	0.028	0.451	0.043	0.298	0.028	0.298	0.028
	4	0.028	0.139	0.043	0.211	0.028	0.139	0.028	0.139
	5	0.298	0.028	0.298	0.028	0.451	0.043	0.298	0.028
	6	0.028	0.139	0.028	0.139	0.043	0.211	0.028	0.139
	7	0.298	0.028	0.298	0.028	0.298	0.028	0.451	0.043
	8	0.028	0.139	0.028	0.139	0.028	0.139	0.043	0.211
2 Corr	1	1.000	0.139	0.660	0.092	0.660	0.092	0.660	0.092
	2	0.139	1.000	0.092	0.660	0.092	0.660	0.092	0.660
	3	0.660	0.092	1.000	0.139	0.660	0.092	0.660	0.092
	4	0.092	0.660	0.139	1.000	0.092	0.660	0.092	0.660
	5	0.660	0.092	0.660	0.092	1.000	0.139	0.660	0.092
	6	0.092	0.660	0.092	0.660	0.139	1.000	0.092	0.660
	7	0.660	0.092	0.660	0.092	0.660	0.092	1.000	0.139
	8	0.092	0.660	0.092	0.660	0.092	0.660	0.139	1.000

It is important at this stage to realize just what the rows and columns of the printed covariance matrix or correlation matrix in Output 3.7 represent. Although the online SAS Help and documentation informs us that the specification TYPE=UN@CS corresponds to the direct product covariance structure shown in (3.23), the estimated covariance matrix created from the ODS OUTPUT R=R statement shown above actually corresponds to the covariance matrix being displayed as $\boldsymbol{\Sigma}^*(\boldsymbol{\theta}) = \boldsymbol{R}(\rho) \otimes \boldsymbol{\Sigma}_{\text{UN}}$. This reversal between model specification and output makes sense in that we are more apt to think about the bivariate aspects of the data from this example when results are presented in blocks of 2 corresponding log salivary and serum estradiol levels rather than in blocks of 4 corresponding to the 4 visits (i.e., as would otherwise be presented if (3.23) were used to display the results). Nonetheless, the reader should be aware of this difference between model specification and output. In any case, results of these preliminary analyses, all of which are based on marginal LM's, suggest that salivary and serum estrogen levels are not all that correlated and that both may be considered as joint covariates for assessing their overall relationship with NAF estrogen levels.

Primary analysis

One of the primary goals of Gann et. al. (2006) was to evaluate the relationship of NAF estrogen levels to those in serum and saliva. Following the above preliminary analyses, a random-effects repeated measures ANCOVA model can be used as a basis for assessing various relationships between hormone levels, particularly the relationship between log serum and salivary estradiol levels versus log NAF estradiol levels. Since NAF samples from women were obtained from both breasts over time, such a model would need to reflect two possible sources of correlation. The first would be the correlation we would expect between any two log NAF estradiol measurements taken from the same woman at different times (i.e., repeated measurements across visits). The second would be the correlation we would expect between NAF estradiol measurements obtained from the left and right breast within the same visit (i.e., "paired" observations associated with the left and right breast). One such model would be the following "split-split plot" type repeated measures ANOVA model used by the authors:

$$\begin{aligned} y_{ijkl} = \mu + \gamma_j + \omega_{i(j)} + \tau_k + (\gamma\tau)_{jk} + b_{i(jk)} \\ + \beta_l + (\gamma\beta)_{jl} + (\tau\beta)_{kl} + (\gamma\tau\beta)_{jkl} + \epsilon_{i(jkl)} \end{aligned} \tag{3.24}$$

where y_{ijkl} is the log NAF estradiol level for the ith woman from the jth treatment group $(j = 1, 2, 3, 4)$ during the kth visit $(k = 1, 2, 3, 4)$ corresponding to the woman's l^{th} breast $(l = 1 \text{ for left breast}, 2 \text{ for right breast})$.

Although strictly speaking the data are not generated from a split-split plot design, the parameters of this LME model are defined in a manner similar to that of a split-split plot ANOVA. Specifically, μ represents an overall mean effect; γ_j represents the jth treatment group effect; $\omega_{i(j)}$ is a random effect due to the ith woman within the jth treatment group; τ_k represents the time effect associated with the kth visit; $(\gamma\tau)_{jk}$ is the treatment group $\times$ time two-way interaction effect; $b_{i(jk)}$ is a random effect due to the ith woman within the jth treatment group during the kth visit; β_l is the effect due to the lth breast $(l = 1$ for left breast, $l = 2$ for right breast); $(\gamma\beta)_{jl}$ and $(\tau\beta)_{kl}$ are the two-way interaction effects of treatment group and visit with breast; $(\gamma\tau\beta)_{jkl}$ is the three-way interaction between treatment group, visit and breast; and $\epsilon_{i(jkl)}$ is within-breast measurement error. The random effect, $\omega_{i(j)}$, is a random intercept associated with the ith woman within the jth treatment group. This random effect yields the usual intraclass correlation coefficient commonly associated with a "split-plot" repeated measures ANOVA where each woman represents a "whole plot" experimental unit randomly assigned to one of the four treatment groups (the levels of a "whole-plot" treatment factor) while the four visits per woman represent the levels of the "split-plot" treatment factor, visit (e.g., see page 71). As such $\omega_{i(j)}$ accounts for overall woman-to-woman variation. The random effect, $b_{i(jk)}$, is a random intercept associated with the ith woman within the jth treatment group *and* k^{th} visit. This term allows us to account for within "sub-plot" variation (i.e., within visit variation within a woman) and as such, accounts for breast-to-breast variation within women across visits. It yields a second intraclass correlation coefficient analogous to that of a split-split plot design. Finally, the error term, $\epsilon_{i(jkl)}$, is simply within-breast measurement error associated with the kth visit from the ith woman within the jth treatment group.

As noted above, the primary feature of this "split-split plot" repeated measures ANOVA model is that it provides two intraclass correlation coefficients for describing different patterns of correlation between observations from the same woman. To see this, one can write model (3.24) as

$$y_{ijkl} = \mu_{jkl} + \omega_{i(j)} + b_{i(jk)} + \epsilon_{i(jkl)}$$

where $\mu_{jkl} = \mu + \gamma_j + \tau_k + (\gamma\tau)_{jk} + \beta_l + (\gamma\beta)_{jl} + (\tau\beta)_{kl} + (\gamma\tau\beta)_{jkl}$ is a cell-means representation that captures the various fixed-effects and their interactions. Assuming that $\omega_{i(j)} \sim$iid $N(0, \sigma^2_\omega)$, $b_{i(jk)} \sim$iid $N(0, \sigma^2_b)$, and $\epsilon_{i(jkl)} \sim$iid $N(0, \sigma^2_\epsilon)$, and assuming these random effects are mutually independent of each other, the overall covariance between any two log NAF estradiol measurements from the same woman but taken at different times (i.e., visits) can be shown to be

$$\begin{aligned}
Cov(y_{ijkl}, y_{ijk'l'}) &= Cov(\omega_{i(j)} + b_{i(jk)} + \epsilon_{i(jkl)}, \omega_{i(j)} + b_{i(jk')} + \epsilon_{i(jk'l')}) \\
&= Var(\omega_{i(j)}) \\
&= \sigma^2_\omega \text{ subject to the restriction that } k \neq k'.
\end{aligned}$$

Likewise, the covariance between the log NAF estradiol levels from the left and right breast from a woman within any given visit can be shown to be

$$\begin{aligned}
Cov(y_{ijk1}, y_{ijk2}) &= Cov(\omega_{i(j)} + b_{i(jk)} + \epsilon_{i(jk1)}, \omega_{i(j)} + b_{i(jk)} + \epsilon_{i(jk2)}) \\
&= Var(\omega_{i(j)}) + Var(b_{i(jk)}) \\
&= \sigma^2_\omega + \sigma^2_b.
\end{aligned}$$

Since, by our assumption of independence, the overall variance at each visit is simply the sum of the variance components,

$$Var(y_{ijkl}) = \sigma^2_\omega + \sigma^2_b + \sigma^2_\epsilon,$$

it follows that the correlation between any two log NAF estradiol measurements from the same woman but taken at different times (i.e., visits) is given by the intraclass correlation

$$Corr(y_{ijkl}, y_{ijk'l'}) = \frac{Cov(y_{ijkl}, y_{ijk'l'})}{\sqrt{Var(y_{ijkl})}\sqrt{Var(y_{ijk'l'})}} = \frac{\sigma^2_\omega}{\sigma^2_\omega + \sigma^2_b + \sigma^2_\epsilon} = \rho_\omega$$

while the correlation between the log NAF estradiol levels from the left and right breast from a woman at any given visit is given by the intraclass correlation

$$Corr(y_{ijk1}, y_{ijk2}) = \frac{Cov(y_{ijk1}, y_{ijk2})}{\sqrt{Var(y_{ijk1})}\sqrt{Var(y_{ijk2})}} = \frac{\sigma^2_\omega + \sigma^2_b}{\sigma^2_\omega + \sigma^2_b + \sigma^2_\epsilon} = \rho_{\omega,b}.$$

We fit this basic "split-split plot" repeated measures ANOVA model to the 47 women using the SAS code shown below.

Program 3.6

```
ods select ModelInfo Dimensions NObs CovParms G V VCorr
           FitStatistics Tests3 ;
proc mixed data=example3_2_3 ic covtest empirical;
 class Subject_ID Group Visit Breast;
 model NAF_Estradiol = Group Visit Group*Visit Breast
       Group*Breast Visit*Breast Group*Visit*Breast /
       solution;
 random Intercept / sub=Subject_ID type=un g;
 random Intercept / sub=Subject_ID(Visit) type=un g v vcorr;
run;
```

Note that in specifying the random effects, we elected to use two RANDOM statements in which we specifically identify the source for each random effect using a SUBJECT= option (or SUB= as shown above). This was done for one primary reason. It allows us to process the data by SUBJECT_ID in such a way as to print the G matrices (i.e., the variance-covariance matrices associated with the two random effects, $\omega_{i(j)}$ and $b_{i(jk)}$) and the V and VCORR matrices for a single woman (which, by default, is the first subject). Of course we could have also specified TYPE=VC rather than TYPE=UN within each RANDOM statement without changing the results. Alternatively, we can use the single RANDOM statement

```
random Intercept Visit / sub=Subject_ID g v vcorr;
```

as a way of specifying the same split-split plot ANOVA model. The options G, V and VCORR will produce exactly the same output (i.e, the same parameter estimates, variance-covariance estimates, and tests of hypotheses) as obtained with the initial SAS code. However, using two RANDOM statements has the advantage of specifically identifying the two nested random effects.

Select output, as determined by the ODS SELECT statement in the above SAS code, are shown in Output 3.8. The correlation matrix for the first subject, displayed as a result of the VCORR option, shows that the correlation between the paired observations (left and right breast) at any one visit is 0.6554 whereas the correlation between any two observations from different visits is 0.5393. Although the estimated variance component, $\widehat{\sigma}_b^2 = 0.1510$ (corresponding to `Subject_ID(Visit)`), is not statistically significant, we elected to keep it in all subsequent models as it does provide further insight into the possible underlying correlation structure. Indeed, given the highly unbalanced nature of the data (e.g., Output 3.5), attempts to fit more complex covariance structures (e.g., allowing for heterogeneity across treatment groups within the above split-split plot specification, or assuming a marginal LM with an unstructured covariance across visits) either failed to improve the fit of the data or the program failed to converge. Thus the "split-split plot" repeated measures ANOVA model (3.24) was selected to serve as the basis for subsequent investigations into the relationship between NAF estradiol levels versus serum and salivary estradiol levels. However, to provide some protection against model misspecification with respect to this "split-split plot" type covariance structure, all inference was done using a robust sandwich estimator of the variance-covariance matrix of the fixed-effects (as indicated by specifying the EMPIRICAL option in the preceding SAS program).

Output 3.8: Select output from the "split-split plot" repeated measures ANOVA model (3.24) as applied to the estrogen hormone data.

```
                    Model Information
Data Set                     WORK.EXAMPLE3_2_3
Dependent Variable           NAF_Estradiol
Covariance Structure         Unstructured
Subject Effects              Subject_ID, Subject_ID(Visit)
Estimation Method            REML
Residual Variance Method     Profile
Fixed Effects SE Method      Empirical
Degrees of Freedom Method    Containment
          Dimensions
Covariance Parameters          3
Columns in X                  75
Columns in Z Per Subject       5
Subjects                      47
Max Obs Per Subject            8
```

Number of Observations	
Number of Observations Read	362
Number of Observations Used	200
Number of Observations Not Used	162

Estimated G Matrix

Row	Effect	Subject_ID	Visit	Col1	Col2	Col3	Col4	Col5
1	Intercept	1		0.7019				
2	Intercept	1	1		0.1510			
3	Intercept	1	2			0.1510		
4	Intercept	1	3				0.1510	
5	Intercept	1	4					0.1510

Estimated V Matrix for Subject_ID 1

Row	Col1	Col2	Col3	Col4	Col5	Col6	Col7	Col8
1	1.3014	0.8529	0.7019	0.7019	0.7019	0.7019	0.7019	0.7019
2	0.8529	1.3014	0.7019	0.7019	0.7019	0.7019	0.7019	0.7019
3	0.7019	0.7019	1.3014	0.8529	0.7019	0.7019	0.7019	0.7019
4	0.7019	0.7019	0.8529	1.3014	0.7019	0.7019	0.7019	0.7019
5	0.7019	0.7019	0.7019	0.7019	1.3014	0.8529	0.7019	0.7019
6	0.7019	0.7019	0.7019	0.7019	0.8529	1.3014	0.7019	0.7019
7	0.7019	0.7019	0.7019	0.7019	0.7019	0.7019	1.3014	0.8529
8	0.7019	0.7019	0.7019	0.7019	0.7019	0.7019	0.8529	1.3014

Estimated V Correlation Matrix for Subject_ID 1

Row	Col1	Col2	Col3	Col4	Col5	Col6	Col7	Col8
1	1.0000	0.6554	0.5393	0.5393	0.5393	0.5393	0.5393	0.5393
2	0.6554	1.0000	0.5393	0.5393	0.5393	0.5393	0.5393	0.5393
3	0.5393	0.5393	1.0000	0.6554	0.5393	0.5393	0.5393	0.5393
4	0.5393	0.5393	0.6554	1.0000	0.5393	0.5393	0.5393	0.5393
5	0.5393	0.5393	0.5393	0.5393	1.0000	0.6554	0.5393	0.5393
6	0.5393	0.5393	0.5393	0.5393	0.6554	1.0000	0.5393	0.5393
7	0.5393	0.5393	0.5393	0.5393	0.5393	0.5393	1.0000	0.6554
8	0.5393	0.5393	0.5393	0.5393	0.5393	0.5393	0.6554	1.0000

Covariance Parameter Estimates

Cov Parm	Subject	Estimate	Standard Error	Z Value	Pr > Z
UN(1,1)	Subject_ID	0.7019	0.2053	3.42	0.0003
UN(1,1)	Subject_ID(Visit)	0.1510	0.1240	1.22	0.1115
Residual		0.4485	0.1067	4.20	<.0001

Fit Statistics	
-2 Res Log Likelihood	506.9
AIC (smaller is better)	512.9
AICC (smaller is better)	513.1
BIC (smaller is better)	518.5

Type 3 Tests of Fixed Effects

Effect	Num DF	Den DF	F Value	Pr > F
Group	3	47	0.87	0.4653
Visit	3	83	0.66	0.5789
Group*Visit	9	47	1.62	0.1382
Breast	1	47	4.82	0.0331
Group*Breast	3	47	2.94	0.0425
Visit*Breast	3	47	2.28	0.0915
Group*Visit*Breast	9	47	0.85	0.5783

Following this base analysis, we fit three "split-split plot" ANCOVA models in an attempt to determine the relationship between NAF estradiol levels versus serum and/or salivary estradiol levels. The first ANCOVA model simply adds log serum estradiol (x_{ijk1}) as a fixed-effect covariate to our basic model as indicated by the following program.

Program 3.7

```
ods select ModelInfo Dimensions NObs CovParms
           FitStatistics SolutionF Tests3;
proc mixed data=example3_2_3 ic covtest empirical;
 class Subject_ID Group Visit Breast;
 model NAF_Estradiol = Group Visit Group*Visit Breast
                      Group*Breast Visit*Breast
                      Group*Visit*Breast
                      Serum_Estradiol / solution;
 random Intercept / sub=Subject_ID type=un g;
 random Intercept / sub=Subject_ID(Visit) type=un g v vcorr;
run;
```

As indicated in Output 3.9, there was no significant association between log NAF estradiol and log serum estradiol levels with the regression coefficient estimated to be 0.2875 ± 0.2392 (p-value= 0.2354). This is the same model as used by Gann et. al. (2006) except that here inference is based on the empirical sandwich estimator of the variance-covariance matrix of the regression parameter estimates. If we exclude the EMPIRICAL option in the above call to MIXED and instead use the default model-based standard errors, we get exactly the same result shown in Table 2 of Gann et. al. (2006) which, when rounded to four decimal places, is 0.2875 ± 0.2318 (p-value= 0.2211).

Output 3.9: Results from a repeated measures ANCOVA model with log serum estradiol as a sole covariate. Other fixed-effects include treatment group, time (visit) and breast and all higher-order interactions.

```
                      Model Information
   Data Set                    WORK.EXAMPLE3_2_3
   Dependent Variable          NAF_Estradiol
   Covariance Structure        Unstructured
   Subject Effects             Subject_ID, Subject_ID(Visit)
   Estimation Method           REML
   Residual Variance Method    Profile
   Fixed Effects SE Method     Empirical
   Degrees of Freedom Method   Containment
            Dimensions
   Covariance Parameters         3
   Columns in X                 76
   Columns in Z Per Subject      5
   Subjects                     47
   Max Obs Per Subject           8
          Number of Observations
   Number of Observations Read        362
   Number of Observations Used        200
   Number of Observations Not Used    162
                      Covariance Parameter Estimates
   Cov Parm   Subject            Estimate  Standard Error  Z Value   Pr > Z
   UN(1,1)    Subject_ID           0.7346          0.2141     3.43   0.0003
   UN(1,1)    Subject_ID(Visit)    0.1373          0.1247     1.10   0.1355
   Residual                        0.4513          0.1084     4.16   <.0001
            Fit Statistics
   -2 Res Log Likelihood      506.5
   AIC (smaller is better)    512.5
   AICC (smaller is better)   512.7
   BIC (smaller is better)    518.1
```

Solution for Fixed Effects

Effect	Breast	Group	Visit	Estimate	Standard Error	DF	t Value	Pr > |t|
Intercept				2.8191	1.0091	38	2.79	0.0081
Group		1		0.5935	0.4394	47	1.35	0.1833
Group		2		0.7728	0.4345	47	1.78	0.0818
Group		3		0.1006	0.6255	47	0.16	0.8729
Group		4		0	.	.	.	.
Visit			1	−0.3478	0.4694	82	−0.74	0.4608
Visit			2	0.4408	0.3926	82	1.12	0.2648
Visit			3	−0.1483	0.2576	82	−0.58	0.5664
Visit			4	0	.	.	.	.
Group*Visit		1	1	0.3027	0.5509	47	0.55	0.5853
Group*Visit		1	2	−0.8887	0.5633	47	−1.58	0.1213
Group*Visit		1	3	−0.1789	0.3534	47	−0.51	0.6151
Group*Visit		1	4	0	.	.	.	.
Group*Visit		2	1	0.05911	0.5540	47	0.11	0.9155
Group*Visit		2	2	−1.1237	0.5295	47	−2.12	0.0391
Group*Visit		2	3	−0.3971	0.3280	47	−1.21	0.2321
Group*Visit		2	4	0	.	.	.	.
Group*Visit		3	1	0.1824	0.6199	47	0.29	0.7699
Group*Visit		3	2	−0.4205	0.5495	47	−0.77	0.4480
Group*Visit		3	3	0.2393	0.5111	47	0.47	0.6418
Group*Visit		3	4	0	.	.	.	.
Group*Visit		4	1	0	.	.	.	.
Group*Visit		4	2	0	.	.	.	.
Group*Visit		4	3	0	.	.	.	.
Group*Visit		4	4	0	.	.	.	.
Breast	Left			−0.1546	0.1550	47	−1.00	0.3239
Breast	Right			0	.	.	.	.
Group*Breast	Left	1		0.07984	0.2330	47	0.34	0.7334
Group*Breast	Right	1		0	.	.	.	.
Group*Breast	Left	2		−0.1099	0.2742	47	−0.40	0.6904
Group*Breast	Right	2		0	.	.	.	.
Group*Breast	Left	3		0.3395	0.5217	47	0.65	0.5184
Group*Breast	Right	3		0	.	.	.	.
Group*Breast	Left	4		0	.	.	.	.
Group*Breast	Right	4		0	.	.	.	.
Visit*Breast	Left		1	−0.1387	0.7165	47	−0.19	0.8473
Visit*Breast	Right		1	0	.	.	.	.
Visit*Breast	Left		2	−0.3147	0.4505	47	−0.70	0.4883
Visit*Breast	Right		2	0	.	.	.	.
Visit*Breast	Left		3	0.7129	0.2237	47	3.19	0.0026
Visit*Breast	Right		3	0	.	.	.	.
Visit*Breast	Left		4	0	.	.	.	.
Visit*Breast	Right		4	0	.	.	.	.
Group*Visit*Breast	Left	1	1	0.3924	0.7347	47	0.53	0.5958
Group*Visit*Breast	Right	1	1	0	.	.	.	.
Group*Visit*Breast	Left	1	2	0.8363	0.6799	47	1.23	0.2248
Group*Visit*Breast	Right	1	2	0	.	.	.	.
Group*Visit*Breast	Left	1	3	0.3624	0.5776	47	0.63	0.5335
Group*Visit*Breast	Right	1	3	0	.	.	.	.
Group*Visit*Breast	Left	1	4	0	.	.	.	.
Group*Visit*Breast	Right	1	4	0	.	.	.	.
Group*Visit*Breast	Left	2	1	0.4522	0.8084	47	0.56	0.5786
Group*Visit*Breast	Right	2	1	0	.	.	.	.
Group*Visit*Breast	Left	2	2	0.8839	0.5831	47	1.52	0.1363
Group*Visit*Breast	Right	2	2	0	.	.	.	.
Group*Visit*Breast	Left	2	3	−0.5305	0.4344	47	−1.22	0.2281
Group*Visit*Breast	Right	2	3	0	.	.	.	.
Group*Visit*Breast	Left	2	4	0	.	.	.	.
Group*Visit*Breast	Right	2	4	0	.	.	.	.
Group*Visit*Breast	Left	3	1	0.4445	0.9592	47	0.46	0.6452
Group*Visit*Breast	Right	3	1	0	.	.	.	.
Group*Visit*Breast	Left	3	2	0.4709	0.8389	47	0.56	0.5772
Group*Visit*Breast	Right	3	2	0	.	.	.	.
Group*Visit*Breast	Left	3	3	−0.4270	0.6746	47	−0.63	0.5298
Group*Visit*Breast	Right	3	3	0	.	.	.	.
Group*Visit*Breast	Left	3	4	0	.	.	.	.
Group*Visit*Breast	Right	3	4	0	.	.	.	.
Group*Visit*Breast	Left	4	1	0	.	.	.	.
Group*Visit*Breast	Right	4	1	0	.	.	.	.
Group*Visit*Breast	Left	4	2	0	.	.	.	.
Group*Visit*Breast	Right	4	2	0	.	.	.	.
Group*Visit*Breast	Left	4	3	0	.	.	.	.
Group*Visit*Breast	Right	4	3	0	.	.	.	.
Group*Visit*Breast	Left	4	4	0	.	.	.	.
Group*Visit*Breast	Right	4	4	0	.	.	.	.
Serum_Estradiol				0.2875	0.2392	47	1.20	0.2354

Type 3 Tests of Fixed Effects

Effect	Num DF	Den DF	F Value	Pr > F
Group	3	47	0.83	0.4864
Visit	3	82	0.81	0.4900
Group*Visit	9	47	1.30	0.2604
Breast	1	47	4.87	0.0323
Group*Breast	3	47	2.59	0.0640
Visit*Breast	3	47	2.31	0.0886
Group*Visit*Breast	9	47	0.99	0.4579
Serum_Estradiol	1	47	1.44	0.2354

We fit a second ANCOVA model to the data by replacing `Serum_Estradiol` with `Saliva_Estradiol` as the covariate in the above call to MIXED. Because of the sparsity of repeated saliva estradiol values, regression estimates from this model are based on 35 women (down from the 47 women with serum estradiol concentrations) and a total of 131 observations (down from 200 in the previous analysis). As with serum estradiol, there was no association between log NAF estradiol and log serum estradiol levels. The regression coefficient was estimated to be -0.1344 ± 0.1545 (p-value= 0.3917) when we use robust standard errors and -0.1344 ± 0.1951 (p-value= 0.4965) when we use model-based standard errors.

We fit a third ANCOVA model to the data by including both log serum estradiol and log salivary estradiol as covariates. This is accomplished using the MODEL statement.

```
model NAF_Estradiol = Group Visit Group*Visit Breast
                      Group*Breast Visit*Breast
                      Group*Visit*Breast Serum_Estradiol
                      Saliva_Estradiol / solution;
```

The results of this third model are displayed in Output 3.10 but with parameter estimates shown for the log serum estradiol and log salivary estradiol covariates only. The results are consistent with results from the previous two ANCOVA models in that there is no significant association between either of these covariates and NAF estradiol levels. Given the reduction in the number of subjects (down from 47 to 35 subjects) and number of observations (down from 200 to 131 observations) as a result of missing salivary estradiol values, one might consider using some form of multiple imputation to determine if the lack of association persists when analyzed using a more complete dataset. Nonetheless, the results indicate that serum and salivary levels of estrogen are relatively poor independent predictors of NAF levels.

Output 3.10: Select results from a repeated measures ANCOVA model with both log serum estradiol and log salivary estradiol as fixed-effect covariates

Model Information

Data Set	WORK.EXAMPLE3_2_3
Dependent Variable	NAF_Estradiol
Covariance Structure	Unstructured
Subject Effects	Subject_ID, Subject_ID(Visit)
Estimation Method	REML
Residual Variance Method	Profile
Fixed Effects SE Method	Empirical
Degrees of Freedom Method	Containment

Dimensions

Covariance Parameters	3
Columns in X	76
Columns in Z Per Subject	5
Subjects	34
Max Obs Per Subject	8

```
          Number of Observations
Number of Observations Read         362
Number of Observations Used         131
Number of Observations Not Used     231
                        Covariance Parameter Estimates
Cov Parm    Subject                Estimate   Standard Error    Z Value    Pr > Z
UN(1,1)     Subject_ID               0.7229           0.2812       2.57    0.0051
UN(1,1)     Subject_ID(Visit)      8.89E-18                .          .         .
Residual                             0.5818          0.09677       6.01    <.0001
            Fit Statistics
-2 Res Log Likelihood         311.9
AIC (smaller is better)       315.9
AICC (smaller is better)      316.0
BIC (smaller is better)       318.9
                                   Solution for Fixed Effects
Effect              Breast   Group   Visit   Estimate   Standard Error   DF   t Value   Pr > |t|
Serum_Estradiol                                0.4432           0.2905   27      1.53     0.1387
Saliva_Estradiol                              -0.1621           0.1432   27     -1.13     0.2675
                 Type 3 Tests of Fixed Effects
Effect                  Num DF    Den DF    F Value    Pr > F
Group                        3        27       1.37    0.2743
Visit                        3        46       0.75    0.5273
Group*Visit                  9        27       1.86    0.1023
Breast                       1        27       5.89    0.0222
Group*Breast                 3        27       4.43    0.0118
Visit*Breast                 3        27       0.52    0.6737
Group*Visit*Breast           8        27       0.93    0.5061
Serum_Estradiol              1        27       2.33    0.1387
Saliva_Estradiol             1        27       1.28    0.2675
```

3.3 Summary

In this chapter, we have extended the marginal LM for correlated data by including random effects into the underlying model. The inclusion of such random effects provides an alternative mechanism for explaining correlation in one's data. In particular, we have demonstrated that correlation is introduced whenever a subject has a set of measurements that share a common set of random effects. Moreover, by introducing subject-specific or cluster-specific random effects into the linear model framework such as we have with the linear mixed-effects (LME) model (3.1), we can proceed with either population-wide or subject-specific inference. We briefly describe methods for conducting population-wide inference as it pertains to tests of hypotheses and confidence intervals on estimable functions of the fixed-effects parameters. We also describe methods for conducting a more narrow base of inference that is subject-specific in nature.

Just as in the last chapter, the methods of estimation and inference presented in this chapter ignore any weighting of the observations one may wish to include in the analysis. For the LME model, one can accommodate weighting by replacing $\boldsymbol{\Lambda}_i(\boldsymbol{\theta}_w)$ with $\boldsymbol{\Lambda}_i^*(\boldsymbol{\theta}_w) = \boldsymbol{W}_i^{-1/2}\boldsymbol{\Lambda}_i(\boldsymbol{\theta}_w)\,\boldsymbol{W}_i^{-1/2}$ where $\boldsymbol{W}_i$ is a diagonal matrix of weights associated with $\boldsymbol{y}_i$, the vector of observations on the ith subject/cluster.

In two of the three examples considered in this chapter, we consider the use of a random coefficient growth curve model (RCGCM) as an alternative to a marginal LM as a means for better accounting for observed correlation. In one case the RCGCM did a better job than various marginal LM's while in the second case, a marginal GMANOVA provided a better fit to the data. In our third example, we illustrate via preliminary analysis how one might conduct a multivariate repeated measures ANOVA. We then consider a multi-level random effects model in the form of a "split-split plot" repeated measures ANCOVA for helping assess relationships between time-dependent covariates and a set of repeated measurements. Although numerous other applications exist which require the use of LME models, it is hoped that this brief introduction will help serve as ground work for understanding how mixed models, in general, can be applied to a host of applications including those that entail non-Gaussian observations and a mean response that is nonlinear in the parameters of interest.

Part II

Nonlinear Models

4 Generalized Linear and Nonlinear Models

While normal-theory linear models find a wide range of use, there are numerous applications where the data are non-Gaussian in nature and, as such, involve distributions where the mean response is likely to be nonlinear in the parameters of interest. In Chapter 1 we considered several such examples including the respiratory disorder data involving repeated binary outcomes over time and the epileptic seizure data in which the primary outcome was a count of the number of epileptic seizures over successive two-week intervals. Nelder and Wedderburn (1972) introduced a class of generalized linear models within the univariate setting that permit regression-type analysis to be applied to data from a broad spectrum of discrete or continuous distributions. Specifically, the generalized linear model (GLIM) of Nelder and Wedderburn (1972) extends the familiar Gaussian-based linear model to models based on a broader class of distributions, namely the exponential family of distributions. This extension combined with the quasi-likelihood estimation approach of Wedderburn (1974) has provided us with a fairly unified approach to modeling either discrete or continuous data. Through the use of a "working" correlation matrix, these likelihood and quasi-likelihood methods have been extended to the repeated measures and longitudinal settings using a non-likelihood generalized estimating equation (GEE) approach to account for correlated response data (e.g., Liang and Zeger, 1986; Zeger, Liang and Albert, 1988).

While GLIM's provide practitioners with one extremely important extension to the normal linear model, they nonetheless carry certain restrictions that may prohibit their use in certain applications. For example, under the GLIM, the mean response, although possibly nonlinear in the parameters, is necessarily expressed as a monotonic invertible function of a linear predictor. In some settings, a more general marginal nonlinear mean response may be needed. Moreover, the allowable distributions under the class of GLIM's are restricted to those from the regular exponential family. This might limit one's ability to adequately model the variance-covariance matrix in the presence of correlated response data. Prentice and Zhao (1991), for example, consider a class of distributions from the quadratic exponential family that includes the multivariate normal distribution as a special case. By allowing the mean response to be a general nonlinear function of parameters and covariates and by allowing for a broader family of distributions, we introduce a class of generalized nonlinear models (GNLM's) that includes the family of marginal normal-theory nonlinear models (NLM's). Estimation under the GNLM is based on a second-order generalized estimating equation (GEE2) approach that, in many instances, corresponds to likelihood-based estimation and is suitable for correlated discrete and/or continuous response data.

In this chapter, we first review the class of marginal generalized linear models (GLIM's) introduced by Nelder and Wedderburn (1972) and Wedderburn (1974) for univariate data. We then extend this model to cover correlated data and present the generalized estimating equations (GEE) approach to analyzing discrete or continuous correlated outcomes as described by Liang and Zeger (1986) and Zeger, Liang and Albert (1988). A number of examples will be presented illustrating the use of GEE's for marginal GLIM's. The primary SAS tools used to illustrate these types of analyses are the GENMOD and GLIMMIX procedures. We then proceed to introduce a class of generalized nonlinear models (GNLM's) and discuss estimation and inference based on the second-order generalized estimating (GEE2) approach of Prentice (1988) and Prentice and Zhao (1991). Examples illustrating this class of GNLM's will be presented along with software code based on the SAS macro %NLINMIX and the SAS procedure NLMIXED.

4.1 The generalized linear model (GLIM)

Nelder and Wedderburn (1972) extended the normal-errors general linear model for univariate data to the generalized linear model (GLIM) by allowing the distribution of the response variable to be in the exponential family with probability density function (pdf)

$$\pi(y;\,\eta,\phi) = \exp\Big\{[y\eta - b(\eta)]/a(\phi) + c(y,\phi)\Big\} \tag{4.1}$$

and by allowing the existence of a monotonic differentiable and invertible function g that relates the mean μ of y to a set of s regressors, $\boldsymbol{x}' = (x_1,\ldots,x_s)$, through a linear predictor $\boldsymbol{x}'\boldsymbol{\beta}$ (i.e., $g(\mu) = \boldsymbol{x}'\boldsymbol{\beta}$). The exponential family as defined by (4.1) is a special case of the regular exponential family obtained by replacing $y\eta$ in (4.1) with $T(y)q(\eta)$ for nontrivial functions $T(y)$ and $q(\eta)$. In general, $a(\phi) = \phi/w$ for known weights w and in most cases $a(\phi) = \phi$. For this reason we will assume that $a(\phi) = \phi$ unless otherwise noted. Under this assumption, the first two moments of y_i are

$$\begin{aligned} E(y_i) &= \mu_i = \mu_i(\boldsymbol{\beta}) = g^{-1}(\boldsymbol{x}_i'\boldsymbol{\beta}),\ i = 1,\ldots,n \text{ observations} \\ Var(y_i) &= a_i(\phi)h(\mu_i) = \phi h(\mu_i) \quad (\text{assuming } a_i(\phi) = \phi/w_i,\ w_i = 1) \end{aligned} \tag{4.2}$$

where

$g(\mu_i) = \boldsymbol{x}_i'\boldsymbol{\beta}$ is an invertible monotonic link function of the marginal mean, μ_i.
$\boldsymbol{x}_i' = (x_{i1},\ldots,x_{is})$ is a set of s covariates or regressors,
$\boldsymbol{x}_i'\boldsymbol{\beta}$ is a linear (in $\boldsymbol{\beta}$) predictor.
$h(\mu_i)$ is a known or specified variance function of the mean
ϕ is a dispersion or possible overdispersion parameter.

GLIM's allow one to model data from the exponential family given by (4.1) which, for known ϕ, includes the following one- and two-parameter families:

Table 4.1 Some common 2-parameter continuous distributions in GENMOD and GLIMMIX. † Note that for the gamma distribution, the GLIMMIX procedure utilizes a default log link function to ensure a positive mean.

	Normal	**Gamma**	**Inverse Gaussian**
	$N(\mu,\sigma^2)$	$G(\mu,\nu)$	$IG(\mu,\sigma^2)$
$\pi(y)$	$\frac{1}{\sqrt{2\pi}\sigma}e^{-\frac{1}{2}\left(\frac{y-\mu}{\sigma}\right)^2}$	$\frac{1}{\Gamma(\nu)y}\left(\frac{y\nu}{\mu}\right)^{\nu}e^{-\frac{y\nu}{\mu}}$	$\frac{1}{\sqrt{2\pi y^3}\sigma}e^{-\frac{1}{2y}\left(\frac{y-\mu}{\mu\sigma}\right)^2}$
$b(\eta)$	$\eta^2/2$	$-\log(-\eta)$	$-(-2\eta)^{1/2}$
ϕ	σ^2	$1/\nu$	σ^2
Link: $\eta(\mu)$	identity (μ)	$-1/\mu$ †	$1/\mu^2$
$\mu(\eta)$	η	$-1/\eta$	$(-2\eta)^{-1/2}$
$h(\mu(\eta))$	1	μ^2	μ^3
Scale	σ	ν	σ
$E(y)$	μ	μ	μ
$V(y)$	$\phi=\sigma^2$	$\phi\mu^2=\mu^2/\nu$	$\phi\mu^3=\sigma^2\mu^3$

- Poisson $P(\mu)$, Binomial $B(m,\mu)$, Binary $B(1,\mu)$ $(\phi\equiv 1)$
- Normal $N(\mu,\sigma^2)$, Gamma $G(\mu,\nu)$, Inverse Gaussian $IG(\mu,\sigma^2)$
- Negative Binomial $NB(\mu,\alpha)$ (one-parameter exponential family for fixed α).

The first generalization of a GLIM over that of the standard linear model is the relaxation of the Gaussian assumption typically made for linear models. By allowing for distributions from the exponential family, we can model data in which the response or outcome variable is either a discrete (e.g., binomial, Poisson) or continuous (e.g., normal, gamma) random variable. Assuming the response variable has density function (4.1), the mean and variance of y are easily shown (see McCullagh and Nelder, 1989, pp. 28-29) to be:

$$E(y)=b'(\eta)=\mu(\eta)$$
$$Var(y)=b''(\eta)a(\phi)=(\partial\mu(\eta)/\partial\eta)a(\phi)=a(\phi)h(\mu(\eta))$$

where $b'(\eta)$ and $b''(\eta)$ denote the first and second derivatives of $b(\eta)$ with respect to canonical parameter η. The second generalization of the GLIM over that of the linear model is the concept of a link function and it is the link function that perhaps best distinguishes generalized linear models from the classical linear model. The link function is any monotonic differentiable (hence invertible) function g that relates the mean of y to a linear predictor $\boldsymbol{x}'\boldsymbol{\beta}$ via $g(\mu(\eta))=\boldsymbol{x}'\boldsymbol{\beta}$. In addition to the link function, GLIM's are also characterized by a variance function denoted by $h(\mu(\eta))$. The variance function merely reflects the fact that the variance of y can depend on the mean $\mu(\eta)$ through the canonical parameter η.

When the canonical parameter η equals the linear predictor $\boldsymbol{x}'\boldsymbol{\beta}$, we refer to g as the canonical link (McCullagh and Nelder, 1989, page 32). To illustrate, let Y be a Poisson random variable with density $\pi(y;\lambda)=(\lambda^y/y!)e^{-\lambda}$. Setting $\lambda=e^{\eta}$, we can write this density in terms of (4.1) by setting $b(\eta)=e^{\eta}, a(\phi)=\phi=1$ and $c(y,\phi)=-\log(y!)$. The mean and variance are therefore $b'(\eta)=b''(\eta)=e^{\eta}$ and the canonical link is simply $g(e^{\eta})=\log(e^{\eta})=\eta$. Shown in Table 4.1 is a list of some of the more common continuous distributions that SAS users have access to via the GENMOD and GLIMMIX procedures. Similarly, shown in Table 4.2 are some of the more common discrete distributions available in GENMOD and GLIMMIX. As suggested in Tables 4.1 - 4.2, rather than specifying the probability density function (pdf) $\pi(y)$ in terms of the natural or canonical parameter, η, the SAS procedures GENMOD and GLIMMIX generally work with a re-parameterization of the model in terms of the mean μ and scale parameter (if present) of the various distributions.

Table 4.2 Some common discrete distributions available in GENMOD and GLIMMIX. †For fixed known α, the negative binomial is a member of the exponential family.

	Binary	**Poisson**	**Negative Binomial**†
	$B(\mu)$	$P(\mu)$	$NB(\mu,\alpha)$
$\pi(y)$	$\mu^y(1-\mu)^{1-y}$	$\frac{\mu^y}{y!}e^{-\mu}$	$\frac{\Gamma(y+1/\alpha)}{\Gamma(1/\alpha)\Gamma(y+1)}\left(\frac{\alpha\mu}{1+\alpha\mu}\right)^y\left(\frac{1}{1+\alpha\mu}\right)^{1/\alpha}$
$b(\eta)$	$\log(1+e^\eta)$	e^η	$-\alpha^{-1}\log(1-e^\eta)$
ϕ	1	1	1
Link: $\eta(\mu)$	logit	log	$\log(\frac{\alpha\mu}{1+\alpha\mu})$
$\mu(\eta)$	$e^\eta/(1+e^\eta)$	e^η	$\alpha^{-1}\{e^\eta/(1-e^\eta)\}$
$h(\mu(\eta))$	$\mu(1-\mu)$	μ	$\mu+\alpha\mu^2$
Scale	1	1	α
$E(y)$	μ	μ	μ
$V(y)$	$\mu(1-\mu)$	μ	$\mu+\alpha\mu^2$

4.1.1 Estimation and inference in the univariate case

Maximum likelihood estimation

Maximum likelihood estimation is the default method of estimation for both GENMOD and GLIMMIX provided the pdf of y is from the exponential family with true dispersion parameter ϕ (i.e., ϕ is not some overdispersion parameter one wishes to use to help explain away excess variation not accounted for in distributions like the Poisson or Binomial where $\phi\equiv 1$). By default, both procedures utilize a variation of the Newton-Raphson algorithm to obtain ML estimates but one can request that the Fisher scoring algorithm be used by specifying the SCORING= option that is available in both procedures. In this case, it can be shown that Fisher's scoring is identical to the Gauss-Newton algorithm applied to a nonlinear regression of the response variable on the mean weighted by the inverse variance function.

Maximum likelihood estimation is also the default method of estimation of the dispersion parameter ϕ in both procedures. The exception to this occurs when one specifies the normal pdf with identity link in GLIMMIX. In this case GLIMMIX defaults to the REML estimate of ϕ and only by specifying the NOREML option within GLIMMIX can one obtain a MLE of ϕ. It should also be noted that the "Scale" parameter displayed in the GENMOD output containing parameter estimates is not the dispersion or overdispersion parameter estimate $\widehat{\phi}$ but rather the square root of $\widehat{\phi}$ (i.e., $\sqrt{\widehat{\phi}}$). The exception to this occurs when one specifies the gamma distribution. In this case the "Scale" parameter shown is $1/\widehat{\phi}=\widehat{\nu}$ (see Table 4.1). GLIMMIX, on the other hand, displays the "Scale" parameter as $\widehat{\phi}$ (in GLIMMIX, the gamma distribution is reparameterized such that $\phi=\nu$). In both procedures, the "Scale" parameter shown for the negative binomial distribution is the MLE of α.

Quasi-likelihood estimation

In many applications one may choose to simply model the mean and variance without making potentially restrictive assumptions regarding the underlying distribution. For example, one may wish to specify an overdispersed GLIM that allows for additional unexplained variation. To illustrate, one may initially model count data assuming a Poisson distribution with mean = variance = μ but examination of the residuals suggests some degree of overdispersion of the form $\phi\mu$ is necessary to explain excess variation in the counts. In cases like this, a quasi-likelihood approach was proposed by Wedderburn (1974) as an alternative means of estimation in the GLIM setting. Given univariate data,

Table 4.3 Quasi-likelihood functions associated with select distributions

Distribution	$Q(\mu, y)$
Binary	$y\log\left(\frac{\mu}{1-\mu}\right)+\log(1-\mu)$
Poisson	$y\log\mu-\mu$
Negative Binomial	$\log\Gamma(y+\frac{1}{\alpha})-\log\Gamma(\frac{1}{\alpha})+y\log\left(\frac{\alpha\mu}{1+\alpha\mu}\right)+\frac{1}{\alpha}\log\left(\frac{1}{1+\alpha\mu}\right)$
Normal	$-(y-\mu)^2/2$
Gamma	$-(y/\mu)-\log\mu$
Inverse Gaussian	$(\mu-\frac{1}{2}y)/\mu^2$

$y_1,\ldots,y_n$, with assumed means $\mu_1,\ldots,\mu_n$ and variances $\phi h(\mu_1),\ldots,\phi h(\mu_n)$, Wedderburn (1974) showed that if the log quasi-likelihood function,

$$Q(\mu_i, y_i) = \int_{y_i}^{\mu_i} \frac{y_i - t}{\phi h(t)} dt, \tag{4.3}$$

exists, then the quasi-score function,

$$\frac{\partial Q(\mu_i, y_i)}{\partial \mu_i} = \frac{y_i - \mu_i}{\phi h(\mu_i)} = U_i \tag{4.4}$$

shares the following properties in common with that of the log-likelihood score function:

$$\begin{aligned} E(U_i) &= 0 \\ Var(U_i) &= 1/\phi h(\mu_i) \\ -E\left(\frac{\partial U_i}{\partial \mu_i}\right) &= 1/\phi h(\mu_i). \end{aligned}$$

For example, suppose that $E(y) = \mu$, and $Var(y) = \phi\mu$ as when y is an overdispersed Poisson variate. Then the log quasi-likelihood function is

$$\begin{aligned} Q(\mu, y) &= \int_y^{\mu} \frac{y-t}{\phi t} dt \\ &= \int_y^{\mu} \phi^{-1}\left(\frac{y}{t} - 1\right) dt \\ &= \phi^{-1}[y\log\mu - \mu] + k(y, \phi) \end{aligned}$$

where $k(y,\phi) = \phi^{-1}[-y\log(y) + y]$. The quasi-score is $U = (y-\mu)/\phi\mu$ which is the same as the log-likelihood score for a Poisson random variable when $\phi = 1$. Shown in Table 4.3 are the quasi-likelihood functions for the select distributions shown in Tables 4.1-4.2 (ignoring constants like $k(y,\phi)$).

Writing $\mu_i = \mu_i(\boldsymbol{\beta})$ to reflect the dependence of the mean on the linear predictor, $\boldsymbol{x}_i'\boldsymbol{\beta}$, the quasi-likelihood estimator $\widehat{\boldsymbol{\beta}}$ of the $s \times 1$ parameter vector $\boldsymbol{\beta}$ will, for any fixed ϕ, be the solution to the set of quasi-likelihood estimating equations

$$U(\boldsymbol{\beta}) = \phi^{-1}\boldsymbol{D}'\boldsymbol{V}^{-1}(\boldsymbol{y} - \boldsymbol{\mu}(\boldsymbol{\beta})) = \boldsymbol{0} \tag{4.5}$$

where $U(\boldsymbol{\beta})$ is the quasi-score function with respect to $\boldsymbol{\beta}$, $\boldsymbol{y}' = (y_1 \;\ldots\; y_n)$ is the vector of observations, $\boldsymbol{\mu}(\boldsymbol{\beta})' = (\mu_1(\boldsymbol{\beta}) \;\ldots\; \mu_n(\boldsymbol{\beta}))$ is the corresponding mean vector,

$$\boldsymbol{V} = \boldsymbol{V}(\boldsymbol{\beta}) = \begin{pmatrix} h(\mu_1(\boldsymbol{\beta})) & 0 & \ldots & 0 \\ 0 & h(\mu_2(\boldsymbol{\beta})) & \ldots & 0 \\ \vdots & & \ddots & \vdots \\ 0 & 0 & \ldots & h(\mu_n(\boldsymbol{\beta})) \end{pmatrix}_{n\times n}$$

is the $n \times n$ matrix of variance functions expressed in terms of $\boldsymbol{\beta}$ and

$$\begin{aligned}
\boldsymbol{D} = \boldsymbol{D}(\boldsymbol{\beta}) = \frac{\partial \boldsymbol{\mu}(\boldsymbol{\beta})}{\partial \boldsymbol{\beta}'} &= \frac{\partial g^{-1}(\boldsymbol{\eta})}{\partial \boldsymbol{\eta}'}\frac{\partial \boldsymbol{\eta}}{\partial \boldsymbol{\beta}'} \\
&= \frac{\partial g^{-1}(\boldsymbol{\eta})}{\partial \boldsymbol{\eta}'}\boldsymbol{X} \\
&= \begin{pmatrix} \frac{\partial g^{-1}(\eta_1)}{\partial \eta_1} & 0 & \cdots & 0 \\ 0 & \frac{\partial g^{-1}(\eta_2)}{\partial \eta_2} & \cdots & 0 \\ \vdots & & \ddots & \\ 0 & 0 & \cdots & \frac{\partial g^{-1}(\eta_n)}{\partial \eta_n} \end{pmatrix} \boldsymbol{X} \\
&= \begin{pmatrix} \frac{\partial g^{-1}(\eta_1)}{\partial \eta_1}\boldsymbol{x}_1' \\ \frac{\partial g^{-1}(\eta_2)}{\partial \eta_2}\boldsymbol{x}_2' \\ \vdots \\ \frac{\partial g^{-1}(\eta_n)}{\partial \eta_n}\boldsymbol{x}_n' \end{pmatrix}
\end{aligned}$$

is the $n \times s$ derivative matrix of the means with respect to the parameters $\boldsymbol{\beta}$, and $\boldsymbol{\eta} = \boldsymbol{X}\boldsymbol{\beta}$ with $\eta_i = \boldsymbol{x}_i'\boldsymbol{\beta}$. In this univariate setting, quasi-likelihood estimates of $\boldsymbol{\beta}$ may be obtained using iteratively reweighted least squares in combination with the Gauss-Newton algorithm.

Inference under the univariate GLIM is based on the fact that the covariance matrix of the quasi-score function, $U(\boldsymbol{\beta})$, is

$$\boldsymbol{I}(\boldsymbol{\beta}) = \phi^{-1}\boldsymbol{D}'\boldsymbol{V}^{-1}\boldsymbol{D}$$

which may be thought of as representing the Fisher expected information matrix for quasi-likelihood estimation in so far as $\boldsymbol{I}(\boldsymbol{\beta}) = -E(\partial U(\boldsymbol{\beta}))/\partial\boldsymbol{\beta}$. Applying likelihood-based asymptotic theory to the quasi-likelihood estimator, it can be shown that inference in the form of hypothesis testing and confidence intervals may be carried out on the basis of the asymptotic normality of $\widehat{\boldsymbol{\beta}}$, namely that $\sqrt{n}(\widehat{\boldsymbol{\beta}} - \boldsymbol{\beta}) \sim N_s(\boldsymbol{0}, n\boldsymbol{I}(\boldsymbol{\beta})^{-1}) + O_p(n^{-1/2})$ (e.g., McCullagh, 1983). Since, for known ϕ, the asymptotic variance-covariance matrix of $\widehat{\boldsymbol{\beta}}$ may be estimated as

$$Var(\widehat{\boldsymbol{\beta}}) = \boldsymbol{\Omega}(\widehat{\boldsymbol{\beta}}) = \boldsymbol{I}(\widehat{\boldsymbol{\beta}})^{-1} = \phi\Big\{\boldsymbol{D}(\widehat{\boldsymbol{\beta}})'\boldsymbol{V}(\widehat{\boldsymbol{\beta}})^{-1}\boldsymbol{D}(\widehat{\boldsymbol{\beta}})\Big\}^{-1},$$

all that remains for valid inference is to obtain a consistent estimate of the dispersion or overdispersion parameter, ϕ. For quasi-likelihood based estimation, the most common approach is to use a method of moments (MM) estimator based on the estimated residuals, $\boldsymbol{y} - \boldsymbol{\mu}(\boldsymbol{\beta})$, namely

$$\widehat{\phi} = \frac{1}{n-s}\sum_{i=1}^{n}\frac{(y_i - \widehat{\mu}_i)^2}{h(\widehat{\mu}_i)} = \frac{X^2}{n-s} \tag{4.6}$$

where X^2 is the generalized Pearson statistic used as a measure of discrepancy in assessing model goodness-of-fit and $n-s$ is the degrees of freedom associated with estimating the $s \times 1$ parameter vector $\boldsymbol{\beta}$ (e.g., McCullagh and Nelder, 1989, pp.33-34). This option is available in the GENMOD procedure by specifying the option SCALE=PEARSON or in GLIMMIX by specifying the statement

```
random _residual_;
```

Alternatively, in GENMOD one may choose an estimate based on the deviance function, namely

$$\widehat{\phi} = \frac{D(\boldsymbol{y}; \widehat{\boldsymbol{\mu}})}{n-s}$$

where

$$D(\boldsymbol{y}; \widehat{\boldsymbol{\mu}}) = \phi\{2l(\boldsymbol{y}; \boldsymbol{y}) - 2l(\boldsymbol{y}; \widehat{\boldsymbol{\mu}})\} \tag{4.7}$$

is the deviance function evaluated at $\widehat{\boldsymbol{\mu}}$, l is the log-likelihood function from the specified distribution expressed as a function of the predicted mean values $\widehat{\boldsymbol{\mu}}$ and the vector $\boldsymbol{y}$ of response values, ϕ is the dispersion parameter from the specified distribution (e.g., normal, gamma, inverse Gaussian, etc.) and $n-s$ is the degrees of freedom defined as the number of observations minus the number of parameters. For example, if $y \sim N(\mu, \sigma^2)$ such that $\phi = \sigma^2$, then the deviance function for this single value is estimated to be

$$\begin{aligned} D(y; \widehat{\mu}) &= \sigma^2\{2l(y; y) - 2l(y; \widehat{\mu})\} \\ &= (y - \widehat{\mu})^2 \end{aligned}$$

where $l(y; y) = \log[\pi(y; y, \sigma^2)] = -\frac{1}{2}\log(2\pi\sigma^2)$ and $l(y; \widehat{\mu}) = \log[\pi(y; \widehat{\mu}, \sigma^2)] = -\frac{1}{2}\log(2\pi\sigma^2) - (y-\widehat{\mu})^2/2\sigma^2$ for probability density function $\pi(y; \mu, \sigma^2)$. In this case the deviance function and the generalized Pearson statistic are the same and the dispersion parameter based on n observations would be estimated as

$$\widehat{\phi} = \widehat{\sigma}^2 = \sum_{i=1}^{n}(y_i - \widehat{\mu}_i)^2/(n-s).$$

It should be noted that estimates of ϕ based on the generalized Pearson statistic or the deviance function must be specified using the SCALE= option of the MODEL statement in GENMOD. As noted above, to estimate an overdispersion parameter ϕ in GLIMMIX, one must specify the statement: `random _residual_;`. The default estimate of ϕ in this case is given by (4.6) while the NOREML option in GLIMMIX results in estimating ϕ by replacing $(n-s)$ in (4.6) with n.

4.2 The GLIM for correlated response data

In Chapter 2 we saw how one can accommodate correlated response data in the normal linear model setting by replacing the single response variable, y_i, with a response vector, $\boldsymbol{y}_i = \begin{pmatrix} y_{i1} & \ldots & y_{ip_i} \end{pmatrix}'$, and by specifying a variance-covariance matrix, $Var(\boldsymbol{y}_i) = \boldsymbol{\Sigma}_i(\boldsymbol{\theta})$, that allows for possible correlation among the components of $\boldsymbol{y}_i$. Following the work of Liang and Zeger (1986), a similar approach may be taken with respect to the GLIM. Specifically, let $\boldsymbol{y}_i = \begin{pmatrix} y_{i1} & \ldots & y_{ip_i} \end{pmatrix}'$ be a vector of p_i responses for the ith subject or cluster $(i = 1, \ldots, n)$ which may or may not be correlated. A marginal (PA) generalized linear model for such clustered or longitudinal data may be obtained from the univariate GLIM defined by (4.1) and (4.2) by assuming the marginal distribution of the individual components, y_{ij}, are in the exponential family (4.1) and by modeling the first two moments of the response vector $\boldsymbol{y}_i$. Specifically, we have the following.

Generalized linear models (GLIM)—correlated response data

The GLIM for correlated response data may be expressed in terms of the mean and variance as

$$E(\boldsymbol{y}_i) = \boldsymbol{\mu}_i = \boldsymbol{\mu}_i(\boldsymbol{\beta}) = \boldsymbol{g}^{-1}(\boldsymbol{X}_i\boldsymbol{\beta}),\ i = 1, \ldots, n \tag{4.8}$$

$$Var(\boldsymbol{y}_i) = \boldsymbol{\Sigma}_i(\boldsymbol{\beta}, \boldsymbol{\theta}) = \boldsymbol{V}_i(\boldsymbol{\mu}_i)^{1/2}\boldsymbol{\Sigma}_i^*(\boldsymbol{\theta})\boldsymbol{V}_i(\boldsymbol{\mu}_i)^{1/2}$$

where

$\boldsymbol{y}_i$ is the $p_i \times 1$ response vector for the ith subject (cluster)
$\boldsymbol{X}_i$ is a $p_i \times s$ matrix of within and between-subject covariates
$\boldsymbol{\beta}$ is an $s \times 1$ vector of fixed-effects parameters
$\boldsymbol{\mu}_i = \boldsymbol{\mu}_i(\boldsymbol{\beta}) = \boldsymbol{g}^{-1}(\boldsymbol{X}_i\boldsymbol{\beta})$ is a $p_i \times 1$ vector of means, $(\mu_{i1} \ldots \mu_{ip_i})'$, expressed as the inverse of the link function $g(\mu_{ij}) = \boldsymbol{x}_{ij}'\boldsymbol{\beta}$ where $\boldsymbol{x}_{ij}'$ is the jth row of $\boldsymbol{X}_i$.
$\boldsymbol{\Sigma}_i(\boldsymbol{\beta}, \boldsymbol{\theta}) = \boldsymbol{V}_i^{1/2}\boldsymbol{\Sigma}_i^*(\boldsymbol{\theta})\boldsymbol{V}_i^{1/2}$ is a $p_i \times p_i$ variance-covariance matrix defined by the square root of the diagonal variance function matrix, $\boldsymbol{V}_i = \boldsymbol{V}_i(\boldsymbol{\mu}_i) = Diag\{h(\mu_{ij})\}$ and by
$$\boldsymbol{\Sigma}_i^*(\boldsymbol{\theta}) = \begin{cases} \boldsymbol{\Sigma}_i(\boldsymbol{\theta}), & \text{if unweighted} \\ \boldsymbol{W}_i^{*-1/2}\boldsymbol{\Sigma}_i(\boldsymbol{\theta})\ \boldsymbol{W}_i^{*-1/2}, & \text{if weighted (e.g., } a_{ij}(\phi) = \phi/w_{ij}) \end{cases}$$
where $\boldsymbol{W}_i^* = Diag\{w_{ij}\}$ is a diagonal matrix of known weights w_{ij} and $\boldsymbol{\Sigma}_i(\boldsymbol{\theta})$ is a "working" covariance matrix with unique parameters defined by a $d \times 1$ vector $\boldsymbol{\theta}$.

Under the "multivariate" GLIM (4.8), the correlated response data are modeled using the same link function and linear predictor setup as defined for the univariate GLIM of §4.1. Likewise, the stochastic distribution of the components of $\boldsymbol{y}_i$ and their variance functions are described by the same distribution and variance function as in the independence case. The primary difference between the univariate GLIM defined by (4.2) and the "multivariate" GLIM defined by (4.8) is the introduction of a "working" covariance matrix, $\boldsymbol{\Sigma}_i(\boldsymbol{\theta})$, used to model correlated discrete or continuous response data.

Within the GENMOD procedure, the "working" covariance matrix takes the form $\boldsymbol{\Sigma}_i(\boldsymbol{\theta}) = \phi\boldsymbol{R}_i(\boldsymbol{\alpha})$ where ϕ represents a possible dispersion or overdispersion parameter, $\boldsymbol{R}_i(\boldsymbol{\alpha})$ is a "working" correlation matrix defined by a vector, $\boldsymbol{\alpha}$, of unique correlation parameters, and $\boldsymbol{\theta} = (\boldsymbol{\alpha}, \phi)$. In this case, the marginal covariance matrix of $\boldsymbol{y}_i$ is $\boldsymbol{\Sigma}_i(\boldsymbol{\beta}, \boldsymbol{\theta}) = \phi\boldsymbol{V}_i(\boldsymbol{\mu}_i)^{1/2}\boldsymbol{R}_i(\boldsymbol{\alpha})\boldsymbol{V}_i(\boldsymbol{\mu}_i)^{1/2}$ provided there is no additional weighting (i.e., when $a_{ij}(\phi) = \phi$). Common choices for a "working" correlation structure in GENMOD include that of working independence with $Corr(y_{ij}, y_{ik}) = 0$ for $j \neq k$, compound symmetry or exchangeable correlation with $Corr(y_{ij}, y_{ik}) = \alpha$ for $j \neq k$, first-order autoregressive with $Corr(y_{ij}, y_{i,j+k}) = \alpha^k\ (k = 0, 1, \ldots p_i - j)$, m-dependent with
$$Corr(y_{ij}, y_{i,j+k}) = \begin{cases} 1 & k = 0 \\ \alpha_k & k = 1, 2, \ldots, m, \\ 0 & k > m \end{cases}$$
and an unstructured correlation structure with
$$Corr(y_{ij}, y_{ik}) = \begin{cases} 1 & j = k \\ \alpha_{jk} & j \neq k \end{cases}.$$
The GLIMMIX procedure provides users with even greater flexibility in specifying a "working" covariance structure including, for example, those listed in Table 2.1. A complete list is displayed in the on-line documentation of GLIMMIX under the TYPE= option of the RANDOM statement.

4.2.1 Estimation

Since specifications of a stochastic distribution, link function, mean and variance function for GLIM's with correlated data are the same as that for GLIM's for independently distributed data, estimation may be carried out by simply extending the quasi-likelihood estimating equations (4.5) to include the "working" covariance matrix. This is precisely what Liang and Zeger (1986) did when they defined a set of generalized estimation equations (GEE) for GLIM's with correlated data. Their approach, which is implemented in the GENMOD procedure, assumes correlation can be accounted for through specification of a "working" correlation matrix, $\boldsymbol{R}_i(\boldsymbol{\alpha})$, while overdispersion is accounted for through ϕ (see above). In GLIMMIX, correlation and overdispersion are accounted for through specification of a more general "working" covariance matrix, $\boldsymbol{\Sigma}_i(\boldsymbol{\theta})$. In both cases estimation is based on the GEE approach of Liang and Zeger. The GEE approach is a non-likelihood semiparametric estimation technique ideally suited to GLIM's for correlated data. This is because marginal GLIM's defined by both (4.1) and (4.8) impose certain constraints on the possible correlation structure that make it difficult to specify a valid joint distribution for the response vector $\boldsymbol{y}_i$. Prentice (1988), for example, noted the constraints and difficulties one faces when attempting to model correlated binary data. For these reasons, the semiparametric GEE approach is the one most often used to analyze the marginal GLIM with correlated data.

Both GENMOD and GLIMMIX use GEE to estimate $\boldsymbol{\beta}$ but they differ in how the covariance parameters $\boldsymbol{\theta}$ are estimated. Consequently, we present the estimation schemes of these two procedures separately.

GEE—Estimation in GENMOD

In accordance with Liang and Zeger (1986), let $\boldsymbol{\Sigma}_i(\boldsymbol{\theta}) = \phi \boldsymbol{R}_i(\boldsymbol{\alpha})$ be a "working" covariance structure with $\boldsymbol{\theta} = (\boldsymbol{\alpha}, \phi)$. Under this setup, the marginal variance-covariance matrix of $\boldsymbol{y}_i$ under (4.8) will be of the form

$$\boldsymbol{\Sigma}_i = \boldsymbol{\Sigma}_i(\boldsymbol{\beta}, \boldsymbol{\theta}) = \boldsymbol{V}_i(\boldsymbol{\mu}_i)^{1/2} \boldsymbol{\Sigma}_i^*(\boldsymbol{\theta}) \boldsymbol{V}_i(\boldsymbol{\mu}_i)^{1/2} \tag{4.9}$$

where $\boldsymbol{\Sigma}_i^*(\boldsymbol{\theta}) = \boldsymbol{W}_i^{*-1/2} \boldsymbol{\Sigma}_i(\boldsymbol{\theta})\ \boldsymbol{W}_i^{*-1/2}$whenever $a_{ij}(\phi) = \phi / w_{ij}$ under (4.1) or $\boldsymbol{\Sigma}_i^*(\boldsymbol{\theta}) = \boldsymbol{\Sigma}_i(\boldsymbol{\theta})$ whenever $a_{ij}(\phi) = \phi$. For fixed $\boldsymbol{\theta}$, the GEE estimator of Liang and Zeger (1986) is that solution $\widehat{\boldsymbol{\beta}}_{\text{GEE}} = \widehat{\boldsymbol{\beta}}(\boldsymbol{\theta})$ to the set of GEE's

$$U_{\boldsymbol{\beta}}(\boldsymbol{\beta}, \boldsymbol{\theta}) = \sum_{i=1}^{n} U_i(\boldsymbol{\beta}, \boldsymbol{\theta}) = \sum_{i=1}^{n} \boldsymbol{D}_i' \boldsymbol{\Sigma}_i^{-1} (\boldsymbol{y}_i - \boldsymbol{\mu}_i(\boldsymbol{\beta})) = \boldsymbol{0} \tag{4.10}$$

where $\boldsymbol{D}_i = \boldsymbol{D}_i(\boldsymbol{\beta}) = \frac{\partial \boldsymbol{\mu}_i(\boldsymbol{\beta})}{\partial \boldsymbol{\beta}'} = \frac{\partial g^{-1}(\boldsymbol{\eta}_i)}{\partial \boldsymbol{\eta}_i'} \frac{\partial \boldsymbol{\eta}_i}{\partial \boldsymbol{\beta}'}$ is the $p_i \times s$ derivative matrix with respect to $\boldsymbol{\beta}$ for the ith subject and $\boldsymbol{\eta}_i = \boldsymbol{X}_i \boldsymbol{\beta}$ is the linear predictor.

The function $U_{\boldsymbol{\beta}}(\boldsymbol{\beta}, \boldsymbol{\theta})$ represents a multivariate quasi-score function suitable for GLIM's for correlated data. It is similar to the quasi-score function for univariate GLIM's except that it depends on the unknown parameter vector $\boldsymbol{\alpha}$ and also on the overdispersion parameter ϕ. However, as shown by Liang and Zeger (1986), one can express (4.10) as a function of $\boldsymbol{\beta}$ alone by letting $\widehat{\boldsymbol{\alpha}}(\boldsymbol{\beta}, \phi)$ be any $\sqrt{n}$-consistent method-of-moment (MM) estimator of $\boldsymbol{\alpha}$ for fixed known $\boldsymbol{\beta}$ and ϕ and letting $\widehat{\phi}(\boldsymbol{\beta})$ be any $\sqrt{n}$ -consistent estimator of ϕ for fixed known $\boldsymbol{\beta}$. We can then re-write (4.10) in terms of the set of first-order generalized estimating equations (GEE)

$$U_{\text{GEE}}(\boldsymbol{\beta}, \widehat{\boldsymbol{\theta}}(\boldsymbol{\beta})) = \sum_{i=1}^{n} U_i(\boldsymbol{\beta}, \widehat{\boldsymbol{\alpha}}(\boldsymbol{\beta}, \widehat{\phi}(\boldsymbol{\beta}))) = \boldsymbol{0} \tag{4.11}$$

which we solve using iteratively reweighted generalized least squares (IRGLS). More specifically, estimation based on GEE can be carried out using a Gauss-Newton algorithm which entails repeated application of Taylor series linearization and weighted linear regression. In our current setting this entails the following steps. Form the nonlinear regression model

$$\boldsymbol{y}_i = \boldsymbol{\mu}_i(\boldsymbol{\beta}) + \boldsymbol{\epsilon}_i \tag{4.12}$$

where $E(\boldsymbol{\epsilon}_i) = \boldsymbol{0}$ and $Var(\boldsymbol{\epsilon}_i) = \boldsymbol{\Sigma}_i(\boldsymbol{\beta}, \boldsymbol{\theta})$. A Gauss-Newton algorithm for estimating $\boldsymbol{\beta}$ under (4.12) via GEE is as follows:

A Gauss-Newton algorithm for estimating $\boldsymbol{\beta}$ via GEE

Step 0: Obtain an initial estimate, say $\tilde{\beta}$, of β assuming $\phi = 1$ and $R_i(\alpha) = I_{p_i}$. This is done by solving (4.5) assuming independence.

Step 1: Let $\widehat{\theta}(\beta)$ be a $\sqrt{n}$-consistent method-of-moment (MM) estimator of $\theta = (\alpha, \phi)$ for fixed β. Compute the estimated residuals, $e_{ij} = (y_{ij} - \tilde{\mu}_{ij})/\sqrt{h(\tilde{\mu}_{ij})}$, at the current estimate $\tilde{\beta}$ of β. Based on e_{ij}, update the MM estimator, $\widehat{\theta} = \widehat{\theta}(\tilde{\beta})$, and evaluate $\Sigma_i^*(\theta)$ at $\widehat{\theta}(\tilde{\beta})$.

Step 2. Perform a Taylor series linearization of $\mu_i(\beta)$ in (4.12) about $\beta = \tilde{\beta}$ yielding

$$y_i = \mu_i(\tilde{\beta}) + \tilde{\Delta}_i X_i(\beta - \tilde{\beta}) + \epsilon_i,$$

where

$$\mu_i(\beta) = \mu_i(\tilde{\beta}) + \tilde{D}_i(\beta - \tilde{\beta}),$$

$$\tilde{D}_i = D_i(\tilde{\beta}) = \left.\frac{\partial \mu_i(\beta)}{\partial \beta'}\right|_{\beta=\tilde{\beta}} = \left.\frac{\partial g^{-1}(\eta_i)}{\partial \eta_i'}\frac{\partial \eta_i}{\partial \beta'}\right|_{\beta=\tilde{\beta}} = \tilde{\Delta}_i X_i,$$

$$\tilde{\Delta}_i = \Delta_i(\tilde{\beta}) = \left.\frac{\partial g^{-1}(\eta_i)}{\partial \eta_i'}\right|_{\eta_i=\tilde{\eta}_i}$$

and $\eta_i = X_i\beta$ is the $p_i \times 1$ vector of linear predictors $x_{ij}'\beta$ for $j = 1, \ldots, p_i$. Form the pseudo weighted LM

$$\tilde{y}_i = X_i\beta + \tilde{W}_i^{-1/2}\epsilon_i^* \tag{4.13}$$

where $\tilde{y}_i = \tilde{\Delta}_i^{-1}(y_i - \mu_i(\tilde{\beta})) + X_i\tilde{\beta}$, $\tilde{W}_i = \tilde{V}_i^{-1}\tilde{\Delta}_i^2 = V_i(\mu_i(\tilde{\beta}))^{-1}\tilde{\Delta}_i^2$, $E(\epsilon_i^*) = \mathbf{0}$, and $Var(\epsilon_i^*) = \Sigma_i^*(\theta)$. Using estimated generalized least squares (EGLS), estimate β under (4.13) as

$$\widehat{\beta} = \widehat{\beta}(\widehat{\theta}) = \left(\sum_{i=1}^{n} X_i'\Sigma_i^*(\tilde{\beta}, \widehat{\theta}(\tilde{\beta}))^{-1}X_i\right)^{-1}\sum_{i=1}^{n} X_i'\Sigma_i^*(\tilde{\beta}, \widehat{\theta}(\tilde{\beta}))^{-1}\tilde{y}_i$$

where $\Sigma_i^*(\tilde{\beta}, \widehat{\theta}(\tilde{\beta})) = \tilde{W}_i^{-1/2}\Sigma_i^*(\widehat{\theta}(\tilde{\beta}))\tilde{W}_i^{-1/2} = \tilde{\Delta}_i^{-1}\Sigma_i(\tilde{\beta}, \widehat{\theta}(\tilde{\beta}))\tilde{\Delta}_i^{-1}$. Continue to iterate between Steps 1-2 by first updating $\tilde{\beta} = \widehat{\beta}(\widehat{\theta})$ and then $\widehat{\theta}(\tilde{\beta})$ in Step 1 and then using the EGLS estimate in Step 2 to obtain an updated estimate of β. This process is repeated until such time that successive differences in updated estimates is negligible.

This is an equivalent description of the fitting algorithm used by GENMOD (see SAS documentation) for fitting models using GEE. It results in a sequence of one-step Gauss-Newton estimators, $\{\widehat{\beta}^{(k)} | k = 1, 2, \ldots\}$, which, as $k \to \infty$, yields a solution to the GEE's (4.11). Typically convergence is determined by some specified criteria such as whenever the maximum absolute difference in successive parameter estimates is below a given tolerance.

The GENMOD procedure uses method of moment (MM) estimates of α and ϕ in step 1. Assuming $a_{ij}(\phi) = \phi/w_{ij}$ under (4.1) where w_{ij} are known weights, the MM estimator of ϕ used in GENMOD is

$$\widehat{\phi} = \frac{1}{N - s}\sum_{i=1}^{n}\sum_{j=1}^{p_i}\frac{(y_{ij} - \widehat{\mu}_{ij})^2}{h(\widehat{\mu}_{ij})/w_{ij}}$$

where $N = \sum_{i=1}^{n} p_i$ is the total number of observations, and $\widehat{\mu}_{ij}$ is the estimate of μ_{ij} at the current estimate of β. The vector of unique correlation parameters, α, is likewise estimated via MM. The MM formulae for estimating α for the various "working" correlation structures available in GENMOD may be found in the documentation of GENMOD.

GEE—Estimation in GLIMMIX

One can specify a marginal GLIM for correlated data in GLIMMIX using a single RANDOM statement with the TYPE= option for specifying a "working" covariance matrix $\boldsymbol{\Sigma}_i(\boldsymbol{\theta})$, and the RESIDUAL or RSIDE option to let SAS know this is a marginal model. As with GENMOD, estimation of $\boldsymbol{\beta}$ will be carried out by solving the set of GEE's (4.11). In fact, GLIMMIX uses a Gauss-Newton type algorithm similar to that used in GENMOD to estimate $\boldsymbol{\beta}$. However, GLIMMIX differs from GENMOD in two key aspects. First, the "working" covariance matrix, $\boldsymbol{\Sigma}_i(\boldsymbol{\theta})$, is much more general in GLIMMIX compared to GENMOD. Choices for $\boldsymbol{\Sigma}_i(\boldsymbol{\theta})$ include those shown in Table 2.1 for the marginal LM as well as generalizations of some of the choices of $\boldsymbol{\Sigma}_i(\boldsymbol{\theta}) = \phi \boldsymbol{R}_i(\boldsymbol{\alpha})$ available in GENMOD. Second, the GLIMMIX procedure actually solves a set of joint "independent" estimating equations in $(\boldsymbol{\beta}, \boldsymbol{\theta})$ that the documentation for GLIMMIX describes as a pseudo-likelihood (PL) estimation technique after Carroll and Ruppert (1988).

The PL estimation technique implemented in GLIMMIX involves computing a GEE estimator $\widehat{\boldsymbol{\beta}}_{\mathrm{GEE}}$ of $\boldsymbol{\beta}$ jointly with either a pseudo-likelihood (PL) estimator $\widehat{\boldsymbol{\theta}}_{\mathrm{PL}}$ or a restricted pseudo-likelihood (REPL) estimator $\widehat{\boldsymbol{\theta}}_{\mathrm{REPL}}$ of $\boldsymbol{\theta}$. It entails modifying the Gauss-Newton algorithm used in GENMOD by simply updating the pseudo-response vector $\tilde{\boldsymbol{y}}_i$ and weight matrix $\tilde{\boldsymbol{W}}_i$ at the current estimate $\tilde{\boldsymbol{\beta}}$ in Step 1 and using normal-theory ML or REML estimation to update $\widehat{\boldsymbol{\theta}}(\tilde{\boldsymbol{\beta}})$ in Step 2 based on the pseudo-weighted LM (4.13). Specifically, if under (4.13), we make the "working" assumption that $\boldsymbol{\epsilon}_i^* \sim N(\boldsymbol{0}, \boldsymbol{\Sigma}_i^*(\boldsymbol{\theta}))$ or, equivalently, that $\tilde{\boldsymbol{y}}_i \sim N(\boldsymbol{X}_i\boldsymbol{\beta}, \boldsymbol{\Sigma}_i^*(\tilde{\boldsymbol{\beta}}, \boldsymbol{\theta}))$ with

$$\boldsymbol{\Sigma}_i^*(\tilde{\boldsymbol{\beta}}, \boldsymbol{\theta}) = \tilde{\boldsymbol{W}}_i^{-1/2} \boldsymbol{\Sigma}_i^*(\boldsymbol{\theta}) \tilde{\boldsymbol{W}}_i^{-1/2} = \tilde{\boldsymbol{\Delta}}_i^{-1} \boldsymbol{V}_i(\tilde{\boldsymbol{\mu}}_i)^{1/2} \boldsymbol{\Sigma}_i^*(\boldsymbol{\theta}) \boldsymbol{V}_i(\tilde{\boldsymbol{\mu}}_i)^{1/2} \tilde{\boldsymbol{\Delta}}_i^{-1}$$

evaluated at a current estimate $\tilde{\boldsymbol{\beta}}$ of $\boldsymbol{\beta}$, then updated estimates of $\boldsymbol{\beta}$ and $\boldsymbol{\theta}$ may be obtained using standard normal-theory ML or REML estimation as described in Chapter 2 (§2.1.1; see also pp. 66). Unlike the Gauss-Newton algorithm for GEE, the modified Gauss-Newton algorithm for PL estimation requires a doubly iterative scheme as described below.

A Gauss-Newton algorithm for estimating $(\boldsymbol{\beta}, \boldsymbol{\theta})$ via PL

Step 0: Obtain an initial estimate, say $\tilde{\boldsymbol{\beta}}$, of $\boldsymbol{\beta}$ assuming $\boldsymbol{\Sigma}_i(\boldsymbol{\theta}) = \boldsymbol{I}_{p_i}$. This is done by solving (4.5) assuming independence.

Step 1: Given a current estimate $\tilde{\boldsymbol{\beta}}$, update the pseudo-response vector,

$$\tilde{\boldsymbol{y}}_i = \tilde{\boldsymbol{\Delta}}_i^{-1}(\boldsymbol{y}_i - \boldsymbol{\mu}_i(\tilde{\boldsymbol{\beta}})) + \boldsymbol{X}_i\tilde{\boldsymbol{\beta}}$$

and weights, $\tilde{\boldsymbol{W}}_i = \tilde{\boldsymbol{V}}_i^{-1}\tilde{\boldsymbol{\Delta}}_i^2$, as defined in Step 2 of the GEE algorithm.

Step 2. Form the pseudo-weighted LM

$$\tilde{\boldsymbol{y}}_i = \boldsymbol{X}_i\boldsymbol{\beta} + \tilde{\boldsymbol{W}}_i^{-1/2}\boldsymbol{\epsilon}_i^*$$

with a "working" assumption that $\boldsymbol{\epsilon}_i^* \sim N(\boldsymbol{0}, \boldsymbol{\Sigma}_i^*(\boldsymbol{\theta}))$ and maximize wrt $\boldsymbol{\theta}$ either the profile log-likelihood (2.12) or restricted profile log-likelihood (2.13) function associated with this weighted LM (see Chapter 2, §2.1.1, and also pp. 66). Let $\widehat{\boldsymbol{\theta}}$ be either the PL estimate, $\widehat{\boldsymbol{\theta}}_{\mathrm{PL}} = \widehat{\boldsymbol{\theta}}_{\mathrm{PL}}(\tilde{\boldsymbol{\beta}})$, or REPL estimate, $\widehat{\boldsymbol{\theta}}_{\mathrm{REPL}} = \widehat{\boldsymbol{\theta}}_{\mathrm{REPL}}(\tilde{\boldsymbol{\beta}})$, and compute the estimated generalized least squares (EGLS) estimator

$$\widehat{\boldsymbol{\beta}} = \widehat{\boldsymbol{\beta}}(\widehat{\boldsymbol{\theta}}) = \Big(\sum_{i=1}^{n} \boldsymbol{X}_i'\boldsymbol{\Sigma}_i^*(\tilde{\boldsymbol{\beta}}, \widehat{\boldsymbol{\theta}}(\tilde{\boldsymbol{\beta}}))^{-1}\boldsymbol{X}_i\Big)^{-1} \sum_{i=1}^{n} \boldsymbol{X}_i'\boldsymbol{\Sigma}_i^*(\tilde{\boldsymbol{\beta}}, \widehat{\boldsymbol{\theta}}(\tilde{\boldsymbol{\beta}}))^{-1}\tilde{\boldsymbol{y}}_i$$

where $\boldsymbol{\Sigma}_i^*(\tilde{\boldsymbol{\beta}}, \widehat{\boldsymbol{\theta}}(\tilde{\boldsymbol{\beta}})) = \tilde{\boldsymbol{W}}_i^{-1/2}\boldsymbol{\Sigma}_i^*(\widehat{\boldsymbol{\theta}}(\tilde{\boldsymbol{\beta}}))\tilde{\boldsymbol{W}}_i^{-1/2} = \tilde{\boldsymbol{\Delta}}_i^{-1}\boldsymbol{\Sigma}_i(\tilde{\boldsymbol{\beta}}, \widehat{\boldsymbol{\theta}}(\tilde{\boldsymbol{\beta}}))\tilde{\boldsymbol{\Delta}}_i^{-1}$. Continue to iterate between Steps 1-2 by updating $\tilde{\boldsymbol{\beta}} = \widehat{\boldsymbol{\beta}}(\widehat{\boldsymbol{\theta}})$ in Step 1 and repeating the iterative procedure in Step 2 to obtain updated joint estimates of $(\boldsymbol{\beta}, \boldsymbol{\theta})$. This process is repeated until such time that successive differences in updated estimates is negligible.

This is adapted from the PL algorithm described by Wolfinger and O'Connell (1993) and is the algorithm implemented in GLIMMIX for marginal GLIM's with correlated data. One can request a GEE estimator of $\boldsymbol{\beta}$ based on the PL estimate of $\boldsymbol{\theta}$ [i.e., $\widehat{\boldsymbol{\beta}}_{\text{GEE}} = \widehat{\boldsymbol{\beta}}(\widehat{\boldsymbol{\theta}}_{\text{PL}})$] by specifying the METHOD=MMPL option within the GLIMMIX statement. Alternatively, a GEE estimator of $\boldsymbol{\beta}$ based on the REPL estimator of $\boldsymbol{\theta}$ [i.e., $\widehat{\boldsymbol{\beta}}_{\text{GEE}} = \widehat{\boldsymbol{\beta}}(\widehat{\boldsymbol{\theta}}_{\text{REPL}})$] may be obtained by specifying the option METHOD=RMPL.

Vonesh and Chinchilli (1997, Chapter 9) and Vonesh et. al. (2001) refer to the PL approach taken in GLIMMIX as quasi-extended least squares (QELS) in that the joint estimate of $(\boldsymbol{\beta}, \boldsymbol{\theta})$ obtained in Step 2 may be viewed as a LS-type estimator obtained by minimizing a quasi-extended least squares (QELS) objective function associated with the pseudo-marginal LM. The term quasi-extended least squares was used by Vonesh and Chinchilli (1997) to describe a LS approach to the joint estimation of $(\boldsymbol{\beta}, \boldsymbol{\theta})$ that does not require the assumption of normality. It is an adaptation of the extended least squares (ELS) approach to joint estimation as first described by Beal and Sheiner (1982, 1988) for nonlinear mixed-effects models. In our current setting, we use the terms pseudo-likelihood (PL) and quasi-extended least squares (QELS) interchangeably. The PL/QELS estimation scheme involves solving two "independent" estimating equations, $U_{\boldsymbol{\beta}}(\boldsymbol{\beta}, \boldsymbol{\theta})$ and $U_{\boldsymbol{\theta}}(\boldsymbol{\beta}, \boldsymbol{\theta})$, that ignore any dependence $\boldsymbol{\Sigma}_i(\boldsymbol{\beta}, \boldsymbol{\theta})$ may have on $\boldsymbol{\beta}$ (Vonesh and Chinchilli, 1997; Chapter 9, §9.2.6). Hall and Severini (1998) refer to these estimating equations as extended GEE (EGEE) which should not to be confused with the EGEE's described by Vonesh et. al. (2001). Chaganty (1997) and Shults and Chaganty (1998) describe a quasi-least-squares (QLS) estimation technique that is based on a partial minimization of a GLS objective function. Their approach is similar in spirit to the PL/QELS technique. Of course these techniques all apply to the Gaussian-based marginal LM of Chapter 2 since it is simply a marginal GLIM with an identity link function (i.e., $\tilde{\boldsymbol{\Delta}}_i^{-1} = \boldsymbol{I}_{p_i}$). In this case, $\tilde{\boldsymbol{y}}_i = \boldsymbol{y}_i$ and $\boldsymbol{\Sigma}_i(\boldsymbol{\beta}, \boldsymbol{\theta}) = \boldsymbol{\Sigma}_i(\boldsymbol{\theta})$ so that the PL approach in GLIMMIX corresponds to exact ML or REML estimation.

GENMOD versus GLIMMIX

Although the PL/QELS approach in GLIMMIX is similar to the GEE approach in GENMOD, the two methods do differ in several aspects. While for any fixed $\boldsymbol{\theta}$ both approaches entail estimating $\boldsymbol{\beta}$ by solving $U_{\boldsymbol{\beta}}(\boldsymbol{\beta}, \boldsymbol{\theta})$ in equation 4.10, the GEE estimator in GENMOD uses method-of-moment estimates to update $\boldsymbol{\theta}$ while the GEE estimator in GLIMMIX uses PL or REPL estimates of $\boldsymbol{\theta}$. To capture this difference, we will denote any generic GEE estimator by $\widehat{\boldsymbol{\beta}}_{\text{GEE}}$ but reserve the notation $\widehat{\boldsymbol{\beta}}(\widehat{\boldsymbol{\theta}}_{\text{MM}})$ when referring to GEE estimators from GENMOD and either $\widehat{\boldsymbol{\beta}}(\widehat{\boldsymbol{\theta}}_{\text{PL}})$ or $\widehat{\boldsymbol{\beta}}(\widehat{\boldsymbol{\theta}}_{\text{REPL}})$ when referring to GEE estimators from GLIMMIX (depending on whether $\boldsymbol{\theta}$ is estimated via PL or REPL). A second difference is that the GEE estimator in GLIMMIX involves simultaneously solving a set of PL estimating equations $U_{\boldsymbol{\theta}}(\boldsymbol{\beta}, \boldsymbol{\theta})$ for $\boldsymbol{\theta}$ (e.g., Vonesh and Chinchilli, 1997, §9.2.6). By treating $U_{\boldsymbol{\beta}}(\boldsymbol{\beta}, \boldsymbol{\theta})$ and $U_{\boldsymbol{\theta}}(\boldsymbol{\beta}, \boldsymbol{\theta})$ as independent estimating equations, we are effectively treating $\boldsymbol{\beta}$ and $\boldsymbol{\theta}$ as though they are orthogonal (see also §4.4.2 of this Chapter). This can result in a less efficient estimate of $\boldsymbol{\beta}$ (Vonesh et. al., 2001). An alternative approach is to jointly estimate $(\boldsymbol{\beta}, \boldsymbol{\theta})$ by solving a set of second-order generalized estimating equations (GEE2) that directly account for any dependence the variance-covariance matrix of $\boldsymbol{y}_i$ has on $\boldsymbol{\beta}$ (see Prentice and Zhao, 1991). As this approach fits better with our discussion of estimation and inference under the generalized nonlinear model, we defer discussion of GEE2 type estimation techniques to §4.4.2.

It should be noted that some authors refer to any estimator of $\boldsymbol{\beta}$ that is a solution to (4.11) as a GEE1-type estimator. This is done to highlight the fact that $\widehat{\boldsymbol{\beta}}_{\text{GEE}}$ only requires correct specification of the *first-order* moments of $\boldsymbol{y}_i$ in order for it to be consistent and asymptotically normal. We shall use GEE to refer to the GEE1/MM based estimation of GENMOD and GEE1 to refer to the GEE1/PL/QELS based estimation of GLIMMIX. We do so with the understanding that such estimators do not require correct specification of the *second-order* moments as do the GEE2-based estimation techniques presented in §4.4.2.

4.2.2 Inference and test statistics

Under the GEE approach for GLIM's, inference centers almost exclusively on the fixed-effects regression coefficients, $\boldsymbol{\beta}$. This is because the overdispersion and correlation parameters are treated more or less as nuisance parameters that are required to improve efficiency resulting in better estimates of precision in the estimated regression coefficients. Indeed, under the GEE approach of GENMOD, the second-order parameters in $\boldsymbol{\theta} = (\boldsymbol{\alpha}, \phi)$ are estimated via method-of-moments (MM) and standard error estimates of such MM estimators are generally not available.

Inference about $\boldsymbol{\beta}$ as estimated using GEE may be based either on the inverse of the Fisher-type expected information matrix evaluated at $\widehat{\boldsymbol{\beta}}_{\text{GEE}}$, i.e.,

$$\widehat{\boldsymbol{\Omega}}(\widehat{\boldsymbol{\beta}}_{\text{GEE}}) = \boldsymbol{I}(\widehat{\boldsymbol{\beta}}_{\text{GEE}})^{-1} = \Big\{\sum_{i=1}^{n} \widehat{\boldsymbol{D}}_i'\widehat{\boldsymbol{\Sigma}}_i^{-1}\widehat{\boldsymbol{D}}_i\Big\}^{-1} \tag{4.14}$$

where $\widehat{\boldsymbol{\Sigma}}_i = \boldsymbol{\Sigma}_i(\widehat{\boldsymbol{\beta}}_{\text{GEE}}, \widehat{\boldsymbol{\theta}}) = \boldsymbol{V}_i(\boldsymbol{\mu}_i(\widehat{\boldsymbol{\beta}}_{\text{GEE}}))^{1/2}\boldsymbol{\Sigma}_i^*(\widehat{\boldsymbol{\theta}})\boldsymbol{V}_i(\boldsymbol{\mu}_i(\widehat{\boldsymbol{\beta}}_{\text{GEE}}))^{1/2}$ and $\widehat{\boldsymbol{D}}_i = \boldsymbol{D}_i(\widehat{\boldsymbol{\beta}}_{\text{GEE}})$, or it may be based on the robust sandwich estimator,

$$\widehat{\boldsymbol{\Omega}}_R(\widehat{\boldsymbol{\beta}}_{\text{GEE}}) = \widehat{\boldsymbol{\Omega}}(\widehat{\boldsymbol{\beta}}_{\text{GEE}})\Big(\sum_{i=1}^{n} \widehat{\boldsymbol{D}}_i'\widehat{\boldsymbol{\Sigma}}_i^{-1}\widehat{\boldsymbol{e}}_i\widehat{\boldsymbol{e}}_i'\widehat{\boldsymbol{\Sigma}}_i^{-1}\widehat{\boldsymbol{D}}_i\Big)\widehat{\boldsymbol{\Omega}}(\widehat{\boldsymbol{\beta}}_{\text{GEE}}), \tag{4.15}$$

where $\widehat{\boldsymbol{e}}_i = \boldsymbol{y}_i - \boldsymbol{\mu}_i(\widehat{\boldsymbol{\beta}}_{\text{GEE}})$. Preference is usually given to inference based on the robust sandwich estimator, $\widehat{\boldsymbol{\Omega}}_R(\widehat{\boldsymbol{\beta}}_{\text{GEE}})$, as it will lead to unbiased inference in large samples provided 1) we have correctly specified the marginal means, $\boldsymbol{\mu}_i(\boldsymbol{\beta}) = g^{-1}(\boldsymbol{X}_i\boldsymbol{\beta})$, regardless of whether we misspecified the "working" covariance matrix, $\boldsymbol{\Sigma}_i(\boldsymbol{\theta})$, and 2) we have a consistent estimator $\widehat{\boldsymbol{\theta}}$ of the working covariance parameters. The default method in GENMOD is to use the robust sandwich estimator (referred to in the GENMOD output as "Empirical Standard Error Estimates") and only by specifying the MODELSE option in the REPEATED statement of GENMOD can one also examine inference based on the inverse information matrix (be it the expected or observed information matrix depending on the options one specifies). Also, GENMOD uses the observed information matrix rather than the expected information matrix as the default method for computing model-based standard errors.

Within GENMOD, all tests of hypotheses and confidence intervals are carried out using large-sample z-tests for individual components of $\boldsymbol{\beta}$ and chi-square tests for general hypotheses of the form

$$H_0 : \boldsymbol{L\beta} = \boldsymbol{0} \tag{4.16}$$

for estimable $r \times s$ hypothesis matrix, $\boldsymbol{L}$. By default, GENMOD uses a generalized score (GS) test as described by Boos (1992) and Rotnitzky and Jewell (1990) for testing $H_0 : \boldsymbol{L\beta} = \boldsymbol{0}$. The GS test statistic is given by

$$T^2_{\text{GS}} = U(\widehat{\boldsymbol{\beta}}^*)'\widehat{\boldsymbol{\Omega}}(\widehat{\boldsymbol{\beta}}^*)\boldsymbol{L}'(\boldsymbol{L}\widehat{\boldsymbol{\Omega}}_R(\widehat{\boldsymbol{\beta}}^*)\boldsymbol{L}')^{-1}\boldsymbol{L}\widehat{\boldsymbol{\Omega}}(\widehat{\boldsymbol{\beta}}^*)U(\widehat{\boldsymbol{\beta}}^*) \tag{4.17}$$

where $\widehat{\boldsymbol{\beta}}^*$ is the vector of regression parameters resulting from solving the GEE under the restricted model $\boldsymbol{L\beta} = \boldsymbol{0}$. The term $U(\widehat{\boldsymbol{\beta}}^*) = U(\widehat{\boldsymbol{\beta}}^*, \widehat{\boldsymbol{\theta}}(\widehat{\boldsymbol{\beta}}^*))$ denotes the GEE score function

(4.11) at $\widehat{\boldsymbol{\beta}}^*$ while $\widehat{\boldsymbol{\Omega}}(\widehat{\boldsymbol{\beta}}^*)$ and $\widehat{\boldsymbol{\Omega}}_R(\widehat{\boldsymbol{\beta}}^*)$ are the model-based and robust variance-covariance estimates at $\widehat{\boldsymbol{\beta}}^*$. The restricted estimate $\widehat{\boldsymbol{\beta}}^*$ satisfies $U(\widehat{\boldsymbol{\beta}}^*) - \boldsymbol{L}'\tilde{\lambda} = \mathbf{0}$, $\boldsymbol{L}\widehat{\boldsymbol{\beta}}^* = \mathbf{0}$ where $\tilde{\lambda}$ is an $r \times 1$ vector of Lagrange multipliers (Boos, 1992). Under H_0, the GS test statistic will be asymptotically distributed as chi-square with r degrees of freedom. Alternatively, by specifying the option WALD along with TYPE3 in the MODEL statement or the WALD option in the CONTRAST statement, the GENMOD procedure will base tests of hypotheses on the Wald chi-square test statistic

$$T_{\mathrm{W}}^2 = (\boldsymbol{L}\widehat{\boldsymbol{\beta}})'\left(\boldsymbol{L}\widehat{\boldsymbol{\Omega}}_R(\widehat{\boldsymbol{\beta}})\boldsymbol{L}'\right)^{-1}(\boldsymbol{L}\widehat{\boldsymbol{\beta}}) \tag{4.18}$$

which is also asymptotically distributed as chi-square with r degrees of freedom (see the documentation for further details).

Inference under marginal GLIM's for correlated data differs in several aspects when one uses GLIMMIX versus GENMOD. First, since GLIMMIX utilizes a PL estimation technique involving a second set of PL-based estimating equations for $\boldsymbol{\theta}$, standard errors of the parameter estimates of $\boldsymbol{\theta}$ are available which one can then use to conduct inference in the form of confidence intervals and certain tests of hypotheses (e.g., see the COVTEST statement of GLIMMIX, page 33, and its application in Example 2.2.2, page 48). Second, with respect to $\boldsymbol{\beta}$, tests of hypotheses and confidence intervals are determined using either the model-based estimated covariance matrix (4.14) or the robust sandwich estimator (4.15) evaluated at $\widehat{\boldsymbol{\beta}}_{\mathrm{GEE1}} = \widehat{\boldsymbol{\beta}}(\widehat{\boldsymbol{\theta}})$ with either $\widehat{\boldsymbol{\theta}} = \widehat{\boldsymbol{\theta}}_{\mathrm{PL}}$ or $\widehat{\boldsymbol{\theta}} = \widehat{\boldsymbol{\theta}}_{\mathrm{REPL}}$. In addition to the classical robust sandwich estimator (4.15) which is the default, GLIMMIX also provides users with an option to specify alternative forms of the robust sandwich estimator (see SAS documentation of GLIMMIX for details). By default, all hypothesis tests of the form (4.16) in GLIMMIX are carried out using the F-test

$$F = T_W^2/r \tag{4.19}$$

with numerator degrees of freedom, r, and denominator degrees of freedom as determined by either the DDF= or DDFM= option of the MODEL statement of GLIMMIX or by the DF= option of the CONTRAST, ESTIMATE or LSMEANS statements. If one specifies DDFM=NONE in the MODEL statement, GLIMMIX carries out tests of hypotheses using the Wald chi-square test statistic T_W^2 with r degrees of freedom.

4.2.3 Model selection and diagnostics

Because the GEE approach for analyzing correlated response data is semiparametric, a number of non-likelihood based methods have been suggested as a means for performing model selection, goodness-of-fit and model diagnostics. In this section, we briefly describe some of the model selection and diagnostic tools available within the GENMOD and GLIMMIX procedures. In addition, we briefly mention some of the tools available within the SAS macros %GOF and %GLIMMIX_GOF both of which are described in Appendix D. Finally, we discuss strategies for selecting an appropriate "working" covariance structure within the marginal GLIM setting and describe how some of the model-fitting tools might be used to help make that selection.

The use of information criterion such Akaike's information criterion (AIC) or Schwarz's Bayesian criterion (SBC) are commonly used for model selection when estimation and inference are likelihood based. Given that GEE is an extension of quasi-likelihood estimation, Pan (2001) developed a quasi-likelihood information criterion (QIC) for GEE that can be used to either select from different regression models (i.e., variable selection) or to select from models with different "working" covariance structures. Let $\boldsymbol{\theta}$ be the vector of parameters associated with "working" covariance matrices, $\boldsymbol{\Sigma}_i(\boldsymbol{\theta})$, and let $\widehat{\boldsymbol{\beta}}(\widehat{\boldsymbol{\theta}})$ be the

GEE1 estimator based on $\mathbf{\Sigma}_i(\widehat{\boldsymbol{\theta}})$. The QIC is defined as

$$QIC(\boldsymbol{\theta}) = -2Q(\widehat{\boldsymbol{\beta}}(\widehat{\boldsymbol{\theta}}); \boldsymbol{I}) + 2\,\text{trace}\left\{\widehat{\mathbf{\Omega}}_I^{-1}\widehat{\mathbf{\Omega}}_R\right\} \tag{4.20}$$

$$= -2\sum_{i=1}^{n}\sum_{j=1}^{p_i} Q(\mu_{ij}(\widehat{\boldsymbol{\beta}}(\widehat{\boldsymbol{\theta}})), y_{ij}) + 2\,\text{trace}\left\{\widehat{\mathbf{\Omega}}_I^{-1}\widehat{\mathbf{\Omega}}_R\right\}$$

where $Q(\mu_{ij}(\widehat{\boldsymbol{\beta}}(\widehat{\boldsymbol{\theta}})), y_{ij})$ is the quasi-likelihood function (4.3) under the "working" independence assumption evaluated at the GEE1 estimator $\widehat{\boldsymbol{\beta}}(\widehat{\boldsymbol{\theta}})$, and

$$\widehat{\mathbf{\Omega}}_I^{-1} = \widehat{\mathbf{\Omega}}_I(\widehat{\boldsymbol{\beta}}(\widehat{\boldsymbol{\theta}}))^{-1} = \sum_{i=1}^{n} \widehat{\boldsymbol{D}}_i'\widehat{\boldsymbol{V}}_i^{-1}\widehat{\boldsymbol{D}}_i$$

is the inverse of the model-based covariance matrix as estimated under "working" independence but evaluated at $\widehat{\boldsymbol{\beta}}(\widehat{\boldsymbol{\theta}})$. Here, $\widehat{\boldsymbol{V}}_i = \widehat{\boldsymbol{V}}_i(\boldsymbol{\mu}_i(\widehat{\boldsymbol{\beta}}(\widehat{\boldsymbol{\theta}})))$ is the diagonal matrix of the variance function, $h(\mu_{ij})$, evaluated at $\widehat{\boldsymbol{\beta}}(\widehat{\boldsymbol{\theta}})$. Finally, $\widehat{\mathbf{\Omega}}_R = \widehat{\mathbf{\Omega}}_R(\widehat{\boldsymbol{\beta}}(\widehat{\boldsymbol{\theta}}))$ is the robust variance-covariance matrix (4.15) of the GEE1 estimate $\widehat{\boldsymbol{\beta}}(\widehat{\boldsymbol{\theta}})$. Model selection in GENMOD may be based either on $QIC(\boldsymbol{\theta})$ or the approximation

$$QIC_u(\boldsymbol{\theta}) = -2Q(\widehat{\boldsymbol{\beta}}(\widehat{\boldsymbol{\theta}}); \boldsymbol{I}) + 2s. \tag{4.21}$$

As noted by Pan, the approximation $QIC_u(\boldsymbol{\theta})$ will be appropriate for variable selection while $QIC(\boldsymbol{\theta})$ may be used for either variable selection or for selecting from different "working" covariance structure candidates. In either case, smaller values of QIC are indicative of a better fit.

Unfortunately, GLIMMIX does not include the quasi-likelihood information criterion of Pan (2001) in its output. Instead, GLIMMIX reports the generalized chi-square statistic

$$\sum_{i=1}^{n} \widehat{\boldsymbol{r}}_i'\mathbf{\Sigma}_i(\widehat{\boldsymbol{\beta}}, \widehat{\boldsymbol{\theta}}(\widehat{\boldsymbol{\beta}}))^{-1}\widehat{\boldsymbol{r}}_i$$

and its degrees of freedom based on the final pseudo-LM with $\widehat{\boldsymbol{r}}_i = \tilde{\boldsymbol{y}}_i - \boldsymbol{X}_i\widehat{\boldsymbol{\beta}}$. This can be useful for detecting possible model misspecifications due to overdispersion, incorrect assumptions regarding the marginal correlation structure or the omission of important covariates from the regression model. It is important to emphasize that the GEE approach in GLIMMIX is based on a PL estimation technique for which no quantifiable objective function of the response vectors $\boldsymbol{y}_i$ and joint parameters $(\boldsymbol{\beta}, \boldsymbol{\theta})$ exists (except in the case of the Gaussian-based LM). Consequently, one should avoid constructing some sort of pseudo-likelihood ratio test for nested GLIM's based on the calculated log pseudo-likelihood function reported in GLIMMIX. This is because the pseudo response vectors, $\tilde{\boldsymbol{y}}_i$, will change from one model to another and explains why GLIMMIX only reports the generalized chi-square statistic.

While one should avoid constructing PL ratio tests for the purposes of variable selection, the GLIMMIX procedure does offer the user the option of performing PL ratio tests for the purposes of conducting inference on the parameters of the "working" covariance matrix. Specifically, inference with respect to the covariance parameters $\boldsymbol{\theta}$ of a GLIM may be carried out using the COVTEST statement in GLIMMIX (e.g., see Output 2.9 of the bone mineral density example, §2.2.2). For a given GLIM with specified "working" covariance structure, GLIMMIX fits the null or reduced model for a test of covariance parameters using the final pseudo-data from the converged optimization of the full model. In this way, the pseudo response vectors $\tilde{\boldsymbol{y}}_i$ are held fixed allowing one to construct an appropriate PL ratio test based on the profile or restricted profile log-likelihood (see the GLIMMIX documentation for details).

The COVB(DETAILS) option of the MODEL statement of GLIMMIX provides additional less formal diagnostic tools for assessing how well an assumed covariance structure fits the data. These tools which are described in the GLIMMIX documentation are based on measuring how close the empirical sandwich estimator $\widehat{\mathbf{\Omega}}_R(\widehat{\boldsymbol{\beta}})$ of the variance-covariance of $\widehat{\boldsymbol{\beta}}$ is to the model-based estimator, $\widehat{\mathbf{\Omega}}(\widehat{\boldsymbol{\beta}})$. In addition, one can use the SAS macro %GOF (see Appendix D) in combination with select output from GENMOD and GLIMMIX to obtain R-square type goodness-of-fit measures useful in assessing model fit and variable selection (see Vonesh et. al. 1996; Vonesh and Chinchilli, 1997). Both the %GOF and %GLIMMIX_GOF macros (the latter being exclusively for use with GLIMMIX), provide an additional test for determining whether the "working" covariance structure may be reasonable. This test has been shown to provide reasonable type I and type II errors for models with Gaussian errors but it can be sensitive to models where the distribution of the residual errors is highly skewed (Vonesh et. al., 1996).

Finally, by using the PLOTS option in GENMOD and GLIMMIX, one can examine various diagnostic plots based on marginal residuals including the raw, studentized, and Pearson-type residuals. These residuals are all based on the pseudo response data evaluated at final convergence. The PLOTS option in GENMOD also provides options for calculating influence and case deletion diagnostics. The ASSESS statement of GENMOD can be used to check the validity of an assumed link function as well as functional forms of select covariates based on work by Lin, Wei, and Ying (2002).

Selecting a working covariance structure

One of the key challenges to fitting marginal GLIM's to correlated data is the selection of an appropriate "working" covariance structure. Indeed there are as many GEE1 estimators as there are choices of the "working" covariance matrix, $\mathbf{\Sigma}_i(\boldsymbol{\theta})$. Liang and Zeger (1986), for example, advocate the use of a "working" correlation structure over that of "working" independence on the basis of improved efficiency. Lipitz et. al. (1994) and Fitzmaurice (1995) also found that GEE estimators computed under "working" correlation structures offered improved efficiencies compared to those computed under "working" independence, particularly with respect to regression parameters linked to within-subject covariates. However, as noted by Crowder (1995) and Sutradhar and Das (1999), there may be occasions where specification of an ambiguous "working" correlation structure may result in a biased and/or inefficient estimate of $\boldsymbol{\beta}$. For example, Crowder (1995) showed that in a balanced GLIM setting involving $p \times 1$ correlated response vectors, $\boldsymbol{y}_1, \ldots, \boldsymbol{y}_n$, if one were to choose a "working" AR(1) structure when in fact the true structure is that of exchangeable correlation, then there will be a unique solution for the working correlation parameter α only when $\rho + a_p^{-1} \geq 0$ where ρ is the true exchangeable correlation and $a_p = \begin{cases} p & \text{when } p \text{ is odd} \\ p-1 & \text{when } p \text{ is even} \end{cases}$. In fact, utilizing the constraint that $\rho \geq -(p-1)^{-1}$, Crowder showed that in the case $p = 3$, there will be no solution to α whenever $-\frac{1}{2} \leq \rho < -\frac{1}{3}$. Thus there can be no guarantee that a solution $\widehat{\alpha}$ exists let alone that it converges to some value as $n \to \infty$ both of which are requirements in order for $\widehat{\boldsymbol{\beta}}_{\text{GEE}}$ to be a consistent estimator of $\boldsymbol{\beta}$. This problem stems from the fact that a moment-based estimating equation for a "working" correlation parameter α does not necessarily coincide with any well-defined objective function involving the first two moments of $\boldsymbol{y}_i$.

As Sutradhar and Das (1999) note, even when an estimate $\widehat{\alpha}$ exists of the working correlation parameter α, it may be that the value to which $\widehat{\alpha}$ converges will result in a less efficient estimate than that obtained assuming a "working" independence structure. Indeed, based on relative efficiencies of the robust variance-covariance matrices as computed under model misspecification of the working correlation structure, Sutradhar and Das (1999) note that use of a misspecified "working" independence structure generally outperformed alternative GEE estimators based on other misspecified "working" correlation structures. They also advocate the use of a Toeplitz type structure for certain kinds of balanced data.

While advocating the use of any one "working" correlation or covariance structure over another is probably ill-advised, there are some guidelines we can follow when selecting a "working" correlation. First, the use of a "working" independence structure will always yield a consistent estimator of $\boldsymbol{\beta}$ provided we have not misspecified the mean. This means that when we couple the use of a "working" independence structure with inference based on a robust variance-covariance matrix, the results will remain asymptotically valid despite some loss in efficiency. Second, when it appears that an analysis would stand to gain substantially in power and efficiency by attempting to correctly model the correlation/covariance structure, we can call on some of the tools described above (e.g., use of QIC in GENMOD or the COVTEST statement in GLIMMIX) to help differentiate between potential candidate covariance structures. However, given the inferential validity associated with the "working" independence structure, it may be a good idea to always compare the regression estimates from a particular "working" correlation structure to that obtained under "working" independence. If there is great disparity between the two sets of estimates, further consideration to model selection and diagnostics is warranted as well as to other underlying assumptions we have not yet touched upon, such as the role of nonignorable missing data (see Chapter 6).

4.3 Examples of GLIM's

As suggested at the beginning of this chapter, the practitioner will no doubt encounter numerous applications where the model is inherently nonlinear in the parameters of interest and the primary outcome variable is not Gaussian. In this section, we present several such applications: two involving count data, one involving binary outcome data, and one involving ordinal response data. We start by first illustrating an application requiring the use of a univariate GLIM for count data taken from the ADEMEX study (page 17). We then consider analyses of the respiratory disorder data described on page 13 involving repeated binary outcomes, and the epileptic seizure data described on page 15 which examines the impact of progabide on the number of partial seizures in patients with epilepsy. We conclude with an analysis of psychiatric data involving repeated ordinal responses taken over time.

4.3.1 ADEMEX peritonitis infection data

Peritonitis is an inflammation of the peritoneum most often due to bacterial infection. It can be a serious complication associated with continuous ambulatory peritoneal dialysis (CAPD) in that patients on CAPD must perform several dialysis exchanges a day requiring the connection and disconnection from a transfer set. Over the course of a year patients must perform a large number of sterile exchanges and, for some patients, bacterial contamination will occur resulting in an episode or peritonitis. Given this background, one of the secondary objectives of the ADEMEX study, described briefly in §1.4, page 17, was to determine if patients randomized to the high dose treated group were at increased risk for peritonitis due to possibly receiving an additional exchange per day compared to control patients (Paniagua et. al., 2002). The key elements of the ADEMEX study as it relates to this objective are outlined below. The name of the SAS variables in the dataset are shown in parentheses (see also author's Web page).

- Response variable:
  - $y_i(t_i)$ =number of peritonitis episodes for the ith patient during the course of follow-up (Episodes). Here t_i represents the patient's months at risk for peritonitis (MonthsAtRisk).
- Six between-patient covariates:
  - Treatment group (standard dose, high dose). The standard dose entails four 2L exchanges a day while the high dose may entail four or five 2.5L or 3L exchanges per day or five 2L exchanges per day (Trt).

$$x_{1i} = \begin{cases} 0, & \text{if control} = \text{standard dose} \\ 1, & \text{if treated} = \text{high dose} \end{cases}$$

  - Gender (Sex)

$$x_{2i} = \begin{cases} 0, & \text{if male} \\ 1, & \text{if female} \end{cases}$$

  - Age at baseline (Age)
    x_{3i} =patient age
  - Presence or absence of diabetes at baseline (Diabetic)

$$x_{4i} = \begin{cases} 0, & \text{if non-diabetic} \\ 1, & \text{if diabetic} \end{cases}$$

  - Baseline value of albumin (Albumin)
    x_{5i} =serum albumin (g/dL)
  - Prior time on dialysis (PriorMonths)
    x_{6i} =prior time on dialysis (months)
- Goal: Determine if patients in the treated group are at increased risk for peritonitis after adjusting for key risk factors

We observe that since peritonitis can occur more than once in a patient, the outcome variable of interest is a count represented by the number of peritonitis episodes observed during the course of follow-up (denoted $y_i(t_i)$ where t_i is the months at risk during follow-up). A partial listing of the data that also includes hospitalization counts (Hosp) are shown in Output 4.1.

Output 4.1: A partial list of ADEMEX peritonitis and hospitalization data

ptid	Trt	Age	Sex	Diabetic	Albumin	PriorMonths	MonthsAtRisk	Episodes	Hosp
01 001	1	34	1	0	2.96	2	1.5132	0	0
01 002	0	60	0	1	3.43	5	32.6316	1	0
01 004	0	46	0	0	3.78	17	22.3355	0	9
01 005	0	57	0	1	2.95	25	9.6382	0	1
01 006	1	20	0	0	4.15	42	4.9342	1	0
01 008	1	43	1	0	3.36	53	10.0000	1	1
01 009	1	42	1	0	3.46	25	29.0132	5	7
01 010	1	34	0	0	2	19	25.1974	0	2
01 011	0	48	1	0	4.02	114	25.9868	1	6
01 012	0	45	0	1	3.65	20	32.7961	0	0

In keeping with goals of the analysis, we fit two different univariate GLIM's that have been previously proposed for the analysis of peritonitis infections: a Poisson model and a negative binomial model (e.g., Vonesh, 1990). Both models require the use of an offset so as to correctly weight the number of infections a patient has by the patient's time at risk (the offset is set at $\log(t_i)$). The basic model, expressed in terms of the mean, is

$$E(y_i(t_i)) = \mu_i(\beta) = \exp(\beta_0 + \beta_1 x_{1i} + \beta_2 x_{2i} + \beta_3 x_{3i} + \beta_4 x_{4i} + \beta_5 x_{5i} + \beta_6 x_{6i})t_i$$

for either the Poisson or negative binomial model. The variance is

$$Var(y_i(t_i)) = \begin{cases} \mu_i(\boldsymbol{\beta}) & \text{if } y_i(t_i) \sim P(\mu_i) \\ \mu_i(\boldsymbol{\beta}) + \alpha\mu_i(\boldsymbol{\beta})^2 & \text{if } y_i(t_i) \sim NB(\mu_i, \alpha) \end{cases}.$$

The SAS code used to generate the partial listing shown in Output 4.1 as well as the GENMOD statements required to run both the Poisson and negative binomial models are shown below.

Program 4.1

```
data example4_3_1;
 set SASdata.ADEMEX_Peritonitis_Data;
 log_time=log(MonthsAtRisk); ** Offset;
run;
proc print data=example4_3_1(obs=10) noobs;
 var ptid Trt Age Sex Diabetic Albumin
     PriorMonths MonthsAtRisk Episodes Hosp;
run;
ods output LSMeans=lsout_P;
ods output LSMeanDiffs=lsdiff_P;
ods exclude ConvergenceStatus ParmInfo LSMeans LSMeanDiffs;
proc genmod data=example4_3_1;
  class Trt;
  model  Episodes=Trt Sex Age Diabetic Albumin PriorMonths
      / dist=Poisson link=log offset=log_time type3 ;
  lsmeans Trt /diff cl;
run;
ods output LSMeans=lsout_NB;
ods output LSMeanDiffs=lsdiff_NB;
ods select ModelInfo Modelfit
             ParameterEstimates Type3;
proc genmod data=example4_3_1;
  class Trt;
  model  Episodes=Trt Sex Age Diabetic Albumin PriorMonths
      / dist=NegBin link=log offset=log_time type3 ;
  lsmeans Trt /diff cl;
run;
```

Note that the ODS EXCLUDE and ODS SELECT statements allow the user to select the desired output to be shown. For example, since much of the information for the negative binomial model is the same as for the Poisson model, we elected to include only output related to the model information, model goodness-of-fit, ML estimates of the parameters and type 3 likelihood ratio chi-square tests of each of the fixed effects. For both models, we excluded printing the least squares means as these will be used later to list adjusted mean peritonitis rates and corresponding rate ratio.

Output 4.2: Analysis of ADEMEX peritonitis infection data assuming a Poisson model. Shown are selected output from GENMOD.

```
            Model Information
Data Set                WORK.EXAMPLE4_3_1
Distribution                      Poisson
Link Function                         Log
Dependent Variable               Episodes
Offset Variable                  log_time
```

Number of Observations Read	965
Number of Observations Used	877
Missing Values	88

Class Level Information

Class	Levels	Values
Trt	2	0 1

Criteria For Assessing Goodness Of Fit

Criterion	DF	Value	Value/DF
Deviance	870	1270.6815	1.4606
Scaled Deviance	870	1270.6815	1.4606
Pearson Chi-Square	870	1655.7619	1.9032
Scaled Pearson X2	870	1655.7619	1.9032
Log Likelihood		−846.6936	
Full Log Likelihood		−1093.2255	
AIC (smaller is better)		2200.4510	
AICC (smaller is better)		2200.5799	
BIC (smaller is better)		2233.8866	

Analysis Of Maximum Likelihood Parameter Estimates

Parameter		DF	Estimate	Standard Error	Wald 95% Confidence Limits		Wald ChiSq	Pr > ChiSq
Intercept		1	−2.3091	0.2770	−2.8519	−1.7663	69.52	<.0001
Trt	0	1	−0.0576	0.0786	−0.2117	0.0966	0.54	0.4642
Trt	1	0	0.0000	0.0000	0.0000	0.0000	.	.
Sex		1	0.0260	0.0806	−0.1319	0.1840	0.10	0.7466
Age		1	−0.0028	0.0034	−0.0095	0.0038	0.70	0.4039
Diabetic		1	0.1585	0.0946	−0.0270	0.3439	2.81	0.0939
Albumin		1	−0.2885	0.0664	−0.4187	−0.1583	18.86	<.0001
PriorMonths		1	−0.0003	0.0016	−0.0035	0.0028	0.04	0.8471
Scale		0	1.0000	0.0000	1.0000	1.0000		

LR Statistics For Type 3 Analysis

Source	DF	Chi-Square	Pr > ChiSq
Trt	1	0.54	0.4642
Sex	1	0.10	0.7467
Age	1	0.70	0.4038
Diabetic	1	2.82	0.0933
Albumin	1	18.69	<.0001
PriorMonths	1	0.04	0.8465

The selected output under an assumed Poisson model is shown in Output 4.2. The first thing to note is that by including adjustment for baseline covariates, the effective sample size is reduced from 965 patients to 877 patients due to missing values in one or more of the covariates. This is the same for both the Poisson and negative binomial models. The next thing we observe is that based on the goodness-of-fit statistics, there appears to be significant overdispersion associated with the number of peritonitis episodes when compared to what one might expect assuming a Poisson distribution. Based on the generalized Pearson chi-square shown in Output 4.2, we estimate the overdispersion to be $\widehat{\phi} = 1.9032$ which is nearly double what one would expect if the underlying distribution was Poisson. What is not clear is whether such overdispersion can be explained on the basis of excluding important explanatory variables or whether it is due to a violation of the Poisson assumption that the mean and variance are the same (i.e., that $\phi \equiv 1$). In terms of our primary goal, the results in Output 4.2 indicate there is no significant difference in the average number of peritonitis episodes over time between control and treated patients (p=0.4642). Although the confidence interval and p-value for comparing the two treatment groups does not account for any overdispersion, we can still assert qualitatively that patients treated at higher doses of CAPD were not at any increased risk for peritonitis (accounting for overdispersion will increase the standard error and p-value).

Output 4.3: Analysis of ADEMEX peritonitis infection data assuming a negative binomial model. Shown are selected output from GENMOD.

Model Information

Data Set	WORK.EXAMPLE4_3_1
Distribution	Negative Binomial
Link Function	Log
Dependent Variable	Episodes
Offset Variable	log_time

Criteria For Assessing Goodness Of Fit

Criterion	DF	Value	Value/DF
Deviance	870	845.1330	0.9714
Scaled Deviance	870	845.1330	0.9714
Pearson Chi-Square	870	1073.2150	1.2336
Scaled Pearson X2	870	1073.2150	1.2336
Log Likelihood		−801.0432	
Full Log Likelihood		−1047.5751	
AIC (smaller is better)		2111.1503	
AICC (smaller is better)		2111.3162	
BIC (smaller is better)		2149.3623	

Analysis Of Maximum Likelihood Parameter Estimates

Parameter		DF	Estimate	Standard Error	Wald 95% Confidence Limits		Wald ChiSq	Pr > ChiSq
Intercept		1	−2.2509	0.3643	−2.9650	−1.5368	38.17	<.0001
Trt	0	1	−0.0497	0.1034	−0.2523	0.1530	0.23	0.6309
Trt	1	0	0.0000	0.0000	0.0000	0.0000	.	.
Sex		1	0.0466	0.1064	−0.1620	0.2551	0.19	0.6618
Age		1	−0.0025	0.0044	−0.0111	0.0061	0.32	0.5711
Diabetic		1	0.1900	0.1233	−0.0517	0.4317	2.37	0.1234
Albumin		1	−0.3043	0.0881	−0.4770	−0.1315	11.92	0.0006
PriorMonths		1	0.0003	0.0021	−0.0039	0.0045	0.01	0.9025
Dispersion		1	0.7779	0.1225	0.5378	1.0181		

LR Statistics For Type 3 Analysis

Source	DF	Chi-Square	Pr > ChiSq
Trt	1	0.23	0.6310
Sex	1	0.19	0.6617
Age	1	0.32	0.5712
Diabetic	1	2.38	0.1232
Albumin	1	11.84	0.0006
PriorMonths	1	0.01	0.9026

We repeated the analysis assuming the number of peritonitis episodes follows a negative binomial model (see the second set of GENMOD statements shown above). The results displayed in Output 4.3 suggest that the overdispersion observed under the Poisson model may be mitigated under the negative binomial model (Output 4.3: Scaled Deviance = 0.9714, Scaled Pearson Chi-square = 1.2336). Indeed, based on the likelihood criteria associated with both models, we see the negative binomial model provides a better fit as indicated by an AIC of 2111.1503 versus 2200.4510. This result is consistent with previous studies that have evaluated different models for analyzing peritonitis rates (e.g., Vonesh, 1985, 1990).

As with the Poisson model, there was no significant difference in the number of peritonitis episodes between control and treated patients (p=0.6309). Moreover, both analyses show that low serum albumin at baseline is a strong predictor of an increased risk of peritonitis. Under the Poisson model, the rate ratio associated with a 1 g/dL *increase* in albumin is estimated to be $\exp(1\times(-0.2885)) = 0.75$ indicating a 1 g/dL lower starting serum albumin is associated with a 33% higher rate of peritonitis ($1/0.75 = 1.33$). Similarly under the negative binomial model, there is an estimated 36% higher risk of peritonitis ($\exp(1\times 0.3043)=1.36$) associated with 1 g/dL lower starting serum albumin.

Finally, we computed the adjusted mean peritonitis rates for the two groups and compared their rate ratio (RR) against a null value of 1 using least squares means. Given its superior fit to the data, this was done using results from the negative binomial model. The SAS programming statements required for this summary are shown below and are based on the SAS datasets created from the ODS OUTPUT statements linked to the GENMOD procedure for the negative binomial model.

Program 4.2

```
data lsout;
 set lsout_NB;
 Rate = exp(LBeta)*12;
run;
data lsdiff;
 set lsdiff_NB;
 RR = exp(-estimate);
 RR_lower = exp(-UpperCL);
 RR_upper = exp(-LowerCL);
 p=ProbChiSq;
 if interval=_interval;
 Trt='1';
run;
proc sort data=lsdiff;
 by Trt;
run;
proc sort data=lsout;
 by Trt;
run;
data summary;
 merge lsout lsdiff;
 by Trt;
run;
proc format;
 value $_trtfmt_ '0'='Control' '1'='Treated';
run;

proc print data=summary split='|' noobs;
 var Trt Rate RR RR_lower RR_upper p;
 label Rate='Peritonitis|(Episodes/Year)'
         RR='Rate Ratio|(Treated:Control)'
   RR_Lower='Lower|95% CL'
   RR_Upper='Upper|95% CL'
   p='p-value';
 format Trt $_trtfmt_.;
run;
```

The above program takes the estimated least squares means on the linear predictor scale (i.e., the log mean count) for the two treatment groups and summarizes them in terms of annualized peritonitis rates. Specifically, by exponentiation of the linear predictor, `LBeta`, the statement

```
Rate = exp(LBeta)*12;
```

computes the monthly rate for each group and then expresses this rate as the number of episodes per year (obtained by multiplying by 12). Similarly, the rate ratio (RR) for assessing whether these rates are significantly different is obtained by exponentiation of the least squares mean difference. The results are shown in Output 4.4 below. The adjusted peritonitis rate for the control group was estimated to be 0.49065 episodes/year while the patients in the intervention group had an estimated peritonitis rate of 0.51564 episodes/year (adjusted rate ratio = 1.05; p-value=0.63095).

Output 4.4: Adjusted mean peritonitis rates (annualized) and their associated rate ratio (RR) based on least squares means under the negative binomial model.

Trt	Peritonitis (Episodes/Year)	Rate Ratio (Treated:Control)	Lower 95% CL	Upper 95% CL	p-value
Control	0.49065	.	.	.	.
Treated	0.51564	1.05093	0.85814	1.28702	0.63095

4.3.2 Respiratory disorder data

The respiratory disorder data, as described in §1.4 (page 13), are data from a randomized controlled trial designed to evaluate patient response to an active treatment versus placebo control among patients with a respiratory disorder. The primary response was patient status which was defined in terms of the ordinal categorical responses: 0 = terrible, 1 = poor, 2 = fair, 3 = good, 4 = excellent. However, for this example, we follow Stokes et. al. (2000) and analyze a collapsed binary response to treatment over time defined by the binary indicator variable

$$y_{ij} = \begin{cases} 0 = \text{negative response if (terrible, poor, or fair)} \\ 1 = \text{positive response if (good, excellent)} \end{cases}$$

where y_{ij} is the binary response from the ith patient on the jth occasion or visit. A partial listing of the data is shown in Output 4.5. The raw data is available from the SAS Sample Library under Example 5 for PROC GENMOD (see also Appendix C, Dataset C.7).

Output 4.5: A partial listing of the respiratory disorder data.

Obs	Center	ID	Treatment	Sex	Age	Baseline	Visit	y
1	1	1	P	M	46	0	1	0
2	1	1	P	M	46	0	2	0
3	1	1	P	M	46	0	3	0
4	1	1	P	M	46	0	4	0
5	1	2	P	M	28	0	1	0
6	1	2	P	M	28	0	2	0
7	1	2	P	M	28	0	3	0
8	1	2	P	M	28	0	4	0
9	1	3	A	M	23	1	1	1
10	1	3	A	M	23	1	2	1
11	1	3	A	M	23	1	3	1
12	1	3	A	M	23	1	4	1
13	1	4	P	M	44	1	1	1
14	1	4	P	M	44	1	2	1
15	1	4	P	M	44	1	3	1
16	1	4	P	M	44	1	4	0
17	1	5	P	F	13	1	1	1
18	1	5	P	F	13	1	2	1
19	1	5	P	F	13	1	3	1
20	1	5	P	F	13	1	4	1

Our primary goal is to fit the data using logistic regression for correlated data and test whether there is a significant treatment effect. Since the logit link function (Table 4.2) is the default link for binary regression in both GENMOD and GLIMMIX, we first performed an analysis to determine if the logit is in fact an appropriate link for this particular dataset. We did so by fitting the data to model (1.3) assuming a "working" independence structure and using the ASSESS statement in GENMOD to determine if the logit link is reasonable. In order to run this analysis we must first convert all character covariates into numeric covariates. This is because the ASSESS statement will not work with models having character-based covariates. The SAS programming statements required for this initial analysis are shown below.

Program 4.3

```
proc sort data=example4_3_2;
 by Center ID;
run;
Data example4_3_2;
 set example4_3_2;
 by Center ID;
 Trt=(Treatment='A');
 Gender=(Sex='F');
 Center_=(Center=2);
 Subject=Compress(Trim(Center_)||'-'||Trim(ID));
 y0=Baseline;
run;
ods graphics on / imagefmt=PS imagename='Resp_Assess' reset=index;
ods select NObs GEEModInfo GEEFitCriteria GEEModPEst GEEEmpPEst
           AssessmentSummary CumulativeResiduals;
proc genmod data=example4_3_2 desc ;
 class ID Center ;
 model y = Trt Center_ Gender Age y0 / dist=bin itprint;
 repeated subject=ID(Center) / corr=ind modelse;
 assess link / resample=1000 seed=60370754 crpanel;
run;
ods graphics off;
quit;
```

In the above program, we use assignment statements like

```
Trt=(Treatment='A');
```

to create numeric binary indicators. We then use these numeric covariates in the MODEL statement of GENMOD in combination with the above ASSESS statement in order to check on the validity of the default logit link (the logit link is the default when one specifies DIST=BIN). Based on methods developed by Lin, Wei, and Ying (2002), the ASSESS statement allows one to test the goodness-of-fit of an assumed link function using a Kolmogorov-type supremum test applied to simulated realizations of the cumulative sum of residuals (as defined by the assumed link function—see the documentation for greater details). The RESAMPLE=1000 option specifies that a p-value for testing the goodness-of-fit of the link be based on 1,000 simulated realizations of the cumulative sum of residuals using a normal random number generator with a specified seed (SEED=60370754). Using ODS GRAPHICS, the option CRPANEL of the ASSESS statement specifies that plots be created comparing the observed cumulative sum of residuals to a few simulated realizations of aggregate residuals. This numerical test and accompanying graphics allows one to detect anomalies reflecting possible misspecifications with respect to the link function, to the functional form of the response variable or to the linear predictor (Lin, Wei and Ying, 2002).

Using ODS SELECT, the output containing model information, QIC fit criterion, the GEE regression estimates using robust standard errors and an assessment summary of the adequacy of the logit link are shown in Output 4.6.

Output 4.6: An initial analysis of the respiratory disorder data assuming a "working" independence structure.

```
Number of Observations Read    444
Number of Observations Used    444
Number of Events               248
Number of Trials               444
                 GEE Model Information
Correlation Structure                          Independent
Subject Effect                      ID(Center) (111 levels)
Number of Clusters                                     111
Correlation Matrix Dimension                             4
Maximum Cluster Size                                     4
Minimum Cluster Size                                     4
GEE Fit Criteria
QIC      512.5723
QICu     499.4873
                       Analysis Of GEE Parameter Estimates
                        Empirical Standard Error Estimates
Parameter   Estimate   Standard Error   95% Confidence Limits       Z   Pr > |Z|
Intercept    -0.8561           0.4564   -1.7506      0.0384     -1.88    0.0607
Trt           1.2654           0.3467    0.5859      1.9448      3.65    0.0003
Center_       0.6495           0.3532   -0.0428      1.3418      1.84    0.0660
Gender        0.1368           0.4402   -0.7261      0.9996      0.31    0.7560
Age          -0.0188           0.0130   -0.0442      0.0067     -1.45    0.1480
y0            1.8457           0.3460    1.1676      2.5238      5.33    <.0001
                            Assessment Summary
                            Maximum
Assessment Variable   Absolute Value   Replications       Seed   Pr>MaxAbsVal
Link Function                 0.5384           1000   60370754         0.7590
```

This initial analysis indicates there is a strong effect associated with the active treatment group (from Output 4.6, the estimated log-odds ratio for treatment effect is 1.2654 with p=0.0003). Moreover, the default logit link function appears reasonable based on the Kolmogorov-type supremum test (p=0.7590). In addition to providing a numerical test of the adequacy of a proposed link function, the CRPANEL option of the ASSESS statement together with the selection of the ODS graph name "CumulativeResiduals" (in the ODS SELECT statement) provides a plot indicating how well the proposed link fits with respect to simulated realizations of the cumulative sum of residuals (see Lin, Wei, and Ying, 2002). Figure 4.1 shows that for this particular application, the observed cumulative residual pattern fits well within the patterns observed from the simulated realizations.

Having established that the logit link is reasonable, one may wish to perform a sensitivity analysis to see what impact different "working" correlation structures have on subsequent inference regarding treatment effect. To that end, we fit the data to the marginal logistic regression model assuming three alternative "working" correlation structures: unstructured, exchangeable, and first-order dependent. Using the ODS SELECT and ODS OUTPUT statements, a SAS macro was written to extract information from GENMOD related to the specified "working" correlation structure, the quasi-likelihood information criterion (QIC), and the estimated odds ratio for the treatment effect (active:placebo) based on the difference in least squares means of the linear predictors. The SAS program is shown below.

Figure 4.1 A plot of cumulative residuals for testing fit of link function

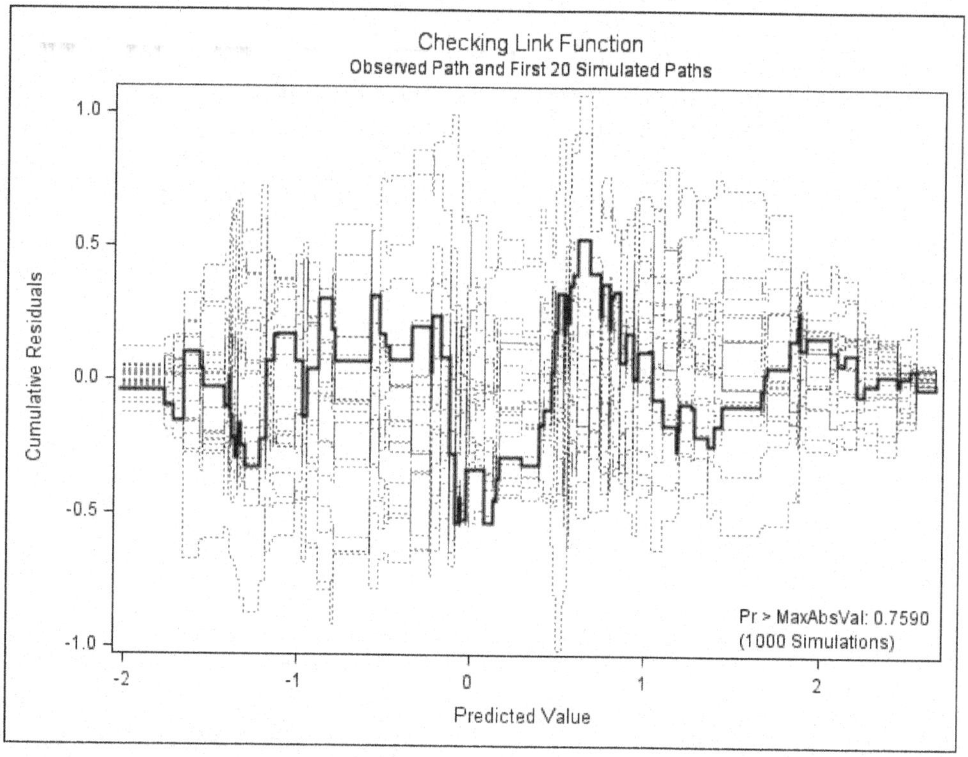

Program 4.4

```
%macro ModelCorr(corr=ind);
  ods listing close;
  ods select GEEModInfo GEEFitCriteria LSMeanDiffs;
  ods output GEEModInfo=info;
  ods output GEEFitCriteria=fit;
  ods output LSMeanDiffs=lsdiff;
  proc genmod data=example4_3_2 desc ;
   class ID Center Treatment(ref=''P'');
   model y = Treatment Center_ Gender Age y0 / dist=bin;
   repeated subject=ID(Center) / corr=&corr;
   lsmeans Treatment / diff cl;
  run;
  data info; set info;
   if Label1='Correlation Structure';
   Structure=cValue1; keep Structure;
  run;
  data fit; set fit;
   if Criterion='QIC';
   QIC=Value; keep QIC;
  run;
  data lsdiff;
   set lsdiff;
   OR=exp(estimate);
   OR_LowerCL=exp(LowerCL);
   OR_UpperCL=exp(UpperCL);
```

```
    p=ProbChiSq;
    keep OR OR_LowerCL OR_UpperCL p;
  run;
  data _summary_;
    merge info fit lsdiff;
  run;
  data summary;  set summary _summary_;
  run;
  ods listing;
%mend ModelCorr;
Data Summary;
 %ModelCorr(corr=ind);
 %ModelCorr(corr=un);
 %ModelCorr(corr=cs);
 %ModelCorr(corr=mdep(1));
proc print data=Summary split='|';
 id Structure;
 var QIC OR OR_LowerCL OR_UpperCL p;
 label OR='Odds Ratio'
       OR_LowerCL='95% LCL'
       OR_UpperCL='95% UCL'
       p='p-value';
 format OR_LowerCL OR_UpperCL 6.3 p 7.5;
run;
```

The results from this analysis, shown in Output 4.7 below, suggest little difference in inference for the four different correlation structures. Indeed, for this balanced design case, inference under the "working" independence and "working" exchangeable correlation structures are identical. Regardless of the specified correlation, patients randomized to the active treatment group had about a three-and-a-half fold higher odds of a positive response to treatment.

Output 4.7: Sensitivity analysis comparing the impact of different "working" correlation structures on inference related to treatment effects for the respiratory disorder data.

Structure	QIC	Odds Ratio	95% LCL	95% UCL	p-value
	.	.	.	.	.
Independent	512.572	3.54435	1.797	6.992	0.00026
Unstructured	512.342	3.47001	1.763	6.831	0.00032
Exchangeable	512.572	3.54435	1.797	6.992	0.00026
1-Dependent	512.859	3.34275	1.693	6.601	0.00051

4.3.3 Epileptic seizure data

Thall and Vail (1990) and Diggle, Liang and Zeger (1994) fit a number of different marginal models to a set of longitudinal count data obtained from a randomized controlled trial of progabide, an anticonvulsant drug. The original study carried out by Leppik et. al. (1985) was a double-blind crossover study designed to compare how effective progabide versus placebo was at reducing the number of partial seizures (counts) in individuals with epilepsy. In the models fit by Thall and Vail (1990), the analyses were restricted to data from the first period of the crossover study. A brief description of the study along with the variables included for analysis can be found in §1.4, page 15. A partial listing of the data is shown in Output 4.8 (see author's Web page for further details).

Output 4.8: A partial listing of the epileptic seizure data

ID	y	Visit	Trt	Bline	Age	y0
101	11	1	1	76	18	2.94444
101	14	2	1	76	18	2.94444
101	9	3	1	76	18	2.94444
101	8	4	1	76	18	2.94444
102	8	1	1	38	32	2.25129
102	7	2	1	38	32	2.25129
102	9	3	1	38	32	2.25129
102	4	4	1	38	32	2.25129
103	0	1	1	19	20	1.55814
103	4	2	1	19	20	1.55814
103	3	3	1	19	20	1.55814
103	0	4	1	19	20	1.55814
104	5	1	0	11	31	1.01160
104	3	2	0	11	31	1.01160
104	3	3	0	11	31	1.01160
104	3	4	0	11	31	1.01160

In the above listing, the variable `y0` is the log of 1/4 the 8-week baseline count (`Bline`). This function of the baseline count is defined such that for Poisson regression with log link, the resulting value of `y0` will be 1) normalized to the same two-week period as defined for the counts recorded during each of the four post-randomization visits, and 2) maintained on the same mean scale as the response variable.

In this section, we fit this longitudinal count data using GEE but with "working" variance-covariance parameters estimated via pseudo-likelihood estimation. We pattern our basic model after that of Thall and Vail in that the counts are modeled assuming an overdispersed Poisson model with possible intra-subject correlation among counts across visits. We fit several such models each having the same mean structure as Thall and Vail, namely

$$E(y_{ij}) = \mu_{ij}(\boldsymbol{\beta}) = g^{-1}(\boldsymbol{x}'_{ij}\boldsymbol{\beta} + \log(t_{ij})) = \exp(\boldsymbol{x}'_{ij}\boldsymbol{\beta} + \log(t_{ij})) \tag{4.22}$$

where $\boldsymbol{x}'_{ij} = \begin{pmatrix} 1 & a_{1i} & \log(a_{2i}) & \log(a_{3i}) & a_{1i}\log(a_{3i}) & v_{i4} \end{pmatrix}$ is the vector of within- and between-subject covariates defined in §1.4 (page 15) and $\boldsymbol{x}'_{ij}\boldsymbol{\beta} + \log(t_{ij}) = \beta_0 + \beta_1 a_{1i} + \beta_2 \log(a_{2i}) + \beta_3 \log(a_{3i}) + \beta_4 a_{1i} \log(a_{3i}) + \beta_5 v_{i4} + \log(t_{ij})$ is the linear predictor. Here a_{1i}, a_{2i}, $\log(a_{3i})$, and v_{i4} are defined as the treatment group indicator (`Trt`), age (`Age`), log normalized baseline count (`y0`), and visit 4 indicator (`Visit4`), respectively. Unlike Thall and Vail (1990), we include an offset, $\log(t_{ij}) = \log 2$, corresponding to the two-week periods over which counts were obtained. This affects the estimated intercepts of each model considered but has no affect on the remaining regression estimates.

As suggested by summary statistics presented by Thall and Vail, there is evidence of significant overdispersion and heterogeneity relative to the Poisson distribution as well as significant intra-subject correlation across visits (see Table 3 of Thall and Vail, 1990). In an attempt to fit the mean response (4.22) accounting for such heterogeneity and correlation, we ran the basic model assuming each of the following five different "working" covariance structures available in GLIMMIX.

Model 1: A working independence or variance components model with heterogeneous variances at each visit (model 11 from Table 1 of Thall and Vail, 1990)—denoted VC,
Model 2: A compound symmetric structure—denoted CS,
Model 3: A compound symmetric structure with heterogeneous variances at each visit (model 13 from Table 1 of Thall and Vail, 1990)—denoted CSH,
Model 4: A Toeplitz structure—denoted Toep,
Model 5: An unstructured covariance—denoted UN.

With the help of the following SAS program and macro along with select ODS OUTPUT statements, one can succinctly summarize results from each of these models.

Program 4.5

```
data thall;
input ID y Visit Trt Bline Age;
cards;
104 5 1  0 11 31
104 3 2  0 11 31
104 3 3  0 11 31
104 3 4  0 11 31
106 3 1  0 11 30
106 5 2  0 11 30
106 3 3  0 11 30
106 3 4  0 11 30
       ⋮
data example4_3_3;
 set thall;
 y0=log(Bline/4);
 LogAge=log(Age);
 LogTime=log(2);
 Visit4=(Visit=4);
run;
proc sort data=example4_3_3;
 by ID Visit;
run;
proc print data=example4_3_3(obs=15) noobs;
 var ID y Visit Trt Bline Age y0;
run;

%macro ModelCov(cov=ind);
  ods listing close;
  ods output ParameterEstimates=pe;
  ods output CovParms=covparms;
  ods output CovBDetails=covb_fit;
  ods output Tests3=tests;
  proc glimmix data=example4_3_3 empirical;
   class ID Visit;
   model y = y0 Trt y0*Trt LogAge Visit4
           / dist=Poisson link=log offset=LogTime
             covb(details) cl s htype=3;
   nloptions maxiter=500;
   random _residual_ / subject=ID type=&cov
    %if %index(&cov, VC) %then %do; group=Visit %end;;
  run;
  data pe;  length Structure $4.;
   set pe;
   Structure=''&cov'';
  run;
  data covparms;  length Structure $4.;
   set covparms;
   Structure=''&cov'';
  run;
```

```
   data covb_fit;  length Structure $4.;
    set covb_fit;
    if Descr IN ('Concordance correlation'
                 'Discrepancy function'
                 'Trace(Adjusted Inv(MBased))' );
    Structure=''&cov'';
    Statistic=Descr;
    Value=Adjusted;
    Keep Structure Statistic Value;
   run;
   data tests;
    length Structure $4.;
    set tests;
    Structure=''&cov'';
   run;
   data covparms_summary;
    set covparms_summary covparms;
   run;
   data covb_summary;
    set covb_summary covb_fit;
   run;
   data pe_summary;
    set pe_summary pe;
   run;
   data tests_summary;
    set tests_summary tests;
   run;
   ods listing;
%mend ModelCov;

Data pe_summary;
Data covparms_summary;
Data covb_summary;
Data tests_summary;
  %ModelCov(cov=VC);
  %ModelCov(cov=CS);
  %ModelCov(cov=CSH);
  %ModelCov(cov=Toep);
  %ModelCov(cov=UN);
run;

proc report data=pe_summary
 HEADLINE HEADSKIP SPLIT='|' nowindows;
 column Effect Structure, (Estimate StdErr);
 define Effect / GROUP ''Effect'' WIDTH=12 ;
 define Structure / Across ''Working Covariance Structure''
                    order=internal WIDTH=8;
 define Estimate / MEAN FORMAT=6.3 'Estimate'
                   width=8 NOZERO spacing=1;
 define StdErr   / MEAN FORMAT=6.3 'SE'
                   width=6 NOZERO spacing=1;
run;
quit;
proc sort data=covparms_summary;
 by Structure;
run;
```

```
proc print data=covparms_summary noobs split='|';
 where Estimate ne .;
 by Structure;
 id Structure;
 var CovParm Group Estimate StdErr;
 label CovParm='Parameter'
       Structure='Working|Covariance|Structure'
       StdErr='SE';
run;
proc report data=covb_summary
 HEADLINE HEADSKIP SPLIT='|' nowindows;
 column Statistic Structure, (Value);
 define Statistic / GROUP ''Goodness-of-Fit Measure''
                    format=$30. WIDTH=30;
 define Structure / Across ''Working Covariance Structure''
                    order=internal WIDTH=8;
 define Value     / MEAN FORMAT=8.4 'Value'
                    width=8 NOZERO spacing=1;
run;
quit;
```

The macro %ModelCov fits the count data to a marginal Poisson model with mean and variance functions both defined by (4.22) and a "working" covariance structure specified by the TYPE= option of the RANDOM _RESIDUAL_ statement using the macro variable &cov. The COVB(DETAILS) option of the MODEL statement displays summary statistics that one can use to assess how well different "working" covariance structures fit the data based on the degree of closeness between the model-based covariance matrix, $\widehat{\Omega}(\widehat{\beta})$, and the empirical or robust sandwich estimator, $\widehat{\Omega}_R(\widehat{\beta})$, (e.g., Vonesh et. al., 1996; Vonesh and Chinchilli, 1997; see also the documentation of GLIMMIX).

Summarized in Output 4.9 are the estimates of the regression parameters and their robust standard errors along with the estimated covariance parameters from each of the five models. Output 4.10 lists three selected summary statistics that measure the goodness-of-fit of the assumed "working" covariance structure. They include the concordance correlation (higher values are better) and the discrepancy function (lower values are better).

Output 4.9: Estimates of the regression parameters (first panel) and covariance parameters (second panel) for the five Poisson models with specified "working" covariance structures.

PANEL 1	Working Covariance Structure					
	CS		CSH		VC	
Effect	Estimate	SE	Estimate	SE	Estimate	SE
Intercept	−3.491	0.958	−3.295	0.918	−3.442	0.909
LogAge	0.908	0.278	0.878	0.271	0.909	0.266
Trt	−1.338	0.430	−1.475	0.419	−1.431	0.416
Visit4	−0.161	0.066	−0.167	0.065	−0.168	0.065
y0	0.951	0.099	0.921	0.083	0.935	0.087
y0*Trt	0.563	0.175	0.603	0.173	0.593	0.170

	Working Covariance Structure			
	Toep		UN	
Effect	Estimate	SE	Estimate	SE
Intercept	−3.517	0.959	−3.870	0.942
LogAge	0.915	0.278	1.051	0.277
Trt	−1.347	0.430	−1.675	0.419
Visit4	−0.166	0.070	−0.163	0.063
y0	0.951	0.099	0.922	0.085
y0*Trt	0.567	0.175	0.689	0.169

PANEL 2

Working Covariance Structure	Parameter	Group	Estimate	SE
CS	CS		1.7430	0.4747
	Residual		2.7774	0.2961
CSH	Var(1)		3.7035	0.7407
	Var(2)		4.2854	0.7960
	Var(3)		7.4901	1.3605
	Var(4)		2.2833	0.4251
	CSH		0.4080	0.07208
Toep	TOEP(2)		1.8660	0.4808
	TOEP(3)		1.4991	0.5368
	TOEP(4)		1.8476	0.8576
	Residual		4.5197	0.5334
UN	UN(1,1)		3.2901	0.6428
	UN(2,1)		1.4062	0.5640
	UN(2,2)		4.7779	0.9297
	UN(3,1)		1.2880	0.6906
	UN(3,2)		3.1237	0.9134
	UN(3,3)		7.7110	1.4281
	UN(4,1)		0.8362	0.4007
	UN(4,2)		1.7602	0.5396
	UN(4,3)		2.2056	0.6490
	UN(4,4)		2.4561	0.4725
VC	Residual (VC)	Visit 1	3.3474	0.6375
	Residual (VC)	Visit 2	4.3679	0.8193
	Residual (VC)	Visit 3	7.5270	1.3880
	Residual (VC)	Visit 4	2.2932	0.4298

Output 4.10: Summary statistics for comparing covariance structures.

	Working Covariance Structure				
	CS	CSH	Toep	UN	VC
Goodness-of-Fit Measure	Value	Value	Value	Value	Value
Concordance correlation	0.9273	0.9237	0.9287	0.9325	0.7201
Discrepancy function	0.9708	0.8285	0.9794	0.7312	2.5094
Trace(Adjusted Inv(MBased))	5.0549	5.4899	5.1494	5.6546	10.6863

In comparing results, we find the regression estimates and their standard errors are fairly similar across the five covariance models (Output 4.9). Moreover, except for the intercepts (due to the inclusion of an offset as noted above), the results are comparable to those obtained by Thall and Vail (1990) for the heterogeneous variance components (VC) structure (model 1 listed above versus model 11 of Thall and Vail) and the heterogeneous compound symmetric (CSH) structure (model 3 above versus model 13 of Thall and Vail). This is not too surprising in that Thall and Vail employ a doubly iterative generalized quasi-likelihood estimation scheme that is very similar to the GLIMMIX-based PL estimation scheme described in §4.2.1. The primary difference is that they solve a set of moments-based estimating equations for $\boldsymbol{\theta}$ as opposed to a set of PL estimating equations corresponding to profiled ML or REML estimation of $\boldsymbol{\theta}$. In comparing the covariance parameter estimates of the two approaches, we again see that the results for models 1 (VC) and 3 (CSH) are similar to those obtained under Thall and Vail's model 11 and model 13, respectively. However, the standard errors of the PL estimates do differ. This is because

Thall and Vail compute robust standard error estimates for both $\widehat{\boldsymbol{\beta}}$ and $\widehat{\boldsymbol{\theta}}$ whereas GLIMMIX computes robust standard errors for $\widehat{\boldsymbol{\beta}}$ only.

Among the five "working" covariance structures considered, the unstructured covariance provided the best fit (Output 4.10). Specifically, it yielded the highest concordance correlation and lowest discrepancy function when compared to the other four structures. The concordance correlation is a standardized measure of the closeness of $\widehat{\boldsymbol{\Omega}}$ and $\widehat{\boldsymbol{\Omega}}_R$ where $\widehat{\boldsymbol{\Omega}}$ is the model-based variance-covariance of $\widehat{\boldsymbol{\beta}}$ (equation 4.14) and $\widehat{\boldsymbol{\Omega}}_R$ is the robust covariance matrix (equation 4.15). It is a slight modification of the covariance concordance correlation defined by Vonesh, et. al. (1996) and Vonesh and Chinchilli (1997, Ch. 8.3) and is defined as

$$r(\widehat{\boldsymbol{\omega}}) = 1 - \frac{||(\widehat{\boldsymbol{\omega}} - \boldsymbol{k})||^2}{||\widehat{\boldsymbol{\omega}}||^2 + ||\boldsymbol{k}||^2} \tag{4.23}$$

where $\widehat{\boldsymbol{\omega}} = Vech(\widehat{\boldsymbol{\Omega}}^{-1/2}\widehat{\boldsymbol{\Omega}}_R\widehat{\boldsymbol{\Omega}}^{-1/2})$ is the lower triangular portion of the matrix $\widehat{\boldsymbol{\Omega}}^{-1/2}\widehat{\boldsymbol{\Omega}}_R\widehat{\boldsymbol{\Omega}}^{-1/2}$ while $\boldsymbol{k} = Vech(\boldsymbol{K}_s)$ is the lower triangular portion of the matrix $\boldsymbol{K}_s$ obtained from the identity matrix of size s by replacing diagonal elements corresponding to singular rows in $\widehat{\boldsymbol{\Omega}}$ with zeros (here we use the notation $||\boldsymbol{x}||$ to represent the usual Euclidian norm of a vector $\boldsymbol{x}$). The concordance correlation measures the agreement between $\widehat{\boldsymbol{\Omega}}$ and $\widehat{\boldsymbol{\Omega}}_R$ on a scale from 0 to 1 with $r(\widehat{\boldsymbol{\omega}}) = 1$ (i.e., perfect agreement) when $\widehat{\boldsymbol{\Omega}} = \widehat{\boldsymbol{\Omega}}_R$ and $r(\widehat{\boldsymbol{\omega}}) = 0$ (i.e., total disagreement) whenever $\widehat{\boldsymbol{\omega}}$ is orthogonal to $\boldsymbol{k}$. The discrepancy function is defined in terms of $\widehat{\boldsymbol{\Omega}}$ and $\widehat{\boldsymbol{\Omega}}_R$ as

$$\widehat{d} = \left\{\log|\widehat{\boldsymbol{\Omega}}| - \log|\widehat{\boldsymbol{\Omega}}_R| + \text{trace}(\widehat{\boldsymbol{\Omega}}_R\widehat{\boldsymbol{\Omega}}^{-1}) - s\right\} \tag{4.24}$$

and will converge in probability to 0 given correct specification of the true covariance structure of $\boldsymbol{y}_i$. Assuming the "working" covariance matrix $\boldsymbol{\Sigma}_i(\boldsymbol{\theta})$ and diagonal variance function $\boldsymbol{V}_i(\boldsymbol{\mu}_i)$ have been correctly specified such that $\boldsymbol{\Sigma}_i(\boldsymbol{\beta}, \boldsymbol{\theta}) = \boldsymbol{V}_i(\boldsymbol{\mu}_i)^{1/2}\boldsymbol{\Sigma}_i^*(\boldsymbol{\theta})\boldsymbol{V}_i(\boldsymbol{\mu}_i)^{1/2}$ is the true covariance of $\boldsymbol{y}_i$, then $\lim_{n\to\infty}\{n\widehat{\boldsymbol{\Omega}}\} = \lim_{n\to\infty}\{n\widehat{\boldsymbol{\Omega}}_R\}$. It is based on this limiting property that we should expect $r(\widehat{\boldsymbol{\omega}})$ to be near 1 and $\widehat{d}$ to be near 0 whenever the specified covariance structure of $\boldsymbol{y}_i$ is not too far from its true structure. In fact, under certain conditions, $\widehat{\lambda} = n\widehat{d}$ can be used to perform a pseudo-likelihood ratio test of the correctness of $\boldsymbol{\Sigma}_i(\boldsymbol{\beta}, \boldsymbol{\theta})$ (Vonesh et. al., 1996).

In terms of comparing treatment effects, each of the models fit resulted in a strong and statistically significant interaction between the baseline count and treatment group as shown in Output 4.11 (SAS code shown below).

Program 4.6

```
proc report data=tests_summary
 HEADLINE HEADSKIP SPLIT='|' nowindows;
 column Effect Structure, (NumDF DenDF FValue ProbF);
 define Effect / GROUP ''Effect'' WIDTH=12 ;
 define Structure / Across ''Working Covariance Structure''
                    order=internal WIDTH=8;
 define NumDF / MEAN FORMAT=6.0 'NumDF' width=6 spacing=1;
 define DenDF / MEAN FORMAT=6.0 'DenDF' width=6 spacing=1;
 define FValue/ MEAN FORMAT=8.2 'FValue' width=6 spacing=1;
 define ProbF / MEAN 'ProbF' width=8 spacing=1;
run;
quit;
```

Output 4.11: Type 3 tests of hypotheses for each of the covariates

Working Covariance Structure

	CS				CSH			
Effect	NumDF	DenDF	FValue	ProbF	NumDF	DenDF	FValue	ProbF
LogAge	1	54	10.70	0.0019	1	54	10.54	0.0020
Trt	1	54	9.68	0.0030	1	54	12.40	0.0009
Visit4	1	176	6.03	0.0150	1	176	6.70	0.0104
y0	1	54	92.15	<.0001	1	54	121.75	<.0001
y0*Trt	1	54	10.36	0.0022	1	54	12.21	0.0010

Working Covariance Structure

	Toep				UN			
Effect	NumDF	DenDF	FValue	ProbF	NumDF	DenDF	FValue	ProbF
LogAge	1	54	10.83	0.0018	1	54	14.34	0.0004
Trt	1	54	9.82	0.0028	1	54	15.96	0.0002
Visit4	1	176	5.59	0.0192	1	54	6.72	0.0123
y0	1	54	91.88	<.0001	1	54	118.10	<.0001
y0*Trt	1	54	10.51	0.0020	1	54	16.57	0.0002

Working Covariance Structure

	VC			
Effect	NumDF	DenDF	FValue	ProbF
LogAge	1	54	11.65	0.0012
Trt	1	54	11.86	0.0011
Visit4	1	176	6.64	0.0108
y0	1	54	115.22	<.0001
y0*Trt	1	54	12.18	0.0010

As noted by Thall and Vail (1990), the strong interaction between baseline counts and treatment suggests there may be a threshold rate in baseline epileptic seizures beyond which there is no benefit of progabide. While this merits further investigation, one might first ask if there are any influential observations that could help explain this strong interaction. Patient 207, for example, had unusually high seizure counts throughout the study and summary statistics presented by Thall and Vail suggest this patient may indeed influence the treatment effect associated with progabide. To that end, we repeated the analysis in GENMOD using an unstructured "working" correlation matrix. Using ODS GRAPHICS, we included case deletion diagnostics plots based on the cluster-specific leverage statistics available within GENMOD to determine the influence of individual patients on the resulting regression estimates. Specifically, using the option `PLOTS(Clusterlabel)=DFBETACS` (see the SAS code below), case deletion diagnostics showing the standardized effect of deleting cluster i (i.e., the ith patient) on the estimated regression parameters is shown in Figure 4.2.

Program 4.7

```
ods graphics on / imagefmt=PS imagename='Epilepsy_influence';
proc genmod data=example4_3_3
             plots(Clusterlabel)=DFBETACS;
 class ID Visit;
 model y = Trt LogAge y0 y0*Trt Visit4
        / dist=Poisson link=log offset=LogTime Type3;
 repeated subject=ID / type=un ;
run;
ods graphics off;
```

Based on selected output shown in Output 4.12, the overall results and conclusions reached from the analysis done in GENMOD are virtually the same as those obtained from GLIMMIX (compare to Output 4.9). However, these results appear to be overly influenced by patient 207. As suggested by the plots in Figure 4.2, this progabide patient has an overwhelming influence on both the estimated treatment effect and its interactive effect with baseline seizure rate. If one repeats the analysis in both GENMOD and GLIMMIX excluding this patient (simply insert a '`where ID ne 207`' clause immediately before the CLASS statement in the above programs), one will find that both treatment and treatment × log baseline count lose their significance.

Output 4.12: Selected output from GENMOD for the epileptic seizure data. The analysis assumes the mean count can be described by a Poisson GLIM with unstructured "working" correlation.

```
Number of Observations Read    236
Number of Observations Used    236
             GEE Model Information
Correlation Structure                Unstructured
Subject Effect                      ID (59 levels)
Number of Clusters                              59
Correlation Matrix Dimension                     4
Maximum Cluster Size                             4
Minimum Cluster Size                             4
 GEE Fit Criteria
QIC     −1324.7231
QICu    −1331.2459
                          Analysis Of GEE Parameter Estimates
                          Empirical Standard Error Estimates
Parameter   Estimate   Standard Error   95% Confidence Limits        Z    Pr > |Z|
Intercept    −3.7788           0.9456   −5.6321     −1.9256    −4.00    <.0001
Trt          −1.5114           0.4228   −2.3402     −0.6827    −3.57    0.0004
LogAge        1.0074           0.2737    0.4710      1.5438     3.68    0.0002
y0            0.9346           0.0924    0.7535      1.1157    10.11    <.0001
Trt*y0        0.6339           0.1698    0.3012      0.9666     3.73    0.0002
Visit4       −0.1559           0.0829   −0.3184      0.0067    −1.88    0.0602
Score Statistics For Type 3 GEE Analysis
Source   DF   Chi-Square        Pr > ChiSq
Trt       1         6.59            0.0103
LogAge    1         6.98            0.0083
y0        1         8.45            0.0036
Trt*y0    1         3.86            0.0494
Visit4    1         2.72            0.0988
```

To illustrate, when we re-ran the analysis in GLIMMIX assuming an unstructured "working" covariance, we found no significant interaction between treatment and baseline count (Output 4.13). While there is some evidence that progabide may result in a mild reduction in seizure counts, it did not reach statistical significance at the 5% level (p=0.0972).

Output 4.13: Parameter estimates from GLIMMIX applied to the epileptic seizure data excluding patient 207 and assuming the marginal mean model (4.22) with an unstructured "working" covariance.

```
                    Solutions for Fixed Effects
Effect       Estimate   Standard Error   DF   t Value   Pr > |t|
Intercept     −3.1087           0.8555   53     −3.63     0.0006
Trt           −0.6539           0.3872   53     −1.69     0.0972
LogAge         0.8273           0.2522   53      3.28     0.0018
y0             0.9151          0.08474   53     10.80     <.0001
Trt*y0         0.1657           0.1763   53      0.94     0.3515
Visit4        −0.1481          0.07396   53     −2.00     0.0504
```

Figure 4.2 Patient deletion diagnostics showing what the standardized effect of removing a patient is on the resulting regression parameter estimates

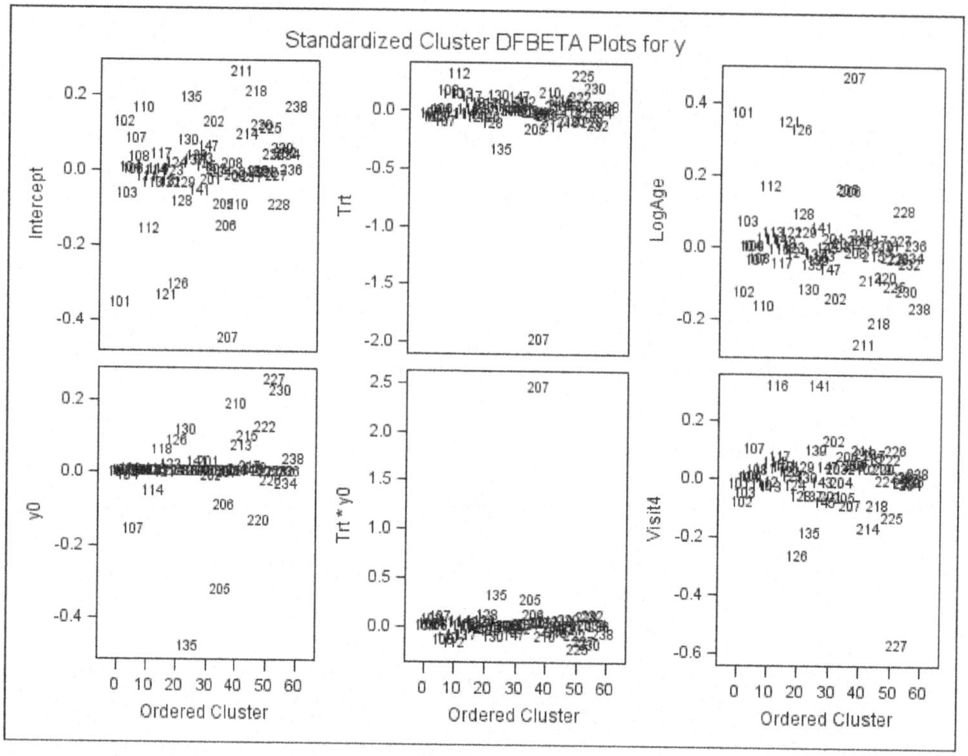

Thall and Vail (1990) correctly point out that there is no clinical basis for removing patient 207 from the final analysis. Nevertheless, the results presented here serve to illustrate some of the dangers associated with the analysis of count data. One of those dangers involves the impact that data from a skewed distribution like the Poisson or negative binomial can have in studies of moderate size. While patient 207 appears to be an "outlier," it is not all that unreasonable to expect such patients to occur, particularly in large studies. In moderate to smaller trials, especially randomized controlled trials, the presence of only one such subject could alter the findings simply because that subject can only represent one group. In the present example, if there had been a similar patient in the placebo control group, it is doubtful we would have seen much of an impact of both subjects on the overall results.

4.3.4 Schizophrenia data

Hedeker and Gibbons (1994) analyzed a set of ordinal longitudinal data from the NIMH Schizophrenia collaborative study on treatment-related changes in overall severity (Cole et. al., 1964). The primary response considered by the authors was a consolidated four-point ordinal score formed from Item 79 of the Inpatient Multidimensional Psychiatric Scale (IMPS), a seven-point severity of illness score defined as: 1 = normal, not at all ill; 2 = borderline mentally ill; 3 = mildly ill; 4 = moderately ill; 5 = markedly ill; 6 = severely ill; and 7 = among the most extremely ill. The study was a randomized controlled trial of 437 patients randomized to one of four medications: placebo, chlorpromazine, fluphenazine, or thioridazine. Since previous analyses revealed similar effects among the three active antipsychotic drugs, the authors confined their analyses to a comparison of treatment effects based on collapsing the three active drug groups into one overall drug group. The primary goal of the study was to determine if there was a treatment-related change in the severity of illness over time. To that end, Hedeker and Gibbons (1994) applied a mixed-effects ordinal regression model to the analysis of the data which we will consider in Chapter 5. In this chapter, we present a marginal GEE analysis of the data assuming different link functions. Some of the key features of the study and a description of variables used are outlined below. The SAS variable names are shown in parentheses.

- National Institute of Mental Health (NIMH) Schizophrenia Collaborative Study—A randomized controlled trial of 437 patients
- Experimental unit or cluster is a patient (ID)
- Response variable (IMPS79o):
  - An ordinal response variable indicating severity of illness from item 79 of Inpatient Multidimensional Psychiatric Scale (IMPS 79)

$$y_{ij} = \begin{cases} 1 & \text{normal or borderline mentally ill,} \\ 2 & \text{mildly or moderately ill,} \\ 3 & \text{markedly ill,} \\ 4 & \text{severely or among the most extremely ill} \end{cases}$$

  is the ordered response of severity of illness for the ith patient during the jth week of follow-up (IMPS79o).
- One within-unit covariate:
  - t_{ij} =time in weeks at which IMPS 79 scores were assessed which were at baseline (week 0), and at weeks 1, 2, 3, 4, 5, 6 (Week)
  - For the primary analysis, the square root of time, $\sqrt{t_{ij}}$, was used to linearize the relationship between IMPS79 scores over time (SWeek)
- One between-patient covariate:
  - Treatment group (Trt) consisting of placebo versus drug and coded as

$$a_{1i} = \begin{cases} 0, & \text{if placebo} \\ 1, & \text{if drug (chlorpromazine, fluphenazine, or thioridazine)} \end{cases}$$

- Goal: Evaluate effectiveness of drug versus placebo on severity of illness over time.

A partial listing of the data are shown in Output 4.14 (see also the author's Web page for a description of the variables). Trends over time in the distribution of patients according to the severity of illness are shown in Figure 4.3 for the two treatment groups. These crude trends suggest there is some improvement over time in the severity of illness for both groups but it is less clear whether these improvements are greater for patients in the active drug group.

Output 4.14: A partial listing of the schizophrenia data.

ID	IMPS79o	Trt	Week	SWeek
1103	4	1	0	0.0000
1103	2	1	1	1.0000
1103	2	1	3	1.7321
1103	2	1	6	2.4495
1104	4	1	0	0.0000
1104	2	1	1	1.0000
1104	1	1	3	1.7321
1104	2	1	6	2.4495
1105	2	1	0	0.0000
1105	2	1	1	1.0000
1105	1	1	3	1.7321
1106	2	1	0	0.0000
1106	1	1	1	1.0000

1106	1	1	3	1.7321
1106	1	1	6	2.4495
1107	3	0	0	0.0000
1107	3	0	1	1.0000
1107	3	0	3	1.7321
1107	4	0	6	2.4495

Hedeker and Gibbons (1994) applied ordinal regression in order to investigate these trends more formally. In the GLIM setting, ordinal regression is based on an underlying assumption that there exists a continuous but unobservable latent random variable that directly corresponds to the levels of the ordered categorical variable. More formally, let $y_{ij} = k\,(k = 1, \ldots, K)$ denote an ordered categorical response variable for subject i at time j (in our example, we have y_{ij} = severity of illness with $k = 1$ = normal, $k = 2$ = mildly ill, $k = 3$ = markedly ill, and $k = 4$ = severely ill). Let $\pi_k(\boldsymbol{x}_{ij}) = \Pr(y_{ij} = k|\boldsymbol{x}_{ij})$, $k = 1, \ldots, K$ denote the K distinct response probabilities relating y_{ij} to a set of predictors $\boldsymbol{x}_{ij}$. Suppose y_{ij} corresponds to a latent (unobserved) continuous random variable, y^*_{ij}, with threshold cutpoints $\alpha_1, \alpha_2, \ldots, \alpha_{K-1}$, corresponding to the K ordered categories. Under an ordinal regression model it is assumed that the y_{ij} are related to threshold cutpoints of y^*_{ij} via:

$$y_{ij} = k \text{ when } \alpha_{k-1} < y^*_{ij} \le \alpha_k \text{ with } \alpha_0 = -\infty,\ \alpha_K = \infty. \tag{4.25}$$

Since the y_{ij} are ordered, we can work with the cumulative response probabilities $\gamma_1 = \pi_1, \gamma_2 = \pi_1 + \pi_2, \ldots, \gamma_K = 1$ where $\gamma_k = \gamma_k(\boldsymbol{x}_{ij}) = \Pr(y_{ij} \le k|\boldsymbol{x}_{ij})$ rather than with the response probabilities, $\pi_k = \pi_k(\boldsymbol{x}_{ij}) = \Pr(y_{ij} = k|\boldsymbol{x}_{ij})$.

Under this framework, we can formulate a generalized linear model that relates the ordered categorical responses y_{ij} to the set of predictors $\boldsymbol{x}_{ij}$ through the standardized cumulative distribution function (cdf) of the latent response variable y^*_{ij}. Specifically, let

Figure 4.3 IMPS 79 Severity of Illness (1=normal, 2=mild/moderate, 3=marked, 4=severe). Observed proportion of patients over time by treatment group

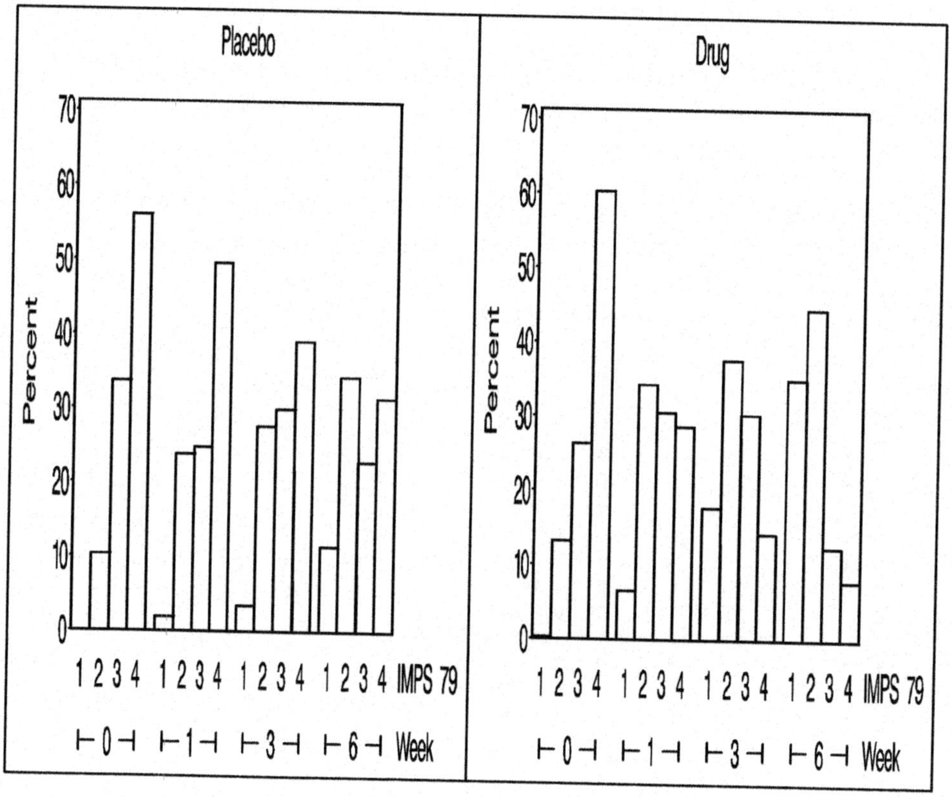

$F(\alpha_k; \mu_{ij}, \sigma) = \Pr(y^*_{ij} \leq \alpha_k)$ denote the cumulative distribution function (cdf) of y^*_{ij} evaluated at α_k where μ_{ij} is a location parameter and σ is a scale parameter. Furthermore, let $G(\alpha_k) = F(\alpha_k; 0, 1)$ denote the standardized cdf obtained by setting $\mu_{ij} = 0$ and $\sigma = 1$. In ordinal regression, the set of predictors, $\boldsymbol{x}_{ij}$, are assumed to be related to a latent response variable, y^*_{ij}, via the linear regression model

$$y^*_{ij} = \boldsymbol{x}'_{ij}\boldsymbol{\beta} + \epsilon_{ij} \tag{4.26}$$

where ϵ_{ij} is distributed according to some standardized cdf $G(\cdot)$ such that

$$\begin{aligned} \gamma_k(\boldsymbol{x}_{ij}) = \Pr(y_{ij} \leq k|\boldsymbol{x}_{ij}) &= \Pr(y^*_{ij} \leq \alpha_k) \\ &= \Pr(\epsilon_{ij} \leq \alpha_k - \boldsymbol{x}'_{ij}\boldsymbol{\beta}) \\ &= G(\alpha_k - \boldsymbol{x}'_{ij}\boldsymbol{\beta}) \end{aligned} \tag{4.27}$$

for $k = 1, \ldots K-1$. One can then form a GLIM for the ordinal response y_{ij} by defining the cumulative link function

$$G^{-1}(\Pr(y_{ij} \leq k|\boldsymbol{x}_{ij})) = G^{-1}(\gamma_k(\boldsymbol{x}_{ij})) = \alpha_k - \boldsymbol{x}'_{ij}\boldsymbol{\beta} \tag{4.28}$$

that maps the cumulative probabilities, $\{\gamma_k(\boldsymbol{x}_{ij}),\ k = 1, \ldots, K-1\}$, onto the real line (note $\alpha_K = \infty \Rightarrow \gamma_K(\boldsymbol{x}_{ij}) = 1$). The class of models defined by (4.26)-(4.28) are sometimes referred to as cumulative link models.

On the scale defined by the cumulative link function [i.e., equation (4.28)], the threshold cutpoints $\alpha_1, \ldots, \alpha_{K-1}$ act as intercepts which, in the absence of any covariates, are merely the quantiles from the cdf of the latent random variable y^* corresponding to the observed cumulative probabilities of the ordered response variable y. Note also that the cumulative link model defined by (4.28) defines a series of $K-1$ parallel regressions. As such the cumulative link model assumes the effects of $\boldsymbol{x}$, as measured by $\boldsymbol{\beta}$, are the same for each cutpoint, $k = 1, \ldots, K-1$. When fitting cumulative link models, some authors recommend excluding an overall intercept term from the linear predictor $\boldsymbol{x}'_{ij}\boldsymbol{\beta}$ because this term will be absorbed by the $K-1$ cutpoints. This is the approach taken in GENMOD. However, one can take the equivalent approach and include an overall intercept parameter, say β_0, within the linear predictor $\boldsymbol{x}'_{ij}\boldsymbol{\beta}$ and simply set $\alpha_1 = 0$ (e.g., Hedeker and Gibbons, 1994). This latter approach is what is implemented in GLIMMIX and the two approaches are equivalent. In fact, if one attempts to fit a cumulative link model in GLIMMIX with the NOINT option, SAS will issue the following error message:

```
ERROR: An intercept can not be suppressed in models for
multinomial data.
```

The GENMOD procedure, on the other hand, will execute when one uses the NOINT option of the MODEL statement. However, the resulting model forces the first threshold cutpoint (labeled as Intercept1 in the output) to be 0. This results in biased estimates of all the parameters as it effectively maps the observed probability from the first ordered category to a quantile value of 0 for the latent random variable. The user should be aware of this problem and avoid specifying the NOINT option when fitting a cumulative link model in GENMOD.

The GENMOD procedure provides three cumulative link functions from which to choose: a cumulative probit link (G corresponds to the standard normal distribution); a cumulative logit link (G corresponds to the standard logistic distribution); and cumulative complementary log-log link (G corresponds to the standard extreme-value distribution). These give rise to three marginal cumulative link models summarized below.

Cumulative Probit Model:
Link: $G^{-1}[\Pr(y_{ij} \le k|\boldsymbol{x}_{ij})] = \Phi^{-1}(\Pr(y_{ij} \le k|\boldsymbol{x}_{ij}))$
SAS specification: LINK=CPROBIT
Standardized CDF: $G(\alpha_k - \boldsymbol{x}'_{ij}\boldsymbol{\beta}) = \Phi(\alpha_k - \boldsymbol{x}'_{ij}\boldsymbol{\beta})$
Cumulative Logit Model (Proportional Odds Model):
Link: $G^{-1}[\Pr(y_{ij} \le k|\boldsymbol{x}_{ij})] = \log\left(\dfrac{\Pr(y_{ij} \le k|\boldsymbol{x}_{ij})}{1-\Pr(y_{ij} \le k|\boldsymbol{x}_{ij})}\right)$
SAS specification: LINK=CLOGIT
Standardized CDF: $G(\alpha_k - \boldsymbol{x}'_{ij}\boldsymbol{\beta}) = \dfrac{\exp\{\alpha_k - \boldsymbol{x}'_{ij}\boldsymbol{\beta}\}}{1+\exp\{\alpha_k - \boldsymbol{x}'_{ij}\boldsymbol{\beta}\}}$
Discrete Proportional Hazards Model (Cumulative complementary log-log link)
Link: $G^{-1}[\Pr(y_{ij} \le k|\boldsymbol{x}_{ij})] = \log[-\log(1-\Pr(y_{ij} \le k|\boldsymbol{x}_{ij}))]$
SAS specification: LINK=CCLL
Standardized CDF: $G(\alpha_k - \boldsymbol{x}'_{ij}\boldsymbol{\beta}) = 1 - \exp\{-\exp(\alpha_k - \boldsymbol{x}'_{ij}\boldsymbol{\beta})\}$

The cumulative probit model is appropriate whenever the underlying latent response distribution is normal or approximately so. The cumulative logit model or proportional odds model will often provide a fit to the data that will be similar to that of the cumulative probit model as both models utilize latent random variables with symmetric distributions. However, the parameters from the two models will differ as they have different interpretations. Under the proportional odds model, the effect of a covariate on the odds of response below the kth category will be the same for all categories. Specifically,

$$\log\left(\frac{\Pr(y \le k|\boldsymbol{x}_1)/\Pr(y > k|\boldsymbol{x}_1)}{\Pr(y \le k|\boldsymbol{x}_2)/\Pr(y > k|\boldsymbol{x}_2)}\right) = (\boldsymbol{x}_1 - \boldsymbol{x}_2)'\boldsymbol{\beta} \text{ for all } k = 1, \ldots K-1.$$

The corresponding odds ratio obtained by exponentiation is referred to as the cumulative odds ratio (e.g., Agresti, 1990). It satisfies the property that the cumulative odds ratio for $(y \le k)$ will remain constant across all $k = 1, \ldots, K-1$ categories.

The discrete proportional hazards model utilizes the cumulative complementary log-log link function which is the inverse cdf of the extreme-value distribution, an asymmetric distribution which is more appropriate in "survival" model settings. In this case the regression parameters satisfy the property that the "survival" function for individuals with covariate vector $\boldsymbol{x}_1$ is related to individuals with covariate vector $\boldsymbol{x}_2$ by the relation $S(k|\boldsymbol{x}_j) = 1 - \Pr(y \le k|\boldsymbol{x}_j) = [1 - \Pr(y \le k|\boldsymbol{x}_k)]^{\exp\{(\boldsymbol{x}_k - \boldsymbol{x}_j)'\boldsymbol{\beta}\}} = S(k|\boldsymbol{x}_k)^{\exp\{(\boldsymbol{x}_k - \boldsymbol{x}_j)'\boldsymbol{\beta}\}}$ where $S(k|\boldsymbol{x}_j)$ is the "survival" function for individuals with covariates $\boldsymbol{x}_j$.

It should be noted that in addition to the above three cumulative link functions, the GLIMMIX procedure also allows one to specify a cumulative log-log link function. However, with GLIMMIX, one can only apply these cumulative link models to data comprised solely of independent observations. Specifically, in SAS 9.2 and later, only GENMOD supports a GEE-based analysis of ordinal multinomial data as GLIMMIX does not allow RSIDE residual variance structures nor does it provide an EMPIRICAL option when DIST=MULTINOMIAL is specified in the MODEL statement.

Before returning to the schizophrenia data, let us illustrate conceptually how the use of a latent random variable can be used to model ordinal response data. Suppose we have an ordinal response variable y having four ordered response categories ($K = 4$) and that y is associated with a single covariate x through a latent response variable y^* satisfying $y^* = x\beta + \epsilon$ with $\epsilon \sim N(0, 1)$. Assuming threshold points at -1, 0 and 1 (corresponding to four ordered categorical regions), Figure 4.4 illustrates what the expected cumulative probabilities would be for the ordinal response variable at each of three levels of x when $\beta = 0.5$. This corresponds to a cumulative probit model and serves to illustrate how the cumulative response probabilities of y vary with a single covariate x according to the

Figure 4.4 A diagram showing how the cumulative response probabilities of an ordinal response variable y with four response categories ($y = 1, 2, 3$ or 4) varies with a single covariate x according to the threshold cutpoints $\alpha_1 = -1, \alpha_2 = 0$, and $\alpha_3 = 1$ from a standard normal distribution.

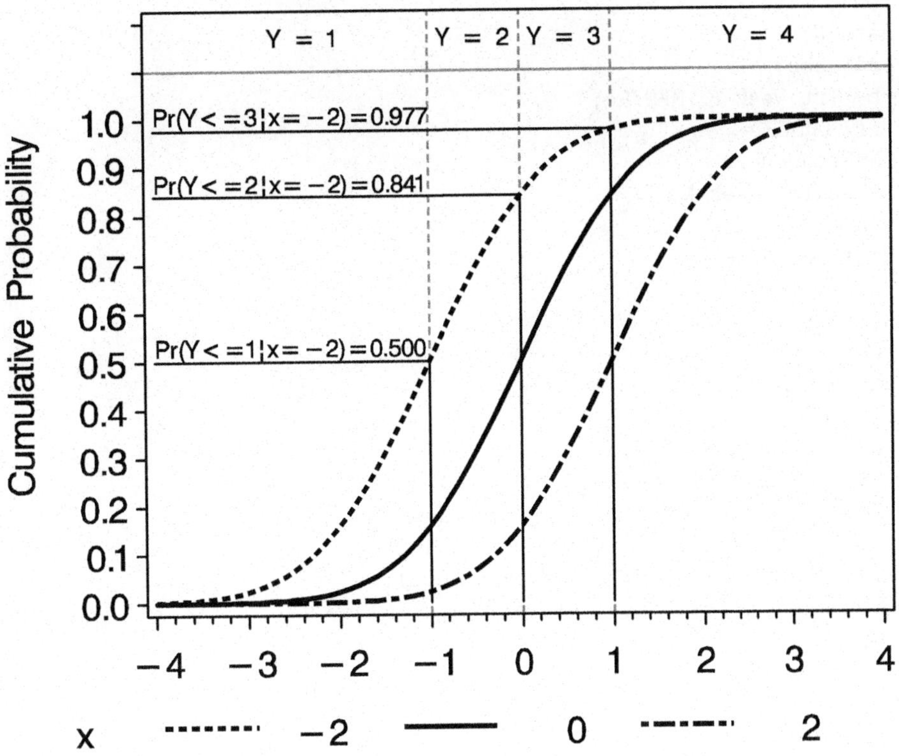

threshold cutpoints $\alpha_1 = -1, \alpha_2 = 0$, and $\alpha_3 = 1$.

Turning our attention back to the schizophrenia data, PROC REPORT was used to summarize trends in both the observed proportions of patients within each ordered category and also the cumulative proportions, cumulative odds and cumulative odds ratios across the first three ordered categories. The programming steps required to produce the necessary data are shown below.

Program 4.8

```
proc sort data=example4_3_4;
 by ID Week;
run;
proc print data=example4_3_4(obs=19) noobs;
run;
ods listing close ;
ods output crosslist=obs;
proc freq data=example4_3_4;
 table Trt*Week*IMPS79o /crosslist nocol nopercent;
run;
data obs;
 set obs;
 _ORDER_=IMPS79o;
 Prob_Obs=RowPercent/100;
 if Prob_Obs>.;
 keep Trt Week F_IMPS79o IMPS79o _ORDER_
      Frequency RowPercent Prob_Obs;;
run;
```

```
ods listing;
proc sort data=obs;
 by Trt Week F_IMPS79o;
run;
data N;
 set obs;
 by Trt Week F_IMPS79o;
 if _ORDER_=.;
 N=Frequency;
 keep Trt Week N;
run;
data obs;
 merge obs N;
 by Trt Week;
run;
data odds;
 set obs;
 by Trt Week F_IMPS79o;
 retain CumProb_Obs 0;
 CumProb_Obs = CumProb_Obs + Prob_Obs;
 if first.week then CumProb_Obs=Prob_Obs;
 CumOdds = CumProb_Obs/(1-CumProb_Obs);
 if CumProb_Obs=2 then delete;
run;
data odds0;
 set odds;
 odds0=CumOdds;
 cprob0=CumProb_Obs;
 if trt=0;
 keep Week IMPS79o cprob0 odds0;
run;
data odds1;
 set odds;
 odds1=CumOdds;
 cprob1=CumProb_Obs;
 if trt=1;
 keep Week IMPS79o cprob1 odds1;
run;
data COR;
 merge odds0 odds1;
 CumOddsRatio = odds1/odds0;
run;
```

PROC REPORT (code not shown) was used to produce the results shown in Output 4.15. The observed cumulative odds and cumulative odds ratios from Output 4.15 support the conjecture drawn from Figure 4.3 that there is a differential reduction in severity of illness over time favoring patients on an active drug. Indeed, the observed cumulative odds ratios (shown for weeks 0,1,3 and 6 only given the sparse data elsewhere) suggest that a proportional odds model may be appropriate for these data. To test whether these observed trends are significant, we fit the data to a cumulative logit model (or proportional odds model) using the default response variable sorting options of GENMOD. In GENMOD and GLIMMIX, the sorting order of an ordinal response variable will determine what level of the response each intercept parameter corresponds to. By default, both GENMOD and GLIMMIX use the external formatted value of the response variable (or the internal value if there is no explicit format).

Output 4.15: Summary of observed proportions of patients by category of illness. Included are the cumulative proportions, their corresponding odds and the observed cumulative odds ratio (drug vs. placebo).

			IMPS 79 Ordinal Score							
			1 = Normal		2 = Mild		3 = Marked		4 = Severe	
Treatment	Week	N	Freq	Prob	Freq	Prob	Freq	Prob	Freq	Prob
Placebo	0	107	0	0.000	11	0.103	36	0.336	60	0.561
	1	105	2	0.019	25	0.238	26	0.248	52	0.495
	2	5	0	0.000	0	0.000	1	0.200	4	0.800
	3	87	3	0.034	24	0.276	26	0.299	34	0.391
	4	2	0	0.000	0	0.000	1	0.500	1	0.500
	5	2	0	0.000	1	0.500	0	0.000	1	0.500
	6	70	8	0.114	24	0.343	16	0.229	22	0.314
Drug	0	327	1	0.003	43	0.131	86	0.263	197	0.602
	1	321	21	0.065	110	0.343	98	0.305	92	0.287
	2	9	3	0.333	4	0.444	1	0.111	1	0.111
	3	287	51	0.178	108	0.376	87	0.303	41	0.143
	4	9	5	0.556	3	0.333	1	0.111	0	0.000
	5	7	3	0.429	3	0.429	0	0.000	1	0.143
	6	265	93	0.351	118	0.445	33	0.125	21	0.079

		Placebo		Drug		
		Cum.	Cum.	Cum.	Cum.	Cumulative
Week	Cumulative IMPS 79 Scores	Prob.	Odds	Prob.	Odds	Odds Ratio
0	Y<=1 (Normal)	0.000	0.000	0.003	0.003	.
	Y<=2 (Normal, Mild)	0.103	0.115	0.135	0.155	1.357
	Y<=3 (Normal, Mild, Marked)	0.439	0.783	0.398	0.660	0.842
1	Y<=1 (Normal)	0.019	0.019	0.065	0.070	3.605
	Y<=2 (Normal, Mild)	0.257	0.346	0.408	0.689	1.992
	Y<=3 (Normal, Mild, Marked)	0.505	1.019	0.713	2.489	2.442
3	Y<=1 (Normal)	0.034	0.036	0.178	0.216	6.051
	Y<=2 (Normal, Mild)	0.310	0.450	0.554	1.242	2.760
	Y<=3 (Normal, Mild, Marked)	0.609	1.559	0.857	6.000	3.849
6	Y<=1 (Normal)	0.114	0.129	0.351	0.541	4.190
	Y<=2 (Normal, Mild)	0.457	0.842	0.796	3.907	4.640
	Y<=3 (Normal, Mild, Marked)	0.686	2.182	0.921	11.619	5.325

The cumulative logit link model (proportional odds model) was fit assuming a "working" independence structure and a linear predictor of the form, $\boldsymbol{x}'_{ij}\boldsymbol{\beta} = \beta_0 + \beta_1 a_{1i} + \beta_2\sqrt{t_{ij}} + \beta_3 a_{1i}\sqrt{t_{ij}}$. The "working" independence structure is the only currently available option in GENMOD for multinomial data. However, all subsequent inference is based on the empirical sandwich estimator so as to correct for possible correlated ordinal responses over time. The SAS code for the cumulative logit model with select output is shown below.

Program 4.9

```
ods select NObs ModelInfo GEEModInfo GEEFitCriteria
           GEEEmpPEst;
ods output Estimates = Estimates;
proc genmod data=example4_3_4;
 class ID;
 model IMPS79o = Trt SWeek Trt*SWeek
                / dist=multinomial link=clogit;
 repeated subject=ID / type=ind modelse;
 estimate 'CumLogOR(Week=0)' Trt 1 Trt*SWeek 0 /exp;
 estimate 'CumLogOR(Week=1)' Trt 1 Trt*SWeek 1 /exp;
 estimate 'CumLogOR(Week=3)' Trt 1 Trt*SWeek 1.7321 /exp;
 estimate 'CumLogOR(Week=3)' Trt 1 Trt*SWeek 2.4495 /exp;
 output out=pred prob=CumProb_Hat;
run;
```

In the preceding code, the ESTIMATE statements are used to compute the cumulative odds ratio at each week as measured by the square root of week (here we set $\sqrt{3} = 1.7321$ and $\sqrt{6} = 2.4495$). These are obtained by exponentiation of the difference in the cumulative log odds for the active drug group versus placebo controls. Using the ODS OUTPUT statement, the results from the ESTIMATE statements are stored in the dataset 'Estimates' which will be merged with the dataset 'COR' (from the previous SAS code) for purposes of plotting the cumulative odds ratios. The call to GENMOD also includes the OUTPUT OUT=PRED statement which will be merged with the dataset 'obs' in order to graphically compare the observed and predicted proportions of patients according to each category of illness. The GENMOD results shown in Output 4.16 shows that there is a strong interaction between treatment group and time with all indications pointing to a differential improvement in outcome in favor of those patients randomized to one of the three active drugs versus those receiving the placebo.

Output 4.16: Summary results from a cumulative logit model comparing the differential effects of active drug versus placebo on severity of illness.

```
              Model Information
Data Set                    WORK.EXAMPLE4_3_4
Distribution                     Multinomial
Link Function               Cumulative Logit
Dependent Variable                   IMPS79o
Number of Observations Read     1603
Number of Observations Used     1603
                GEE Model Information
Correlation Structure                Independent
Subject Effect                   ID (437 levels)
Number of Clusters                           437
Correlation Matrix Dimension                   5
Maximum Cluster Size                           5
Minimum Cluster Size                           2
GEE Fit Criteria
QIC      3774.2961
QICu     3768.1953
```

Analysis Of GEE Parameter Estimates
Empirical Standard Error Estimates

Parameter	Estimate	Standard Error	95% Confidence Limits		Z	Pr > |Z|
Intercept1	−3.8073	0.2027	−4.2046	−3.4099	−18.78	<.0001
Intercept2	−1.7602	0.1898	−2.1322	−1.3881	−9.27	<.0001
Intercept3	−0.4221	0.1880	−0.7905	−0.0537	−2.25	0.0247
Trt	0.0006	0.2126	−0.4160	0.4173	0.00	0.9977
SWeek	0.5366	0.1001	0.3404	0.7329	5.36	<.0001
Trt*SWeek	0.7510	0.1198	0.5161	0.9858	6.27	<.0001

To illustrate the fit of the data, we first plot the observed response proportions and compared them to the predicted proportions computed from the cumulative logit model. The observed proportions are captured in the SAS dataset 'obs' and this data is merged with the predicted proportions in dataset 'pred' and the results displayed in Figure 4.5. This plot suggests the cumulative logit model provides a reasonably good fit to the proportions observed across the four categories of illness.

We repeated the above analysis but with the cumulative logit link replaced by the cumulative probit link in one case and by the cumulative complementary log-log link in another. Using the same SAS code as shown above but without the ESTIMATE statements, the cumulative probit model was run by specifying the model option LINK=CPROBIT (rather than LINK=CLOGIT) while the cumulative complementary log-log link model was run by specifying LINK=CCLL. Both models gave qualitatively

Figure 4.5 Cumulative logit (proportional odds) model of IMPS 79 severity of illness scores. Observed versus predicted proportion of patients over time by treatment group. Given the sparse data at weeks 2, 4 and 5, the observed values are plotted for weeks 0, 1, 3 and 6.

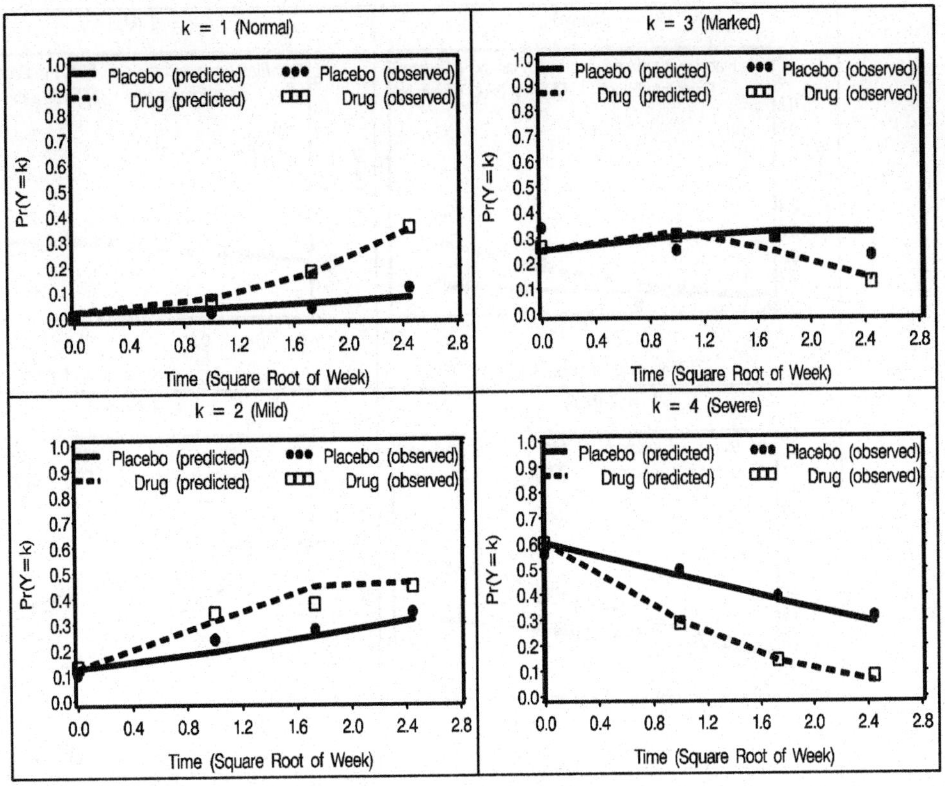

similar results as suggested by the estimates shown in Output 4.17 (cumulative probit model) and Output 4.18 (cumulative complementary log-log link model). The QIC of 3763.7450 for the cumulative probit model (Output 4.17) suggests a slightly better fit compared to the cumulative logit model (a QIC of 3774.2961, Output 4.16) but a graph (not shown) comparing the observed proportions with the predicted proportions was very similar to that shown in Figure 4.5. This is not unexpected given that both models utilize symmetric distributions that frequently overlap each other.

Output 4.17: Results from the cumulative probit link model.

```
GEE Fit Criteria
QIC      3763.7450
QICu     3757.8408
                    Analysis Of GEE Parameter Estimates
                    Empirical Standard Error Estimates
                        Standard
Parameter   Estimate     Error   95% Confidence Limits       Z   Pr > |Z|
Intercept1   -2.2802    0.1137   -2.5031     -2.0572    -20.05   <.0001
Intercept2   -1.0865    0.1087   -1.2996     -0.8734     -9.99   <.0001
Intercept3   -0.2878    0.1098   -0.5029     -0.0727     -2.62   0.0087
Trt           0.0308    0.1228   -0.2099      0.2716      0.25   0.8018
SWeek         0.3426    0.0595    0.2259      0.4593      5.76   <.0001
Trt*SWeek     0.4202    0.0702    0.2826      0.5579      5.98   <.0001
```

Figure 4.6 Cumulative complementary log-log link model of IMPS 79 severity of illness scores. Observed versus predicted proportion of patients over time by treatment group. Given the sparse data at weeks 2, 4 and 5, the observed values are plotted for weeks 0, 1, 3 and 6.

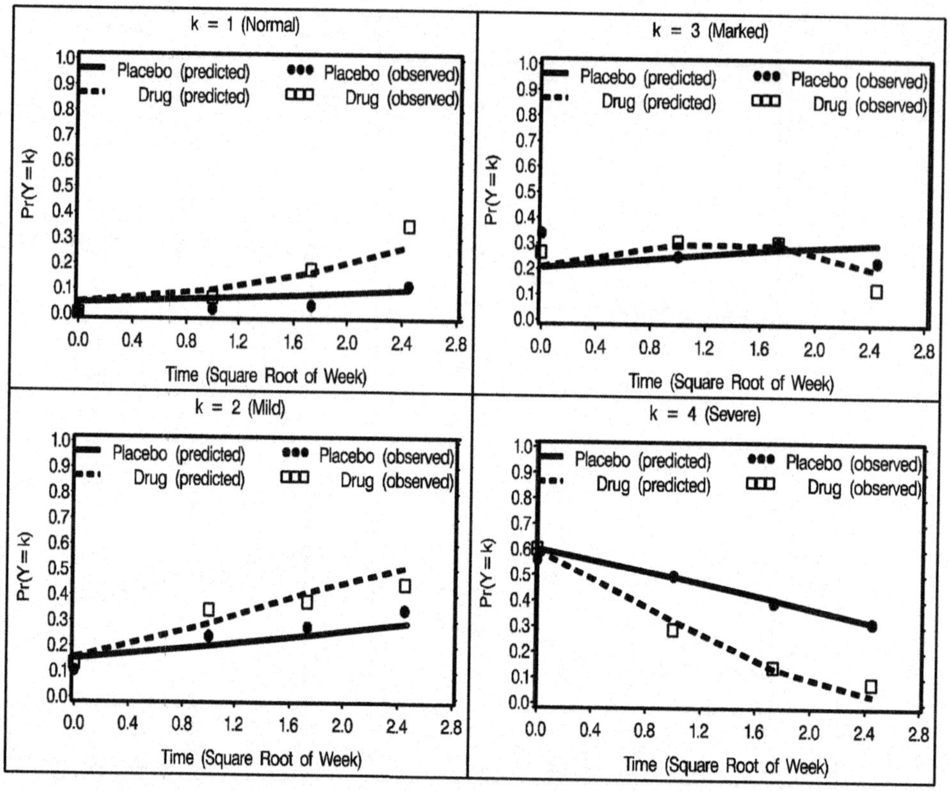

Figure 4.7 Observed and predicted cumulative odds ratios (drug:placebo) over time. The predicted cumulative odds ratios and 95 percent confidence intervals are estimated from the cumulative logit model and presented on a log-10 scale. Given the sparse data at weeks 2, 4 and 5, the observed values are plotted for weeks 0, 1, 3 and 6 only.

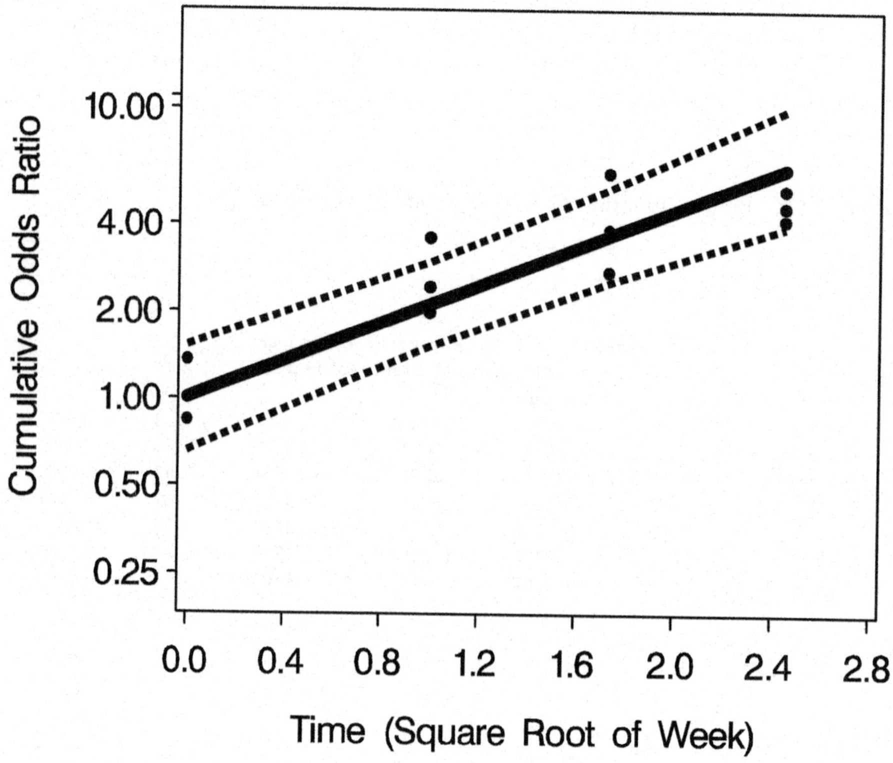

Output 4.18: Results from the cumulative complementary log-log link model.

```
GEE Fit Criteria
QIC     3839.7458
QICu    3833.1205
```

Analysis Of GEE Parameter Estimates
Empirical Standard Error Estimates

Parameter	Estimate	Standard Error	95% Confidence Limits		Z	Pr > |Z|
Intercept1	−3.0926	0.1597	−3.4056	−2.7795	−19.36	<.0001
Intercept2	−1.5253	0.1509	−1.8210	−1.2296	−10.11	<.0001
Intercept3	−0.6719	0.1487	−0.9633	−0.3804	−4.52	<.0001
Trt	0.0236	0.1660	−0.3019	0.3490	0.14	0.8872
SWeek	0.3324	0.0727	0.1900	0.4748	4.58	<.0001
Trt*SWeek	0.4442	0.0858	0.2759	0.6124	5.17	<.0001

However, there are some clear indications that the cumulative complementary log-log link model does not provide nearly as good a fit as either the cumulative logit or probit models. The first indication follows from the fact that it had the highest QIC of the three models (a QIC of 3839.7458, Output 4.18). The second indication can be found when we compare the observed proportions within each category of illness against the predicted proportions (Figure 4.6). Here we see that predicted proportions from the cumulative complementary log-log link model tend to have a poorer fit to the observed proportions compared with the cumulative logit link (Figure 4.5). This is particularly noticeable for patients in the active drug group.

As can be seen by the preceding analysis, the cumulative logit and probit models both provide a reasonably good fit to the data and all three models yield qualitatively similar results relating to the differential impact the use of an active drug has on reducing the severity of illness over time. The cumulative logit model also allows one to compare trends over time based on a cumulative odds ratio (drug:placebo) that is common across the first three categories of illness. To illustrate, Figure 4.7 compares the observed versus predicted cumulative odds ratios across the first three categories of illness (normal, mild, marked) at each period of follow-up. This graph was obtained by taking the output dataset 'Estimates' created from the ODS OUTPUT ESTIMATES= statement from the GENMOD procedure used to run the cumulative logit model (see above SAS code) and merging select portions of this dataset with the 'COR' dataset (see above previous SAS code related to Output 4.15) containing the observed cumulative odds ratios. The complete SAS code for generating Figure 4.7 is shown below.

Program 4.10

```
data COR_Pred;
 set Estimates;
 if Label IN ('Exp(CumLogOR(Week=0))' 'Exp(CumLogOR(Week=1))'
              'Exp(CumLogOR(Week=3))' 'Exp(CumLogOR(Week=6))');
 if Label = 'Exp(CumLogOR(Week=0))' then do;
    Week=0; COR_Pred=LBetaEstimate;
    Lower=LBetaLowerCL; Upper=LBetaUpperCL;
 end;
 if Label = 'Exp(CumLogOR(Week=1))' then do;
    Week=1; COR_Pred=LBetaEstimate;
    Lower=LBetaLowerCL; Upper=LBetaUpperCL;
 end;
 if Label = 'Exp(CumLogOR(Week=3))' then do;
    Week=3; COR_Pred=LBetaEstimate;
    Lower=LBetaLowerCL; Upper=LBetaUpperCL;
 end;
```

```
  if Label = 'Exp(CumLogOR(Week=6))' then do;
     Week=6; COR_Pred=LBetaEstimate;
     Lower=LBetaLowerCL; Upper=LBetaUpperCL;
  end;
  SWeek=Sqrt(Week);
  keep Week SWeek COR_Pred Lower Upper;
run;
proc sort data=COR_Pred;
 by Week;
run;
data COR_all;
 merge COR COR_Pred;
 by Week;
 if IMPS79o< 4 and Week IN (0 1 3 6);
run;

goptions reset rotate=landscape gsfmode=replace ftext=swiss
         htext=2.0 hby=2.0 device=sasemf;
proc gplot data=COR_all;
 plot CumOddsRatio*SWeek=1 COR_Pred*SWeek=2
      Lower*SWeek=3 Upper*SWeek=4 / overlay
      vaxis=axis1 haxis=axis2;
 axis1 label=(a=90 'Cumulative Odds Ratio')
       order=0.25 0.5 1 2 4 10 11
       value=(t=7 ' ')
       interval=uneven
       logbase=10
       logstyle=expand
       minor=none;
 axis2 label=('Time (Square Root of Week)')
       order=0 to 2.8 by .4
       minor=none;
 symbol1 c=black v=dot i=none;
 symbol2 c=black v=none i=join l=1 w=5;
 symbol3 c=black v=none i=join l=2 w=3;
 symbol4 c=black v=none i=join l=2 w=3;
run;
quit;
```

The cumulative odds ratios displayed in Figure 4.7 and summarized in Output 4.19 clearly show that patients in the active drug group improve at a rate proportionately higher over time compared to those in the placebo control group. For example, at baseline (week 0) the odds of response at or below any of the first three IMPS79 severity scores (normal, mildly ill, markedly ill) is essentially the same for patients in both groups with an estimated cumulative odds ratio of 1.001. This merely reflects the fact that patients in the two groups have similar response at baseline as would be expected with a randomized controlled trial. In contrast, at week 1 the odds of being classified as normal ($y \leq 1$), normal or mildly ill ($y \leq 2$), or normal, mildly ill or markedly ill ($y \leq 3$) are estimated to be 2.120 times higher among patients receiving an active drug compared to those receiving placebo. From this analysis we can conclude that while patients in both groups showed some improvement over time, patients randomized to one of the three active drugs showed significantly greater improvement over the 6 week course of treatment.

Output 4.19: A summary comparison of observed versus predicted cumulative odds ratios across time.

Week	Cumulative IMPS 79 Scores	Placebo Cum. Prob.	Drug Cum. Prob.	Observed Cumulative Odds Ratio	Predicted Cumulative Odds Ratio
0	Y<=1 (Normal)	0.000	0.003	.	1.001
	Y<=2 (Normal, Mild)	0.103	0.135	1.357	1.001
	Y<=3 (Normal, Mild, Marked)	0.439	0.398	0.842	1.001
1	Y<=1 (Normal)	0.019	0.065	3.605	2.120
	Y<=2 (Normal, Mild)	0.257	0.408	1.992	2.120
	Y<=3 (Normal, Mild, Marked)	0.505	0.713	2.442	2.120
3	Y<=1 (Normal)	0.034	0.178	6.051	3.674
	Y<=2 (Normal, Mild)	0.310	0.554	2.760	3.674
	Y<=3 (Normal, Mild, Marked)	0.609	0.857	3.849	3.674
6	Y<=1 (Normal)	0.114	0.351	4.190	6.297
	Y<=2 (Normal, Mild)	0.457	0.796	4.640	6.297
	Y<=3 (Normal, Mild, Marked)	0.686	0.921	5.325	6.297

4.4 The generalized nonlinear model (GNLM)

While GLIM's offer practitioners a wide range of models and options with which to analyze correlated response data, they do have certain limitations as noted at the beginning of this chapter. The first and most notable limitation deals with the fact that under a GLIM, the means of cluster-specific observations y_{ij} are restricted to having the form $E(y_{ij}) = \mu_{ij} = \mu_{ij}(\boldsymbol{\beta}) = \boldsymbol{g}^{-1}(\boldsymbol{x}'_{ij}\boldsymbol{\beta})$ where $\boldsymbol{g}^{-1}(\boldsymbol{x}'_{ij}\boldsymbol{\beta})$ is a monotonic invertible function of a linear predictor $\boldsymbol{x}'_{ij}\boldsymbol{\beta}$ of within- and between-subject covariates $\boldsymbol{x}'_{ij}$. While $\boldsymbol{g}^{-1}(\boldsymbol{x}'_{ij}\boldsymbol{\beta})$ may be nonlinear, it is nevertheless strictly a function of linear predictors (neither GENMOD nor GLIMMIX allow nonlinear predictors). A second limitation is that the marginal distribution of individual observations, when specified, will generally be restricted to the regular exponential family of distributions and correlation will be induced through specification of a "working" covariance structure. This prevents one from specifying a truly multivariate distribution, like the multivariate normal, that can be used to directly model correlation via a stochastic mechanism. Finally, when one does specify a univariate marginal distribution for individual observations, y_{ij}, the GLIM restricts the variance function to that of the assumed distribution.

Generalized nonlinear models—correlated response data

In this section we consider a class of generalized nonlinear models that 1) allow for more flexible modeling of the mean and variance-covariance structure and 2) allow for a broader class of multivariate distributions if and when the need arises. As in the preceding section on GLIM's, let $\boldsymbol{y}_i = \begin{pmatrix} y_{i1} & \cdots & y_{ip_i} \end{pmatrix}'$ denote a $p_i \times 1$ response vector for the ith subject or cluster ($i = 1, \ldots, n$) which may or may not be correlated. Taking a quasi-likelihood approach in which we simply specify the first two moments of $\boldsymbol{y}_i$, a marginal population-averaged generalized nonlinear model (GNLM) for correlated response data may be written as follows:

$$\begin{aligned} E(\boldsymbol{y}_i) &= \boldsymbol{\mu}_i = \boldsymbol{\mu}_i(\boldsymbol{\beta}) = f(\boldsymbol{X}_i, \boldsymbol{\beta});\ i = 1, \ldots, n \\ Var(\boldsymbol{y}_i) &= \boldsymbol{\Sigma}_i(\boldsymbol{\beta}, \boldsymbol{\theta}) \end{aligned} \tag{4.29}$$

where

$\boldsymbol{y}_i$ is the $p_i \times 1$ response vector for the ith subject (cluster)
$\boldsymbol{X}_i$ is a $p_i \times u$ matrix of within- and between-subject covariates

$\boldsymbol{\beta}$ is an $s \times 1$ vector of fixed-effects parameters
$\boldsymbol{\mu}_i(\boldsymbol{\beta}) = f(\boldsymbol{X}_i, \boldsymbol{\beta})$ is a general nonlinear function of $\boldsymbol{\beta}$
$\boldsymbol{\Sigma}_i(\boldsymbol{\beta}, \boldsymbol{\theta})$ is a $p_i \times p_i$ variance-covariance matrix that depends on an unknown $d \times 1$ vector of parameters, $\boldsymbol{\theta}$ and possibly on $\boldsymbol{\beta}$.

The first thing to note regarding this GNLM is that the mean $\boldsymbol{\mu}_i$ is a general nonlinear function of $\boldsymbol{\beta}$. As such, the GNLM is not restricted to having a design matrix $\boldsymbol{X}_i$ that necessarily conforms to the dimension of $\boldsymbol{\beta}$ (i.e., $\boldsymbol{X}_i$ need not have dimension $p_i \times s$ as would otherwise occur in a GLIM involving linear predictors $\boldsymbol{x}_{ij}'\boldsymbol{\beta}$). Second, the variance-covariance structure is not restricted to matrices having the form $\boldsymbol{\Sigma}_i(\boldsymbol{\beta}, \boldsymbol{\theta}) = \boldsymbol{V}_i(\boldsymbol{\mu}_i)^{1/2}\boldsymbol{\Sigma}_i^*(\boldsymbol{\theta})\boldsymbol{V}_i(\boldsymbol{\mu}_i)^{1/2}$ where $\boldsymbol{V}_i(\boldsymbol{\mu}_i)$ is a diagonal matrix whose elements correspond to some specified variance function of the mean [i.e., $\boldsymbol{V}_i(\boldsymbol{\mu}_i) = \boldsymbol{V}_i(\boldsymbol{\mu}_i(\boldsymbol{\beta})) = Diag\{h(\mu_{ij})\}$]. Implicit under (4.29), is the assumption that the parameter space of $(\boldsymbol{\beta}, \boldsymbol{\theta})$ is a compact subspace of Euclidean space $\Re^{s+d}$ which is determined by the underlying distribution function of $\boldsymbol{y}_i$ and which may depend directly on the dimension of $\boldsymbol{y}_i$ for unstructured correlation. Finally, the GNLM specified by (4.29) is semiparametric in that no explicit assumptions regarding the distribution of $\boldsymbol{y}_i$ are made. This means that likelihood-based estimation and inference will not be possible based solely on the assumptions in (4.29).

The semiparametric GNLM (4.29) can be extended to a fully parametric multivariate GNLM by specifying a joint probability density function (pdf) for $\boldsymbol{y}_i$. One important class of pdf's that allow for continuous or discrete outcomes are those pdf's from the quadratic exponential family in which the pdf of $\boldsymbol{y}_i$ may be written as

$$\pi(\boldsymbol{y}_i) = \Delta_i^{-1} \exp\{\boldsymbol{y}_i'\boldsymbol{\lambda}_i + \boldsymbol{w}_i'\boldsymbol{\gamma}_i + c_i(\boldsymbol{y}_i)\} \tag{4.30}$$

where $\boldsymbol{y}_i'$ is the response vector, and $\boldsymbol{w}_i' = (y_{i1}^2, y_{i1}y_{i2}, ..., y_{i1}y_{ip_i}, y_{i2}^2, y_{i2}y_{i3}, ...)$ is the vector of unique second-order components of $\boldsymbol{y}_i$ which one can write as $\boldsymbol{w}_i = Vech(\boldsymbol{y}_i\boldsymbol{y}_i')$. The $\boldsymbol{\lambda}_i$ and $\boldsymbol{\gamma}_i$ are canonical parameters expressed in terms of vector functions of the marginal moments as

$$\boldsymbol{\lambda}_i' = \boldsymbol{\lambda}_i'(\boldsymbol{\mu}_i, \boldsymbol{\sigma}_i) = (\lambda_{i1}, \lambda_{i2}, ..., \lambda_{ip_i})$$
$$\boldsymbol{\gamma}_i' = \boldsymbol{\gamma}_i'(\boldsymbol{\mu}_i, \boldsymbol{\sigma}_i) = (\gamma_{i11}, \gamma_{i12}...\gamma_{i1p_i}, \gamma_{i22}, \gamma_{i23}, ...).$$

where $\boldsymbol{\sigma}_i = \boldsymbol{\sigma}_i(\boldsymbol{\beta}, \boldsymbol{\theta}) = Vech(\boldsymbol{\Sigma}_i(\boldsymbol{\beta}, \boldsymbol{\theta}))$ denotes the lower diagonal elements of the variance-covariance matrix in vector form. The function $c_i(\boldsymbol{y}_i)$ is a "shape" function, and $\Delta_i = \Delta_i(\boldsymbol{\lambda}_i, \boldsymbol{\gamma}_i, c_i)$ is a normalization constant required to make (4.30) a proper pdf (e.g., Prentice and Zhao, 1991).

By combining the marginal first- and second-order moment specifications of (4.29) with the distributional assumptions of (4.30), we end up with a class of GNLM's that will include the normal-theory multivariate linear model (2.1) and the univariate generalized linear model (4.1)-(4.2) as special cases. For example, the univariate GLIM is obtained by setting $\boldsymbol{\mu}_i = \mu_i(\boldsymbol{\beta}) = g^{-1}(\boldsymbol{x}_i'\boldsymbol{\beta})$, $\boldsymbol{\sigma}_i = a_i(\phi)h(\mu_i)$ with $\boldsymbol{\theta} = \phi$, $\boldsymbol{y}_i = y_i$, $\boldsymbol{\lambda}_i = \eta_i/a_i(\phi) = \boldsymbol{x}_i'\boldsymbol{\beta}/a_i(\phi)$, $\gamma_i \equiv 0$, $\Delta_i^{-1} = \exp\{-b(\eta_i)/a_i(\phi)\}$ and $c_i(\boldsymbol{y}_i) = c(y_i, \phi)$. The marginal multivariate LM (2.1) is obtained by setting $\boldsymbol{\mu}_i = \boldsymbol{\mu}_i(\boldsymbol{\beta}) = \boldsymbol{X}_i\boldsymbol{\beta}$, $\boldsymbol{\Sigma}_i(\boldsymbol{\beta}, \boldsymbol{\theta}) = \boldsymbol{\Sigma}_i(\boldsymbol{\theta}) = \boldsymbol{\Sigma}_i$, $\Delta_i^{-1} = (2\pi)^{-p/2}|\boldsymbol{\Sigma}_i|^{1/2}\exp\{-\frac{1}{2}\boldsymbol{\mu}_i'\boldsymbol{\Sigma}_i^{-1/2}\boldsymbol{\mu}_i\}$, $\boldsymbol{\lambda}_i(\boldsymbol{\mu}_i, \boldsymbol{\sigma}_i) = \boldsymbol{\Sigma}_i^{-1}\boldsymbol{\mu}_i$, $\boldsymbol{\gamma}_i(\boldsymbol{\mu}_i, \boldsymbol{\sigma}_i) = -\frac{1}{2}\boldsymbol{C}_i'\boldsymbol{C}_i Vech\{\boldsymbol{\Sigma}_i^{-1}\}$, $c_i(\boldsymbol{y}_i) = 0$, and $\boldsymbol{C}_i$ is a $p_i^2 \times p_i(p_i+1)/2$ matrix of 0's and 1's satisfying $Vec(\boldsymbol{y}_i\boldsymbol{y}_i') = \boldsymbol{C}_i Vech(\boldsymbol{y}_i\boldsymbol{y}_i')$ (see Appendix A for definitions of the $Vec(\cdot)$ and $Vech(\cdot)$ matrix operators). A number of authors have considered applications involving both discrete and continuous outcomes that arise from the quadratic exponential family including correlated binary data (Zhao and Prentice, 1988; Prentice and Zhao, 1991). One important set of applications involves the normal-theory multivariate nonlinear model which we present in the following section.

4.4.1 Normal-theory nonlinear model (NLM)

Marginal nonlinear models for correlated normally distributed data play an important role in numerous disciplines and as such represent an important sub-class of models under the GNLM defined by (4.29)-(4.30). The normal-theory nonlinear model (NLM) for correlated responses may be written as

$$y_{ij} = f(\boldsymbol{x}'_{ij}, \boldsymbol{\beta}) + \epsilon_{ij},\ i = 1, \ldots, n;\ j = 1, \ldots, p_i \tag{4.31}$$

where

y_{ij} is the jth measurement on the ith subject or cluster
$\boldsymbol{x}'_{ij}$ are $1 \times u$ vectors of within- and between-subject covariates of the form $\boldsymbol{x}'_{ij} = (\boldsymbol{x}^{*\prime}_{ij}, \boldsymbol{a}'_i)$ where $\boldsymbol{x}^*_{ij}$ is a $t \times 1$ vector of within-subject covariates and $\boldsymbol{a}_i$ is a $q \times 1$ vector of between-subject covariates $(u = t + q)$
$\boldsymbol{\beta}$ is an $s \times 1$ unknown vector of regression parameters
$\epsilon_i = (\epsilon_{i1}, \epsilon_{i2}, \ldots, \epsilon_{ip_i})'$ are $p_i \times 1$ random error vectors that are mutually independent and distributed as $N_{p_i}(\mathbf{0}, \Sigma_i(\boldsymbol{\beta}, \boldsymbol{\theta}))$
$\boldsymbol{\Sigma}_i(\boldsymbol{\beta}, \boldsymbol{\theta})$ is a $p_i \times p_i$ variance-covariance matrix that depends on an unknown $d \times 1$ vector of parameters, $\boldsymbol{\theta}$ and possibly on $\boldsymbol{\beta}$.

Following the general notation of §1.3.2, we can write this model in matrix form as:

$$\begin{aligned} \boldsymbol{y}_i &= f(\boldsymbol{X}^*_i, \boldsymbol{A}_i, \boldsymbol{\beta}) + \epsilon_i \\ &= f(\boldsymbol{X}_i, \boldsymbol{\beta}) + \epsilon_i,\quad i = 1, \ldots, n \text{ subjects (clusters)} \end{aligned} \tag{4.32}$$

where

$\boldsymbol{y}_i$ is the $p_i \times 1$ response vector, $\boldsymbol{y}_i = (y_{i1} \ldots y_{ip_i})'$
$\boldsymbol{X}^*_i = \begin{pmatrix} \boldsymbol{x}^{*\prime}_{i1} \\ \vdots \\ \boldsymbol{x}^{*\prime}_{ip_i} \end{pmatrix}$ is a $p_i \times t$ within-subject design matrix formed from the row vectors, $\boldsymbol{x}^{*\prime}_{ij}$
and $\boldsymbol{A}_i = \mathbf{1}_{p_i \times 1} \otimes \boldsymbol{a}'_i$ is a $p_i \times q$ between-subject design matrix formed from the row vector $\boldsymbol{a}'_i$
$\boldsymbol{X}_i = \begin{pmatrix} \boldsymbol{X}^*_i & \mathbf{1}_{p_i \times 1} \otimes \boldsymbol{a}'_i \end{pmatrix}_{p_i \times (t+q)}$ is a $p_i \times u$ design matrix of within- and between-subject covariates formed from the row vectors $\boldsymbol{x}'_{ij}$ $(u = t + q)$
$\epsilon_i \sim \text{ind } N_{p_i}(\mathbf{0}, \boldsymbol{\Sigma}_i(\boldsymbol{\beta}, \boldsymbol{\theta}))$.

As is the case with the marginal LM, the marginal NLM (4.32) may be expressed in terms of the GNLM by setting $\boldsymbol{\mu}_i = f(\boldsymbol{X}_i, \boldsymbol{\beta})$, $\boldsymbol{\lambda}_i(\boldsymbol{\mu}_i, \boldsymbol{\sigma}_i) = \boldsymbol{\Sigma}_i^{-1}\boldsymbol{\mu}_i$, $\gamma_i(\boldsymbol{\mu}_i, \boldsymbol{\sigma}_i) = -\frac{1}{2}\boldsymbol{C}'_i\boldsymbol{C}_i Vech\{\boldsymbol{\Sigma}_i^{-1}\}$, $\Delta_i^{-1} = (2\pi)^{-p/2}|\boldsymbol{\Sigma}_i|^{1/2} exp\{-\frac{1}{2}\boldsymbol{\mu}'_i\boldsymbol{\Sigma}_i^{-1/2}\boldsymbol{\mu}_i\}$, $c_i(\boldsymbol{y}_i) = 0$, and $\boldsymbol{C}_i$ is the matrix of 0's and 1's as defined previously.

The marginal NLM (4.32) has received considerable attention in the field of econometrics as it applies to the analysis of panel data. Gallant (1987, Chapter 5) provides an excellent discourse on multivariate NLM's including estimation and inference based on both least squares (LS) and maximum likelihood (ML) principles as well as the asymptotic theory associated with these methods. In the following section, we present a general estimation procedure for the multivariate GNLM that includes the LS and ML estimation procedures for the marginal NLM (4.32) as well as the GEE-based estimation methods for the GLIM. The approach involves solving a set of estimating equations that formally take into account any additional information about regression parameters $\boldsymbol{\beta}$ that may occur in the second-order moments determined by $\boldsymbol{\Sigma}_i(\boldsymbol{\beta}, \boldsymbol{\theta})$.

4.4.2 Estimation

In this section we first consider maximum likelihood estimation under the GNLM (4.29) assuming the distribution of $\boldsymbol{y}_i$ is in the quadratic exponential family (4.30). This requires specification of the third- and fourth-order moments of $\boldsymbol{y}_i$ either directly or through a doubly iterative procedure as described by Prentice and Zhao (1991). The ML estimates of $(\boldsymbol{\beta}, \boldsymbol{\theta})$ are obtained by solving a set of second-order generalized estimating equations (GEE2) that are the ML score equations under any special case of (4.30). As shown by Gourieroux et. al. (1984), Prentice and Zhao (1991) and Vonesh et. al. (2001), the resulting ML estimates will be consistent for $(\boldsymbol{\beta}, \boldsymbol{\theta})$ even under distributional misspecification provided the first- and second-order moments in (4.29) have not been misspecified. Based on this result, we discuss how, through the specification of "working" third- and fourth-order moments, one can adopt a method-of-moments approach as an alternative to ML-based estimation. This yields a second-order generalized estimating equations (GEE2) approach that generalizes Liang and Zeger's (1986) GEE1 approach for estimating the parameters, $\boldsymbol{\beta}$, of the mean to estimating the parameters, $(\boldsymbol{\beta}, \boldsymbol{\theta})$, of both the mean and variance. Both the ML and GEE2 approach are available using PROC NLMIXED. Alternatively, PROC NLMIXED may be used to obtain ML estimates for GNLM's with joint density functions that are not in the quadratic exponential family but this does require the user to specify the pdf (see §4.5.4 for an example of fitting a multivariate Poisson model).

Maximum likelihood (ML) estimation

Gourieroux et. al. (1984) and Prentice and Zhao (1991) derive ML estimates of $(\boldsymbol{\beta}, \boldsymbol{\theta})$ under the fully parametric GNLM defined by (4.29)-(4.30). Following Prentice and Zhao (1991), let $\boldsymbol{s}_i = \boldsymbol{s}_i(\boldsymbol{\beta}) = Vech\{[\boldsymbol{y}_i - \boldsymbol{\mu}_i(\boldsymbol{\beta})][\boldsymbol{y}_i - \boldsymbol{\mu}_i(\boldsymbol{\beta})]'\}$ be the empirical covariance vector associated with $\boldsymbol{y}_i$ and let $\boldsymbol{\Upsilon}_i = Cov(\boldsymbol{y}_i, \boldsymbol{s}_i)$ and $\boldsymbol{\Gamma}_i = Var(\boldsymbol{s}_i)$ be the matrices containing the third- and fourth-order moments of $\boldsymbol{y}_i$, respectively. Further, let

$$\boldsymbol{D}_i(\boldsymbol{\beta}) = \frac{\partial \boldsymbol{\mu}_i(\boldsymbol{\beta})}{\partial \boldsymbol{\beta}'},\ \boldsymbol{E}_i(\boldsymbol{\beta}) = \frac{\partial \boldsymbol{\sigma}_i(\boldsymbol{\beta}, \boldsymbol{\theta})}{\partial \boldsymbol{\beta}'}, \boldsymbol{E}_i(\boldsymbol{\theta}) = \frac{\partial \boldsymbol{\sigma}_i(\boldsymbol{\beta}, \boldsymbol{\theta})}{\partial \boldsymbol{\theta}'}$$

denote the first-order derivative matrices of the first- and second-order moments $\boldsymbol{\mu}_i(\boldsymbol{\beta})$ and $\boldsymbol{\sigma}_i(\boldsymbol{\beta}, \boldsymbol{\theta})$, respectively, and let $\boldsymbol{\tau} = (\boldsymbol{\beta}, \boldsymbol{\theta})$ be the joint vector of mean and covariance parameters. Prentice and Zhao (1991) show the log-likelihood score equations under any specific form of (4.30) will be given by

$$\begin{aligned} U(\boldsymbol{\tau}) &= \sum_{i=1}^{n} \begin{pmatrix} \boldsymbol{D}_i(\boldsymbol{\beta}) & \boldsymbol{0} \\ \boldsymbol{E}_i(\boldsymbol{\beta}) & \boldsymbol{E}_i(\boldsymbol{\theta}) \end{pmatrix}' \begin{pmatrix} \boldsymbol{\Sigma}_i(\boldsymbol{\beta}, \boldsymbol{\theta}) & \boldsymbol{\Upsilon}_i \\ \boldsymbol{\Upsilon}_i' & \boldsymbol{\Gamma}_i \end{pmatrix}^{-1} \begin{pmatrix} \boldsymbol{y}_i - \boldsymbol{\mu}_i(\boldsymbol{\beta}) \\ \boldsymbol{s}_i - \boldsymbol{\sigma}_i(\boldsymbol{\beta}, \boldsymbol{\theta}) \end{pmatrix} \\ &= \sum_{i=1}^{n} \boldsymbol{X}_i(\boldsymbol{\tau})' \boldsymbol{V}_i(\boldsymbol{\tau})^{-1} \{\boldsymbol{Y}_i - \boldsymbol{F}_i(\boldsymbol{\tau})\} = \boldsymbol{0} \end{aligned} \tag{4.33}$$

where $\boldsymbol{Y}_i = \begin{pmatrix} \boldsymbol{y}_i \\ \boldsymbol{s}_i \end{pmatrix}$ is an extended "response" vector, $\boldsymbol{F}_i(\boldsymbol{\tau}) = \begin{pmatrix} \boldsymbol{\mu}_i(\boldsymbol{\beta}) \\ \boldsymbol{\sigma}_i(\boldsymbol{\beta}, \boldsymbol{\theta}) \end{pmatrix}$ is the nonlinear mean "response" function, $\boldsymbol{X}_i(\boldsymbol{\tau}) = \begin{pmatrix} \boldsymbol{D}_i(\boldsymbol{\beta}) & \boldsymbol{0} \\ \boldsymbol{E}_i(\boldsymbol{\beta}) & \boldsymbol{E}_i(\boldsymbol{\theta}) \end{pmatrix}$ is the matrix of first-order derivatives of $\boldsymbol{F}_i(\boldsymbol{\tau})$, and $\boldsymbol{V}_i(\boldsymbol{\tau}) = \begin{pmatrix} \boldsymbol{\Sigma}_i(\boldsymbol{\beta}, \boldsymbol{\theta}) & \boldsymbol{\Upsilon}_i \\ \boldsymbol{\Upsilon}_i' & \boldsymbol{\Gamma}_i \end{pmatrix}$ is the variance-covariance matrix of $\boldsymbol{Y}_i$.

The score estimating equations (4.33) may be derived as the estimating equations of $\boldsymbol{\tau}$ under the multivariate nonlinear regression model

$$\begin{aligned}\boldsymbol{Y}_i &= \boldsymbol{F}_i(\boldsymbol{X}_i, \boldsymbol{\tau}) + \boldsymbol{e}_i = \boldsymbol{F}_i(\boldsymbol{\tau}) + \boldsymbol{e}_i,\ i = 1, \ldots, n \\ E(\boldsymbol{e}_i) &= \boldsymbol{0}, \\ Var(\boldsymbol{e}_i) &= \boldsymbol{V}_i(\boldsymbol{\tau})\end{aligned} \tag{4.34}$$

where $\boldsymbol{F}_i(\boldsymbol{X}_i, \boldsymbol{\tau}) = \begin{pmatrix} f(\boldsymbol{X}_i, \boldsymbol{\beta}) \\ \boldsymbol{\sigma}_i(\boldsymbol{\beta}, \boldsymbol{\theta}) \end{pmatrix}$ and $f(\boldsymbol{X}_i, \boldsymbol{\beta}) = \boldsymbol{\mu}_i(\boldsymbol{\beta})$ is a general nonlinear function of $\boldsymbol{\beta}$. In fact, it is under a generalization of this nonlinear regression setup in which one replaces $\boldsymbol{V}_i(\boldsymbol{\tau})$ by $\boldsymbol{V}_i(\boldsymbol{\tau}, \boldsymbol{\nu})$ where $\boldsymbol{\nu}$ are possible nuisance parameters associated with "working" third- and fourth-order moments of $\boldsymbol{y}_i$ that one can view the ML score estimating equations (4.33) as being a special case of the set of GEE2 described in the following section.

As shown by Prentice and Zhao (1991) and Vonesh et. al. (2001), derivation of the score equations (4.33) may be expressed in terms of the extended "response" vector $\boldsymbol{Y}_i^* = \begin{pmatrix} \boldsymbol{y}_i \\ \boldsymbol{w}_i \end{pmatrix}$ rather than $\boldsymbol{Y}_i = \begin{pmatrix} \boldsymbol{y}_i \\ \boldsymbol{s}_i \end{pmatrix}$ as follows. Let $E(\boldsymbol{w}_i) = \boldsymbol{\eta}_i(\boldsymbol{\beta}, \boldsymbol{\theta}) = Vech[\boldsymbol{\Sigma}_i(\boldsymbol{\beta}, \boldsymbol{\theta}) + \boldsymbol{\mu}_i(\boldsymbol{\beta})\boldsymbol{\mu}_i(\boldsymbol{\beta})']$ and let $\boldsymbol{B}_i$ be a $p_i(p_i+1)/2 \times p_i^2$ matrix of 0's and 1's such that $Vech(\boldsymbol{y}_i\boldsymbol{y}_i') = \boldsymbol{B}_i Vec(\boldsymbol{y}_i\boldsymbol{y}_i')$ and $\boldsymbol{B}_i\boldsymbol{C}_i = \boldsymbol{I}_{p_i(p_i+1)/2}$ where $\boldsymbol{C}_i$ is defined above. If we further let

$$\boldsymbol{W}_i(\boldsymbol{\beta}) = \begin{pmatrix} \boldsymbol{I}_{p_i} & \boldsymbol{0} \\ \boldsymbol{T}_i(\boldsymbol{\beta}) & \boldsymbol{I}_{p_i(p_i+1)/2} \end{pmatrix}$$

where $\boldsymbol{T}_i(\boldsymbol{\beta}) = \boldsymbol{B}_i\{[\boldsymbol{\mu}_i(\boldsymbol{\beta}) \otimes \boldsymbol{I}_{p_i}] + [\boldsymbol{I}_{p_i} \otimes \boldsymbol{\mu}_i(\boldsymbol{\beta})]\}$, then the score equations (4.33) can be written as

$$\boldsymbol{U}(\boldsymbol{\tau}) = \sum_{i=1}^{n} \boldsymbol{X}_i^*(\boldsymbol{\tau})'\boldsymbol{V}_i^*(\boldsymbol{\tau})^{-1}\{\boldsymbol{Y}_i^* - \boldsymbol{F}_i^*(\boldsymbol{\tau})\} = \boldsymbol{0} \tag{4.35}$$

where

$$\begin{aligned}\boldsymbol{X}_i^*(\boldsymbol{\tau}) &= \begin{pmatrix} \boldsymbol{D}_i(\boldsymbol{\beta}) & \boldsymbol{0} \\ \boldsymbol{T}_i(\boldsymbol{\beta})\boldsymbol{D}_i(\boldsymbol{\beta}) + \boldsymbol{E}_i(\boldsymbol{\beta}) & \boldsymbol{E}_i(\boldsymbol{\theta}) \end{pmatrix} \\ &= \begin{pmatrix} \partial\boldsymbol{\mu}_i(\boldsymbol{\beta})/\partial\boldsymbol{\beta}' & \boldsymbol{0} \\ \partial\boldsymbol{\eta}_i(\boldsymbol{\beta}, \boldsymbol{\theta})/\partial\boldsymbol{\beta}' & \partial\boldsymbol{\eta}_i(\boldsymbol{\beta}, \boldsymbol{\theta})/\partial\boldsymbol{\theta}' \end{pmatrix} \\ \boldsymbol{V}_i^*(\boldsymbol{\tau}) &= \boldsymbol{W}_i(\boldsymbol{\beta}) \begin{pmatrix} \boldsymbol{\Sigma}_i(\boldsymbol{\beta}, \boldsymbol{\theta}) & \boldsymbol{\Upsilon}_i \\ \boldsymbol{\Upsilon}_i' & \boldsymbol{\Gamma}_i \end{pmatrix} \boldsymbol{W}_i(\boldsymbol{\beta})' \\ &= \begin{pmatrix} \boldsymbol{\Sigma}_i & \boldsymbol{\Sigma}_i\boldsymbol{T}_i' + \boldsymbol{\Upsilon}_i \\ \boldsymbol{T}_i\boldsymbol{\Sigma}_i + \boldsymbol{\Upsilon}_i' & \boldsymbol{T}_i\boldsymbol{\Sigma}_i\boldsymbol{T}_i' + \boldsymbol{\Upsilon}_i'\boldsymbol{T}_i' + \boldsymbol{T}_i\boldsymbol{\Upsilon}_i + \boldsymbol{\Gamma}_i \end{pmatrix} \\ &= \begin{pmatrix} \boldsymbol{\Sigma}_i(\boldsymbol{\beta}, \boldsymbol{\theta}) & \boldsymbol{\Upsilon}_i^* \\ \boldsymbol{\Upsilon}_i^{*\prime} & \boldsymbol{\Gamma}_i^* \end{pmatrix} \\ &= \begin{pmatrix} Var(\boldsymbol{y}_i) & Cov(\boldsymbol{y}_i, \boldsymbol{w}_i) \\ Cov(\boldsymbol{y}_i, \boldsymbol{w}_i)' & Var(\boldsymbol{w}_i) \end{pmatrix} \\ \boldsymbol{Y}_i^* &= \begin{pmatrix} \boldsymbol{y}_i \\ \boldsymbol{w}_i \end{pmatrix} \\ \boldsymbol{F}_i^*(\boldsymbol{\tau}) &= \begin{pmatrix} \boldsymbol{\mu}_i(\boldsymbol{\beta}) \\ \boldsymbol{\eta}_i(\boldsymbol{\beta}, \boldsymbol{\theta}) \end{pmatrix}.\end{aligned} \tag{4.36}$$

Under the GNLM (4.29), maximum likelihood estimation for any specific form of (4.30) requires the correct specification of the third- and fourth-order moments via $\boldsymbol{\Upsilon}_i$ and $\boldsymbol{\Gamma}_i$. This is often difficult to do and estimation will usually require a doubly iterative procedure that can be computationally intensive (Prentice and Zhao, 1991). A notable exception to this occurs when the data are normally distributed as the GNLM then reduces to the NLM (4.32).

Under the NLM (4.32), the third-order moments will satisfy $\boldsymbol{\Upsilon}_i = \mathbf{0}$ while the fourth-order moments will satisfy

$$\begin{aligned} \boldsymbol{\Gamma}_i = \boldsymbol{\Gamma}_i(\boldsymbol{\beta},\boldsymbol{\theta}) &= Var(\boldsymbol{s}_i(\boldsymbol{\beta})) \\ &= \boldsymbol{B}_i[\boldsymbol{I}_{p_i^2} + \boldsymbol{I}_{(p_i,p_i)}][\boldsymbol{\Sigma}_i(\boldsymbol{\beta},\boldsymbol{\theta}) \otimes \boldsymbol{\Sigma}_i(\boldsymbol{\beta},\boldsymbol{\theta})]\boldsymbol{B}_i' \end{aligned} \tag{4.37}$$

where $\boldsymbol{I}_{(p_i,p_i)}$ is the $p_i^2 \times p_i^2$ permuted identity matrix (e.g., Henderson and Searle, 1981). In this case, the score equations (4.35) based on the extended "response" vector $\boldsymbol{Y}_i^*$ are weighted by the inverse of the variance-covariance of $\boldsymbol{Y}_i^*$ which we can express in closed form as

$$Var(\boldsymbol{Y}_i^*) = \begin{pmatrix} \boldsymbol{\Sigma}_i(\boldsymbol{\beta},\boldsymbol{\theta}) & \boldsymbol{\Upsilon}_i^*(\boldsymbol{\beta},\boldsymbol{\theta}) \\ \boldsymbol{\Upsilon}_i^*(\boldsymbol{\beta},\boldsymbol{\theta})' & \boldsymbol{\Gamma}_i^*(\boldsymbol{\beta},\boldsymbol{\theta}) \end{pmatrix} \tag{4.38}$$

where $\boldsymbol{\Upsilon}_i^*(\boldsymbol{\beta},\boldsymbol{\theta}) = Cov(\boldsymbol{y}_i, \boldsymbol{w}_i) = \boldsymbol{\Sigma}_i(\boldsymbol{\beta},\boldsymbol{\theta})\boldsymbol{T}_i(\boldsymbol{\beta})'$ and $\boldsymbol{\Gamma}_i^*(\boldsymbol{\beta},\boldsymbol{\theta}) = Var(\boldsymbol{w}_i) = \boldsymbol{T}_i(\boldsymbol{\beta})\boldsymbol{\Sigma}_i(\boldsymbol{\beta},\boldsymbol{\theta})\boldsymbol{T}_i(\boldsymbol{\beta})' + \boldsymbol{\Gamma}_i(\boldsymbol{\beta},\boldsymbol{\theta})$. Utilizing this closed form expression for $Var(\boldsymbol{Y}_i^*)$ one can apply a Gauss-Newton or Newton-Raphson–type algorithm to obtain the normal-theory ML estimates of $(\boldsymbol{\beta},\boldsymbol{\theta})$. Since the Gauss-Newton algorithm also applies to solving GEE2 under the more general method-of-moment scheme of Prentice and Zhao (1991), a description of this algorithm is presented in the following section on GEE2.

Second-order generalized estimating equations (GEE2)

As indicated in the previous section, ML estimation under the GNLM (4.29) requires specification of the third- and fourth-order moments of $\boldsymbol{y}_i$ assuming the pdf is from the quadratic exponential family. When possible, this results in an estimator of $(\boldsymbol{\beta},\boldsymbol{\theta})$ that is strongly consistent and asymptotically efficient (Gourieroux et. al., 1984; and Prentice and Zhao, 1991). However, even if the third- and fourth-order moments are misspecified, Gourieroux et. al. (1984) showed that if the assumed but possibly incorrect pdf of $\boldsymbol{y}_i$ is in the quadratic exponential family and the first two moments, $\boldsymbol{\mu}_i(\boldsymbol{\beta})$ and $\boldsymbol{\Sigma}_i(\boldsymbol{\beta},\boldsymbol{\theta})$, are correctly specified, then the resulting estimator of $(\boldsymbol{\beta},\boldsymbol{\theta})$, which the authors refer to as a pseudo-maximum likelihood (PML) estimator, will still be strongly consistent and asymptotically normal.

Prentice and Zhao (1991) utilize this result to extend the first-order GEE approach of Liang and Zeger (1986) which requires correct specification of the mean of $\boldsymbol{y}_i$ to a second-order GEE approach which requires the correct specification of both the mean and variance-covariance of $\boldsymbol{y}_i$. By specifying closed form expressions of "working" third- and fourth-order moments of $\boldsymbol{y}_i$ in terms of $\boldsymbol{\mu}_i(\boldsymbol{\beta})$, $\boldsymbol{\sigma}_i(\boldsymbol{\beta},\boldsymbol{\theta})$ and a set of possible nuisance parameters $\boldsymbol{\nu}$, one can solve a set of second-order generalized estimating equations (GEE2) using a Gauss-Newton type algorithm suitable for the GNLM (4.29) without making specific distributional assumptions. As shown by Prentice and Zhao (1991) the resulting GEE2 estimates of $(\boldsymbol{\beta},\boldsymbol{\theta})$ will remain consistent and asymptotically normal.

Using a setup based on the extended "response" vector, $\boldsymbol{Y}_i^* = \begin{pmatrix} \boldsymbol{y}_i \\ \boldsymbol{w}_i \end{pmatrix}$, we can formulate a semiparametric method-of-moments GEE2 approach within the context of the general nonlinear regression model:

$$\begin{aligned} \boldsymbol{Y}_i^* &= \boldsymbol{F}_i^*(\boldsymbol{X}_i, \boldsymbol{\tau}) + \boldsymbol{e}_i^* = \boldsymbol{F}_i^*(\boldsymbol{\tau}) + \boldsymbol{e}_i^*,\ i = 1, \ldots, n \\ E(\boldsymbol{e}_i^*) &= \boldsymbol{0}, \\ Var(\boldsymbol{e}_i^*) &= \boldsymbol{V}_i^*(\boldsymbol{\tau}, \boldsymbol{\nu}) \end{aligned} \tag{4.39}$$

where

$\boldsymbol{Y}_i^* = \begin{pmatrix} \boldsymbol{y}_i \\ \boldsymbol{w}_i \end{pmatrix}$ is the extended "response" vector,

$\boldsymbol{F}_i^*(\boldsymbol{X}_i, \boldsymbol{\tau}) = \boldsymbol{F}_i^*(\boldsymbol{\tau}) = \begin{pmatrix} E(\boldsymbol{y}_i) \\ E(\boldsymbol{w}_i) \end{pmatrix} = \begin{pmatrix} \boldsymbol{\mu}_i(\boldsymbol{\beta}) \\ \boldsymbol{\eta}_i(\boldsymbol{\beta}, \boldsymbol{\theta}) \end{pmatrix}$ is a specified nonlinear function of the mean and covariance parameters $\boldsymbol{\tau} = (\boldsymbol{\beta}, \boldsymbol{\theta})$ with $E(\boldsymbol{y}_i) = \boldsymbol{\mu}_i(\boldsymbol{\beta})$ and $E(\boldsymbol{w}_i) = \boldsymbol{\eta}_i(\boldsymbol{\beta}, \boldsymbol{\theta}) = Vech[\boldsymbol{\Sigma}_i(\boldsymbol{\beta}, \boldsymbol{\theta}) + \boldsymbol{\mu}_i(\boldsymbol{\beta})\boldsymbol{\mu}_i(\boldsymbol{\beta})']$

$\boldsymbol{e}_i^*$ is an extended "error" term with $E(\boldsymbol{e}_i^*) = \boldsymbol{0}$ and $Var(\boldsymbol{e}_i^*) = \boldsymbol{V}_i^*(\boldsymbol{\tau}, \boldsymbol{\nu})$

$\boldsymbol{V}_i^*(\boldsymbol{\tau}, \boldsymbol{\nu}) = \begin{pmatrix} \boldsymbol{\Sigma}_i(\boldsymbol{\beta}, \boldsymbol{\theta}) & \boldsymbol{\Upsilon}_i^*(\boldsymbol{\beta}, \boldsymbol{\theta}, \boldsymbol{\nu}) \\ \boldsymbol{\Upsilon}_i^*(\boldsymbol{\beta}, \boldsymbol{\theta}, \boldsymbol{\nu})' & \boldsymbol{\Gamma}_i^*(\boldsymbol{\beta}, \boldsymbol{\theta}, \boldsymbol{\nu}) \end{pmatrix}$ is a "working" covariance matrix of $\boldsymbol{Y}_i^*$ but with $\boldsymbol{\Sigma}_i(\boldsymbol{\beta}, \boldsymbol{\theta})$ assumed to be the true variance-covariance of $\boldsymbol{y}_i$. The matrix $\boldsymbol{V}_i^*$ depends on $\boldsymbol{\tau}$ and possibly on additional nuisance parameters $\boldsymbol{\nu}$ that one might include when specifying "working" third- and fourth-order moments, $\boldsymbol{\Upsilon}_i^*$ and $\boldsymbol{\Gamma}_i^*$.

Based on this nonlinear regression setup, joint estimates of $(\boldsymbol{\beta}, \boldsymbol{\theta})$ can be carried out by solving a set of GEE2 using an adaptation of the Gauss-Newton algorithm described by Vonesh et. al. (2001) which entails repeated application of Taylor series linearization and weighted linear regression. The basic algorithm is defined as follows.

A Gauss-Newton algorithm for estimating $(\boldsymbol{\beta}, \boldsymbol{\theta})$ via GEE2

Step 0: Obtain an initial estimate, say $\tilde{\boldsymbol{\tau}}$, of $\boldsymbol{\tau}$

Step 1: Given $\tilde{\boldsymbol{\tau}}$, estimate $\boldsymbol{\nu}$ (if needed) by method of moments such that $\widehat{\boldsymbol{\nu}} = \widehat{\boldsymbol{\nu}}(\boldsymbol{\tau})$ is $\sqrt{n}$-consistent for $\boldsymbol{\nu}$ under the "working" covariance matrix $\boldsymbol{V}_i^*(\boldsymbol{\tau}, \boldsymbol{\nu})$. Update $\boldsymbol{V}_i^*$ at $\tilde{\boldsymbol{V}}_i^* = \boldsymbol{V}_i^*(\tilde{\boldsymbol{\tau}}, \widehat{\boldsymbol{\nu}}(\tilde{\boldsymbol{\tau}}))$ and hold this fixed.

Step 2. Perform a Taylor series linearization of $\boldsymbol{F}_i^*(\boldsymbol{\tau})$ in (4.39) about $\boldsymbol{\tau} = \tilde{\boldsymbol{\tau}}$ yielding

$$\boldsymbol{Y}_i^* = \boldsymbol{F}_i^*(\tilde{\boldsymbol{\tau}}) + \tilde{\boldsymbol{X}}_i^*\boldsymbol{\tau} - \tilde{\boldsymbol{X}}_i^*\tilde{\boldsymbol{\tau}} + \boldsymbol{e}_i^*,$$

where $\tilde{\boldsymbol{X}}_i^* = \boldsymbol{X}_i^*(\tilde{\boldsymbol{\tau}}) = \partial \boldsymbol{F}_i^*(\boldsymbol{\tau})/\partial\boldsymbol{\tau}\Big|_{\boldsymbol{\tau}=\tilde{\boldsymbol{\tau}}}$. Form the pseudo-marginal linear model

$$\tilde{\boldsymbol{Y}}_i^* = \tilde{\boldsymbol{X}}_i^*\boldsymbol{\tau} + \boldsymbol{e}_i^*$$

where $\tilde{\boldsymbol{Y}}_i^* = \boldsymbol{Y}_i^* - \boldsymbol{F}_i^*(\tilde{\boldsymbol{\tau}}) + \tilde{\boldsymbol{X}}_i^*\tilde{\boldsymbol{\tau}}$ is a pseudo-response vector and estimate $\boldsymbol{\tau}$ via estimated generalized least squares (EGLS)

$$\widehat{\boldsymbol{\tau}} = \widehat{\boldsymbol{\tau}}(\widehat{\boldsymbol{\nu}}) = \Big(\sum_{i=1}^n \tilde{\boldsymbol{X}}_i^{*\prime}\boldsymbol{V}_i^*(\tilde{\boldsymbol{\tau}}, \widehat{\boldsymbol{\nu}}(\tilde{\boldsymbol{\tau}}))^{-1}\tilde{\boldsymbol{X}}_i^*\Big)^{-1} \sum_{i=1}^n \tilde{\boldsymbol{X}}_i^{*\prime}\boldsymbol{V}_i^*(\tilde{\boldsymbol{\tau}}, \widehat{\boldsymbol{\nu}}(\tilde{\boldsymbol{\tau}}))^{-1}\tilde{\boldsymbol{Y}}_i^*.$$

Continue to iterate between Steps 1-2 by updating $\tilde{\boldsymbol{\tau}} = \widehat{\boldsymbol{\tau}}(\widehat{\boldsymbol{\nu}})$ in Step 1 and using the EGLS estimate in Step 2 to obtain an updated estimate of $\boldsymbol{\tau}$. This process is repeated until such time that successive differences in updated estimates is negligible.

This algorithm corresponds to solving the set of GEE2

$$\boldsymbol{U}(\boldsymbol{\tau}) = \sum_{i=1}^n \boldsymbol{X}_i^*(\boldsymbol{\tau})'\boldsymbol{V}_i^*(\boldsymbol{\tau}, \widehat{\boldsymbol{\nu}}(\boldsymbol{\tau}))^{-1}(\boldsymbol{Y}_i^* - \boldsymbol{F}_i^*(\boldsymbol{\tau})) = \boldsymbol{0} \tag{4.40}$$

where $\boldsymbol{X}_i^*(\boldsymbol{\tau}) = \partial \boldsymbol{F}_i^*(\boldsymbol{\tau})/\partial \boldsymbol{\tau}$ and $\widehat{\boldsymbol{\nu}}(\boldsymbol{\tau})$ is a consistent estimate of $\boldsymbol{\nu}$ given $\boldsymbol{\tau}$. As shown in the previous section, this will be equivalent to solving the set of GEE2

$$\boldsymbol{U}(\boldsymbol{\tau}) = \sum_{i=1}^{n} \boldsymbol{X}_i(\boldsymbol{\tau})'\boldsymbol{V}_i(\boldsymbol{\tau}, \widehat{\boldsymbol{\nu}}(\boldsymbol{\tau}))^{-1}(\boldsymbol{Y}_i - \boldsymbol{F}_i(\boldsymbol{\tau})) = \boldsymbol{0} \tag{4.41}$$

obtained when one specifies the "working" third- and fourth-order moments based on $\boldsymbol{y}_i$ and $\boldsymbol{s}_i$ [i.e., $Cov(\boldsymbol{y}_i, \boldsymbol{s}_i)$ and $Var(\boldsymbol{s}_i)$] rather than $\boldsymbol{y}_i$ and $\boldsymbol{w}_i$.

Since unbiased inference under GEE2 only requires correct specification of the first two moments of $\boldsymbol{y}_i$, one can specify "working" third- and fourth-order moments assuming normality as suggested by Prentice and Zhao (1991) and Vonesh et. al. (2001). Under this "working" assumption, the resulting GEE2 estimator of $\boldsymbol{\tau} = (\boldsymbol{\beta}, \boldsymbol{\theta})$ is obtained by setting $\boldsymbol{\Upsilon}_i = \boldsymbol{0}$ and $\boldsymbol{\Gamma}_i = \boldsymbol{B}_i[\boldsymbol{I}_{p_i^2} + \boldsymbol{I}_{(p_i,p_i)}][\boldsymbol{\Sigma}_i(\boldsymbol{\beta}, \boldsymbol{\theta}) \otimes \boldsymbol{\Sigma}_i(\boldsymbol{\beta}, \boldsymbol{\theta})]\boldsymbol{B}_i'$ as defined for the NLM in (4.37). In this case, the GEE2 estimator will remain consistent and asymptotically normally distributed even if $\boldsymbol{y}_i$ is not normally distributed. One explanation for this is that these are the second-order estimating equations of an extended least squares (ELS) estimator obtained by minimizing the objective function

$$Q_{\text{ELS}}(\boldsymbol{\beta}, \boldsymbol{\theta}) = \sum_{i=1}^{n} \Big\{ (\boldsymbol{y}_i - \boldsymbol{\mu}_i(\boldsymbol{\beta}))'\boldsymbol{\Sigma}_i(\boldsymbol{\beta}, \boldsymbol{\theta})^{-1}(\boldsymbol{y}_i - \boldsymbol{\mu}_i(\boldsymbol{\beta})) + \log|\boldsymbol{\Sigma}_i(\boldsymbol{\beta}, \boldsymbol{\theta})| \Big\}. \tag{4.42}$$

Here the term extended least squares, as first used by Beal and Sheiner (1982, 1988), reflects the fact that the objective function (4.42) extends the usual generalized least square (GLS) objective function by including a "penalty" term, $\log|\boldsymbol{\Sigma}_i(\boldsymbol{\beta}, \boldsymbol{\theta})|$. Such a "penalty" term is required to ensure the estimating equations required to minimize a GLS-based objective function are unbiased (Vonesh et. al., 2001). Although this ELS estimator is equivalent to the ML estimator under normality, it may also be classified as a least means difference estimator (Gallant, 1987) and does not require the assumption of normality in order for it to achieve consistency and asymptotically normality (Gourieroux et. al., 1984; Gallant, 1987; Prentice and Zhao, 1991). Conceptually, the GEE2/ELS estimator is similar to the GEE1/PL/QELS estimator that GLIMMIX uses to estimate the parameters from a GLIM. However, because it directly takes into account the assumed dependency the marginal covariance structure of $\boldsymbol{y}_i$ may have on $\boldsymbol{\beta}$, the GEE2/ELS estimator can result in an inconsistent estimate of $\boldsymbol{\beta}$ when $\boldsymbol{\Sigma}_i(\boldsymbol{\beta}, \boldsymbol{\theta})$ is misspecified.

Alternative methods of estimation

The ML and GEE2/ELS estimation techniques just described utilize information about $\boldsymbol{\beta}$ within the assumed marginal covariance matrix of $\boldsymbol{y}_i$ to improve on the efficiency of GEE1-type estimators. One can, however, estimate $\boldsymbol{\beta}$ and $\boldsymbol{\theta}$ under the GNLM (4.29) using alternative least squares (LS) or GEE1 methods such as estimated generalized least squares (EGLS), or a GEE1/PL/QELS approach like that used in GLIMMIX.

If we treat $\boldsymbol{\theta}$ as a nuisance parameter to be estimated via method-of-moments (MM), an EGLS estimate of $\boldsymbol{\beta}$ may be obtained by first computing a consistent estimate of $\boldsymbol{\beta}$ by OLS, say $\tilde{\boldsymbol{\beta}}$, and based on this estimate, compute a MM estimate $\tilde{\boldsymbol{\theta}}$ of $\boldsymbol{\theta}$. We then minimize the EGLS objective function

$$Q_{\text{EGLS}}(\boldsymbol{\beta}) = \sum_{i=1}^{n} \Big\{ (\boldsymbol{y}_i - \boldsymbol{\mu}_i(\boldsymbol{\beta}))'\boldsymbol{\Sigma}_i(\tilde{\boldsymbol{\beta}}, \tilde{\boldsymbol{\theta}})^{-1}(\boldsymbol{y}_i - \boldsymbol{\mu}_i(\boldsymbol{\beta})) \Big\} \tag{4.43}$$

obtained by holding the marginal covariance matrix fixed at these initial estimates. Although seldom used, the EGLS approach may be implemented in certain cases using the SAS macro %MIXNLIN (Vonesh and Chinchilli, 1997). For GNLM's where $Var(\boldsymbol{y}_i) = \boldsymbol{\Sigma}_i(\boldsymbol{\theta})$ does not depend on $\boldsymbol{\beta}$, an EGLS estimator of $\boldsymbol{\beta}$ may be computed via a two-step approach using the %NLINMIX macro or the NLMIXED procedure. However, as we shall demonstrate in the next chapter, the real value of EGLS lies in its ability to provide good starting values for more complex nonlinear mixed-effects models.

Alternatively, one can jointly estimate $\boldsymbol{\beta}$ and $\boldsymbol{\theta}$ by simply adapting the PL/QELS algorithm described in §4.2.1 to the current GNLM setting. For a GNLM in which $\boldsymbol{\Sigma}_i(\boldsymbol{\beta},\boldsymbol{\theta}) = \boldsymbol{\Sigma}_i(\boldsymbol{\theta})$, the resulting GEE1/PL/QELS estimator will be equivalent to the GEE2/ELS estimator and joint estimates of $\boldsymbol{\beta}$ and $\boldsymbol{\theta}$ may be obtained using the SAS macro %NLINMIX in combination with the REPEATED statement (rather than a RANDOM statement). An example of this is the analysis of the LDH enzyme leakage data presented in §4.5.1. In other cases where $Var(\boldsymbol{y}_i) = \boldsymbol{\Sigma}_i(\boldsymbol{\beta},\boldsymbol{\theta})$ does depend on $\boldsymbol{\beta}$, one may be able to express the GNLM as a random effects model which is strictly linear in the random effects. In this case, one can use %NLINMIX in combination with the RANDOM statement and the option EXPAND=ZERO to obtain GEE1/PL/QELS estimates of the model parameters. This is illustrated with the analysis of the orange tree data in §4.5.2.

In either case, under the GNLM, a PL/QELS based estimation scheme for $\boldsymbol{\tau} = (\boldsymbol{\beta},\boldsymbol{\theta})$ will correspond to solving the set of GEE2 (4.41) but with the derivative matrix set at $\boldsymbol{X}_i(\boldsymbol{\tau}) = \begin{pmatrix} \boldsymbol{D}_i(\boldsymbol{\beta}) & \boldsymbol{0} \\ \boldsymbol{0} & \boldsymbol{E}_i(\boldsymbol{\theta}) \end{pmatrix}$ and the variance-covariance matrix set at $\boldsymbol{V}_i(\boldsymbol{\tau},\boldsymbol{\nu}) = \boldsymbol{V}_i(\boldsymbol{\tau})$ $= \begin{pmatrix} \boldsymbol{\Sigma}_i(\boldsymbol{\beta},\boldsymbol{\theta}) & \boldsymbol{0} \\ \boldsymbol{0} & \boldsymbol{\Gamma}_i(\boldsymbol{\beta},\boldsymbol{\theta}) \end{pmatrix}$ with $\boldsymbol{\Gamma}_i(\boldsymbol{\beta},\boldsymbol{\theta})$ defined by (4.37). Because the matrices $\boldsymbol{X}_i(\boldsymbol{\tau})$ and $\boldsymbol{V}_i(\boldsymbol{\tau})$ are both set to be diagonal, the PL/QELS estimator coincides with a normal-theory pseudo-likelihood estimator of $(\boldsymbol{\beta},\boldsymbol{\theta})$ obtained by expressing the GNLM in terms of the NLM (4.32). It is a pseudo-likelihood estimator in that it ignores any dependence $Var(\boldsymbol{y}_i)$ may have on $\boldsymbol{\beta}$ just as is done for GLIM's using GLIMMIX. This results in a GEE1-type estimator of $\boldsymbol{\beta}$ and either a pseudo-likelihood (PL) estimator $\widehat{\boldsymbol{\theta}}_{\text{PL}}$ or a restricted pseudo-likelihood (REPL) estimator $\widehat{\boldsymbol{\theta}}_{\text{REPL}}$ of $\boldsymbol{\theta}$. This GEE1/PL/QELS approach to estimation under the GNLM can be implemented with the SAS macro %NLINMIX using the following doubly iterative PL/QELS algorithm (see the %NLINMIX syntax).

A Gauss-Newton algorithm for estimating $(\boldsymbol{\beta},\boldsymbol{\theta})$ via PL/QELS

Step 0: Specify starting values $\tilde{\boldsymbol{\beta}}$ of $\boldsymbol{\beta}$ (for example, by using the macro PARMS option of %NLINMIX) and, optionally, $\tilde{\boldsymbol{\theta}}$ of $\boldsymbol{\theta}$ (for example, by using the PARMS statement of PROC MIXED within the macro STMTS option of %NLINMIX).

Step 1: Given the current estimate $\tilde{\boldsymbol{\beta}}$, compute the pseudo-response vector,

$$\begin{aligned}\tilde{\boldsymbol{y}}_i &= \boldsymbol{y}_i - \boldsymbol{\mu}_i(\tilde{\boldsymbol{\beta}}) + \tilde{\boldsymbol{X}}_i\tilde{\boldsymbol{\beta}} \\ &= \boldsymbol{y}_i - f(\boldsymbol{X}_i,\tilde{\boldsymbol{\beta}}) + \tilde{\boldsymbol{X}}_i\tilde{\boldsymbol{\beta}}\end{aligned}$$

where $\tilde{\boldsymbol{X}}_i = \tilde{\boldsymbol{X}}_i(\tilde{\boldsymbol{\beta}}) = \partial f(\boldsymbol{X}_i,\boldsymbol{\beta})/\partial\boldsymbol{\beta}'\Big|_{\boldsymbol{\beta}=\tilde{\boldsymbol{\beta}}}$.

Step 2. Form the pseudo-marginal LM

$$\tilde{\boldsymbol{y}}_i = \tilde{\boldsymbol{X}}_i\boldsymbol{\beta} + \boldsymbol{\epsilon}_i$$

with a "working" assumption that $\boldsymbol{\epsilon}_i \sim N(\boldsymbol{0},\boldsymbol{\Sigma}_i(\tilde{\boldsymbol{\beta}},\boldsymbol{\theta}))$ and maximize wrt $\boldsymbol{\theta}$ either the profile log-likelihood (2.12) or restricted profile log-likelihood (2.13) function associated with this LM (see Chapter 2, §2.1.1). Let $\widehat{\boldsymbol{\theta}}$ be either the PL estimate, $\widehat{\boldsymbol{\theta}}_{\text{PL}} = \widehat{\boldsymbol{\theta}}_{\text{PL}}(\tilde{\boldsymbol{\beta}})$, or REPL estimate, $\widehat{\boldsymbol{\theta}}_{\text{REPL}} = \widehat{\boldsymbol{\theta}}_{\text{REPL}}(\tilde{\boldsymbol{\beta}})$, and compute the estimated generalized least squares (EGLS) estimator

$$\widehat{\boldsymbol{\beta}} = \widehat{\boldsymbol{\beta}}(\widehat{\boldsymbol{\theta}}) = \Big(\sum_{i=1}^{n}\tilde{\boldsymbol{X}}_i'\boldsymbol{\Sigma}_i(\tilde{\boldsymbol{\beta}},\widehat{\boldsymbol{\theta}}(\tilde{\boldsymbol{\beta}}))^{-1}\tilde{\boldsymbol{X}}_i\Big)^{-1}\sum_{i=1}^{n}\tilde{\boldsymbol{X}}_i'\boldsymbol{\Sigma}_i(\tilde{\boldsymbol{\beta}},\widehat{\boldsymbol{\theta}}(\tilde{\boldsymbol{\beta}}))^{-1}\tilde{\boldsymbol{y}}_i.$$

Continue to iterate between Steps 1-2 by updating $\tilde{\boldsymbol{\beta}} = \widehat{\boldsymbol{\beta}}(\widehat{\boldsymbol{\theta}})$ in Step 1 and repeating the iterative procedure in Step 2 to obtain updated joint estimates of $(\boldsymbol{\beta}, \boldsymbol{\theta})$. This process is repeated until such time that successive differences in updated estimates is negligible.

As noted previously, whenever $Var(\boldsymbol{y}_i) = \boldsymbol{\Sigma}_i(\boldsymbol{\theta})$ such that $\boldsymbol{E}_i(\boldsymbol{\beta}) = \boldsymbol{0}$, the PL/QELS algorithm implemented in %NLINMIX will be equivalent to the Gauss-Newton GEE2 algorithm under the "working" assumption of third- and fourth-order Gaussian moments for $\boldsymbol{y}_i$. In this case, the PL/QELS estimator of $\boldsymbol{\tau} = (\boldsymbol{\beta}, \boldsymbol{\theta})$ solves a set of joint "independent" estimating equations which are GEE2. More generally, this algorithm is equivalent to the PL/QELS algorithm implemented in GLIMMIX for GLIM's (see page 113).

Summary

The use of GEE, MLE, GEE1/PL/QELS and GEE2/ELS provides users with an array of estimators from which to choose when estimating the parameters from either a GLIM or GNLM. It can be shown that by using the basic model formulation shown in (4.34) but with $\boldsymbol{V}_i(\boldsymbol{\tau})$ replaced by the more general $\boldsymbol{V}_i(\boldsymbol{\tau}, \boldsymbol{\nu})$, one can retrieve any of the estimators discussed in this chapter through proper specification of $\boldsymbol{Y}_i$, $\boldsymbol{F}_i(\boldsymbol{\tau})$, $\boldsymbol{\tau}$, $\boldsymbol{V}_i(\boldsymbol{\tau}, \boldsymbol{\nu})$, $\widehat{\boldsymbol{\nu}}(\boldsymbol{\tau})$ and $\boldsymbol{X}_i(\boldsymbol{\tau})$. To better understand the relationship between these estimators and the SAS procedures used to obtain them, Table 4.4 displays the different estimators presented in this chapter along with the key model specifications. Of course not all SAS procedures and macros share the same options and flexibility and in some applications, one might need to perform additional programming in order to implement a particular method. For example, estimation based on GEE2 requires specification of the third- and fourth-order moments of $\boldsymbol{y}_i$ while ML estimation requires specification of the marginal log-likelihood function. Under the assumption of "working normality" the former is easily accomplished using the NLMIXED procedure as shown for the respiratory disorder data (§4.5.3) and the epileptic seizure data (§4.5.4). In the latter case, special programming would be required as is illustrated for the epileptic seizure data assuming a multivariate Poisson distribution (§4.5.4). The examples in §4.5 illustrate how one can implement GEE1, GEE2 and ML estimation using currently available tools within SAS.

4.4.3 Inference and test statistics

Inference under the GNLM essentially mimics that of the GLIM except for two key distinctions. First, under the GNLM, a more deliberate attempt is made to allow for joint inference on the mean and covariance parameters $\boldsymbol{\tau} = (\boldsymbol{\beta}, \boldsymbol{\theta})$. Second, given its more general nonlinear framework, inference under the GNLM may target both linear and nonlinear hypotheses of the regression parameters $\boldsymbol{\beta}$. With regards to *joint* inference, since estimation of $(\boldsymbol{\beta}, \boldsymbol{\theta})$ may be based on either ML, GEE2/ELS or GEE1/PL/QELS, one need consider inference under these methods separately. Assuming the distribution of $\boldsymbol{y}_i$ is in the quadratic exponential family, all three methods entail solving a set of joint estimating equations of the form (4.41) but with different specifications of $\boldsymbol{X}_i(\boldsymbol{\tau})$ and $\boldsymbol{V}_i(\boldsymbol{\tau}, \boldsymbol{\nu})$ as indicated in Table 4.4. Consequently, a general theory of inference exists for any estimator $\widehat{\boldsymbol{\tau}} = (\widehat{\boldsymbol{\beta}}, \widehat{\boldsymbol{\theta}})$ that is a solution to a set of joint estimating equations (4.41). Specific inference may be recovered by simply substituting in the correct specifications of $\boldsymbol{X}_i(\boldsymbol{\tau})$ and $\boldsymbol{V}_i(\boldsymbol{\tau}, \boldsymbol{\nu})$.

With respect to SAS, the procedure GLIMMIX and the macro %NLINMIX "jointly" estimate $(\boldsymbol{\beta}, \boldsymbol{\theta})$ using PL/QELS which essentially entails solving a set of independent estimating equations for $\boldsymbol{\beta}$ and $\boldsymbol{\theta}$ separately. Perhaps as a consequence, joint inference in terms of hypothesis testing is not possible within GLIMMIX or %NLINMIX in that tests of hypotheses are carried out separately for the regression parameters $\boldsymbol{\beta}$ (via a CONTRAST

Table 4.4 Estimation techniques available in SAS for GLIM's and GNLM's. [1.] GENMOD and GLIMMIX are restricted to GLIM's. [2.] The nuisance parameter $\boldsymbol{\nu}$, if present, is estimated by any $\sqrt{n}$-consistent estimator $\widehat{\boldsymbol{\nu}}(\boldsymbol{\tau})$ given $\boldsymbol{\tau}$. [3.] Both EGLS and GEE treat $\boldsymbol{\theta}$ as a nuisance parameter to be estimated via MM. [4.] Both GEE1 and GEE2 jointly estimate $\boldsymbol{\beta}$ and $\boldsymbol{\theta}$ under a "working normality" assumption, i.e., $\boldsymbol{\Gamma}_i = \boldsymbol{B}_i[\boldsymbol{I}_{p_i^2} + \boldsymbol{I}_{(p_i,p_i)}](\boldsymbol{\Sigma}_i(\boldsymbol{\beta},\boldsymbol{\theta}) \otimes \boldsymbol{\Sigma}_i(\boldsymbol{\beta},\boldsymbol{\theta}))\boldsymbol{B}_i'$. GEE1 uses PL/QELS to jointly estimate $(\boldsymbol{\beta},\boldsymbol{\theta})$ while GEE2 uses ELS. [5.] MLE, in general, requires specification of third- and fourth-order moments or specification of the marginal log-likelihood. In special cases, it may be possible to program MLE using NLMIXED.

Method	SAS PROC [1.]	$\boldsymbol{Y}_i$	$\boldsymbol{F}_i(\boldsymbol{\tau})$	$\boldsymbol{\tau}$	$\boldsymbol{V}_i(\boldsymbol{\tau},\widehat{\boldsymbol{\nu}})$	$\widehat{\boldsymbol{\nu}}(\boldsymbol{\tau})$ [2.]	$\boldsymbol{X}_i(\boldsymbol{\tau})$
OLS	NLIN	$\boldsymbol{y}_i$	$\boldsymbol{\mu}_i(\boldsymbol{\beta})$	$\boldsymbol{\beta}$	$\widehat{\sigma}^2\boldsymbol{I}_{p_i}$	$\widehat{\sigma}^2$	$\partial\boldsymbol{\mu}_i/\partial\boldsymbol{\beta}'$
EGLS [3.]	NLMIXED %NLINMIX	$\boldsymbol{y}_i$	$\boldsymbol{\mu}_i(\boldsymbol{\beta})$	$\boldsymbol{\beta}$	$\boldsymbol{\Sigma}_i(\widehat{\boldsymbol{\theta}})$	$\widehat{\boldsymbol{\theta}}(\boldsymbol{\beta})$	$\partial\boldsymbol{\mu}_i/\partial\boldsymbol{\beta}'$
GEE [3.]	GENMOD	$\boldsymbol{y}_i$	$\boldsymbol{\mu}_i(\boldsymbol{\beta})$	$\boldsymbol{\beta}$	$\boldsymbol{\Sigma}_i(\boldsymbol{\beta},\widehat{\boldsymbol{\theta}}(\boldsymbol{\beta}))$	$\widehat{\boldsymbol{\theta}}(\boldsymbol{\beta})$	$\partial\boldsymbol{\mu}_i/\partial\boldsymbol{\beta}'$
GEE1 [4.]	GLIMMIX %NLINMIX	$\begin{pmatrix}\boldsymbol{y}_i\\ \boldsymbol{s}_i\end{pmatrix}$	$\begin{pmatrix}\boldsymbol{\mu}_i(\boldsymbol{\beta})\\ \boldsymbol{\sigma}_i(\boldsymbol{\beta},\boldsymbol{\theta})\end{pmatrix}$	$(\boldsymbol{\beta},\boldsymbol{\theta})$	$\begin{pmatrix}\boldsymbol{\Sigma}_i & \boldsymbol{0}\\ \boldsymbol{0}' & \boldsymbol{\Gamma}_i\end{pmatrix}$	–	$\begin{pmatrix}\frac{\partial\boldsymbol{\mu}_i}{\partial\boldsymbol{\beta}'} & \boldsymbol{0}\\ \boldsymbol{0}' & \frac{\partial\boldsymbol{\sigma}_i}{\partial\boldsymbol{\theta}'}\end{pmatrix}$
GEE2 [4.]	NLMIXED %NLINMIX	$\begin{pmatrix}\boldsymbol{y}_i\\ \boldsymbol{s}_i\end{pmatrix}$	$\begin{pmatrix}\boldsymbol{\mu}_i(\boldsymbol{\beta})\\ \boldsymbol{\sigma}_i(\boldsymbol{\beta},\boldsymbol{\theta})\end{pmatrix}$	$(\boldsymbol{\beta},\boldsymbol{\theta})$	$\begin{pmatrix}\boldsymbol{\Sigma}_i & \boldsymbol{0}\\ \boldsymbol{0}' & \boldsymbol{\Gamma}_i\end{pmatrix}$	–	$\begin{pmatrix}\frac{\partial\boldsymbol{\mu}_i}{\partial\boldsymbol{\beta}'} & \boldsymbol{0}\\ \frac{\partial\boldsymbol{\sigma}_i}{\partial\boldsymbol{\beta}'} & \frac{\partial\boldsymbol{\sigma}_i}{\partial\boldsymbol{\theta}'}\end{pmatrix}$
MLE [5.]	NLMIXED	$\begin{pmatrix}\boldsymbol{y}_i\\ \boldsymbol{s}_i\end{pmatrix}$	$\begin{pmatrix}\boldsymbol{\mu}_i(\boldsymbol{\beta})\\ \boldsymbol{\sigma}_i(\boldsymbol{\beta},\boldsymbol{\theta})\end{pmatrix}$	$(\boldsymbol{\beta},\boldsymbol{\theta})$	$\begin{pmatrix}\boldsymbol{\Sigma}_i & \boldsymbol{\Upsilon}_i\\ \boldsymbol{\Upsilon}_i' & \boldsymbol{\Gamma}_i\end{pmatrix}$	–	$\begin{pmatrix}\frac{\partial\boldsymbol{\mu}_i}{\partial\boldsymbol{\beta}'} & \boldsymbol{0}\\ \frac{\partial\boldsymbol{\sigma}_i}{\partial\boldsymbol{\beta}'} & \frac{\partial\boldsymbol{\sigma}_i}{\partial\boldsymbol{\theta}'}\end{pmatrix}$

statement within GLIMMIX or a CONTRAST statement under MIXED as an option within %NLINMIX) and the covariance parameters $\boldsymbol{\theta}$ (via the COVTEST statement in GLIMMIX or the COVTEST option in MIXED). Moreover the CONTRAST statements used to test components of $\boldsymbol{\beta}$ within GLIMMIX and %NLINMIX are restricted to comparisons based on the general linear hypothesis, H_0: $\boldsymbol{L\beta} = \boldsymbol{0}$, as described in §2.1.2 and §4.2.2. Alternatively, the procedure NLMIXED jointly estimates $(\boldsymbol{\beta},\boldsymbol{\theta})$ using either ML estimation or GEE2/ELS. Consequently joint inference is possible in so far as the CONTRAST statement of NLMIXED allows one to specify general nonlinear hypotheses that involve components of both $\boldsymbol{\beta}$ and $\boldsymbol{\theta}$. Since inferential techniques based on PL/QELS have already been described in detail for LM's and GLIM's in §2.1.2 and §4.2.2 and since these techniques also apply to the GNLM and are easily implemented within %NLINMIX through optional CONTRAST and ESTIMATE statements (see the STMTS macro option of %NLINMIX), we shall focus our attention here on the types of inference available within NLMIXED when applied to the GNLM.

Joint inference on $(\boldsymbol{\beta},\boldsymbol{\theta})$

Assume the nuisance parameter $\boldsymbol{\nu}$, when present, is either known or may be replaced by some consistent estimator $\widehat{\boldsymbol{\nu}}(\boldsymbol{\tau})$. Let $\widehat{\boldsymbol{\tau}} = (\widehat{\boldsymbol{\beta}},\widehat{\boldsymbol{\theta}})$ be any solution to the set of estimating equations (4.41) that entails specifications of $\boldsymbol{X}_i(\boldsymbol{\tau})$ and $\boldsymbol{V}_i(\boldsymbol{\tau},\boldsymbol{\nu})$ consistent with either the GEE1 or GEE2 approach as detailed in Table 4.4. Based on Fisher's scoring algorithm, one can choose between model-based inference using the inverse Fisher information matrix to estimate the asymptotic variance-covariance of $\widehat{\boldsymbol{\tau}}$ or robust inference using an empirical sandwich estimator of the asymptotic variance-covariance of $\widehat{\boldsymbol{\tau}}$.

Under the GNLM, valid model-based inference, whether it be based on ML, GEE1/PL/QELS, or GEE2/ELS, requires that the first four moments of the response vector $\boldsymbol{y}_i$ be correctly specified. When this assumption is satisfied along with other regularity conditions (e.g., Gourieroux et. al., 1984; and Prentice and Zhao, 1991), inference may be based on the following asymptotic results:

Model-based inference (correct 1st-4th order-moments): (4.44)

$$\sqrt{n}(\widehat{\boldsymbol{\tau}}-\boldsymbol{\tau})\xrightarrow{p}\mathrm{N}(\mathbf{0},\boldsymbol{\Omega}_0)$$

$$Var(\widehat{\boldsymbol{\tau}})\simeq\widehat{\boldsymbol{\Omega}}(\widehat{\boldsymbol{\tau}})=\Big(\sum_{i=1}^{n}\boldsymbol{X}_i(\widehat{\boldsymbol{\tau}})'\boldsymbol{V}_i(\widehat{\boldsymbol{\tau}},\widehat{\boldsymbol{\nu}}(\widehat{\boldsymbol{\tau}}))^{-1}\boldsymbol{X}_i(\widehat{\boldsymbol{\tau}})\Big)^{-1}$$

$$n\widehat{\boldsymbol{\Omega}}(\widehat{\boldsymbol{\tau}})\xrightarrow{p}\boldsymbol{\Omega}_0=\lim_{n\to\infty}\Big\{n\Big(\sum_{i=1}^{n}\boldsymbol{X}_i(\boldsymbol{\tau})'\boldsymbol{V}_i(\boldsymbol{\tau},\boldsymbol{\nu})^{-1}\boldsymbol{X}_i(\boldsymbol{\tau})\Big)^{-1}\Big\}.$$

These results simply state that when $\boldsymbol{\mu}_i(\boldsymbol{\beta})$, $\boldsymbol{\Sigma}_i(\boldsymbol{\beta},\boldsymbol{\theta})$ and $\boldsymbol{V}_i(\boldsymbol{\tau},\boldsymbol{\nu})$ are all correctly specified, the joint estimator $\widehat{\boldsymbol{\tau}}=(\widehat{\boldsymbol{\beta}},\widehat{\boldsymbol{\theta}})$ will be a consistent asymptotically normal (CAN) estimator of $\boldsymbol{\tau}=(\boldsymbol{\beta},\boldsymbol{\theta})$. If, in addition, the pdf of $\boldsymbol{y}_i$ $(i=1,\ldots,n)$ is in the quadratic exponential family and $\widehat{\boldsymbol{\tau}}$ is the GEE2 estimator, then $\widehat{\boldsymbol{\tau}}$ will also be the ML estimator in which case $\widehat{\boldsymbol{\tau}}$ is fully asymptotically efficient. Under these conditions, valid large sample tests of hypotheses and confidence intervals may be formulated using the model-based estimated variance-covariance matrix $\widehat{\boldsymbol{\Omega}}(\widehat{\boldsymbol{\tau}})$.

Since the assumption that all four moments are correctly specified is extremely difficult to verify in practice, preference is usually given to robust inference in which the model-based covariance estimator $\widehat{\boldsymbol{\Omega}}(\widehat{\boldsymbol{\tau}})$ is replaced by an empirical sandwich estimator $\widehat{\boldsymbol{\Omega}}_R(\widehat{\boldsymbol{\tau}})$ that only requires the first two moments of $\boldsymbol{y}_i$ to be correctly specified. Under this assumption, namely that $E(\boldsymbol{y}_i)=\boldsymbol{\mu}_i(\boldsymbol{\beta})$ and $Var(\boldsymbol{y}_i)=\boldsymbol{\Sigma}_i(\boldsymbol{\beta},\boldsymbol{\theta})$ have been correctly specified, Prentice and Zhao (1991) showed that valid robust inference may be based on the following results:

Robust inference (correct 1st-2nd order moments) : (4.45)

$$\sqrt{n}(\widehat{\boldsymbol{\tau}}-\boldsymbol{\tau})\xrightarrow{p}\mathrm{N}(\mathbf{0},\boldsymbol{\Omega}_1)$$

$$Var(\widehat{\boldsymbol{\tau}})\simeq\widehat{\boldsymbol{\Omega}}_R(\widehat{\boldsymbol{\tau}})=\widehat{\boldsymbol{\Omega}}(\widehat{\boldsymbol{\tau}})\Big(\sum_{i=1}^{n}\boldsymbol{U}_i(\widehat{\boldsymbol{\tau}})\boldsymbol{U}_i(\widehat{\boldsymbol{\tau}})'\Big)^{-1}\widehat{\boldsymbol{\Omega}}(\widehat{\boldsymbol{\tau}})$$

$$n\widehat{\boldsymbol{\Omega}}_R(\widehat{\boldsymbol{\tau}})\xrightarrow{p}\boldsymbol{\Omega}_1=\lim_{n\to\infty}\Big\{n\Big(\sum_{i=1}^{n}\boldsymbol{X}_i(\boldsymbol{\tau})'\boldsymbol{V}_i^{-1}Var(\boldsymbol{Y}_i)\boldsymbol{V}_i^{-1}\boldsymbol{X}_i(\boldsymbol{\tau})\Big)^{-1}\Big\}$$

where $\boldsymbol{U}_i(\widehat{\boldsymbol{\tau}})=\boldsymbol{X}_i(\widehat{\boldsymbol{\tau}})'\boldsymbol{V}_i(\widehat{\boldsymbol{\tau}},\widehat{\boldsymbol{\nu}}(\widehat{\boldsymbol{\tau}}))^{-1}(\boldsymbol{Y}_i-\boldsymbol{F}_i(\widehat{\boldsymbol{\tau}}))$ and $\boldsymbol{V}_i=\boldsymbol{V}_i(\boldsymbol{\tau},\boldsymbol{\nu})$. These results indicate that even when the third- and fourth-order moments, $\boldsymbol{\Upsilon}_i$ and $\boldsymbol{\Gamma}_i$, of $\boldsymbol{V}_i(\boldsymbol{\tau},\boldsymbol{\nu})$ are misspecified, the estimator $\widehat{\boldsymbol{\tau}}$ will remain a consistent asymptotically normal (CAN) estimator of $\boldsymbol{\tau}$. Note that if all four moments are correctly specified, then $\boldsymbol{\Omega}_0=\boldsymbol{\Omega}_1$ and $\widehat{\boldsymbol{\tau}}$ will remain asymptotically efficient provided $\boldsymbol{y}_i$ is in the quadratic exponential family.

Regardless of whether one bases inference on the model-based covariance estimator, $\widehat{\boldsymbol{\Omega}}$, or the robust empirical sandwich estimator, $\widehat{\boldsymbol{\Omega}}_R$, it will be convenient to express these matrices in partitioned form as follows:

$$\widehat{\boldsymbol{\Omega}}=\widehat{\boldsymbol{\Omega}}(\widehat{\boldsymbol{\tau}})=\begin{pmatrix}\widehat{\boldsymbol{\Omega}}(\widehat{\boldsymbol{\beta}}) & \widehat{\boldsymbol{C}}(\widehat{\boldsymbol{\beta}},\widehat{\boldsymbol{\theta}})\\ \widehat{\boldsymbol{C}}(\widehat{\boldsymbol{\beta}},\widehat{\boldsymbol{\theta}})' & \widehat{\boldsymbol{\Omega}}(\widehat{\boldsymbol{\theta}})\end{pmatrix} \tag{4.46}$$

$$\widehat{\boldsymbol{\Omega}}_R=\widehat{\boldsymbol{\Omega}}_R(\widehat{\boldsymbol{\tau}})=\begin{pmatrix}\widehat{\boldsymbol{\Omega}}_R(\widehat{\boldsymbol{\beta}}) & \widehat{\boldsymbol{C}}_R(\widehat{\boldsymbol{\beta}},\widehat{\boldsymbol{\theta}})\\ \widehat{\boldsymbol{C}}_R(\widehat{\boldsymbol{\beta}},\widehat{\boldsymbol{\theta}})' & \widehat{\boldsymbol{\Omega}}_R(\widehat{\boldsymbol{\theta}})\end{pmatrix}$$

where $\widehat{\boldsymbol{\Omega}}(\widehat{\boldsymbol{\beta}})$ and $\widehat{\boldsymbol{\Omega}}_R(\widehat{\boldsymbol{\beta}})$ are, respectively, the model-based and robust variance-covariance matrices of $\widehat{\boldsymbol{\beta}}$ (i.e., $Cov(\widehat{\beta}_l, \widehat{\beta}_m)$; $l, m = 1, \ldots, s$); $\widehat{\boldsymbol{\Omega}}(\widehat{\boldsymbol{\theta}})$ and $\widehat{\boldsymbol{\Omega}}_R(\widehat{\boldsymbol{\theta}})$ are the corresponding estimated variance-covariance matrices of $\widehat{\boldsymbol{\theta}}$ (i.e., $Cov(\widehat{\theta}_l, \widehat{\theta}_m); l, m = 1, \ldots, d$); and $\widehat{\boldsymbol{C}}(\widehat{\boldsymbol{\beta}}, \widehat{\boldsymbol{\theta}})$ and $\widehat{\boldsymbol{C}}_R(\widehat{\boldsymbol{\beta}}, \widehat{\boldsymbol{\theta}})$ are the corresponding estimated covariances of the crossed components of $\widehat{\boldsymbol{\beta}}$ and $\widehat{\boldsymbol{\theta}}$ (i.e., $Cov(\widehat{\beta}_l, \widehat{\theta}_m)$; $l = 1, \ldots, s$; $m = 1, \ldots, d$). Note that although we write the partitioned covariance matrices of $\widehat{\boldsymbol{\beta}}$ and $\widehat{\boldsymbol{\theta}}$ as $\widehat{\boldsymbol{\Omega}}(\widehat{\boldsymbol{\beta}})$ and $\widehat{\boldsymbol{\Omega}}(\widehat{\boldsymbol{\theta}})$, respectively, this does not mean these sub-matrices do not also depend on other elements of the joint vector $\widehat{\boldsymbol{\tau}}$. This notation is used simply to designate what the variance-covariance matrix of the sub-component $\widehat{\boldsymbol{\beta}}$ of $\widehat{\boldsymbol{\tau}}$ is or what the variance-covariance matrix of the sub-component $\widehat{\boldsymbol{\theta}}$ of $\widehat{\boldsymbol{\tau}}$ is. The specific forms that these partitioned matrices take will depend on both the SAS procedure one uses and estimation scheme specified within the procedure.

Under the GNLM, joint inference in terms of hypothesis testing will typically take on a more general structure than what we find for linear models. In the most general case, inference with respect to the $(s+d) \times 1$ parameter vector $\boldsymbol{\tau} = (\boldsymbol{\beta}, \boldsymbol{\theta})$ will entail testing general nonlinear hypotheses of the form

$$H_0 : \boldsymbol{l}(\boldsymbol{\tau}) = \mathbf{0} \text{ versus } H_1 : \boldsymbol{l}(\boldsymbol{\tau}) \neq \mathbf{0} \tag{4.47}$$

where $\boldsymbol{l}(\boldsymbol{\tau})$ is a once continuously differentiable vector-valued function of order $r \times 1$ $(r \leq s+d)$. If we let $\boldsymbol{L}(\boldsymbol{\tau}) = \partial \boldsymbol{l}(\boldsymbol{\tau})/\partial \boldsymbol{\tau}'$ be the $r \times (s+d)$ derivative matrix of $\boldsymbol{l}(\boldsymbol{\tau})$, a robust test of the general nonlinear hypothesis (4.47) may be based on the Wald chi-square test statistic

$$T^2(\widehat{\boldsymbol{\tau}}) = \boldsymbol{l}(\widehat{\boldsymbol{\tau}})' \left\{ \boldsymbol{L}(\widehat{\boldsymbol{\tau}}) \widehat{\boldsymbol{\Omega}}_R(\widehat{\boldsymbol{\tau}}) \boldsymbol{L}(\widehat{\boldsymbol{\tau}})' \right\}^{-1} \boldsymbol{l}(\widehat{\boldsymbol{\tau}}) \tag{4.48}$$

in conjunction with the robust covariance matrix $\widehat{\boldsymbol{\Omega}}_R(\widehat{\boldsymbol{\tau}})$. Alternatively, one could replace $\widehat{\boldsymbol{\Omega}}_R(\widehat{\boldsymbol{\tau}})$ with the model-based covariance matrix $\widehat{\boldsymbol{\Omega}}(\widehat{\boldsymbol{\tau}})$. In either case, $T^2(\widehat{\boldsymbol{\tau}})$ will have an approximate chi-square distribution with r degrees of freedom provided the appropriate assumptions are met. A more conservative test would be one based on an F-approximation of the form

$$F(\widehat{\boldsymbol{\tau}}) = T^2(\widehat{\boldsymbol{\tau}})/r \tag{4.49}$$

which would be evaluated against the F distribution with numerator degrees of freedom r and denominator degrees of freedom v_e where v_e is an error degrees of freedom that one must specify. Currently only the NLMIXED procedure supports tests of the general nonlinear hypothesis (4.47). It uses a default denominator degrees of freedom of $v_e = n - v$ where n is the number of subjects and v is the number of random effects or $v_e = N$ where N is the total number of observations when no random effects are specified.

Typically, inference will target either functions of the regression coefficients $\boldsymbol{\beta}$ or functions of the variance-covariance parameters $\boldsymbol{\theta}$ but seldom on joint functions of both. For example, inference with respect to the regression parameters $\boldsymbol{\beta}$ would entail testing general nonlinear hypotheses of the form

$$H_0 : \boldsymbol{l}(\boldsymbol{\beta}) = \mathbf{0} \text{ versus } H_1 : \boldsymbol{l}(\boldsymbol{\beta}) \neq \mathbf{0} \tag{4.50}$$

where $\boldsymbol{l}(\boldsymbol{\beta})$ is a once continuously differentiable vector-valued function of order $r \times 1$ $(r \leq s)$. While in some applications $\boldsymbol{l}(\boldsymbol{\beta})$ will assume the familiar linear form $\boldsymbol{l}(\boldsymbol{\beta}) = \boldsymbol{L}\boldsymbol{\beta}$, there will be many applications where $\boldsymbol{l}(\boldsymbol{\beta})$ is a nonlinear function of $\boldsymbol{\beta}$ such as in pharmacokinetic applications where $\boldsymbol{l}(\boldsymbol{\beta})$ may represent the area under a curve, a peak concentration value or the biological half-life of a drug. In such cases, the test would be carried out using the Wald chi-square test statistic (4.48) or the F-statistic (4.49) but with

$\boldsymbol{\tau}$ replaced by $\boldsymbol{\beta}$. In cases where $\boldsymbol{l}(\boldsymbol{\beta})$ is an estimable scalar function of $\boldsymbol{\beta}$, inference may also be based on confidence interval coverage of the form

$$\boldsymbol{l}(\widehat{\boldsymbol{\beta}}) \pm t_{\alpha/2}(v_e)\sqrt{\boldsymbol{L}(\widehat{\boldsymbol{\beta}})\widehat{\boldsymbol{\Omega}}_R(\widehat{\boldsymbol{\beta}})\boldsymbol{L}(\widehat{\boldsymbol{\beta}})'} \tag{4.51}$$

where, again, the user specifies an appropriate degrees of freedom, v_e. Alternative model-based F-statistics that include different choices for v_e have been suggested by Gallant (1975, 1987) and Vonesh and Chinchilli (1997) for tests of hypotheses about $\boldsymbol{\beta}$.

It should be pointed out that inference carried out with NLMIXED will be based on Fisher's observed information matrix (i.e., the Hessian matrix) rather than Fisher's expected information matrix. This does not alter the asymptotic behavior of the estimates as described above. Moreover, there are a number of NLMIXED options that the user can specify to control covariance matrix tolerances and other features.

Model selection and goodness-of-fit

The number of tools available within NLMIXED for model selection and goodness-of-fit are currently limited to likelihood-based information criteria. The SAS macro %GOF described in Appendix D can be used with output from NLMIXED to provide alternative R-square type goodness-of-fit measures useful in assessing model fit and variable selection (see Vonesh et. al. 1996; Vonesh and Chinchilli, 1997).

4.5 Examples of GNLM's

In this section, we present several examples illustrating the use of NLMIXED and %NLINMIX to analyze data from a GNLM. We start by considering an extension of the GMANOVA model that allows for a nonlinear mean response and illustrate its utility with an application involving LDH enzyme leakage data. In this example, the variance-covariance matrix of $\boldsymbol{y}_i$ is of the form $\boldsymbol{\Sigma}_i(\boldsymbol{\theta})$ which does not depend on $\boldsymbol{\beta}$. In our second example, we consider the orange tree data of Draper and Smith (1981, p. 524) as analyzed by Lindstrom and Bates (1990) but within the context of a marginal nonlinear model. In this example, the variance-covariance matrix of $\boldsymbol{y}_i$ will depend on the unknown components of $\boldsymbol{\beta}$. In our third example, we revisit the respiratory disorder data of §4.3.2 by comparing binary logistic regression results obtained from our GEE analysis with results obtained using GEE2 in combination with "working" third- and fourth-order moments assuming normality (i.e., ELS). Finally, we fit the epileptic seizure data of §4.3.3 assuming both a "working" independence structure and a "working" exchangeable correlation structure. We do so using both the GEE approach as implemented in the GENMOD procedure and the GEE2 approach as implemented with ELS using NLMIXED. In addition, we illustrate how one can analyze this count data under a true multivariate Poisson model using either GEE2 or ML estimation.

4.5.1 LDH enzyme leakage data.

Gennings, Chinchilli and Carter (1989) analyzed a set of data from an in vitro toxicity study of isolated hepatocyte suspensions using a nonlinear GMANOVA. The data consists of the percentage of lactic dehydrogenase (LDH) enzyme leakage obtained as a response of hepatocyte cell toxicity to the effects of different combinations of carbon tetrachloride (CCl_4) and chloroform ($CHCl_3$) as determined from a 4×4 factorial design. Specific features and assumptions of the data include (see also the author's Web page):

- In vitro toxicity study of isolated hepatocyte suspensions
- The experimental unit is a flask of isolated hepatocyte suspensions

Figure 4.8 Mean profiles of proportion LDH leakage for the 16 groups formed by the various dose combinations of CCl4 and CHCl3

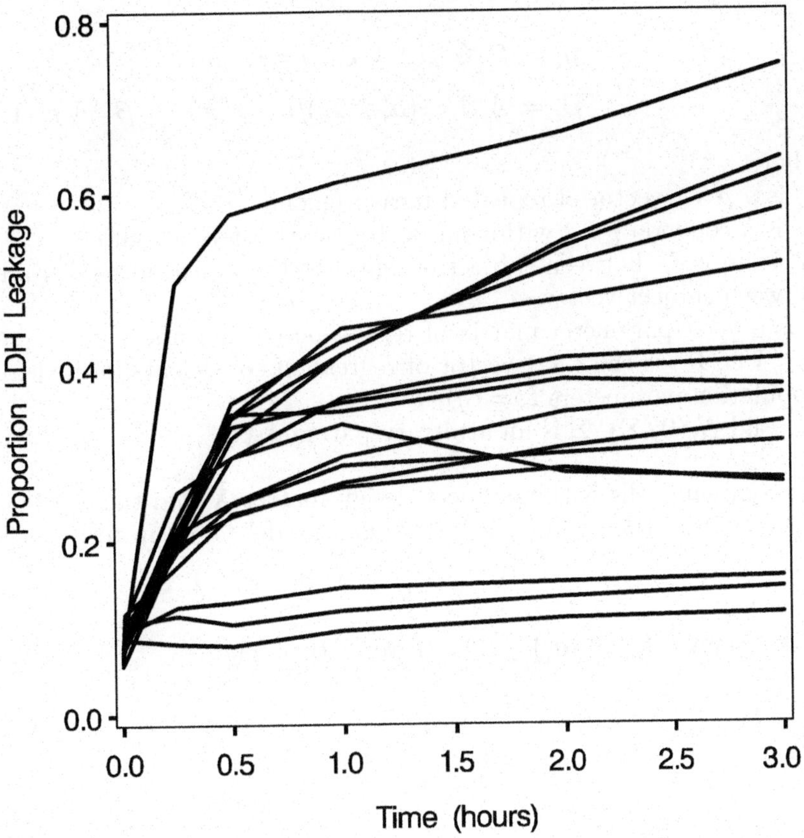

- Response variable:
  - y_{ij} = proportion of lactic dehydrogenase (LDH) enzyme leakage, a surrogate for cell toxicity, in the ith flask at j^{th} hour $(i = 1, \ldots, 64; j = 1, \ldots, 7)$.
- One within-unit covariate:
  - $t_{ij} = t_j$ =time 0, 0.01, 0.25, 0.50, 1.00, 2.00 and 3.00 hours $(j = 1, \ldots, 7)$
- Two between-unit covariates:
  - a_{i1}=carbon tetrachloride (CCl_4) at 0, 1, 2.5, 5 mM of CCl_4
  - a_{i2}=chloroform ($CHCl_3$) at 0, 5, 10, 25 mM of $CHCl_3$
- Structural Model: $\%\ LDH = 1 - \beta_0 \exp\Big(1 - \beta_1 t^{\beta_2}\Big)$
- Objective: Evaluate, via a nonlinear response surface analysis, the effects of CCl_4, $CHCl_3$ and their interactions on % LDH.

Shown in Figure 4.8 are the mean profiles of the proportion of LDH enzyme leakage according to the 16 groups formed by the 16 different combinations of carbon tetrachloride (CCl_4) and chloroform ($CHCl_3$). As suggested by the graph, the proportion LDH leakage appears to follow a sigmoidal growth curve response over time. Since the data are balanced and complete, Gennings, Chinchilli and Carter (1989) fit the data to a multivariate nonlinear logistic growth curve model which is a special case of a nonlinear GMANOVA model. The nonlinear GMANOVA model is itself a special case of the GNLM (4.29) as shown below.

Nonlinear GMANOVA Model

$$\begin{aligned} \boldsymbol{y}_i &= f(\boldsymbol{X}, \boldsymbol{\beta}_i) + \boldsymbol{\epsilon}_i; \;\; i=1,\ldots,n \\ \boldsymbol{\beta}_i &= \boldsymbol{A}_i\boldsymbol{\beta} = (\boldsymbol{a}_i' \otimes \boldsymbol{I}_r)\boldsymbol{\beta} = \boldsymbol{\Gamma}'\boldsymbol{a}_i \quad (\, \boldsymbol{\beta} = Vec(\boldsymbol{\Gamma}')\,) \end{aligned} \tag{4.52}$$

where
$\boldsymbol{y}_i$ is a $p \times 1$ vector of repeated measurements
$\boldsymbol{X}$ is a common $p \times t$ within-subject design matrix for all i
$\boldsymbol{A}_i$ is an $r \times s$ between-subject design matrix ($s = rq$) formed from a $q \times 1$ between-subject vector, $\boldsymbol{a}_i$
$\boldsymbol{\Gamma}$ is a $q \times r$ parameter matrix of regression coefficients
$\boldsymbol{\beta} = Vec(\boldsymbol{\Gamma}')$ is the $s \times 1$ vector of regression coefficients formed from the matrix of population parameters ($s = rq$)
$\boldsymbol{\epsilon}_i \sim$ iid $N_p(\boldsymbol{0}, \boldsymbol{\Sigma})$, $\boldsymbol{\Sigma}$ is an arbitrary p.d. matrix.

By combining the within-subject design matrix $\boldsymbol{X}$ and the $q \times 1$ between-subject vector, $\boldsymbol{a}_i$, we can write the nonlinear GMANOVA model (4.52) in terms of the GNLM

$$\boldsymbol{y}_i = f(\boldsymbol{X}_i, \boldsymbol{\beta}) + \boldsymbol{\epsilon}_i$$

with $\boldsymbol{X}_i = \begin{pmatrix}\boldsymbol{X} & \boldsymbol{1}_{p\times 1} \otimes \boldsymbol{a}_i'\end{pmatrix}_{p\times(t+q)}$. If we further set

$$\boldsymbol{Y}_{n\times p} = \begin{pmatrix}\boldsymbol{y}_1' \\ \vdots \\ \boldsymbol{y}_n'\end{pmatrix}, \quad \boldsymbol{A}_{n\times q} = \begin{pmatrix}\boldsymbol{a}_1' \\ \vdots \\ \boldsymbol{a}_n'\end{pmatrix}, \quad \boldsymbol{\epsilon}_{n\times p} = \begin{pmatrix}\boldsymbol{\epsilon}_1' \\ \vdots \\ \boldsymbol{\epsilon}_n'\end{pmatrix}$$

then we can write the general multivariate nonlinear growth curve model in the nonlinear GMANOVA format of Gennings et. al. (1989) as

$$\begin{aligned} \boldsymbol{Y} &= \boldsymbol{f}(\boldsymbol{\Gamma}; \boldsymbol{A}, \boldsymbol{X}') + \boldsymbol{\epsilon} \\ &= \boldsymbol{f}(\boldsymbol{X}', \boldsymbol{A\Gamma}) + \boldsymbol{\epsilon} \end{aligned} \tag{4.53}$$

where

$$\boldsymbol{f}(\boldsymbol{X}', \boldsymbol{A\Gamma}) = \begin{pmatrix}\boldsymbol{f}(\boldsymbol{X}', \boldsymbol{a}_1'\boldsymbol{\Gamma}) \\ \vdots \\ \boldsymbol{f}(\boldsymbol{X}', \boldsymbol{a}_n'\boldsymbol{\Gamma})\end{pmatrix}$$

and $\boldsymbol{a}_i'\boldsymbol{\Gamma} = \boldsymbol{\beta}_i'$.

It should be noted that although we have assumed an error structure that is multivariate normal with unstructured variance-covariance matrix $\boldsymbol{\Sigma}$, the assumption of normality is not necessary in order to achieve consistent and asymptotically normal estimators of the regression coefficients $\boldsymbol{\beta}$ (Gennings, Chinchilli and Carter, 1989). Moreover, one can replace the unstructured covariance matrix $\boldsymbol{\Sigma}$ with any structured covariance matrix, $\boldsymbol{\Sigma}(\boldsymbol{\theta})$, where $\boldsymbol{\theta}$ is a vector of unique variance-covariance parameters of dimension less than $p \times (p+1)/2$ [e.g., $\boldsymbol{\Sigma}(\boldsymbol{\theta}) = \sigma^2(1-\rho)\boldsymbol{I}_p + \sigma^2\rho\boldsymbol{J}_p$ with $\boldsymbol{\theta} = (\sigma^2, \rho)$].

A partial listing of the LDH leakage data is shown in Output 4.20. In seeking to fit a sigmoidal dose-response model, Gennings et. al. (1989) performed a nonlinear response surface analysis in which the data were fit to the multivariate logistic growth curve model:

$$y_{ij} = 1/\{1 + \exp(-\boldsymbol{a}_i'\boldsymbol{\Gamma}\boldsymbol{x}_j)\} + \epsilon_{ij} \quad (i = 1, \ldots, 64; j = 1, \ldots, 7) \tag{4.54}$$

where $\boldsymbol{\Gamma}$ is a matrix of regression coefficients, $\boldsymbol{a}_i' = (1, a_{i1}, a_{i2}, a_{i1}a_{i2})$ is a between-flask design vector, $\boldsymbol{x}_j = \begin{pmatrix} 1 \\ t_j \end{pmatrix}$ is the jth column of a within-flask design matrix $\boldsymbol{X}'$, and a_{i1}, a_{i2} and t_j are the covariates defined above.

Output 4.20: A partial listing of the LDH enzyme leakage data. The variables LDH0 LDH001 LDH025 LDH05 LDH1 LDH2 LDH3 represent the percentage LDH leakage (i.e., 100×proportion) at times $t_j = 0, 0.01, 0.25, 0.5, 1.0, 2.0$ and 3.0 hours, respectively. The data are reprinted with permission from *The Journal of the American Statistical Association*. Copyright 1989 by the American Statistical Association. All rights reserved.

Flask	CCl4	CHCl3	LDH0	LDH001	LDH025	LDH05	LDH1	LDH2	LDH3
1	0	0	8	9	9	8	10	10	12
2	0	0	8	10	10	9	12	15	13
3	0	0	7	8	8	8	9	9	10
4	0	0	6	8	6	7	8	10	11
5	0	5	6	11	14	12	14	13	14
6	0	5	11	14	16	18	20	21	21
7	0	5	5	7	13	13	14	15	16
8	0	5	6	6	7	9	11	12	12
9	0	10	6	11	20	36	46	44	46
10	0	10	8	14	24	27	29	32	34
11	0	10	6	7	17	18	21	22	22
12	0	10	5	5	15	16	19	22	23
13	0	25	7	10	25	51	65	66	70
14	0	25	11	11	33	39	48	52	55
15	0	25	7	7	17	24	34	37	41
16	0	25	7	6	16	24	31	36	41
17	1	0	6	11	13	9	10	11	11
18	1	0	8	14	15	14	16	19	21
19	1	0	5	8	10	10	11	12	13
20	1	0	5	9	8	9	11	12	13
21	1	5	5	13	18	37	41	42	46
22	1	5	10	16	22	22	29	30	21
23	1	5	6	10	14	16	16	20	18
24	1	5	5	8	15	18	19	21	21
25	1	10	6	10	25	61	57	60	63
26	1	10	11	14	26	30	30	35	29
27	1	10	5	7	24	27	29	32	32
28	1	10	5	6	16	21	24	27	27
29	1	25	7	9	23	39	58	53	67
30	1	25	8	11	28	40	42	75	72
31	1	25	6	6	15	22	30	44	56
32	1	25	6	5	15	27	36	43	55

Under this logistic response curve, the proportion of LDH leakage is assumed to be related to the levels of CCl_4, $CHCl_3$ and time via the inverse logit function of the linear predictor

$$\begin{aligned}\eta_{ij} &= \boldsymbol{a}_i'\boldsymbol{\Gamma}\boldsymbol{x}_j \\ &= \boldsymbol{\beta}_i'\boldsymbol{x}_j \\ &= \beta_{i0} + \beta_{i1}t_j\end{aligned}$$

where

$$\begin{aligned}\beta_{i0} &= \beta_{00} + \beta_{01}a_{i1} + \beta_{02}a_{i2} + \beta_{03}a_{i1}a_{i2} \\ \beta_{i1} &= \beta_{10} + \beta_{11}a_{i1} + \beta_{12}a_{i2} + \beta_{13}a_{i1}a_{i2}\end{aligned}$$

are the intercept and slope parameters of the linear predictor expressed as linear functions of CCl_4 (i.e., a_{i1}) and $CHCl_3$ (i.e., a_{i2}) and their interaction. The authors fit this model using a two-stage EGLS estimation scheme assuming the error terms $\boldsymbol{\epsilon}_i = (\epsilon_{i1}, \ldots, \epsilon_{i7})$ are independent and identically distributed with mean $\mathbf{0}$ and unstructured variance-covariance matrix $\boldsymbol{\Sigma}$.

This logistic growth curve model seems reasonable insofar as it allows one to model a sigmoidal dose-response relationship for a response variable that is a proportion. However, it does appear that in this application, the predicted response under (4.54) grossly underestimates the observed response, particularly at the higher levels of $CHCl_3$. For example, when one compares the observed proportion of LDH leakage shown in Table 1 of Gennings, et. al. (1989) to that predicted from the model (see Figure 1 of Gennings et. al.), the proportion of LDH leakage predicted at a 25 mM dose of $CHCl_3$ is consistently and substantially lower than that observed at 1, 2 and 3 hours. This may be due to model misspecification with respect to the assumed form of the linear predictor $\eta_{ij} = \boldsymbol{a}_i'\boldsymbol{\Gamma}\boldsymbol{x}_j$ or to misspecification of the assumed structural form of the nonlinear model or possibly to both.

The mean profiles shown in Figure 4.8 suggest the proportion of LDH leakage increases with time very rapidly for some dose combinations of CCl_4 and $CHCl_3$ and very slowly for other combinations. Since the logistic growth model (4.54) relies solely on the single linear predictor, $\eta_{ij} = \boldsymbol{a}_i'\boldsymbol{\Gamma}\boldsymbol{x}_j$, to describe the shape of the response curve, it may simply be inadequate to describe these widely different trends. As an alternative, we consider a three-parameter Weibull growth curve model as a means for describing the heterogenous shapes seen in the mean response profiles across the different combinations of CCl_4 and $CHCl_3$. To ensure the parameters of the model conform with a response variable satisfying $0 \leq y_{ij} \leq 1$ (recall y_{ij} is the proportion of LDH leakage in the ith flask at time j), the three parameter Weibull growth curve model may be written as:

$$y_{ij} = 1 - \eta_{i0}\exp\{-\eta_{i1}t_j^{\eta_{i2}}\} + \epsilon_{ij} \quad (i = 1, \ldots, 64; j = 1, \ldots, 7) \tag{4.55}$$

where

$$\begin{aligned}\eta_{i0} &= 1/(1 + \exp(-\beta_{i0}))\\ \eta_{i1} &= \exp(\beta_{i1})\\ \eta_{i2} &= \beta_{i2}\end{aligned}$$

are "canonical" parameters expressed as functions of model parameters $\boldsymbol{\beta}_i = \begin{pmatrix}\beta_{i0}\\ \beta_{i1}\\ \beta_{i2}\end{pmatrix} = \boldsymbol{a}_i'\boldsymbol{\Gamma}$ where $\boldsymbol{a}_i' = (1, a_{i1}, a_{i2}, a_{i1}a_{i2})$ is the vector of between-flask covariates,

$$\boldsymbol{\Gamma} = \begin{pmatrix}\beta_{00} & \beta_{10} & \beta_{20}\\ \beta_{01} & \beta_{11} & \beta_{21}\\ \beta_{02} & \beta_{12} & \beta_{22}\\ \beta_{03} & \beta_{13} & \beta_{23}\end{pmatrix} \tag{4.56}$$

is the matrix of regression coefficients and

$$\begin{aligned}\beta_{i0} &= \beta_{00} + \beta_{01}a_{i1} + \beta_{02}a_{i2} + \beta_{03}a_{i1}a_{i2}\\ \beta_{i1} &= \beta_{10} + \beta_{11}a_{i1} + \beta_{12}a_{i2} + \beta_{13}a_{i1}a_{i2}\\ \beta_{i2} &= \beta_{20} + \beta_{21}a_{i1} + \beta_{22}a_{i2} + \beta_{23}a_{i1}a_{i2}\end{aligned} \tag{4.57}$$

are the "intercept," "rate" and "shape" parameters, respectively, of a Weibull growth curve model expressed in terms of the same linear functions of the carbon tetrachloride (CCl_4) levels and chloroform ($CHCl_3$) levels as modeled by Gennings et. al. (1989). Here the "canonical" parameter η_{i0} is expressed in terms of the inverse-logit of the "intercept" parameter β_{i0} while η_{i1} is expressed in terms of the exponential of the "rate" parameter β_{i1}. This parameterization is required to ensure the response function has the same range as the response variable.

The Weibull model (4.55) may be written in terms of the nonlinear GMANOVA model (4.52) by simple substitution; i.e.,

$$\begin{aligned} y_{ij} &= f(\boldsymbol{x}_j, \boldsymbol{\beta}_i) + \epsilon_{ij} \\ &= 1 - [1/(1+\exp(-\beta_{i0}))]\exp\{-\exp(\beta_{i1})t_j^{\beta_{i2}}\} + \epsilon_{ij} \\ &= 1 - \eta_{i0}\exp\{-\eta_{i1}t_j^{\eta_{i2}}\}. \end{aligned}$$

Following Gennings et. al. (1989), we assume the vector of within-flask error terms $\boldsymbol{\epsilon}_i = (\epsilon_{i1}, \ldots, \epsilon_{i7})'$ are independent and identically distributed with mean $\mathbf{0}$ and variance-covariance $\boldsymbol{\Sigma}$.

Prior to fitting the data under the response surface structure defined by (4.56) and (4.57), we fit the data to a cell-means Weibull growth curve model in which separate curves were fit to each of the 16 combinations of carbon tetrachloride (CCl_4) and chloroform ($CHCl_3$) levels assuming a "working" independence structure. This was accomplished with the NLMIXED procedure. To do so, we first arranged the response data vertically using the TRANSPOSE procedure and defined the time points at which LDH leakage was assessed. The SAS code used to arrange the data and fit the data to a cell-means Weibull growth curve model is shown below.

Program 4.11

```
/* Arrange the data vertically and define the time points */
proc sort data=LDH_data out=example4_5_1;
 by Flask CCl4 CHCl3;
run;
proc transpose data=example4_5_1 out=example4_5_1T
               name=timechar prefix=LDH;
 by Flask CCl4 CHCl3;
run;
proc sort data=example4_5_1T;
 by CCl4 CHCl3;
run;
data example4_5_1T;
 retain Group 0;
 set example4_5_1T;
 by CCl4 CHCl3;
 if first.CHCl3 then do;
  Group+1;
 end;
 Time=0 + 10*('.'||substr(timechar,4));
 LDHpct=LDH1;
 LDH=LDH1/100;
 _Time_=Time;
 keep Flask CCl4 CHCl3 Group Time _Time_ LDHpct LDH;
run;
```

```
/* Multivariate Weibull growth model by dose group */
proc sort data=example4_5_1T;
 by Group Time;
run;
ods listing close;
proc nlmixed data=example4_5_1T;
 by Group;
 parms b0=2.2 b1=-.693 b2=1 sigsq=0.1;
 eta0 = 1/(1+exp(-b0));
 eta1 = exp(b1);
 eta2 = b2;
 if time=0 then predv = 1 - eta0;
 if time> 0 then predv = 1 - eta0*exp(-eta1*(time**eta2));
 model LDH ~ normal(predv, sigsq);
 id b0 b1 b2 sigsq predv;
 predict predv out=pred ;
run;
ods listing;
```

The TRANSPOSE procedure is used to arrange the data vertically rather than horizontally. This is usually required when using NLMIXED and always required when using the %NLINMIX macro. Under the transposed dataset, example4_5_1T, we converted the percentage LDH leakage into a proportion (the variable `LDH`) and we created two time variables, `Time` and `_Time_`, coinciding with the time points, in hours, at which LDH leakage was assessed. The second time variable, `_Time_`, is identical with the first but is created for later use with the %NLINMIX macro as will be explained when we describe the %NLINMIX syntax required to fit the Weibull response surface model. The NLMIXED code simply implements an OLS analysis in which parameter estimates under the structural model (4.55) are obtained for each of the 16 combinations of carbon tetrachloride (CCl_4) and chloroform ($CHCl_3$). Initial parameter estimates were set at $\beta_0 = 2.2, \beta_1 = -0.693$, and $\beta_2 = 1$ corresponding to an overall intercept of $0.10 = 1 - [1/(1 + \exp(-\beta_0))] = 1 - \eta_0$ at time 0, a "rate" parameter of $0.50 = \exp(\beta_1) = \eta_1$, and a "shape" parameter of $\beta_2 = \eta_2 = 1.0$, respectively. We define the predicted response based on the time at which the proportion LDH leakage is measured, i.e.,

```
if time=0 then predv = 1 - eta0;
if time> 0 then predv = 1 - eta0*exp(-eta1*(time**eta2));
```

This is done so as to avoid the problem of having an undefined derivative at time 0 [Note: the derivative with respect to `eta2` of the above structural model, `1 - eta0*exp(eta1*(time**eta2))`, requires evaluating the term log(time) which, at time=0, results in an execution error message within NLMIXED]. The output generated from the PREDICT statement allows us to plot the predicted curves under this structural model. As shown in Figure 4.9, the individual mean profiles and corresponding predicted mean profiles agree reasonably well across all 16 combinations of $CHCl_3$ and CCl_4.

Having established that a structural model based on the Weibull growth curve provides an adequate fit across the different combinations of $CHCl_3$ and CCl_4, we proceeded to fit a nonlinear GMANOVA response surface model in which the "intercept," "rate" and "shape" parameters are defined by the linear predictors of carbon tetrachloride (CCl_4) and chloroform ($CHCl_3$) shown in (4.57). We did so using the macro %NLINMIX in combination with an unstructured variance-covariance matrix. The SAS code used to carry out this response surface analysis is shown below.

Figure 4.9 Observed mean and predicted response curves obtained by fitting each CCl4 and CHCl3 dose combination separately to a multivariate Weibull growth curve model

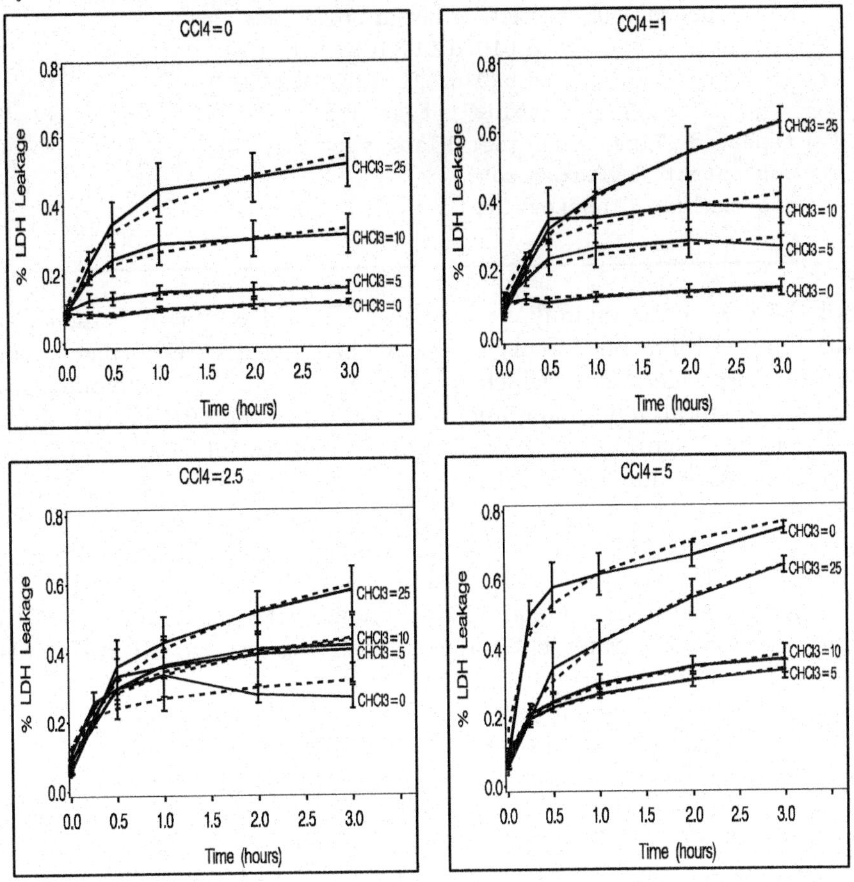

Program 4.12

```
/* Fit the multivariate Weibull response surface model */
%nlinmix(data=example4_5_1T,
    procopt=method=ml covtest,
    maxit=150,
    model=%str(
       beta0 = b00 + b01*ccl4 + b02*chcl3 + b03*ccl4*chcl3;
       beta1 = b10 + b11*ccl4 + b12*chcl3 + b13*ccl4*chcl3;
       beta2 = b20 + b21*ccl4 + b22*chcl3 + b23*ccl4*chcl3;
       eta0 = 1/(1+exp(-beta0));
       eta1 = exp(beta1);
       eta2 = beta2;
       if time=0 then predv = 1-eta0;
       if time> 0 then predv = 1-eta0*exp(-eta1*(time**eta2));
    ),
    parms=%str(b00=2.2 b01=0 b02=0 b03=0
               b10=-0.693 b11=0 b12=0 b13=0
               b20=1 b21=0 b22=0 b23=0),
```

```
     stmts=%str(
       class Flask _Time_;
       model pseudo_LDH = d_b00 d_b01 d_b02 d_b03
                          d_b10 d_b11 d_b12 d_b13
                          d_b20 d_b21 d_b22 d_b23 /
                          noint notest s cl;
       repeated _Time_ / subject=Flask type=un;
       ods output CovParms=_cov;
       ods exclude ClassLevels; )
);
```

For those who are unfamiliar with the %NLINMIX macro, some explanation of its syntax follows. First, one should be aware that %NLINMIX utilizes a Gauss-Newton type algorithm (see page 159) in which PL/QELS estimates of the model parameters are computed by iteratively fitting linearized models using PROC MIXED. As such, much of the %NLINMIX syntax involves specifying macro variable arguments that involve options and statements available under PROC MIXED. The DATA= argument specifies the SAS dataset to be used for the analysis; in this case `data=example4_5_1T`. The PROCOPT= argument allows users to specify any number of options available under the PROC MIXED statement. In this example, we have selected the options METHOD=ML and COVTEST which instruct the macro to compute and summarize ML estimates of the variance-covariance parameters under normality or pseudo-ML estimates under non-Gaussian assumptions.

The MODEL= argument of %NLINMIX provides information regarding the nonlinear response function being used to fit the data. Its syntax typically involves several data step type programming statements, each separated by a semicolon, that one can use to define various auxiliary functions of the parameters (much like one does when working with the NLIN or NLMIXED procedures). These assignment statements, which must be enclosed with the %STR() macro to allow for semicolons, are used to first specify the linear predictor parameters, $\boldsymbol{\beta}_i$, in (4.57) which are in turn expressed as functions of the "canonical" parameters, $\boldsymbol{\eta}_i = (\eta_{i1}, \eta_{i2}, \eta_{i3})'$, of the structural model. We then define a single predictor variable that describes the response function. This predictor variable *MUST* be assigned to the PREDV variable in order for %NLINMIX to successfully execute. In this example, we define the linear predictors BETA0, BETA1 and BETA2 corresponding to β_{i0}, β_{i1} and β_{i2} of (4.57). We then define the "canonical" parameters ETA0, ETA1, and ETA2 corresponding to η_{i0}, η_{i1} and η_{i2} of the structural model (4.55) which, in turn, is defined by the statements,

```
if time=0 then predv = 1 - eta0;
if time> 0 then predv = 1 - eta0*exp(-eta1*(time**eta2));
```

for the same reasons as explained above.

With %NLINMIX, starting values are specified using the PARMS= argument. In this example, the parameters $\beta_{00}, \beta_{01}, \ldots, \beta_{23}$ of the parameter matrix $\boldsymbol{\Gamma}$ in (4.56) are assigned starting values using the PARMS= argument

```
parms=%str(b00=2.2 b01=0 b02=0 b03=0
           b10=-0.693 b11=0 b12=0 b13=0
           b20=1 b21=0 b22=0 b23=0),
```

where we use the same starting values for the linear intercept parameters β_{00}, β_{10} and β_{20} as done for the individual fits above. We then set the remaining parameters to 0 (corresponding to no effects due to CCl_4 or $CHCl_3$).

The STMTS= argument provides the necessary PROC MIXED syntax needed to iteratively fit linearized models in accordance with Step 2 of the Gauss-Newton algorithm described on page 159. At a minimum (for example, when running an OLS nonlinear regression), this entails specification of the MODEL statement of PROC MIXED. However, for most applications one would also include other PROC MIXED statements such as the CLASS statement, the RANDOM and/or REPEATED statements, and possibly the WEIGHT, ESTIMATE or CONTRAST statements along with ODS OUTPUT statements. The response variable of the MODEL statement *MUST* be of the form "pseudo_y" where y is the name of the response variable in the dataset. In this example, the response variable in the dataset example4_5_1T is LDH and so the response variable in the MODEL statement of PROC MIXED is `pseudo_LDH`. In accordance with Step 2 of the Gauss-Newton algorithm shown on page 159, the 'fixed-effects' variables of the MODEL statement are simply the derivatives with respect to each of the fixed-effects parameters listed in the PARMS= argument. These derivatives are computed in %NLINMIX by either directly specifying some or all of the derivatives analytically using the DERIVS= argument of %NLINMIX or by letting %NLINMIX compute the derivatives automatically using numerical differentiation. In either case, the syntax calls for specifying the 'fixed-effects' by listing the parameters in the PARMS= argument preceded by "d_" to indicate that these 'fixed-effects' are in fact the derivatives of the parameters. In addition, the MODEL statement *MUST* include the NOINT, NOTEST, SOLUTION and CL options. In our example, the STMTS= argument is

```
stmts=%str(
      class Flask _Time_;
      model pseudo_LDH = d_b00 d_b01 d_b02 d_b03
                         d_b10 d_b11 d_b12 d_b13
                         d_b20 d_b21 d_b22 d_b23 /
                         noint notest s cl;
      repeated _Time_ / subject=Flask type=un;
      ods output CovParms=_cov;
      ods exclude ClassLevels; )
```

It includes a CLASS statement identifying `Flask` as the SUBJECT= variable for the REPEATED statement and the variable `_Time_`. Recall that we defined the second time variable, `_Time_`, to be identical with the variable `Time`. This was done so that we can include it as a categorical variable within the REPEATED statement of MIXED in order to identify the levels of the repeated measurement for each flask. The REPEATED statement then instructs the macro to fit an unstructured variance-covariance matrix according to the levels of the repeated measures factor, `_Time_`.

The macro produces output in both the SAS log and SAS output. In particular, %NLINMIX displays an iteration history in the SAS log (a partial listing is shown below) and then displays the results of the final call to PROC MIXED in the SAS output (unless otherwise instructed by the OPTIONS= argument of %NLINMIX).

SAS Log

```
                          The NLINMIX Macro
          Data Set                      : example4_5_1T
          Response                      : LDH
          Fixed-Effect Parameters       : b00 b01 b02 b03
                                           b10 b11 b12 b13
                                           b20 b21 b22 b23

b00 = b00
```

```
b01 = b01
b02 = b02
b03 = b03
b10 = b10
b11 = b11
b12 = b12
b13 = b13
b20 = b20
b21 = b21
b22 = b22
b23 = b23
Iteratively calling PROC MIXED.
   PROC MIXED call 0
iteration = 0
convergence criterion = .
b00=3.4643744417 b01=0.2186439935 b02=-0.024424689 b03=-0.017961727
b10=-1.99813815 b11=0.3068500843 b12=0.0543983706
b13=-0.01797385 b20=0.1291494164 b21=0.0223362995 b22=0.0135589823
b23=0.0019677251
   PROC MIXED call 1
iteration = 1
convergence criterion = 0.1253383443
b00=2.301668173 b01=-0.035085182 b02=0.0190658438 b03=-0.001191178
b10=-2.29331585 b11=0.2725989708 b12=0.0557032305
b13=-0.01258906 b20=0.176183393 b21=0.0288506594 b22=0.0160732556
b23=-0.000235474
   PROC MIXED call 2
iteration = 2
convergence criterion = 0.0190911168
b00=2.6314954565 b01=0.0031813271 b02=0.0030435492 b03=-0.000071061
b10=-2.392422142 b11=0.2559136129 b12=0.0569802475
b13=-0.011579564 b20=0.2112325934 b21=0.0269092966 b22=0.0159060621
b23=-0.000354747
   PROC MIXED call 3
iteration = 3
convergence criterion = 0.0023307956
b00=2.6797944554 b01=0.0222724439 b02=0.0000132543 b03=-0.000326136
b10=-2.407268558 b11=0.2556813287 b12=0.057845371
b13=-0.011848469 b20=0.2153904202 b21=0.0241539756 b22=0.015544971
b23=-0.000101674
   PROC MIXED call 4
iteration = 4
convergence criterion = 0.0006484942
b00=2.6808851982 b01=0.0235957565 b02=-0.000134001 b03=-0.000348098
b10=-2.404340258 b11=0.2533505331 b12=0.057545848
b13=-0.011604234 b20=0.2146491459 b21=0.0249523561 b22=0.0156691841
b23=-0.000194842
   PROC MIXED call 5
iteration = 5
convergence criterion = 0.0003916228
b00=2.6808300407 b01=0.023710676 b02=-0.000145429 b03=-0.000345173
b10=-2.405567048 b11=0.2539990272 b12=0.0576722885
```

```
b13=-0.011703189 b20=0.214934383 b21=0.0245401592 b22=0.0156128671
b23=-0.000146047
:
  PROC MIXED call 21
iteration = 21
convergence criterion = 5.0164859E-9
b00=2.6808527216 b01=0.0236747337 b02=-0.000144569 b03=-0.000345069
b10=-2.405165191 b11=0.2537251811 b12=0.0576302819
b13=-0.011667812 b20=0.2148428656 b21=0.0246712593 b22=0.0156307527
b23=-0.000161549
NLINMIX convergence criteria met.
```

The convergence criteria used in %NLINMIX is based on a relative parameter convergence criterion in which consecutive parameter estimates do not differ by some specified criterion which by default is 1E-8. Several datasets are generated from %NLINMIX including an output working dataset which one can name using the OUTDATA= argument of the macro (the default is _NLINMIX) as well as datasets _SOLN, _FIT and _COV although _COV is only generated automatically when one specifies a RANDOM statement (in this example we generate this dataset using the ODS statement: `ODS Output COVTEST=_COV;` within the STMTS= argument). The results of the analysis are shown in Output 4.21.

Output 4.21: Final output from %NLINMIX summarizing parameter estimates for the multivariate Weibull response surface model defined by (4.55)-(4.57).

Model Information

Data Set	WORK._NLINMIX
Dependent Variable	pseudo_LDH
Covariance Structure	Unstructured
Subject Effect	Flask
Estimation Method	ML
Residual Variance Method	None
Fixed Effects SE Method	Model-Based
Degrees of Freedom Method	Between-Within

Dimensions

Covariance Parameters	28
Columns in X	12
Columns in Z	0
Subjects	64
Max Obs Per Subject	7

Number of Observations

Number of Observations Read	448
Number of Observations Used	448
Number of Observations Not Used	0

Iteration History

Iteration	Evaluations	-2 Log Like	Criterion
0	1	−770.59278037	
1	2	−1519.29040286	6592844.7890
2	1	−1547.24263121	2673250.2981
3	1	−1563.43594743	0.12011309
4	1	−1573.95637890	0.00942995
5	1	−1579.27864987	0.00489438
6	1	−1586.41486536	0.00097737
7	1	−1587.76673941	0.00007640
8	1	−1587.86448414	0.00000078
9	1	−1587.86543337	0.00000000

Convergence criteria met.

Covariance Parameter Estimates

Cov Parm	Subject	Estimate	Standard Error	Z Value	Pr Z
UN(1,1)	Flask	0.000369	0.000070	5.28	<.0001
UN(2,1)	Flask	0.000255	0.000071	3.58	0.0003
UN(2,2)	Flask	0.000651	0.000125	5.22	<.0001
UN(3,1)	Flask	0.000548	0.000270	2.03	0.0425
UN(3,2)	Flask	0.000198	0.000355	0.56	0.5774
UN(3,3)	Flask	0.007648	0.001642	4.66	<.0001
UN(4,1)	Flask	0.000563	0.000440	1.28	0.2005
UN(4,2)	Flask	0.000254	0.000583	0.43	0.6636
UN(4,3)	Flask	0.009924	0.002414	4.11	<.0001
UN(4,4)	Flask	0.02303	0.004417	5.21	<.0001
UN(5,1)	Flask	0.000502	0.000438	1.15	0.2518
UN(5,2)	Flask	0.000381	0.000582	0.65	0.5128
UN(5,3)	Flask	0.009391	0.002427	3.87	0.0001
UN(5,4)	Flask	0.02250	0.004442	5.07	<.0001
UN(5,5)	Flask	0.02327	0.004572	5.09	<.0001
UN(6,1)	Flask	0.000529	0.000381	1.39	0.1644
UN(6,2)	Flask	0.000578	0.000504	1.15	0.2516
UN(6,3)	Flask	0.009406	0.002289	4.11	<.0001
UN(6,4)	Flask	0.01893	0.003959	4.78	<.0001
UN(6,5)	Flask	0.01859	0.003983	4.67	<.0001
UN(6,6)	Flask	0.01855	0.003727	4.98	<.0001
UN(7,1)	Flask	−4.99E−6	0.000377	−0.01	0.9894
UN(7,2)	Flask	0.000102	0.000509	0.20	0.8405
UN(7,3)	Flask	0.009089	0.002321	3.92	<.0001
UN(7,4)	Flask	0.01773	0.003949	4.49	<.0001
UN(7,5)	Flask	0.01785	0.003992	4.47	<.0001
UN(7,6)	Flask	0.01805	0.003716	4.86	<.0001
UN(7,7)	Flask	0.01994	0.003887	5.13	<.0001

Fit Statistics

-2 Log Likelihood	−1587.9
AIC (smaller is better)	−1507.9
AICC (smaller is better)	−1499.8
BIC (smaller is better)	−1421.5

Null Model Likelihood Ratio Test

DF	Chi-Square	Pr > ChiSq
27	817.27	<.0001

Solution for Fixed Effects

Effect	Estimate	Standard Error	DF	t Value	Pr > |t|	Alpha	Lower	Upper
d_b00	2.6809	0.06824	64	39.29	<.0001	0.05	2.5445	2.8172
d_b01	0.02367	0.02506	64	0.94	0.3484	0.05	−0.02639	0.07374
d_b02	−0.00014	0.004921	64	−0.03	0.9767	0.05	−0.00998	0.009687
d_b03	−0.00035	0.001799	64	−0.19	0.8485	0.05	−0.00394	0.003248
d_b10	−2.4052	0.1622	64	−14.83	<.0001	0.05	−2.7291	−2.0812
d_b11	0.2537	0.04300	64	5.90	<.0001	0.05	0.1678	0.3396
d_b12	0.05763	0.008568	64	6.73	<.0001	0.05	0.04051	0.07475
d_b13	−0.01167	0.002580	64	−4.52	<.0001	0.05	−0.01682	−0.00651
d_b20	0.2148	0.03246	64	6.62	<.0001	0.05	0.1500	0.2797
d_b21	0.02467	0.008744	64	2.82	0.0064	0.05	0.007203	0.04214
d_b22	0.01563	0.002169	64	7.21	<.0001	0.05	0.01130	0.01996
d_b23	−0.00016	0.000754	64	−0.21	0.8309	0.05	−0.00167	0.001344

The results shown in Output 4.21 include basic model information, dimensions and numbers of observations. An "Iteration History" table reveals that 9 sub-iterations were required in the last call of Step 2 of the Gauss-Newton algorithm to obtain sub-convergence with respect to the variance-covariance parameters, $\boldsymbol{\theta}$. Also shown are the likelihood fit statistics assuming normality and the final parameter estimates $\widehat{\beta}_{00}, \widehat{\beta}_{01}, \ldots, \widehat{\beta}_{23}$ associated with the "fixed-effects" d_b00, d_b01,. . .,d_b23, respectively. These fit statistics reflect the fit of the pseudo-linear model based on the final call to MIXED and do not reflect how well the data fit the nonlinear model specified by %NLINMIX.

The results from our analysis of the data using a multivariate Weibull response surface model along with PL/QELS differs substantially from results reported by Gennings et. al. (1989) using a multivariate logistic response surface model. Computationally, Gennings et. al. (1989) utilize an EGLS procedure in which a single non-iterative estimate of the variance-covariance matrix $\mathbf{\Sigma}$ is used to minimize an EGLS objective function with respect to the regression parameters in $\mathbf{\Gamma}$ whereas the PL/QELS procedure implemented in %NLINMIX is fully iterative. However, the differences in results are not attributable to the use of PL/QELS versus EGLS but rather to the structural model itself. For example, the model used by Gennings et. al. suggest there is a strong interaction between CCl_4 and $CHCl_3$ in terms of both the logistic intercept and slope parameters. In contrast, under the Weibull response surface model, only the "rate" parameter interaction, $\widehat{\beta}_{13} = -0.01167$ (corresponding to `d_b13` in the output), was found to be significant indicating a significant interaction between CCl_4 and $CHCl_3$ exists with respect to the rate of LDH leakage. Moreover, Figure 4.10 illustrates that the three-dimensional response surface predicted using the Weibull growth model predicts higher proportions of LDH leakage at the higher levels of $CHCl_3$ as compared to the three-dimensional plots shown in Figure 1 of Gennings et. al. (1989). As indicated earlier, this may reflect the ability of the Weibull "shape" parameter to better reflect the rapid increase in the proportion of LDH leakage seen at different levels of CCl_4 and $CHCl_3$.

Figure 4.10 Three-dimensional dose-response relationship between the combinations of CCl4 and CHCl3 and the proportion of LDH leakage over time under a multivariate Weibull growth curve model with an unstructured variance-covariance matrix

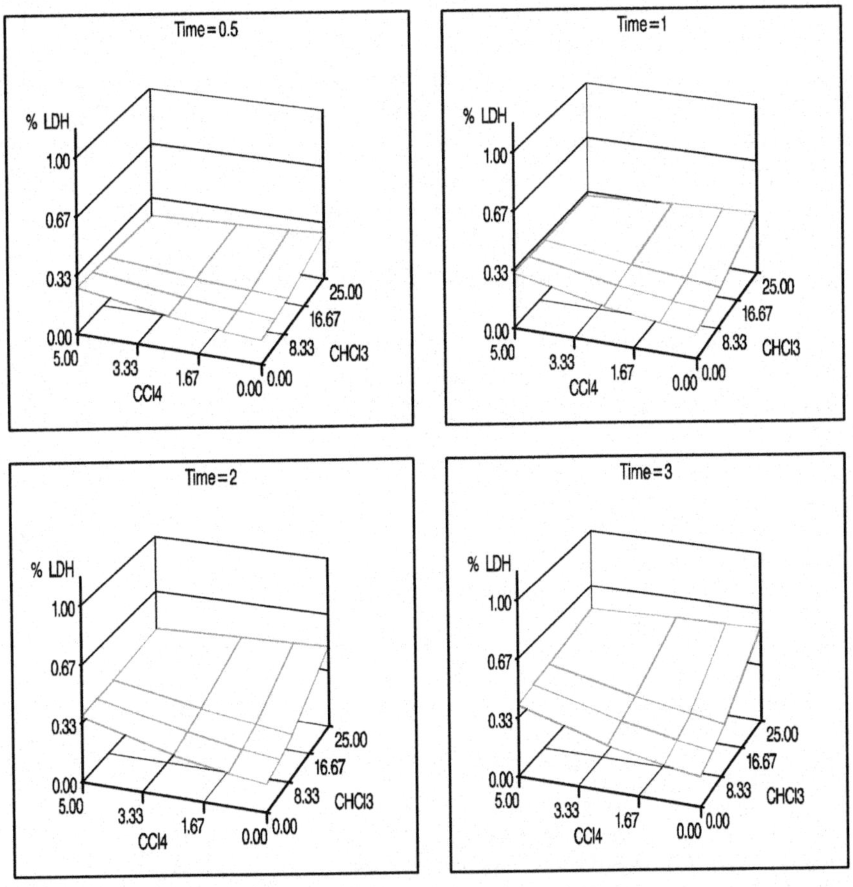

To gain further insight into how well the model fits the data, we ran the SAS macro %GOF (see Appendix D) as shown below.

Program 4.13

```
%GOF(proc=mixed,
     parms=_soln,
     covparms=_cov,
     data=_nlinmix,
     subject=Flask,
     response=LDH,
     pred_ind=predv,
     pred_avg=predv);
```

The %GOF macro takes output generated from MIXED, GENMOD, GLIMMIX, or NLMIXED and summarizes various goodness-of-fit measures using R-square type criteria described by Vonesh et. al. (1996) and Vonesh and Chinchilli (1997). The PROC= argument informs the macro which procedure was used to obtain parameter estimates. In the case of the macro %NLINMIX, the MIXED procedure is used to iteratively estimate the parameters and %GOF uses final parameter estimates from the last call to MIXED. The PARMS= and COVPARMS= arguments instruct the %GOF macro as to where the model parameter estimates may be located—in this example, they are found in the datasets _SOLN (corresponding to the regression parameter estimates) and _COV (corresponding to the covariance parameter estimates) as determined by our call to the %NLINMIX macro above. The DATA= argument identifies the SAS dataset containing the observed and predicted response from the model (here DATA=_NLINMIX) while the FITSTATS= argument specifies the SAS dataset containing the final fit statistics as generated under the SAS procedure specified by the PROC= argument. Here we ignore the FITSTATS= option as it contains information on how well the pseudo-linear model from MIXED fits rather than the actual nonlinear model of interest. Finally, the RESPONSE= argument specifies the name of the response variable while the PRED_IND= argument and PRED_AVG= argument specify, respectively, the individual's predicted mean response and the average individual's predicted mean response. The former is a function of both the fixed-effects and random-effects parameters (as determined when one has a random-effects model) and relates to the subject-specific means while the latter is a function of only the fixed-effects (i.e., is the individual's mean response obtained by setting all random effects equal to 0). For marginal models such as the GNLM, the value of PRED_IND= and PRED_AVG= are the same and in this example are simply the values of the %NLINMIX variable, PREDV. Shown in Output 4.22 are the goodness-of-fit statistics for this example.

Output 4.22: Summary goodness-of-fit statistics describing how well the multivariate Weibull response surface model fits the observed LDH leakage.

```
R-Square Type Goodness-of-Fit Information
Results based on predicted values from SAS procedure MIXED
MODEL FITTING INFORMATION
DESCRIPTION                                                    VALUE
Total Observations                                               448
N (number of subjects)                                            64
Number of Total Parameters                                        40
Number of Fixed-Effects Regression Parameters                     12
Average Model R-Square:                                      0.60125
Average Model Adjusted R-Square:                            0.562157
Average Model Concordance Correlation:                      0.735991
Average Model Adjusted Concordance Correlation:             0.710108
Conditional Model R-Square:                                  0.60125
Conditional Model Adjusted R-Square:                        0.562157
Conditional Model Concordance Correlation:                  0.735991
Conditional Model Adjusted Concordance Correlation:         0.710108
```

The unadjusted R-square for this marginal nonlinear model is 0.601 and the unadjusted concordance correlation is 0.736 suggesting that perhaps some improvement in the predicted response surface may be possible. To examine this further, we plotted the mean profiles against the predicted means based on the response surface model for each combination of CCl_4 and $CHCl_3$. The results, shown in Figure 4.11, suggest the response surface model may be reasonable for CCl_4 levels below 5 nM. However, at a CCl_4 dose of 5 nM, we find the observed mean response at 0 nM of $CHCl_3$ to be significantly higher than that observed at dose levels of 5, 10 and 25 nM of $CHCl_3$. This is in stark contrast with the ordered responses seen at the lower levels of CCl_4 where mean LDH leakage increases with each increase in dose of $CHCl_3$. The "outlying" behavior of LDH leakage at 5 nM CCl_4 and 0 nM $CHCl_3$ suggests that a well-behaved response surface model may only be possible for this dataset if one were to exclude this particular CCl_4-$CHCl_3$ dose combination.

Figure 4.11 Observed mean and predicted response curves obtained from a response surface analysis in which CCl4/CHCl3 dose combinations were fit to a multivariate Weibull growth curve model assuming an unstructured variance-covariance matrix over time

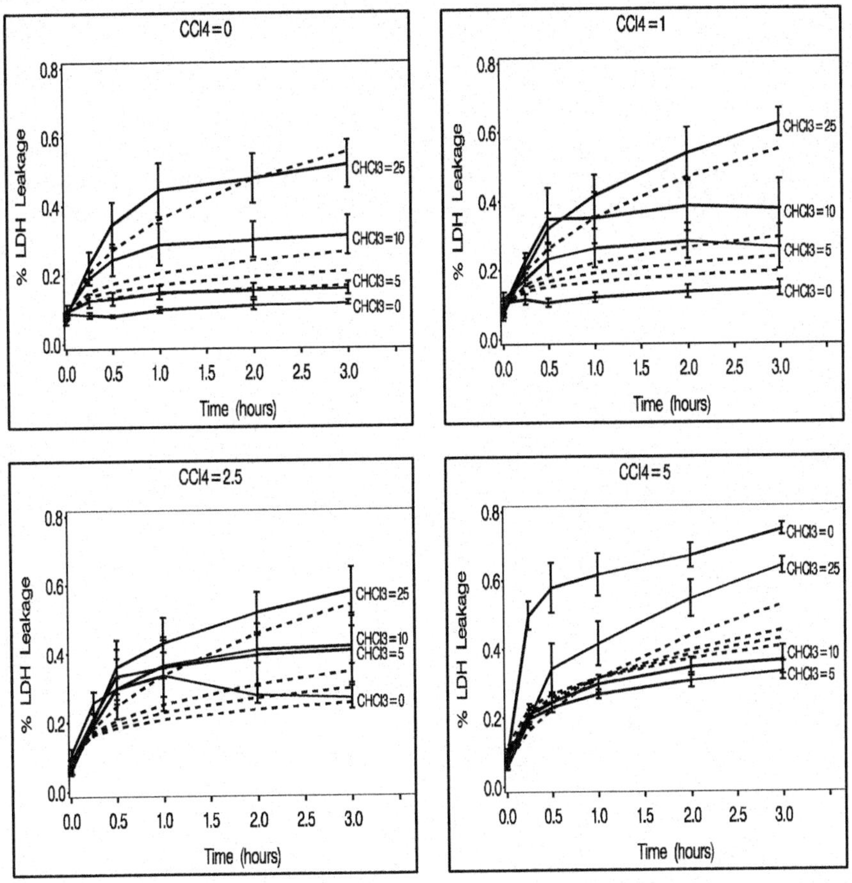

4.5.2 Orange tree data

Lindstrom and Bates (1990) applied a logistic model to a set of growth curve data presented in Draper and Smith (1981, p. 524) in which the trunk circumference (millimeters) of five different orange trees were measured over 7 different time points. The structure of the data and a description of a re-parameterized version of the logistic growth curve model of Lindstrom and Bates (1990) which was proposed by Pinheiro and Bates (2000) follow (see also the author's Web page).

- Experimental unit is an orange tree
- Each tree was measured at 7 different time points
- Response variable:
  - y_{ij} = trunk circumference (mm) of ith tree at time t_{ij}
- One within-unit covariate:
  - Time: $t_{ij} = t_j$ =days (118, 484, 664, 1004, 1231, 1372, 1582)
- Structural Model:
  - $y = \beta_1/(1 + exp\{-(t - \beta_2)/\beta_3\})$
- Goal: Estimate the orange tree growth parameters

The data, listed in Output 4.23 and plotted in Figure 1.1 of Chapter 1, were analyzed by Pinheiro and Bates (2000) using nonlinear mixed-effects regression. Based on prior analyses by Lindstrom and Bates (1990) in which only the asymptote β_1 showed any appreciable variation between trees, the data were fit to the following logistic growth curve model:

$$y_{ij} = (\beta_1 + b_i)/\{1 + \exp(-(t_j - \beta_2)/\beta_3)\} + \epsilon_{ij} \; (i = 1, \ldots, 5; j = 1, \ldots, 7) \tag{4.58}$$

where the b_i are subject-specific random effects assumed to be iid $N(0, \psi)$ and ϵ_{ij} ~iid $N(0, \sigma^2)$ independent of b_i. Although this model is in the class of normal-theory nonlinear mixed-effects models which we consider in Chapter 5, it may also be expressed strictly in terms of a marginal nonlinear model for correlated normally distributed data (see §4.4.1). Specifically, since the model is linear in b_i, one can express (4.58) in terms of the marginal NLM

$$\begin{aligned} y_{ij} &= f(\boldsymbol{x}_j', \boldsymbol{\beta}) + \epsilon_{ij}^* \\ &= \beta_1/\{1 + \exp(-(t_j - \beta_2)/\beta_3)\} + \epsilon_{ij}^* \end{aligned} \tag{4.59}$$

where $Var(\epsilon_{ij}^*) = \psi g_{ij}^2(\boldsymbol{\beta}) + \sigma^2$, $Cov(\epsilon_{ij}^*, \epsilon_{ij'}^*) = \psi g_{ij}(\boldsymbol{\beta}) g_{ij'}(\boldsymbol{\beta})$ and $g_{ij}(\boldsymbol{\beta}) = 1/\{1 + \exp(-(t_j - \beta_2)/\beta_3)\}$. Here $\boldsymbol{\beta} = (\beta_1 \;\; \beta_2 \;\; \beta_3)'$ is the vector of regression coefficients and $\boldsymbol{x}_j' = (1, \, t_j)$ is the design vector common for each tree.

Output 4.23: Listing of the orange tree data.

Tree	Days	y
1	118	30
1	484	58
1	664	87
1	1004	115
1	1231	120
1	1372	142
1	1582	145
2	118	33
2	484	69
2	664	111
2	1004	156
2	1231	172
2	1372	203
2	1582	203
3	118	30
3	484	51
3	664	75
3	1004	108
3	1231	115
3	1372	139
3	1582	140

4	118	32
4	484	62
4	664	112
4	1004	167
4	1231	179
4	1372	209
4	1582	214
5	118	30
5	484	49
5	664	81
5	1004	125
5	1231	142
5	1372	174
5	1582	177

Under the assumption of normality, ML estimates of the model parameters can be obtained using the SAS procedure NLMIXED. Unlike the %NLINMIX macro which utilizes PL/QELS estimation, NLMIXED directly minimizes a negative log-likelihood objective function using any one of several optimization techniques available to the user. To do so, NLMIXED first evaluates the marginal log-likelihood by numerically integrating over the conditional likelihood given a set of random effects. Specifically, if $\pi(\boldsymbol{y}_i|\boldsymbol{b}_i)$ is the conditional pdf of $\boldsymbol{y}_i$ given a set of random effects $\boldsymbol{b}_i$ that are assumed to normally distributed, then NLMIXED minimizes the negative log-likelihood function

$$-L(\boldsymbol{\beta},\boldsymbol{\theta};\boldsymbol{y}) = -\sum_{i=1}^{n}\log\pi(\boldsymbol{y}_i) = -\sum_{i=1}^{n}\log\int\pi(\boldsymbol{y}_i|\boldsymbol{b}_i)\pi(\boldsymbol{b}_i)d\boldsymbol{b}_i \tag{4.60}$$

where $\pi(\boldsymbol{b}_i)$ is the assumed Gaussian pdf of $\boldsymbol{b}_i$. When the model is linear in the random effects as in this example, one can employ any of the numerical integration methods available in NLMIXED and they will each yield the exact marginal log-likelihood assuming normality (we defer discussion of these integration methods to Chapter 5). Presented below is the SAS code required to obtain ML estimates of the parameters of the marginal NLM (4.59).

Program 4.14

```
ods output CovMatParmEst=Omega;
proc nlmixed data=example4_5_2 cov;
 parms b1=150 b2=700  b3=350  sigma_sq=60;
 num = b1+u1;
 den = 1 + exp(-(Days-b2)/b3);
 predmean = (num/den);
 predvar = sigma_sq;
 model y ~ normal(predmean, predvar);
 random u1 ~ normal(0, psi) subject=Tree;
run;
```

Some comments regarding the NLMIXED syntax is required. Unlike the SAS procedure NLIN and the macro %NLINMIX, the NLMIXED procedure does not require one to have a PARMS statement whereby one specifies initial starting values for the model parameters. However, it is strongly recommended that one use the PARMS statement in order to provide reasonably good starting values as this will help speed up the convergence of the algorithm used. It is also important to scale the parameters of the model so that they are all within one to two orders of magnitude of each other. Each of these issues is addressed in more detail in §4.6. In this example, starting values for the regression parameters $\boldsymbol{\beta}$ and the within-tree variance component, σ^2, are specified by the PARMS= statement and are based on results reported by Pinheiro and Bates (2000). The random-effects variance component, ψ, is assigned a default starting value of 1.0. This is the default initial value of any parameter in the model that is not assigned a starting value in a PARMS statement. Chapter 5 provides a more in-depth treatment of how to specify good starting values of the model parameters, especially the variance-covariance parameters of random effects.

In the above code, we used the default method of estimation which is to apply a dual quasi-Newton optimization algorithm (TECH=QUANEW) to the problem of minimizing the negative log-likelihood function (4.60). The marginal log-likelihood is evaluated via numerical integration using adaptive Gaussian quadrature which is the default method of integration in NLMIXED. Since model (4.58) is linear in b_i, one could also evaluate the marginal log-likelihood by specifying METHOD=FIRO, a first-order Taylor series approximation to the integrated log-likelihood. We chose the default method because it also enables one to compute an empirical sandwich estimator of the variance-covariance matrix of the estimated model parameters (the EMPIRICAL option of NLMIXED requires the use of a RANDOM statement in combination with METHOD=GAUSS (default) or METHOD=ISAMP). We specified the NLMIXED option COV along with the ODS statement,

```
ODS OUTPUT CovMatParmEst=Omega;
```

in order to save the model-based variance-covariance matrix of the estimated parameters for later use when evaluating the overall model goodness-of-fit.

The results of this model fit are shown in Output 4.24. In addition to the model specifications, a summary of the dimensions of the problem and a display of the starting values as specified by the PARMS= statement, the default NLMIXED output also summarizes the iteration history, likelihood-based goodness-of-fit statistics and the final parameter estimates including inferential statistics in the form of t-tests and confidence intervals as described in §4.4.3. In addition, because we specified the NLMIXED option COV, the output includes the model-based covariance matrix of the parameter estimates, $\widehat{\boldsymbol{\Omega}}(\widehat{\boldsymbol{\tau}}_{\mathrm{ML}})$.

Output 4.24: ML estimates of the orange tree growth parameters. The standard errors are based on the model-based covariance matrix, $\widehat{\boldsymbol{\Omega}}(\widehat{\boldsymbol{\tau}}_{\mathrm{ML}})$.

```
                         Specifications
Data Set                                  WORK.EXAMPLE4_5_2
Dependent Variable                        y
Distribution for Dependent Variable       Normal
Random Effects                            u1
Distribution for Random Effects           Normal
Subject Variable                          Tree
Optimization Technique                    Dual Quasi-Newton
Integration Method                        Adaptive Gaussian Quadrature
            Dimensions
Observations Used          35
Observations Not Used       0
Total Observations         35
Subjects                    5
Max Obs Per Subject         7
Parameters                  5
Quadrature Points           1
                    Parameters
 b1     b2     b3   sigma_sq   psi   NegLogLike
150    700    350         60     1   434.522063
```

Iteration History

Iter	Calls	NegLogLike	Diff	MaxGrad	Slope
1	4	249.859067	184.663	4.074741	−116.881
2	6	174.815992	75.04308	1.253194	−37.5903
3	8	155.142876	19.67312	0.367729	−14.1912
4	10	153.852867	1.290008	0.430311	−1.34072
5	13	147.927114	5.925753	0.494522	−0.09807
6	14	139.905725	8.021389	0.123389	−4.52754
7	16	138.222308	1.683417	0.112809	−1.26484
8	18	137.946677	0.275631	0.053735	−0.25622
9	20	137.881736	0.064942	0.053863	−0.07867
10	21	137.826007	0.055729	0.05471	−0.04335
11	23	137.819975	0.006031	0.054631	−0.00948
12	25	137.816906	0.003069	0.054542	−0.00162
13	31	136.325253	1.491654	0.041836	−0.00487
14	35	135.387365	0.937888	0.037703	−4.3384
15	37	132.970547	2.416818	0.03264	−3.40294
16	39	132.820571	0.149976	0.048541	−0.31664
17	41	132.747403	0.073168	0.055385	−0.10896
18	43	132.742674	0.004729	0.05793	−0.00667
19	44	132.735095	0.007579	0.060201	−0.0018
20	46	132.663577	0.071518	0.06382	−0.01581
21	48	132.236835	0.426742	0.048444	−0.10786
22	49	131.764909	0.471926	0.023202	−0.27557
23	51	131.68883	0.076079	0.009665	−0.08376
24	53	131.667124	0.021706	0.008942	−0.01898
25	55	131.660003	0.00712	0.009951	−0.00981
26	57	131.659405	0.000599	0.009435	−0.00078
27	59	131.659313	0.000092	0.009745	−0.00006
28	62	131.657234	0.002079	0.008603	−0.0001
29	64	131.624066	0.033167	0.009907	−0.00298
30	65	131.576858	0.047208	0.004194	−0.04487
31	67	131.571898	0.00496	0.000174	−0.00877
32	69	131.571885	0.000013	0.000016	−0.00002
33	71	131.571885	9.396E−8	4.878E−7	−2.02E−7

NOTE: GCONV convergence criterion satisfied.

Fit Statistics

−2 Log Likelihood	263.1
AIC (smaller is better)	273.1
AICC (smaller is better)	275.2
BIC (smaller is better)	271.2

Parameter Estimates

Parameter	Estimate	Standard Error	DF	t Value	Pr > |t|
b1	192.05	15.6578	4	12.27	0.0003
b2	727.91	35.2486	4	20.65	<.0001
b3	348.07	27.0798	4	12.85	0.0002
sigma_sq	61.5129	15.8826	4	3.87	0.0179
psi	1001.51	649.50	4	1.54	0.1979

Parameter	Alpha	Lower	Upper	Gradient
b1	0.05	148.58	235.53	−4.88E−7
b2	0.05	630.04	825.77	−2.25E−7
b3	0.05	272.89	423.26	−1.26E−7
sigma_sq	0.05	17.4157	105.61	1.148E−7
psi	0.05	−801.80	2804.82	5.593E−8

Covariance Matrix of Parameter Estimates

Row	Parameter	b1	b2	b3	sigma_sq	psi
1	b1	245.17	217.27	158.22	0.05025	420.83
2	b2	217.27	1242.46	739.42	0.06451	2259.50
3	b3	158.22	739.42	733.31	0.4336	1605.72
4	sigma_sq	0.05025	0.06451	0.4336	252.26	−81.9164
5	psi	420.83	2259.50	1605.72	−81.9164	421855

As an additional check on model goodness-of-fit, we repeated the above analysis but with standard errors computed based on the robust covariance matrix, $\widehat{\boldsymbol{\Omega}}_R(\widehat{\boldsymbol{\tau}}_{\text{ML}})$. This is accomplished by specifying the EMPIRICAL option in the NLMIXED statement. We then requested additional goodness-of-fit statistics using the SAS macro %GOF including a pseudo-likelihood ratio test for evaluating the goodness-of-fit of the marginal variance-covariance matrix of $\boldsymbol{y}_i$. The SAS code required for this analysis is shown below.

Program 4.15

```
ods output ParameterEstimates=PE;
ods output CovMatParmEst=Omega_R;
ods select ParameterEstimates CovMatParmEst;
proc nlmixed data=example4_5_2 cov empirical;
 parms b1=150 b2=700  b3=350  sigma_sq=60;
 num = b1+u1;
 den = 1 + exp(-(Days-b2)/b3);
 predmean = (num/den);
 predvar = sigma_sq;
 model y ~ normal(predmean, predvar);
 random u1 ~ normal(0, psi) subject=Tree;
 predict predmean out=pred;
run;

data PE_beta;
 set PE;
 if Parameter IN ('b1' 'b2' 'b3');
run;

data PE_cov;
 set PE;
 if Parameter IN ('sigma_sq' 'psi');
run;

%GOF(proc=nlmixed,
     parms=PE_beta,
     covparms=PE_cov,
     data=_pred_,
     subject=Tree,
     omega=Omega,
     omega_r=Omega_R,
     response=y,
     pred_ind=pred,
     pred_avg=pred
    );
```

The ODS OUTPUT statements allow us to save the parameter estimates and the covariance matrix of these estimates into the SAS datasets `PE` and `Omega_R`, respectively. Since the model specifications, dimensions and starting values remain the same, the ODS SELECT statement instructs the Output Delivery System to only print results pertaining to the parameter estimates and their estimated covariance matrix. The NLMIXED option EMPIRICAL requests that inference be based on robust standard errors computed from the empirical covariance matrix, $\widehat{\boldsymbol{\Omega}}_R(\widehat{\boldsymbol{\tau}})$ rather than the model-based covariance matrix $\widehat{\boldsymbol{\Omega}}(\widehat{\boldsymbol{\tau}})$. We then call on the SAS macro %GOF to compute goodness-of-fit statistics based on R-square and concordance correlation coefficients as described by Vonesh et. al. (1996)

[see also the author's Web page]. The macro also computes a pseudo-likelihood ratio test based on the discrepancy function (4.24) to determine if the assumed covariance structure under (4.59) is reasonable (see Appendix D and also page 135). The input for running %GOF requires specifying SAS datasets containing

1) the regression parameter estimates only (i.e., dataset `PE_beta`),
2) the variance-covariance parameters (i.e., dataset `PE_cov`),
3) the model-based covariance matrix $\widehat{\Omega}(\widehat{\beta})$ (i.e., dataset `Omega`),
4) the robust covariance matrix $\widehat{\Omega}_R(\widehat{\beta})$ (i.e., dataset `Omega_R`),
5) and a dataset containing the predicted mean response (i.e., dataset `PRED` as generated from the `PREDICT` statement shown above).

The %GOF macro automatically determines that $\widehat{\Omega}(\widehat{\beta})$ and $\widehat{\Omega}_R(\widehat{\beta})$ are formed from the upper 3×3 sub-matrix of $\widehat{\Omega}(\widehat{\tau})$ and $\widehat{\Omega}_R(\widehat{\tau})$ (corresponding to the first three rows and columns of the datasets `Omega` and `Omega_R`, respectively) provided one has correctly listed the regression parameters β_1, β_2, and β_3 as the first three parameters in the PARMS= statement (see the self-contained instructions for running %GOF). In those cases where one does not wish to carry out the pseudo-likelihood ratio test, one can skip this test by merely leaving the macro argument OMEGA= or the macro argument OMEGA_R= out of the syntax when calling the macro %GOF.

Output 4.25: ML estimates of the orange tree growth parameters. The standard errors are based on the robust covariance matrix, $\widehat{\Omega}_R(\widehat{\tau}_{\mathrm{ML}})$.

Parameter Estimates

Parameter	Estimate	Standard Error	DF	t Value	Pr > |t|
b1	192.05	13.0064	4	14.77	0.0001
b2	727.91	28.8732	4	25.21	<.0001
b3	348.07	17.6379	4	19.73	<.0001
sigma_sq	61.5129	7.4135	4	8.30	0.0012
psi	1001.51	219.07	4	4.57	0.0102

Parameter	Alpha	Lower	Upper	Gradient
b1	0.05	155.94	228.16	−4.88E−7
b2	0.05	647.74	808.07	−2.25E−7
b3	0.05	299.10	397.04	−1.26E−7
sigma_sq	0.05	40.9296	82.0961	1.148E−7
psi	0.05	393.28	1609.74	5.593E−8

Empirical Covariance Matrix of Parameter Estimates

Row	Parameter	b1	b2	b3	sigma_sq	psi
1	b1	169.17	65.8104	−152.51	48.5407	−154.05
2	b2	65.8104	833.66	265.67	67.4218	−4119.31
3	b3	−152.51	265.67	311.10	−38.4200	−2317.76
4	sigma_sq	48.5407	67.4218	−38.4200	54.9602	−639.04
5	psi	−154.05	−4119.31	−2317.76	−639.04	47991

Shown in Output 4.25 are the parameter estimates along with their robust standard errors, t-tests and confidence intervals, and the empirical covariance matrix, $\widehat{\Omega}_R(\widehat{\tau})$. When compared with results based on the model-based covariance matrix shown in Output 4.24, we see that the robust standard errors for the regression parameters are somewhat lower than the model-based standard errors. This suggests the possibility that the assumed marginal variance-covariance structure under model (4.59) may be misspecified. One means for checking this assumption is to carry out the pseudo-likelihood ratio test comparing $\widehat{\Omega}(\widehat{\beta})$ with $\widehat{\Omega}_R(\widehat{\beta})$ as described by Vonesh et. al. (1996). Summarized in Output 4.26 are the results of that test along with R-square measures of goodness-of-fit as implemented in the %GOF macro. The overall fit of the marginal logistic growth curve model (4.59) to the observed data is extremely good as judged by an adjusted R-square measure of 0.981 and adjusted concordance correlation of 0.990. The variance-covariance concordance correlation between the elements of $\widehat{\mathbf{\Omega}}(\widehat{\beta})$ and $\widehat{\mathbf{\Omega}}_R(\widehat{\beta})$ is estimated to be 0.865 and although the pseudo-likelihood ratio chi-square test was not significant at the 0.05 significance level (p=0.088092), the near-borderline significance suggests the possibility that an alternative covariance structure may provide a better fit.

Output 4.26: Goodness-of-fit results comparing the model-based covariance matrix to the robust covariance matrix of the regression parameter estimates

```
R-Square Type Goodness-of-Fit Information
Results based on predicted values from SAS procedure NLMIXED
DESCRIPTION                                                  VALUE
Total Observations                                              35
N (number of subjects)                                           5
Number of Total Parameters                                       5
Number of Fixed-Effects Regression Parameters                    3
Average Model R-Square:                                   0.983523
Average Model Adjusted R-Square:                          0.980777
Average Model Concordance Correlation:                    0.991723
Average Model Adjusted Concordance Correlation:           0.990344
Conditional Model R-Square:                               0.983523
Conditional Model Adjusted R-Square:                      0.980777
Conditional Model Concordance Correlation:                0.991723
Conditional Model Adjusted Concordance Correlation:       0.990344
Pseudo-Likelihood Ratio Test of Variance-Covariance Structure
DESCRIPTION                                                              VALUE
Variance-Covariance Concordance Correlation:                          0.864686
Discrepancy Function                                                  2.201844
s = Rank of robust sandwich estimator, OmegaR                                3
s1 = Number of unique non-zero off-diagonal elements of OmegaR               3
Approx. Chi-Square for H0: Covariance Structure is Correct            11.00922
DF1 = s(s+1)/2, per Vonesh et al (Biometrics 52:572-587, 1996)               6
Pr > Chi Square based on degrees of freedom, DF1                      0.088092
DF2 = s+s1, a modified degress of freedom                                    6
Pr > Chi Square based on modified degrees of freedom, DF2             0.088092
```

Since ML estimation under model (4.59) is equivalent to solving a set of GEE2 assuming the first- and second-order moments are correctly specified, one may wish to protect against any second-order misspecifications by estimating the regression parameters via PL/QELS. To that end, the following code shows how one can obtain PL/QELS estimates of the model parameters using the %NLINMIX macro.

Program 4.16

```
%nlinmix(data=example4_5_2,
    procopt=method=ml covtest,
    model=%str(
       num = b1+u1;
       den = 1 + exp(-(Days-b2)/b3);
       predv=num/den;
    ),
```

```
    parms=%str(b1=150 b2=700 b3=350),
    stmts=%str(class Tree;
      model pseudo_y = d_b1 d_b2 d_b3 /
                       noint notest s cl covb;
      random d_u1 / subject=Tree type=un s cl;),
    expand=zero
  );
run;
```

The PL estimates along with their model-based standard errors are summarized in Output 4.27. When compared with the corresponding ML estimates from Output 4.24 we find the results are very similar. By observing that the log-likelihood based profile traces of $\boldsymbol{\theta}$ and $\boldsymbol{\beta}$ meet almost perpendicularly, Pinheiro and Bates (2000) suggest such agreement is indicative of a local orthogonality between the variance components $\boldsymbol{\theta}$ and fixed effects $\boldsymbol{\beta}$. Indeed, when the covariance matrix of $\boldsymbol{y}_i$ strongly depends on $\boldsymbol{\beta}$, then evidence of misspecification with respect to the covariance structure would likely manifest itself whenever the ML/GEE2 estimator of $\boldsymbol{\beta}$ differs substantially from the corresponding PL/GEE1 estimator (Vonesh et. al., 2001).

Output 4.27: PL/QELS estimates of the orange tree logistic growth parameters using %NLINMIX. The standard error estimates are based on the model-based covariance matrix, $\widehat{\Omega}(\widehat{\tau}_{\mathrm{ML}})$, as computed using PL/QELS.

```
              Model Information
Data Set                        WORK._NLINMIX
Dependent Variable              pseudo_y
Covariance Structure            Unstructured
Subject Effect                  Tree
Estimation Method               ML
Residual Variance Method        Profile
Fixed Effects SE Method         Model-Based
Degrees of Freedom Method       Containment
  Class Level Information
Class    Levels    Values
Tree          5    1 2 3 4 5
          Dimensions
Covariance Parameters           2
Columns in X                    3
Columns in Z Per Subject        1
Subjects                        5
Max Obs Per Subject             7
        Number of Observations
Number of Observations Read           35
Number of Observations Used           35
Number of Observations Not Used        0
                  Iteration History
Iteration    Evaluations     -2 Log Like      Criterion
        0              1    316.79742544
        1              1    263.22701631     0.00000000
Convergence criteria met.
                  Covariance Parameter Estimates
Cov Parm    Subject    Estimate   Standard Error   Z Value    Pr > Z
UN(1,1)     Tree        1007.37           649.96      1.55    0.0606
Residual                61.6906          15.9284      3.87    <.0001
          Fit Statistics
-2 Log Likelihood              263.2
AIC (smaller is better)        273.2
AICC (smaller is better)       275.3
BIC (smaller is better)        271.3
```

Null Model Likelihood Ratio Test

DF	Chi-Square	Pr > ChiSq
1	53.57	<.0001

Solution for Fixed Effects

Effect	Estimate	Standard Error	DF	t Value	Pr > |t|	Alpha	Lower	Upper
d_b1	192.69	15.7403	28	12.24	<.0001	0.05	160.45	224.93
d_b2	728.76	36.0582	28	20.21	<.0001	0.05	654.89	802.62
d_b3	353.53	27.3790	28	12.91	<.0001	0.05	297.45	409.62

Covariance Matrix for Fixed Effects

Row	Effect	Col1	Col2	Col3
1	d_b1	247.76	226.12	161.94
2	d_b2	226.12	1300.19	760.02
3	d_b3	161.94	760.02	749.61

4.5.3 Respiratory disorder data—continued

In §4.3.2, we applied GEE to the analysis of the respiratory disorder data in which the proportion of patients having a positive response (a binary outcome) to either an active or placebo control drug was compared at each of four visits using logistic regression with adjustment for gender, age and center-to-center differences. For a detailed description of the data, the reader is referred to §1.4, page 13. Here we apply the semiparametric GEE2 approach of Prentice and Zhao (1991) to analyze the data by specifying "working" third- and fourth-order moments assuming normality (i.e., ELS). We do so under two different assumptions about the second-order moments of the repeated binary outcomes. The first assumption is that of an independence structure and the second assumption is that of an exchangeable correlation structure. Our goal here is to compare results from this GEE2-based analysis with that of our GEE-based analysis in §4.3.2. The SAS code for structuring the data and performing a GEE analysis using GENMOD is presented in §4.3.2. Presented below is the PROC NLMIXED code needed to implement a GEE2/ELS analysis under the assumption that the second-order moments satisfy an independence structure.

Program 4.17

```
ods output parameterestimates=GEE2pe_ind;
proc nlmixed data=Example4_3_2 empirical;
  parms b0=-0.8561 b1=1.2654 b2=0.6495 b3=0.1368
        b4=-0.0188 b5=1.8457;
  X_b = b0 + b1*Trt + b2*Center_ + b3*Gender + b4*Age + b5*y0;
  pi = exp(X_b)/(1+exp(X_b));
  Sigma_i = pi*(1-pi);
  Mu_i = pi + u;
  Var_i = Sigma_i;
  model y ~ normal(Mu_i, Var_i);
  random u ~ normal(0, 0) subject=Subject;
run;
```

In the above code, we specify the mean and variance of the binary response variable, y_{ij}, based on the logistic model (1.3). In accordance with an assumed independence structure, the above code reflects our assumption that y_{ij} and $y_{ij'}$ are mutually independent for all $j \neq j'$. By then specifying "working" third- and fourth-order moments for $\boldsymbol{y}_i = (y_{i1}, \ldots, y_{i4})'$ that coincide with those obtained under normality, we can carry out a GEE2-based analysis by simply treating the binary response variable y as though it were normally distributed and proceed to minimize a Gaussian-based negative log-likelihood. This is precisely what is implemented when we specify the model statement

```
model y ~ normal(Mu_i, Var_i);
```

where `Mu_i` and `Var_i` are the mean and variance under model (1.3). To safeguard against model misspecification with respect to the "working" third- and fourth-order moments, the usual approach with GEE2 is to base inference on a robust variance-covariance matrix of the model parameters. To do this, we must "trick" NLMIXED into computing robust standard errors via the EMPIRICAL option. The problem we face is that NLMIXED will ignore the EMPIRICAL option unless 1) we specify a random effect within the model and 2) we integrate over the random effect using either adaptive Gaussian quadrature (METHOD=GAUSS) or importance sampling (METHOD=ISAMP). Therefore, in order to maintain our marginal binary logistic regression model, we "trick" NLMIXED by simply specifying an additive Gaussian random effect having a fixed mean of 0 and a fixed variance of 0 which we add to the assumed marginal mean and then request that this "random effect" be integrated out using adaptive Gaussian quadrature (which is the default method) [Note: the marginal mean is specified in the assignment statement: `pi = exp(X_b)/(1+exp(X_b));`]. This is accomplished by first adding in an additive random effect to the mean via the code

```
Mu_i = pi + u;
```

and then specifying this random effect has mean 0 and variance 0 using the RANDOM statement

```
random u ~ normal(0, 0) subject=Subject;
```

in which the `subject=Subject` option is needed to identify subjects according to their repeated binary outcomes.

The results from this GEE2-based analysis using NLMIXED are shown in Output 4.28. When compared with results from a GEE analysis assuming "working" independence (see Output 4.6), we find that the results are qualitatively similar although the center effects are nearly significant in the GEE analysis (p-value of 0.0660). Likewise, the regression parameter estimates for both GEE and GEE2 are quantitatively similar although the robust standard errors under GEE2 tend to be more conservative than the GEE-based robust standard errors.

Output 4.28: GEE2 results from the respiratory disorder data assuming the second-order moments satisfy an independence structure and that "working" third- and fourth-order moments coincide with those obtained under normality.

```
                              Specifications
Data Set                                    WORK.EXAMPLE4_3_2
Dependent Variable                          y
Distribution for Dependent Variable         Normal
Random Effects                              u
Distribution for Random Effects             Normal
Subject Variable                            Subject
Optimization Technique                      Dual Quasi-Newton
Integration Method                          Adaptive Gaussian Quadrature
              Dimensions
Observations Used            444
Observations Not Used          0
Total Observations           444
Subjects                     111
Max Obs Per Subject            4
Parameters                     6
Quadrature Points              1
                                  Parameters
      b0         b1        b2         b3          b4        b5    NegLogLike
 -0.8561     1.2654    0.6495     0.1368     -0.0188    1.8457    242.484489
```

Iteration History

Iter	Calls	NegLogLike	Diff	MaxGrad	Slope
1	5	242.474255	0.010234	21.09154	−308.875
2	7	242.147311	0.326944	13.04167	−2.07421
3	9	241.931238	0.216073	4.451172	−0.81369
4	11	241.871769	0.05947	5.485537	−0.27999
5	12	241.869269	0.0025	0.663006	−0.01176
6	14	241.867995	0.001274	0.033745	−0.00256
7	16	241.867994	5.59E−7	0.000876	−1.34E−6

NOTE: GCONV convergence criterion satisfied.

Fit Statistics

−2 Log Likelihood	483.7
AIC (smaller is better)	495.7
AICC (smaller is better)	495.9
BIC (smaller is better)	512.0

Parameter Estimates

Parameter	Estimate	Standard Error	DF	t Value	Pr > |t|
b0	−1.1362	0.5885	110	−1.93	0.0561
b1	1.3787	0.3499	110	3.94	0.0001
b2	0.6112	0.4083	110	1.50	0.1372
b3	0.1159	0.5107	110	0.23	0.8209
b4	−0.01117	0.01745	110	−0.64	0.5235
b5	1.8423	0.3628	110	5.08	<.0001

Parameter	Alpha	Lower	Upper	Gradient
b0	0.05	−2.3025	0.03003	0.000035
b1	0.05	0.6853	2.0720	−0.00011
b2	0.05	−0.1979	1.4204	−2.2E−6
b3	0.05	−0.8962	1.1280	−0.00066
b4	0.05	−0.04574	0.02341	0.000876
b5	0.05	1.1233	2.5613	0.000137

Next, we fit the data using GEE2/ELS under the assumption that the second-order moments satisfy the assumption of an exchangeable correlation structure. The PROC NLMIXED code required for this analysis is shown below.

Program 4.18

```
ods output parameterestimates=GEE2pe_cs;
proc nlmixed data=Example4_3_2 empirical;
  parms b0=-0.8561 b1=1.2654 b2=0.6495 b3=0.1368
        b4=-0.0188 b5=1.8457 rho=.327;
  bounds -1<rho< 1;
  X_b = b0 + b1*Trt + b2*Center_ + b3*Gender + b4*Age + b5*y0;
  pi = exp(X_b)/(1+exp(X_b));
  Sigma_i = pi*(1-pi);
  Mu_i = pi + sqrt(Sigma_i)*u;
  Var_i = Sigma_i*(1-rho);
  model y ~ normal(Mu_i, Var_i);
  random u ~ normal(0, rho) subject=Subject;
run;
```

The code is very similar to that specified for the independence structure with one important exception. In the above code, we specify an additive random effect, `sqrt(Sigma_i)*u`, and a within-subject variance, `Var_i = Sigma_i*(1-rho)`, that together induce an exchangeable correlation structure for our marginal repeated measures logistic regression model. The NLMIXED syntax specifies that the mean and variance-covariance structure for the marginal logistic regression model is of the form

$$E(\boldsymbol{y}_i) = \mu_i(\boldsymbol{X}_i, \boldsymbol{\beta}) = g^{-1}(\boldsymbol{X}_i\boldsymbol{\beta})$$

$$Var(\boldsymbol{y}_i) = \boldsymbol{\Sigma}_i(\boldsymbol{\beta}, \boldsymbol{\theta}) = \boldsymbol{H}_i(\boldsymbol{\mu}_i)^{1/2}\boldsymbol{R}_i(\boldsymbol{\alpha})\boldsymbol{H}_i(\boldsymbol{\mu}_i)^{1/2}$$

where $\boldsymbol{H}_i(\boldsymbol{\mu}_i)$ is the 4×4 diagonal variance matrix with diagonal elements $Var(y_{ij}) = \mu_{ij}(1-\mu_{ij})$, $\boldsymbol{H}_i(\boldsymbol{\mu}_i)^{1/2}$ is the square root of $\boldsymbol{H}_i(\boldsymbol{\mu}_i)$ and $\boldsymbol{\theta} = \boldsymbol{\alpha}$ is the correlation parameter coded as `rho` above. We use the bounds statement, `bounds -1<rho< 1;` to ensure the correlation coefficient is properly bounded between -1 and 1. Using ODS OUTPUT statements, we capture the parameter estimates and their robust standard errors from both of the above GEE2-based analyses. We also capture the corresponding results from the GEE analysis performed in §4.3.2 under the "working" independence structure and "working" exchangeable correlation structure. The GEE and GEE2 results are displayed, side by side, in Output 4.29. In referring to this output, we see that the GEE-based estimates are identical under the "working" independence and "working" exchangeable correlation structures for this balanced data. However, the results under GEE2 do vary according to the type of second-order moments one is willing to assume. This is because the GEE2 approach incorporates the second-order moments directly into the set of estimating equations for $\boldsymbol{\beta}$ resulting in some changes to the model parameter estimates.

Output 4.29: A comparison of GEE and GEE2 parameter estimates, their robust standard errors and associated p-values for the respiratory disorder data. Tests (i.e., p-values) are based on the default test criteria in GENMOD (a z-test) and NLMIXED (a t-test), respectively.

		Method of Estimation					
		GEE			GEE2/ELS		
Structure	Effect	Estimate	SE	p	Estimate	SE	p
1.Independence	Intercept	−0.856	0.456	0.061	−1.136	0.589	0.056
	Trt	1.265	0.347	0.000	1.379	0.350	0.000
	Center_	0.649	0.353	0.066	0.611	0.408	0.137
	Gender	0.137	0.440	0.756	0.116	0.511	0.821
	Age	−0.019	0.013	0.148	−0.011	0.017	0.524
	y0	1.846	0.346	0.000	1.842	0.363	0.000
2.Exchangeable	Intercept	−0.856	0.456	0.061	−1.348	0.749	0.075
	Trt	1.265	0.347	0.000	1.441	0.413	0.001
	Center_	0.649	0.353	0.066	0.443	0.547	0.420
	Gender	0.137	0.440	0.756	0.201	0.733	0.784
	Age	−0.019	0.013	0.148	−0.005	0.026	0.853
	y0	1.846	0.346	0.000	1.776	0.444	0.000
	Rho	0.327	.	.	0.356	0.058	0.000

The results for GEE and GEE2 are qualitatively and quantitatively similar. This is in keeping with a similar analysis conducted by Vonesh et. al. (2001) in which they applied both GEE and GEE2 to a different set of binary data, the respiratory infection data of Zeger, Liang and Albert (1988). In that analysis, the results were remarkably similar between GEE and GEE2 both in terms of the regression parameter estimates and the robust standard errors. Under the assumption of exchangeable correlation, both the moment estimate and GEE2 estimate of the common correlation coefficient are similar. In our current example, the GEE2 method tended to yield more conservative standard error estimates but this is not always the case as illustrated with the respiratory infection data (see Vonesh et. al., 2001).

4.5.4 Epileptic seizure data—continued

In §4.3.3, we fit the number of partial seizures measured repeatedly over time in patients with epilepsy to a marginal Poisson regression model with different "working" covariance matrices. Using the GEE1-based PL/QELS estimation scheme of GLIMMIX, we examined the goodness-of-fit of five different "working" covariance structures. The data is from a randomized controlled trial comparing an active drug, progabide, to a placebo control over the course of four bi-weekly visits (see Thall and Vail, 1990). A description of the study and data is presented in §1.4 (page 15).

In this section, we analyze the data using GEE and GEE2/ELS assuming a marginal Poisson model having mean structure given by (4.22) of §4.3.3 but excluding the offset, $\log(t_{ij}) = \log(2)$. As in the previous example, we do so under each of two "working" correlation structures: independence and exchangeable correlation. The SAS code required to estimate the model parameters via GEE using GENMOD and GEE2/ELS using NLMIXED is similar to that of the previous example. For example, below is the SAS code for implementing the GEE and GEE2/ELS analysis under the assumption of an exchangeable correlation structure.

Program 4.19

```
proc genmod data=example4_3_3;
 where ID ne 207;
 class ID Visit;
 model y = Trt LogAge y0 y0*Trt Visit4 /
           dist=Poisson noscale scale=1 type3;
 repeated subject=id /type=cs modelse ECOVB MCOVB;
 output out=pred pred=yhat;
run;

proc nlmixed data=example4_3_3 empirical;
 where ID ne 207;
 parms b0 = -3 b1 = -1 b2 = 1 b3 = 1 b4 = .5 b5 = 0 rho=0;
 bounds -1<rho< 1;
 xbeta = b0 + b1*Trt + b2*LogAge + b3*y0 + b4*y0*Trt +
         b5*Visit4;
 Sigma_i = exp(xbeta);
 Mu_i = exp(xbeta) + sqrt(Sigma_i)*u;
 Var_i = Sigma_i*(1-rho);
 model y ~ normal(Mu_i, Var_i);
 random u ~ normal(0, rho) subject=ID;
run;
```

Here we use the same dataset and structure as described in §4.3.3. However, we exclude patient 207 from the analysis given the strong influence this patient exerts on both the treatment effect and its interaction with baseline seizure rates (see Figure 4.2). The results for both correlation structures are shown in Output 4.30. When compared with GEE, the GEE2/ELS parameter estimates are slightly attenuated toward zero (except for Visit4). Using the macro %GOF, the adjusted R-square under the exchangeable correlation structure was 0.447 versus 0.365 in favor of the GEE-based fit to the data. This suggests some bias may be present with the GEE2-based analysis.

Output 4.30: A comparison of GEE and GEE2 parameter estimates, their robust standard errors and associated p-values for the epileptic seizure data. Results reflect the exclusion of patient 207 as an outlier. Also, the estimated intercept reflects our exclusion of the offset, $\log(2)$, from this analysis.

		Method of Estimation					
		GEE			GEE2/ELS		
Structure	Effect	Estimate	SE	p	Estimate	SE	p
1.Independence	Intercept	−2.327	0.871	0.008	−2.026	0.844	0.020
	Trt	−0.521	0.415	0.209	−0.422	0.506	0.408
	LogAge	0.769	0.254	0.002	0.742	0.246	0.004
	y0	0.950	0.096	0.000	0.942	0.137	0.000
	y0*Trt	0.138	0.194	0.477	0.092	0.229	0.690
	Visit4	−0.148	0.076	0.053	−0.284	0.092	0.003
2.Exchangeable	Intercept	−2.358	0.884	0.008	−1.880	0.868	0.035
	Trt	−0.520	0.418	0.214	−0.424	0.568	0.458
	LogAge	0.777	0.257	0.002	0.716	0.250	0.006
	y0	0.951	0.098	0.000	0.948	0.160	0.000
	y0*Trt	0.139	0.195	0.476	0.076	0.254	0.765
	Visit4	−0.148	0.076	0.053	−0.252	0.084	0.004
	Rho	0.336	.	.	0.226	0.036	0.000

In this particular example and, in general, for any GLIM for correlated data, we can perform a GEE2/ELS analysis assuming a "working" independence structure for the second-order moments and a "working" Gaussian assumption for the third- and fourth-order moments using GLIMMIX rather than NLMIXED. The GLIMMIX code for performing the same analysis under the "working" independence structure is shown below.

Program 4.20

```
proc glimmix data=example4_3_3 empirical method=laplace;
 where ID ne 207;
 class ID;
 _variance_=_mu_;
 parms (0) / hold=1;
 model y = Trt LogAge y0 y0*Trt Visit4 / link=log s;
 random intercept / subject=id;
 output out=GEE2out pred=yhat;
      id id _xbeta_ _mu_ _variance_ ;
run;
```

Here we use the same idea as with NLMIXED in that we utilize a dummy random effect to "trick" GLIMMIX into performing a GEE2/ELS analysis under a "working" independence structure. To implement GEE2/ELS, we first must specify the appropriate link function either through the LINK=option or through programming statements. (Note: one cannot use the DIST= option to assign a default link function when implementing GEE2/ELS.) In this example, we specify a log link function via the MODEL option, `link=log`, so as to be consistent with our prior GEE-based analysis. We then specify the variance function via the programming statement

```
_variance_=_mu_;
```

where `_variance_` and `_mu_` are GLIMMIX automatic variables that define the variance and mean function, respectively (see the GLIMMIX documentation for a complete definition of automatic variables). We then specify a dummy random intercept effect with a variance fixed at 0. This is achieved with the above RANDOM and PARMS statements. (Note: the statement, `parms (0) / hold=1;` instructs GLIMMIX to hold the variance of the random intercept to be 0.) Finally, by specifying METHOD=LAPLACE in combination with our user-defined variance function, we are instructing GLIMMIX to apply maximum likelihood estimation under normality (the DIST= option defaults to normality in this case) assuming the mean and variance are those defined by the inverse link function and specified variance function. This results in exactly the same GEE2/ELS estimates shown in Output 4.30 for the "working" independence structure. The advantage of using GLIMMIX for a GEE2 analysis is that we can now utilize the full set of features available in GLIMMIX to conduct our analysis—features such as the use of a CLASS statement to define the model and the use of the ESTIMATE, CONTRAST and LSMEANS statements to conduct specific tests of interest, etc.

In addition to the preceding analysis, we also fit the data to the following multivariate Poisson model as described by Vonesh et. al. (2001):

$$y_{ij} = u_i + z_{ij} \quad (i = 1, \ldots, n;\ j = 1, \ldots, 4) \tag{4.61}$$

where $u_i \sim$iid $P(\mu)$ with $\mu = \exp(\theta)$, and $z_{ij} \sim$ind $P(\mu_{ij}(\boldsymbol{\beta}))$ such that, conditional on u_i, $E(y_{ij}|u_i) = u_i + \mu_{ij}(\boldsymbol{\beta}) = u_i + \exp(\boldsymbol{x}'_{ij}\boldsymbol{\beta})$ with the linear predictor $\boldsymbol{x}'_{ij}\boldsymbol{\beta}$ being the same as that corresponding to the marginal mean (4.22) of §4.3.3 but excluding the common offset, $\log(t_{ij}) = \log(2)$ [i.e., $\boldsymbol{x}'_{ij}\boldsymbol{\beta} = \beta_0 + \beta_1 a_{1i} + \beta_2 \log(a_{2i}) + \beta_3 \log(a_{3i}) + \beta_4 a_{1i} \log(a_{3i}) + \beta_5 v_{i4}$]. Under this model, the marginal means, variances, covariances and correlations are given by

$$\begin{aligned} E(y_{ij}) &= \mu + \mu_{ij}(\boldsymbol{\beta}) = \exp(\theta) + \exp(\boldsymbol{x}'_{ij}\boldsymbol{\beta}) \\ Var(y_{ij}) &= \mu + \mu_{ij}(\boldsymbol{\beta}) = \exp(\theta) + \exp(\boldsymbol{x}'_{ij}\boldsymbol{\beta}) \\ Cov(y_{ij}, y_{ij'}) &= \mu = \exp(\theta) \\ Corr(y_{ij}, y_{ij'}) &= \frac{\mu}{\sqrt{\mu + \mu_{ij}(\boldsymbol{\beta})}\sqrt{\mu + \mu_{ij'}(\boldsymbol{\beta})}} \\ &= \frac{\exp(\theta)}{\sqrt{\exp(\theta) + \exp(\boldsymbol{x}'_{ij}\boldsymbol{\beta})}\sqrt{\exp(\theta) + \exp(\boldsymbol{x}'_{ij'}\boldsymbol{\beta})}}. \end{aligned} \tag{4.62}$$

Unlike with the marginal Poisson model (4.22) where an exchangeable correlation assumption implies the correlation between repeated measurements remains constant and the covariance changes with the mean, under the multivariate Poisson model (4.61), the correlation varies with the mean while the covariance remains constant.

We can fit this multivariate Poisson model to the epileptic seizure data using either GEE2 with "working" third- and fourth-order moments assuming normality (i.e., ELS) or we can directly maximize the log-likelihood function as specified by Johnson and Kotz (1969). If we let $y_i^* = \min\{y_{ij}; j = 1, \ldots, 4\}$, then under (4.61), the multivariate pdf of $\boldsymbol{y}_i = (y_{i1}, y_{i2}, y_{i3}, y_{i4})'$ is

$$\pi(\boldsymbol{y}_i; \boldsymbol{\beta}, \theta) = e^{-\left(\mu + \sum_{j=1}^{4} \mu_{ij}(\boldsymbol{\beta})\right)} \sum_{k=0}^{y_i^*} \left\{ \frac{\mu^k}{k!} \prod_{j=1}^{4} \frac{\mu_{ij}(\boldsymbol{\beta})^{(y_{ij}-k)}}{(y_{ij}-k)!} \right\} \tag{4.63}$$

where $\mu = \exp(\theta) = E(u_i)$ and $\mu_{ij}(\boldsymbol{\beta}) = \exp(\boldsymbol{x}'_{ij}\boldsymbol{\beta}) = E(z_{ij})$. From (4.63), we can write the log-likelihood function as

$$L(\boldsymbol{\beta},\theta;\boldsymbol{y}) = \sum_{i=1}^{n}\left(-\Big(\mu + \sum_{j=1}^{4}\mu_{ij}(\boldsymbol{\beta})\Big) + \log\left(\sum_{k=0}^{y_i^*}\left\{\frac{\mu^k}{k!}\prod_{j=1}^{4}\frac{\mu_{ij}(\boldsymbol{\beta})^{(y_{ij}-k)}}{(y_{ij}-k)!}\right\}\right)\right) \tag{4.64}$$

$$= \sum_{i=1}^{n} -\Big(\exp(\theta) + \sum_{j=1}^{4}\exp(\boldsymbol{x}'_{ij}\boldsymbol{\beta})\Big) +$$

$$\sum_{i=1}^{n}\log\left(\sum_{k=0}^{y_i^*}\left\{\frac{\exp(\theta k)}{k!}\frac{\exp\left\{\sum_{j=1}^{4}(\boldsymbol{x}'_{ij}\boldsymbol{\beta})(y_{ij}-k)\right\}}{\prod_{j=1}^{4}(y_{ij}-k)!}\right\}\right).$$

One can carry out both GEE2/ELS and ML estimation using the SAS procedure NLMIXED. Below is the SAS code and NLMIXED syntax required to fit the epileptic seizure data to the multivariate Poisson model (4.61) using GEE2/ELS in combination with the first two moments specified in (4.62). The code is based on defining a "Gaussian" linear random effects model, $y_{ij} = \mu_{ij} + u_i + \epsilon_{ij}$, with $u_i \sim\text{iid}\, N(\mu,\mu) \perp \epsilon_{ij} \sim\text{ind}\, N(0,\mu_{ij})$ which yields an ELS analysis. We first compute mean values of the covariates, `LogAge` and `y0`, which we merge into a new dataset, `example4_5_4, in order to` compute least squares means using ESTIMATE statements in NLMIXED.

Program 4.21

```
proc means data=example4_3_3 noprint mean;
 where Visit=1 and ID ne 207;
 var y0 LogAge;
 output out=MeanCovariates mean=Avg_y0 Avg_LogAge;
run;
data example4_5_4;
 set example4_3_3;
 by ID;
 if _n_=1 then set MeanCovariates;
run;
ods output ParameterEstimates = pe;
ods output AdditionalEstimates = ae;
ods listing close;
proc nlmixed data=example4_5_4 qpoints=1 cov empirical;
 where ID ne 207;
 parms b0 = -3 b1 = 1 b2 = 1 b3 = -1 b4 = .5
       b5 = -.30 theta=0;
 /* Define time-dependent conditional linear predictor */
 xbeta = b0 + b1*Trt + b2*LogAge + b3*y0 + b4*y0*Trt +
         b5*Visit4;
 /* Define Mu_ij=E(Zij) and Mu = E(Ui) under the */
 /* multivariate Poisson model: Yij=Ui+Zij       */
 Mu_ij = exp(xbeta);
 Mu = exp(theta);
 /* Define an exchangeable random effect Ui */
 /* and the conditional moments of Yij | Ui */
```

```
  Ui =  exp(theta) + sqrt(exp(theta))*u;
  Mean_ij = exp(xbeta) + Ui;
  Var_ij = exp(xbeta);
  /* Define least squares mean linear predictors  */
  /* for visits 1 and 4 for both treatment groups */
  xbeta1_0 = b0 + b1*0 + b2*Avg_LogAge + b3*Avg_y0 +
             b4*Avg_y0*0;
  xbeta4_0 = b0 + b1*0 + b2*Avg_LogAge + b3*Avg_y0 +
             b4*Avg_y0*0 + b5;
  xbeta1_1 = b0 + b1*1 + b2*Avg_LogAge + b3*Avg_y0 +
             b4*Avg_y0*1;
  xbeta4_1 = b0 + b1*1 + b2*Avg_LogAge + b3*Avg_y0 +
             b4*Avg_y0*1 + b5;

  /* Define predicted mean count per individual */
  E_Yij = Mu + Mu_ij;
  /* Define first- and second-order moments evaluated at */
  /* the least squares mean linear predictors for both   */
  /* treatment groups at visits 1 and 4 noting that the  */
  /* moments are all the same for visits 1-3             */
  Mu1_0 = exp(xbeta1_0) + exp(theta);
  Mu4_0 = exp(xbeta4_0) + exp(theta);
  Mu1_1 = exp(xbeta1_1) + exp(theta);
  Mu4_1 = exp(xbeta4_1) + exp(theta);
  Var1_0 = Mu1_0;
  Var4_0 = Mu4_0;
  Var1_1 = Mu1_1;
  Var4_1 = Mu4_1;
  Cov_jk =  exp(theta);
  Corr_jk_0 = Cov_jk/Var1_0;
  Corr_j4_0 = Cov_jk/( sqrt(Var1_0)*sqrt(Var4_0) );
  Corr_jk_1 = Cov_jk/Var1_1;
  Corr_j4_1 = Cov_jk/( sqrt(Var1_1)*sqrt(Var4_1) );
  estimate 'Mean Rate Trt(0): Visit< 4' Mu1_0;
  estimate 'Mean Rate Trt(1): Visit< 4' Mu1_1;
  estimate 'Mean Difference : Visit< 4' Mu1_0 - Mu1_1;
  estimate 'Rate Ratio(1:0)  : Visit< 4' Mu1_1/Mu1_0;
  estimate 'Mean Rate Trt(0): Visit=4' Mu4_0;
  estimate 'Mean Rate Trt(1): Visit=4' Mu4_1;
  estimate 'Mean Difference : Visit=4' Mu4_0 - Mu4_1;
  estimate 'Rate Ratio(1:0)  : Visit=4' Mu4_1/Mu4_0;
  estimate 'Cov(Yij,Yik): All Visits' Cov_jk;
  estimate 'Corr(Yij,Yik) (j<k, k< 4) Trt(0):' Corr_jk_0;
  estimate 'Corr(Yij,Yik) (j<k, k=4) Trt(0):' Corr_j4_0;
  estimate 'Corr(Yij,Yik) (j<k, k< 4) Trt(1):' Corr_jk_1;
  estimate 'Corr(Yij,Yik) (j<k, k=4) Trt(1):' Corr_j4_1;
  model y ~ Normal(Mean_ij, Var_ij);
  random u ~ normal(0,1) subject=ID;
  predict E_Yij out=pred;
run;
ods listing;
data pe;
 set pe;
 Effect=Parameter;
run;
```

```
proc print data=pe noobs split='|' ;
 var Parameter Effect Estimate StandardError DF tValue
     Probt Lower Upper;
 label tValue='t Test' Probt='p-value' ;
 format Effect $effectf. Estimate StandardError tValue
        Probt Lower Upper 5.3;
run;
proc print data=ae noobs split='|' ;
 var Label Estimate StandardError DF Lower Upper;
 label Label='Statistic' StandardError='SE';
 format Estimate StandardError Lower Upper 5.3;
run;
```

The NLMIXED syntax is heavily commented so that one can follow the various programming assignment statements needed to define the various parameter estimates, etc. The ESTIMATE statements compute summary statistics on the marginal moments for visits 1-3 and for visit 4 evaluated at the mean values of the two covariates, `LogAge` and `y0`. Select results are shown in Output 4.31.

Output 4.31: Select GEE2/ELS results for the epileptic seizure data under the multivariate Poisson model (4.61) with first- and second-order moments defined by (4.62). Inference is based on the robust covariance matrix, $\widehat{\Omega}_R(\widehat{\tau})$.

Parameter	Effect	Estimate	SE	DF	t Value	p	Lower	Upper
b0	Intercept	−2.75	0.961	57	−2.86	0.006	−4.67	−.822
b1	Trt	−.545	0.703	57	−.776	0.441	−1.95	0.862
b2	LogAge	0.859	0.263	57	3.265	0.002	0.332	1.386
b3	y0	1.038	0.163	57	6.382	0.000	0.712	1.363
b4	y0*Trt	0.120	0.299	57	0.402	0.689	−.479	0.720
b5	Visit4	−.297	0.101	57	−2.94	0.005	−.500	−.095
theta	Theta	0.459	0.267	57	1.718	0.091	−.076	0.994

Statistic	Estimate	SE	DF	Lower	Upper
Mean Rate Trt(0): Visit<4	8.342	1.395	57	5.550	11.14
Mean Rate Trt(1): Visit<4	6.412	1.110	57	4.189	8.634
Mean Difference : Visit<4	1.931	1.572	57	−1.22	5.079
Rate Ratio(1:0) : Visit<4	0.769	0.163	57	0.443	1.094
Mean Rate Trt(0): Visit=4	6.604	0.845	57	4.911	8.297
Mean Rate Trt(1): Visit=4	5.169	0.839	57	3.489	6.850
Mean Difference : Visit=4	1.434	1.124	57	−.817	3.685
Rate Ratio(1:0) : Visit=4	0.783	0.153	57	0.476	1.089
Cov(Yij,Yik): All Visits	1.583	0.423	57	0.736	2.430
Corr(Yij,Yik) (j<k, k<4) Trt(0):	0.190	0.046	57	0.097	0.282
Corr(Yij,Yik) (j<k, k=4) Trt(0):	0.213	0.050	57	0.113	0.314
Corr(Yij,Yik) (j<k, k<4) Trt(1):	0.247	0.055	57	0.136	0.358
Corr(Yij,Yik) (j<k, k=4) Trt(1):	0.275	0.061	57	0.152	0.397

The application of GEE2/ELS to the multivariate Poisson model with exchangeable covariance structure yields regression parameter estimates, $\widehat{\boldsymbol{\beta}}$, that are in closer agreement with the GEE estimates than the GEE2/ELS estimates from the Poisson model with exchangeable correlation (see Output 4.30). Moreover when we exclude patient 207, we find no evidence of a treatment effect (parameter b1 in above output) nor of any significant interaction between baseline count and treatment groups (parameter b4 above). In addition to inference based strictly on the model parameters, the results in Output 4.31 also summarize what the marginal moments are predicted to be for the two treatment groups at the average of the covariates (i.e., `Avg_y0` and `Avg_LogAge` in the above SAS code). The visit-specific marginal means at the average baseline log(count) and average log(age) are similar to the overall summary means reported by Thall and Vail when one excludes patient 207. We also estimated the mean difference in the seizure counts as well as the ratio of the seizure rates (i.e., the rate ratio) evaluated at the average of the modeled covariates. Neither the mean difference between the two groups (the 95% confidence interval covers 0) nor the rate ratio (the 95% confidence interval covers 1.00) provides any evidence of a significant treatment effect. The estimated correlation coefficients at visits 1-4 for the two treatment groups tend to agree more closely with the GEE2/ELS based exchangeable correlation coefficient than with its GEE counterpart (see Output 4.30). This raises the possibility that the multivariate Poisson model may be a good approximation to the true but unknown stochastic model for this particular application. To check this, we applied the macro %GOF to the GEE2/ELS predicted means and found the adjusted R-square to be 0.389 which is in-between that of the GEE (0.447) and GEE2/ELS (0.365) predicted means from the Poisson model with exchangeable correlation.

Given these inconclusive results from the application of GEE2/ELS to model (4.61), we proceed to estimate the parameters of the multivariate Poisson model using ML estimation. To do so, we first re-arrange the data on the horizontal scale using the TRANSPOSE procedure as shown below. This will make programming the log-likelihood (4.64) within NLMIXED much easier than if we had kept the data vertically arrayed. The variable, `Yij`, is a dummy variable created as a place mark for specifying the log-likelihood function (see the NLMIXED syntax below). As with the GEE2/ELS approach, we apply a dummy random effect, `u`, which we specify as being normally distributed with mean 0 and variance 0. This allows us to compute an empirical sandwich estimator of the variance-covariance matrix of the ML estimate $\widehat{\tau}_{\text{MLE}} = (\widehat{\beta}_{\text{MLE}}, \widehat{\theta}_{\text{MLE}})$ using the EMPIRICAL option of NLMIXED. After having defined the log-likelihood function via the assignment statement, `loglike = logexpon + log(Summ_ij);` the remainder of the NLMIXED syntax is the same as that used for GEE2/ELS estimation. The additional code yields the ML-based estimates of the marginal moments evaluated at the mean of the covariates.

Program 4.22

```
proc transpose data=example4_5_4 out=new prefix=y;
 by ID;
 var y;
 copy y0 LogAge Avg_y0 Avg_LogAge Trt LogTime;
run;
data new;
 set new;
 if _NAME_='y';
 Yij=1;
 Drop _NAME_;
run;
proc print data=new;
run;
ods output ParameterEstimates = pe_ML;
ods output AdditionalEstimates = ae_ML;
ods listing close;
proc nlmixed data=new qpoints=1 cov empirical;
 where ID ne 207;
 parms b0 = -3 b1 = -1 b2 = 0 b3 = 0 b4 = 0 b5 = 0
       theta=0;
 /* Define xbeta for visits 1-4 */
 xbeta1  = (b0 + b1*Trt + b2*LogAge + b3*y0 + b4*y0*Trt);
 xbeta2  = (b0 + b1*Trt + b2*LogAge + b3*y0 + b4*y0*Trt);
 xbeta3  = (b0 + b1*Trt + b2*LogAge + b3*y0 + b4*y0*Trt);
 xbeta4  = (b0 + b1*Trt + b2*LogAge + b3*y0 + b4*y0*Trt + b5);
```

```
/* Define joint likelihood function via Johnson and Kotz */
logexpon = -( exp(xbeta1) + exp(xbeta2) +
              exp(xbeta3) + exp(xbeta4) + exp(theta) + u );
array y[4] y1-y4;
array xbeta[4] xbeta1-xbeta4;
Min_y=min(of y1-y4);
Summ_ij = 0;
do j=0 to Min_y by 1;
  xbeta_j = xbeta[1]*(y[1]-j);
  gamma_j = gamma(y[1]-j+1);
 do k=2 to 4;
  xbeta_j = xbeta_j + xbeta[k]*(y[k]-j);
  gamma_j = gamma_j*gamma(y[k]-j+1);
 end;
 Summ_ij = Summ_ij +
         (exp(theta*j)/gamma(j+1))*exp(xbeta_j)/gamma_j;
end;
random u ~ normal(0,0) subject=ID;
loglike = logexpon + log(Summ_ij);
/* Define least squares mean linear predictors  */
/* for visits 1 and 4 for both treatment groups */
xbeta1_0 = b0 + b1*0 + b2*Avg_LogAge + b3*Avg_y0 +
           b4*Avg_y0*0;
xbeta4_0 = b0 + b1*0 + b2*Avg_LogAge + b3*Avg_y0 +
           b4*Avg_y0*0 + b5;
xbeta1_1 = b0 + b1*1 + b2*Avg_LogAge + b3*Avg_y0 +
           b4*Avg_y0*1;
xbeta4_1 = b0 + b1*1 + b2*Avg_LogAge + b3*Avg_y0 +
           b4*Avg_y0*1 + b5;
/* Define first- and second-order moments evaluated at */
/* the least squares mean linear predictors for both   */
/* treatment groups at visits 1 and 4 noting that the  */
/* moments are all the same for visits 1-3             */
Mu1_0 = exp(xbeta1_0) + exp(theta);
Mu4_0 = exp(xbeta4_0) + exp(theta);
Mu1_1 = exp(xbeta1_1) + exp(theta);
Mu4_1 = exp(xbeta4_1) + exp(theta);
Var1_0 = Mu1_0;
Var4_0 = Mu4_0;
Var1_1 = Mu1_1;
Var4_1 = Mu4_1;
Cov_jk =  exp(theta);
Corr_jk_0 = Cov_jk/Var1_0;
Corr_j4_0 = Cov_jk/( sqrt(Var1_0)*sqrt(Var4_0) );
Corr_jk_1 = Cov_jk/Var1_1;
Corr_j4_1 = Cov_jk/( sqrt(Var1_1)*sqrt(Var4_1) );
estimate 'Mean Rate Trt(0): Visit< 4' Mu1_0;
estimate 'Mean Rate Trt(1): Visit< 4' Mu1_1;
estimate 'Mean Difference : Visit< 4' Mu1_0 - Mu1_1;
estimate 'Rate Ratio(1:0)  : Visit< 4' Mu1_1/Mu1_0;
estimate 'Mean Rate Trt(0): Visit=4' Mu4_0;
estimate 'Mean Rate Trt(1): Visit=4' Mu4_1;
estimate 'Mean Difference : Visit=4' Mu4_0 - Mu4_1;
estimate 'Rate Ratio(1:0)  : Visit=4' Mu4_1/Mu4_0;
```

```
 estimate 'Cov(Yij,Yik): All Visits' Cov_jk;
 estimate 'Corr(Yij,Yik) (j<k, k< 4) Trt(0):' Corr_jk_0;
 estimate 'Corr(Yij,Yik) (j<k, k=4) Trt(0):' Corr_j4_0;
 estimate 'Corr(Yij,Yik) (j<k, k< 4) Trt(1):' Corr_jk_1;
 estimate 'Corr(Yij,Yik) (j<k, k=4) Trt(1):' Corr_j4_1;
 model Yij ~ general(loglike);
run;
ods listing;
data pe_ML;
 set pe_ML;
 Effect=Parameter;
run;
proc print data=pe_ML noobs split='|' ;
 var Parameter Effect Estimate StandardError DF
     tValue Probt Lower Upper;
 label StandardError='SE' tValue='t Value' Probt='p' ;
 format Effect $effectf. Estimate StandardError
        tValue Probt Lower Upper 5.3;
run;
proc print data=ae_ML noobs split='|' ;
 var Label Estimate StandardError DF Lower Upper;
 ark Label='Statistic' StandardError='SE';
 format Estimate StandardError Lower Upper 5.3;
run;
```

Output 4.32: Select MLE results for the epileptic seizure data under the multivariate Poisson model (4.61) with first- and second-order moments defined by (4.62). Inference is based on the robust covariance matrix, $\widehat{\Omega}_R(\widehat{\tau})$.

Parameter	Effect	Estimate	SE	DF	t Value	p	Lower	Upper
b0	Intercept	−2.64	1.013	57	−2.61	0.012	−4.67	−.612
b1	Trt	−.670	0.521	57	−1.28	0.204	−1.71	0.374
b2	LogAge	0.822	0.279	57	2.950	0.005	0.264	1.380
b3	y0	0.981	0.118	57	8.286	0.000	0.744	1.218
b4	y0*Trt	0.180	0.232	57	0.776	0.441	−.285	0.645
b5	Visit4	−.163	0.084	57	−1.94	0.057	−.331	0.005
theta	Theta	−.456	0.527	57	−.866	0.390	−1.51	0.599

Statistic	Estimate	SE	DF	Lower	Upper
Mean Rate Trt(0): Visit<4	6.652	0.806	57	5.039	8.265
Mean Rate Trt(1): Visit<4	4.845	0.659	57	3.525	6.165
Mean Difference : Visit<4	1.807	1.005	57	−.206	3.820
Rate Ratio(1:0) : Visit<4	0.728	0.128	57	0.472	0.985
Mean Rate Trt(0): Visit=4	5.747	0.572	57	4.602	6.891
Mean Rate Trt(1): Visit=4	4.212	0.585	57	3.039	5.384
Mean Difference : Visit=4	1.535	0.824	57	−.114	3.185
Rate Ratio(1:0) : Visit=4	0.733	0.126	57	0.480	0.985
Cov(Yij,Yik): All Visits	0.634	0.334	57	−.035	1.302
Corr(Yij,Yik) (j<k, k<4) Trt(0):	0.095	0.052	57	−.008	0.199
Corr(Yij,Yik) (j<k, k=4) Trt(0):	0.102	0.055	57	−.008	0.213
Corr(Yij,Yik) (j<k, k<4) Trt(1):	0.131	0.069	57	−.007	0.268
Corr(Yij,Yik) (j<k, k=4) Trt(1):	0.140	0.073	57	−.006	0.286

Shown in Output 4.32 are the ML estimates of the model parameters under the multivariate Poisson model (4.61). The estimated regression parameters are quantitatively more closely aligned with the GEE versus GEE2 estimates under the two "working" Poisson models (Output 4.30). Likewise, inference based on the empirical covariance matrix shown in Output 4.32 is in better agreement with GEE-based inference under either the "working" independence structure or "working" exchangeable correlation structure

(Output 4.30). Indeed, the adjusted R-square is estimated to be 0.443 for the ML-based predicted means under the multivariate Poisson model which is close to the 0.447 adjusted R-square for the GEE-based predicted means. Interestingly, under the multivariate Poisson model, the ML and GEE2 estimates of θ differ ($\widehat{\theta}_{\mathrm{ML}} = -0.456$ versus $\widehat{\theta}_{\mathrm{GEE2}} = 0.459$) resulting in lower likelihood-based covariance and correlation estimates. These likelihood-based correlations are also lower than the common correlation under the Poisson model with exchangeable correlation (Output 4.30). It is not clear whether the close agreement between the GEE-based analysis assuming either "working" independence or "working" exchangeable correlation and the likelihood-based analysis under a multivariate Poisson model is indicative of a possible underlying stochastic mechanism. However, this close agreement does suggest that fully parametric models like the multivariate Poisson model can provide users with a reasonable alternative to a strictly moments-based approach that is GEE. When applying a parametric ML approach to the analysis as done here, it may still prove useful to conduct inference based on a robust covariance matrix of $\widehat{\boldsymbol{\tau}}_{\mathrm{MLE}}$ so as to safeguard against possible model misspecification. For example, rather than assuming u_i and z_{ij} are each distributed as Poisson as is done under model (4.61), it may be that the underlying multivariate model corresponds to pdf's of u_i and z_{ij} that are negative binomial with overdispersion. In our current example, the use of a robust covariance matrix, $\widehat{\boldsymbol{\Omega}}_R(\widehat{\boldsymbol{\tau}})$, might safeguard against such misspecification.

4.6 Computational considerations

In this chapter and the following, we are faced with the problem of estimating parameters from nonlinear models which in turn entails the general problem of nonlinear optimization. As a result, there are a number of practical issues to be considered when applying the various nonlinear optimization algorithms available with NLMIXED, %NLINMIX and, to a lesser extent, GLIMMIX. In this section, we briefly consider three such issues: model parameterization, scaling of parameters and starting values.

4.6.1 Model parameterization and scaling

One of the key aspects to consider when fitting data to a nonlinear model is the parameterization one uses. By carefully choosing the parameterization, one can minimize problems with convergence as well as ensure the parameter estimates make sense for a given application. We recall from the LDH leakage data (§4.5.1) that the underlying model should yield predicted values that have the same range as the response variable which, in this example, is the measured proportion of LDH enzyme leakage. The logistic growth curve model fit by Gennings et. al. (1989) is parameterized to ensure the predicted response lies in an interval between 0 and 1 which is in the range of LDH values measured. However, their model provided a poor fit to the observed proportion of LDH leakage at the higher dose levels of $CHCl_3$. Consequently, we fit the data using a Weibull growth curve model as a means for providing greater flexibility in modeling the observed proportions over time. The structural form of the Weibull growth curve model is given by

$$LDH = 1 - \eta_0 \exp\left(\eta_1 t^{\eta_2}\right)$$

where LDH is the proportion of LDH leakage at a given time point. The problem with this parameterization is there is no guarantee that at time $t = 0$, the "intercept" $1 - \eta_0$ will lie in the range between 0 and 1. In fact, there are no guarantees that the predicted response at any value of t will lie between 0 and 1. Consequently, we reparameterized this model to enforce the constraint that the predicted response lies between 0 and 1 by setting $\eta_0 = 1/(1+\exp(-\beta_0))$ and $\eta_1 = \exp(\beta_1)$. The reparameterization based on the inverse-logit, $\eta_0 = 1/(1+\exp(-\beta_0))$, ensures us that for any value of β_0 on the real line, η_0

will lie between 0 and 1. When combined with the reparameterization $\eta_1 = \exp(\beta_1)$ which forces η_1 to be positive for any real value β_1, the resulting reparameterized model will be constrained to have a predicted response that lies between 0 and 1. In some cases, one may be able to impose such constraints directly, for example, through the use of the BOUNDS statement in NLMIXED.

The use of the BOUNDS statement is also useful for problems where, during optimization, the specified algorithm iterates to a point where either the objective function or one of its derivatives cannot be evaluated. One can use a BOUNDS statement to direct the algorithm away from such problem values. However, such problems may also be due to the scale one uses when specifying parameters. The scaling of parameters is important if one is to avoid floating-point errors and overflows. As noted in the on-line documentation for NLMIXED, if the scaling of parameters varies by more than a few orders of magnitude, the numerical stability of the optimization problem can be seriously reduced resulting in computational difficulties. One suggestion is to rescale each of the parameters so that the final estimated values will all be within one or two orders of magnitude of each other. Examples of this will be forthcoming in Chapter 5.

4.6.2 Starting values

As with any nonlinear optimization problem, poor starting values can lead to problems with convergence. Indeed, without good starting values, there is no guarantee that the algorithm used will converge and even if it does converge, it may converge to a wrong set of values. There are any number of ways one can go about obtaining good starting values depending on the model being fit. One approach available with PROC NLMIXED is to specify a grid of starting values using the PARMS statement. PROC NLMIXED computes the objective function value at each possible grid point and chooses the best grid point as the initial point for the optimization process.

In some cases, good starting values can be obtained by simply "looking" at the data. For example, if there is an "intercept" parameter and/or an "asymptote" parameter, one can usually determine a set of reasonable starting values by plotting the data and eyeballing what values appear to be reasonable. In other cases, one may have a basic structural model in which a set of parameters describes the intrinsic response function of a "control" group to which is amended additional parameters describing the effects of various covariates. One may specify 0 starting values for the additional parameters knowing that a value of 0 corresponds to the null hypothesis of no effect due to each particular covariate. For example, in the LDH leakage data of §4.5.1, we specify starting values for the intrinsic parameters of the "control" group (i.e., the group corresponding to a 0 mM dose of CCl_4 and 0 mM dose of $CHCl_3$) with all other regression parameters set to 0. Specifically, we set the "intercept" parameter β_{00} to 2.2 which corresponds to an initial proportion of LDH leakage of 10% at time 0 and we set the "rate" parameter β_{10} to -0.693 and the "shape" parameter β_{20} to 1 corresponding to a 0.50 constant leakage rate. We then set all remaining parameters to 0 which would correspond to the overall null hypothesis of no dose-response relationship with CCl_4 or $CHCl_3$ (see the NLMIXED syntax of this example for details).

While there have been a number of recommendations made for choosing good starting values, most of the emphasis has been on selecting starting values for the regression parameters, of interest. For a more thorough treatment of how one might go about choosing good starting values of the regression parameters, $\boldsymbol{\beta}$, the reader is referred to texts by Bates and Watts (1988) and Ratkowsky (1983). In Chapter 5, we return to the issue of starting values as it pertains to the variance-covariance parameters of nonlinear random effects. There we will demonstrate just how crucial it is to have good starting values.

4.7 Summary

In this chapter we have presented two classes of marginal models for correlated response data—the generalized linear model (GLIM) and the generalized nonlinear model (GNLM). Both have important applications as has been illustrated with a number of unique and challenging examples. The GLIM is most useful in applications involving discrete and/or continuous outcomes for which the assumption of normality is unlikely to hold. It is applicable to distributions from the regular exponential family and one can easily account for correlated response data through specification of a "working" covariance structure. Its principal drawback is that it requires the mean response to be a monotonic function of a linear predictor. The GNLM extends the GLIM by 1) allowing for a broader class of distributions including those from the quadratic exponential family and 2) allowing for a more general nonlinear structure that does not require the specification of a linear predictor. The GNLM includes, as a special case, the important class of marginal nonlinear models, the normal-theory nonlinear model (NLM).

We have also described a number of different estimation schemes for GLIM's and GNLM's each within the context of available software by SAS. These include the use of generalized estimating equations which, with respect to $\boldsymbol{\beta}$, may be classified as either first-order (GEE1) or second-order (GEE2) depending on whether the estimating equations account for the presence of the regression parameters, $\boldsymbol{\beta}$, within the marginal variance-covariance matrix. Specifically, the class of GEE1 estimators are characterized by estimating equations for $\boldsymbol{\beta}$ that are based solely on the first-order moments (i.e., marginal means). This includes the familiar GEE approach of Liang and Zeger (1986) in which the variance-covariance parameters $\boldsymbol{\theta}$ are estimated via method-of moments and the

Table 4.5 Estimation techniques available within SAS for marginal GLIM's and GNLM's for correlated response data.

Procedure or Macro	Class of Models	Key Statements and/or Options	Estimation of $\boldsymbol{\beta}$	Remarks
GENMOD	GLIM	REPEATED/TYPE=	GEE	$\boldsymbol{\theta}$ estimated via MM
GLIMMIX	GLIM	RANDOM/RSIDE	GEE1	$\boldsymbol{\theta}$ estimated via PL/REPL
%NLINMIX	GNLM	REPEATED/TYPE=	GEE1	$\boldsymbol{\theta}$ estimated via PL/REPL
NLMIXED GLIMMIX	GNLM	MODEL statement: y ~ normal(mu, sigsq); A dummy RANDOM statement combined with the EMPIRICAL option	GEE2	$\boldsymbol{\tau} = (\boldsymbol{\beta}, \boldsymbol{\theta})$ estimated via GEE2/ELS
		MODEL statement: y ~ general(loglike); where loglike defines the log-likelihood	MLE	$\boldsymbol{\tau} = (\boldsymbol{\beta}, \boldsymbol{\theta})$ estimated via MLE as specified by loglike

pseudo-likelihood/quasi-extended least squares (PL/QELS) approach in which $\boldsymbol{\theta}$ is estimated via normal-theory ML or REML estimation applied to a pseudo-linear model. The former is the approach implemented in GENMOD while the latter is what is implemented in both GLIMMIX and %NLINMIX. Under the assumption that the marginal means, $\boldsymbol{\mu}_i(\boldsymbol{\beta})$, are correctly specified, GEE1 estimators of $\boldsymbol{\beta}$ will be consistent and asymptotically normally distributed as $n \to \infty$.

The class of GEE2 estimators are characterized by estimating equations for $\boldsymbol{\beta}$ that are based on both the first- and second-order moments (i.e., the marginal means, variances and covariances). This class includes the extended least squares (ELS) estimator obtained by simply treating the response vector, $\boldsymbol{y}_i$, as though it were normally distributed with mean $\boldsymbol{\mu}_i(\boldsymbol{\beta})$ and variance-covariance $\boldsymbol{\Sigma}_i(\boldsymbol{\beta}, \boldsymbol{\theta})$ regardless of what the underlying distribution of $\boldsymbol{y}_i$ is. As illustrated in the preceding examples, estimation based on GEE2/ELS can often be accomplished using NLMIXED. It is also possible to use GLIMMIX to obtain GEE2/ELS estimates assuming a "working" independence structure for the second-order moments and "working" Gaussian assumptions for the third- and fourth-order moments (see Example 4.3.3). When both the marginal mean $\boldsymbol{\mu}_i(\boldsymbol{\beta})$ and variance-covariance $\boldsymbol{\Sigma}_i(\boldsymbol{\beta}, \boldsymbol{\theta})$ are correctly specified, GEE2 estimators of both $\boldsymbol{\beta}$ and $\boldsymbol{\theta}$ will be consistent and asymptotically normally distributed as $n \to \infty$. However, if $\boldsymbol{\Sigma}_i(\boldsymbol{\beta}, \boldsymbol{\theta})$ is misspecified then the GEE2 estimator of $\boldsymbol{\beta}$ may be inconsistent even when $\boldsymbol{\mu}_i(\boldsymbol{\beta})$ is correctly specified.

Finally, we also describe a general likelihood-based estimation scheme suitable for distributions from the quadratic exponential family which happens to include the multivariate normal distribution. In this case, ML estimation is possible using NLMIXED and in limited cases (e.g., when $Var(\boldsymbol{y}_i) = \boldsymbol{\Sigma}_i(\boldsymbol{\theta})$ does not depend on $\boldsymbol{\beta}$) with GLIMMIX and %NLINMIX. Table 4.5 summarizes the estimation methods and features available in SAS for marginal nonlinear models.

5

Generalized Linear and Nonlinear Mixed-Effects Models

In the preceding chapter we considered two classes of models for correlated response data, generalized linear models (GLIM's) and generalized nonlinear models (GNLM's). Both classes focus on population-averaged (PA) inference in which model parameters describe various characteristics of an underlying marginal distribution. For example, the vector $\boldsymbol{\beta}$ of regression parameters is used to predict the average response in a population of individuals and also to describe what effect different covariates have on the mean response of a given population. In this chapter, we turn our attention to subject-specific (SS) inference as it relates to the inclusion of subject-specific random effects within the same basic structural framework of models considered in Chapter 4. Recall from Chapter 1 that with SS inference, the regression parameters describe what the "average" or "typical" individual's mean response is rather than what the average response is for a population of individuals. When the model is strictly linear in the random effects, both interpretations are possible as is the case with the LME model of Chapter 3. Moreover, the introduction of random effects allows one to easily account for observed heterogeneity as demonstrated with the analysis of the orange tree data shown in Figure 1.1 and analyzed in §4.5.2. The model used for the orange tree data is nonlinear in the fixed-effects but linear in the random effect. Hence while it can be formally classified as a nonlinear mixed-effects model, the fact that it is linear in the random effect allows us to also write the model as a population-averaged marginal nonlinear model. However, when a model is nonlinear in one or more random effects, we have a very distinct difference in interpretation of results. The marginal means, for example, are no longer a function of simply the regression parameters $\boldsymbol{\beta}$ as would be the case under the marginal models of Chapter 4. Instead, the marginal means will likely be a function of both the regression parameters $\boldsymbol{\beta}$ and the random-effect variance components as was demonstrated with the exponential decay model (1.1) of Chapter 1 (see also the multivariate Poisson model (4.61) which one may classify as a mixed-effects model with a non-Gaussian random effect). Hence, while nonlinear random-effects models provide a natural mechanism for describing both heterogeneity and correlation in the response variable, they can lead to certain difficulties in interpretation and are certainly more challenging when it comes to estimating the population parameters. These issues will become more apparent as we consider various applications involving models with nonlinear random effects.

We start then by first presenting a class of generalized linear mixed-effects (GLME) models. The GLME model represents a natural extension of the GLIM of the previous chapter and is obtained by simply including random effects within the linear predictor of a GLIM. The resulting model may be classified as a quasi-likelihood model in that only the conditional moments are specified or it may be a likelihood-based model in that a conditional pdf is specified given a set of random effects. In either case the GLME model will generally be nonlinear in both the fixed- and random-effects which means that we can not, in general, express the likelihood function or even the marginal moments in closed form. This makes estimating the parameters much more difficult and various estimation techniques have been proposed as a means for overcoming this problem. In this chapter, we will focus on those estimation techniques that involve solving either a set of conditional generalized estimating equations (CGEE) or a set of log-likelihood score equations that maximize an integrated log-likelihood function. Both approaches are available within GLIMMIX and several examples will be presented illustrating these techniques. In a manner similar to what was done in Chapter 4, we extend the class of GLME models to a class of generalized nonlinear mixed-effects (GNLME) models by 1) expressing the conditional moments (conditional on the random effects) in terms of a general nonlinear function of fixed- and random-effect parameters and 2) allowing the conditional distribution to be from a broader family of distributions. The basic methods of estimation described for the GLME model are extended to the GNLME model and several examples illustrating these methods are presented using both the NLMIXED procedure and the %NLINMIX macro.

5.1 The generalized linear mixed-effects (GLME) model

In the mid-1980s and early 1990s, a number of publications extended the GLIM of Nelder and Wedderburn (1972) to a generalized linear mixed-effects (GLME) model by introducing random effects into the linear predictor of the GLIM. Most notable among these were the papers by Stiratelli, Laird and Ware (1984); Zeger, Liang and Albert (1988); Schall (1991); and Breslow and Clayton (1993). These articles all considered the case where observations within clusters or subjects are conditionally independent given a set of random effects. Wolfinger and O'Connell (1993) extended this basic formulation by also allowing for within-cluster or within-subject correlation. We shall present methods of estimation and inference under each formulation of the GLME model. Following Wolfinger and O'Connell (1993), the GLME model may be written quite generally as follows.

Generalized linear mixed-effects (GLME) models

The GLME model for correlated response data may be cast strictly within the framework of a moments-based mixed-effects regression model of the form

$$\begin{aligned} \boldsymbol{y}_i &= \boldsymbol{\mu}_i(\boldsymbol{\beta}, \boldsymbol{b}_i) + \boldsymbol{\epsilon}_i \\ &= \boldsymbol{g}^{-1}(\boldsymbol{X}_i\boldsymbol{\beta} + \boldsymbol{Z}_i\boldsymbol{b}_i) + \boldsymbol{\epsilon}_i, \quad i = 1, \ldots, n \end{aligned} \tag{5.1}$$

where

$\boldsymbol{y}_i$ is the $p_i \times 1$ response vector for the ith subject (cluster)

$\boldsymbol{\mu}_i(\boldsymbol{\beta}, \boldsymbol{b}_i) = \boldsymbol{g}^{-1}(\boldsymbol{X}_i\boldsymbol{\beta} + \boldsymbol{Z}_i\boldsymbol{b}_i) = E(\boldsymbol{y}_i|\boldsymbol{b}_i)$ is a $p_i \times 1$ vector of conditional means with $\boldsymbol{\mu}_i = (\mu_{i1} \dots \mu_{ip_i})'$ expressed as the inverse of a differentiable monotonic link function satisfying $g(\mu_{ij}) = \boldsymbol{x}'_{ij}\boldsymbol{\beta} + \boldsymbol{z}'_{ij}\boldsymbol{b}_i$ where $\boldsymbol{x}'_{ij}$ is the jth row of a design matrix $\boldsymbol{X}_i$ and $\boldsymbol{z}'_{ij}$ is the jth row of a within-subject design matrix $\boldsymbol{Z}_i$.
$\boldsymbol{X}_i$ is a $p_i \times s$ matrix of within and between-subject covariates
$\boldsymbol{\beta}$ is an $s \times 1$ vector of fixed-effects regression parameters
$\boldsymbol{Z}_i$ is a $p_i \times v$ matrix of within-subject covariates
$\boldsymbol{b}_i$ is a $v \times 1$ vector of subject-specific random-effects
$\boldsymbol{\epsilon}_i$ is a $p_i \times 1$ within-subject error vector satisfying

$$\begin{aligned} E(\boldsymbol{\epsilon}_i|\boldsymbol{\mu}_i) &= \boldsymbol{0}, \\ Var(\boldsymbol{\epsilon}_i|\boldsymbol{\mu}_i) &= \boldsymbol{\Lambda}_i(\boldsymbol{\beta}, \boldsymbol{b}_i, \boldsymbol{\theta}_w) \\ &= \boldsymbol{V}_i(\boldsymbol{\mu}_i)^{1/2}\boldsymbol{\Lambda}^*_i(\boldsymbol{\theta}_w)\boldsymbol{V}_i(\boldsymbol{\mu}_i)^{1/2} \\ &= Var(\boldsymbol{y}_i|\boldsymbol{b}_i) \end{aligned}$$

$\boldsymbol{\Lambda}_i(\boldsymbol{\beta}, \boldsymbol{b}_i, \boldsymbol{\theta}_w)$ is a $p_i \times p_i$ conditional variance-covariance matrix defined by the square root of a diagonal matrix containing a specified variance function of the means, $\boldsymbol{V}_i = \boldsymbol{V}_i(\boldsymbol{\mu}_i) = \boldsymbol{V}_i(\boldsymbol{\mu}_i(\boldsymbol{\beta}, \boldsymbol{b}_i)) = Diag\{h(\mu_{ij})\}$, and by

$$\boldsymbol{\Lambda}^*_i(\boldsymbol{\theta}_w) = \begin{cases} \boldsymbol{\Lambda}_i(\boldsymbol{\theta}_w), & \text{if unweighted} \\ \boldsymbol{W}_i^{*-1/2}\boldsymbol{\Lambda}_i(\boldsymbol{\theta}_w)\ \boldsymbol{W}_i^{*-1/2}, & \text{if weighted} \end{cases}$$

where $\boldsymbol{W}^*_i = Diag\{w_{ij}\}$ is a diagonal matrix of known weights w_{ij} and $\boldsymbol{\Lambda}_i(\boldsymbol{\theta}_w)$ is a "working" covariance matrix with unique within-subject covariance parameters defined by a $d_w \times 1$ vector $\boldsymbol{\theta}_w$.
The random effects vector $\boldsymbol{b}_i$ has finite first- and second-order moments

$$\begin{aligned} E(\boldsymbol{b}_i) &= \boldsymbol{0}, \\ Var(\boldsymbol{b}_i) &= \boldsymbol{\Psi}(\boldsymbol{\theta}_b) \end{aligned}$$

with $\boldsymbol{\Psi}(\boldsymbol{\theta}_b)$ representing a $v \times v$ between-subject variance-covariance matrix that depends on a $d_b \times 1$ vector $\boldsymbol{\theta}_b$ of unique between-subject covariance parameters.

This model, considered by Wolfinger and O'Connell (1993), is based strictly on specifications of the conditional first- and second-order moments. No distributional assumptions are made other than the tacit assumption that the first four moments of the otherwise unknown conditional distribution of $\boldsymbol{y}_i|\boldsymbol{b}_i$ exist and that the parameters, $\boldsymbol{\tau} = (\boldsymbol{\beta}, \boldsymbol{\theta})$, lie in a compact subspace of Euclidean space $\Re^{s+d}$ where $\boldsymbol{\theta} = (\boldsymbol{\theta}_b, \boldsymbol{\theta}_w)$ is a $d \times 1$ vector of between- and within-subject covariance parameters, $(d = d_w + d_b)$. For those familiar with the GLIMMIX procedure, the matrix $\boldsymbol{\Lambda}_i(\boldsymbol{\theta}_w)$ plays the role of the R matrix while $\boldsymbol{\Psi}(\boldsymbol{\theta}_b)$ plays the role of the G matrix.

Fully parametric versions of the GLME model have been considered by a number of authors including Breslow and Clayton (1993) and McCulloch (1997). Most frequently, a parametric GLME model is formed by 1) assuming the elements of $\boldsymbol{y}_i$ are conditionally independent given a set of normally distributed random effects, $\boldsymbol{b}_i \sim N(\boldsymbol{0}, \boldsymbol{\Psi}(\boldsymbol{\theta}_b))$, and 2) assuming the conditional distribution of $y_{ij}|\boldsymbol{b}_i$ is from the linear exponential family with conditional probability density function (pdf)

$$\pi(y_{ij}; \boldsymbol{\beta}, \phi|\boldsymbol{b}_i) = \exp\Big\{[y_{ij}\eta_{ij} - b(\eta_{ij})]/a_{ij}(\phi) + c(y_{ij}, \phi)\Big\}. \tag{5.2}$$

Under (5.2), $\eta_{ij} = \boldsymbol{x}'_{ij}\boldsymbol{\beta} + \boldsymbol{z}'_{ij}\boldsymbol{b}_i$ denotes a linear predictor of both fixed and random effects, while $a_{ij}(\phi) = \phi/w_{ij}$ is a function of a dispersion parameter ϕ and known weights, w_{ij}.

Under this formulation, we take g to be the canonical link function $g(\mu_{ij}) = \eta_{ij} = \boldsymbol{x}'_{ij}\boldsymbol{\beta} + \boldsymbol{z}'_{ij}\boldsymbol{b}_i$ satisfying the relationship $E(y_{ij}|\boldsymbol{b}_i) = \mu_{ij}(\boldsymbol{\beta}, \boldsymbol{b}_i) = g^{-1}(\boldsymbol{x}'_{ij}\boldsymbol{\beta} + \boldsymbol{z}'_{ij}\boldsymbol{b}_i) = b'(\eta_{ij})$. Under model (5.2) the conditional first- and second-order moments are the same as those under model (5.1) except that the variance function has the specific form $h(\mu_{ij}) = h(\mu_{ij}(\boldsymbol{\beta}, \boldsymbol{b}_i)) = b''(\eta_{ij})$ in which case $\boldsymbol{\Lambda}_i(\boldsymbol{\beta}, \boldsymbol{b}_i, \boldsymbol{\theta}_w) = \boldsymbol{V}_i^{1/2}\boldsymbol{\Lambda}_i^*(\boldsymbol{\theta}_w)\boldsymbol{V}_i^{1/2}$with $\boldsymbol{V}_i^{1/2} = \boldsymbol{V}_i(\boldsymbol{\mu}_i(\boldsymbol{\beta}, \boldsymbol{b}_i))^{1/2} = Diag\{\sqrt{b''(\eta_{ij})}\}$ and $\boldsymbol{\Lambda}_i(\boldsymbol{\theta}_w) = \phi \boldsymbol{I}_{p_i}$ with $\boldsymbol{\theta}_w = \phi$. If $a_{ij}(\phi) = \phi$ then $\boldsymbol{\Lambda}_i^*(\boldsymbol{\theta}_w) = \phi \boldsymbol{I}_{p_i}$ while $\boldsymbol{\Lambda}_i^*(\boldsymbol{\theta}_w) = \phi \boldsymbol{W}_i^{*-1}$ whenever $a_{ij}(\phi) = \phi/w_{ij}$. For known ϕ, conditional distributions within the linear exponential family (5.2) include the normal, gamma, inverse Gaussian, binomial, Poisson and negative binomial distributions (see Tables 4.1 and 4.2).

The family of GLME models considered in this chapter may be classified according to one of the following three cases:

Case 1: The GLME model formed from the conditional moment specifications of model (5.1). In this case the within-subject covariance matrix, $\boldsymbol{\Lambda}_i(\boldsymbol{\theta}_w)$, is allowed to be quite general and estimation is based on solving a set of first-order conditional generalized estimating equations (CGEE1) with respect to $\boldsymbol{\beta}$ and $\boldsymbol{b} = (\boldsymbol{b}'_1, \ldots \boldsymbol{b}'_n)'$ and pseudo-likelihood or restricted pseudo-likelihood estimation with respect to $\boldsymbol{\theta}$.

Case 2: The GLME model formed from the conditional moment specifications of model (5.1) but with the added assumption of conditional independence given $\boldsymbol{b}_i \sim$iid $N(\boldsymbol{0}, \boldsymbol{\Psi}(\boldsymbol{\theta}_b))$. In this case $\boldsymbol{\Lambda}_i(\boldsymbol{\theta}_w) = \phi \boldsymbol{I}_{p_i}$ and estimation may be based on maximizing an approximation to the integrated quasi-likelihood function

$$e^{q(\boldsymbol{\beta},\boldsymbol{\theta};\boldsymbol{y})} = \prod_{i=1}^{n} |\boldsymbol{\Psi}(\boldsymbol{\theta}_b)|^{-1/2} \int \exp\Big[-\frac{1}{2}\sum_{j=1}^{p_i} Q(\mu_{ij}; y_{ij}) - \frac{1}{2}\boldsymbol{b}'_i\boldsymbol{\Psi}(\boldsymbol{\theta}_b)^{-1}\boldsymbol{b}_i\Big] d\boldsymbol{b}_i \tag{5.3}$$

where

$$Q(\mu_{ij}, y_{ij}) = -2\int_{y_{ij}}^{\mu_{ij}} \frac{y_{ij} - t}{a_{ij}(\phi)h(t)} dt.$$

Following Breslow and Clayton (1993) and Wolfinger and O'Connell (1993) this can be accomplished by again solving a set of first-order conditional generalized estimating equations (CGEE1) with respect to $\boldsymbol{\beta}$ and $\boldsymbol{b}$ and pseudo-likelihood or restricted pseudo-likelihood estimation with respect to $\boldsymbol{\theta} = (\boldsymbol{\theta}_b, \phi)$.

Case 3: The GLME model formed from the conditional distribution specifications of model (5.2) and the assumption of conditional independence given $\boldsymbol{b}_i \sim$iid $N(\boldsymbol{0}, \boldsymbol{\Psi}(\boldsymbol{\theta}_b))$. In this case $\boldsymbol{\Lambda}_i(\boldsymbol{\theta}_w) = \phi \boldsymbol{I}_{p_i}$ and ML estimation may be carried out by maximizing the integrated log-likelihood function

$$\begin{aligned} L(\boldsymbol{\beta}, \boldsymbol{\theta}; \boldsymbol{y}) &= \sum_{i=1}^{n} \log\{\pi(\boldsymbol{y}_i; \boldsymbol{\beta}, \boldsymbol{\theta})\} \qquad (5.4) \\ &= \sum_{i=1}^{n} \log \int \pi(\boldsymbol{y}_i; \boldsymbol{\beta}, \phi|\boldsymbol{b}_i)\pi(\boldsymbol{b}_i;\boldsymbol{\theta}_b) d\boldsymbol{b}_i \\ &= \sum_{i=1}^{n} \log \int \exp\Big[\sum_{j=1}^{p_i} l_{ij}(\boldsymbol{\beta}, \boldsymbol{b}_i, \phi; y_{ij}) - \frac{v}{2}\log(2\pi) \\ &\qquad - \frac{1}{2}\log|\boldsymbol{\Psi}(\boldsymbol{\theta}_b)| - \frac{1}{2}\boldsymbol{b}'_i\boldsymbol{\Psi}(\boldsymbol{\theta}_b)^{-1}\boldsymbol{b}_i\Big] d\boldsymbol{b}_i \end{aligned}$$

where $\boldsymbol{\theta} = (\boldsymbol{\theta}_b, \phi)$ and $l_{ij}(\boldsymbol{\beta}, \boldsymbol{b}_i, \phi; y_{ij}) = \log\{\pi(y_{ij}; \boldsymbol{\beta}, \phi|\boldsymbol{b}_i)\}$. This is accomplished using either the Laplace approximation or adaptive Gaussian quadrature to evaluate the marginal pdf of $\boldsymbol{y}_i$ via numerical integration.

Case 1 is the GLME model setup considered by Wolfinger and O'Connell (1993) while Case 2 entails a GLME model that may be cast within the context of a conditional quasi-likelihood modeling framework as was done by Breslow and Clayton (1993). Case 3 is a fully parametric GLME model. It is easy to verify that the GLME model (5.1) will encompass all three cases provided one has properly designated the components of $\boldsymbol{\mu}_i(\boldsymbol{\beta}, \boldsymbol{b}_i) = \boldsymbol{g}^{-1}(\boldsymbol{X}_i\boldsymbol{\beta} + \boldsymbol{Z}_i\boldsymbol{b}_i)$ and $\boldsymbol{\Lambda}_i(\boldsymbol{\beta}, \boldsymbol{b}_i, \boldsymbol{\theta}_w) = \boldsymbol{V}_i^{1/2}\boldsymbol{\Lambda}_i^*(\boldsymbol{\theta}_w)\boldsymbol{V}_i^{1/2}$ as well as any assumptions regarding the distribution of $\boldsymbol{b}_i$. Regardless of which set of assumptions one makes, the resulting class of GLME models will allow one to model discrete, ordinal or continuous outcomes as a function of both between- and within-subject covariates. The GLIMMIX procedure provides the necessary tools for analyzing such data including options for conducting moment-based GEE methods of estimation as well as methods based on the quasi-likelihood approach of Breslow and Clayton (1993) or methods based on a fully parametric maximum likelihood approach.

5.1.1 Estimation

In general, there are no closed form expressions for the marginal log-likelihood function or the marginal moments of a GLME model. This is because GLME models are nonlinear in the random effects making it virtually impossible to obtain closed-form expressions except in special cases. While a number of different estimation techniques have been proposed for estimating the parameters of a GLME model, the most frequently cited are those based on some form of linearization via Taylor series expansion or those that utilize numerical integration techniques. Taylor series linearization techniques are used to either 1) approximate the marginal moments of an approximate marginal quasi-likelihood function or 2) approximate an integrated quasi-likelihood function corresponding to specified first- and second-order conditional moments (Cases 1 and 2 above). Numerical integration techniques are used exclusively in cases when estimation is based on maximizing an integrated log-likelihood function (Case 3 above). Both approaches are available in the GLIMMIX procedure and are described below.

Methods based on linearization

In this section, an estimation scheme based on the pseudo-likelihood (PL) approach for the marginal GLIM (see page 113 of Chapter 4) is extended to the GLME model (5.1). This scheme was developed by Wolfinger and O'Connell (1993) and is applicable to the entire class of GLME models considered here. It is based on taking a first-order Taylor series expansion of $\boldsymbol{g}^{-1}(\boldsymbol{X}_i\boldsymbol{\beta} + \boldsymbol{Z}_i\boldsymbol{b}_i)$ about current values of $\boldsymbol{\beta}$ and $\boldsymbol{b}_i$ and forming a linear mixed-effects model based on pseudo data. Two different Taylor series expansions are available in GLIMMIX, a first-order subject-specific expansion and a first-order population-averaged (marginal or mean) expansion. We consider each separately.

First-order subject-specific expansion:

Within GLIMMIX, the default method of estimation based on linearization involves expanding the mean $\boldsymbol{\mu}_i(\boldsymbol{\beta}, \boldsymbol{b}_i) = \boldsymbol{g}^{-1}(\boldsymbol{X}_i\boldsymbol{\beta} + \boldsymbol{Z}_i\boldsymbol{b}_i)$ about current estimates $\tilde{\boldsymbol{\beta}}$ of $\boldsymbol{\beta}$ and $\tilde{\boldsymbol{b}}_i$ of $\boldsymbol{b}_i$ and forming a linear mixed-effects model based on pseudo data from which new estimates of $\boldsymbol{\theta} = (\boldsymbol{\theta}_b, \boldsymbol{\theta}_w)$, $\boldsymbol{\beta}$ and $\boldsymbol{b} = (\boldsymbol{b}_1', \ldots, \boldsymbol{b}_n')'$ can be obtained. Overall estimation is therefore based on a doubly iterative scheme in which, for fixed $\boldsymbol{\theta}$, a set of first-order conditional generalized estimating equations (CGEE1) are used to estimate $\boldsymbol{\beta}$ and $\boldsymbol{b}$ while normal-theory likelihood is used to update either a pseudo-likelihood (PL) estimator $\widehat{\boldsymbol{\theta}}_{\text{PL}}$ of $\boldsymbol{\theta} = (\boldsymbol{\theta}_b, \boldsymbol{\theta}_w)$ or a restricted pseudo-likelihood (REPL) estimator $\widehat{\boldsymbol{\theta}}_{\text{REPL}}$. A modified Gauss-Newton type algorithm for this PL estimation scheme is described below. It entails forming a pseudo-linear mixed-effects (LME) model from a Taylor series expansion of the conditional mean and applying ML or REML estimation to the variance-covariance parameters $\boldsymbol{\theta}$ assuming normality. This is followed by updating current estimates of $\boldsymbol{\beta}$ and $\boldsymbol{b}$ under the pseudo LME model and repeating the process.

A Gauss-Newton algorithm for estimating (β, b, θ) via PL/CGEE1

Step 0: Obtain an initial estimate, say $\tilde{\boldsymbol{\beta}}$, of $\boldsymbol{\beta}$ using, for example, fixed-effects estimates from a GLIM fit (see the INITGLM option of GLIMMIX). Likewise, initially set $\tilde{\boldsymbol{b}}_i = \boldsymbol{0}$ for all $i = 1, \ldots, n$.

Step 1: Given a current value $\widehat{\boldsymbol{\theta}}$ of $\boldsymbol{\theta}$, let $\tilde{\boldsymbol{\beta}} = \tilde{\boldsymbol{\beta}}(\widehat{\boldsymbol{\theta}})$ and $\tilde{\boldsymbol{b}}_i = \tilde{\boldsymbol{b}}_i(\widehat{\boldsymbol{\theta}})$ be estimates of $\boldsymbol{\beta}$ and $\boldsymbol{b}_i$. Perform a first-order Taylor series expansion of $\boldsymbol{\mu}_i(\boldsymbol{\beta}, \boldsymbol{b}_i)$ about $\boldsymbol{\beta} = \tilde{\boldsymbol{\beta}}, \boldsymbol{b}_i = \tilde{\boldsymbol{b}}_i$ yielding

$$\boldsymbol{y}_i = \boldsymbol{\mu}_i(\tilde{\boldsymbol{\beta}}, \tilde{\boldsymbol{b}}_i) + \tilde{\boldsymbol{\Delta}}_i\{\boldsymbol{X}_i(\boldsymbol{\beta} - \tilde{\boldsymbol{\beta}}) + \boldsymbol{Z}_i(\boldsymbol{b}_i - \tilde{\boldsymbol{b}}_i)\} + \boldsymbol{\epsilon}_i \tag{5.5}$$

where

$$\begin{aligned}
\boldsymbol{\mu}_i(\boldsymbol{\beta}, \boldsymbol{b}_i) &= \boldsymbol{\mu}_i(\tilde{\boldsymbol{\beta}}, \tilde{\boldsymbol{b}}_i) + \tilde{\boldsymbol{X}}_i(\boldsymbol{\beta} - \tilde{\boldsymbol{\beta}}) + \tilde{\boldsymbol{Z}}_i(\boldsymbol{b}_i - \tilde{\boldsymbol{b}}_i)\\
\tilde{\boldsymbol{X}}_i &= \frac{\partial \boldsymbol{\mu}_i(\boldsymbol{\beta}, \boldsymbol{b}_i)}{\partial \boldsymbol{\beta}'}\bigg|_{\boldsymbol{\beta}=\tilde{\boldsymbol{\beta}}, \boldsymbol{b}_i=\tilde{\boldsymbol{b}}_i} = \frac{\partial g^{-1}(\boldsymbol{\eta}_i)}{\partial \boldsymbol{\eta}_i'}\frac{\partial \boldsymbol{\eta}_i}{\partial \boldsymbol{\beta}'}\bigg|_{\boldsymbol{\beta}=\tilde{\boldsymbol{\beta}}, \boldsymbol{b}_i=\tilde{\boldsymbol{b}}_i} = \tilde{\boldsymbol{\Delta}}_i\boldsymbol{X}_i\\
\tilde{\boldsymbol{Z}}_i &= \frac{\partial \boldsymbol{\mu}_i(\boldsymbol{\beta}, \boldsymbol{b}_i)}{\partial \boldsymbol{b}_i'}\bigg|_{\boldsymbol{\beta}=\tilde{\boldsymbol{\beta}}, \boldsymbol{b}_i=\tilde{\boldsymbol{b}}_i} = \frac{\partial g^{-1}(\boldsymbol{\eta}_i)}{\partial \boldsymbol{\eta}_i'}\frac{\partial \boldsymbol{\eta}_i}{\partial \boldsymbol{b}_i'}\bigg|_{\boldsymbol{\beta}=\tilde{\boldsymbol{\beta}}, \boldsymbol{b}_i=\tilde{\boldsymbol{b}}_i} = \tilde{\boldsymbol{\Delta}}_i\boldsymbol{Z}_i\\
\tilde{\boldsymbol{\Delta}}_i &= \boldsymbol{\Delta}_i(\tilde{\boldsymbol{\beta}}, \tilde{\boldsymbol{b}}_i) = \frac{\partial g^{-1}(\boldsymbol{\eta}_i)}{\partial \boldsymbol{\eta}_i'}\bigg|_{\boldsymbol{\eta}_i=\tilde{\boldsymbol{\eta}}_i}
\end{aligned}$$

and $\boldsymbol{\eta}_i = \boldsymbol{X}_i\boldsymbol{\beta} + \boldsymbol{Z}_i\boldsymbol{b}_i$ is the $p_i \times 1$ vector of linear predictors $\eta_{ij} = \boldsymbol{x}_{ij}'\boldsymbol{\beta} + \boldsymbol{z}_{ij}'\boldsymbol{b}_i$ for $j = 1, \ldots, p_i$ and $\tilde{\boldsymbol{\eta}}_i = \boldsymbol{X}_i\tilde{\boldsymbol{\beta}} + \boldsymbol{Z}_i\tilde{\boldsymbol{b}}_i$. From this, compute the pseudo-response vector,

$$\begin{aligned}
\tilde{\boldsymbol{y}}_i &= \tilde{\boldsymbol{\Delta}}_i^{-1}(\boldsymbol{y}_i - \boldsymbol{\mu}_i(\tilde{\boldsymbol{\beta}}, \tilde{\boldsymbol{b}}_i)) + \boldsymbol{X}_i\tilde{\boldsymbol{\beta}} + \boldsymbol{Z}_i\tilde{\boldsymbol{b}}_i\\
&= \tilde{\boldsymbol{\Delta}}_i^{-1}(\boldsymbol{y}_i - \boldsymbol{\mu}_i(\tilde{\boldsymbol{\beta}}, \tilde{\boldsymbol{b}}_i)) + \tilde{\boldsymbol{\eta}}_i
\end{aligned}$$

and weight matrix

$$\tilde{\boldsymbol{W}}_i = \tilde{\boldsymbol{V}}_i^{-1}\tilde{\boldsymbol{\Delta}}_i^2$$

where $\tilde{\boldsymbol{W}}_i$ is a diagonal matrix of weights obtained by evaluating the variance function $h(\mu_{ij}(\boldsymbol{\beta}, \boldsymbol{b}_i))$ at $\tilde{\boldsymbol{\beta}}$ and $\tilde{\boldsymbol{b}}_i$, i.e., $\tilde{\boldsymbol{V}}_i = \boldsymbol{V}_i(\boldsymbol{\mu}_i(\tilde{\boldsymbol{\beta}}, \tilde{\boldsymbol{b}}_i)) = Diag\{h(\mu_{ij}(\tilde{\boldsymbol{\beta}}, \tilde{\boldsymbol{b}}_i))\}$ and by evaluating the derivative matrix, $\boldsymbol{\Delta}_i(\boldsymbol{\beta}, \boldsymbol{b}_i)$ at $\tilde{\boldsymbol{\beta}}$ and $\tilde{\boldsymbol{b}}_i$. Note that for canonical link functions, $\tilde{\boldsymbol{W}}_i = \tilde{\boldsymbol{V}}_i = \tilde{\boldsymbol{\Delta}}_i$.

Step 2. Form the pseudo-weighted LME model

$$\tilde{\boldsymbol{y}}_i = \boldsymbol{X}_i\boldsymbol{\beta} + \boldsymbol{Z}_i\boldsymbol{b}_i + \tilde{\boldsymbol{W}}_i^{-1/2}\boldsymbol{\epsilon}_i^* \tag{5.6}$$

together with the "working" assumption that $\boldsymbol{\epsilon}_i^* \sim N(\boldsymbol{0}, \boldsymbol{\Lambda}_i^*(\boldsymbol{\theta}_w))$. Note that $\tilde{\boldsymbol{V}}_i$ and $\tilde{\boldsymbol{\Delta}}_i$ are both diagonal matrices such that $\tilde{\boldsymbol{W}}_i = \tilde{\boldsymbol{W}}_i'$ and $\tilde{\boldsymbol{W}}_i^{-1/2} = \tilde{\boldsymbol{\Delta}}_i^{-1}\tilde{\boldsymbol{V}}_i^{1/2} = \tilde{\boldsymbol{V}}_i^{1/2}\tilde{\boldsymbol{\Delta}}_i^{-1}$. Under (5.6), the marginal variance-covariance of $\tilde{\boldsymbol{y}}_i$ is

$$\tilde{\boldsymbol{\Sigma}}_i(\boldsymbol{\theta}) = \boldsymbol{Z}_i\boldsymbol{\Psi}(\boldsymbol{\theta}_b)\boldsymbol{Z}_i' + \tilde{\boldsymbol{W}}_i^{-1/2}\boldsymbol{\Lambda}_i^*(\boldsymbol{\theta}_w)\tilde{\boldsymbol{W}}_i^{-1/2}$$

which depends on $\tilde{\boldsymbol{\beta}}$ and $\tilde{\boldsymbol{b}}_i$ (and hence $\boldsymbol{\theta}$) through the weight matrix $\tilde{\boldsymbol{W}}_i$ which we treat as a fixed known matrix. Using normal-theory likelihood estimation, obtain either a PL estimate, $\widehat{\boldsymbol{\theta}}_{\text{PL}}$, or REPL estimate, $\widehat{\boldsymbol{\theta}}_{\text{REPL}}$, of $\boldsymbol{\theta} = (\boldsymbol{\theta}_b, \boldsymbol{\theta}_w)$ by maximizing either the profile log-likelihood function 2.12 or the restricted profile log-likelihood function 2.13 evaluated at the pseudo-residual vector, $\tilde{\boldsymbol{r}}_i = \tilde{\boldsymbol{y}}_i - \boldsymbol{X}_i\widehat{\boldsymbol{\beta}}(\boldsymbol{\theta})$. Letting $\widehat{\boldsymbol{\theta}}$ be either the PL or REPL estimator of $\boldsymbol{\theta}$, we then update the profiled regression parameters $\boldsymbol{\beta}$ and the random effects $\boldsymbol{b}$ based on the one-step Gauss-Newton estimates

$$\widehat{\boldsymbol{\beta}} = \widehat{\boldsymbol{\beta}}(\widehat{\boldsymbol{\theta}}) = \Big(\sum_{i=1}^{n} \boldsymbol{X}_i'\tilde{\boldsymbol{\Sigma}}_i(\widehat{\boldsymbol{\theta}})^{-1}\boldsymbol{X}_i\Big)^{-1} \sum_{i=1}^{n} \boldsymbol{X}_i'\tilde{\boldsymbol{\Sigma}}_i(\widehat{\boldsymbol{\theta}})^{-1}\tilde{\boldsymbol{y}}_i \tag{5.7}$$

and

$$\widehat{\boldsymbol{b}}_i = \widehat{\boldsymbol{b}}_i(\widehat{\boldsymbol{\theta}}) = \boldsymbol{\Psi}(\widehat{\boldsymbol{\theta}}_b)\boldsymbol{Z}_i'\tilde{\boldsymbol{\Sigma}}_i(\widehat{\boldsymbol{\theta}})^{-1}(\tilde{\boldsymbol{y}}_i - \boldsymbol{X}_i\widehat{\boldsymbol{\beta}}(\widehat{\boldsymbol{\theta}}));\ i = 1, \ldots, n \tag{5.8}$$

where $\tilde{\boldsymbol{\Sigma}}_i(\widehat{\boldsymbol{\theta}}) = \boldsymbol{Z}_i\boldsymbol{\Psi}(\widehat{\boldsymbol{\theta}}_b)\boldsymbol{Z}_i' + \tilde{\boldsymbol{W}}_i^{-1/2}\boldsymbol{\Lambda}_i^*(\widehat{\boldsymbol{\theta}}_w)\tilde{\boldsymbol{W}}_i^{-1/2}$. The estimates (5.7) and (5.8) are solutions to the LME model equations coinciding with the pseudo-weighted LME model (5.6) and as such correspond to one-step Gauss-Newton estimates associated with a set of CGEE1 for $\boldsymbol{\beta}$ and $\boldsymbol{b}$. Also $\widehat{\boldsymbol{\beta}}(\widehat{\boldsymbol{\theta}})$ is the EGLS estimator of $\boldsymbol{\beta}$ and $\widehat{\boldsymbol{b}}_i(\widehat{\boldsymbol{\theta}})$ is the empirical best linear unbiased predictor (EBLUP) of $\boldsymbol{b}_i$ under (5.6) given the current PL estimator of $\boldsymbol{\theta}$. Continue to iterate between Steps 1-2 by updating $\tilde{\boldsymbol{\beta}} = \widehat{\boldsymbol{\beta}}(\widehat{\boldsymbol{\theta}})$ and $\tilde{\boldsymbol{b}}_i = \widehat{\boldsymbol{b}}_i(\widehat{\boldsymbol{\theta}})$ in Step 1 and repeating the iterative procedure in Step 2 to obtain updated joint estimates of $(\boldsymbol{\beta}, \boldsymbol{\theta})$ and EBLUP's of $\boldsymbol{b}_i$. This process is repeated until such time that successive differences in updated estimates of $\boldsymbol{\theta}$ are negligible.

This algorithm is adapted from the PL algorithm described by Wolfinger and O'Connell (1993) and is the algorithm implemented in GLIMMIX whenever one specifies METHOD=MSPL (yields a PL estimate of $\boldsymbol{\theta}$ and CGEE1 estimates of $\boldsymbol{\beta}$ and $\boldsymbol{b}$) or METHOD=RSPL (yields a REPL estimate of $\boldsymbol{\theta}$ and CGEE1 estimates of $\boldsymbol{\beta}$ and $\boldsymbol{b}$). The latter is the default method of estimation in GLIMMIX.

The CGEE1 estimators of $\boldsymbol{\beta}$ and $\boldsymbol{b}_i$ $(i = 1, \ldots, n)$ combined with either the PL or REPL estimator of $\boldsymbol{\theta}$ are based on a subject-specific Taylor series expansion of the conditional mean, $\boldsymbol{\mu}_i(\boldsymbol{\beta}, \boldsymbol{b}_i)$, about current estimates $\tilde{\boldsymbol{\beta}}$ and $\tilde{\boldsymbol{b}}_i$, respectively. For fixed $\boldsymbol{\theta} = (\boldsymbol{\theta}_b, \boldsymbol{\theta}_w)$, the CGEE1 estimates of $\boldsymbol{\beta}$ and $\boldsymbol{b}_i$ will be those solutions $\widehat{\boldsymbol{\beta}}_{\text{CGEE1}} = \widehat{\boldsymbol{\beta}}(\boldsymbol{\theta})$ and $\widehat{\boldsymbol{b}}_i(\boldsymbol{\theta})$ to the set of first-order conditional generalized estimating equations (CGEE1):

$$\begin{aligned}
U_{\boldsymbol{\beta}}(\boldsymbol{\beta}, \boldsymbol{b}, \boldsymbol{\theta}) &= \sum_{i=1}^{n} \tilde{\boldsymbol{X}}_i'\boldsymbol{\Lambda}_i(\boldsymbol{\beta}, \boldsymbol{b}_i, \boldsymbol{\theta}_w)^{-1}[\boldsymbol{y}_i - \boldsymbol{\mu}_i(\boldsymbol{\beta}, \boldsymbol{b}_i)] = \mathbf{0} \\
U_{\boldsymbol{b}_1}(\boldsymbol{\beta}, \boldsymbol{b}_1, \boldsymbol{\theta}) &= \tilde{\boldsymbol{Z}}_1'\boldsymbol{\Lambda}_i(\boldsymbol{\beta}, \boldsymbol{b}_1, \boldsymbol{\theta}_w)^{-1}[\boldsymbol{y}_1 - \boldsymbol{\mu}_1(\boldsymbol{\beta}, \boldsymbol{b}_1)] = \boldsymbol{\Psi}(\boldsymbol{\theta}_b)^{-1}\boldsymbol{b}_1 \\
&\ \vdots \\
U_{\boldsymbol{b}_n}(\boldsymbol{\beta}, \boldsymbol{b}_n, \boldsymbol{\theta}) &= \tilde{\boldsymbol{Z}}_n'\boldsymbol{\Lambda}_i(\boldsymbol{\beta}, \boldsymbol{b}_n, \boldsymbol{\theta}_w)^{-1}[\boldsymbol{y}_n - \boldsymbol{\mu}_n(\boldsymbol{\beta}, \boldsymbol{b}_n)] = \boldsymbol{\Psi}(\boldsymbol{\theta}_b)^{-1}\boldsymbol{b}_n
\end{aligned} \tag{5.9}$$

where

$$\begin{aligned}
\tilde{\boldsymbol{X}}_i &= \tilde{\boldsymbol{X}}_i(\boldsymbol{\beta}, \boldsymbol{b}_i) = \partial\boldsymbol{\mu}_i(\boldsymbol{\beta}, \boldsymbol{b}_i)/\partial\boldsymbol{\beta}' = \boldsymbol{\Delta}_i\boldsymbol{X}_i, \\
\tilde{\boldsymbol{Z}}_i &= \tilde{\boldsymbol{Z}}_i(\boldsymbol{\beta}, \boldsymbol{b}_i) = \partial\boldsymbol{\mu}_i(\boldsymbol{\beta}, \boldsymbol{b}_i)/\partial\boldsymbol{b}_i' = \boldsymbol{\Delta}_i\boldsymbol{Z}_i
\end{aligned}$$

and $\boldsymbol{\Delta}_i = \boldsymbol{\Delta}_i(\boldsymbol{\beta}, \boldsymbol{b}_i) = \frac{\partial g^{-1}(\boldsymbol{\eta}_i)}{\partial \boldsymbol{\eta}_i'}$. The set of CGEE1 are first-order conditional in that they yield estimates of $\boldsymbol{\beta}$ and $\boldsymbol{b}_i$ $(i = 1, \ldots, n)$ that are based strictly on the first-order conditional moments, $E(\boldsymbol{y}_i|\boldsymbol{b}_i)$, as compared with conditional second-order generalized estimating equations (see §5.3.4). The PL or REPL estimating equations for $\boldsymbol{\theta}$ are those of the profile log-likelihood or restricted profile log-likelihood of the pseudo LME model in which the dependence of $\tilde{\boldsymbol{W}}_i$ on $\boldsymbol{\theta}$ through $\tilde{\boldsymbol{\beta}}(\boldsymbol{\theta})$ and $\tilde{\boldsymbol{b}}_i(\boldsymbol{\theta})$ is ignored.

For fixed $\boldsymbol{\theta}$, the CGEE1 estimators obtained by jointly solving (5.9) are equivalent to the penalized quasi-likelihood (PQL) estimates of Breslow and Clayton (1993) insofar as the solutions, $\widehat{\boldsymbol{\beta}}(\boldsymbol{\theta})$ and $\widehat{\boldsymbol{b}}_i(\boldsymbol{\theta})$, $i = 1, \ldots, n$, jointly satisfy $U_{\boldsymbol{\beta}}(\boldsymbol{\beta}, \boldsymbol{b}, \boldsymbol{\theta}) = \partial l^*/\partial \boldsymbol{\beta}' = \boldsymbol{0}$ and $U_{\boldsymbol{b}_i}(\boldsymbol{\beta}, \boldsymbol{b}_i, \boldsymbol{\theta}) = \partial l^*/\partial \boldsymbol{b}_i' = \boldsymbol{0}$, where

$$l^* = l^*(\boldsymbol{\beta}, \boldsymbol{b}, \boldsymbol{\theta}; \boldsymbol{y}) = \sum_{i=1}^{n} \Big[\sum_{j=1}^{p_i} l_{ij}^*(\boldsymbol{\beta}, \boldsymbol{b}_i, \boldsymbol{\theta}_w; y_{ij}) - \frac{1}{2} \boldsymbol{b}_i' \boldsymbol{\Psi}(\boldsymbol{\theta}_b)^{-1} \boldsymbol{b}_i \Big] \tag{5.10}$$

is either Green's (1987) log penalized likelihood which is proportional to the unnormalized log-posterior density $\log\{\pi(\boldsymbol{y}_i|\boldsymbol{b}_i)\pi(\boldsymbol{b}_i)\}$ or Breslow and Clayton's (1993) log penalized quasi-likelihood. In the former case $l_{ij}^* = l_{ij}(\boldsymbol{\beta}, \boldsymbol{b}_i, \phi; y_{ij}) = \log\{\pi(y_{ij}; \boldsymbol{\beta}, \phi|\boldsymbol{b}_i)\}$ while in the latter case $l_{ij}^* = -\frac{1}{2} Q(\mu_{ij}; y_{ij})$.

First-order population-averaged (marginal) expansion:

Alternatively, GLIMMIX offers the user the choice of expanding about the mean of the random effects, i.e., about $\boldsymbol{b}_i = \boldsymbol{0}$. In this case, one would simply modify the above algorithm by setting $\tilde{\boldsymbol{b}}_i = \boldsymbol{0}$ at each stage in the iteration history. This is referred to as a population-averaged or marginal expansion in that it yields an approximation to the marginal moments by expanding about the average random effect, i.e., about $E(\boldsymbol{b}_i) = \boldsymbol{0}$. Under this marginal expansion, we replace the pseudo-response vector under (5.6) with

$$\begin{aligned} \tilde{\boldsymbol{y}}_i &= \tilde{\boldsymbol{\Delta}}_i^{-1} (\boldsymbol{y}_i - \boldsymbol{\mu}_i(\tilde{\boldsymbol{\beta}}, \boldsymbol{0})) + \boldsymbol{X}_i \tilde{\boldsymbol{\beta}} \\ &= \tilde{\boldsymbol{\Delta}}_i^{-1} (\boldsymbol{y}_i - \boldsymbol{\mu}_i(\tilde{\boldsymbol{\beta}}, \boldsymbol{0})) + \tilde{\boldsymbol{\eta}}_i \end{aligned}$$

where $\tilde{\boldsymbol{\Delta}}_i = \boldsymbol{\Delta}_i(\tilde{\boldsymbol{\beta}}, \boldsymbol{0}) = \frac{\partial g^{-1}(\boldsymbol{\eta}_i)}{\partial \boldsymbol{\eta}_i'}\Big|_{\boldsymbol{\eta}_i = \tilde{\boldsymbol{\eta}}_i}$ and $\boldsymbol{\eta}_i = \boldsymbol{X}_i \boldsymbol{\beta} + \boldsymbol{Z}_i \boldsymbol{b}_i$ is the $p_i \times 1$ vector of linear predictors $\eta_{ij} = \boldsymbol{x}_{ij}' \boldsymbol{\beta} + \boldsymbol{z}_{ij}' \boldsymbol{b}_i$ evaluated at $\boldsymbol{\beta} = \tilde{\boldsymbol{\beta}}$ and $\boldsymbol{b}_i = \boldsymbol{0}$ (i.e., $\tilde{\boldsymbol{\eta}}_i = \boldsymbol{X}_i \tilde{\boldsymbol{\beta}}$). Likewise the weight matrix, $\tilde{\boldsymbol{W}}_i = \tilde{\boldsymbol{V}}_i^{-1} \tilde{\boldsymbol{\Delta}}_i^2$, is evaluated at the variance function $h(\mu_{ij}(\tilde{\boldsymbol{\beta}}, \boldsymbol{0}))$, i.e., $\tilde{\boldsymbol{V}}_i = \boldsymbol{V}_i(\boldsymbol{\mu}_i(\tilde{\boldsymbol{\beta}}, \boldsymbol{0})) = Diag\{h(\mu_{ij}(\tilde{\boldsymbol{\beta}}, \boldsymbol{0}))\}$.

By expanding $\boldsymbol{\mu}_i(\boldsymbol{\beta}, \boldsymbol{b}_i)$ about $\boldsymbol{\beta} = \tilde{\boldsymbol{\beta}}$ and $\boldsymbol{b}_i = \boldsymbol{0}$, the above PL/CGEE1 algorithm can be shown to be equivalent to the PL/QELS algorithm (see page 159) applied to a marginal GNLM with

$$\begin{aligned} E(\boldsymbol{y}_i) &= \boldsymbol{\mu}_i(\boldsymbol{\beta}) = g^{-1}(\boldsymbol{X}_i \boldsymbol{\beta} + \boldsymbol{Z}_i \boldsymbol{0}) = g^{-1}(\boldsymbol{X}_i \boldsymbol{\beta}) \\ Var(\boldsymbol{y}_i) &= \boldsymbol{\Sigma}_i(\boldsymbol{\beta}, \boldsymbol{\theta}) \\ &= \boldsymbol{Z}_i(\boldsymbol{\beta}) \boldsymbol{\Psi}(\boldsymbol{\theta}_b) \boldsymbol{Z}_i(\boldsymbol{\beta})' + \boldsymbol{V}_i(\boldsymbol{\mu}_i)^{1/2} \boldsymbol{\Lambda}_i^*(\boldsymbol{\theta}_w) \boldsymbol{V}_i(\boldsymbol{\mu}_i)^{1/2} \end{aligned} \tag{5.11}$$

where $\boldsymbol{Z}_i(\boldsymbol{\beta}) = \boldsymbol{\Delta}_i(\boldsymbol{\beta}, \boldsymbol{0}) \boldsymbol{Z}_i$. By setting $\boldsymbol{b}_i = \boldsymbol{0}$ in (5.9), the CGEE1 estimator for $\boldsymbol{\beta}$ will reduce to the usual GEE1 estimator for any fixed value of $\boldsymbol{\theta}$ since $U_{\boldsymbol{\beta}}(\boldsymbol{\beta}, \boldsymbol{b}, \boldsymbol{\theta})|_{\boldsymbol{b}=\boldsymbol{0}}$ will be equivalent in form to the GEE, $U_{\boldsymbol{\beta}}(\boldsymbol{\beta}, \boldsymbol{\theta})$, of equation (4.10). Under GLIMMIX, one can implement this marginal expansion and obtain a GEE1 estimator of $\boldsymbol{\beta}$ and either a PL or REPL estimator of $\boldsymbol{\theta}$ by specifying either METHOD=MMPL or METHOD=RMPL, respectively. The resulting estimates are equivalent to the marginal quasi-likelihood (MQL) estimates of Breslow and Clayton (1993). When employing this marginal first-order expansion, one can obtain estimates of the random effects $\boldsymbol{b}_i$ by applying equation (5.8) using the final estimates of $(\boldsymbol{\beta}, \boldsymbol{\theta})$ obtained at convergence.

Table 5.1 shows the different Taylor series expansion options available in GLIMMIX. The METHOD= option of GLIMMIX allows the user to specify one of four estimation

Table 5.1 Methods of estimation in GLIMMIX based on linearization using a first-order Taylor series expansion. The default method in GLIMMIX is RSPL. When $\mathbf{\Lambda}_i(\boldsymbol{\theta}_w) = \phi \boldsymbol{I}_{p_i}$, the PL/CGEE1 estimator based on RSPL or MSPL will be equivalent to the penalized quasi-likelihood (PQL) estimator of Breslow and Clayton (1993) and the PL/QELS estimator based on RMPL or MMPL will be equivalent to the marginal quasi-likelihood (MQL) estimator of Breslow and Clayton (1993). In all cases, estimation of $\boldsymbol{\theta}$ depends on whether one uses a profile or restricted profile log-likelihood approach.

	Taylor Series Expansion of $\boldsymbol{\mu}_i(\boldsymbol{\beta}, \boldsymbol{b}_i)$ in (5.5)	
Method for estimating $\boldsymbol{\theta} = (\boldsymbol{\theta}_b, \boldsymbol{\theta}_w)$ under the pseudo LME model (5.6)	$\boldsymbol{\beta} = \tilde{\boldsymbol{\beta}}, \boldsymbol{b}_i = \tilde{\boldsymbol{b}}_i$	$\boldsymbol{\beta} = \tilde{\boldsymbol{\beta}}, \boldsymbol{b}_i = \boldsymbol{0}$
REML (REPL)	RSPL	RMPL
ML (PL)	MSPL	MMPL

methods based on a Taylor series expansion of the conditional mean: RSPL, MSPL, RMPL and MMPL. The last two letters, PL, simply indicates that the **P**seudo-**L**ikelihood algorithm of Wolfinger and O'Connell (1993) is to be used. The first letter determines whether the variance-covariance parameters of a pseudo LME model are to be estimated via normal-theory **R**estricted Maximum Likelihood (the RSPL and RMPL options) or normal-theory **M**aximum Likelihood (the MSPL and MMPL options). The second letter determines whether the Taylor series expansion of the conditional mean is about the current **S**ubject-specific estimate of the random effect, i.e., about $\boldsymbol{b}_i = \tilde{\boldsymbol{b}}_i$ (the RSPL and MSPL options) or about the **M**ean random effect, i.e., about $\boldsymbol{b}_i = \boldsymbol{0}$ (the RMPL and MMPL options).

If the underlying model includes a scale parameter $\phi \neq 1$, then by default, GLIMMIX profiles ϕ from the pseudo-likelihood function in the above algorithm and replaces $\boldsymbol{\Psi}(\boldsymbol{\theta}_b)$ with $\boldsymbol{\Psi}(\boldsymbol{\theta}_b^*)$ and $\mathbf{\Lambda}_i^*(\boldsymbol{\theta}_w)$ with $\mathbf{\Lambda}_i^*(\boldsymbol{\theta}_w^*)$ where $\boldsymbol{\theta}^* = (\boldsymbol{\theta}_b^*, \boldsymbol{\theta}_w^*) = (\phi^{-1}\boldsymbol{\theta}_b, \phi^{-1}\boldsymbol{\theta}_w)$ resulting in $d^* = d - 1$ parameters in $\boldsymbol{\theta}^*$. In this case, the profile log-likelihood functions are based on $\phi\tilde{\boldsymbol{\Sigma}}_i(\boldsymbol{\theta}^*)$ with ϕ estimated as

$$\widehat{\phi}_{\mathrm{PL}} = \sum_{i=1}^{n} \tilde{\boldsymbol{r}}_i' \tilde{\boldsymbol{\Sigma}}_i(\widehat{\boldsymbol{\theta}}^*)^{-1} \tilde{\boldsymbol{r}}_i / N$$

$$\widehat{\phi}_{\mathrm{REPL}} = \sum_{i=1}^{n} \tilde{\boldsymbol{r}}_i' \tilde{\boldsymbol{\Sigma}}_i(\widehat{\boldsymbol{\theta}}^*)^{-1} \tilde{\boldsymbol{r}}_i / (N - s)$$

depending on whether one specifies PL or REPL estimation of $\boldsymbol{\theta}^*$. Here, $\tilde{\boldsymbol{r}}_i = (\tilde{\boldsymbol{y}}_i - \boldsymbol{X}_i\widehat{\boldsymbol{\beta}})$ and N is either the total number of observations across subjects, $N = \sum_{i=1}^n p_i$, or it is the total frequency across subjects, $N = \sum_{i=1}^n \sum_{j=1}^{p_i} f_{ij}$, where f_{ij} is the frequency associated with j^{th} observation from the ith subject. One can override this default by specifying the NOPROFILE option in GLIMMIX. For further details and options on the PL algorithm, see the documentation for GLIMMIX.

Methods based on numerical integration

When the conditional pdf of y given $\boldsymbol{b}$ is from the exponential family (5.2) and one assumes the random effects are normally distributed, maximum likelihood estimation may be carried by directly minimizing $-2L(\boldsymbol{\beta}, \boldsymbol{\theta}; \boldsymbol{y})$ where $L(\boldsymbol{\beta}, \boldsymbol{\theta}; \boldsymbol{y})$ is the integrated log-likelihood function (5.4) with $l_{ij}(\boldsymbol{\beta}, \boldsymbol{b}_i, \phi; y_{ij}) = [y_{ij}\eta_{ij} - b(\eta_{ij})]/a_{ij}(\phi) + c(y_{ij}, \phi)$. By utilizing some form of numerical integration to approximate $\pi(\boldsymbol{y}_i; \boldsymbol{\beta}, \boldsymbol{\theta})$, ML estimation may be achieved using a singly iterative algorithm rather than a doubly iterative algorithm as required for the preceding methods based on linearization. The GLIMMIX procedure offers two numerical approaches to approximating the integrated log-likelihood function: the Laplace approximation and Gauss-Hermite quadrature. We consider each separately.

Laplace approximation:

The Laplace approximation has long been utilized as a means for approximating integrals and is frequently employed in Bayesian analyses for approximating posterior means, marginal densities, etc. (Tierney and Kadane, 1986). Within the context of the GLME model, the Laplace approximation is used to approximate the marginal pdf, $\pi(\boldsymbol{y}_i;\boldsymbol{\beta},\boldsymbol{\theta})$, which we then use to obtain ML estimates of $(\boldsymbol{\beta},\boldsymbol{\theta})$ by minimizing the resulting approximation to the objective function, $-2L(\boldsymbol{\beta},\boldsymbol{\theta};\boldsymbol{y})$.

A second-order Laplace approximation to $\pi(\boldsymbol{y}_i;\boldsymbol{\beta},\boldsymbol{\theta})$ may be obtained as follows. Let $\boldsymbol{\tau}=(\boldsymbol{\beta},\boldsymbol{\theta})$ where $\boldsymbol{\theta}=(\boldsymbol{\theta}_b,\phi)$ and let

$$p_iL_i(\boldsymbol{\tau},\boldsymbol{b}_i)=l_i(\boldsymbol{\beta},\boldsymbol{b}_i,\phi;\boldsymbol{y}_i)-\frac{1}{2}\log|\boldsymbol{\Psi}(\boldsymbol{\theta}_b)|-\frac{1}{2}\boldsymbol{b}_i'\boldsymbol{\Psi}(\boldsymbol{\theta}_b)^{-1}\boldsymbol{b}_i \tag{5.12}$$

be the unnormalized log-posterior density of the ith subject/cluster where $l_i(\boldsymbol{\beta},\boldsymbol{b}_i,\phi;\boldsymbol{y}_i)=\sum_{j=1}^{p_i}l_{ij}(\boldsymbol{\beta},\boldsymbol{b}_i,\phi;y_{ij})=\sum_{j=1}^{p_i}\Big\{[y_{ij}\eta_{ij}-b(\eta_{ij})]/a_{ij}(\phi)+c(y_{ij},\phi)\Big\}$ is the log probability density function, $\log[\pi(\boldsymbol{y}_i;\boldsymbol{\beta},\phi|\boldsymbol{b}_i)]$, assuming conditional independence. Let $\tilde{\boldsymbol{b}}_i(\boldsymbol{\tau})$ be the posterior mode of $\boldsymbol{b}_i$ obtained by maximizing the log-posterior (5.12) holding $\boldsymbol{\tau}$ fixed. It can be shown that under the GLME model, $\tilde{\boldsymbol{b}}_i(\boldsymbol{\tau})$ satisfies the relation

$$\tilde{\boldsymbol{b}}_i(\boldsymbol{\tau})=\boldsymbol{\Psi}(\boldsymbol{\theta}_b)\boldsymbol{Z}_i'\tilde{\boldsymbol{\Sigma}}_i(\boldsymbol{\tau})^{-1}(\tilde{\boldsymbol{y}}_i-\boldsymbol{X}_i\boldsymbol{\beta}) \tag{5.13}$$

where $\tilde{\boldsymbol{y}}_i=\tilde{\boldsymbol{\Delta}}_i^{-1}(\boldsymbol{y}_i-\tilde{\boldsymbol{\mu}}_i)+\tilde{\boldsymbol{\eta}}_i$ is the linearized pseudo-response vector, $\tilde{\boldsymbol{\mu}}_i=\boldsymbol{\mu}_i(\boldsymbol{\beta},\tilde{\boldsymbol{b}}_i(\boldsymbol{\tau}))$, $\tilde{\boldsymbol{\eta}}_i=\boldsymbol{X}_i\boldsymbol{\beta}+\boldsymbol{Z}_i\tilde{\boldsymbol{b}}_i(\boldsymbol{\tau})$, and $\tilde{\boldsymbol{\Sigma}}_i(\boldsymbol{\tau})=\boldsymbol{Z}_i\boldsymbol{\Psi}(\boldsymbol{\theta}_b)\boldsymbol{Z}_i'+\tilde{\boldsymbol{\Delta}}_i^{-1/2}\boldsymbol{\Lambda}_i^*(\boldsymbol{\theta}_w)\tilde{\boldsymbol{\Delta}}_i^{-1/2}=\boldsymbol{Z}_i\boldsymbol{\Psi}(\boldsymbol{\theta}_b)\boldsymbol{Z}_i'+\phi\tilde{\boldsymbol{W}}_i^{-1}$ with

$$\tilde{\boldsymbol{W}}_i=\begin{cases}\tilde{\boldsymbol{\Delta}}_i, & \text{if unweighted}\\ \tilde{\boldsymbol{\Delta}}_i^{1/2}\boldsymbol{W}_i^*\tilde{\boldsymbol{\Delta}}_i^{1/2}, & \text{if weighted}\end{cases}$$

and $\tilde{\boldsymbol{\Delta}}_i=\boldsymbol{\Delta}_i(\boldsymbol{\beta},\tilde{\boldsymbol{b}}_i(\boldsymbol{\tau}))=\frac{\partial g^{-1}(\boldsymbol{\eta}_i)}{\partial\boldsymbol{\eta}_i'}\Big|_{\boldsymbol{\eta}_i=\tilde{\boldsymbol{\eta}}_i}=\frac{\partial\boldsymbol{\mu}_i(\boldsymbol{\eta}_i)}{\partial\boldsymbol{\eta}_i'}\Big|_{\boldsymbol{\eta}_i=\tilde{\boldsymbol{\eta}}_i}$. Upon setting $\tilde{\boldsymbol{b}}_i=\tilde{\boldsymbol{b}}_i(\boldsymbol{\tau})$ and applying a multivariate Taylor series expansion of $p_iL_i(\boldsymbol{\tau},\boldsymbol{b}_i)$ about $\tilde{\boldsymbol{b}}_i$, the marginal pdf of $\boldsymbol{y}_i$ may be expressed in terms of the second-order Laplace approximation as

$$\begin{aligned}\pi(\boldsymbol{y}_i;\boldsymbol{\beta},\boldsymbol{\theta})=\pi(\boldsymbol{y}_i;\boldsymbol{\tau})&=\frac{1}{(2\pi)^{v/2}}\int\exp\{p_iL_i(\boldsymbol{\tau},\boldsymbol{b}_i)\}d\boldsymbol{b}_i \\ &=\exp\Big\{p_iL_i(\boldsymbol{\tau},\tilde{\boldsymbol{b}}_i)\Big\}\frac{1}{\left|-p_iL_i''\right|^{1/2}}\left(1+\frac{A_i(\boldsymbol{\tau},\tilde{\boldsymbol{b}}_i)}{p_i}\right) \\ &=\exp\Big\{p_iL_i(\boldsymbol{\tau},\tilde{\boldsymbol{b}}_i)\Big\}\frac{1}{\left|-p_iL_i''\right|^{1/2}}\Big(1+O(p_i^{-1})\Big)\end{aligned} \tag{5.14}$$

where $A_i(\boldsymbol{\tau},\tilde{\boldsymbol{b}}_i)$ is a uniformly bounded function in p_i, and

$$\begin{aligned}-p_iL_i''=\ -p_iL_i''(\boldsymbol{\tau},\tilde{\boldsymbol{b}}_i)&=\frac{\partial^2}{\partial\boldsymbol{b}_i\partial\boldsymbol{b}_i'}\Big\{-p_iL_i(\boldsymbol{\tau},\boldsymbol{b}_i)\Big\}\Big|_{\boldsymbol{b}_i=\tilde{\boldsymbol{b}}_i(\boldsymbol{\tau})} \\ &=\phi^{-1}\boldsymbol{Z}_i'\tilde{\boldsymbol{W}}_i\boldsymbol{Z}_i+\boldsymbol{\Psi}(\boldsymbol{\theta}_b)^{-1}.\end{aligned} \tag{5.15}$$

Ignoring the remainder term $A_i(\boldsymbol{\tau}, \tilde{\boldsymbol{b}}_i)/p_i = O(p_i^{-1})$, a ML estimate of $\boldsymbol{\tau} = (\boldsymbol{\beta}, \boldsymbol{\theta})$ may be obtained by minimizing minus twice the Laplace-approximated log-likelihood function $-2L_{\text{LA}}(\boldsymbol{\tau}, \tilde{\boldsymbol{b}}(\boldsymbol{\tau}); \boldsymbol{y})$ where

$$\begin{aligned}
-2L(\boldsymbol{\beta}, \boldsymbol{\theta}; \boldsymbol{y}) &= -2L(\boldsymbol{\tau}; \boldsymbol{y}) \\
&\approx -2\sum_{i=1}^{n}\left[p_i L_i(\boldsymbol{\tau}, \tilde{\boldsymbol{b}}_i(\boldsymbol{\tau})) - \frac{1}{2}\log\left(\left|-p_i L_i''(\boldsymbol{\tau}, \tilde{\boldsymbol{b}}_i(\boldsymbol{\tau}))\right|\right)\right] \\
&= -2\sum_{i=1}^{n}\left[p_i L_i(\boldsymbol{\tau}, \tilde{\boldsymbol{b}}_i(\boldsymbol{\tau})) - \frac{1}{2}\log\left(\left|\phi^{-1}\boldsymbol{Z}_i'\tilde{\boldsymbol{W}}_i\boldsymbol{Z}_i + \boldsymbol{\Psi}(\boldsymbol{\theta}_b)^{-1}\right|\right)\right] \\
&= -2L_{\text{LA}}(\boldsymbol{\tau}, \tilde{\boldsymbol{b}}(\boldsymbol{\tau}); \boldsymbol{y}).
\end{aligned} \tag{5.16}$$

One can minimize the objective function (5.16) by applying a Fisher scoring algorithm. Raudenbush et. al. (2000), for example, apply Fisher scoring to minimize a sixth-order Laplace approximation to $\pi(\boldsymbol{y}_i; \boldsymbol{\tau})$ which includes the first two terms of the second-order Laplace approximation. They provide formulas for the set of joint score estimating equations for $(\boldsymbol{\beta}, \boldsymbol{\theta})$ which one can adapt to the problem of minimizing the second-order Laplace approximation. The GLIMMIX procedure does not use Fisher scoring to minimize (5.16). Rather it offers users several nonlinear optimization techniques including a Newton-Raphson algorithm and a quasi-Newton algorithm. The default approach is to apply a dual quasi-Newton algorithm which does not require second-order derivatives and which should suffice for most applications. For further details on the different nonlinear optimization routines available, see the documentation for GLIMMIX.

Gauss-Hermite quadrature:

For highly discrete data such as with paired binary outcomes, the second-order Laplace approximation will generally lead to biased estimation due to the small number of observations per cluster (e.g., Breslow and Lin, 1995; Lin and Breslow, 1996; Vonesh, 1996; Vonesh et. al., 2002). Alternative numerical integration techniques based on Gaussian quadrature can overcome this deficiency as demonstrated by Pinheiro and Chao (2006). In this section, we briefly describe the use of Gauss-Hermite quadrature as a means for evaluating the integrated log-likelihood function.

For ease of presentation, let us assume we have a single random effect $b_i \sim N(0, \psi)$ and let $z = \psi^{-1/2}b_i$ such that $z \sim N(0, 1)$. The Gauss-Hermite quadrature with q quadrature points, $(z_k; k = 1, \ldots, q)$ in b_i centered about 0 is of the form

$$\begin{aligned}
\pi(\boldsymbol{y}_i; \boldsymbol{\tau}) &= \frac{1}{\sqrt{2\pi}}|\psi|^{-1/2}\int \pi(\boldsymbol{y}_i; \boldsymbol{\tau}|b_i)\exp(-\frac{1}{2}b_i^2/\psi)db_i \\
&= \int \pi(\boldsymbol{y}_i; \boldsymbol{\tau}|\psi^{1/2}z)\exp\{-z^2/2\}dz \\
&\approx \sum_{k=1}^{q}\pi(\boldsymbol{y}_i; \boldsymbol{\tau}|a_k)w_k
\end{aligned}$$

where $a_k = \psi^{1/2}z_k$ reflects the scaling of z by $\psi^{1/2}$ and where z_k and w_k are Gauss-Hermite abscissas and weights based on the Gaussian kernel $\exp\{-z^2/2\}$ (Pinheiro and Chao, 2006; see also (5.66) of §5.3.4 and ensuing discussion).

When one specifies METHOD=QUAD under GLIMMIX, the procedure approximates the marginal pdf of $\boldsymbol{y}_i$ with an adaptive Gaussian quadrature (AGQ). To apply this to the GLME model defined by (5.1) and (5.2), let $\tilde{b}_i = \tilde{b}_i(\boldsymbol{\tau})$ be the posterior mode that maximizes the log-posterior (5.11) holding $\boldsymbol{\tau} = (\boldsymbol{\beta}, \psi, \phi)$ fixed, i.e., $\tilde{b}_i$ maximizes

$p_iL_i(\boldsymbol{\tau}, b_i) = l_i(\boldsymbol{\beta}, b_i, \phi; \boldsymbol{y}_i) - \frac{1}{2}log|\psi| - \frac{1}{2}b_i^2/\psi$. Adaptive Gaussian quadrature with q quadrature points in b_i centered at $\tilde{b}_i$ and scaled to $(-p_iL_i'')^{-1/2}$ is then given by

$$\begin{aligned}\pi(\boldsymbol{y}_i; \boldsymbol{\tau}) &= \frac{1}{\sqrt{2\pi}}|\psi|^{-1/2}\int \pi(\boldsymbol{y}_i; \boldsymbol{\tau}|b_i)\exp(-\frac{1}{2}b_i^2/\psi)db_i \\ &= \frac{1}{\sqrt{2\pi}}\int \exp\{p_iL_i(\boldsymbol{\tau}, b_i)\}\, db_i \\ &\approx \left|-p_iL_i''\right|^{-1/2}\sum_{k=1}^{q}\exp\{p_iL_i(\boldsymbol{\tau}, a_k)\}w_k\exp(z_k^2/2)\end{aligned} \tag{5.17}$$

where $a_k = \tilde{b}_i + (-p_iL_i'')^{-1/2}z_k$ and $p_iL_i'' = \frac{\partial^2}{\partial b_i^2}\left\{p_iL_i(\boldsymbol{\tau}, b_i)\right\}\Big|_{b_i=\tilde{b}_i}$ is the second-order derivative matrix as defined in (5.15). The log-likelihood objective function is then approximated by $L_{\text{AGQ}}(\boldsymbol{\tau}, \tilde{\boldsymbol{b}}(\boldsymbol{\tau}); \boldsymbol{y})$ where

$$\begin{aligned}L(\boldsymbol{\tau}; \boldsymbol{y}) &\approx \sum_{i=1}^{n}\log\left\{\sum_{k=1}^{q}\exp\{p_iL_i(\boldsymbol{\tau}, a_k)\}w_k\exp(z_k^2/2)\right\} \\ &\quad -\frac{1}{2}\sum_{i=1}^{n}\log\left|-p_iL_i''(\boldsymbol{\tau}, \tilde{b}_i(\boldsymbol{\tau}))\right| \\ &= L_{\text{AGQ}}(\boldsymbol{\tau}, \tilde{\boldsymbol{b}}(\boldsymbol{\tau}); \boldsymbol{y}).\end{aligned} \tag{5.18}$$

It is worth noting that when $q = 1$, we have $z_1 = 0$, $w_1 = 1$,and $a_1 = \tilde{b}_i$ in which case (5.18) reduces to the Laplace approximation (5.16),

$$L(\boldsymbol{\tau}; \boldsymbol{y}) \approx \sum_{i=1}^{n}\left\{p_iL_i(\boldsymbol{\tau}, \tilde{b}_i) - \frac{1}{2}\log|-p_iL_i''(\boldsymbol{\tau}, \tilde{b}_i(\boldsymbol{\tau}))|\right\}.$$

A more general description of Gaussian quadrature is presented in §5.3.4 where we address ML estimation for generalized nonlinear mixed models.

Adaptive Gaussian quadrature is the default approach in GLIMMIX when one specifies METHOD=QUAD. It utilizes the above formulation but with standard Gauss-Hermite abscissas and weights based on the kernel $\exp\{-z^2\}$ and with the number of quadrature points, q, selected adaptively by evaluating the log-likelihood function at the starting values of the parameters until two successive evaluations have a relative difference less that a set tolerance level (as specified by the option QTOL). Alternatively, one can specify a fixed-point Gaussian quadrature by specifying the number of quadrature points using the option QPOINTS= under METHOD=QUAD (see the GLIMMIX syntax).

5.1.2 Comparing different estimators

Given the fact that the GLME model is nonlinear in the random effects, the different estimation techniques available in GLIMMIX all rely on some approximation to either the first- and second-order moments of the model (linearization methods) or to the marginal log likelihood (numerical integration). Consequently, in terms of asymptotic bias and efficiency, the performance of the different estimators available in GLIMMIX depends, in large part, on how well the different approximations perform. We briefly review some of the findings in the literature with regards to the consistency and asymptotic bias associated with the PL/CGEE1, Laplace and Gaussian quadrature based estimators.

Breslow and Lin (1995) and Lin and Breslow (1996) investigate the asymptotic bias associated with the Laplace-based ML estimator (LMLE) versus the PQL or PL/CGEE1 estimator for GLME models having a single random intercept effect (Breslow and Lin, 1995) and GLME models with variance component random effects (Lin and Breslow, 1996). For $\boldsymbol{\theta}$ fixed, the LMLE has a lower order of bias when compared to the PQL estimator with the order of bias being $o(\|\boldsymbol{\theta}\|^2)$ for LMLE versus $o(\|\boldsymbol{\theta}\|)$ for PQL as $\boldsymbol{\theta} \downarrow \mathbf{0}$ (see Appendix B for a brief description of o and O notation). However, both the LMLE and PQL estimates of $\boldsymbol{\theta}$ are shown to be severely biased for sparse within-subject data and this source of bias affects the bias of the LMLE and PQL estimators of $\boldsymbol{\beta}$ (Breslow and Lin, 1995; Lin and Breslow 1996). Breslow and Lin (1995) showed that the degree of bias for paired binary data can be quite severe for either LMLE or PQL confirming what was demonstrated via simulation by Breslow and Clayton (1993). In terms of the effect of cluster size p_i, the Laplace-based ML estimator of $\boldsymbol{\beta}$ was shown by Vonesh (1996) to be consistent upon order $O_p\{\max[n^{-1/2}, \min(p_i)^{-1}]\}$. Similarly, Vonesh et. al. (2002) showed that for fixed $\boldsymbol{\theta}$, the PQL estimator of $\boldsymbol{\beta}$ is consistent upon order $O_p\{\max[n^{-1/2}, \min(p_i)^{-1/2}]\}$. This result holds even for discrete binary data as these authors demonstrate via simulation. Specifically, they show that the bias associated with PQL decreases with increasing cluster size under a mixed-effects binary logistic regression model. This was further confirmed by Bellamy et. al. (2005) who showed that for fixed n and $\boldsymbol{\theta}$, the asymptotic bias (i.e., $E(\widehat{\boldsymbol{\beta}} - \boldsymbol{\beta})$) of the PQL estimator was on order $O_p\{\min(p_i)^{-1}\}$.

The PQL estimator of $\boldsymbol{\beta}$ is closely related to the Laplace-based ML estimator in that, for fixed $\boldsymbol{\theta}$, the PQL estimator maximizes the objective function

$$\begin{aligned}\sum_{i=1}^n p_i L_i(\boldsymbol{\tau}, \tilde{\boldsymbol{b}}_i) &= \sum_{i=1}^n p_i L_i(\boldsymbol{\beta}, \boldsymbol{\theta}, \tilde{\boldsymbol{b}}_i) \\ &= \sum_{i=1}^n \Big\{ l_i(\boldsymbol{\beta}, \tilde{\boldsymbol{b}}_i, \phi; \boldsymbol{y}_i) - \frac{1}{2}\log|\boldsymbol{\Psi}(\boldsymbol{\theta}_b)| - \frac{1}{2}\tilde{\boldsymbol{b}}_i'\boldsymbol{\Psi}(\boldsymbol{\theta}_b)^{-1}\tilde{\boldsymbol{b}}_i \Big\}.\end{aligned} \tag{5.19}$$

Specifically, for fixed $\boldsymbol{\theta} = (\boldsymbol{\theta}_b, \phi)$, the PQL estimates of $\boldsymbol{\beta}$ and $\boldsymbol{b}_1, \ldots \boldsymbol{b}_n$ are obtained by jointly solving the set of CGEE1 (5.9) which are the score estimating equations for jointly maximizing the PQL objective function (5.10) or, equivalently, the log posterior (5.12) given that $\frac{1}{2}\log|\boldsymbol{\Psi}(\boldsymbol{\theta}_b)|$ is constant. Indeed, for fixed $\boldsymbol{\theta}$, the Laplace-based MLE of $\boldsymbol{\beta}$ solves the estimating equations:

$$\begin{aligned}U_{\text{LMLE}}(\boldsymbol{\beta}) &= \frac{\partial}{\partial\boldsymbol{\beta}}\sum_{i=1}^n \log\left[\exp\Big\{p_i L_i(\boldsymbol{\tau}, \tilde{\boldsymbol{b}}_i)\Big\}\frac{1}{\left|-p_i L_i''(\boldsymbol{\tau}, \tilde{\boldsymbol{b}}_i)\right|^{1/2}}\right] \\ &= \sum_{i=1}^n \left[\frac{\partial}{\partial\boldsymbol{\beta}}\Big\{p_i L_i(\boldsymbol{\beta}, \boldsymbol{\theta}, \tilde{\boldsymbol{b}}_i)\Big\} - \frac{1}{2}\frac{\partial}{\partial\boldsymbol{\beta}}\log\Big(\left|-p_i L_i''(\boldsymbol{\beta}, \boldsymbol{\theta}, \tilde{\boldsymbol{b}}_i)\right|\Big)\right] \\ &= U_{\text{CGEE1}}(\boldsymbol{\beta}, \tilde{\boldsymbol{b}}, \boldsymbol{\theta}) - \frac{1}{2}\sum_{i=1}^n \left\{\frac{\partial}{\partial\boldsymbol{\beta}}\log\Big(\left|-p_i L_i''(\boldsymbol{\beta}, \boldsymbol{\theta}, \tilde{\boldsymbol{b}}_i)\right|\Big)\right\} \\ &= 0\end{aligned} \tag{5.20}$$

where

$$\begin{aligned}U_{\text{CGEE1}}(\boldsymbol{\beta}, \tilde{\boldsymbol{b}}, \boldsymbol{\theta}) &= \sum_{i=1}^n \frac{\partial}{\partial\boldsymbol{\beta}}\Big\{p_i L_i(\boldsymbol{\tau}, \tilde{\boldsymbol{b}}_i)\Big\} \\ &= \sum_{i=1}^n \frac{\partial}{\partial\boldsymbol{\beta}}\Big\{l_i(\boldsymbol{\beta}, \tilde{\boldsymbol{b}}_i, \phi; \boldsymbol{y}_i)\Big\} \\ &= \sum_{i=1}^n \boldsymbol{X}_i'\boldsymbol{W}_i^*[\boldsymbol{y}_i - \boldsymbol{\mu}_i(\boldsymbol{\beta}, \tilde{\boldsymbol{b}}_i)]/\phi \\ &= \mathbf{0}\end{aligned} \tag{5.21}$$

are the set of CGEE1, $U_\beta(\beta, \tilde{b}, \theta)$, for estimating β given θ (Breslow and Lin, 1995; Lin and Breslow, 1996). One will observe that the estimating equations in (5.21) are equivalent to those in (5.9) whenever g is a canonical link function since under (5.2),
$\Lambda_i(\beta, b_i, \theta_w) = \phi V_i^{1/2} W_i^{*-1} V_i^{1/2} = \phi V_i W_i^{*-1} = \phi W_i^{*-1} V_i$ and $\tilde{X}_i = \Delta_i X_i = V_i X_i$ imply $\tilde{X}_i' \Lambda_i(\beta, b_i, \theta_w)^{-1} = X_i' W_i^* / \phi$.

Pinheiro and Chao (2006) show how one can overcome the small sample bias associated with LMLE and PQL by estimating (β, θ) using adaptive Gaussian quadrature (AGQ). Consequently, it is recommended that for highly discrete data, one should avoid the use of LMLE or PQL in favor of AGQ unless one has large cluster sizes or small random-effects dispersion. For continuous outcomes and outcomes based on count data, the use of Laplace-based ML estimation or PQL will, in many cases, compare quite favorably with the more computer-intensive methods based on AGQ. One approach we have not presented but which is closely related to LMLE and PQL is the use of penalized extended least squares (PELS). This approach, presented by Vonesh et. al. (2002), entails solving a set of conditional second-order generalized estimating equations (CGEE2) that minimize a PELS objective function formed from the conditional first- and second-order moments. A similar approach would be to minimize an augmented PELS objective function that coincides with a Laplace approximation to an integrated log-likelihood assuming normality but with the specified conditional mean and variance of a GLME model. As both approaches fit more appropriately under our treatment of estimation for generalized nonlinear mixed-effects (GNLME) models, we defer discussion of these methods to §5.3.4.

Finally, estimates based on the MQL or PL/QELS approach in which we approximate the marginal moments via a first-order marginal Taylor series expansion of $\mu_i(\beta, b_i)$ about $b_i = 0$ will generally be biased. The order of bias will depend on the magnitude of the random effects variance-covariance parameters, θ_b. For example, if there is a single random intercept effect with variance component θ_b, then estimates based on MQL or PL/QELS will be biased with the order of the bias depending on θ_b as shown by Breslow and Lin (1995) and Demidenko (2004, pp. 456-459). As $\theta_b \downarrow 0$, the bias disappears and the first-order PA approximation implemented under GLIMMIX with METHOD=MMPL or RMPL will provide reasonable estimates of the population parameters. The real value of the MQL approach, as discussed by Breslow and Clayton (1993) and Vonesh and Chinchilli (1997, pp. 441-443), lies in its ability to provide a marginal or PA interpretation to the effects of covariates on population averages. For example, one may be willing to assume that the true marginal mean is of the form $E(y_i) = g^{-1}(X_i\beta)$ but may seek greater flexibility in defining a marginal "working" covariance structure. One approach to this would be to apply MQL estimation to an "intermediate" GLME model with conditional mean $\mu_i(\beta, b_i) = g^{-1}(X_i\beta + Z_i b_i)$ as a means for retrieving the desired marginal mean structure $\mu_i(\beta, 0) = g^{-1}(X_i\beta)$ while also providing a potentially more flexible "working" covariance matrix based on the random-effects type structure shown in (5.11).

5.1.3 Inference and test statistics

In general, inference under GLME models (or indeed any model with nonlinear random effects) is restricted to SS inference as outlined in §1.3.1. This can make the interpretation of results difficult for investigators who are accustomed to inferential statements made based on results from marginal models. This is particularly true in a repeated measures/longitudinal setting where we have individuals followed through time. For example, consider the following logistic regression model with random intercept terms

$$\begin{aligned} E(y_{ij}|b_i) &= \mu_{ij} = \mu_{ij}(\beta, b_i) = g^{-1}(x_{ij}'\beta + b_i) \qquad (5.22)\\ &= \frac{\exp\{x_{ij}'\beta + b_i\}}{1 + \exp\{x_{ij}'\beta + b_i\}};\ (i = 1, \ldots, n; j = 1, \ldots, p)\\ x_{ij}'\beta + b_i &= \beta_0 + \beta_1 X_i + \beta_2 t_j + b_i\\ Var(y_{ij}|b_i) &= \mu_{ij}(1 - \mu_{ij}). \end{aligned}$$

Figure 5.1 Subject-specific probability responses (dashed lines) under the mixed-effects logistic model (5.22) for both small and large inter-subject variation. The probability response for the average subject is shown as a bold solid line.

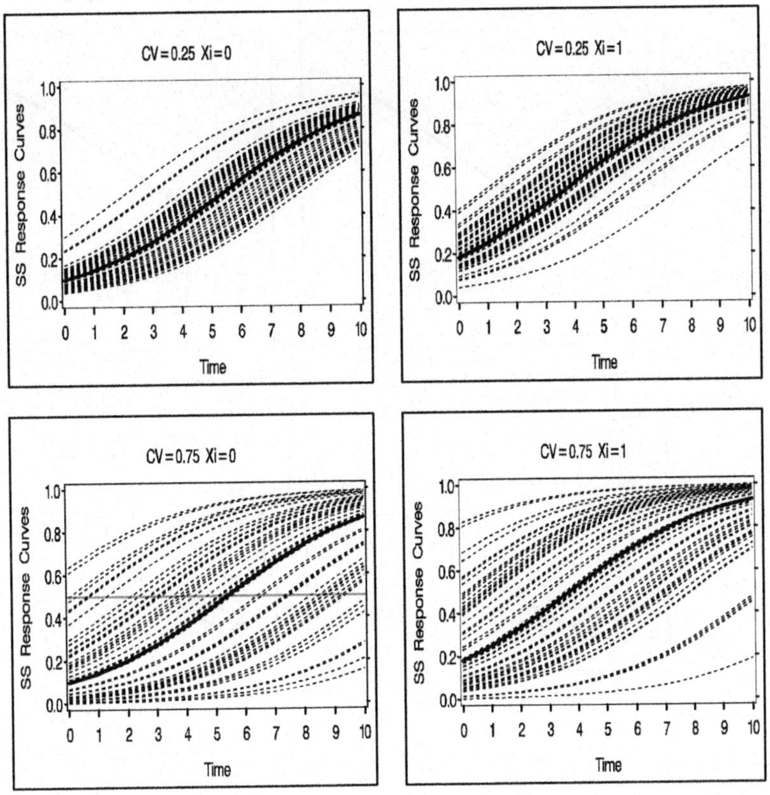

Here y_{ij} is a 0-1 indicator variable defining whether a binary event took place on the ith subject at the jth visit. The variable X_i is a known risk factor for the event with $X_i = 1$ indicating exposure to the risk factor and $X_i = 0$ indicating no exposure. Finally, t_j is the jth time point corresponding to visit j and $b_i \sim N(0, \psi)$ is a subject-specific random intercept term. Under this mixed-effects logistic regression model, we cannot interpret the regression coefficients as population-based log-odds ratios. Specifically, we cannot interpret $\exp(\beta_1)$ as an odds ratio depicting the relative increase or decrease in the odds of an event occurring among subjects exposed to the risk factor relative to subjects unexposed to the risk factor. Rather, as noted by Zeger, Liang and Albert (1988), the coefficients from a mixed-effects logistic model are log-odds ratios for an individual subject. One might therefore interpret $\exp(\beta_1)$ as indicating how one individual's risk (odds) will change if the individual were exposed to the risk factor. A slightly alternative interpretation is that $\exp(\beta_1)$ represents the odds ratio describing the average exposed subject's risk relative to the average unexposed subject's risk.

To illustrate the difference in interpretation, shown in Figure 5.1 are the subject-specific logistic response curves over time for the exposed and unexposed groups assuming the following parameter values for the logistic model (5.22): $\beta_0 = -\log(9)$ corresponding to a baseline probability of 0.10 in the average unexposed individual, $\beta_1 = \log(2)$ corresponding to an odds ratio of 2.0 for the average exposed versus unexposed individual, $\beta_3 = \log(1.5)$ corresponding to an odds ratio of 1.5 per unit increase in time per subject, and values of $\sqrt{\psi} = 0.25 \times |\beta_0|$ and $\sqrt{\psi} = 0.75 \times |\beta_0|$ corresponding to a random intercept coefficient of variation (CV) of 25% and 75%, respectively. Also shown is the average subject's probability response curve. Conversely, shown in Figure 5.2 are the marginal response curves for exposed and unexposed individuals obtained via numerical integration using the

Figure 5.2 Marginal probability response (dashed line) versus the probability response for the average subject (solid line) under the mixed-effects logistic model (5.22) for both small and large inter-subject variation.

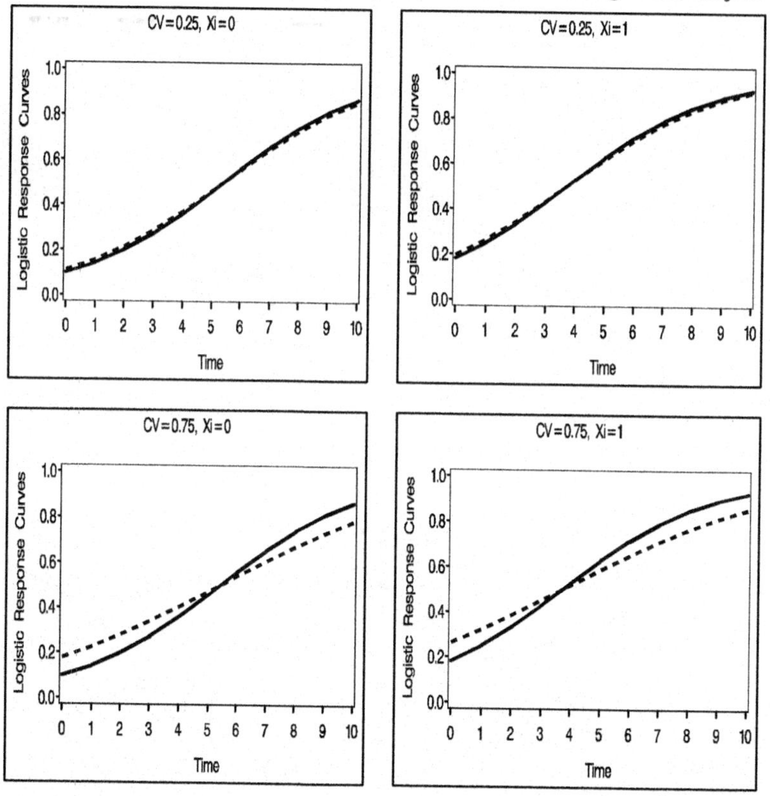

QUAD subroutine of IML. For comparison, included are the average subject response curves for the small and large inter-subject variances. As one might expect, when the CV is relatively small such as the case with a CV of 25%, the marginal probability of response across subjects is similar to the average subject's probability response in both groups. However, as illustrated in Figure 5.2, this similarity disappears when the CV increases to 75%. Moreover, when we compute the marginal odds ratio at each time point based on the marginal probability of response in the exposed and unexposed groups, we find there is a noticeable difference between the marginal odds ratios and the odds ratio for the average exposed versus unexposed individual (Figure 5.3). The marginal odds ratio, for example, varies with time whereas the odds ratio between the average exposed versus unexposed individual is constant at 2.0. Moreover, the marginal odds ratio will be higher for smaller inter-subject variation (CV=25%) and lower for larger inter-subject variation (CV=75%).

As with the LME model, inference under the GLME model is not restricted to estimable functions of the fixed-effects regression parameters, $\boldsymbol{\beta}$, but may also involve tests of hypotheses and confidence interval coverage on predictable functions involving linear combinations of both fixed and random effects parameters, $\boldsymbol{\beta}$ and $\boldsymbol{b}$. By default, the GLIMMIX procedure conducts such inference under the linked scale (i.e., the scale of the linear predictors $\boldsymbol{x}_{ij}'\boldsymbol{\beta} + \boldsymbol{z}_{ij}'\boldsymbol{b}_i$) wherein the model effects are additive. For example, under a Poisson model with log link, inference in the form of hypotheses tests and confidence intervals are on the scale of the log means. However, one may also be interested in tests of hypotheses and/or confidence intervals involving estimable functions on the mean scale (i.e., inverse linked scale). The GLIMMIX procedure provides this option for single degree of freedom tests of hypotheses. In addition one may be interested in tests of hypotheses and/or confidence interval coverage on select components of the variance-covariance parameters $\boldsymbol{\theta} = (\boldsymbol{\theta}_b, \boldsymbol{\theta}_w)$. In both cases, inference is based on large-sample theory. Below we

Figure 5.3 Marginal odds ratios from Figure 5.2 for exposed versus unexposed subjects. The subject-specific odds ratio of 2.0 remains constant with time under the mixed-effects logistic model (5.22) whereas the population-averaged odds ratio varies with time.

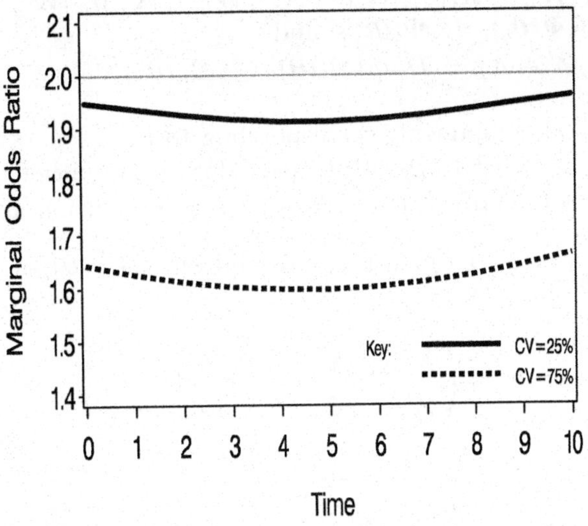

first consider inference with respect to the fixed effects parameters $\boldsymbol{\beta}$ and random effects, $\boldsymbol{b}$. We then consider inference with respect to the variance-covariance parameters, $\boldsymbol{\theta}$.

Inference about fixed effects parameters $\boldsymbol{\beta}$ and random effects $\boldsymbol{b}$

Under GLIMMIX, inference with respect to estimable functions of $\boldsymbol{\beta}$ or predictable functions of $\boldsymbol{\beta}$ and $\boldsymbol{b}$ are carried out in one of two ways depending on whether one uses pseudo-likelihood estimation based on linearization (METHOD=RSPL, RMPL, MSPL or MMPL) or likelihood estimation based on numerical integration (METHOD=LAPLACE or METHOD=QUAD). We consider each case separately.

Case 1. Inference based on pseudo-likelihood estimation: Recall from Chapters 2 and 3 that estimable functions of $\boldsymbol{\beta}$ involve linear combinations, $\boldsymbol{l}'\boldsymbol{\beta}$, of fixed effects whereas predictable functions of $\boldsymbol{\beta}$ and $\boldsymbol{b}$ involve linear combinations, $\boldsymbol{l}'\boldsymbol{\beta}+\boldsymbol{m}'\boldsymbol{b}$, of both the fixed and random effects. To carry out tests of hypotheses and/or confidence intervals on either requires we first obtain an estimate of the precision of $\widehat{\boldsymbol{\beta}}$, i.e., $Var(\widehat{\boldsymbol{\beta}})$, and an estimate of the prediction variability for $\widehat{\boldsymbol{b}}$, i.e., $Var(\widehat{\boldsymbol{b}}-\boldsymbol{b})$. Under the PL approach to estimation, both are readily available using the LME model equations associated with the pseudo-weighted LME model (5.6). Specifically, under the GLME model, the one-step Gauss-Newton solution to the set of CGEE1 (5.9) used to jointly estimate $\boldsymbol{\beta}$ and $\boldsymbol{b}$ for fixed $\boldsymbol{\theta}$ may be obtained by solving the pseudo-linear mixed-model equations

$$\begin{pmatrix} \boldsymbol{X}'\tilde{\boldsymbol{\Lambda}}_D^{-1}\boldsymbol{X} & \boldsymbol{X}'\tilde{\boldsymbol{\Lambda}}_D^{-1}\boldsymbol{Z} \\ \boldsymbol{Z}'\tilde{\boldsymbol{\Lambda}}_D^{-1}\boldsymbol{X} & \boldsymbol{Z}'\tilde{\boldsymbol{\Lambda}}_D^{-1}\boldsymbol{Z}+\boldsymbol{\Psi}_D^{-1} \end{pmatrix} \begin{pmatrix} \widehat{\boldsymbol{\beta}}(\boldsymbol{\theta}) \\ \widehat{\boldsymbol{b}}(\boldsymbol{\theta}) \end{pmatrix} = \begin{pmatrix} \boldsymbol{X}'\tilde{\boldsymbol{\Lambda}}_D^{-1}\tilde{\boldsymbol{y}} \\ \boldsymbol{Z}'\tilde{\boldsymbol{\Lambda}}_D^{-1}\tilde{\boldsymbol{y}} \end{pmatrix} \quad (5.23)$$

where

$\widehat{\boldsymbol{\beta}}(\boldsymbol{\theta}) = (\boldsymbol{X}'\tilde{\boldsymbol{\Sigma}}_D^{-1}\boldsymbol{X})^{-1}\boldsymbol{X}'\tilde{\boldsymbol{\Sigma}}_D^{-1}\tilde{\boldsymbol{y}}$ is the GLS estimate (BLUE) of the fixed-effects parameters $\boldsymbol{\beta}$ under (5.6),

$\widehat{\boldsymbol{b}}(\boldsymbol{\theta}) = \boldsymbol{\Psi}_D \boldsymbol{Z}' \tilde{\boldsymbol{\Sigma}}_D^{-1}(\tilde{\boldsymbol{y}} - \boldsymbol{X}\widehat{\boldsymbol{\beta}}(\boldsymbol{\theta}))$ is the best linear unbiased predictor (BLUP) of the random-effects $\boldsymbol{b}' = \begin{bmatrix}\boldsymbol{b}_1' \cdots \boldsymbol{b}_n'\end{bmatrix}$ under (5.6),

$\boldsymbol{X}' = \begin{bmatrix}\boldsymbol{X}_1' \cdots \boldsymbol{X}_n'\end{bmatrix}$, $\boldsymbol{Z}' = Diag\{\boldsymbol{Z}_1', \ldots, \boldsymbol{Z}_n'\}$, $\tilde{\boldsymbol{y}}' = \begin{bmatrix}\tilde{\boldsymbol{y}}_1' \cdots \tilde{\boldsymbol{y}}_n'\end{bmatrix}$,

$\tilde{\boldsymbol{\Lambda}}_D = Diag\{\tilde{\boldsymbol{W}}_1^{-1/2} \boldsymbol{\Lambda}_1^*(\boldsymbol{\theta}_w) \tilde{\boldsymbol{W}}_1^{-1/2}, \ldots, \tilde{\boldsymbol{W}}_n^{-1/2} \boldsymbol{\Lambda}_n^*(\boldsymbol{\theta}_w) \tilde{\boldsymbol{W}}_n^{-1/2}\}$

$\boldsymbol{\Psi}_D = Diag\{\boldsymbol{\Psi}(\boldsymbol{\theta}_b), \ldots, \boldsymbol{\Psi}(\boldsymbol{\theta}_b)\}$, and

$\tilde{\boldsymbol{\Sigma}}_D = \boldsymbol{Z}\boldsymbol{\Psi}_D\boldsymbol{Z}' + \tilde{\boldsymbol{\Lambda}}_D = Diag\{\tilde{\boldsymbol{\Sigma}}_1(\boldsymbol{\theta}), \ldots, \tilde{\boldsymbol{\Sigma}}_n(\boldsymbol{\theta})\}$

are the individual components corresponding to the pseudo-weighted LME model (5.6). By applying the PL/CGEE1 algorithm (page 358) using either the SS expansion (i.e., expansion about $\boldsymbol{b}_i = \tilde{\boldsymbol{b}}_i$) or PA expansion (i.e., expansion about $\boldsymbol{b}_i = \boldsymbol{0}$), one can estimate $Var\begin{pmatrix}\widehat{\boldsymbol{\beta}}(\boldsymbol{\theta}) \\ \widehat{\boldsymbol{b}}(\boldsymbol{\theta}) - \boldsymbol{b}\end{pmatrix} = \boldsymbol{\Upsilon}(\widehat{\boldsymbol{\beta}}, \widehat{\boldsymbol{b}})$ at the final solutions $(\widehat{\boldsymbol{\beta}}, \widehat{\boldsymbol{b}}, \widehat{\boldsymbol{\theta}})$ as

$$\widehat{\boldsymbol{\Upsilon}}(\widehat{\boldsymbol{\beta}}, \widehat{\boldsymbol{b}}) = \begin{pmatrix} \boldsymbol{X}'\widehat{\boldsymbol{\Lambda}}_D^{-1}\boldsymbol{X} & \boldsymbol{X}'\widehat{\boldsymbol{\Lambda}}_D^{-1}\boldsymbol{Z} \\ \boldsymbol{Z}'\widehat{\boldsymbol{\Lambda}}_D^{-1}\boldsymbol{X} & \boldsymbol{Z}'\widehat{\boldsymbol{\Lambda}}_D^{-1}\boldsymbol{Z} + \widehat{\boldsymbol{\Psi}}_D^{-1} \end{pmatrix}^{-1} \tag{5.24}$$
$$= \begin{pmatrix} \widehat{\boldsymbol{\Omega}} & -\widehat{\boldsymbol{\Omega}}\boldsymbol{X}'\widehat{\boldsymbol{\Sigma}}_D^{-1}\boldsymbol{Z}\widehat{\boldsymbol{\Psi}}_D \\ -\widehat{\boldsymbol{\Psi}}_D\boldsymbol{Z}'\widehat{\boldsymbol{\Sigma}}_D^{-1}\boldsymbol{X}\widehat{\boldsymbol{\Omega}} & \widehat{\boldsymbol{Q}} + \widehat{\boldsymbol{\Psi}}_D\boldsymbol{Z}'\widehat{\boldsymbol{\Sigma}}_D^{-1}\boldsymbol{X}\widehat{\boldsymbol{\Omega}}\boldsymbol{X}'\widehat{\boldsymbol{\Sigma}}_D^{-1}\boldsymbol{Z}\widehat{\boldsymbol{\Psi}}_D \end{pmatrix}$$

where $\widehat{\boldsymbol{\Omega}} = \widehat{\boldsymbol{\Omega}}(\widehat{\boldsymbol{\beta}}(\widehat{\boldsymbol{\theta}})) = (\boldsymbol{X}'\widehat{\boldsymbol{\Sigma}}_D^{-1}\boldsymbol{X})^{-1}$ is the estimated variance-covariance matrix of $\widehat{\boldsymbol{\beta}}(\widehat{\boldsymbol{\theta}})$ while $\widehat{\boldsymbol{Q}} = \boldsymbol{Q}(\widehat{\boldsymbol{\theta}}) = (\boldsymbol{Z}'\widehat{\boldsymbol{\Lambda}}_D^{-1}\boldsymbol{Z} + \widehat{\boldsymbol{\Psi}}_D^{-1})^{-1}$ with $\widehat{\boldsymbol{\Lambda}}_D$ and $\widehat{\boldsymbol{\Psi}}_D$ representing estimates of $\tilde{\boldsymbol{\Lambda}}_D$ and $\boldsymbol{\Psi}_D$ at the final estimates $(\widehat{\boldsymbol{\beta}}, \widehat{\boldsymbol{b}}, \widehat{\boldsymbol{\theta}})$. As with the MIXED procedure, when one specifies the EMPIRICAL option in GLIMMIX, $\widehat{\boldsymbol{\Omega}}$ is replaced by the robust variance-covariance matrix $\widehat{\boldsymbol{\Omega}}_R = \widehat{\boldsymbol{\Omega}}_R(\widehat{\boldsymbol{\beta}}(\widehat{\boldsymbol{\theta}}))$.

Because the PL approach utilizes model linearization to estimate the parameters, inference with respect to predictable linear functions of $\boldsymbol{\beta}$ and $\boldsymbol{b}$ for the GLME model may be based on the same methods described for the LME model of Chapter 3. For example, tests of hypotheses and confidence interval coverage involving the fixed-effects parameters $\boldsymbol{\beta}$ are the same as described in §2.1.2 (page 29) and §3.1.3 (page 78) with inference based either on the model-based covariance matrix, $\widehat{\boldsymbol{\Omega}}$, or the robust covariance matrix, $\widehat{\boldsymbol{\Omega}}_R$, as computed under a final optimization call to the pseudo-weighted LME model (5.6). Likewise, inference with respect to predictable linear functions of $\boldsymbol{\beta}$ and $\boldsymbol{b}$ will be based on the same methods described in §3.1.3 (page 78) for the LME model. Specifically under the GLME model, tests of the general linear hypothesis $H_0 : \boldsymbol{L\beta} = \boldsymbol{0}$ would be carried out using the same Wald chi-square test statistic $T^2(\widehat{\boldsymbol{\beta}})$ defined in (2.15) while "subject-specific" tests of hypotheses on predictable functions of the form $H_0 : \boldsymbol{L}\begin{pmatrix}\boldsymbol{\beta} \\ \boldsymbol{b}\end{pmatrix} = \boldsymbol{0}$ would be carried out using the Wald chi-square test statistic $T^2(\widehat{\boldsymbol{\beta}}, \widehat{\boldsymbol{b}})$ defined in (3.21) but with variance-covariance matrix $\widehat{\boldsymbol{\Upsilon}}(\widehat{\boldsymbol{\beta}}, \widehat{\boldsymbol{b}})$ based on (5.24). In both cases, one can replace the chi-square test with the corresponding F-test as discussed in §2.1.2 and §3.1.3, respectively. Narrow or subject-specific inference based on confidence interval coverage for a single linear combination would be of the form

$$\boldsymbol{L}\begin{pmatrix}\widehat{\boldsymbol{\beta}} \\ \widehat{\boldsymbol{b}}\end{pmatrix} \pm t_{\alpha/2}(\widehat{v}_e)\sqrt{\boldsymbol{L}\widehat{\boldsymbol{\Upsilon}}(\widehat{\boldsymbol{\beta}}, \widehat{\boldsymbol{b}})\boldsymbol{L}'} \tag{5.25}$$

where $\boldsymbol{L}_{1 \times snv} = (\boldsymbol{l}_{1\times s}' \;\; \boldsymbol{m}_{1\times nv}')$ with $\boldsymbol{l}'$ satisfying the estimability criteria for the fixed-effects parameter $\boldsymbol{\beta}$ under the pseudo-weighted LME model (5.6). Here $\widehat{v}_e$ is an estimated denominator degrees of freedom as determined by the DDFM option of the MODEL statement of GLIMMIX. Note that when $\boldsymbol{m}' = \mathbf{0}$, then (5.25) yields the same confidence interval coverage for fixed effects as defined for the LME model (§3.1.3).

While the preceding discussion focuses on inference with respect to estimable functions of $\boldsymbol{\beta}$ or predictable functions of $\boldsymbol{\beta}$ and $\boldsymbol{b}$ on the linear predictor scale (i.e., link scale), GLIMMIX also enables one to conduct limited inference on the mean scale (i.e., inverse link scale). This is further addressed below where we examine the case of likelihood-based inference.

Case 2: Inference based on likelihood estimation: When the population parameters, $\boldsymbol{\tau} = (\boldsymbol{\beta}, \boldsymbol{\theta})$, of the GLME model are estimated via maximum likelihood using the Laplace approximation or numerical quadrature to approximate the integrated log-likelihood (5.4), a slightly different approach to inference with respect to $\boldsymbol{\beta}$ and $\boldsymbol{b}$ is required. Specifically, if we let $\widehat{\boldsymbol{\tau}} = (\widehat{\boldsymbol{\beta}}, \widehat{\boldsymbol{\theta}})$ denote the MLE of $\boldsymbol{\tau}$ obtained by minimizing $-2L(\boldsymbol{\beta}, \boldsymbol{\theta}; \boldsymbol{y})$ where $L(\boldsymbol{\beta}, \boldsymbol{\theta}; \boldsymbol{y})$ is the integrated log-likelihood function (5.4), inference with respect to the population parameters $\boldsymbol{\tau} = (\boldsymbol{\beta}, \boldsymbol{\theta})$ may be based on standard ML theory. This implies that tests of hypotheses with respect to $\boldsymbol{\tau}$ may be based either on a likelihood ratio test or a Wald chi-square test. The former approach requires one to fit nested models using GLIMMIX and then use the output to manually perform the likelihood ratio test. The latter approach is based on large sample ML theory which states that

$$\sqrt{n}(\widehat{\boldsymbol{\tau}} - \boldsymbol{\tau}) \xrightarrow{p} \mathrm{N}(\mathbf{0}, \boldsymbol{\Omega}_0) \tag{5.26}$$

$$Var(\widehat{\boldsymbol{\tau}}) \simeq \widehat{\boldsymbol{\Omega}}(\widehat{\boldsymbol{\tau}}) = \left(\frac{\partial^2 L^*(\boldsymbol{\tau}, \tilde{\boldsymbol{b}}(\boldsymbol{\tau}); \boldsymbol{y})}{\partial \boldsymbol{\tau} \partial \boldsymbol{\tau}'} \bigg|_{\boldsymbol{\tau} = \widehat{\boldsymbol{\tau}}} \right)^{-1}$$

$$n\widehat{\boldsymbol{\Omega}}(\widehat{\boldsymbol{\tau}}) \xrightarrow{p} \boldsymbol{\Omega}_0$$

where $L^*(\boldsymbol{\tau}, \tilde{\boldsymbol{b}}(\boldsymbol{\tau}); \boldsymbol{y})$ is the approximate integrated log-likelihood function $L_{\mathrm{AGQ}}(\boldsymbol{\tau}, \tilde{\boldsymbol{b}}(\boldsymbol{\tau}); \boldsymbol{y})$ based on adaptive Gaussian quadrature (5.18). When $L^*(\boldsymbol{\tau}, \tilde{\boldsymbol{b}}(\boldsymbol{\tau}); \boldsymbol{y})$ is the approximate log-likelihood function $L_{\mathrm{LA}}(\boldsymbol{\tau}, \tilde{\boldsymbol{b}}(\boldsymbol{\tau}); \boldsymbol{y})$ from the Laplace approximation (5.16), it was shown by Vonesh (1996) that the Laplace-based ML estimate, $\widehat{\boldsymbol{\tau}}_{\mathrm{LMLE}}$, will be asymptotically equivalent to the unconditional MLE but that $\widehat{\boldsymbol{\tau}}_{\mathrm{LMLE}} - \boldsymbol{\tau} = O_p(n^{-1/2}) + O_p\{\min(p_i)^{-1}\}$. In many cases we can effectively ignore this additional error term $O_p\{\min(p_i)^{-1}\}$ in which case inference based on the results in (5.26) will remain valid for the Laplace-based MLE (Vonesh, 1996).

The matrix $\widehat{\boldsymbol{\Omega}}(\widehat{\boldsymbol{\tau}})$ is the model-based estimated asymptotic covariance matrix of $\widehat{\boldsymbol{\tau}}$ which is simply the inverse Hessian matrix associated with the approximated integrated log-likelihood function. It is computed in GLIMMIX using finite forward differences based on the analytic gradient of $L^*(\boldsymbol{\tau}, \tilde{\boldsymbol{b}}(\boldsymbol{\tau}); \boldsymbol{y})$ (see the on-line documentation of GLIMMIX). Alternatively, one may request inference based on the robust sandwich estimator

$$\widehat{\boldsymbol{\Omega}}_R(\widehat{\boldsymbol{\tau}}) = \widehat{\boldsymbol{\Omega}}(\widehat{\boldsymbol{\tau}}) \left(\sum_{i=1}^n U_i^*(\widehat{\boldsymbol{\tau}}) U_i^*(\widehat{\boldsymbol{\tau}})' \right) \widehat{\boldsymbol{\Omega}}(\widehat{\boldsymbol{\tau}}) \tag{5.27}$$

where $U_i^*(\widehat{\boldsymbol{\tau}}) = \frac{\partial}{\partial \boldsymbol{\tau}} L_i^*(\boldsymbol{\tau}, \tilde{\boldsymbol{b}}(\boldsymbol{\tau}); \boldsymbol{y}_i)$ is the gradient of the approximated log-likelihood with respect to $\boldsymbol{\tau}$ for the ith of the n independent experimental units (i.e., subjects or clusters). For example, under the Laplace approximation (5.16),

$$\begin{aligned} U_i^*(\widehat{\boldsymbol{\tau}}) &= \frac{\partial}{\partial \boldsymbol{\tau}} L_i^*(\boldsymbol{\tau}, \tilde{\boldsymbol{b}}(\boldsymbol{\tau}); \boldsymbol{y}_i) \bigg|_{\boldsymbol{\tau} = \widehat{\boldsymbol{\tau}}} \\ &= \frac{\partial}{\partial \boldsymbol{\tau}} \left[p_i L_i(\boldsymbol{\tau}, \tilde{\boldsymbol{b}}_i(\boldsymbol{\tau})) - \frac{1}{2} \log\left(\left| \phi^{-1} \boldsymbol{Z}_i' \tilde{\boldsymbol{W}}_i \boldsymbol{Z}_i + \boldsymbol{\Psi}(\boldsymbol{\theta}_b)^{-1} \right| \right) \right] \bigg|_{\boldsymbol{\tau} = \widehat{\boldsymbol{\tau}}} . \end{aligned}$$

The robust sandwich estimator (5.27) is the default estimator used in GLIMMIX whenever one specifies the EMPIRICAL option together with METHOD =LAPLACE or METHOD=QUAD. Alternatively, one can specify EMPIRICAL=MBN to obtain an adjusted version of (5.27) as described in the GLIMMIX documentation.

For inferential purposes, the GLIMMIX procedure partitions $\widehat{\boldsymbol{\Omega}}(\widehat{\boldsymbol{\tau}})$ or $\widehat{\boldsymbol{\Omega}}_R(\widehat{\boldsymbol{\tau}})$ as

$$\widehat{\boldsymbol{\Omega}} = \widehat{\boldsymbol{\Omega}}(\widehat{\boldsymbol{\tau}}) = \begin{pmatrix} \widehat{\boldsymbol{\Omega}}(\widehat{\boldsymbol{\beta}}) & \widehat{\boldsymbol{C}}(\widehat{\boldsymbol{\beta}},\widehat{\boldsymbol{\theta}}) \\ \widehat{\boldsymbol{C}}(\widehat{\boldsymbol{\beta}},\widehat{\boldsymbol{\theta}})' & \widehat{\boldsymbol{\Omega}}(\widehat{\boldsymbol{\theta}}) \end{pmatrix} \quad (5.28)$$

$$\widehat{\boldsymbol{\Omega}}_R = \widehat{\boldsymbol{\Omega}}_R(\widehat{\boldsymbol{\tau}}) = \begin{pmatrix} \widehat{\boldsymbol{\Omega}}_R(\widehat{\boldsymbol{\beta}}) & \widehat{\boldsymbol{C}}_R(\widehat{\boldsymbol{\beta}},\widehat{\boldsymbol{\theta}}) \\ \widehat{\boldsymbol{C}}_R(\widehat{\boldsymbol{\beta}},\widehat{\boldsymbol{\theta}})' & \widehat{\boldsymbol{\Omega}}_R(\widehat{\boldsymbol{\theta}}) \end{pmatrix},$$

where $\widehat{\boldsymbol{\Omega}}(\widehat{\boldsymbol{\beta}})$ and $\widehat{\boldsymbol{\Omega}}_R(\widehat{\boldsymbol{\beta}})$ are, respectively, the model-based and robust variance-covariance matrices of $\widehat{\boldsymbol{\beta}}$; $\widehat{\boldsymbol{\Omega}}(\widehat{\boldsymbol{\theta}})$ and $\widehat{\boldsymbol{\Omega}}_R(\widehat{\boldsymbol{\theta}})$ are the corresponding estimated covariance matrices of $\widehat{\boldsymbol{\theta}}$; and $\widehat{\boldsymbol{C}}(\widehat{\boldsymbol{\beta}},\widehat{\boldsymbol{\theta}})$ and $\widehat{\boldsymbol{C}}_R(\widehat{\boldsymbol{\beta}},\widehat{\boldsymbol{\theta}})$ are the corresponding estimated covariances of the crossed components of $\widehat{\boldsymbol{\beta}}$ and $\widehat{\boldsymbol{\theta}}$. By partitioning $\widehat{\boldsymbol{\Omega}}(\widehat{\boldsymbol{\tau}})$ or $\widehat{\boldsymbol{\Omega}}_R(\widehat{\boldsymbol{\tau}})$ in this manner, inference with respect to predictable functions of $\boldsymbol{\beta}$ and $\boldsymbol{b}$ may be based on the extended covariance matrix (Booth and Hobert, 1998)

$$\widehat{\boldsymbol{\Upsilon}}(\widehat{\boldsymbol{\tau}},\widehat{\boldsymbol{b}}) = \begin{pmatrix} \widehat{\boldsymbol{\Omega}}^* & \widehat{\boldsymbol{\Omega}}^*\left(\frac{\partial \tilde{\boldsymbol{b}}(\boldsymbol{\tau})}{\partial \boldsymbol{\tau}}\Big|_{\widehat{\boldsymbol{\tau}}}\right) \\ \left(\frac{\partial \tilde{\boldsymbol{b}}(\boldsymbol{\tau})}{\partial \boldsymbol{\tau}'}\Big|_{\widehat{\boldsymbol{\tau}}}\right)\widehat{\boldsymbol{\Omega}}^* & \widehat{\boldsymbol{\Gamma}}^{-1} + \left(\frac{\partial \tilde{\boldsymbol{b}}(\boldsymbol{\tau})}{\partial \boldsymbol{\tau}'}\Big|_{\widehat{\boldsymbol{\tau}}}\right)\widehat{\boldsymbol{\Omega}}^*\left(\frac{\partial \tilde{\boldsymbol{b}}(\boldsymbol{\tau})}{\partial \boldsymbol{\tau}}\Big|_{\widehat{\boldsymbol{\tau}}}\right) \end{pmatrix} \quad (5.29)$$

where $\widehat{\boldsymbol{\Omega}}^* = \widehat{\boldsymbol{\Omega}}$ or $\widehat{\boldsymbol{\Omega}}^* = \widehat{\boldsymbol{\Omega}}_R$ (depending on whether one invokes the EMPIRICAL option of GLIMMIX), and where $\widehat{\boldsymbol{\Gamma}}^{-1} = \widehat{\phi}^{-1}\boldsymbol{Z}'\tilde{\boldsymbol{W}}_D\boldsymbol{Z} + \widehat{\boldsymbol{\Psi}}_D^{-1}$ is the second-order derivative matrix associated with maximizing the log posterior (5.12) across all subjects. As noted by Booth and Hobert (1998), $\widehat{\boldsymbol{\Gamma}}^{-1}$ may be used to approximate the posterior covariance matrix, $Var(\boldsymbol{b}|\boldsymbol{y})$, while the posterior mean, $E(\boldsymbol{b}|\boldsymbol{y})$, may be approximated by the estimated posterior mode $\tilde{\boldsymbol{b}}(\widehat{\boldsymbol{\tau}}) = \begin{pmatrix} \tilde{\boldsymbol{b}}_1(\widehat{\boldsymbol{\tau}}) \\ \vdots \\ \tilde{\boldsymbol{b}}_n(\widehat{\boldsymbol{\tau}}) \end{pmatrix}$. The matrix $\widehat{\boldsymbol{\Gamma}}^{-1}$ is simply an $nv \times nv$ block diagonal matrix obtained by grouping the individual second-order derivative matrices in (5.15) along the main diagonal. In particular,

$$\begin{aligned} \widehat{\boldsymbol{\Gamma}}^{-1} &= \begin{pmatrix} \widehat{\phi}^{-1}\boldsymbol{Z}_1'\tilde{\boldsymbol{W}}_1\boldsymbol{Z}_1 + \boldsymbol{\Psi}(\widehat{\boldsymbol{\theta}}_b)^{-1} & \ldots & \boldsymbol{0} \\ \boldsymbol{0} & \ddots & \boldsymbol{0} \\ \boldsymbol{0} & \ldots & \widehat{\phi}^{-1}\boldsymbol{Z}_n'\tilde{\boldsymbol{W}}_n\boldsymbol{Z}_n + \boldsymbol{\Psi}(\widehat{\boldsymbol{\theta}}_b)^{-1} \end{pmatrix} \\ &= \widehat{\phi}^{-1}\boldsymbol{Z}'\tilde{\boldsymbol{W}}_D\boldsymbol{Z} + \widehat{\boldsymbol{\Psi}}_D^{-1} \end{aligned}$$

where $\boldsymbol{Z}' = Diag\{\boldsymbol{Z}_1', \ldots, \boldsymbol{Z}_n'\}$, $\tilde{\boldsymbol{W}}_D = Diag\{\tilde{\boldsymbol{W}}_1, \ldots, \tilde{\boldsymbol{W}}_n\}$ with $\tilde{\boldsymbol{W}}_i$ evaluated at $\widehat{\boldsymbol{\tau}}$ and $\widehat{\boldsymbol{\Psi}}_D = Diag\{\boldsymbol{\Psi}(\widehat{\boldsymbol{\theta}}_b), \ldots, \boldsymbol{\Psi}(\widehat{\boldsymbol{\theta}}_b)\}$. Note that for canonical link functions, $\widehat{\boldsymbol{\Gamma}}^{-1}$ will be identical to $\widehat{\boldsymbol{Q}}$ in (5.24) provided $\boldsymbol{\Lambda}_i^*(\boldsymbol{\theta}_w) = \phi\boldsymbol{I}$. However, the prediction variance submatrix for the random effects, $\widehat{\boldsymbol{\Gamma}}^{-1} + \left(\frac{\partial \tilde{\boldsymbol{b}}(\boldsymbol{\tau})}{\partial \boldsymbol{\tau}'}\Big|_{\widehat{\boldsymbol{\tau}}}\right)\widehat{\boldsymbol{\Omega}}^*\left(\frac{\partial \tilde{\boldsymbol{b}}(\boldsymbol{\tau})}{\partial \boldsymbol{\tau}}\Big|_{\widehat{\boldsymbol{\tau}}}\right)$, will differ from its counterpart under pseudo-likelihood estimation, i.e., $\widehat{\boldsymbol{Q}} + \widehat{\boldsymbol{\Psi}}_D\boldsymbol{Z}'\widehat{\boldsymbol{\Sigma}}_D^{-1}\boldsymbol{X}\widehat{\boldsymbol{\Omega}}\boldsymbol{X}'\widehat{\boldsymbol{\Sigma}}_D^{-1}\boldsymbol{Z}\widehat{\boldsymbol{\Psi}}_D$, even for LME models.

When one specifies METHOD=LAPLACE or METHOD=QUAD, inference with respect to predictable linear functions of $\boldsymbol{\beta}$ and $\boldsymbol{b}$ will have the same structural form (5.25) as described for PL-based estimation but with the estimated asymptotic covariance matrix $\widehat{\boldsymbol{\Upsilon}}(\widehat{\boldsymbol{\beta}},\widehat{\boldsymbol{b}})$ formed from the appropriate submatrix components of $\widehat{\boldsymbol{\Upsilon}}(\widehat{\boldsymbol{\tau}},\widehat{\boldsymbol{b}})$ such as the submatrix $\widehat{\Omega}(\widehat{\beta})$ of $\widehat{\Omega}(\widehat{\tau})$ corresponding to the fixed-effects parameter estimates, $\widehat{\beta}$. Likewise, inference with respect to the fixed effects parameters β will be based solely on the submatrix $\widehat{\Omega}(\widehat{\beta})$ of $\widehat{\Omega}(\widehat{\tau})$ or its robust counterpart, $\widehat{\Omega}_R(\widehat{\beta})$ of $\widehat{\Omega}_R(\widehat{\tau})$, depending on whether one requests inference be based on the robust sandwich estimator using the EMPIRICAL option of GLIMMIX.

A key additional feature available with GLIMMIX is the option to also summarize results based on the mean scale (or inverse link scale) for those estimable linear functions of β or predictable linear functions of β and b resulting from the ESTIMATE, LSMEANS, and LSMESTIMATE statements. This feature, which is implemented using the ILINK option under one of these statements, entails the use of the delta method for computing the standard error of the inverse linked estimate while confidence intervals on the mean scale are obtained by taking the inverse link of the confidence intervals computed under the link scale. This feature is available with either method of estimation, PL or ML.

Inference about covariance parameters θ

Under the GLME model, inference with respect to variance-covariance parameters θ will be based on the submatrix $\widehat{\Omega}(\widehat{\theta})$ or its robust counterpart $\widehat{\Omega}_R(\widehat{\theta})$ in (5.28). By default, GLIMMIX will print the estimated components $\widehat{\theta}_k$ of $\widehat{\theta}$ $(k = 1, \ldots, d)$ along with their estimated standard errors. Moreover, as described briefly in §2.1.2, one can conduct tests of hypotheses of the form

$$H_0 : C\theta = 0$$

where C is a contrast matrix that tests whether select linear combinations of the variance-covariance parameters are zero. This is accomplished using the COVTEST statement of GLIMMIX as was illustrated in the analysis of the bone mineral density data of §2.2.2 (see, for example, Output 2.9).

5.1.4 Model selection, goodness-of-fit and diagnostics

In §4.2.3 of Chapter 4, we described a number of model selection and diagnostic tools available with GLIMMIX for marginal GLIM's for correlated data. These same tools are also available for GLME models. In terms of model selection, GLIMMIX reports a generalized chi-square statistic and its degrees of freedom based on the final pseudo-LME whenever one requests one of the pseudo-likelihood-based estimation techniques. Alternatively, one may prefer to use the SAS macro %GOF or %GLIMMIX_GOF (see Appendix D) in combination with select output from GLIMMIX to obtain R-square type goodness-of-fit measures useful in assessing model fit and variable selection (see Vonesh et. al. 1996; Vonesh and Chinchilli, 1997). For mixed-effects models like the GLME model, these macros summarize how well the predicted values fit the observed response values based on both the average individual's fit (i.e., predicted values based on $\mu_i(\widehat{\beta}, \mathbf{0})$ which is identified in the output as Average Model R-Square, etc.) and the subject-specific fit for each individual (i.e., predicted values based on $\mu_i(\widehat{\beta}, \widehat{b}_i)$ which is identified in the output as Conditional Model R-Square, etc.). These macros also provide an approximate test for determining whether the assumed covariance structure is reasonable (see, for example, Output 2.11 from the bone mineral density example).

When one uses ML estimation based on numerical integration, GLIMMIX reports various information criteria such as the AIC, AICC, BIC, CAIC, and HQIC fit statistics described in the on-line documentation of GLIMMIX. These can be used to help discriminate between the fit of various GLME models including the detection of possible model misspecification due to overdispersion, incorrect assumptions regarding the random effect structure or the omission of important covariates from the overall regression model. Moreover, one can conduct a likelihood ratio test for select hypotheses regarding the assumed random-effects covariance structure using the COVTEST statement. This is likely to be a more powerful test than that offered with either of the SAS macros, %GOF or %GLIMMIX_GOF, given that GLIMMIX carries out a likelihood ratio test as opposed to the pseudo-likelihood ratio test implemented in the macros.

The GLIMMIX procedure also provides a number of diagnostic tools for assessing how sensitive a particular model is to influential observations and/or random effect misspecifications. For example, the COVB(DETAILS) option of the MODEL statement of GLIMMIX provides additional less formal diagnostic tools for assessing how well an assumed covariance structure fits the data. As noted in §4.2.3, these tools are based on measuring how close the empirical sandwich estimator $\widehat{\boldsymbol{\Omega}}_R(\widehat{\boldsymbol{\beta}})$ of the variance-covariance of $\widehat{\boldsymbol{\beta}}$ is to the model-based estimator, $\widehat{\boldsymbol{\Omega}}(\widehat{\boldsymbol{\beta}})$. GLIMMIX also provides a PLOTS option which one can use to examine various diagnostic plots based on marginal and/or conditional residuals including the raw, studentized, and Pearson-type residuals. These can be useful in identifying influential observations and/or influential subjects. For example, by specifying PLOTS = STUDENTPANEL (NOBLUP), the GLIMMIX procedure will provide a panel of plots based on studentized residuals on the linear scale (i.e., on the link scale). The suboption (NOBLUP) instructs GLIMMIX to not use the predictors of the random effects in computing the studentized residuals. Alternatively, by specifying the suboption (BLUP), GLIMMIX will produce a panel of studentized residuals that do incorporate the predictors of the random effects. Not all suboptions are available with the PLOTS option and one will need to refer to the GLIMMIX documentation for details on the available combination of options and suboptions.

5.2 Examples of GLME models

In the previous chapter, several examples were given wherein the response variable of interest was non-Gaussian and the mean response was nonlinear with respect to the regression parameters of interest. In this section, we re-analyze three of these examples in which random effects are introduced as a means for accounting for correlation among the repeated responses over time. These include the respiratory disorder data involving repeated binary outcomes, the epileptic seizure data involving repeated count data and the schizophrenia data involving repeated ordinal data. We also consider an example from the ADEMEX study in which we compare the rate of hospital admissions between control and treated patients using a cluster-specific random-effects Poisson model in which the number of hospital admissions among patients within the same center (i.e., cluster) are assumed correlated.

5.2.1 Respiratory disorder data—continued

In §4.3.2 and §4.5.3, we fit repeated binary outcomes from the respiratory disorder data described in §1.4 (page 13) to a marginal logistic regression model. Correlation among repeated binary responses was accounted for through the use of a "working" correlation structure. Both GEE1 and GEE2 were used to estimate the regression parameters of interest and to test whether patients receiving the active drug had an improved response over time compared to those on the placebo control drug.

In this section, we apply a mixed-effects logistic regression model with a random intercept term as an alternative approach to modeling correlated binary outcomes among patients with respiratory disorder. Specifically, we fit the GLME model

$$E(y_{ij}|b_i) = \mu_{ij} = \mu_{ij}(\boldsymbol{x}'_{ij}, \boldsymbol{\beta}, b_i) = g^{-1}(\boldsymbol{x}'_{ij}\boldsymbol{\beta} + b_i) \tag{5.30}$$

$$= \frac{\exp\{\boldsymbol{x}'_{ij}\boldsymbol{\beta} + b_i\}}{1+\exp\{\boldsymbol{x}'_{ij}\boldsymbol{\beta} + b_i\}}$$

$$\boldsymbol{x}'_{ij}\boldsymbol{\beta} = \beta_0 + \beta_1 a_{1i} + \beta_2 a_{2i} + \beta_3 a_{3i} + \beta_4 a_{4i} + \beta_5 a_{5i}$$

$$Var(y_{ij}|b_i) = \mu_{ij}(1-\mu_{ij})$$

where $\mu_{ij} = \mu_{ij}(\boldsymbol{x}_{ij}, \boldsymbol{\beta}, b_i) = \Pr(y_{ij} = 1|\boldsymbol{x}_{ij}, b_i)$ is the *conditional* probability of a positive response for the ith subject on the jth visit, $\boldsymbol{x}'_{ij} = \begin{pmatrix}1 & a_{1i} & a_{2i} & a_{3i} & a_{4i} & a_{5i}\end{pmatrix}$ is the design vector of within- and between-subject covariates as described in §1.4, $\boldsymbol{\beta}' = \begin{pmatrix}\beta_0 & \beta_1 & \beta_2 & \beta_3 & \beta_4 & \beta_5\end{pmatrix}$ is the vector of fixed-effects regression parameters associated with the conditional mean response and b_i is a subject-specific random intercept term assumed to be $N(0, \psi)$. As in Chapter 4, we do not model a visit effect but rather assume a common visit effect that is reflected in the overall intercept parameter, β_0. We can express this model in terms of the GLME model (5.1) by setting

$$\boldsymbol{y}'_i = (y_{i1}, \ldots, y_{ip_i}),\ \boldsymbol{X}_i = \begin{pmatrix}\boldsymbol{x}'_{i1} \\ \vdots \\ \boldsymbol{x}'_{ip_i}\end{pmatrix},\ \boldsymbol{\mu}_i(\boldsymbol{\beta}, \boldsymbol{b}_i) = g^{-1}(\boldsymbol{X}_i\boldsymbol{\beta} + \boldsymbol{Z}_i b_i),\ \boldsymbol{Z}_i = \boldsymbol{1}_{p_i},$$

$\boldsymbol{V}_i(\boldsymbol{\mu}_i) = Diag\{h(\mu_{ij})\}$, $h(\mu_{ij}) = \mu_{ij}(1-\mu_{ij})$ and $\boldsymbol{\Lambda}^*_i(\boldsymbol{\theta}_w) = \boldsymbol{I}_{p_i}$. In addition, we assume the conditional pdf of $y_{ij}|b_i$ is that of the binary distribution with logit link function (see Table 4.2).

Under these assumptions, we fit the respiratory disorder data to model (5.30) using GLIMMIX. We did so using both the PL approach based on linearization and the ML approach based on numerical integration. To illustrate, shown below is the SAS code used to fit the data using the GLIMMIX option METHOD=MMPL. This option implements the PL/QELS estimation approach to GLME models and is based on a first-order population-averaged Taylor series expansion of the conditional mean about $b_i = 0$ together with normal-theory maximum likelihood estimation of the random effects variance, ψ. In fact, this option combined with the MODEL /DIST=BIN option and the RANDOM statement represent the required GLIMMIX syntax needed to apply the marginal quasi-likelihood (MQL) estimation procedure of Breslow and Clayton (1993) to model (5.30) using pseudo ML estimation to estimate ψ. Included in the syntax is the EMPIRICAL option which requests that inference be based on robust standard errors.

Program 5.1

```
ods select Dimensions ModelInfo FitStatistics CovParms
            ParameterEstimates;
proc glimmix data=example5_2_1 method=MMPL empirical;
 class ID Center ;
 model y = Trt Center_ Gender Age y0 /
             dist=bin solution covb(details);
 random intercept / subject=ID(Center) ;
run;
quit;
```

Shown in Output 5.1 are select results from GLIMMIX based on the ODS SELECT statement.

Output 5.1: Regression estimates based on MQL (METHOD=MMPL).

```
                 Model Information
Data Set                       WORK.EXAMPLE5_2_1
Response Variable              y
Response Distribution          Binomial
Link Function                  Logit
Variance Function              Default
Variance Matrix Blocked By     ID(Center)
Estimation Technique           MPL
Degrees of Freedom Method      Containment
Fixed Effects SE Adjustment    Sandwich - Classical
             Dimensions
G-side Cov. Parameters          1
Columns in X                    6
Columns in Z per Subject        1
Subjects (Blocks in V)        111
Max Obs per Subject             4
             Fit Statistics
-2 Log Pseudo-Likelihood     2000.60
Generalized Chi-Square        333.24
Gener. Chi-Square / DF          0.75
             Covariance Parameter Estimates
Cov Parm     Subject       Estimate    Standard Error
Intercept    ID(Center)     1.3520             0.3649
                  Solutions for Fixed Effects
Effect        Estimate   Standard Error    DF    t Value    Pr > |t|
Intercept      -0.9187           0.4440   105      -2.07      0.0410
Trt             1.2947           0.3410   333       3.80      0.0002
Center_         0.6396           0.3512   333       1.82      0.0694
Gender          0.1221           0.4354   333       0.28      0.7794
Age           -0.01705          0.01243   333      -1.37      0.1712
y0              1.8420           0.3417   333       5.39      <.0001
```

Output 5.2: Regression estimates under four GLIMMIX estimation schemes.

Method	Effect	Estimate	SE	DF	p-value	Odds Ratio	Lower CL	Upper CL
1. MMPL	Intercept	−0.919	0.444	105	0.0410	.	.	.
	Trt	1.295	0.341	333	0.0002	3.65	1.87	7.14
	Center_	0.640	0.351	333	0.0694	.	.	.
	Gender	0.122	0.435	333	0.7794	.	.	.
	Age	−0.017	0.012	333	0.1712	.	.	.
	y0	1.842	0.342	333	0.0000	.	.	.
2. MSPL	Intercept	−1.040	0.523	105	0.0495	.	.	.
	Trt	1.453	0.380	333	0.0002	4.28	2.03	9.03
	Center_	0.710	0.399	333	0.0760	.	.	.
	Gender	0.169	0.499	333	0.7349	.	.	.
	Age	−0.020	0.015	333	0.1792	.	.	.
	y0	2.100	0.389	333	0.0000	.	.	.
3. Laplace	Intercept	−1.501	0.760	105	0.0509	.	.	.
	Trt	2.048	0.554	333	0.0003	7.76	2.61	23.04
	Center_	0.981	0.583	333	0.0932	.	.	.
	Gender	0.247	0.701	333	0.7253	.	.	.
	Age	−0.027	0.021	333	0.1941	.	.	.
	y0	2.948	0.640	333	0.0000	.	.	.
4. Quad	Intercept	−1.486	0.749	105	0.0499	.	.	.
	Trt	2.028	0.539	333	0.0002	7.60	2.63	21.97
	Center_	0.980	0.577	333	0.0902	.	.	.
	Gender	0.244	0.697	333	0.7261	.	.	.
	Age	−0.027	0.021	333	0.1971	.	.	.
	y0	2.920	0.607	333	0.0000	.	.	.

Not surprisingly, the results in Output 5.1 are quite similar to the GEE-based results shown in Output 4.6 in which a marginal logistic regression model with "working" independence was fit to the data. As explained previously, this is because the PL/QELS approach implemented with METHOD= MMPL uses a first-order PA expansion of the conditional mean about $b_i = 0$. This is equivalent to MQL in that it yields an approximation to the marginal mean that is structurally equivalent to the marginal mean one would specify when fitting the data directly to a marginal GLIM. Shown in Output 5.2 are the regression parameter estimates obtained when we also fit the data to model (5.30) using METHOD=MSPL, METHOD=Laplace and METHOD=Quad corresponding to PQL, Laplace-based MLE and adaptive quadrature based MLE, respectively. Included are the estimated odds ratios for the active versus placebo control treatment effect. The output was generated using repeated calls to a SAS macro that invokes GLIMMIX and the use of PROC REPORT to summarize the GLIMMIX output. Except for the METHOD= option, the GLIMMIX syntax for all four methods is the same as that shown for MQL (i.e., METHOD=MMPL). We elected to compute the MQL and PQL estimates of $\boldsymbol{\beta}$ based on the PL estimate of ψ (i.e., normal-theory MLE) as implemented with the METHOD=MMPL or METHOD=MSPL option. We did so in order to compare the different estimates of $\boldsymbol{\beta}$ when a ML type estimation scheme is used to estimate ψ.

In comparing parameter estimates across the four estimation schemes, we find there is a much stronger treatment effect when we apply the ML approach using METHOD=Laplace or METHOD=Quad. This may be due in part to an underestimation of the variance component ψ. As shown by Breslow and Lin (1995), some of the bias in estimating $\boldsymbol{\beta}$ may be attributed to bias in estimating ψ. In this example, the Laplace and quadrature based ML estimates of ψ are of two-fold higher magnitude than the PL based estimates (see Output 5.3). Such differences can explain the attenuation of treatment effects associated with the MQL and PQL estimation procedures (see Appendix B for further results on the asymptotic bias of MQL, PQL, Laplace and ML estimators under GLME models for binary outcomes).

Output 5.3: Variance component estimates under the four GLIMMIX estimation schemes.

Method	CovParm	Subject	Estimate	StdErr
1. MMPL	Intercept	ID(Center)	1.3520	0.3649
2. MSPL	Intercept	ID(Center)	1.7215	0.4572
3. Laplace	Intercept	ID(Center)	3.6424	1.1854
4. Quad	Intercept	ID(Center)	3.9395	1.1751

Despite some differences in the magnitude of the parameter estimates, the results from all four analyses are qualitatively similar when it comes to assessing the significance of the active drug treatment effect (p< 0.001 in all four cases). However, choosing which set of parameter estimates "best" predicts a patient's response to treatment is far less clear. For example, one cannot use information criteria to help select between PL and ML estimates since the former do not correspond to any well-defined objective function. For this example we chose to compare goodness-of-fit using the R-square and concordance correlation coefficient criteria described by Vonesh et. al. (1996) and Vonesh and Chinchilli (1997) as implemented in the %GLIMMIX_GOF macro (Appendix D). This macro takes output generated from GLIMMIX and summarizes various goodness-of-fit measures including an average model R^2 and a conditional model R^2 statistic (Vonesh and Chinchilli, 1997). The average model R^2 computed in %GLIMMIX_GOF (as well as in %GOF) is

$$R^2_{Avg} = 1 - \frac{\sum_{i=1}^{n}(y_{ij} - \widehat{y}_{ij})^2}{\sum_{i=1}^{n}(y_{ij} - \bar{y})^2} \tag{5.31}$$

where $\widehat{y}_{ij} = E(y_{ij}|\boldsymbol{b}_i = \boldsymbol{0}) = \mu_{ij}(\boldsymbol{x}'_{ij}, \widehat{\boldsymbol{\beta}}, \boldsymbol{z}_{ij}, \boldsymbol{0}) = \mu_{ij}(\widehat{\boldsymbol{\beta}}, \boldsymbol{0})$ is the ith subject's predicted value under the GLME model but evaluated at the average random effect, $\boldsymbol{b}_i = \boldsymbol{0}$, and $\bar{y} = \sum_{i=1}^{n} y_{ij}/N$ is the overall mean of the response variable. The conditional model R^2, denoted by R^2_{Cond}, is defined in the same way as in (5.31) but with $\widehat{y}_{ij} = E(y_{ij}|\widehat{\boldsymbol{b}}_i) = \mu_{ij}(\boldsymbol{x}'_{ij}, \widehat{\boldsymbol{\beta}}, \boldsymbol{z}_{ij}, \widehat{\boldsymbol{b}}_i) = \mu_{ij}(\widehat{\boldsymbol{\beta}}, \widehat{\boldsymbol{b}}_i)$ representing the ith individual's predicted value evaluated at the individual's estimated random effect, $\boldsymbol{b}_i = \widehat{\boldsymbol{b}}_i$. The average model R^2 measures a reduction in residual variation explained by the fixed-effects alone while the conditional model R^2 measures a reduction in explained variation due to fitting both fixed and random effects versus explained variation due to fitting an overall mean.

The macro also computes a goodness-of-fit measure based on a concordance correlation coefficient developed by Lin (1989) and adapted to goodness-of-fit by Vonesh et. al. (1996) and Vonesh and Chinchilli (1997). The average model concordance correlation is defined as

$$\widehat{\rho}_{c,Avg} = 1 - \frac{\sum_{i=1}^{n}(y_{ij} - \widehat{y}_{ij})^2}{\sum_{i=1}^{n}(y_{ij} - \bar{y})^2 + \sum_{i=1}^{n}(\widehat{y}_{ij} - \bar{\widehat{y}})^2 + N(\bar{y} - \bar{\widehat{y}})^2} \tag{5.32}$$

where $\bar{\widehat{y}} = \sum_{i=1}^{n} \widehat{y}_{ij}/N$ is the overall mean of the individual predicted values evaluated at the average random effect, i.e., $\widehat{y}_{ij} = \mu_{ij}(\widehat{\boldsymbol{\beta}}, \boldsymbol{0})$. The conditional concordance correlation, $\widehat{\rho}_{c,Cond}$, is also given by (5.32) but with the predicted values evaluated at the subject-specific random effects, i.e., $\widehat{y}_{ij} = \mu_{ij}(\widehat{\boldsymbol{\beta}}, \widehat{\boldsymbol{b}}_i)$. The concordance correlation is a measure of how well the predicted values agree with the observed values. As a measure of agreement, it reflects how well a scatter plot of observed versus predicted values falls about the line of identity. Consequently, it does not require specification of a null model (e.g., an overall intercept model with no fixed effects) that is otherwise required for R^2 type goodness-of-fit measures (Vonesh and Chinchilli, 1997).

Shown below is the syntax required to run %GLIMMIX_GOF in combination with GLIMMIX for the ML based analysis using the Laplace approximation to the integrated log-likelihood.

Program 5.2

```
ods select Dimensions ModelInfo FitStatistics CovParms
              CovBDetails ParameterEstimates;
proc glimmix data=example5_2_1 method=Laplace empirical;
 class ID Center ;
 model y = Trt Center_ Gender Age y0 /
              dist=bin solution cl rs,
 random intercept / subject=ID(Center) ;
 ods output ParameterEstimates=pe_lmle;
 ods output CovBDetails=gof_lmle;
 ods output dimensions=n_lmle;
 output out=pred_lmle /allstats;
run;
quit;
%GLIMMIX_GOF(dimension=n_lmle,
             parms=pe_lmle,
             covb_gof=gof_lmle,
             output=pred_lmle,
             response=y,
             pred_ind=PredMu,
             pred_avg=PredMuPA,
             opt=noprint);
```

With respect to the %GLIMMIX_GOF syntax (see the author's Web page), the DIMENSION= argument of the macro provides information on the number of subjects. The PARMS= argument instructs %GLIMMIX_GOF as to where the regression parameter estimates may be located—in this example, they are found in the dataset PE_LMLE as determined by the ODS OUTPUT statement above. The COVB_GOF= argument provides the macro with necessary statistics computed from the MODEL option COVB(DETAILS) which the macro uses to compute a pseudo-likelihood ratio goodness-of-fit test of the assumed model covariance structure (see the example in §2.2.2 for an explanation). Finally, the RESPONSE= argument specifies the name of the response variable from GLIMMIX while the PRED_IND= argument and PRED_AVG= argument specify, respectively, the SAS variables representing the individual's predicted mean response and the predicted average mean response. The former estimates the individual SS mean responses evaluated at both $\widehat{\boldsymbol{\beta}}$ and $\widehat{\boldsymbol{b}}_i$ while the latter is an average mean response evaluated at $\widehat{\boldsymbol{\beta}}$ alone (i.e., is the average conditional mean response obtained by setting $\widehat{\boldsymbol{b}}_i = 0$). Using the SAS macro %GLIMMIX_GOF, we obtained estimates of both the average model R^2 and conditional model R^2 for the MQL and PQL estimation procedures (corresponding to MMPL and MSPL, see Output 5.4) and for ML estimation based on the Laplace and quadrature-based integral approximations (corresponding to Laplace and Quad, see Output 5.5). Also included are the corresponding concordance correlations as well as an informal pseudo-likelihood ratio test for assessing whether the assumed covariance structure is reasonable.

Output 5.4: Goodness-of-fit estimates for PL-based estimation.

DESCRIPTION	MMPL	MSPL
Total Observations	444.000	444.000
N (number of subjects)	111.000	111.000
Number of Fixed-Effects Parameters	6.000	6.000
Average Model R-Square:	0.252	0.250
Average Model Adjusted R-Square:	0.242	0.240
Average Model Concordance Correlation:	0.403	0.422
Average Model Adjusted Concordance Correlation:	0.394	0.414
Conditional Model R-Square:	0.557	0.572
Conditional Model Adjusted R-Square:	0.551	0.566
Conditional Model Concordance Correlation:	0.667	0.691
Conditional Model Adjusted Concordance Correlation:	0.663	0.686
Variance-Covariance Concordance Correlation:	0.974	0.982
Discrepancy Function	0.242	0.178
s = Rank of robust sandwich estimator, OmegaR	6.000	6.000
s1 = Number of unique non-zero off-diagonal elements of OmegaR	15.000	15.000
Approx. Chi-Square for H0: Covariance Structure is Correct	26.890	19.808
DF1 = s(s+1)/2, per Vonesh et al (Biometrics 52:572-587, 1996)	21.000	21.000
Pr > Chi Square based on degrees of freedom, DF1	0.175	0.533
DF2 = s+s1, a modified degress of freedom	21.000	21.000
Pr > Chi Square based on modified degrees of freedom, DF2	0.175	0.533

In comparing the goodness-of-fit statistics between the two PL estimation schemes, there is a better fit to the data when one uses the PQL (MSPL) estimation scheme versus the MQL (MMPL) scheme. Between the two ML estimation schemes, the ML fit based on adaptive quadrature (Quad) provided a slight improvement over the Laplace-based ML fit (Laplace). The fit of the data to both the predicted average conditional mean response and the predicted conditional mean responses favors the use of the ML-based estimation schemes over the PL schemes. This is particularly true with respect to predicting the individual SS responses wherein the conditional adjusted concordance correlation (adjusted or corrected for the number of parameters) is 0.734 and 0.737 for the Laplace and quadrature-based ML estimates, respectively, versus ≤ 0.686 for the two PL-based methods. There was no evidence of misspecification with respect to the assumed model covariance structure for any of the four estimation schemes although the quadrature-based ML scheme provided the lowest discrepancy function value of the four schemes. These results are not surprising in consideration of our discussion comparing the asymptotic behavior of PL and ML based estimates (see §5.1.2).

Output 5.5: Goodness-of-fit estimates for ML-based estimation.

DESCRIPTION	Laplace	Quad
Total Observations	444.000	444.000
N (number of subjects)	111.000	111.000
Number of Fixed-Effects Parameters	6.000	6.000
Average Model R-Square:	0.228	0.229
Average Model Adjusted R-Square:	0.217	0.218
Average Model Concordance Correlation:	0.458	0.457
Average Model Adjusted Concordance Correlation:	0.451	0.450
Conditional Model R-Square:	0.611	0.614
Conditional Model Adjusted R-Square:	0.605	0.609
Conditional Model Concordance Correlation:	0.738	0.741
Conditional Model Adjusted Concordance Correlation:	0.734	0.737
Variance-Covariance Concordance Correlation:	0.978	0.984
Discrepancy Function	0.184	0.149
s = Rank of robust sandwich estimator, OmegaR	6.000	6.000
s1 = Number of unique non-zero off-diagonal elements of OmegaR	15.000	15.000
Approx. Chi-Square for H0: Covariance Structure is Correct	20.406	16.494
DF1 = s(s+1)/2, per Vonesh et al (Biometrics 52:572-587, 1996)	21.000	21.000
Pr > Chi Square based on degrees of freedom, DF1	0.496	0.741
DF2 = s+s1, a modified degress of freedom	21.000	21.000
Pr > Chi Square based on modified degrees of freedom, DF2	0.496	0.741

If we therefore base inference on the quadrature-based ML results in Output 5.2, we find that the average patient receiving active drug has a 7.60-fold higher odds of experiencing a positive response versus the average patient receiving the placebo control. Note, however, that this is not a population-based odds ratio in that it does not depict what the relative odds of a favorable response is among a population of patients receiving active drug versus a similar population of patients receiving the placebo control drug. Under the GLME model (5.30), a population-based odds ratio for treated and control patients each having a common set of characteristics (i.e., center, gender, age and baseline response) would require numerically evaluating the marginal means (i.e., probabilities of a positive response) in the two groups and forming the odds ratio from those marginal probabilities. Alternatively, one can apply the approximation described by Zeger, Liang and Albert (1988) to this problem and estimate the marginal probability of a positive response as

$$\mu_{ij}(\boldsymbol{x}'_{ij}, \boldsymbol{\beta}, \psi) = \Pr(y_{ij} = 1 | \boldsymbol{x}'_{ij}, \boldsymbol{\beta}, \psi) = \frac{\exp\{\boldsymbol{x}'_{ij}\boldsymbol{\beta} \times a_l(\psi)\}}{1 + \exp\{\boldsymbol{x}'_{ij}\boldsymbol{\beta} \times a_l(\psi)\}} \tag{5.33}$$

where $a_l(\psi) = (c^2\psi + 1)^{-1/2}$ and $c = 16\sqrt{3}/(15\pi)$. Based on this approximation, the marginal or PA odds ratio for active drug versus control is

$$OR(Active : Control) = \exp(\beta_1 \times a_l(\psi)) = \exp(\beta_1)^{a_l(\psi)} \tag{5.34}$$

where β_1 is the regression parameter associated with the treatment group indicator variable, a_{1i}, of $\boldsymbol{x}_{ij}$ where $a_{1i} = 1$ if active drug and $a_{1i} = 0$ if placebo control. One can compute a point estimate of this marginal odds ratio by simply substituting estimates $\widehat{\beta}_1$ and $\widehat{\psi}$ from GLIMMIX into (5.34). Unfortunately, there are no options available within GLIMMIX that will allow us to form a confidence interval or test the significance of this approximate PA odds ratio. However, such inference is possible using the ESTIMATE statement of the NLMIXED procedure. To illustrate, the NLMIXED syntax required to fit the GLME model (5.30) to the respiratory disorder data and also to estimate the marginal odds ratio together with a 95% confidence interval is shown below.

Program 5.3

```
ods listing close;
ods output ParameterEstimates=pe;
ods output AdditionalEstimates=ae;
proc nlmixed data=example5_2_1 empirical;
 parms b0=-1.5  b1=2  b2=1 b3=0 b4=0 b5=3 psi=4;
 Pi=Constant('PI');
 c=16*sqrt(3)/(15*Pi);
 A_psi = ((c**2)*psi + 1)**(-1/2);
 logit_pij = b0 + b1*Trt + b2*Center_ + b3*Gender +
                  b4*Age + b5*y0 + bi;
 pij = exp(logit_pij)/(1+exp(logit_pij));
 OddsRatioPA = exp(b1)**A_psi;
 OddsRatioSS = exp(b1);
 LogOddsRatioPA = b1*A_psi;
 LogOddsRatioSS = b1;
 model y ~ binary(pij);
 random bi ~ normal(0,psi) subject=id;
 estimate 'PA OR(Active:Placebo)' OddsRatioPA;
 estimate 'SS OR(Active:Placebo)' OddsRatioSS;
 estimate 'PA Log OR(Active:Placebo) ' LogOddsRatioPA;
 estimate 'SS Log OR(Active:Placebo) ' LogOddsRatioSS;
run;
quit;
data ae;
 set ae;
 output;
 If Label='PA Log OR(Active:Placebo) ' then do;
    Label='Alt. PA OR(Active:Placebo)';
    Estimate=exp(Estimate);
    StandardError=Estimate*StandardError;
    Lower=exp(Lower);
    Upper=exp(Upper);
    output;
 end;
 If Label='SS Log OR(Active:Placebo) ' then do;
    Label='Alt. SS OR(Active:Placebo)';
    Estimate=exp(Estimate);
    StandardError=Estimate*StandardError;
    Lower=exp(Lower);
    Upper=exp(Upper);
    output;
 end;
run;
ods listing;
proc print data=pe noobs;
 var Parameter Estimate StandardError DF Alpha Lower Upper;
run;
proc print data=ae noobs;
 var Label Estimate StandardError Lower Upper;
run;
```

The SAS variable, `A_psi`, is the correction factor $a_l(\psi)$ required to approximate the marginal response probabilities as well as estimate the approximate PA odds ratio (5.34) associated with the active versus placebo control drug. For comparative purposes, we also computed the estimated SS odds ratio which is simply $\exp(\widehat{\beta}_1)$. So as to compare results with GLIMMIX, we estimated both the PA and SS odds ratios directly using the following ESTIMATE statements

```
estimate 'PA OR(Active:Placebo)' OddsRatioPA;
estimate 'SS OR(Active:Placebo)' OddsRatioSS;
```

We also estimated the odds ratios indirectly by first estimating the log odds ratios and their confidence bounds using the ESTIMATE statements

```
estimate 'PA Log OR(Active:Placebo)' LogOddsRatioPA;
estimate 'SS Log OR(Active:Placebo)' LogOddsRatioSS;
```

and then back transforming these results to the odds ratio scale by exponentiating both the estimates and their lower and upper 95% confidence bounds in the ensuing data step. This latter approach is what is done in GLIMMIX when estimating odds ratios from a logistic regression model. It is the preferred approach because it maintains a valid range of values for both the lower and upper confidence bounds of an odds ratio. As with the analysis done using GLIMMIX, we specified the EMPIRICAL option of NLMIXED so that inference would be based on the robust sandwich estimator $\widehat{\boldsymbol{\Omega}}_R(\widehat{\boldsymbol{\tau}})$ of the covariance matrix of $\widehat{\boldsymbol{\tau}} = (\widehat{\boldsymbol{\beta}}, \widehat{\psi})$. Results of this additional analysis are shown in Output 5.6.

We see from the results that except for minor differences due to rounding, the estimated regression parameters from NLMIXED are identical with those from GLIMMIX when one uses METHOD=QUAD (Output 5.2). One difference of note between the output of the two procedures is the default degrees of freedom used for inference. For example, the GLIMMIX default degrees of freedom for treatment effect is 333 while the NLMIXED default is 110. While not done here, one can control the degrees of freedom used for either procedure by specifying the DF option available with the ESTIMATE statement.

Output 5.6: Select output from NLMIXED including the parameter estimates $\widehat{\boldsymbol{\tau}} = (\widehat{\boldsymbol{\beta}}, \widehat{\psi})$ from Model (5.30) along with the SS odds ratio and the approximate PA odds ratio for assessing the effectiveness of the active drug.

Parameter	Estimate	StandardError	DF	Alpha	Lower	Upper
b0	−1.4862	0.7490	110	0.05	−2.9706	−0.00181
b1	2.0292	0.5395	110	0.05	0.9601	3.0983
b2	0.9799	0.5768	110	0.05	−0.1632	2.1230
b3	0.2435	0.6965	110	0.05	−1.1368	1.6239
b4	−0.02685	0.02078	110	0.05	−0.06804	0.01434
b5	2.9197	0.6071	110	0.05	1.7165	4.1230
psi	3.9395	1.1752	110	0.05	1.6106	6.2685

Label	Estimate	StandardError	Lower	Upper
PA OR(Active:Placebo)	3.7442	1.2920	1.1837	6.3046
SS OR(Active:Placebo)	7.6079	4.1041	−0.5256	15.7413
PA Log OR(Active:Placebo)	1.3202	0.3451	0.6363	2.0040
Alt. PA OR(Active:Placebo)	3.7442	1.2920	1.8896	7.4190
SS Log OR(Active:Placebo)	2.0292	0.5395	0.9601	3.0983
Alt. SS OR(Active:Placebo)	7.6079	4.1041	2.6119	22.1595

In comparing estimates of the odds ratio, we note that the estimated PA odds ratio is about half the size of the SS odds ratio. This is not surprising in lieu of the shrinkage effect the correction factor $a_l(\psi)$ has when approximating the marginal or PA-based linear predictor. Specifically, as ψ increases, the correction factor $a_l(\psi)$ decreases causing the approximate PA-based linear predictor, $\boldsymbol{x}'_{ij}\boldsymbol{\beta}^* = \boldsymbol{x}'_{ij}\boldsymbol{\beta} \times a_l(\psi)$, to shrink toward 0 as compared to the SS-based linear predictor $\boldsymbol{x}'_{ij}\boldsymbol{\beta}$. Here $\boldsymbol{\beta}^* = \boldsymbol{\beta} \times a_l(\psi)$ is an approximation to the marginal fixed-effects regression parameters from a logistic regression model obtained by setting $\psi = 0$ (Zeger, Liang and Albert; 1988). In our example, $\widehat{\psi} = 3.94$ which means the approximate PA regression parameters are all shrunk by a factor of $a_l(\widehat{\psi}) = 0.65$. Hence the PA log odds ratio estimate is $\widehat{\beta}_1^* = \widehat{\beta}_1 \times 0.65 = 2.029 \times 0.65 = 1.319$ which yields a PA odds ratio of 3.74 rounded to two decimals. We also obtain the same point estimate of the SS odds ratio, namely 7.6079, as was obtained with GLIMMIX. However, we find that the confidence interval for the direct estimated SS odds ratio contains a lower bound that is outside the valid range of values for an odds ratio. As noted above, this is a result of applying the delta method directly to the odds ratio as opposed to applying it to the log odds ratio and then back transforming to the odds ratio scale.

Finally, we observe that the approximate PA odds ratio of 3.74 computed under the GLME model (5.30) using numerical quadrature is similar to the odds ratio estimate of 3.65 computed under GLIMMIX using MQL estimation (METHOD=MMPL, see Output 5.2). This should not be surprising since the MQL approach essentially fits the data to a marginal GLIM with a "working" covariance structure. The advantage of fitting the data to the GLME model (5.30) using numerical quadrature and then estimating the PA odds ratio using (5.34) is that one can estimate the effect of the active drug versus placebo control on both an individual patient basis and on a population basis. Under either interpretation, the active drug is superior to the placebo control in improving patient response over time.

5.2.2 Epileptic seizure data—continued

The epileptic seizure data described in §1.4 (page 15) was fit to several different marginal GLIM's in Chapter 4 (see §4.3.3 and §4.5.4). In this section, we fit the data to a mixed-effects Poisson model in which we allow for intra-subject overdispersion and where we account for correlation in seizure counts across visits by introducing a subject-specific random intercept into the model. The inclusion of an intra-subject dispersion parameter requires that we fit the data to the moments-based GLME model

$$\begin{aligned} E(y_{ij}|b_i) &= \mu_{ij}(\boldsymbol{\beta}, b_i) = g^{-1}(\boldsymbol{x}'_{ij}\boldsymbol{\beta} + b_i) = \exp(\boldsymbol{x}'_{ij}\boldsymbol{\beta} + b_i) \\ Var(y_{ij}|b_i) &= \phi\mu_{ij}(\boldsymbol{\beta}, b_i) \end{aligned} \tag{5.35}$$

where $\boldsymbol{x}'_{ij} = \begin{pmatrix}1 & a_{1i} & \log(a_{2i}) & \log(a_{3i}) & a_{1i}\log(a_{3i}) & v_{i4}\end{pmatrix}$ is the vector of within- and between-subject covariates defined in §1.4 (page 15) and $\boldsymbol{x}'_{ij}\boldsymbol{\beta} + b_i = \beta_0 + \beta_1 a_{1i} + \beta_2 \log(a_{2i}) + \beta_3 \log(a_{3i}) + \beta_4 a_{1i} \log(a_{3i}) + \beta_5 v_{i4} + b_i$ is the conditional linear predictor with $b_i \sim$iid $N(0, \psi)$. As in §4.3.3, the covariates a_{1i}, a_{2i}, $\log(a_{3i})$, and v_{i4} represent the treatment group indicator (`Trt`), age (`Age`), log normalized baseline count (`y0`), and visit 4 indicator (`Visit4`). We exclude the offset term $\log(t_{ij}) = \log 2$ in the same manner as in §4.5.4 so as to more closely mirror the analysis of Thall and Vail (1990). By including the overdispersion parameter, ϕ, model (5.35) no longer coincides with a Poisson conditional probability density function. Consequently, estimation under (5.35) is carried out using PL/CGEE1 with restricted pseudo-likelihood estimation of $\boldsymbol{\theta} = (\boldsymbol{\theta}_b, \boldsymbol{\theta}_w) = (\psi, \phi)$ and PQL estimation of $\boldsymbol{\beta}$.

The GLIMMIX code to run an initial analysis with all patients is shown below. The syntax is similar to that used in Example 4.3.3 except that we exclude the offset option and include a LSMEANS statement and a second random statement (i.e., `random intercept / subject=ID type=vc;`). The second random statement allows us to include the random intercept effect, b_i, in the linear predictor. To request least squares means for the treatment effect via the LSMEANS statement, we must specify the treatment variable, Trt, as a class variable. In doing so, we define formatted values of the Trt variable with the FORMAT procedure and use the GLIMMIX option ORDER=FORMATTED in order to obtain fixed-effects parameter estimates consistent with those modeled under the marginal GLIM's of Example 4.3.3.

Program 5.4

```
proc format;
 value Trtfmt 0='Placebo' 1='Active';
data example5_2_2; set example4_3_3;
 by ID Trt;
 noobs=_n_;
run;
ods graphics on;
ods exclude IterHistory NObs SolutionR;
ods output SolutionR=re;
proc glimmix data=example5_2_2 noclprint=3 order=formatted
             plots(obsno)=(Boxplot(Fixed Observed ILINK)
                           PearsonPanel(ILINK) ) empirical;
 class ID Trt;
 model y = Trt LogAge y0 Trt y0*Trt Visit4 / dist=Poisson s;
 random _residual_ / subject=ID;
 random intercept / subject=ID type=vc s;
 lsmeans Trt / ilink diff e;
 output out=out1  Pearson(blup ilink)=residual;
 format Trt Trtfmt.;
run;
ods graphics off;
```

Estimation was carried out using the default METHOD=RSPL together with the EMPIRICAL option. The ILINK option of the LSMEANS statement allows us to compute both the log mean count and the mean count per patient visit (i.e., per two-week period) for the two treatment groups evaluated at the mean values of the remaining covariates (as displayed by the LSMEANS option E). Finally, we specified a PLOTS(OBSNO)= option in order to examine the set of conditional Pearson residuals on the mean scale for possible outliers. The conditional Pearson residuals on the mean scale are defined to be $r_{ij} = \left(y_{ij} - g^{-1}(\boldsymbol{x}'_{ij}\widehat{\beta} + \widehat{b}_i)\right)/\sqrt{Var(y_{ij}|\widehat{b}_i)}$. In order to identify influential observations according to patient and visit, we created a variable NOOBS=_N_ in the dataset EXAMPLE5_2_2 above that will correspond directly with the observation numbers shown in the graphical displays. The results of this analysis are shown in Output 5.7 and Figures 5.4-5.6.

Output 5.7: Parameter estimates and least squares means from the GLME model (5.35) using PL/CGEE1 as implemented in GLIMMIX using the default option METHOD=RSPL (corresponding to PQL estimation for β and REPL estimation for θ). Inference is based on robust standard errors.

```
                    Model Information
Data Set                        WORK.EXAMPLE5_2_2
Response Variable               y
Response Distribution           Poisson
Link Function                   Log
Variance Function               Default
Variance Matrix Blocked By      ID
Estimation Technique            Residual PL
Degrees of Freedom Method       Containment
Fixed Effects SE Adjustment     Sandwich - Classical
      Class Level Information
Class    Levels    Values
ID           59    not printed
Trt           2    Active Placebo
```

Dimensions

G-side Cov. Parameters	1
R-side Cov. Parameters	1
Columns in X	8
Columns in Z per Subject	1
Subjects (Blocks in V)	59
Max Obs per Subject	4

Optimization Information

Optimization Technique	Dual Quasi−Newton
Parameters in Optimization	1
Lower Boundaries	1
Upper Boundaries	0
Fixed Effects	Profiled
Residual Variance	Profiled
Starting From	Data

Convergence criterion (PCONV=1.11022E−8) satisfied.

Fit Statistics

-2 Res Log Pseudo-Likelihood	519.24
Generalized Chi-Square	453.19
Gener. Chi-Square / DF	1.97

Covariance Parameter Estimates

Cov Parm	Subject	Estimate	Standard Error
Intercept	ID	0.2242	0.06089
Residual (VC)		1.9704	0.2068

Solutions for Fixed Effects

Effect	Trt	Estimate	Standard Error	DF	t Value	Pr > |t|
Intercept		−1.4437	0.9610	55	−1.50	0.1388
Trt	Active	−0.9116	0.3978	175	−2.29	0.0231
Trt	Placebo	0	.	.	.	.
LogAge		0.5285	0.2844	175	1.86	0.0648
y0		0.8810	0.1096	175	8.04	<.0001
y0*Trt	Active	0.3389	0.1985	175	1.71	0.0896
y0*Trt	Placebo	0	.	.	.	.
Visit4		−0.1611	0.06558	175	−2.46	0.0150

Type III Tests of Fixed Effects

Effect	Num DF	Den DF	F Value	Pr > F
Trt	1	175	5.25	0.0231
LogAge	1	175	3.45	0.0648
y0	1	175	114.43	<.0001
y0*Trt	1	175	2.91	0.0896
Visit4	1	175	6.03	0.0150

Coefficients for Trt Least Squares Means

Effect	Trt	Row1	Row2
Intercept		1	1
Trt	Active	1	
Trt	Placebo		1
LogAge		3.3198	3.3198
y0		1.768	1.768
y0*Trt	Active	1.768	
y0*Trt	Placebo		1.768
Visit4		0.25	0.25

Trt Least Squares Means

Trt	Estimate	Standard Error	DF	t Value	Pr > |t|	Mean	Standard Error Mean
Active	1.5157	0.1140	175	13.30	<.0001	4.5525	0.5189
Placebo	1.8281	0.08763	175	20.86	<.0001	6.2223	0.5453

Differences of Trt Least Squares Means

Trt	_Trt	Estimate	Standard Error	DF	t Value	Pr > |t|
Active	Placebo	−0.3125	0.1447	175	−2.16	0.0322

As with the marginal GLIM's of Example 4.3.3, there is evidence of significant overdispersion as estimated by the within-subject dispersion parameter $\widehat{\phi} = 1.9704 \pm 0.2068$ (see the Residual (VC) estimate under the output labeled "Covariance

Parameter Estimates" in Output 5.7). There is also evidence of significant between-subject variability with $\widehat{\psi} = 0.2242 \pm 0.06089$. Except for the intercept, the fixed-effects parameter estimates are similar to those obtained from the various marginal GLIM's of Example 4.3.3 (see Output 4.9). For example, while the treatment effect is somewhat weaker, it does remain statistically significant (p=0.0231). Interestingly, the treatment $\times$ log baseline count is no longer significant although it does retain the same directional effect.

The fact that the estimated intercept in Output 5.7 differs substantially from those shown for the marginal GLIM's of Output 4.9 follows from the fact that under model (5.35), the regression parameters for the marginal mean are directly related to those of the conditional mean via

$$\begin{aligned} E(y_{ij}) &= E_b\Big\{\exp(\boldsymbol{x}_{ij}'\boldsymbol{\beta} + b_i)\Big\} \\ &= \exp\Big\{\boldsymbol{x}_{ij}'\boldsymbol{\beta} + \frac{1}{2}\psi\Big\}. \end{aligned} \tag{5.36}$$

If we partition $\boldsymbol{\beta}$ in model (5.35) as $\boldsymbol{\beta} = (\beta_0, \boldsymbol{\beta}_1)$ where β_0 is the intercept and $\boldsymbol{\beta}_1 = (\beta_1, \beta_2, \beta_3, \beta_4, \beta_5)'$ are the regression coefficients for the fixed effects, then the marginal means may be expressed as

$$E(y_{ij}) = \exp(\beta_0^* + \beta_1 a_{1i} + \beta_2 \log(a_{2i}) + \beta_3 \log(a_{3i}) + \beta_4 a_{1i} \log(a_{3i}) + \beta_5 v_{i4})$$

where $\beta_0^* = \beta_0 + \frac{1}{2}\psi$ is the marginal intercept parameter expressed as a simple additive offset to the conditional intercept. Since we also excluded the offset term, log(2), from model (5.35), the estimated intercept, $\widehat{\beta}_0 = -1.4437$, will differ from the marginal intercepts in Output 4.9 by a factor of $\log(2) + \frac{1}{2}\psi$.

Based on the relationship in (5.36), we should see little difference between parameter estimates obtained by fitting data to a Poisson GLME with a random intercept like model (5.35) versus those obtained by fitting the same data to a corresponding marginal GLIM. However, since the marginal and conditional generalized estimating equations used to fit the two models use different weighting schemes, actual parameter estimates will differ particularly with smaller sample sizes. For a more general mixed-effects Poisson regression model with $E(y_{ij}|\boldsymbol{b}_i) = \exp\{\boldsymbol{x}_{ij}'\boldsymbol{\beta} + \boldsymbol{z}_{ij}'\boldsymbol{b}_i\}$, the marginal mean will be related to the conditional mean via

$$\begin{aligned} E(y_{ij}) &= E_b\Big\{\exp(\boldsymbol{x}_{ij}'\boldsymbol{\beta} + \boldsymbol{z}_{ij}'\boldsymbol{b}_i)\Big\} \\ &= \exp\Big\{\boldsymbol{x}_{ij}'\boldsymbol{\beta} + \frac{1}{2}\boldsymbol{z}_{ij}'\boldsymbol{\Psi}\boldsymbol{z}_{ij}\Big\}. \end{aligned} \tag{5.37}$$

Under this more general setting, the parameter estimates may all differ but the random effects lead to a simple offset, $\frac{1}{2}\boldsymbol{z}_{ij}'\boldsymbol{\Psi}\boldsymbol{z}_{ij}$, in the marginal means.

It should be noted that because the marginal and conditional means of model (5.35) differ by the multiplicative factor, $\exp(\psi/2)$, the population regression coefficients $\boldsymbol{\beta}_1 = (\beta_1, \beta_2, \beta_3, \beta_4, \beta_5)$ retain both a PA and SS interpretation. For example, because $\exp(\psi/2)$ cancels when we take the rate ratio,

$$\begin{aligned} RR(Active : Control) &= \frac{\exp\{\beta_1 \times 1 + \beta_4 \times 1 \times \log(\bar{y}_0) + \psi/2\}}{\exp\{\beta_1 \times 0 + \beta_4 \times 0 \times \log(\bar{y}_0) + \psi/2\}} \\ &= \exp\{\beta_1 + \beta_4 \times \log(\bar{y}_0)\}, \end{aligned}$$

we may interpret this rate ratio as a PA rate ratio depicting the relative increase or decrease in the average seizure rate among a population of patients receiving the active drug, progabide, versus a similar population of patients receiving the placebo control. The two populations would have the same set of baseline characteristics including a mean baseline count of $\bar{y}_0$. For these same two populations, this PA rate ratio may also be interpreted as a SS rate ratio depicting the relative increase or decrease in the seizure rate of the average patient from the active drug population versus the average patient from the placebo control population. In other words, under model (5.35) we have

$$\frac{E(y|\boldsymbol{x}_1)}{E(y|\boldsymbol{x}_0)} = \frac{E_b[E_{y|b}(y|\boldsymbol{x}_1, b)]}{E_b[E_{y|b}(y|\boldsymbol{x}_0, b)]} = \frac{E_{y|b}(y|\boldsymbol{x}_1, b = 0)}{E_{y|b}(y|\boldsymbol{x}_0, b = 0)}$$

where $\boldsymbol{x}_1$ and $\boldsymbol{x}_0$ represent two populations of subjects having the same set of covariates except that those with $\boldsymbol{x}_1$ receive the active drug (i.e., $a_{i1} = 1$) while those with $\boldsymbol{x}_0$ receive the placebo control (i.e., $a_{i1} = 0$). Note that because we excluded the two-week time period as an offset, the mean seizure count, $E(y|\boldsymbol{x})$, is also the seizure rate per two-week period.

Included in Output 5.7 are the least squares means for the two treatment groups presented on both the link (log mean) and inverse link (mean) scales. They are SS least squares means in that they represent the average subject's log mean seizure count and mean seizure count, respectively, per two-week period. They are obtained by setting the SS random effects to 0 and by setting all other covariates to their mean value. (Note: these values are displayed in Output 5.7 under the heading, "Coefficients for Trt Least Squares Means.") Based on (5.36), one can easily convert these to PA least squares means by adding $\widehat{\psi}/2 = 0.1121$ to the least squares log mean counts and multiplying $\exp(\widehat{\psi}/2) = 1.1186$ to the least squares mean counts. For example, the PA least squares mean count per two-week period for patients receiving the active drug, progabide, would be $1.1186 \times 4.5525 = 5.0924$ while the PA least squares mean count for the placebo control group would be $1.1186 \times 6.2223 = 6.9603$. Additional programming would be required if one wished to compute the corresponding standard errors of these PA least squares means by applying the delta method in conjunction with the estimated covariance matrices of $\widehat{\boldsymbol{\beta}}$ and $\widehat{\boldsymbol{\theta}}$ which one can access with ODS OUTPUT AsyCov= and ODS OUTPUT CovB= statements. In this particular example, the difference between the SS and PA least squares means is not too great due to the fact that the random-effects variance component $\widehat{\psi} = 0.2242$ is relatively small in magnitude. Moreover, on the link scale (i.e., the log mean count scale), the difference in least squares means between active versus control drug may also be interpreted as either a PA or SS least squares mean difference.

Finally, shown in Figures 5.4-5.6 are plots showing the distribution of seizure counts by treatment group (Figure 5.4), the corresponding distribution of conditional Pearson residuals by treatment group (Figure 5.5), and a paneled display of the conditional Pearson residuals across both treatment groups (Figure 5.6). Based on the distribution of observed seizure counts (Figure 5.4), we observe the overt influence patient 207 might have on both the treatment effect and its interaction with baseline count as indicated by Thall and Vail (1990). One can identify this patient by simply printing the output dataset OUT1 created by the OUTPUT OUT statement shown above in which the value of the variable NOOBS corresponds to the observation numbers shown in the plot. While this patient appears to be overly influential, we find that when we examine the conditional Pearson residuals in each treatment group (Figure 5.5) as well as overall (Figure 5.6), the influence that this patient exerts is relatively minimal. Indeed, the panel of overall residuals, which consists of a plot of the conditional Pearson residuals versus the linear predictor, a histogram with a normal density overlaid, a Q-Q plot, and a box plot of the residuals, suggests the only influential observation (identified as observation 211) is the seizure count corresponding to visit 3 of patient 227. This observation was also identified as being highly influential in analyses conducted by Thall and Vail (1990) and Breslow and Clayton (1993) based on slightly different models.

Figure 5.4 Box plots showing the distribution of seizure counts according to treatment group. Observations 145-148 correspond to patient 207 at visits 1-4 respectively while observation 211 correpsonds to patient 227 at visit 3

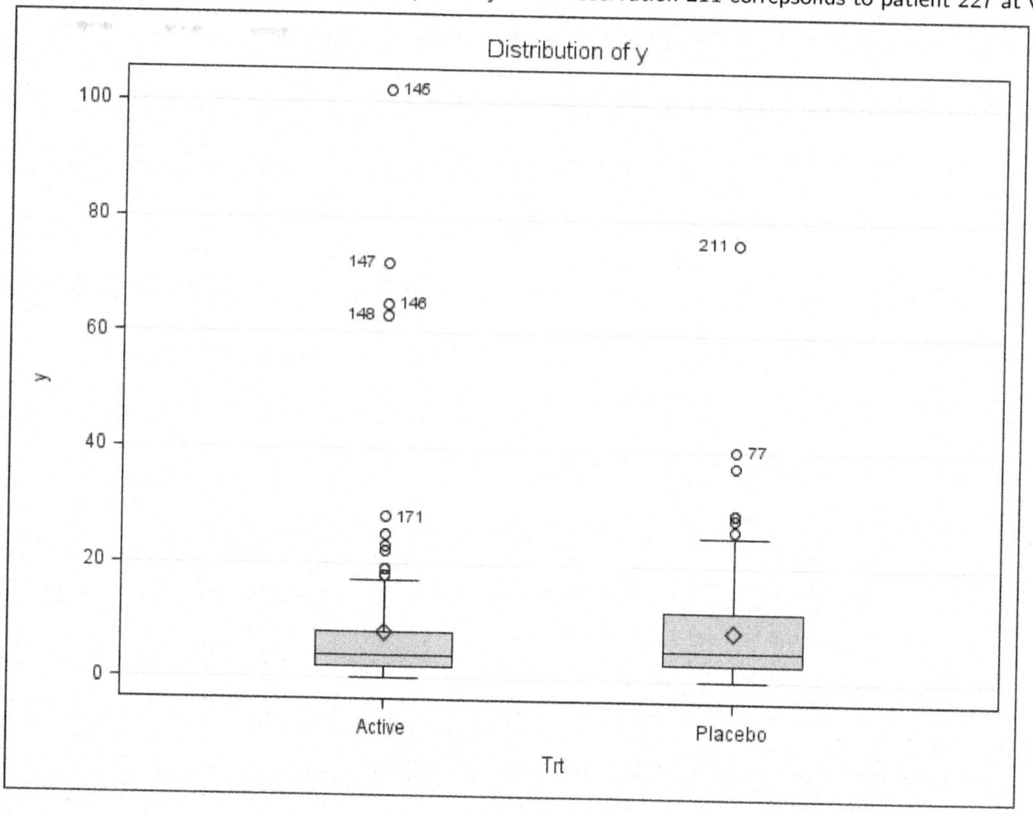

Figure 5.5 Box plots of the conditional Pearson residuals on the mean scale by treatment group showing whether there any highly influential observations

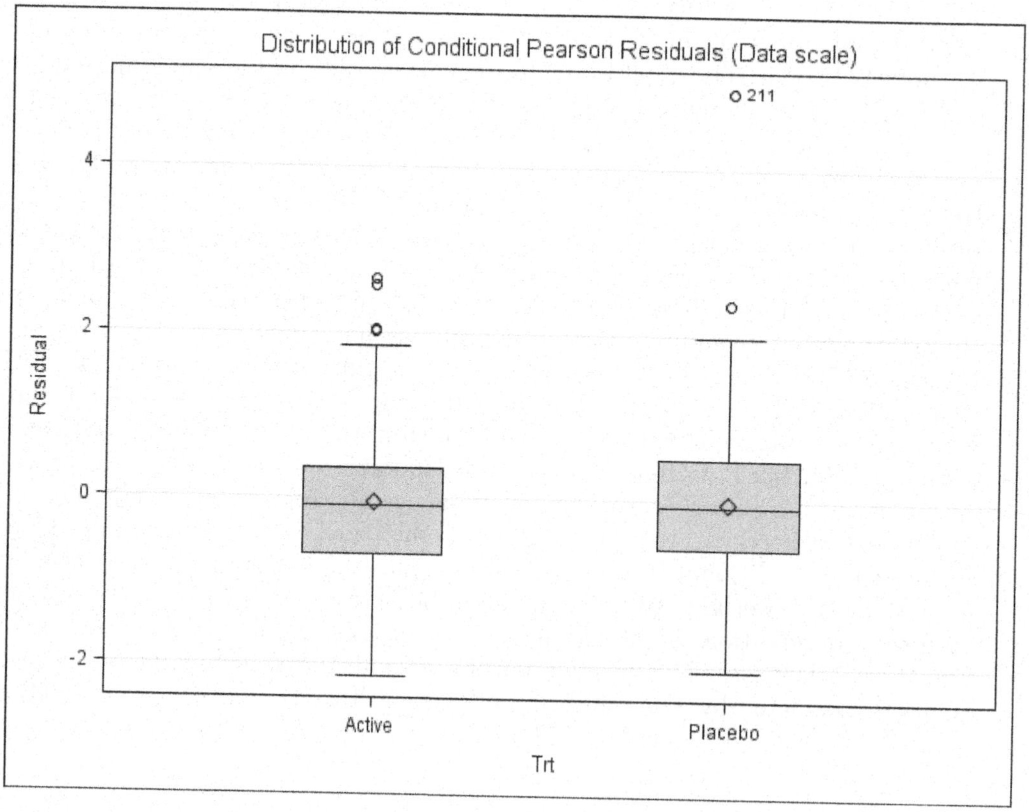

Figure 5.6 Overall panel of conditional Pearson residuals on the mean scale showing whether there any highly influential observations

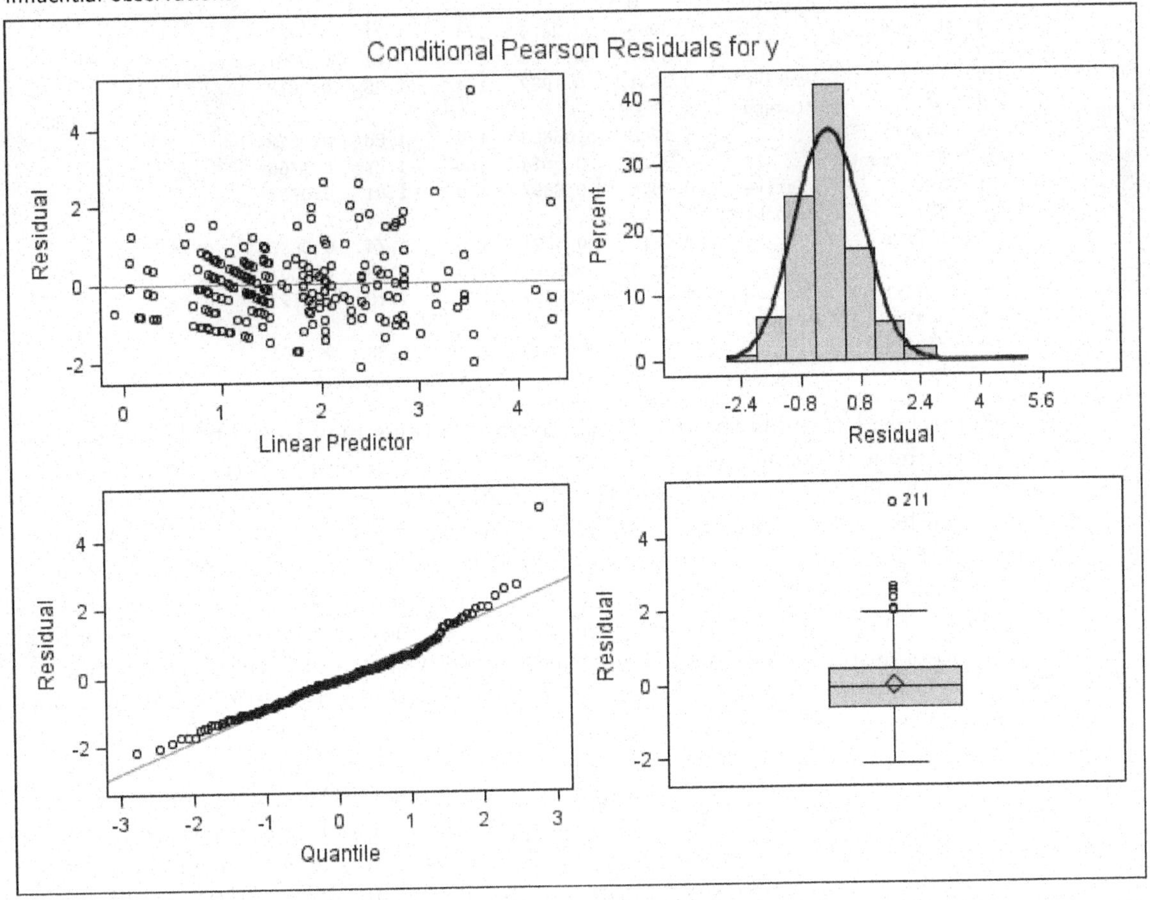

Output 5.8: Results from the GLME model (5.35) when select observations are excluded. Case 1 includes all the data. Case 2 excludes all four observations from patient 207. Case 3 excludes the observation from visit 3 of patient 227. Case 4 excludes all five of these observations.

Case	Effect	Trt	Estimate	StdErr	DF	tValue	Probt
1	Intercept	_	−1.4437	0.9610	55	−1.50	0.1388
1	Trt	Active	−0.9116	0.3978	175	−2.29	0.0231
1	Trt	Placebo	0	.	.	.	.
1	LogAge	_	0.5285	0.2844	175	1.86	0.0648
1	y0	_	0.8810	0.1096	175	8.04	<.0001
1	y0*Trt	Active	0.3389	0.1985	175	1.71	0.0896
1	y0*Trt	Placebo	0	.	.	.	.
1	Visit4	_	−0.1611	0.06558	175	−2.46	0.0150
2	Intercept	_	−1.3828	0.9425	55	−1.47	0.1480
2	Trt	Active	−0.6348	0.3741	171	−1.70	0.0916
2	Trt	Placebo	0	.	.	.	.
2	LogAge	_	0.5092	0.2787	171	1.83	0.0694
2	y0	_	0.8819	0.1096	171	8.04	<.0001
2	y0*Trt	Active	0.1737	0.1883	171	0.92	0.3577
2	y0*Trt	Placebo	0	.	.	.	.
2	Visit4	_	−0.1479	0.07634	171	−1.94	0.0543

3	Intercept	_	−1.3593	0.9690	55	−1.40	0.1663
3	Trt	Active	−0.9367	0.3925	174	−2.39	0.0181
3	Trt	Placebo	0	.	.	.	.
3	LogAge	_	0.5095	0.2876	174	1.77	0.0783
3	y0	_	0.8517	0.1000	174	8.51	<.0001
3	y0*Trt	Active	0.3631	0.1932	174	1.88	0.0618
3	y0*Trt	Placebo	0	.	.	.	.
3	Visit4	_	−0.1249	0.06653	174	−1.88	0.0622
4	Intercept	_	−1.3025	0.9514	55	−1.37	0.1765
4	Trt	Active	−0.6676	0.3697	170	−1.81	0.0727
4	Trt	Placebo	0	.	.	.	.
4	LogAge	_	0.4910	0.2821	170	1.74	0.0836
4	y0	_	0.8523	0.09997	170	8.53	<.0001
4	y0*Trt	Active	0.2017	0.1831	170	1.10	0.2722
4	y0*Trt	Placebo	0	.	.	.	.
4	Visit4	_	−0.1046	0.07595	170	−1.38	0.1701

To further determine the impact these observations may have on the results, we repeated the analysis for four different cases. Case 1 includes all the data while case 2 excludes the four observations from patient 207 and case 3 excludes the observation corresponding to visit 3 for patient 227. In case 4 we exclude all five of these potentially influential observations. The results obtained by excluding patient 207 are similar to those in Example 4.3.3 in that we no longer have an overall significant treatment effect nor a significant treatment ×log baseline count effect as shown in Output 5.8. Excluding the single observation for patient 227 corresponding to visit 3 had virtually no effect on the results when compared to case 1 while excluding all five observations reveals a borderline treatment effect (p=0.0727) but no treatment × log baseline count effect.

Finally, to evaluate how excluding these observations might impact results from the GLME model (5.35) versus a corresponding marginal GLIM and to determine how a potential treatment by log baseline count interaction might alter treatment comparisons, we computed least squares means of the treatment effects at log baseline counts of 0.56, 1.18, 1.79, 2.25 and 3.63 corresponding to quantile values for y_0 of 0% (minimum value), 25% (lower quartile), 50% (median), 75% (upper quartile) and 100% (maximum value). This was done using the PA or marginal GLIM of Example 4.3.3 with a "working" compound symmetric covariance structure and the SS GLME model (5.35). The GLIMMIX syntax is essentially the same for the two models except that we replace the two RANDOM statements from the GLME model with the single RANDOM statement

```
random _residual_ / subject=ID type=cs;
```

when running the PA GLIM. In addition, we include the following LSMEANS statements for both models,

```
lsmeans Trt / ilink diff cl at y0=0.56; ** 0% quantile;
lsmeans Trt / ilink diff cl at y0=1.18; ** 25% quantile;
lsmeans Trt / ilink diff cl at y0=1.79; ** 50% quantile;
lsmeans Trt / ilink diff cl at y0=2.25; ** 75% quantile;
lsmeans Trt / ilink diff cl at y0=3.63; ** 100% quantile;
```

Using ODS OUTPUT statements, results from the least squares mean statements were consolidated and summarized using PROC REPORT. The results are displayed in Output 5.9 for the PA GLIM with "working" compound symmetry and the SS GLME model (5.35).

Output 5.9: Least squares treatment means by select log baseline counts according to type of model (PA versus SS) and data exclusion cases (Cases 1-4). Included are the mean counts and corresponding rate ratios for the two types of models. As noted previously, the rate ratio for the PA Poisson model is directly comparable to the rate ratio for the SS Poisson model (5.35).

Marginal Model
PA

Case	y0	Trt	LS Mean	Log Link Scale Diff.	Std Err	p-value	Mean Count	Rate Ratio
1	0.56	Active	−0.316	.	.	.	0.73	.
		Placebo	0.707	−1.023	0.341	0.0040	2.03	0.36
	1.18	Active	0.623	.	.	.	1.86	.
		Placebo	1.297	−0.673	0.249	0.0091	3.66	0.51
	1.79	Active	1.547	.	.	.	4.70	.
		Placebo	1.876	−0.330	0.179	0.0704	6.53	0.72
	2.25	Active	2.243	.	.	.	9.42	.
		Placebo	2.314	−0.071	0.156	0.6523	10.11	0.93
	3.63	Active	4.332	.	.	.	76.11	.
		Placebo	3.625	0.707	0.281	0.0150	37.54	2.03
2	0.56	Active	0.278	.	.	.	1.32	.
		Placebo	0.719	−0.441	0.321	0.1745	2.05	0.64
	1.18	Active	0.953	.	.	.	2.59	.
		Placebo	1.309	−0.355	0.224	0.1188	3.70	0.70
	1.79	Active	1.618	.	.	.	5.04	.
		Placebo	1.889	−0.271	0.164	0.1047	6.61	0.76
	2.25	Active	2.120	.	.	.	8.33	.
		Placebo	2.326	−0.207	0.167	0.2220	10.24	0.81
	3.63	Active	3.624	.	.	.	37.48	.
		Placebo	3.639	−0.015	0.357	0.9670	38.04	0.99
3	0.56	Active	−0.312	.	.	.	0.73	.
		Placebo	0.710	−1.021	0.341	0.0041	2.03	0.36
	1.18	Active	0.626	.	.	.	1.87	.
		Placebo	1.277	−0.651	0.247	0.0110	3.59	0.52
	1.79	Active	1.548	.	.	.	4.70	.
		Placebo	1.835	−0.287	0.170	0.0965	6.27	0.75
	2.25	Active	2.244	.	.	.	9.43	.
		Placebo	2.256	−0.013	0.136	0.9269	9.55	0.99
	3.63	Active	4.331	.	.	.	75.99	.
		Placebo	3.519	0.811	0.246	0.0018	33.76	2.25
4	0.56	Active	0.283	.	.	.	1.33	.
		Placebo	0.722	−0.439	0.320	0.1764	2.06	0.64
	1.18	Active	0.957	.	.	.	2.60	.
		Placebo	1.290	−0.333	0.222	0.1394	3.63	0.72
	1.79	Active	1.620	.	.	.	5.05	.
		Placebo	1.848	−0.228	0.154	0.1441	6.35	0.80
	2.25	Active	2.120	.	.	.	8.33	.
		Placebo	2.270	−0.150	0.149	0.3204	9.68	0.86
	3.63	Active	3.619	.	.	.	37.32	.
		Placebo	3.534	0.086	0.331	0.7963	34.25	1.09

Mixed-Effects Model
SS

Case	y0	Trt	LS Mean	Log Link Scale Diff.	Std Err	p-value	Mean Count	Rate Ratio
1	0.56	Active	0.042	.	.	.	1.04	.
		Placebo	0.764	−0.722	0.297	0.0160	2.15	0.49
	1.18	Active	0.798	.	.	.	2.22	.
		Placebo	1.310	−0.512	0.198	0.0106	3.71	0.60
	1.79	Active	1.543	.	.	.	4.68	.
		Placebo	1.848	−0.305	0.144	0.0358	6.34	0.74
	2.25	Active	2.104	.	.	.	8.20	.
		Placebo	2.253	−0.149	0.162	0.3587	9.51	0.86
	3.63	Active	3.787	.	.	.	44.13	.
		Placebo	3.469	0.319	0.378	0.4003	32.09	1.38

2	0.56	Active	0.229	.	.	.	1.26	.
		Placebo	0.767	−0.538	0.280	0.0562	2.15	0.58
	1.18	Active	0.884	.	.	.	2.42	.
		Placebo	1.313	−0.430	0.189	0.0242	3.72	0.65
	1.79	Active	1.528	.	.	.	4.61	.
		Placebo	1.851	−0.324	0.143	0.0251	6.37	0.72
	2.25	Active	2.013	.	.	.	7.49	.
		Placebo	2.257	−0.244	0.163	0.1363	9.55	0.78
	3.63	Active	3.470	.	.	.	32.13	.
		Placebo	3.474	−0.004	0.367	0.9907	32.27	1.00
3	0.56	Active	0.044	.	.	.	1.05	.
		Placebo	0.778	−0.733	0.294	0.0134	2.18	0.48
	1.18	Active	0.797	.	.	.	2.22	.
		Placebo	1.306	−0.508	0.196	0.0104	3.69	0.60
	1.79	Active	1.538	.	.	.	4.66	.
		Placebo	1.825	−0.287	0.140	0.0414	6.20	0.75
	2.25	Active	2.097	.	.	.	8.14	.
		Placebo	2.217	−0.120	0.153	0.4363	9.18	0.89
	3.63	Active	3.774	.	.	.	43.54	.
		Placebo	3.392	0.381	0.361	0.2923	29.73	1.46
4	0.56	Active	0.226	.	.	.	1.25	.
		Placebo	0.780	−0.555	0.277	0.0470	2.18	0.57
	1.18	Active	0.879	.	.	.	2.41	.
		Placebo	1.309	−0.430	0.187	0.0231	3.70	0.65
	1.79	Active	1.522	.	.	.	4.58	.
		Placebo	1.829	−0.307	0.139	0.0284	6.23	0.74
	2.25	Active	2.007	.	.	.	7.44	.
		Placebo	2.221	−0.214	0.154	0.1671	9.21	0.81
	3.63	Active	3.461	.	.	.	31.86	.
		Placebo	3.397	0.065	0.350	0.8539	29.87	1.07

With regard to model sensitivity to outlying observations, we find that the marginal GLIM appears to be more sensitive to the inclusion or exclusion of patient 207 when compared with the GLME model. When either all the data are included (Case 1) or we simply exclude the visit 3 observation from patient 227 (Case 3), the log mean counts and mean counts are reasonably similar for the two models. The major exception to this involves the predicted mean count and rate ratio at the maximum baseline count of 3.63 which corresponds to patient 207. Under the marginal GLIM, the predicted mean count in the active drug treatment group is substantially higher when compared to the GLME model and this is reflected in the rate ratios of the two types of models (Cases 1 and 3). This difference between models may well reflect the fact that the least squares means under the GLME model are evaluated by setting the random effects to 0. However, the rate ratio under the GLME model has both a PA and SS interpretation and so it would appear that by including a SS random intercept effect, the GLME model yields parameter estimates that are less sensitive to the influence of patient 207. Further evidence illustrating the sensitivity of the marginal GLIM to patient 207 may be found when we compare results between the two models when patient 207 is excluded (Cases 2 and 4). Here we find that under the marginal GLIM, there is no significant treatment effect due to the active compound progabide at any of the five selected log baseline counts for which least squares means are computed. In contrast, under the GLME model, we find that among patients with log baseline counts at or below the median of 1.79, those who received progabide showed a significant reduction in their two-week seizure rates compared with those who received the placebo control.

In summary, the results from fitting the GLME model (5.35) to the data across all four data exclusion cases are fairly consistent with one another while results from fitting the marginal GLIM to the data change depending on whether patient 207 is included or excluded from the analysis. Since, as pointed out by Thall and Vail (1990), there is no clinical basis for dropping patient 207 from the analysis and since the results from the fitting the GLME model to the data is reasonably robust to the inclusion or exclusion of this patient, it seems reasonable to conclude that progabide is effective in reducing seizure rates among individuals with a mean baseline count at or below 6 seizures per two-week period.

5.2.3 Schizophrenia data—continued

Following Hedeker and Gibbons (1994), we fit a mixed-effects ordinal regression model to the four-point ordinal response data on overall severity of illness from the NIMH Schizophrenia collaborative study (see §4.3.4). The objective here is to determine if the average patient's response to one of three active medications for the treatment of schizophrenia: chlorpromazine, fluphenazine, or thioridazine, differs from that of the average patient receiving the placebo control drug. As in Example 4.3.4, we grouped the three active antipsychotic drugs into one active treatment arm. Using the same general notation as before, we can extend the class of marginal GLIM's for ordinal data described in §4.3.4 to a class of GLME models by simply including a set of random effects into the linear predictor of an underlying latent response variable, y^*_{ij}, corresponding to the ordinal response variable y_{ij} of interest. In particular, a set of fixed-effects predictors, $\boldsymbol{x}_{ij}$, and random-effects predictors, $\boldsymbol{z}_{ij}$, are assumed to be related to a latent response variable, y^*_{ij}, via the linear mixed-effects regression model

$$y^*_{ij} = \boldsymbol{x}'_{ij}\boldsymbol{\beta} + \boldsymbol{z}'_{ij}\boldsymbol{b}_i + \epsilon_{ij} \tag{5.38}$$

where ϵ_{ij} is distributed according to some standardized cdf $G(\cdot)$ such that

$$\begin{aligned}\gamma_k(\boldsymbol{x}_{ij}, \boldsymbol{z}_{ij}, \boldsymbol{b}_i) = \Pr(y_{ij} \le k|\boldsymbol{x}_{ij}, \boldsymbol{z}_{ij}, \boldsymbol{b}_i) &= \Pr(y^*_{ij} \le \alpha_k) \\ &= \Pr(\epsilon_{ij} \le \alpha_k - \boldsymbol{x}'_{ij}\boldsymbol{\beta} - \boldsymbol{z}'_{ij}\boldsymbol{b}_i) \\ &= G(\alpha_k - \boldsymbol{x}'_{ij}\boldsymbol{\beta} - \boldsymbol{z}'_{ij}\boldsymbol{b}_i)\end{aligned}$$

and α_k are threshold cutpoints, $\alpha_1, \alpha_2, \ldots, \alpha_{K-1}$, corresponding to K ordered categories of the ordinal response variable y_{ij}.

Given this representation, one can form a GLME model for the ordinal response y_{ij} by defining the cumulative link function

$$\begin{aligned}G^{-1}(\Pr(y_{ij} \le k|\boldsymbol{x}_{ij}, \boldsymbol{z}_{ij}, \boldsymbol{b}_i)) &= G^{-1}(\gamma_k(\boldsymbol{x}_{ij}, \boldsymbol{z}_{ij}, \boldsymbol{b}_i)) \\ &= \alpha_k - \boldsymbol{x}'_{ij}\boldsymbol{\beta} - \boldsymbol{z}'_{ij}\boldsymbol{b}_i\end{aligned} \tag{5.39}$$

which maps the cumulative probabilities, $\{\gamma_k(\boldsymbol{x}_{ij}, \boldsymbol{z}_{ij}, \boldsymbol{b}_i),\ k = 1, \ldots, K-1\}$, onto the real line (note $\alpha_K = \infty \Rightarrow \gamma_K(\boldsymbol{x}_{ij}, \boldsymbol{z}_{ij}, \boldsymbol{b}_i) = 1$). Choices for the cumulative link function include the cumulative probit link, $\Phi^{-1}(\Pr(y_{ij} \le k|\boldsymbol{x}_{ij}, \boldsymbol{z}_{ij}, \boldsymbol{b}_i))$, the cumulative logit link (corresponding to a proportional odds model), $\log\left(\frac{\Pr(y_{ij} \le k|\boldsymbol{x}_{ij}, \boldsymbol{z}_{ij}, \boldsymbol{b}_i)}{1-\Pr(y_{ij} \le k|\boldsymbol{x}_{ij}, \boldsymbol{z}_{ij}, \boldsymbol{b}_i)}\right)$, and the discrete proportional hazards link (i.e., cumulative complementary log-log link), $\log[-\log(1 - \Pr(y_{ij} \le k|\boldsymbol{x}_{ij}, \boldsymbol{z}_{ij}, \boldsymbol{b}_i))]$, all of which are described at length in §4.3.4.

In this example, we fit the ordinal response variable, `IMPS79o`, from the schizophrenia data to two mixed-effects ordinal regression models: one with a random intercept and one with a random intercept and random slope. Fitting the data using the same fixed effects as in Example 4.3.4, the conditional linear predictor for the two models would therefore be

$$\text{Model 1: } \boldsymbol{x}'_{ij}\boldsymbol{\beta} + \boldsymbol{z}'_{ij}\boldsymbol{b}_i = \beta_0 + \beta_1 a_{1i} + \beta_2\sqrt{t_{ij}} + \beta_3 a_{1i}\sqrt{t_{ij}} + b_{i0}$$

$$\text{Model 2: } \boldsymbol{x}'_{ij}\boldsymbol{\beta} + \boldsymbol{z}'_{ij}\boldsymbol{b}_i = \beta_0 + \beta_1 a_{1i} + \beta_2\sqrt{t_{ij}} + \beta_3 a_{1i}\sqrt{t_{ij}} + b_{i0} + b_{i1}\sqrt{t_{ij}}.$$

where a_{1i} is the treatment drug indicator (SAS variable `Trt`) and $\sqrt{t_{ij}}$ is the square root of time in weeks (SAS variable `SWeek`) as described in §4.3.4. Unlike Hedeker and Gibbons

(1994) who fit these two GLME models assuming a cumulative probit link function, we fit both models assuming a cumulative logit link function. We also did not exclude intermediate observations obtained at weeks 2, 4 and 5 nor did we restrict the analysis to those who completed the study as did Hedeker and Gibbons (1994).

Shown below is the GLIMMIX syntax required to fit these two models including ESTIMATE statements for estimating the cumulative log odds ratios and cumulative odds ratios across time for the average patient in the active drug group versus the average patient in the placebo control group. The ESTIMATE statements are identical to those used in Example 4.3.4 when we fit a corresponding marginal GLIM with "working" independence in GENMOD. We base inference on robust standard errors (the `empirical` option) but we also use the COVTEST statement to perform likelihood ratio tests about the covariance structure since the assumed random-effects covariance structure can affect the fixed-effects estimates of a GLME model.

Program 5.5

```
ods select CovTests FitStatistics CondFitStatistics
           Covparms ParameterEstimates Estimates;
proc glimmix data=example5_2_3 method=Quad maxopt=50 empirical;
 class ID;
 model IMPS79o = Trt SWeek Trt*SWeek
              / dist=multinomial link=clogit s;
 random intercept / subject=ID;
 estimate 'CumLogOR(Week=0)' Trt 1 Trt*SWeek 0 /exp;
 estimate 'CumLogOR(Week=1)' Trt 1 Trt*SWeek 1 /exp;
 estimate 'CumLogOR(Week=3)' Trt 1 Trt*SWeek 1.7321 /exp;
 estimate 'CumLogOR(Week=6)' Trt 1 Trt*SWeek 2.4495 /exp;
 covtest 'H0: No random intercept variance' 0;
run;

ods select CovTests FitStatistics CondFitStatistics
           Covparms ParameterEstimates Estimates;
proc glimmix data=example5_2_3 method=Quad maxopt=50 empirical;
 class ID ;
 model IMPS79o = Trt SWeek Trt*SWeek
              / dist=multinomial link=clogit s;
 random intercept Sweek / subject=ID type=un;
 estimate 'CumLogOR(Week=0)' Trt 1 Trt*SWeek 0 /exp;
 estimate 'CumLogOR(Week=1)' Trt 1 Trt*SWeek 1 /exp;
 estimate 'CumLogOR(Week=3)' Trt 1 Trt*SWeek 1.7321 /exp;
 estimate 'CumLogOR(Week=6)' Trt 1 Trt*SWeek 2.4495 /exp e;
 covtest 'H0: No covariance component' . 0 .;
 covtest 'H0: No random intercept variance' 0 0 .;
 covtest 'H0: No random slope variance' . 0 0;
 covtest 'H0: No random-effects' 0 0 0;
 covtest DiagG;
 covtest ZeroG;
 output out=pred pred(ilink blup)=pred
                 pred(ilink noblup)=predmean;
run;
```

For the GLME model with random intercepts only, the COVTEST statement

```
covtest 'H0: No random intercept variance component' 0;
```

is equivalent to specifying the statement '`Covtest ZeroG;`' which tests whether the $\boldsymbol{\Psi}$ matrix (or G matrix in SAS) can be reduced to a zero matrix. This eliminates all G-side random effects from the model. In specifying the second GLME model with random intercepts and random slopes, the TYPE=UN option of the RANDOM statement instructs GLIMMIX to fit the data assuming an unstructured variance-covariance matrix for the random effects $\boldsymbol{b}_i' = (b_{i0}, b_{i1})$. In this case, the COVTEST statements are

```
covtest 'H0: No covariance component' . 0 .;
covtest 'H0: No random intercept variance' 0 0 .;
covtest 'H0: No random slope variance' . 0 0;
covtest 'H0: No random-effects' 0 0 0;
```

The first COVTEST statement tests whether the second component of $\boldsymbol{\Psi}$ which is the covariance between b_{i0} and b_{i1} is zero. The second and third COVTEST statements test whether the random intercept variance component is zero and whether the random slope variance component is zero, respectively (in both cases the covariance between b_{i0} and b_{i1} must also be zero). The syntax in all three cases places a period for those variance component(s) that are to be re-estimated in accordance with fitting a reduced log-likelihood. The fourth COVTEST statement simply tests whether the entire $\boldsymbol{\Psi}$ matrix is zero. The additional COVTEST statements

```
covtest DiagG;
covtest ZeroG;
```

are alternative but equivalent ways of testing whether the covariance between b_{i0} and b_{i1} is zero and whether the entire $\boldsymbol{\Psi}$ matrix is zero.

Results of fitting the random intercept model are shown in Output 5.10. The fixed effects regression coefficients from this model are considerably different from that of the marginal cumulative logit model shown in Output 4.16. For example, the treatment group by time interaction effect (`Trt*SWeek`) is considerably greater than with the marginal model. This helps to partly explain why the estimated SS cumulative odds ratio at week 6 for the average treated patient versus average placebo control patient is over threefold higher than the PA cumulative odds ratio obtained under the marginal GLIM of Example 4.3.4 (20.2898 from Output 5.10 versus 6.297 from Output 4.16).

Output 5.10: Parameter estimates and other select results (obtained using the ODS SELECT statement) from fitting the schizophrenia data to a mixed-effects cumulative logit model with random intercepts. Included are the SS cumulative log odds ratios and SS cumulative odds ratios (Exponentiated Estimates) at weeks 0, 1, 3 and 6. They represent the cumulative log odds ratios and cumulative odds ratios of the average treated patient versus the average placebo control patient over time.

```
                 Fit Statistics
-2 Log Likelihood              3402.92
AIC (smaller is better)        3416.92
AICC (smaller is better)       3416.99
BIC (smaller is better)        3445.48
CAIC (smaller is better)       3452.48
HQIC (smaller is better)       3428.19
```

Fit Statistics for Conditional Distribution

-2 log L(IMPS79o | r. effects)	2476.35

Covariance Parameter Estimates

Cov Parm	Subject	Estimate	Standard Error
Intercept	ID	3.7641	0.4486

Solutions for Fixed Effects

Effect	IMPS79o	Estimate	Standard Error	DF	t Value	Pr > |t|
Intercept	1	−5.8567	0.3247	435	−18.04	<.0001
Intercept	2	−2.8249	0.2882	435	−9.80	<.0001
Intercept	3	−0.7077	0.2846	435	−2.49	0.0133
Trt		0.05812	0.3201	1162	0.18	0.8560
SWeek		0.7660	0.1459	1162	5.25	<.0001
Trt*SWeek		1.2051	0.1764	1162	6.83	<.0001

Estimates

Label	Estimate	Standard Error	DF	t Value	Pr > |t|	Exponentiated Estimate
CumLogOR(Week=0)	0.05812	0.3201	1162	0.18	0.8560	1.0598
CumLogOR(Week=1)	1.2633	0.2635	1162	4.79	<.0001	3.5369
CumLogOR(Week=3)	2.1455	0.2911	1162	7.37	<.0001	8.5467
CumLogOR(Week=6)	3.0101	0.3634	1162	8.28	<.0001	20.2898

Tests of Covariance Parameters Based on the Likelihood

Label	DF	-2 Log Like	ChiSq	Pr > ChiSq	Note
H0: No random intercept variance	1	3756.20	353.27	<.0001	MI

The magnitude and significance of the random intercept variance component also help account for this discrepancy. In general, the degree of disparity in parameter estimates and hence in the interpretation of results when fitting data to a marginal GLIM versus a corresponding GLME model will increase as the random effect variance components increase in magnitude.

When we fit a mixed-effects cumulative logit model assuming both random intercepts and random slopes (Output 5.11 below), we find an even greater effect associated with the treatment by time interaction (an effect of

Output 5.11: Parameter estimates and other select results when we fit the schizophrenia data to a mixed-effects cumulative logit model with random intercepts and random slopes.

Fit Statistics

-2 Log Likelihood	3326.47
AIC (smaller is better)	3344.47
AICC (smaller is better)	3344.58
BIC (smaller is better)	3381.19
CAIC (smaller is better)	3390.19
HQIC (smaller is better)	3358.96

Fit Statistics for Conditional Distribution

-2 log L(IMPS79o | r. effects)	1808.23

Covariance Parameter Estimates

Cov Parm	Subject	Estimate	Standard Error
UN(1,1)	ID	6.8471	1.2527
UN(2,1)	ID	−1.4470	0.5173
UN(2,2)	ID	1.9488	0.3732

Solutions for Fixed Effects

Effect	IMPS79o	Estimate	Standard Error	DF	t Value	Pr > |t|
Intercept	1	−7.2832	0.4655	435	−15.65	<.0001
Intercept	2	−3.3996	0.3867	435	−8.79	<.0001
Intercept	3	−0.8057	0.3601	435	−2.24	0.0258
Trt		−0.05630	0.3874	726	−0.15	0.8845
SWeek		0.8785	0.2050	436	4.28	<.0001
Trt*SWeek		1.6840	0.2382	726	7.07	<.0001

Estimates

Label	Estimate	Standard Error	DF	t Value	Pr > |t|	Exponentiated Estimate
CumLogOR(Week=0)	−0.05630	0.3874	726	−0.15	0.8845	0.9453
CumLogOR(Week=1)	1.6277	0.3303	726	4.93	<.0001	5.0921
CumLogOR(Week=3)	2.8605	0.3886	726	7.36	<.0001	17.4710
CumLogOR(Week=6)	4.0686	0.5010	726	8.12	<.0001	58.4772

Tests of Covariance Parameters Based on the Likelihood

Label	DF	-2 Log Like	ChiSq	Pr > ChiSq	Note
H0: No covariance component	1	3339.10	12.63	0.0004	DF
H0: No random intercept variance	2	3505.75	179.28	<.0001	MI
H0: No random slope variance	2	3402.92	76.45	<.0001	MI
H0: No random−effects	3	3756.20	429.73	<.0001	−−
Diagonal G	1	3339.10	12.63	0.0004	DF
No G−side effects	3	3756.20	429.73	<.0001	−−

1.2051±0.1764 versus 1.6840±0.2382). This further inflates the SS cumulative odds ratio at week 6 (58.4772) by nine fold compared with the marginal cumulative odds ratio of 6.297. Indeed, even at week 1, we find the SS cumulative odds ratio of 5.0921 is over twofold greater than the marginal estimate of 2.120 (Output 4.16). The increased effect associated with the treatment by time interaction mirrors the significant patient-to-patient variation in slopes (random slope variance of 1.9488±0.3732 - Output 5.11).

Given the disparities in parameter estimates and resulting interpretation of results between the marginal cumulative logit model of §4.3.4 and the two mixed-effects cumulative logit models presented here, it is natural to ask which model one should choose from. While we cannot formally conduct goodness-of-fit tests comparing the marginal GLIM of Example 4.3.4 with the GLME models presented here, we can compare the two GLME models. From the 'Fit Statistics' shown in Output 5.10 and 5.11, we see a significant reduction in Akaike's information criteria (AIC) when we include the random slope effect along with an unstructured variance-covariance matrix for the random intercepts and slopes. Indeed, from Output 5.11, the formal likelihood ratio tests carried out using the COVTEST statements described above indicate that an unstructured covariance matrix, $\boldsymbol{\Psi}$, between b_{i0} and b_{i1} is preferred over any of the possibly reduced covariance structures considered.

Having determined that a SS cumulative logit model with random intercepts and random slopes provides a better fit than the random-intercepts-only model, one may then wonder whether this GLME model also offers a reasonable fit to the observed percentage of severity of illness over time as was illustrated in Figure 4.5 for the PA cumulative logit model. To that end, Figure 5.7 compares the observed versus empirically estimated marginal percentages of illness severity over time. The marginal percentages predicted from the GLME model were computed as simple arithmetic averages of the predicted patient-specific probabilities of illness severity per treatment group and time. That is, for fixed vectors $\boldsymbol{x}$ and $\boldsymbol{z}$ corresponding to a specific treatment group and time, the marginal probabilities were estimated as

$$\Pr(y=k|\boldsymbol{x},\boldsymbol{z}) = E_{\boldsymbol{b}}\{\Pr(y=k|\boldsymbol{x},\boldsymbol{z},\boldsymbol{b})\} \approx \frac{\sum_{i=1}^{n}\sum_{j=1}^{p_i}\delta_{ij}\Pr(y_{ij}=k|\boldsymbol{x}_{ij}'\widehat{\boldsymbol{\beta}}+\boldsymbol{z}_{ij}'\widehat{\boldsymbol{b}}_i)}{\sum_{i=1}^{n}\sum_{j=1}^{p_i}\delta_{ij}} \tag{5.40}$$

where δ_{ij} are 0-1 indicator variables such that

$$\delta_{ij} = \begin{cases} 1, & \text{if } \boldsymbol{x}_{ij}=\boldsymbol{x} \text{ and } \boldsymbol{z}_{ij}=\boldsymbol{z} \\ 0, & \text{otherwise} \end{cases}.$$

For example, if we are interested in estimating the marginal probability of response for patients in the active drug group at week 6, then

Figure 5.7 A GLME cumulative logit (proportional odds) model of IMPS 79 severity of illness scores. Observed versus PA predicted proportion of patients over time by treatment group. The PA predicted proportions are the arithmetic averages of the SS predicted probabilities over time per treatment group. Given the sparse data at weeks 2,4 and 5, the observed values are plotted for weeks 0,1,3 and 6.

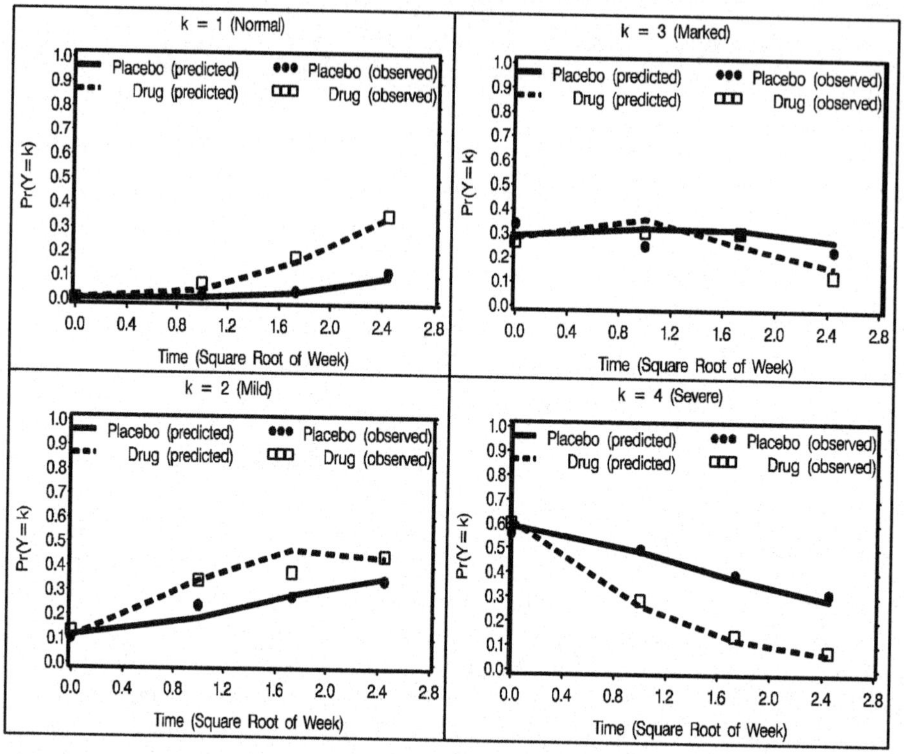

$$x' = (1 \quad a_{1i} \quad \sqrt{t_{ij}} \quad a_{1i}\sqrt{t_{ij}})\Big|_{a_{1i}=1, t_{ij}=6} = (1 \quad 1 \quad \sqrt{6} \quad \sqrt{6}),\ z' = (1 \quad \sqrt{t_{ij}})\Big|_{t_{ij}=6} = (1 \quad \sqrt{6})$$

and $\delta_{ij} = 1$ for those patients in the active drug group who had their severity of illness measured at week 6. The predicted proportion of patients in the active drug group having illness severity of k at week 6 can then be estimated by simply computing the mean of the individual (SS) predicted probabilities of illness severity for that treatment group and time. The SS probabilities are computed from the SS predicted cumulative probabilities as

$$\Pr(y_{ij} = k|x'_{ij}\widehat{\beta} + z'_{ij}\widehat{b}_i) = \Pr(y_{ij} \le k|x'_{ij}\widehat{\beta} + z'_{ij}\widehat{b}_i) - \Pr(y_{ij} \le k-1|x'_{ij}\widehat{\beta} + z'_{ij}\widehat{b}_i)$$

where $\Pr(y_{ij} \le k|x'_{ij}\widehat{\beta} + z'_{ij}\widehat{b}_i)$ is generated with the PRED(ILINK BLUP)= option of the OUTPUT OUT statement (see SAS code above). A complete listing of the SAS code used to generate Figure 5.7 is available at the author's Web page.

In comparing Figure 5.7 with Figure 4.5, it appears as though both the mixed-effects and marginal cumulative logit models provide a reasonable fit to the observed proportions. Selecting which model to use on the basis of goodness-of-fit is therefore problematic. It is more important that the criteria for selecting a model be guided by the primary goal of the study. Is the focus of the study to draw inference at the population level (PA) or the patient level (SS)? In this example, we see just how different the interpretation of a drug-related treatment effect can be when analyzed with a marginal versus mixed-effects model. Under the marginal cumulative logit model of Example 4.3.4, for example, it is estimated that at week 6, patients receiving the active drug will be 6.3 times more likely to be at or below a given illness severity threshold (as measured by the IMPS 79 ordinal score) than patients receiving placebo control. In contrast, under the cumulative logit model with random intercepts and slopes, it is estimated that at week 6, the average treated patient will be 58.5 times more likely to be at or below a given illness severity

Figure 5.8 A profile plot of individual (SS) cumulative probability estimates at week 6. The cumulative probability profile for the average patient is shown as a thick solid line.

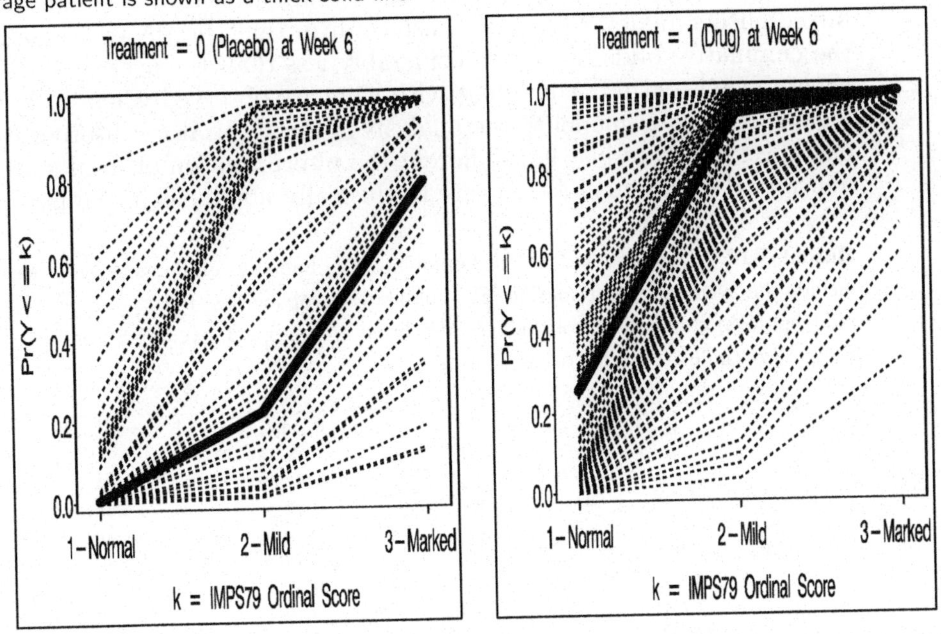

threshold compared with the average placebo control patient (Output 5.11). The treatment effect, as measured by the cumulative odds ratio, is highly significant in both cases but what that effect represents is quite different. The cumulative odds ratio is a measure of relative treatment effect. For the marginal GLIM, it is a measure of the relative treatment effect in a population of individuals while for the GLME model, it is a measure of the relative treatment in the average or "typical" patient.

To better understand the difference between the PA and SS treatment effects, we first plot the subject-specific cumulative probability profiles predicted from the mixed-effects cumulative logit model. As this plot in Figure 5.8 illustrates, there is considerable variation in the individual predicted profiles making a visual comparison of the population treatment effect difficult to ascertain. However, by highlighting what the cumulative probability response is predicted to be at week 6 for the average patient, we see just how effective the active drug can be at the patient level. Indeed, if we were to compute the corresponding odds for the two average patients, we would predict that the average individual's odds of having an ordinal response at or below a given level of illness to be 58.5 higher if that patient received one of the active drugs versus the placebo control.

To gain some insight into what impact the patient-to-patient variation has on the response in the population, we compared the "typical" SS cumulative probabilities of illness at weeks 0, 1, 3 and 6 to the corresponding observed cumulative proportions of patients at those times. We also computed population-averaged (PA) cumulative probability estimates at those times. The SS cumulative probabilities are those predicted from the model for the average or "typical" subject (i.e., $\boldsymbol{b}_i = \mathbf{0}$) while the PA cumulative probabilities are estimated from the model by simply averaging over the SS cumulative probability estimates in Figure 5.8 in a manner analogous to (5.40). Based on these estimates, we then compared the average SS cumulative odds ratios to the PA cumulative odds ratios. The results, shown in Output 5.12, reveal that the PA cumulative probability estimates are similar to the observed cumulative proportions. Moreover, unlike the SS estimates, the PA cumulative probability estimates are not exactly proportional to one another within any given week. This is because the proportional odds assumption under the mixed-effects model does not carry through when we average over individuals. That is, if we were to integrate out the random effects from the mixed-effects model, the ensuing marginal model would not satisfy the proportional odds assumption that is conditionally met in the mixed-effects model. The results also confirm that the PA treatment effect, as estimated by the cumulative odds ratio, is considerably less than the average patient-specific treatment effect. This is entirely consistent with what we observed when we applied a GLME binary logistic regression model to the analysis of the respiratory disorder data (see §5.2.1). Both examples illustrate just how different the interpretation of results are when fitting marginal and mixed-effects models that are intrinsically nonlinear in the parameters of interest.

Output 5.12: Observed and modeled cumulative probabilities of illness severity over time. Modeled probabilities include the average subject-specific (SS) cumulative probabilities, $\Pr(y_{ij} \le k|\boldsymbol{x}_{ij}, \boldsymbol{z}_{ij}, \widehat{\boldsymbol{\beta}}, \boldsymbol{b}_i = \mathbf{0})$ and estimated marginal (PA) cumulative probabilities. Also listed are observed, average SS and PA cumulative odds as well as SS and PA cumulative odds ratios by time.

		Treatment Group					
		Placebo			Active Drug		
Week	IMPS 79 Scores	Observed Cum. Prob.	SS Cum. Prob.	PA Cum. Prob.	Observed Cum. Prob.	SS Cum. Prob.	PA Cum. Prob.
0	Normal	0.000	0.001	0.004	0.003	0.001	0.005
	Mild	0.103	0.032	0.119	0.135	0.031	0.112
	Marked	0.439	0.309	0.409	0.398	0.297	0.391
1	Normal	0.019	0.002	0.010	0.065	0.008	0.040
	Mild	0.257	0.074	0.196	0.408	0.290	0.382
	Marked	0.505	0.518	0.514	0.713	0.846	0.738
3	Normal	0.034	0.003	0.028	0.178	0.052	0.157
	Mild	0.310	0.133	0.310	0.554	0.728	0.626
	Marked	0.609	0.672	0.625	0.857	0.973	0.876
6	Normal	0.114	0.006	0.092	0.351	0.257	0.342
	Mild	0.457	0.223	0.444	0.796	0.944	0.778
	Marked	0.686	0.794	0.712	0.921	0.996	0.936

		Treatment Group							
		Placebo			Active Drug			Odds Ratios	
Week	IMPS 79 Scores	Obs. Cum. Odds	SS Cum. Odds	PA Cum. Odds	Obs. Cum. Odds	SS Cum. Odds	PA Cum. Odds	SS Cum. OR	PA Cum. OR
0	Normal	0.000	0.001	0.004	0.003	0.001	0.005	0.95	1.13
	Mild	0.115	0.033	0.135	0.155	0.032	0.126	0.95	0.94
	Marked	0.783	0.447	0.691	0.660	0.422	0.641	0.95	0.93
1	Normal	0.019	0.002	0.010	0.070	0.008	0.042	5.09	4.11
	Mild	0.346	0.080	0.243	0.689	0.409	0.617	5.09	2.54
	Marked	1.019	1.076	1.056	2.489	5.477	2.823	5.09	2.67
3	Normal	0.036	0.003	0.029	0.216	0.055	0.186	17.47	6.49
	Mild	0.450	0.153	0.449	1.242	2.671	1.673	17.47	3.73
	Marked	1.559	2.046	1.667	6.000	35.749	7.080	17.47	4.25
6	Normal	0.129	0.006	0.101	0.541	0.346	0.519	58.48	5.13
	Mild	0.842	0.287	0.797	3.907	16.793	3.496	58.48	4.39
	Marked	2.182	3.843	2.477	11.619	224.72	14.562	58.48	5.88

In summary, application of a cumulative logit model with random intercepts and slopes revealed that for individual patients with acute schizophrenia, there was a strong and significant benefit of using antipsychotic medication over time on the reduction of the severity of illness. A population-based treatment effect, although lower in magnitude, was also apparent when we averaged across individuals. Finally, Hedeker and Gibbons (1994) obtained results from a mixed-effects ordinal regression analysis that are qualitatively similar to those presented here. Using a cumulative probit link function applied to a balanced subset of these data, they demonstrated a strong treatment by time interaction effect that favored the active drugs. They, too, demonstrated significant patient-to-patient variation in the trends over time (i.e., random slope effect) and found that their model also provides a reasonable fit to the data.

5.2.4 ADEMEX hospitalization data

In §4.3.1, we presented and summarized results from two univariate GLIM's used to compare peritonitis rates among treated and control patients participating in the ADEMEX trial (see §1.4, §2.2.3 and §4.3.1 for details about the ADEMEX trial). In this section, we consider a similar set of analyses based on hospitalization data from the ADEMEX study. Specifically, it was thought that patients randomized to a higher dose of dialysis (treated group) would achieve a lower hospitalization rate compared to those randomized to the standard dose of dialysis (control group). Consequently, a key secondary goal of the ADEMEX study was to compare hospitalization rates among patients randomized to the intervention and control arms of the trial.

As was done with the peritonitis data, we can express the hospitalization data as a single outcome variable by defining $y_i(t_i)$ to be the number of hospital admissions (Hosp) per time at risk t_i (MonthsAtRisk) for the ith patient. The key independent variables are the same as those used in the analysis of peritonitis rates (see §4.3.1 for a description). A partial listing of the hospitalization data is included in Output 4.1 of §4.3.1. In the original analysis published by Paniagua et. al. (2002), this data was fit to an overdispersed Poisson model using GENMOD. The SAS code required to perform this analysis is not shown here because it is identical with that used for the analysis of the peritonitis data in Example 4.3.1 with the exception that the dependent variable EPISODE is replaced by HOSP and the option PSCALE is included in the MODEL statement so as to base comparisons on an overdispersed Poisson model. The results, shown in Output 5.13, indicate that the hospitalization rates are not significantly different between treated and control patients and are consistent with the results reported by Paniagua et. al. (2002). However, based on the generalized Pearson chi-square statistic (4.6), the overdispersion parameter is estimated to be $\widehat{\phi} = 3.0293$ which may artificially

Output 5.13: Summary comparison of ADEMEX hospitalization rates based on a univariate Poisson model with overdispersion.

```
              Model Information
Data Set                  WORK.EXAMPLE4_3_1
Distribution                        Poisson
Link Function                           Log
Dependent Variable                     Hosp
Offset Variable                    log_time
          Criteria For Assessing Goodness Of Fit
Criterion                  DF        Value     Value/DF
Deviance                  870    2180.7596       2.5066
Scaled Deviance           870     719.8795       0.8274
Pearson Chi-Square        870    2635.5258       3.0293
Scaled Pearson X2         870     870.0000       1.0000
Log Likelihood                   -220.4331
Full Log Likelihood              -614.8787
AIC (smaller is better)          1243.7573
AICC (smaller is better)         1243.8862
BIC (smaller is better)          1277.1929
```

Analysis Of Maximum Likelihood Parameter Estimates

Parameter		DF	Estimate	Standard Error	Wald 95% Confidence Limits		Wald ChiSq	Pr > ChiSq
Intercept		1	−1.2107	0.3112	−1.8206	−0.6007	15.13	0.0001
Trt	0	1	−0.1220	0.0886	−0.2956	0.0517	1.89	0.1687
Trt	1	0	0.0000	0.0000	0.0000	0.0000	.	.
Sex		1	0.1369	0.0903	−0.0400	0.3139	2.30	0.1293
Age		1	−0.0082	0.0040	−0.0159	−0.0004	4.30	0.0382
Diabetic		1	0.4893	0.1089	0.2758	0.7028	20.18	<.0001
PriorMonths		1	0.0022	0.0017	−0.0012	0.0055	1.60	0.2055
Albumin		1	−0.3553	0.0747	−0.5018	−0.2089	22.61	<.0001
Scale		0	1.7405	0.0000	1.7405	1.7405		

LR Statistics For Type 3 Analysis

Source	Num DF	Den DF	F Value	Pr > F	Chi-Square	Pr > ChiSq
Trt	1	870	1.90	0.1687	1.90	0.1684
Sex	1	870	2.29	0.1304	2.29	0.1300
Age	1	870	4.30	0.0384	4.30	0.0381
Diabetic	1	870	20.61	<.0001	20.61	<.0001
PriorMonths	1	870	1.53	0.2165	1.53	0.2161
Albumin	1	870	22.34	<.0001	22.34	<.0001

Trt	Hospitalization Rate (Admissions/Year)	Rate Ratio (Trt:Cntrl)	Lower 95% CL	Upper 95% CL	p-value
Control	1.03283	.	.	.	.
Treated	1.16679	1.12970	0.94960	1.34396	0.16872

inflate the p-value depending on what the root cause of overdispersion is. For example, overdispersion can be the result of excluding important explanatory variables or it may be the result of positive correlation among the observations. It is difficult to rule out the former even when results are from a randomized controlled trial. However, we can investigate the possibility that some level of correlation is present among the observations and that such correlation may account for the extra-Poisson variation.

One possible source of correlation may be variation in hospitalization rates between participating centers in that patients within centers are more likely to receive a similar standard of care resulting in a more homogeneous group of patients. This suggests fitting a cluster-specific random effects Poisson model wherein patients are clustered according to their participating center. Below is the SAS code required to run this analysis.

Program 5.6

```
data example5_2_4; set SASdata.ADEMEX_Peritonitis_Data;
 log_time=log(MonthsAtRisk); ** Offset;
 Center=scan(ptid, 1);
run;
proc sort data=example5_2_4;
 by Center ptid;
run;
ods select ModelInfo FitStatistics CondFitStatistics
           CovParms CovTests ParameterEstimates Tests3;
ods output LSMeans = LSmeans;
ods output Diffs = LSdiffs;
ods output CovBDetails=gof;
ods output ParameterEstimates=pe;
ods output CovParms=cov;
ods output dimensions=n;
```

```
proc glimmix data=example5_2_4 noclprint=3 order=formatted
                                   method=Quad maxopt=50 empirical;
  class Center Trt;
  model  Hosp=Trt Sex Age Diabetic PriorMonths Albumin
       / dist=Poisson link=log offset=log_time s covb(details);
  lsmeans Trt /diff cl ilink;
  random intercept / subject=Center type=vc;
  covtest ZeroG;
  output out=pred /allstats;
run;
data lsout;
 set LSmeans;
 Rate = exp(estimate)*12;
run;
data lsdiff;
 set LSdiffs;
 RR = exp(-estimate);
 RR_lower = exp(-Upper);
 RR_upper = exp(-Lower);
 p=Probt;
 if interval=_interval;
 Trt='1';
run;
proc sort data=lsdiff;
 by Trt;
run;
proc sort data=lsout;
 by Trt;
run;
data summary;
 merge lsout lsdiff;
 by Trt;
run;
proc format;
 value _trtfmt_ 0='Control' 1='Treated';
run;
proc print data=summary split='|' noobs;
 var Trt Rate RR RR_lower RR_upper p;
 label Rate='Hospitalization|Rate|(Admissions/Year)'
       RR='Rate Ratio|(Trt:Cntrl)'
       RR_Lower='Lower|95% CL'
       RR_Upper='Upper|95% CL'
       p='p-value';
 format Trt _trtfmt_.;
run;
%GLIMMIX_GOF(dimension=n,
             parms=pe,
             covb_gof=gof,
             output=pred,
             response=HOSP,
             pred_ind=PredMu,
             pred_avg=PredMuPA,
             printopt=NOPRINT);
proc print data=_fitting noobs;
run;
```

The first task is to identify each patient's participating center which we can do based on the value of the patient ID variable, `ptid`. Each patient's ID has two sets of numbers separated by a blank. The first two digits identify the patient's participating center and the last three digits identify the patient. Hence the above SAS statement, "`Center=scan(ptid, 1);`" defines the center ID based on the first two digits of the variable `ptid`. Next, we sort the data by center and patient so as to maintain a proper ordering of the random effects as specified by the SUBJECT=CENTER option of the RANDOM statement. Using GLIMMIX, we fit exactly the same model as used when fitting the univariate Poisson model in GENMOD except that we have included a RANDOM statement that identifies centers as a cluster-specific random effect. We also included a test of whether the random center effect is present using the COVTEST statement, `covtest ZeroG;` shown above. Parameter estimates are based on ML estimation using numerical quadrature (METHOD=QUAD). All inference is based on the empirical sandwich estimator as specified by the EMPIRICAL option of GLIMMIX. The remaining code is similar to that in Example 4.3.1 in that we use ODS OUTPUT statements along with the LSMEANS statement to create and summarize annualized hospitalization rates using PROC PRINT. We included the macro %GLIMMIX_GOF in order to examine alternative goodness-of-fit measures.

With this code, patients from the same center are assumed to share a common random effect. This in turn induces correlation among the hospitalization counts from those patients. We can express the underlying model for this analysis in terms of a cluster-specific random effects Poisson model by writing the conditional mean and variance as

$$\begin{aligned} E(y_{ij}|b_i) &= \mu_{ij}(\boldsymbol{\beta}, b_i) = \exp\{\boldsymbol{x}_{ij}'\boldsymbol{\beta} + b_i + \log(t_{ij})\} \\ Var(y_{ij}|b_i) &= \mu_{ij}(\boldsymbol{\beta}, b_i) \end{aligned} \tag{5.41}$$

where y_{ij} is the number of hospital admissions for the jth patient within the ith center, b_i is the center-specific random effect, and

$$\boldsymbol{x}_{ij}'\boldsymbol{\beta} + b_i = \beta_0 + \beta_1 x_{1ij} + \beta_2 x_{2ij} + \beta_3 x_{3ij} + \beta_4 x_{4ij} + \beta_5 x_{5ij} + \beta_6 x_{6ij} + b_i$$

is the linear predictor (excluding the offset $\log(t_{ij})$) based on the patient-specific covariates $x_{1ij}, x_{2ij}, \ldots, x_{6ij}$ defined in §4.3.1.

The results of fitting the data to this random-effects Poisson model are shown in Output 5.14. While results from the COVTEST statement reveal a significant center random effect, this had little impact on the results. As suggested by the summary table on hospitalization rates at the bottom of Output 5.14, the results remain virtually unchanged with the addition of the center random effect. However, when we examine the conditional fit statistics, we find that there is still a significant source of overdispersion among patients within a center as reflected by an overdispersion parameter of $\widehat{\phi} = 2.78$ (as determined from the `Pearson Chi-Square / DF` value in the output). If we repeat the analysis using the model-based standard errors (i.e., re-run the above code without the EMPIRICAL option), the p-value for the treatment group effect becomes 0.025 indicating a statistically significant effect.

Output 5.14: Summary comparison of ADEMEX hospitalization rates based on a cluster-specific random effects Poisson model with centers representing the cluster-specific random effect.

Model Information

Data Set	WORK.EXAMPLE5_2_4
Response Variable	Hosp
Response Distribution	Poisson
Link Function	Log
Variance Function	Default
Offset Variable	log_time
Variance Matrix Blocked By	Center
Estimation Technique	Maximum Likelihood
Likelihood Approximation	Gauss–Hermite Quadrature
Degrees of Freedom Method	Containment
Fixed Effects SE Adjustment	Sandwich – Classical

Fit Statistics

-2 Log Likelihood	3645.64
AIC (smaller is better)	3661.64
AICC (smaller is better)	3661.81
BIC (smaller is better)	3671.07
CAIC (smaller is better)	3679.07
HQIC (smaller is better)	3664.14

Fit Statistics for Conditional Distribution

-2 log L(Hosp | r. effects)	3583.74
Pearson Chi-Square	2434.77
Pearson Chi-Square / DF	2.78

Covariance Parameter Estimates

Cov Parm	Subject	Estimate	Standard Error
Intercept	Center	0.08617	0.02351

Solutions for Fixed Effects

Effect	Trt	Estimate	Standard Error	DF	t Value	Pr > |t|
Intercept		−1.3539	0.2998	23	−4.52	0.0002
Trt	0	−0.1155	0.08890	847	−1.30	0.1943
Trt	1	0	.	.	.	.
Sex		0.1037	0.07573	847	1.37	0.1714
Age		−0.00777	0.003459	847	−2.25	0.0249
Diabetic		0.4883	0.1156	847	4.22	<.0001
PriorMonths		0.003355	0.001569	847	2.14	0.0327
Albumin		−0.3322	0.07765	847	−4.28	<.0001

Type III Tests of Fixed Effects

Effect	Num DF	Den DF	F Value	Pr > F
Trt	1	847	1.69	0.1943
Sex	1	847	1.87	0.1714
Age	1	847	5.05	0.0249
Diabetic	1	847	17.85	<.0001
PriorMonths	1	847	4.57	0.0327
Albumin	1	847	18.31	<.0001

Tests of Covariance Parameters Based on the Likelihood

Label	DF	-2 Log Like	ChiSq	Pr > ChiSq	Note
No G–side effects	1	3725.35	79.71	<.0001	MI

Trt	Hospitalization Rate (Admissions/Year)	Rate Ratio (Trt:Cntrl)	Lower 95% CL	Upper 95% CL	p-value
Control	0.98509	.	.	.	.
Treated	1.10569	1.12242	0.94271	1.33640	0.19426

To assess whether there truly is a significant treatment effect, we must first assess whether the overdispersion is significant or not. One means for doing this is to run the SAS macro %GLIMMIX_GOF as shown above using the created output datasets as input. On the basis of a discrepancy function of 5.365, a pseudo-likelihood ratio test of the assumed covariance structure suggests this model provides a poor fit to the data (approximate Chi-Square of 128.751 with 28 degrees of freedom and a p-value < 0.0001, see Output 5.15

below). This suggests that assuming conditional Poisson variation for the individual patient-level counts is not a viable option.

Output 5.15: Goodness-of-fit results from a random-effects Poisson model with centers serving as a cluster-specific random effect.

DESCRIPTION	VALUE
Total Observations	877.000
N (number of subjects)	24.000
Number of Fixed-Effects Parameters	7.000
Average Model R-Square:	0.012
Average Model Adjusted R-Square:	0.004
Average Model Concordance Correlation:	0.201
Average Model Adjusted Concordance Correlation:	0.194
Conditional Model R-Square:	0.073
Conditional Model Adjusted R-Square:	0.066
Conditional Model Concordance Correlation:	0.310
Conditional Model Adjusted Concordance Correlation:	0.304
Variance-Covariance Concordance Correlation:	0.573
Discrepancy Function	5.365
s = Rank of robust sandwich estimator, OmegaR	7.000
s1 = Number of unique non-zero off-diagonal elements of OmegaR	21.000
Approx. Chi-Square for H0: Covariance Structure is Correct	128.751
DF1 = s(s+1)/2, per Vonesh et al (Biometrics 52:572-587, 1996)	28.000
Pr > Chi Square based on degrees of freedom, DF1	0.000
DF2 = s+s1, a modified degress of freedom	28.000
Pr > Chi Square based on modified degrees of freedom, DF2	0.000

To accommodate extra-Poisson variation at the patient level, we fit the data to a multi-level random effects Poisson model assuming both center-specific and patient-specific random effects. The SAS code is identical to that shown above except that some modifications to the GLIMMIX syntax is required. Specifically, we need to specify two random-effect statements with patient-level random intercept effects nested within center-specific random intercept effects. In addition, we used Laplace-based ML estimation because of insufficient resources as indicated in the SAS log message,

```
ERROR: Insufficient resources to perform adaptive quadrature with 3
quadrature points. METHOD=LAPLACE, corresponding to a single point, may
provide a computationally less intensive possibility.
```

one receives if one attempts to fit the model using adaptive Gaussian quadrature. Moreover, precisely because there is only a single observation per patient, we would get 0 degrees of freedom for all fixed-effects F-tests carried out in GLIMMIX. Consequently, we specified the options CHISQ and DDFM=NONE within the MODEL statement in order to compute approximate p-values. This two-level random-effects Poisson model may be written in terms of the conditional moments as

$$\begin{aligned} E(y_{ij}|b_i, b_{ij}) &= \mu_{ij}(\boldsymbol{\beta}, b_i, b_{ij}) = \exp\{\boldsymbol{x}_{ij}'\boldsymbol{\beta} + b_i + b_{ij} + \log(t_{ij})\} \\ Var(y_{ij}|b_i, b_{ij}) &= \mu_{ij}(\boldsymbol{\beta}, b_i, b_{ij}) \end{aligned} \quad (5.42)$$

where $b_i \sim$iid $N(0, \psi_c)$ and $b_{ij} \sim$iid $N(0, \psi_p)$ are the center-specific and patient-specific random effects, respectively, with b_{ij} nested within b_i. The conditional pdf of the y_{ij} are assumed to be Poisson, i.e., $y_{ij}|b_i, b_{ij} \sim$ind $P(\mu_{ij}|b_i, b_{ij})$ with b_i and b_{ij} assumed independent of one another. The GLIMMIX code required to fit this model is modified as follows.

Program 5.7

```
proc glimmix data=example5_2_4 noclprint=3 order=formatted
                         method=Laplace maxopt=50 empirical;
  class Center ptid Trt;
  model  Hosp=Trt Sex Age Diabetic PriorMonths Albumin
       / dist=Poisson link=log offset=log_time s covb(details)
         chisq ddfm=none;
  lsmeans Trt /diff cl ilink;
  random intercept / subject=Center type=vc;
  random intercept / subject=ptid(Center) type=vc;
  covtest 'No Center effect' 0 .;
  covtest 'No ptid(Center) effect' . 0;
  covtest ZeroG;
  output out=pred /allstats;
run;
```

As suggested by the covariance parameter estimates themselves as well as the COVTEST results displayed in Output 5.16, there appears to be significant overdispersion due to patient-to-patient variation within centers (a variance component of 0.8580, $p< 0.0001$) as well as some modest center-to-center variation (a variance component of 0.02924, $p=0.0420$). The p-values for all three tests performed with the COVTEST statements are based on a mixture of chi-squares as indicated by the MI listed under the Note column.

Output 5.16: Summary comparison of ADEMEX hospitalization rates based on a multi-level random effects Poisson model with centers representing a cluster-specific random effect and patients within centers representing a patient-level random effect.

```
                    Model Information
    Data Set                      WORK.EXAMPLE5_2_4
    Response Variable             Hosp
    Response Distribution         Poisson
    Link Function                 Log
    Variance Function             Default
    Offset Variable               log_time
    Variance Matrix Blocked By    Center
    Estimation Technique          Maximum Likelihood
    Likelihood Approximation      Laplace
    Degrees of Freedom Method     None
    Fixed Effects SE Adjustment   Sandwich - Classical
               Fit Statistics
    -2 Log Likelihood          3224.05
    AIC (smaller is better)    3242.05
    AICC (smaller is better)   3242.25
    BIC (smaller is better)    3252.65
    CAIC (smaller is better)   3261.65
    HQIC (smaller is better)   3244.86
    Fit Statistics for Conditional Distribution
    -2 log L(Hosp | r. effects)         2085.19
    Pearson Chi-Square                   423.55
    Pearson Chi-Square / DF                0.48
              Covariance Parameter Estimates
    Cov Parm    Subject        Estimate   Standard Error
    Intercept   Center          0.02924          0.02505
    Intercept   ptid(Center)     0.8580           0.1055
```

Solutions for Fixed Effects

Effect	Trt	Estimate	Standard Error	DF	t Value	Pr > |t|
Intercept		−1.2490	0.3526	Infty	−3.54	0.0004
Trt	0	−0.1400	0.1065	Infty	−1.31	0.1885
Trt	1	0	.	.	.	.
Sex		0.1506	0.08513	Infty	1.77	0.0770
Age		−0.00776	0.004197	Infty	−1.85	0.0645
Diabetic		0.5073	0.1298	Infty	3.91	<.0001
PriorMonths		0.001455	0.001435	Infty	1.01	0.3106
Albumin		−0.4468	0.09313	Infty	−4.80	<.0001

Type III Tests of Fixed Effects

Effect	Num DF	Den DF	Chi-Square	F Value	Pr > ChiSq	Pr > F
Trt	1	Infty	1.73	1.73	0.1885	0.1885
Sex	1	Infty	3.13	3.13	0.0770	0.0770
Age	1	Infty	3.42	3.42	0.0645	0.0645
Diabetic	1	Infty	15.27	15.27	<.0001	<.0001
PriorMonths	1	Infty	1.03	1.03	0.3106	0.3106
Albumin	1	Infty	23.02	23.02	<.0001	<.0001

Tests of Covariance Parameters Based on the Likelihood

Label	DF	-2 Log Like	ChiSq	Pr > ChiSq	Note
No Center effect	1	3227.03	2.99	0.0420	MI
No ptid(Center) effect	1	3645.64	421.59	<.0001	MI
No G−side effects	2	3725.35	501.31	<.0001	MI

Trt	Hospitalization Rate (Admissions/Year)	Rate Ratio (Trt:Cntrl)	Lower 95% CL	Upper 95% CL	p-value
Control	0.76647	.	.	.	.
Treated	0.88167	1.15030	0.93362	1.41728	0.18853

Indeed, when we evaluate model goodness-of-fit using the macro %GLIMMIX_GOF, we find that with respect to the assumed covariance structure, the multi-level random-effects Poisson model provides a slightly better fit to the data than the single-level random-effects Poisson model (a variance-covariance concordance correlation of 0.830 versus 0.573 from Output 5.17 and Output 5.15, respectively). However, this appears to be at the expense of underestimating the average hospitalization rate (average model concordance correlation of 0.181 versus 0.310 from Output 5.17 and Output 5.15, respectively). This may be due to a continued poor fit to the assumed covariance structure as indicated by a discrepancy function of 1.963 and a pseudo-likelihood ratio chi-square statistic of 47.104 (p=0.013, Output 5.17). Alternatively, it may be due to misspecifying the conditional distribution as being Poisson. From the fit statistics for the conditional distribution, we estimate

Output 5.17: Goodness-of-fit results from a two-level random-effects Poisson model with centers serving as a cluster-specific random effect and patients within centers serving as a patient-level random effect.

DESCRIPTION	VALUE
Total Observations	877.000
N (number of subjects)	24.000
Number of Fixed−Effects Parameters	7.000
Average Model R−Square:	0.010
Average Model Adjusted R−Square:	0.002
Average Model Concordance Correlation:	0.181
Average Model Adjusted Concordance Correlation:	0.175
Conditional Model R−Square:	0.887
Conditional Model Adjusted R−Square:	0.886
Conditional Model Concordance Correlation:	0.930
Conditional Model Adjusted Concordance Correlation:	0.929
Variance−Covariance Concordance Correlation:	0.830
Discrepancy Function	1.963

```
s = Rank of robust sandwich estimator, OmegaR                          7.000
s1 = Number of unique non-zero off-diagonal elements of OmegaR        21.000
Approx. Chi-Square for H0: Covariance Structure is Correct            47.104
DF1 = s(s+1)/2, per Vonesh et al (Biometrics 52:572-587, 1996)        28.000
Pr > Chi Square based on degrees of freedom, DF1                       0.013
DF2 = s+s1, a modified degress of freedom                             28.000
Pr > Chi Square based on modified degrees of freedom, DF2              0.013
```

a dispersion parameter of 0.48 (Pearson Chi-Square/DF, Output 5.16) indicating an underdispersed conditional distribution relative to that of a Poisson.

These considerations lead to fitting an alternative model that accommodates some form of extra-Poisson variation at the patient level while continuing to account for correlation resulting from a center-specific random effect. Specifically, we fit the data assuming that, conditional on the center-specific random effect, the number of hospital admissions for a patient within a center follows a negative binomial distribution. One can fit such a model in GLIMMIX using the following syntax.

Program 5.8

```
proc glimmix data=example5_2_4 noclprint=3 order=formatted
                         method=Quad maxopt=50 empirical;
  class Center Trt;
  model  Hosp=Trt Sex Age Diabetic PriorMonths Albumin
       / dist=NB link=log offset=log_time s covb(details);
  lsmeans Trt /diff cl ilink;
  random intercept / subject=Center type=vc;
  covtest ZeroG;
  output out=pred /allstats;
run;
```

This random effects negative binomial model is defined by the conditional moments

$$\begin{aligned} E(y_{ij}|b_i) &= \mu_{ij}(\boldsymbol{\beta}, b_i) = \exp\{\boldsymbol{x}'_{ij}\boldsymbol{\beta} + b_i + \log(t_{ij})\} \\ Var(y_{ij}|b_i) &= \mu_{ij}(\boldsymbol{\beta}, b_i) + \alpha\mu_{ij}(\boldsymbol{\beta}, b_i)^2 \end{aligned} \tag{5.43}$$

where $\boldsymbol{x}'_{ij}\boldsymbol{\beta} + b_i$ is the conditional linear predictor (excluding the offset) with the same fixed effects $\boldsymbol{x}_{ij}$ as above and α is a scale or dispersion parameter. This single-level random-effects model can be derived from a multi-level random-effects model by noting that the negative binomial distribution is the marginal distribution of a conditional Poisson distribution having a random rate parameter that is distributed according to a gamma distribution. Specifically, let b_{ij} be a random effect for the jth patient within the ith center and suppose that the conditional distribution of counts, y_{ij}, given b_i and b_{ij} is Poisson with conditional mean and variance

$$\begin{aligned} E(y_{ij}|b_i, b_{ij}) = \mu^*_{ij}(\boldsymbol{\beta}^*, b_i)\exp(b_{ij}) &= \exp\{\boldsymbol{x}'_{ij}\boldsymbol{\beta}^* + b_i + \log(t_{ij})\}\exp(b_{ij}) \\ &= \exp\{\boldsymbol{x}'_{ij}\boldsymbol{\beta}^* + b_i + b_{ij} + \log(t_{ij})\} \\ &= Var(y_{ij}|b_i, b_{ij}). \end{aligned}$$

This has the same form as the conditional mean and variance under the two-level random-effects Poisson model but the fixed-effects portion of the conditional linear predictor is defined by

$$\boldsymbol{x}'_{ij}\boldsymbol{\beta}^* = \beta^*_0 + \beta_1 x_{1ij} + \beta_2 x_{2ij} + \beta_3 x_{3ij} + \beta_4 x_{4ij} + \beta_5 x_{5ij} + \beta_6 x_{6ij}$$

where we represent the intercept in the parameter vector $\boldsymbol{\beta}^*$ by β_0^* rather than β_0 for reasons that will become clear shortly. Now, rather than assume b_{ij} ∼iid $N(0, \psi_p)$ as was done under the two-level random-effects Poisson model, suppose b_{ij} has a log-gamma distribution or, equivalently, $\lambda_{ij} = \exp(b_{ij})$ has a gamma distribution, $G(\gamma, 1/\alpha)$, with mean $1/\alpha\gamma$, variance $1/\alpha\gamma^2$ and a reparameterized probability density function (pdf)

$$\pi(\lambda_{ij}) = \frac{\gamma^{\frac{1}{\alpha}}}{\Gamma(\frac{1}{\alpha})}\lambda_{ij}^{(\frac{1}{\alpha}-1)}\exp\{-\gamma\lambda_{ij}\} \tag{5.44}$$

that differs from the pdf given in Table 4.1. Then we can write the conditional pdf of $y_{ij}|b_i, b_{ij}$ as

$$\begin{aligned}\pi(y_{ij}|b_i, b_{ij}) &= \Pr[y_{ij} = y|b_i, b_{ij}] \\ &= \frac{[\lambda_{ij}\mu_{ij}^*(\boldsymbol{\beta}^*, b_i)]^y}{y!}\exp\{-\lambda_{ij}\mu_{ij}^*(\boldsymbol{\beta}^*, b_i)\}.\end{aligned} \tag{5.45}$$

Given that the gamma distribution (5.44) is the conjugate prior for the Poisson, it is easy to verify by integration that the conditional distribution of $y_{ij}|b_i$ will be distributed according to a negative binomial distribution with probability density function,

$$\begin{aligned}\pi(y_{ij}|b_i) &= \int_0^\infty \frac{[\lambda_{ij}\mu_{ij}^*(\boldsymbol{\beta}^*, b_i)]^y}{y!}\exp\{-\lambda_{ij}\mu_{ij}^*(\boldsymbol{\beta}^*, b_i)\}\pi(\lambda_{ij})d\lambda_{ij} \\ &= \frac{\Gamma(y+\frac{1}{\alpha})}{\Gamma(\frac{1}{\alpha})\Gamma(y+1)}\left(\frac{\mu_{ij}^*}{\mu_{ij}^*+\gamma}\right)^y\left(\frac{\gamma}{\mu_{ij}^*+\gamma}\right)^{1/\alpha} \\ &= \frac{\Gamma(y+\frac{1}{\alpha})}{\Gamma(\frac{1}{\alpha})\Gamma(y+1)}\left(\frac{\alpha\mu_{ij}}{1+\alpha\mu_{ij}}\right)^y\left(\frac{1}{1+\alpha\mu_{ij}}\right)^{1/\alpha}.\end{aligned} \tag{5.46}$$

where $\mu_{ij} = \mu_{ij}(\boldsymbol{\beta}, b_i) = \frac{1}{\alpha\gamma}\mu_{ij}^*(\boldsymbol{\beta}^*, b_i)$ and $\boldsymbol{\beta}$ is the parameter vector obtained from $\boldsymbol{\beta}^*$ by replacing the intercept β_0^* with $\beta_0 = -\log(\alpha\gamma) + \beta_0^*$. From Table 4.2, the mean and variance under (5.46) are defined by

$$\begin{aligned}E(y_{ij}|b_i) &= \frac{1}{\alpha\gamma}\mu_{ij}^*(\boldsymbol{\beta}^*, b_i) \\ &= \exp\{-\log(\alpha\gamma) + \boldsymbol{x}_{ij}'\boldsymbol{\beta}^* + b_i + \log(t_{ij})\} \\ &= \exp\{\boldsymbol{x}_{ij}'\boldsymbol{\beta} + b_i + \log(t_{ij})\} \\ &= \mu_{ij}(\boldsymbol{\beta}, b_i) \\ Var(y_{ij}|b_i) &= \mu_{ij}(1+\alpha\mu_{ij})\end{aligned}$$

as specified directly in (5.43). For identifiability, we set $\gamma = 1/\alpha$ such that $E(\lambda_{ij}) = 1$, $Var(\lambda_{ij}) = \alpha$ and $\mu_{ij}^*(\boldsymbol{\beta}^*, b_i) = \mu_{ij}(\boldsymbol{\beta}, b_i)$ (Nelson et. al., 2006).

Shown in Output 5.18 are the results of fitting the hospitalization data to this random-effects negative binomial model. When we examine the marginal fit of this model versus either of the two random-effects Poisson models, we find a much lower AIC of 3225.95 suggesting this provides a better fit to the marginal distribution. We also see from the conditional fit statistics that the dispersion parameter estimated by Pearson Chi-Square/DF is 1.10 which is further indicative that, conditional on the center-specific random effects, the distribution of patient-level hospitalizations is best described by a negative binomial rather than a Poisson distribution. When we examine

Output 5.18: Summary comparison of ADEMEX hospitalization rates based on a cluster-specific random effects negative binomial model with centers representing the cluster-specific random effect.

Model Information

Data Set	WORK.EXAMPLE5_2_4
Response Variable	Hosp
Response Distribution	Negative Binomial
Link Function	Log
Variance Function	Default
Offset Variable	log_time
Variance Matrix Blocked By	Center
Estimation Technique	Maximum Likelihood
Likelihood Approximation	Gauss−Hermite Quadrature
Degrees of Freedom Method	Containment
Fixed Effects SE Adjustment	Sandwich − Classical

Fit Statistics

-2 Log Likelihood	3207.95
AIC (smaller is better)	3225.95
AICC (smaller is better)	3226.16
BIC (smaller is better)	3236.55
CAIC (smaller is better)	3245.55
HQIC (smaller is better)	3228.77

Fit Statistics for Conditional Distribution

-2 log L(Hosp | r. effects)	3178.70
Pearson Chi-Square	966.49
Pearson Chi-Square / DF	1.10

Covariance Parameter Estimates

Cov Parm	Subject	Estimate	Standard Error
Intercept	Center	0.05037	0.02313
Scale		0.8882	0.1014

Solutions for Fixed Effects

Effect	Trt	Estimate	Standard Error	DF	t Value	Pr > |t|
Intercept		−1.0072	0.3724	23	−2.71	0.0126
Trt	0	−0.1307	0.09545	847	−1.37	0.1713
Trt	1	0	.	.	.	.
Sex		0.1120	0.08560	847	1.31	0.1912
Age		−0.00695	0.003969	847	−1.75	0.0801
Diabetic		0.4812	0.1230	847	3.91	<.0001
PriorMonths		0.002934	0.001447	847	2.03	0.0429
Albumin		−0.4023	0.08808	847	−4.57	<.0001

Type III Tests of Fixed Effects

Effect	Num DF	Den DF	F Value	Pr > F
Trt	1	847	1.87	0.1713
Sex	1	847	1.71	0.1912
Age	1	847	3.07	0.0801
Diabetic	1	847	15.31	<.0001
PriorMonths	1	847	4.11	0.0429
Albumin	1	847	20.86	<.0001

Tests of Covariance Parameters Based on the Likelihood

Label	DF	-2 Log Like	ChiSq	Pr > ChiSq	Note
No G−side effects	1	3215.70	7.75	0.0027	MI

Trt	Hospitalization Rate (Admissions/Year)	Rate Ratio (Trt:Cntrl)	Lower 95% CL	Upper 95% CL	p-value
Control	1.15682	.	.	.	.
Treated	1.31833	1.13962	0.94492	1.37443	0.17129

the variance components, we find that both the scale parameter, $\widehat{\alpha} = 0.8882$, and random-effects variance component, $\widehat{\psi} = 0.05037$, are significant (a COVTEST likelihood ratio test p-value of 0.0027 for testing no G-side random effect). Likewise we fail to reject the hypothesis that the covariance structure incorporating both α and ψ is correct (a pseudo-likelihood ratio chi-square of 34.712 on 28 degrees of freedom, p-value = 0.178, Output 5.19).

Output 5.19: Goodness-of-fit results from a random-effects negative binomial model with centers serving as a cluster-specific random effect.

DESCRIPTION	VALUE
Total Observations	877.000
N (number of subjects)	24.000
Number of Fixed−Effects Parameters	7.000
Average Model R−Square:	−0.056
Average Model Adjusted R−Square:	−0.064
Average Model Concordance Correlation:	0.227
Average Model Adjusted Concordance Correlation:	0.221
Conditional Model R−Square:	−0.014
Conditional Model Adjusted R−Square:	−0.022
Conditional Model Concordance Correlation:	0.297
Conditional Model Adjusted Concordance Correlation:	0.291
Variance−Covariance Concordance Correlation:	0.892
Discrepancy Function	1.446
s = Rank of robust sandwich estimator, OmegaR	7.000
s1 = Number of unique non−zero off−diagonal elements of OmegaR	21.000
Approx. Chi−Square for H0: Covariance Structure is Correct	34.712
DF1 = s(s+1)/2, per Vonesh et al (Biometrics 52:572−587, 1996)	28.000
Pr > Chi Square based on degrees of freedom, DF1	0.178
DF2 = s+s1, a modified degress of freedom	28.000
Pr > Chi Square based on modified degrees of freedom, DF2	0.178

In summary, under the random-effects negative binomial model, the mean rate of hospitalization for the typical center is 1.157 and 1.318 admissions per year, respectively, for control and treated patients. This corresponds to a rate ratio (treated:control) of 1.139 which is not statistically significant (p-value = 0.1713 based on robust standard errors). When we make this same comparison using model-based standard errors (obtained when we exclude the EMPIRICAL option), the ensuing p-value is 0.1415 which provides further evidence that this model is preferred to the two random-effects Poisson models. While we have gone through great pains to arrive at the same basic conclusion as obtained in the original analysis of Paniagua et. al. (2002), we have done so with a purpose of identifying possible sources of overdispersion seen in the original analysis. By specifying a center-specific random effect into a model that allows for overdispersed counts at the patient-level, we have accounted for two sources of overdispersion that, in addition to any differences due to treatment, gender, age, diabetic status, prior months on dialysis and baseline serum albumin, appear to explain most of the variability in the observed number of hospitalizations per patient. This provides some assurance that the results are not biased by other factors like the exclusion of key confounding variables that might otherwise explain the extra-Poisson variation.

5.3 The generalized nonlinear mixed-effects (GNLME) model

Recall from Chapter 4 that the class of marginal generalized linear models (GLIM's) for correlated data can be extended to a class of generalized nonlinear models (GNLM's) by allowing the marginal moments to be expressed in terms of more general nonlinear functions of the parameters and, when applicable, by allowing for a broader family of distributions. In this section, one can apply this same strategy to extend the class of generalized linear mixed-effects (GLME) models to a class of generalized nonlinear mixed-effects (GNLME) models. Using our basic notation (page 8; see also Table 1.1), a GNLME model may be cast in terms of conditional first- and second-order moments as follows:

$$E(\boldsymbol{y}_i|\boldsymbol{b}_i) = \boldsymbol{\mu}_i(\boldsymbol{\beta}, \boldsymbol{b}_i) = f(\boldsymbol{X}_i, \boldsymbol{\beta}, \boldsymbol{b}_i);\ i = 1, \ldots, n \tag{5.47}$$
$$Var(\boldsymbol{y}_i|\boldsymbol{b}_i) = \boldsymbol{\Lambda}_i(\boldsymbol{\beta}, \boldsymbol{b}_i, \boldsymbol{\theta}_w)$$

where

$\boldsymbol{y}_i$ is the $p_i \times 1$ response vector for the ith subject (cluster)
$\boldsymbol{\mu}_i(\boldsymbol{\beta}, \boldsymbol{b}_i) = f(\boldsymbol{X}_i, \boldsymbol{\beta}, \boldsymbol{b}_i)$ is a general nonlinear function of $\boldsymbol{\beta}$ and $\boldsymbol{b}_i$
$\boldsymbol{X}_i$ is a $p_i \times u$ matrix of within- and between-subject covariates
$\boldsymbol{\beta}$ is an $s \times 1$ vector of fixed-effects parameters
$\boldsymbol{b}_i$ are $v \times 1$ vectors of subject-specific random-effects with finite first- and second-order moments

$$E(\boldsymbol{b}_i) = \boldsymbol{0},$$
$$Var(\boldsymbol{b}_i) = \boldsymbol{\Psi}(\boldsymbol{\theta}_b)$$

with $\boldsymbol{\Psi}(\boldsymbol{\theta}_b)$ representing a $v \times v$ between-subject variance-covariance matrix that depends on a $d_b \times 1$ vector $\boldsymbol{\theta}_b$ of unique between-subject covariance parameters.
$\boldsymbol{\Lambda}_i(\boldsymbol{\beta}, \boldsymbol{b}_i, \boldsymbol{\theta}_w)$ is a $p_i \times p_i$ conditional variance-covariance matrix that depends on a $d_w \times 1$ vector $\boldsymbol{\theta}_w$ of unique within-subject covariance parameters and possibly on $\boldsymbol{\beta}$ and $\boldsymbol{b}_i$. We assume $\boldsymbol{\Lambda}_i(\boldsymbol{\beta}, \boldsymbol{b}_i, \boldsymbol{\theta}_w)$ has the specific form

$$\boldsymbol{\Lambda}_i(\boldsymbol{\beta}, \boldsymbol{b}_i, \boldsymbol{\theta}_w) = \boldsymbol{W}_i(\boldsymbol{X}_i, \boldsymbol{\beta}, \boldsymbol{b}_i)^{-1/2}\boldsymbol{\Lambda}_i(\boldsymbol{\theta}_w)\boldsymbol{W}_i(\boldsymbol{X}_i, \boldsymbol{\beta}, \boldsymbol{b}_i)^{-1/2}$$

where $\boldsymbol{W}_i(\boldsymbol{X}_i, \boldsymbol{\beta}, \boldsymbol{b}_i) = Diag\{w_{ij}\}$ is some $p_i \times p_i$ diagonal weight matrix having weights w_{ij} that may depend on $\boldsymbol{X}_i$ as well as $\boldsymbol{\beta}$ and $\boldsymbol{b}_i$ and $\boldsymbol{\Lambda}_i(\boldsymbol{\theta}_w)$ is some specified $p_i \times p_i$ covariance matrix with parameters $\boldsymbol{\theta}_w$.
$\boldsymbol{\theta} = (\boldsymbol{\theta}_b, \boldsymbol{\theta}_w)$ is a $d \times 1$ vector of between- and within-subject variance-covariance parameters ($d = d_b + d_w$).

Note that, as with GNLM's, the conditional moments from a GNLME model do not require design matrices $\boldsymbol{X}_i$ that necessarily conform to the dimension of $\boldsymbol{\beta}$ (i.e., $\boldsymbol{X}_i$ need not have dimension $p_i \times s$ as would otherwise occur in a GLME model involving linear predictors $\boldsymbol{x}'_{ij}\boldsymbol{\beta} + \boldsymbol{z}'_{ij}\boldsymbol{b}_i$). Moreover, as the weight matrix $\boldsymbol{W}_i(\boldsymbol{X}_i, \boldsymbol{\beta}, \boldsymbol{b}_i)$ is quite arbitrary, the conditional variance-covariance structure is not restricted to matrices having the form $\boldsymbol{\Lambda}_i(\boldsymbol{\beta}, \boldsymbol{b}_i, \boldsymbol{\theta}_w) = \boldsymbol{V}_i(\boldsymbol{\mu}_i)^{1/2}\boldsymbol{\Lambda}_i^*(\boldsymbol{\theta}_w)\boldsymbol{V}_i(\boldsymbol{\mu}_i)^{1/2}$ as described for the GLME model (5.1). The parameter space of $(\boldsymbol{\beta}, \boldsymbol{\theta})$ is assumed to be a compact subspace of Euclidean space $\Re^{s+d}$ which is determined by the underlying distribution of $\boldsymbol{y}_i$. Under this assumption and the specification of the conditional moments alone, the GNLME model (5.47) may be classified as a semiparametric mixed-effects regression model in that no explicit assumptions regarding the conditional distribution of $\boldsymbol{y}_i$ are made. This means that under the assumptions of (5.47) alone, estimation and inference are restricted to moments-based semiparametric methods like the PL/CGEE1 methods described for the GLME model.

5.3.1 Fully parametric GNLME models

The semiparametric GNLME model (5.47) can be extended to a fully parametric GNLME model for continuous or discrete outcomes by simply specifying a conditional pdf of $\boldsymbol{y}_i$, say $\pi(\boldsymbol{y}_i; \boldsymbol{\beta}, \boldsymbol{\theta}_w|\boldsymbol{b}_i)$, and a pdf for $\boldsymbol{b}_i$, say $\pi(\boldsymbol{b}_i; \boldsymbol{\theta}_b)$, where $\boldsymbol{\beta}$ and $\boldsymbol{\theta}_w$ are location and scale parameters associated with the conditional distribution of $\boldsymbol{y}_i|\boldsymbol{b}_i$ and $\boldsymbol{\theta}_b$ is a vector of parameters for the distribution of $\boldsymbol{b}_i$. For the remainder of this chapter, we shall assume the random effects are continuous multivariate normal random vectors satisfying $\boldsymbol{b}_i \sim$iid $N(\boldsymbol{0}, \boldsymbol{\Psi}(\boldsymbol{\theta}_b))$. Following the work of Nelson et. al. (2006), we relax this assumption in

Chapter 7 by considering alternative continuous distributions for $\boldsymbol{b}_i$. While the conditional pdf, $\pi(\boldsymbol{y}_i;\boldsymbol{\beta},\boldsymbol{\theta}_w|\boldsymbol{b}_i)$, can be any general parametric joint density, we confine ourselves to any conditional pdf to which the ML estimation techniques available in the GLIMMIX and NLMIXED procedures can be applied or to which the pseudo-likelihood estimation techniques available in the SAS macro %NLINMIX can be applied.

Case 1: Parametric models assuming conditional independence

For fully parametric ML estimation, we consider those cases where the y_{ij} are assumed to be conditionally independent given a set of random effects. Specifically, we assume the joint conditional pdf of $\boldsymbol{y}_i|\boldsymbol{b}_i$ may be written as

$$\pi(\boldsymbol{y}_i;\boldsymbol{\beta},\boldsymbol{\theta}_w|\boldsymbol{b}_i) = \prod_{j=1}^{p_i} \pi(y_{ij};\boldsymbol{\beta},\boldsymbol{\theta}_w|\boldsymbol{b}_i) \tag{5.48}$$

where $\pi(y_{ij};\boldsymbol{\beta},\boldsymbol{\theta}_w|\boldsymbol{b}_i)$ is any general parametric density function with location and scale parameters $\boldsymbol{\beta}$ and $\boldsymbol{\theta}_w$, respectively. This assumption allows us to apply likelihood-based inference to a full array of GNLME models using the general log-likelihood function that one can construct using SAS programming statements together with the *general(ll)* distribution option available with the MODEL statement of NLMIXED. This assumption also allows us to specify GLME models and GNLME models that utilize any of the built-in conditional distributions available in GLIMMIX and NLMIXED. The parameters $\boldsymbol{\beta}$ and $\boldsymbol{\theta}_w$ in (5.48) will typically coincide with a conditional mean and covariance structure as defined by (5.47).

Case 2: Parametric models with conditional intra-subject correlation

Alternatively, for those cases where one wishes to explicitly model the conditional mean and covariance structure defined by (5.47) but within the context of a fully parametric conditional distribution for $\boldsymbol{y}_i$, we consider the class of GNLME models within the quadratic exponential family in which $\boldsymbol{y}_i$ has the joint conditional pdf

$$\pi(\boldsymbol{y}_i|\boldsymbol{b}_i) = \Delta_i^{-1}\exp\{\boldsymbol{y}_i'\boldsymbol{\lambda}_i + \boldsymbol{w}_i'\boldsymbol{\gamma}_i + \boldsymbol{c}_i(\boldsymbol{y}_i)\} \tag{5.49}$$

where $\boldsymbol{y}_i'$ is the response vector, and $\boldsymbol{w}_i' = (y_{i1}^2, y_{i1}y_{i2}, ..., y_{i1}y_{ip_i}, y_{i2}^2, y_{i2}y_{i3}, ...)$ is the vector of unique second-order components of $\boldsymbol{y}_i$ which one can write as $\boldsymbol{w}_i = Vech(\boldsymbol{y}_i\boldsymbol{y}_i')$. Under this family of distributions, the $\boldsymbol{\lambda}_i$ and $\boldsymbol{\gamma}_i$ are canonical parameters expressed in terms of vector functions of the conditional moments as

$$\boldsymbol{\lambda}_i' = \boldsymbol{\lambda}_i'(\boldsymbol{\mu}_i,\boldsymbol{\sigma}_i) = (\lambda_{i1},\lambda_{i2},...,\lambda_{ip_i})$$
$$\boldsymbol{\gamma}_i' = \boldsymbol{\gamma}_i'(\boldsymbol{\mu}_i,\boldsymbol{\sigma}_i) = (\gamma_{i11}, \gamma_{i12}...\gamma_{i1p_i}, \gamma_{i22}, \gamma_{i23}, ...).$$

where $\boldsymbol{\mu}_i$ is the conditional mean and $\boldsymbol{\sigma}_i = \boldsymbol{\sigma}_i(\boldsymbol{\beta},\boldsymbol{b}_i,\boldsymbol{\theta}_w) = Vech(\boldsymbol{\Lambda}_i(\boldsymbol{\beta},\boldsymbol{b}_i,\boldsymbol{\theta}_w))$ is the conditional covariance matrix expressed in vector form as the lower diagonal elements of $\boldsymbol{\Lambda}_i(\boldsymbol{\beta},\boldsymbol{b}_i,\boldsymbol{\theta}_w)$. The function $\boldsymbol{c}_i(\boldsymbol{y}_i)$ is a "shape" function, and $\Delta_i = \Delta_i(\boldsymbol{\lambda}_i,\boldsymbol{\gamma}_i, c_i)$ is a normalization constant required to make (5.49) a proper pdf (e.g., Prentice and Zhao, 1991).

The family of conditional pdf's under (5.49) include those of the linear exponential family shown in Tables 4.1 and 4.2 (simply set $\boldsymbol{\gamma}_i$ to $\boldsymbol{0}$ and redefine $\boldsymbol{\Delta}_i$ in terms of $b(\eta)$ and $a(\phi)$). In this case, the GNLME model jointly satisfies the assumptions of (5.47), (5.48) and (5.49) since $\boldsymbol{\gamma}_i = \boldsymbol{0}$ implies conditional independence. The weight matrix will then assume the form $\boldsymbol{W}_i(\boldsymbol{X}_i,\boldsymbol{\beta},\boldsymbol{b}_i) = \boldsymbol{V}_i(\boldsymbol{\mu}_i)^{-1}$. Estimation and inference in this case can be conducted based on methods for the GLME model described in §5.1.1 provided the mean is a monotonic function of a linear predictor, i.e., $\boldsymbol{\mu}_i(\boldsymbol{\beta},\boldsymbol{b}_i) = g^{-1}(\boldsymbol{\eta}_i)$ where $\boldsymbol{\eta}_i = \boldsymbol{X}_i\boldsymbol{\beta} + \boldsymbol{Z}_i\boldsymbol{b}_i$. In cases where one wishes to model the data assuming a conditional pdf from the linear exponential family but with the canonical parameter $\boldsymbol{\eta}_i$ allowed to be intrinsically nonlinear in one or more parameters, say $\boldsymbol{\eta}_i = f^*(\boldsymbol{X}_i, \boldsymbol{\beta}, \boldsymbol{b}_i)$, one would have to fit the data in NLMIXED in order to define the generalized nonlinear mean $\boldsymbol{\mu}_i(\boldsymbol{\beta}, \boldsymbol{b}_i) = g^{-1}(\boldsymbol{\eta}_i) = g^{-1}(f^*(\boldsymbol{X}_i,\boldsymbol{\beta}, \boldsymbol{b}_i))$ [e.g., McCullagh and Nelder, 1989; Chapter 11, §11.4-11.5]. Thus the family of GNLME models defined by (5.47)-(5.49) extends the GLME models defined by (5.1)-(5.2) of §5.1 by allowing for nonlinear predictors $\eta_{ij} = f^*(\boldsymbol{x}_{ij}, \boldsymbol{\beta}, \boldsymbol{b}_i)$ in the link function [i.e., $E(g(\mu_{ij})) = f^*(\boldsymbol{x}_{ij}, \boldsymbol{\beta}, \boldsymbol{b}_i)$]. Perhaps more importantly, the family of conditional pdf's under (5.49) includes the multivariate normal distribution which is not a member of the linear exponential family. Applications requiring a conditional pdf that is multivariate normal head up an important sub-class of GNLME models, namely the normal-theory nonlinear mixed-effects (NLME) model which we present in the following section.

5.3.2 Normal-theory nonlinear mixed-effects (NLME) model

The normal-theory nonlinear mixed-effects (NLME) model for correlated responses may be written as

$$y_{ij} = f(\boldsymbol{x}_{ij}', \boldsymbol{\beta}, \boldsymbol{b}_i) + w_{ij}(\boldsymbol{x}_{ij}', \boldsymbol{\beta}, \boldsymbol{b}_i)^{-1/2}\epsilon_{ij},\ i = 1, \ldots, n;\ j = 1, \ldots, p_i \tag{5.50}$$

where

y_{ij} is the jth measurement on the ith subject or cluster $(i = 1, \ldots, n)$
$\boldsymbol{x}_{ij}'$ are $1 \times u$ vectors of within- and between-subject covariates of the form $\boldsymbol{x}_{ij}' = (\boldsymbol{x}_{ij}^{*\prime}, \boldsymbol{a}_i')$ where $\boldsymbol{x}_{ij}^*$ is a $t \times 1$ vector of within-subject covariates and $\boldsymbol{a}_i$ is a $q \times 1$ vector of between-subject covariates $(u = t + q)$
$\boldsymbol{\beta}$ is an $s \times 1$ unknown vector of regression parameters
$\boldsymbol{b}_i$ are $v \times 1$ vectors of random-effects assumed to be iid $N(\boldsymbol{0}, \boldsymbol{\Psi}(\boldsymbol{\theta}_b))$
$w_{ij}(\boldsymbol{x}_{ij}', \boldsymbol{\beta}, \boldsymbol{b}_i)$ are scalar weights that may depend on $\boldsymbol{\beta}$ and $\boldsymbol{b}_i$
$\boldsymbol{\epsilon}_i = (\epsilon_{i1}, \epsilon_{i2}, \ldots, \epsilon_{ip_i})'$ are $p_i \times 1$ random error vectors that are mutually independent and distributed as $N_{p_i}(\boldsymbol{0}, \boldsymbol{\Lambda}_i(\boldsymbol{\theta}_w))$
$\boldsymbol{\Lambda}_i(\boldsymbol{\beta}, \boldsymbol{b}_i, \boldsymbol{\theta}_w) = \boldsymbol{W}_i(\boldsymbol{X}_i, \boldsymbol{\beta}, \boldsymbol{b}_i)^{-1/2}\boldsymbol{\Lambda}_i(\boldsymbol{\theta}_w)\boldsymbol{W}_i(\boldsymbol{X}_i, \boldsymbol{\beta}, \boldsymbol{b}_i)^{-1/2}$ is a $p_i \times p_i$ variance-covariance matrix that depends on an unknown $d_w \times 1$ vector of parameters, $\boldsymbol{\theta}_w$ and possibly on $\boldsymbol{\beta}$ and $\boldsymbol{b}_i$ as defined in (5.47) above.

As was done for the NLM in §4.4.1, we can write this model in matrix form as follows:

$$\begin{aligned}\boldsymbol{y}_i &= f(\boldsymbol{X}_i^*, \boldsymbol{A}_i, \boldsymbol{\beta}, \boldsymbol{b}_i) + \boldsymbol{W}_i(\boldsymbol{X}_i, \boldsymbol{\beta}, \boldsymbol{b}_i)^{-1/2}\boldsymbol{\epsilon}_i \\ &= f(\boldsymbol{X}_i, \boldsymbol{\beta}, \boldsymbol{b}_i) + \boldsymbol{W}_i(\boldsymbol{X}_i, \boldsymbol{\beta}, \boldsymbol{b}_i)^{-1/2}\boldsymbol{\epsilon}_i,\end{aligned} \tag{5.51}$$

where

$\boldsymbol{y}_i$ is the $p_i \times 1$ response vector, $\boldsymbol{y}_i = (y_{i1} \cdots y_{ip_i})'$
$\boldsymbol{X}_i^* = \begin{pmatrix} \boldsymbol{x}_{i1}^{*\prime} \\ \vdots \\ \boldsymbol{x}_{ip_i}^{*\prime} \end{pmatrix}$ is a $p_i \times t$ within-subject design matrix formed from the row vectors, $\boldsymbol{x}_{ij}^{*\prime}$
and $\boldsymbol{A}_i = \boldsymbol{1}_{p_i \times 1} \otimes \boldsymbol{a}_i'$ is a $p_i \times q$ between-subject design matrix formed from the row vector $\boldsymbol{a}_i'$
$\boldsymbol{X}_i = \begin{pmatrix}\boldsymbol{X}_i^* & \boldsymbol{1}_{p_i \times 1} \otimes \boldsymbol{a}_i'\end{pmatrix}_{p_i \times (t+q)}$ is a $p_i \times u$ design matrix of within- and between-subject covariates formed from the row vectors $\boldsymbol{x}_{ij}'$ $(u = t + q)$
$\boldsymbol{\epsilon}_i \sim$ind $N_{p_i}(\boldsymbol{0}, \boldsymbol{\Lambda}_i(\boldsymbol{\theta}_w))$.

The NLME model (5.51) may be expressed in terms of the GNLME model defined by (5.47) and (5.49) by setting $\boldsymbol{\mu}_i = f(\boldsymbol{X}_i, \boldsymbol{\beta}, \boldsymbol{b}_i)$, $\boldsymbol{\lambda}_i(\boldsymbol{\mu}_i, \boldsymbol{\sigma}_i) = \boldsymbol{\Lambda}_i^{-1}\boldsymbol{\mu}_i$,

$\gamma_i(\boldsymbol{\mu}_i, \boldsymbol{\sigma}_i) = -\frac{1}{2}\boldsymbol{C}_i'\boldsymbol{C}_i Vech\{\boldsymbol{\Lambda}_i^{-1}\}, \Delta_i^{-1} = (2\pi)^{-p/2}|\boldsymbol{\Lambda}_i|^{1/2}\exp\{-\frac{1}{2}\boldsymbol{\mu}_i'\boldsymbol{\Lambda}_i^{-1/2}\boldsymbol{\mu}_i\}$, $\boldsymbol{\Lambda}_i = \boldsymbol{\Lambda}_i(\boldsymbol{\beta}, \boldsymbol{b}_i, \boldsymbol{\theta}_w)$, $c_i(\boldsymbol{y}_i) = 0$, and $\boldsymbol{C}_i$ is a matrix of 0's and 1's as defined for the GNLM.

Nonlinear mixed-effects models for normally distributed data play a key role in the analysis of pharmacokinetic (PK) data (e.g., Sheiner and Beal, 1980, 1981; Beal and Sheiner, 1982; Yuh, et. al., 1994). In addition to providing a fairly comprehensive treatment of the NLME model, the book by Davidian and Giltinan (1995) provides a number of examples from the field of pharmacokinetics illustrating not only the use of NLME models for PK data but also a brief background on some of the basic terminology and methodology of population pharmacokinetics. Given the importance of NLME models to this area, we devote an entire section of Chapter 7 to examples illustrating the analysis of PK data using SAS. Other applications requiring the use of NLME models may be found in the fields of agronomy (e.g., the soybean growth data example described in §1.4 and analyzed in §5.4.2) and econometrics (e.g., Mátyás and Sevestre, 2008). An important subclass of NLME models are those nonlinear models that are linear in the random effects. Such models are considered separately in the following subsection.

Nonlinear mixed-effects models with linear random effects

Here we consider normal-theory NLME models that are nonlinear in the fixed effects regression parameters $\boldsymbol{\beta}$ but linear in the random effects $\boldsymbol{b}_i$. Following Vonesh and Carter (1992), one class of such models may be written as follows:

$$\boldsymbol{y}_i = f(\boldsymbol{X}_i, \boldsymbol{\beta}) + \boldsymbol{Z}_i(\boldsymbol{\beta})\boldsymbol{b}_i + \boldsymbol{\epsilon}_i, \quad i = 1, \ldots, n \text{ subjects (clusters)} \tag{5.52}$$

where

$\boldsymbol{X}_i$ is a $p_i \times u$ matrix of within and between-subject covariates
$\boldsymbol{Z}_i(\boldsymbol{\beta})$ is a design matrix linked to the *linear* random effects, $\boldsymbol{b}_i$, and which may depend on $\boldsymbol{\beta}$
$\boldsymbol{\beta}$ is an $s \times 1$ vector of fixed-effects parameters
$\boldsymbol{b}_i \sim$iid $N_v(\boldsymbol{0}, \boldsymbol{\Psi}(\boldsymbol{\theta}_b))$ and $\boldsymbol{\epsilon}_i \sim$ind $N_{p_i}(\boldsymbol{0}, \sigma^2\boldsymbol{I}_{p_i})$

This class of models admits both PA and SS inference with respect to $\boldsymbol{\beta}$. Moreover, as was illustrated with the orange tree example of §4.5.2, estimation may be based on the marginal GNLM with mean $f(\boldsymbol{X}_i, \boldsymbol{\beta})$, covariance matrix $\Sigma_i(\boldsymbol{\beta}, \boldsymbol{\theta}) = \boldsymbol{Z}_i(\boldsymbol{\beta})\boldsymbol{\Psi}(\boldsymbol{\theta}_b)\boldsymbol{Z}_i(\boldsymbol{\beta})' + \sigma^2\boldsymbol{I}_{p_i}$ and $\boldsymbol{\theta} = (\boldsymbol{\theta}_b, \sigma^2)$.

Model (5.52) extends split-plot and nested ANOVA models to a nonlinear regression setting (e.g., Gumpertz and Pantula, 1992). For example, consider the following split-plot ANOVA model

$$\begin{aligned} y_{ijk} &= \mu + \gamma_j + b_{i(j)} + \tau_k + (\gamma\tau)_{jk} + \epsilon_{ijk} \\ &= \mu_{jk} + b_{i(j)} + \epsilon_{ijk} \end{aligned}$$

where y_{ijk} is some measured response from the ith subject in the jth group at time k, and

$\mu_{jk} = \mu + \gamma_j + \tau_k + (\gamma\tau)_{jk}$ is the mean for group j at time k,
$\mu + \gamma_j$ is the overall mean + the added effect of the jth group,
$b_{i(j)} \sim N(0, \psi)$ is the random effect of subject i within group j,
τ_k is the kth time effect,
$(\gamma\tau)_{jk}$ is group×time interaction
$\epsilon_{ijk} \sim$iid $N(0, \sigma^2)$ are random measurement errors assumed independent of the random effects, $b_{i(j)}$.

Suppose we model the means via a response surface analysis as

$$\mu_{jk} = f(t_k, \boldsymbol{\beta}_j) = \beta_{j1}\exp\{-\beta_{j2}t_k\}$$

where $\boldsymbol{\beta}_j = (\beta_{j1}, \beta_{j2})$. Then we have a NLME model in the form of model (5.52) which we can write as

$$y_{ijk} = f(t_k, \boldsymbol{\beta}_j) + b_{i(j)} + \epsilon_{ijk}.$$

We can classify this as a split-plot nonlinear regression model in which the marginal covariance structure satisfies the assumption of compound symmetry. One can easily extend this to a class of multi-level mixed-effects nonlinear models involving multiple levels of nesting. Such models offer a natural approach to incorporating nonlinear regression into a response surface analysis suitable for a variety of experimental designs including randomized block designs, split-plot type designs, and nested factorial designs. Depending on the underlying design, such models allow for multiple sources of variation within a nonlinear regression framework without having the added complexity of dealing with nonlinear random effects. This, in turn, makes estimation and inference available at both a population-average level and a unit-specific level.

Another class of NLME models with linear random effects but one which can account for both within-subject and between-subject heterogeneity assuming a constant coefficient of variation would be the following class of NLME models for repeated measurements:

$$y_{ij} = f(\boldsymbol{x}'_{ij}, \boldsymbol{\beta})(1 + b_i + \epsilon_{ij}), \;\; i = 1, \ldots, n; \, j = 1, \ldots, p_i$$

where $b_i \sim$iid $N(0, \psi)$ independent of $\epsilon_{ij} \sim$iid $N(0, \sigma^2)$. Under this class of NLME models, the marginal means are given by $E(y_{ij}) = \mu_{ij}(\boldsymbol{\beta}) = f(\boldsymbol{x}'_{ij}, \boldsymbol{\beta})$ and the marginal variance-covariance matrix of the vector of repeated measurements, $\boldsymbol{y}_i = (y_{i1}, \ldots, y_{ip_i})'$, will satisfy the assumption of proportional compound symmetry. These models offer an attractive alternative to more complex NLME models as discussed in Vonesh and Chinchilli (1997, §8.2 and Figure 8.2.1).

5.3.3 Overcoming modeling limitations in SAS

While the class of GNLME models defined by (5.47) in possible combination with (5.48) and/or (5.49) is quite broad, not all models within this class are easily analyzed using SAS. Both GLIMMIX and NLMIXED, for example, assume the random effects are normally distributed and this would "apparently" place severe limitations on the kinds of random effect models one can analyze. For now we will continue with the assumption that the random effects are normally distributed. Later, in Chapter 7, we will return to this problem and provide a way for handling non-Gaussian random effects.

Another limitation might be the inability to specify a conditional distribution from the quadratic exponential family that meets the needs of a particular dataset. Other than the multivariate normal distribution or distributions in the linear exponential family, specification of a distribution in the quadratic exponential family may require one to iteratively calculate the canonical parameters $\boldsymbol{\lambda}_i$ and $\boldsymbol{\gamma}_i$ from which the third- and fourth-order moments of $\boldsymbol{y}_i$ would need to be determined via summation or integration in order to estimate the model parameters (Prentice and Zhao, 1991). Even then, likelihood-based inference may not be possible given that both GLIMMIX and NLMIXED restrict ML estimation to mixed models satisfying the conditional independence assumption (5.48). This would "apparently" limit one's ability to introduce an intra-subject covariance matrix, $\boldsymbol{\Lambda}_i(\boldsymbol{\theta}_w)$, into a model to account for intra-subject correlation. For example, the introduction of an R-side covariance structure that takes into account intra-subject correlation is not directly possible in GLIMMIX or NLMIXED even for a normal-theory NLME model. However, it might be possible to account for such an R-side structure indirectly by defining an additional set of orthogonal random effects as described below.

Overcoming certain limitations using linear random effects

Consider a class of nonlinear mixed-effects models where an intra-subject variance-covariance matrix, $\boldsymbol{\Lambda}_i(\boldsymbol{\theta}_w)$, can be specified based on a group of linear random effects. By combining $\boldsymbol{\Lambda}_i(\boldsymbol{\theta}_w)$ with a diagonal weight matrix, $\boldsymbol{W}_i(\boldsymbol{X}_i,\boldsymbol{\beta},\boldsymbol{b}_i)$, one can form a conditional variance-covariance matrix $\boldsymbol{\Lambda}_i(\boldsymbol{\beta},\boldsymbol{b}_i,\boldsymbol{\theta}_w)=\boldsymbol{W}_i(\boldsymbol{X}_i,\boldsymbol{\beta},\boldsymbol{b}_i)^{-1/2}\boldsymbol{\Lambda}_i(\boldsymbol{\theta}_w)\boldsymbol{W}_i(\boldsymbol{X}_i,\boldsymbol{\beta},\boldsymbol{b}_i)^{-1/2}$ such as specified under the GNLME (5.47). It might then be possible to define a joint conditional pdf having a true multivariate parametric distribution. We illustrate how one can accomplish this for special cases of $\boldsymbol{\Lambda}_i(\boldsymbol{\theta}_w)$ within the NLME model framework defined by (5.51).

Let $\boldsymbol{b}_i=\begin{pmatrix}\boldsymbol{b}_{i1}\\ \boldsymbol{b}_{i2}\end{pmatrix}$ be a $v\times 1$ partitioned vector of random effects with $v_1\times 1$ and $v_2\times 1$ component vectors $\boldsymbol{b}_{i1}$ and $\boldsymbol{b}_{i2}$, respectively, such that $\begin{pmatrix}\boldsymbol{b}_{i1}\\ \boldsymbol{b}_{i2}\end{pmatrix}\sim N\left[\begin{pmatrix}\boldsymbol{0}\\ \boldsymbol{0}\end{pmatrix},\begin{pmatrix}\boldsymbol{\Psi}_{11}(\boldsymbol{\theta}_{b_1}) & \boldsymbol{0}\\ \boldsymbol{0}' & \boldsymbol{\Psi}_{22}(\boldsymbol{\theta}_{b_2})\end{pmatrix}\right]$. Since $\boldsymbol{b}_{i1}$ and $\boldsymbol{b}_{i2}$ are orthogonal (independent) to each other, we can form a NLME model which is nonlinear in $\boldsymbol{b}_{i1}$ but strictly linear in $\boldsymbol{b}_{i2}$. This in turn can be used to induce intra-subject correlation into a NLME model using NLMIXED. Specifically, consider the class of NLME models of the form

$$\boldsymbol{y}_i=f(\boldsymbol{X}_i,\boldsymbol{\beta},\boldsymbol{b}_{i1})+\boldsymbol{W}_i(\boldsymbol{X}_i,\boldsymbol{\beta},\boldsymbol{b}_{i1})^{-1/2}\Big(\boldsymbol{Z}_i\boldsymbol{b}_{i2}+\boldsymbol{\epsilon}_i^*\Big) \tag{5.53}$$

where

$f(\boldsymbol{X}_i,\boldsymbol{\beta},\boldsymbol{b}_{i1})$ is a general nonlinear function of $\boldsymbol{\beta}$ and $\boldsymbol{b}_{i1}$,
$\boldsymbol{W}_i(\boldsymbol{X}_i,\boldsymbol{\beta},\boldsymbol{b}_{i1})=Diag\{w_{ij}\}$ is some $p_i\times p_i$ diagonal weight matrix having weights w_{ij} that may depend on $\boldsymbol{X}_i$ as well as $\boldsymbol{\beta}$ and $\boldsymbol{b}_{i1}$,
$\boldsymbol{Z}_i$ is a $p_i\times v_2$ matrix of fixed known constants
$\boldsymbol{\epsilon}_i^*$ ~iid $N(\boldsymbol{0},\sigma_w^2\boldsymbol{I}_{p_i})$.

We can rewrite this in terms of the NLME model (5.51) as

$$\boldsymbol{y}_i=f(\boldsymbol{X}_i,\boldsymbol{\beta},\boldsymbol{b}_{i1})+\boldsymbol{W}_i(\boldsymbol{X}_i,\boldsymbol{\beta},\boldsymbol{b}_{i1})^{-1/2}\boldsymbol{\epsilon}_i$$

where $\boldsymbol{\epsilon}_i=\boldsymbol{Z}_i\boldsymbol{b}_{i2}+\boldsymbol{\epsilon}_i^*$ is an "intra-subject" error term assumed to be independent of $\boldsymbol{b}_{i1}$. Under this formulation, we have

$$\begin{aligned}Var(\boldsymbol{y}_i|\boldsymbol{b}_{i1})&=\boldsymbol{\Lambda}_i(\boldsymbol{\beta},\boldsymbol{b}_{i1},\boldsymbol{\theta}_w)\\ &=\boldsymbol{W}_i(\boldsymbol{X}_i,\boldsymbol{\beta},\boldsymbol{b}_{i1})^{-1/2}\boldsymbol{\Lambda}_i(\boldsymbol{\theta}_w)\boldsymbol{W}_i(\boldsymbol{X}_i,\boldsymbol{\beta},\boldsymbol{b}_{i1})^{-1/2}\end{aligned}$$

where

$$\begin{aligned}Var(\boldsymbol{\epsilon}_i)&=\boldsymbol{\Lambda}_i(\boldsymbol{\theta}_w)\\ &=\boldsymbol{Z}_i\boldsymbol{\Psi}_{22}(\boldsymbol{\theta}_{b_2})\boldsymbol{Z}_i'+\sigma_w^2\boldsymbol{I}_{p_i}\end{aligned}$$

and $\boldsymbol{\theta}_w=(\boldsymbol{\theta}_{b_2},\sigma_w^2)$ represents a vector of "within-subject" covariance parameters. For example, if we set $\boldsymbol{Z}_i=\boldsymbol{1}_{p_i}$ and $\boldsymbol{\Psi}_{22}(\boldsymbol{\theta}_{b_2})=\sigma_{b_2}^2$ then the intra-subject covariance matrix $\boldsymbol{\Lambda}_i(\boldsymbol{\theta}_w)$ satisfies the assumption of compound symmetry in that $\boldsymbol{\Lambda}_i(\boldsymbol{\theta}_w)=\sigma^2(1-\rho)\boldsymbol{I}_{p_i}+\sigma^2\rho\boldsymbol{J}_{p_i}$ with $\sigma^2=(\sigma_w^2+\sigma_{b_2}^2)$ and $\rho=\sigma_{b_2}^2/(\sigma_w^2+\sigma_{b_2}^2)$. Depending on the data, other choices are also possible. For example, if one has balanced data with $p_i=p$, one could set $\boldsymbol{Z}_i=\boldsymbol{I}_p$ and define $\boldsymbol{\Psi}_{22}(\boldsymbol{\theta}_{b_2})$ to be any structured $p\times p$ matrix. One would then have to set σ_w^2 to some arbitrarily small value, say $\sigma_w^2=\delta\le$ 1E-8 so that $\boldsymbol{\Psi}_{22}(\boldsymbol{\theta}_{b_2})+\delta\boldsymbol{I}_p=\boldsymbol{\Lambda}(\boldsymbol{\theta}_w)$ is the desired $p\times p$ intra-subject variance-covariance matrix with $\boldsymbol{\theta}_{b_2}=\boldsymbol{\theta}_w$. Examples of this approach are given in §5.4.1 and §5.4.3. Depending on the assumed covariance structure (e.g., compound symmetry), one might be able to extend this to unbalanced data where $p_i \leq p$ for all $i = 1, \ldots, n$ by simply forming $\boldsymbol{Z}_i$ from those rows of $\boldsymbol{I}_p$ that correspond to the time points at which measurements are obtained for the ith subject.

We note that while model (5.53) satisfies the conditional independence assumption (5.48) required to run NLMIXED, this approach of constructing a set of linear random effects that are orthogonal to nonlinear random effects enables us to model intra-subject correlation in NLMIXED using ML estimation. Now rather than assume $\boldsymbol{\epsilon}_i \sim N(\boldsymbol{0}, \boldsymbol{\Lambda}_i(\boldsymbol{\theta}_w))$, suppose $\boldsymbol{\epsilon}_i$ has finite first- through fourth-order moments and that the first two moments are $E(\boldsymbol{\epsilon}_i) = \boldsymbol{0}$ and $Var(\boldsymbol{\epsilon}_i) = \boldsymbol{\Lambda}_i(\boldsymbol{\theta}_w)$, respectively. Under these broader assumptions, model (5.53) reduces to the semiparametric GNLME model (5.47). In this case, estimation may be carried out by solving a set of conditional second-order generalized estimating equations (CGEE2) under a "working" normality assumption as will be discussed in the following section on estimation.

5.3.4 Estimation

There are two procedures and a macro available in SAS for estimating the parameters of a GNLME model. These include methods that iteratively solve either a set of conditional first-order generalized estimating equations (CGEE1) or a set of conditional second-order generalized estimating equations (CGEE2). The former is a simple extension of the PL/CGEE1 methods described for the GLME model while the latter involves solving a set of extended estimating equations that take into account any additional information regarding $\boldsymbol{\beta}$ and $\boldsymbol{b}_i$ that may occur in the second-order conditional moments specified by $\boldsymbol{\Lambda}_i(\boldsymbol{\beta}, \boldsymbol{b}_i, \boldsymbol{\theta}_w)$. We also consider maximum likelihood estimation techniques that rely on numerical integration to maximize an integrated log-likelihood. Of course these techniques all apply to estimating the parameters from NLME models as well. Moreover, as we shall see, all of these estimation techniques are, for the most part, equivalent to the techniques used to estimate parameters from the class of GLME models.

We start first with a description of the pseudo-likelihood (PL) estimation techniques available with the SAS macro %NLINMIX. As with GLIMMIX, these are linearization-based methods that rely on a first-order Taylor expansion of the conditional moments about either current estimates of the individual random effects (i.e., a SS expansion about $\boldsymbol{b}_i = \widehat{\boldsymbol{b}}_i$) or about the mean random effect (i.e., a PA expansion about $\boldsymbol{b}_i = \boldsymbol{0}$). Both estimation techniques are semiparametric in that they do not require any distributional assumptions beyond that of the moments-based GNLME model (5.47) although they may also be used assuming either of the parametric models (5.48) or (5.49). We also consider a first-order approximation technique available in NLMIXED that is closely related to the first-order PA expansion approach within %NLINMIX. Following this, we examine the parametric ML estimation techniques available in NLMIXED. These techniques all maximize an integrated log-likelihood function but differ with respect to the numerical methods used to evaluate the integrated log-likelihood. Finally, we consider estimation based on solving a modified set of conditional second-order generalized estimating equations (CGEE2). This approach is a modification of the CGEE2 approach described by Vonesh et. al. (2002) but differs in that it entails using a singly iterative algorithm which one can easily implement in either GLIMMIX or NLMIXED using Laplace's approximation. The modified set of CGEE2 are the estimating equations required to minimize an augmented penalized least squares objective function suitable for GNLME models where one need only specify the conditional mean and variance of $\boldsymbol{y}_i$ rather than a conditional pdf.

Other estimation techniques which we will not present include various two-stage estimation techniques as well as nonparametric and smooth nonparametric techniques. Two-stage estimation involves special cases of the NLME model and an extensive overview of these methods can be found in Davidian and Giltinan (Chapter 5, 1995), Vonesh and

Chinchilli (Chapter 7, §7.4.1, 1997) and Demidenko (Chapter 8, §8.5, 2004). Nonparametric (Mallet, 1986; Mallet et. al., 1988) and smooth nonparametric (Davidian and Gallant, 1992, 1993) estimation are aimed at relaxing the assumption of normality with respect to the random effects of a NLME model. Likewise, we forego discussion of Bayesian inference under a full hierarchical NLME model in which one introduces a hyperprior distribution on the fixed-effects parameters of $\boldsymbol{\beta}$ and $\boldsymbol{\theta} = (\boldsymbol{\theta}_b, \boldsymbol{\theta}_w)$. Material related to Bayesian inference for the hierarchical NLME model may be found in Wakefield et. al. (1994) as well as the book by Davidian and Giltinan (1995) and references therein.

Methods based on linearization

Linearization-based methods are available in both %NLINMIX and PROC NLMIXED and we consider each separately. With some minor modifications, the SAS macro %NLINMIX implements an estimation scheme for GNLME models based on the same pseudo-likelihood (PL) approach used for the GLME model (see page 358). This scheme follows that of Lindstrom and Bates (1990), Breslow and Clayton (1993), Wolfinger (1993), Wolfinger and O'Connell (1993) and Wolfinger and Lin (1997) and is applicable to the entire class of GNLME models considered here. It entails iteratively solving a set of linear-mixed model equations corresponding to a pseudo-LME model formed from the residuals of a first-order Taylor series expansion of the conditional mean, $\boldsymbol{\mu}_i(\boldsymbol{\beta}, \boldsymbol{b}_i) = f(\boldsymbol{X}_i, \boldsymbol{\beta}, \boldsymbol{b}_i)$, about values of $\boldsymbol{\beta}$ and $\boldsymbol{b}_i$. Two different Taylor series expansions are available in %NLINMIX, a first-order subject-specific expansion and a first-order population-averaged (marginal or mean) expansion. We consider each separately.

First-order subject-specific expansion:

Recall from §5.1.1 that a pseudo-likelihood (PL) estimation procedure based on linearization may be implemented by expanding the conditional means, $\boldsymbol{\mu}_i(\boldsymbol{\beta}, \boldsymbol{b}_i)$, about current estimates $\tilde{\boldsymbol{\beta}}$ of $\boldsymbol{\beta}$ and $\tilde{\boldsymbol{b}}_i$ of $\boldsymbol{b}_i$ and forming a LME model from pseudo data. Based on this pseudo-LME model, one can then estimate the variance-covariance parameters, $\boldsymbol{\theta} = (\boldsymbol{\theta}_b, \boldsymbol{\theta}_w)$, using normal-theory ML or REML methods and then update estimates of $\boldsymbol{\beta}$ and $\boldsymbol{b} = (\boldsymbol{b}_1', \ldots, \boldsymbol{b}_n')'$ using a one-step Gauss-Newton solution to a set of conditional first-order generalized estimating equations (CGEE1). This one-step Gauss-Newton solution is, in turn, the solution to the mixed-model equations of the pseuodo-LME model. Since the assumption of normality is simply a "working" assumption, we continue to refer to this as PL estimation in that normal-theory likelihood is used solely as a means for obtaining a pseudo-likelihood (PL) estimate $\widehat{\boldsymbol{\theta}}_{\mathrm{PL}}$ or restricted pseudo-likelihood (REPL) estimate $\widehat{\boldsymbol{\theta}}_{\mathrm{REPL}}$ of $\boldsymbol{\theta}$. In either case, overall estimation is based on a doubly iterative scheme which entails repeated application of Taylor series linearization and weighted linear mixed-effects regression. It entails first expressing the conditional moments (5.47) in terms of the following generalized nonlinear mixed-effects regression model

$$\boldsymbol{y}_i = \boldsymbol{\mu}_i(\boldsymbol{\beta}, \boldsymbol{b}_i) + \boldsymbol{W}_i(\boldsymbol{X}_i, \boldsymbol{\beta}, \boldsymbol{b}_i)^{-1/2}\boldsymbol{\epsilon}_i \tag{5.54}$$

where $E(\boldsymbol{\epsilon}_i) = \boldsymbol{0}$ and $Var(\boldsymbol{\epsilon}_i) = \boldsymbol{\Lambda}_i(\boldsymbol{\theta}_w)$. We note that while no distributional assumptions regarding $\boldsymbol{\epsilon}_i$ are made, this model will belong to the class of NLME models (5.51) assuming normality. Under model (5.54), estimates of $\boldsymbol{\tau} = (\boldsymbol{\beta}, \boldsymbol{\theta})$ and $\boldsymbol{b}' = (\boldsymbol{b}_1', \ldots, \boldsymbol{b}_n')$ may be obtained by applying the following modified Gauss-Newton type algorithm.

A Gauss-Newton algorithm for estimating $(\boldsymbol{\beta}, \boldsymbol{b}, \boldsymbol{\theta})$ via PL/CGEE1

Step 0: Obtain an initial estimate $\tilde{\boldsymbol{\beta}}$ of $\boldsymbol{\beta}$ using, for example, nonlinear OLS or by using some form of grid search. Likewise, initially set $\tilde{\boldsymbol{b}}_i = \boldsymbol{0}$ for all $i = 1, \ldots, n$. Optionally, define a set of starting values for $\boldsymbol{\theta}$.

Step 1: Given a current value $\widehat{\boldsymbol{\theta}}$ of $\boldsymbol{\theta}$, let $\tilde{\boldsymbol{\beta}} = \tilde{\boldsymbol{\beta}}(\widehat{\boldsymbol{\theta}})$ and $\tilde{\boldsymbol{b}}_i = \tilde{\boldsymbol{b}}_i(\widehat{\boldsymbol{\theta}})$ be estimates of $\boldsymbol{\beta}$ and $\boldsymbol{b}_i$ (initially, $\tilde{\boldsymbol{\beta}}$ and $\tilde{\boldsymbol{b}}_i$ are from Step 0 above). Perform a first-order SS Taylor series expansion of $\boldsymbol{\mu}_i(\boldsymbol{\beta}, \boldsymbol{b}_i) + \boldsymbol{W}_i(\boldsymbol{X}_i, \boldsymbol{\beta}, \boldsymbol{b}_i)^{-1/2}\boldsymbol{\epsilon}_i$ about $\boldsymbol{\beta} = \tilde{\boldsymbol{\beta}}, \boldsymbol{b}_i = \tilde{\boldsymbol{b}}_i$ and $\boldsymbol{\epsilon}_i = \boldsymbol{0}$ yielding

$$\boldsymbol{y}_i = \boldsymbol{\mu}_i(\tilde{\boldsymbol{\beta}}, \tilde{\boldsymbol{b}}_i) + \tilde{\boldsymbol{X}}_i(\boldsymbol{\beta} - \tilde{\boldsymbol{\beta}}) + \tilde{\boldsymbol{Z}}_i(\boldsymbol{b}_i - \tilde{\boldsymbol{b}}_i) + \boldsymbol{W}_i(\boldsymbol{X}_i, \tilde{\boldsymbol{\beta}}, \tilde{\boldsymbol{b}}_i)^{-1/2}\boldsymbol{\epsilon}_i \tag{5.55}$$

where

$$\boldsymbol{\mu}_i(\boldsymbol{\beta}, \boldsymbol{b}_i) = \boldsymbol{\mu}_i(\tilde{\boldsymbol{\beta}}, \tilde{\boldsymbol{b}}_i) + \tilde{\boldsymbol{X}}_i(\boldsymbol{\beta} - \tilde{\boldsymbol{\beta}}) + \tilde{\boldsymbol{Z}}_i(\boldsymbol{b}_i - \tilde{\boldsymbol{b}}_i)$$

$$\tilde{\boldsymbol{X}}_i = \frac{\partial \boldsymbol{\mu}_i(\boldsymbol{\beta}, \boldsymbol{b}_i)}{\partial \boldsymbol{\beta}'}\bigg|_{\boldsymbol{\beta}=\tilde{\boldsymbol{\beta}}, \boldsymbol{b}_i=\tilde{\boldsymbol{b}}_i} = \frac{\partial f(\boldsymbol{X}_i, \boldsymbol{\beta}, \boldsymbol{b}_i)}{\partial \boldsymbol{\beta}'}\bigg|_{\boldsymbol{\beta}=\tilde{\boldsymbol{\beta}}, \boldsymbol{b}_i=\tilde{\boldsymbol{b}}_i}$$

$$\tilde{\boldsymbol{Z}}_i = \frac{\partial \boldsymbol{\mu}_i(\boldsymbol{\beta}, \boldsymbol{b}_i)}{\partial \boldsymbol{b}_i'}\bigg|_{\boldsymbol{\beta}=\tilde{\boldsymbol{\beta}}, \boldsymbol{b}_i=\tilde{\boldsymbol{b}}_i} = \frac{\partial f(\boldsymbol{X}_i, \boldsymbol{\beta}, \boldsymbol{b}_i)}{\partial \boldsymbol{b}_i'}\bigg|_{\boldsymbol{\beta}=\tilde{\boldsymbol{\beta}}, \boldsymbol{b}_i=\tilde{\boldsymbol{b}}_i}$$

From this, compute the pseudo-response vector,

$$\tilde{\boldsymbol{y}}_i = (\boldsymbol{y}_i - \boldsymbol{\mu}_i(\tilde{\boldsymbol{\beta}}, \tilde{\boldsymbol{b}}_i)) + \tilde{\boldsymbol{X}}_i\tilde{\boldsymbol{\beta}} + \tilde{\boldsymbol{Z}}_i\tilde{\boldsymbol{b}}_i$$

and note that the weight matrix $\tilde{\boldsymbol{W}}_i = \boldsymbol{W}_i(\boldsymbol{X}_i, \tilde{\boldsymbol{\beta}}, \tilde{\boldsymbol{b}}_i)$ is automatically updated at values of $\tilde{\boldsymbol{\beta}}$ and $\tilde{\boldsymbol{b}}_i$ when we include in our expansion the term $\boldsymbol{W}_i(\boldsymbol{X}_i, \boldsymbol{\beta}, \boldsymbol{b}_i)^{-1/2}\boldsymbol{\epsilon}_i$ expanded about $\boldsymbol{\beta} = \tilde{\boldsymbol{\beta}}, \boldsymbol{b}_i = \tilde{\boldsymbol{b}}_i$ **and** $\boldsymbol{\epsilon}_i = \boldsymbol{0}$ (this same argument can also be applied to formally justify updating the weight matrix in the PL/CGEE1 algorithm used for GLME models).

Step 2. Form the pseudo-weighted LME model

$$\tilde{\boldsymbol{y}}_i = \tilde{\boldsymbol{X}}_i\boldsymbol{\beta} + \tilde{\boldsymbol{Z}}_i\boldsymbol{b}_i + \tilde{\boldsymbol{W}}_i^{-1/2}\boldsymbol{\epsilon}_i \tag{5.56}$$

together with the "working" assumption that $\boldsymbol{\epsilon}_i \sim N(\boldsymbol{0}, \boldsymbol{\Lambda}_i(\boldsymbol{\theta}_w))$. Under (5.56), the marginal variance-covariance of $\tilde{\boldsymbol{y}}_i$ is

$$\tilde{\boldsymbol{\Sigma}}_i(\boldsymbol{\theta}) = \tilde{\boldsymbol{Z}}_i\boldsymbol{\Psi}(\boldsymbol{\theta}_b)\tilde{\boldsymbol{Z}}_i' + \tilde{\boldsymbol{W}}_i^{-1/2}\boldsymbol{\Lambda}_i(\boldsymbol{\theta}_w)\tilde{\boldsymbol{W}}_i^{-1/2}$$

which depends on $\tilde{\boldsymbol{\beta}}$ and $\tilde{\boldsymbol{b}}_i$ (and hence the current value of $\boldsymbol{\theta}$) through the derivative matrix $\tilde{\boldsymbol{Z}}_i$ and weight matrix $\tilde{\boldsymbol{W}}_i$ both of which we treat as fixed known matrices. Using normal-theory likelihood estimation, obtain either a PL estimate, $\widehat{\boldsymbol{\theta}}_{\mathrm{PL}}$, or REPL estimate, $\widehat{\boldsymbol{\theta}}_{\mathrm{REPL}}$, of $\boldsymbol{\theta} = (\boldsymbol{\theta}_b, \boldsymbol{\theta}_w)$ by maximizing either the profile log-likelihood function 2.12 or the restricted profile log-likelihood function 2.13 evaluated at the pseudo-residual vector, $\tilde{\boldsymbol{r}}_i = \tilde{\boldsymbol{y}}_i - \tilde{\boldsymbol{X}}_i\widehat{\boldsymbol{\beta}}(\boldsymbol{\theta})$. Letting $\widehat{\boldsymbol{\theta}}$ be either the updated PL or REPL estimator of $\boldsymbol{\theta}$, proceed to update the profiled regression parameters $\boldsymbol{\beta}$ and the random effects $\boldsymbol{b}$ based on the one-step Gauss-Newton estimates

$$\widehat{\boldsymbol{\beta}} = \widehat{\boldsymbol{\beta}}(\widehat{\boldsymbol{\theta}}) = \Big(\sum_{i=1}^{n} \tilde{\boldsymbol{X}}_i'\tilde{\boldsymbol{\Sigma}}_i(\widehat{\boldsymbol{\theta}})^{-1}\tilde{\boldsymbol{X}}_i\Big)^{-1} \sum_{i=1}^{n} \tilde{\boldsymbol{X}}_i'\tilde{\boldsymbol{\Sigma}}_i(\widehat{\boldsymbol{\theta}})^{-1}\tilde{\boldsymbol{y}}_i \tag{5.57}$$

and

$$\widehat{\boldsymbol{b}}_i = \widehat{\boldsymbol{b}}_i(\widehat{\boldsymbol{\theta}}) = \boldsymbol{\Psi}(\widehat{\boldsymbol{\theta}}_b)\tilde{\boldsymbol{Z}}_i'\tilde{\boldsymbol{\Sigma}}_i(\widehat{\boldsymbol{\theta}})^{-1}(\tilde{\boldsymbol{y}}_i - \tilde{\boldsymbol{X}}_i\widehat{\boldsymbol{\beta}}(\widehat{\boldsymbol{\theta}})); \; i = 1, \ldots, n \tag{5.58}$$

where $\tilde{\boldsymbol{\Sigma}}_i(\widehat{\boldsymbol{\theta}}) = \tilde{\boldsymbol{Z}}_i\boldsymbol{\Psi}(\widehat{\boldsymbol{\theta}}_b)\tilde{\boldsymbol{Z}}_i' + \tilde{\boldsymbol{W}}_i^{-1/2}\boldsymbol{\Lambda}_i(\widehat{\boldsymbol{\theta}}_w)\tilde{\boldsymbol{W}}_i^{-1/2}$. The estimates (5.57) and (5.58) are solutions to the LME model equations coinciding with the pseudo-weighted LME model (5.56) and as such correspond to one-step Gauss-Newton estimates associated with a set of CGEE1 for $\boldsymbol{\beta}$ and $\boldsymbol{b}$. Also $\widehat{\boldsymbol{\beta}}(\widehat{\boldsymbol{\theta}})$ is the EGLS estimator of $\boldsymbol{\beta}$ and $\widehat{\boldsymbol{b}}_i(\widehat{\boldsymbol{\theta}})$ is the empirical best linear unbiased predictor (EBLUP) of $\boldsymbol{b}_i$ under (5.56) given the current PL estimator of $\boldsymbol{\theta}$. Continue to iterate between Steps 1-2 by updating $\tilde{\boldsymbol{\beta}} = \widehat{\boldsymbol{\beta}}(\widehat{\boldsymbol{\theta}})$ and $\tilde{\boldsymbol{b}}_i = \widehat{\boldsymbol{b}}_i(\widehat{\boldsymbol{\theta}})$ in Step 1 and repeating the iterative procedure in Step 2 to obtain updated joint estimates of $(\boldsymbol{\beta}, \boldsymbol{\theta})$ and EBLUP's of $\boldsymbol{b}_i$. This process is repeated until such time that successive differences in updated estimates of $\boldsymbol{\theta}$ are negligible.

This algorithm can be shown to be identical to PL/CGEE1 algorithm for the GLME model (page 210) and is the algorithm implemented whenever one specifies the argument, EXPAND=EBLUP, within a macro call to %NLINMIX. To obtain a PL or REPL estimate of $\boldsymbol{\theta}$ one need specify either METHOD=ML or METHOD=REML, respectively, as one of the options to the macro argument PROCOPT of %NLINMIX (see the %NLINMIX syntax for a complete description of macro arguments and options). The CGEE1 estimators of $\boldsymbol{\beta}$ and $\boldsymbol{b}$ combined with either the PL or REPL estimator of $\boldsymbol{\theta}$ are based on a subject-specific Taylor series expansion of the conditional means, $\boldsymbol{\mu}_i(\boldsymbol{\beta}, \boldsymbol{b}_i)$, about current estimates $\tilde{\boldsymbol{\beta}}$ and $\tilde{\boldsymbol{b}}_i$, respectively. For fixed $\boldsymbol{\theta} = (\boldsymbol{\theta}_b, \boldsymbol{\theta}_w)$, the estimates of $\boldsymbol{\beta}$ and $\boldsymbol{b}_i$ are those solutions $\widehat{\boldsymbol{\beta}}_{\text{CGEE1}} = \widehat{\boldsymbol{\beta}}(\boldsymbol{\theta})$ and $\widehat{\boldsymbol{b}}_i(\boldsymbol{\theta})$ to the set of first-order conditional generalized estimating equations (CGEE1) given in (5.9) but with $\tilde{\boldsymbol{X}}_i = \tilde{\boldsymbol{X}}_i(\boldsymbol{\beta}, \boldsymbol{b}_i) = \partial f(\boldsymbol{X}_i, \boldsymbol{\beta}, \boldsymbol{b}_i)/\partial \boldsymbol{\beta}'$ and $\tilde{\boldsymbol{Z}}_i = \tilde{\boldsymbol{Z}}_i(\boldsymbol{\beta}, \boldsymbol{b}_i) = \partial f(\boldsymbol{X}_i, \boldsymbol{\beta}, \boldsymbol{b}_i)/\partial \boldsymbol{b}_i'$. We classify these as CGEE1 estimates in that no additional information beyond what is contained in the mean $\boldsymbol{\mu}_i(\boldsymbol{\beta}, \boldsymbol{b}_i)$ are used in estimating $\boldsymbol{\beta}$ and $\boldsymbol{b}$ (i.e., any additional information about $\boldsymbol{\beta}$ and $\boldsymbol{b}_i$ that may be contained in the weight matrix $\boldsymbol{W}_i(\boldsymbol{X}_i, \boldsymbol{\beta}, \boldsymbol{b}_i)$ is ignored). Finally, we note that for GLME models, this algorithm yields the same PQL estimates as would be obtained using GLIMMIX.

First-order population-averaged (marginal) expansion:
When one specifies the %NLINMIX option EXPAND=ZERO, a pseudo-likelihood (PL) estimation procedure based on a first-order PA expansion is implemented. In this case, we apply the same PL/CGEE1 algorithm as described above but with the expansion of $\boldsymbol{\mu}_i(\boldsymbol{\beta}, \boldsymbol{b}_i) + \boldsymbol{W}_i(\boldsymbol{X}_i, \boldsymbol{\beta}, \boldsymbol{b}_i)^{-1/2}\boldsymbol{\epsilon}_i$ carried about $\boldsymbol{\beta} = \tilde{\boldsymbol{\beta}}, \boldsymbol{b}_i = \boldsymbol{0}$ and $\boldsymbol{\epsilon}_i = \boldsymbol{0}$. In this case, the above PL/CGEE1 algorithm will be equivalent to the PL/QELS algorithm (see page 159) applied to a GNLM with

$$\begin{aligned} E(\boldsymbol{y}_i) &= \boldsymbol{\mu}_i(\boldsymbol{\beta}) = f(\boldsymbol{X}_i, \boldsymbol{\beta}, \boldsymbol{0}) \\ Var(\boldsymbol{y}_i) &= \boldsymbol{\Sigma}_i(\boldsymbol{\beta}, \boldsymbol{\theta}) \\ &= \boldsymbol{Z}_i(\boldsymbol{\beta})\boldsymbol{\Psi}(\boldsymbol{\theta}_b)\boldsymbol{Z}_i(\boldsymbol{\beta})' + \boldsymbol{W}_i(\boldsymbol{\beta})^{-1/2}\boldsymbol{\Lambda}_i(\boldsymbol{\theta}_w)\boldsymbol{W}_i(\boldsymbol{\beta})^{-1/2} \end{aligned} \tag{5.59}$$

where $\boldsymbol{Z}_i(\boldsymbol{\beta}) = \tilde{\boldsymbol{Z}}_i(\boldsymbol{\beta}, \boldsymbol{0}) = \partial f(\boldsymbol{X}_i, \boldsymbol{\beta}, \boldsymbol{b}_i)/\partial \boldsymbol{b}_i' \Big|_{\boldsymbol{\beta}, \boldsymbol{b}_i = \boldsymbol{0}}$ and $\boldsymbol{W}_i(\boldsymbol{\beta})$ is the weight matrix $\boldsymbol{W}_i(\boldsymbol{X}_i, \boldsymbol{\beta}, \boldsymbol{0})^{-1/2}$. For GLME models, this algorithm yields MQL estimates of the model parameters and is equivalent to the PL algorithm implemented in GLIMMIX using METHOD=RMPL or METHOD=MMPL.

An alternative first-order marginal approximation is available with PROC NLMIXED using the option METHOD=FIRO. This approach, developed by Beal and Sheiner (1982, 1988), is for normal-theory NLME models only (i.e., one must specify the NORMAL distribution in the MODEL statement of NLMIXED). Under this approach, $\boldsymbol{\beta}$ is held fixed and a one-time first-order Taylor series approximation to the NLME model is obtained by expanding $\boldsymbol{\mu}_i(\boldsymbol{\beta}, \boldsymbol{b}_i) + \boldsymbol{W}_i(\boldsymbol{X}_i, \boldsymbol{\beta}, \boldsymbol{b}_i)^{-1/2}\boldsymbol{\epsilon}_i$ about $\boldsymbol{b}_i = \boldsymbol{0}$ and $\boldsymbol{\epsilon}_i = \boldsymbol{0}$. Estimation is then based on maximizing the log-likelihood of a normal-theory NLM having mean and variance-covariance defined by (5.59) with derivative matrix $\boldsymbol{Z}_i(\boldsymbol{\beta}) = \tilde{\boldsymbol{Z}}_i(\boldsymbol{\beta}, \boldsymbol{0}) = \partial f(\boldsymbol{X}_i, \boldsymbol{\beta}, \boldsymbol{b}_i)/\partial \boldsymbol{b}_i' \Big|_{\boldsymbol{\beta}, \boldsymbol{b}_i = \boldsymbol{0}}$ and weight matrix $\boldsymbol{W}_i(\boldsymbol{X}_i, \boldsymbol{\beta}, \boldsymbol{0})$. While the vector parameter $\boldsymbol{\tau} = (\boldsymbol{\beta}, \boldsymbol{\theta})$ is estimated by maximizing a Gaussian-based log-likelihood function, one can also view the resulting estimate of $\boldsymbol{\tau}$ as minimizing an extended least squares (ELS) objective function based solely on the assumption that the marginal mean and variance-covariance of $\boldsymbol{y}_i$ may be approximated by (5.59). As discussed in Chapter 4 (§4.4.2), minimizing the ELS objective function (4.42) corresponds to solving a set of GEE2 under the "working" assumption of Gaussian-based 1st-4th order moments. The resulting estimate of $\boldsymbol{\tau}$ can be shown to be consistent and asymptotically normally distributed provided the approximate moments in (5.59) actually correspond to the true marginal mean and variance-covariance of $\boldsymbol{y}_i$. If true, then the assumption of normality may be relaxed by assuming the pdf of $\boldsymbol{y}_i$ is from the broader quadratic exponential family. However, the degree of bias associated with this first-order approximation depends on the magnitude of $\boldsymbol{\Psi}$. Solomon and Cox (1992), for example, show that for a single random effect, the degree of bias is $O(\psi^2)$ suggesting that there is minimal bias as $\psi \to 0$. Indeed, for "small" $\boldsymbol{\Psi}$, the random effects will all be concentrated around $\mathbf{0}$ in which case this first-order approximation will yield minimally biased estimates of $\boldsymbol{\tau} = (\boldsymbol{\beta}, \boldsymbol{\theta})$.

Methods based on numerical integration

With both GLIMMIX and NLMIXED, maximum likelihood estimation is achieved by maximizing the integrated log-likelihood function

$$L(\boldsymbol{\beta},\boldsymbol{\theta};\boldsymbol{y}) = \sum_{i=1}^{n} \log \int \pi(\boldsymbol{y}_i;\boldsymbol{\beta},\boldsymbol{\theta}_w|\boldsymbol{b}_i)\pi(\boldsymbol{b}_i;\boldsymbol{\theta}_b)d\boldsymbol{b}_i \tag{5.60}$$

$$= \sum_{i=1}^{n} \log \int \exp\Big[\sum_{j=1}^{p_i} l_{ij}(\boldsymbol{\beta},\boldsymbol{b}_i,\boldsymbol{\theta}_w;y_{ij}) - \frac{v}{2}\log(2\pi)$$

$$- \frac{1}{2}\log|\boldsymbol{\Psi}(\boldsymbol{\theta}_b)| - \frac{1}{2}\boldsymbol{b}_i'\boldsymbol{\Psi}(\boldsymbol{\theta}_b)^{-1}\boldsymbol{b}_i\Big]d\boldsymbol{b}_i$$

where $L(\boldsymbol{\beta},\boldsymbol{\theta};\boldsymbol{y})$ is the integrated log-likelihood function, $\pi(\boldsymbol{y}_i;\boldsymbol{\beta},\boldsymbol{\theta}_w|\boldsymbol{b}_i) = \exp\Big[\log\Big\{\prod_{j=1}^{p_i}\pi(y_{ij};\boldsymbol{\beta},\boldsymbol{\theta}_w|\boldsymbol{b}_i)\Big\}\Big] = \exp\Big[\sum_{j=1}^{p_i} l_{ij}(\boldsymbol{\beta},\boldsymbol{b}_i,\boldsymbol{\theta}_w;y_{ij})\Big]$ is the pdf of $\boldsymbol{y}_i|\boldsymbol{b}_i$ assuming independence and $l_{ij}(\boldsymbol{\beta},\boldsymbol{b}_i,\boldsymbol{\theta}_w;y_{ij}) = \log\{\pi(y_{ij};\boldsymbol{\beta},\boldsymbol{\theta}_w|\boldsymbol{b}_i)\}$.

To maximize the log-likelihood objective function (5.60), one must first evaluate the marginal pdf of $\boldsymbol{y}_i$ be evaluating the integral

$$\pi(\boldsymbol{y}_i;\boldsymbol{\beta},\boldsymbol{\theta}) = \int \pi(\boldsymbol{y}_i;\boldsymbol{\beta},\boldsymbol{\theta}_w|\boldsymbol{b}_i)\pi(\boldsymbol{b}_i;\boldsymbol{\theta}_b)d\boldsymbol{b}_i \tag{5.61}$$

$$= \int \exp\Big[\sum_{j=1}^{p_i} l_{ij}(\boldsymbol{\beta},\boldsymbol{b}_i,\boldsymbol{\theta}_w;y_{ij})\Big]\pi(\boldsymbol{b}_i;\boldsymbol{\theta}_b)d\boldsymbol{b}_i.$$

Like GLIMMIX, the NLMIXED procedure offers two approaches for evaluating integrals of the form (5.61), a second-order Laplace approximation and Gaussian-based quadrature. As these are the same techniques described for the GLME model, we provide a brief description of each below.

Laplace approximation:

A second-order Laplace approximation to $\pi(\boldsymbol{y}_i;\boldsymbol{\beta},\boldsymbol{\theta})$ may be obtained as follows. Let $\boldsymbol{\tau} = (\boldsymbol{\beta},\boldsymbol{\theta})$ where $\boldsymbol{\theta} = (\boldsymbol{\theta}_b, \boldsymbol{\theta}_w)$ and let

$$p_i L_i(\boldsymbol{\tau},\boldsymbol{b}_i) = l_i(\boldsymbol{\beta},\boldsymbol{b}_i,\boldsymbol{\theta}_w;\boldsymbol{y}_i) - \frac{1}{2}\log|\boldsymbol{\Psi}(\boldsymbol{\theta}_b)| - \frac{1}{2}\boldsymbol{b}_i'\boldsymbol{\Psi}(\boldsymbol{\theta}_b)^{-1}\boldsymbol{b}_i \tag{5.62}$$

be the unnormalized log-posterior density of the ith subject/cluster where $l_i(\boldsymbol{\beta},\boldsymbol{b}_i,\boldsymbol{\theta}_w;\boldsymbol{y}_i) = \sum_{j=1}^{p_i} l_{ij}(\boldsymbol{\beta},\boldsymbol{b}_i,\boldsymbol{\theta}_w;y_{ij})$. Fix $\boldsymbol{\tau}$ and let $\tilde{\boldsymbol{b}}_i = \tilde{\boldsymbol{b}}_i(\boldsymbol{\tau})$ be the posterior mode obtained by maximizing (5.62) with respect to $\boldsymbol{b}_i$. By applying a multivariate Taylor series expansion of $p_i L_i(\boldsymbol{\tau}, \boldsymbol{b}_i)$ about $\tilde{\boldsymbol{b}}_i$, the marginal pdf of $\boldsymbol{y}_i$ may be expressed in terms of the second-order Laplace approximation as

$$\begin{aligned}\pi(\boldsymbol{y}_i; \boldsymbol{\beta}, \boldsymbol{\theta}) = \pi(\boldsymbol{y}_i; \boldsymbol{\tau}) &= \frac{1}{(2\pi)^{v/2}} \int \exp\{p_i L_i(\boldsymbol{\tau}, \boldsymbol{b}_i)\} d\boldsymbol{b}_i \qquad (5.63)\\ &= \exp\Big\{p_i L_i(\boldsymbol{\tau}, \tilde{\boldsymbol{b}}_i)\Big\} \frac{1}{\big|-p_i L_i''\big|^{1/2}} \Big(1 + \frac{A_i(\boldsymbol{\tau}, \tilde{\boldsymbol{b}}_i)}{p_i}\Big)\\ &= \exp\Big\{p_i L_i(\boldsymbol{\tau}, \tilde{\boldsymbol{b}}_i)\Big\} \frac{1}{\big|-p_i L_i''\big|^{1/2}} \Big(1 + O(p_i^{-1})\Big)\end{aligned}$$

where $A_i(\boldsymbol{\tau}, \tilde{\boldsymbol{b}}_i(\boldsymbol{\tau}))$ is a uniformly bounded function in p_i, and

$$\begin{aligned}-p_i L_i'' = -p_i L_i''(\boldsymbol{\tau}, \tilde{\boldsymbol{b}}_i(\boldsymbol{\tau})) &= \frac{\partial^2}{\partial \boldsymbol{b}_i \partial \boldsymbol{b}_i'} \Big\{-p_i L_i(\boldsymbol{\tau}, \boldsymbol{b}_i)\Big\}\Big|_{\boldsymbol{b}_i = \tilde{\boldsymbol{b}}_i(\boldsymbol{\tau})} \qquad (5.64)\\ &= -\frac{\partial^2}{\partial \boldsymbol{b}_i \partial \boldsymbol{b}_i'} l_i(\boldsymbol{\beta}, \boldsymbol{b}_i, \boldsymbol{\theta}_w; \boldsymbol{y}_i)\Big|_{\boldsymbol{b}_i = \tilde{\boldsymbol{b}}_i(\boldsymbol{\tau})} + \boldsymbol{\Psi}(\boldsymbol{\theta}_b)^{-1}\\ &= -\frac{\partial^2}{\partial \boldsymbol{b}_i \partial \boldsymbol{b}_i'} \sum_{j=1}^{p_i} l_{ij}(\boldsymbol{\beta}, \boldsymbol{b}_i, \boldsymbol{\theta}_w; y_{ij})\Big|_{\boldsymbol{b}_i = \tilde{\boldsymbol{b}}_i(\boldsymbol{\tau})} + \boldsymbol{\Psi}(\boldsymbol{\theta}_b)^{-1}\\ &= -\sum_{j=1}^{p_i} l_{ij}''(\boldsymbol{\beta}, \tilde{\boldsymbol{b}}_i(\boldsymbol{\tau}), \boldsymbol{\theta}_w; y_{ij}) + \boldsymbol{\Psi}(\boldsymbol{\theta}_b)^{-1}\\ &= -l_i''(\boldsymbol{\beta}, \tilde{\boldsymbol{b}}_i(\boldsymbol{\tau}), \boldsymbol{\theta}_w) + \boldsymbol{\Psi}(\boldsymbol{\theta}_b)^{-1}.\end{aligned}$$

where $-l_i''(\boldsymbol{\beta}, \tilde{\boldsymbol{b}}_i(\boldsymbol{\tau}), \boldsymbol{\theta}_w) = -\sum_{j=1}^{p_i} l_{ij}''(\boldsymbol{\beta}, \tilde{\boldsymbol{b}}_i(\boldsymbol{\tau}), \boldsymbol{\theta}_w; y_{ij})$. By ignoring the remainder term $A_i(\boldsymbol{\tau}, \tilde{\boldsymbol{b}}_i(\boldsymbol{\tau}))/p_i = O(p_i^{-1})$, we can obtain a ML estimate of $\boldsymbol{\tau} = (\boldsymbol{\beta}, \boldsymbol{\theta})$ by minimizing minus twice the Laplace approximated (LA) log-likelihood function $-2L_{\text{LA}}(\boldsymbol{\tau}, \tilde{\boldsymbol{b}}(\boldsymbol{\tau}); \boldsymbol{y})$ where

$$\begin{aligned}-2L(\boldsymbol{\beta}, \boldsymbol{\theta}; \boldsymbol{y}) &= -2L(\boldsymbol{\tau}; \boldsymbol{y}) \qquad (5.65)\\ &\approx -2\sum_{i=1}^{n} \Big[p_i L_i(\boldsymbol{\tau}, \tilde{\boldsymbol{b}}_i(\boldsymbol{\tau})) - \frac{1}{2}\log\Big(\big|-p_i L_i''(\boldsymbol{\tau}, \tilde{\boldsymbol{b}}_i(\boldsymbol{\tau}))\big|\Big)\Big]\\ &= -2L_{\text{LA}}(\boldsymbol{\tau}, \tilde{\boldsymbol{b}}(\boldsymbol{\tau}); \boldsymbol{y}).\end{aligned}$$

One can minimize the objective function (5.65) by applying any one of several nonlinear optimization algorithms available in NLMIXED including a Newton-Raphson algorithm and a quasi-Newton algorithm. The default approach is to apply a dual quasi-Newton algorithm which does not require second-order derivatives and which should suffice for most applications. For further details on the different nonlinear optimization routines available, the reader is referred to the NLMIXED documentation.

Gauss-Hermite quadrature:

The default method in PROC NLMIXED for computing integrals of the form (5.61) is adaptive Gaussian quadrature as described in Pinheiro and Bates (1995). Quadrature methods are a means for approximating a given integral by a weighted sum over predefined abscissas (nodes) for the random effects. By taking an adequate number of quadrature points as well as appropriate centering and scaling of the abscissas, one can usually obtain a good approximation to the given integral. A brief description of this approach was presented in §5.1.1 for GLME models involving a single random effect. Here we extend this to GNLME models satisfying conditional independence (5.48).

Following Liu and Pierce (1994), one-dimensional integrals of the form $\int f(z)\exp\{-z^2/2\}dz$ may be approximated by Gaussian quadrature as

$$\int_{-\infty}^{\infty} f(z)\exp\{-z^2/2\}dz = \sqrt{2\pi}\int_{-\infty}^{\infty} f(z)\phi(z)dz \approx \sqrt{2\pi}\sum_{k=1}^{q} f(z_k)w_k \tag{5.66}$$

where $\phi(z) = \frac{1}{\sqrt{2\pi}}\exp\{-z^2/2\}$ is the standard normal density, $z_k = \sqrt{2}z_k^*$, $w_k = w_k^*/\sqrt{\pi}$, and $\{z_k^*, w_k^*; k = 1, \ldots q\}$ are the quadrature abscissas (nodes) and weights, respectively, for a one-dimensional Gaussian quadrature rule with q quadrature points based on the kernel $\exp\{-z^2\}$; that is, based on the standard Gauss-Hermite quadrature approximation $\int f(z)\exp\{-z^2\}dz \approx \sum_{k=1}^{q} f(z_k^*)w_k^*$ (Golub and Welsch 1969; see also Table 25.10 of Abramowitz and Stegun 1972 for select values of z_k^* and w_k^*).

Let $\{z_k, w_k; k = 1, \ldots q\}$ denote, respectively, the quadrature abscissas and weights for the one-dimensional Gaussian quadrature rule (5.66) with q points based on the Gaussian kernel, $\exp\{-z^2/2\}$. By making a change in variable, $\boldsymbol{z} = \boldsymbol{\Psi}(\boldsymbol{\theta}_b)^{-1/2}\boldsymbol{b}_i$, and noting that the Jacobian of the inverse transformation is $|\boldsymbol{\Psi}(\boldsymbol{\theta}_b)|^{1/2}$, we can, through successive applications of the simple one-dimensional Gaussian quadrature rule (5.66), approximate (5.61) as

$$\begin{aligned}
\pi(\boldsymbol{y}_i;\boldsymbol{\tau}) &= \int \pi(\boldsymbol{y}_i;\boldsymbol{\beta},\boldsymbol{\theta}_w|\boldsymbol{b}_i)\pi(\boldsymbol{b}_i;\boldsymbol{\theta}_b)d\boldsymbol{b}_i \\
&= (2\pi)^{-v/2}|\boldsymbol{\Psi}(\boldsymbol{\theta}_b)|^{-1/2}\int \pi(\boldsymbol{y}_i;\boldsymbol{\beta},\boldsymbol{\theta}_w|\boldsymbol{b}_i)\exp\{-\frac{1}{2}\boldsymbol{b}_i'\boldsymbol{\Psi}(\boldsymbol{\theta}_b)^{-1}\boldsymbol{b}_i\}d\boldsymbol{b}_i \\
&= (2\pi)^{-v/2}\int \pi(\boldsymbol{y}_i;\boldsymbol{\beta},\boldsymbol{\theta}_w|\boldsymbol{\Psi}(\boldsymbol{\theta}_b)^{1/2}\boldsymbol{z})\exp\{-\frac{1}{2}\boldsymbol{z}'\boldsymbol{z}\}d\boldsymbol{z} \\
&\approx \sum_{k_1=1}^{q}\cdots\sum_{k_v=1}^{q}\left[\pi(\boldsymbol{y}_i;\boldsymbol{\beta},\boldsymbol{\theta}_w|\boldsymbol{a}_{\boldsymbol{k}})\times\prod_{l=1}^{v} w_{k_l}\right]
\end{aligned}$$

where $\boldsymbol{a}_{\boldsymbol{k}} = \boldsymbol{\Psi}(\boldsymbol{\theta}_b)^{1/2}\boldsymbol{z}_{\boldsymbol{k}}$ is a vector of scaled abscissas based on a point $\boldsymbol{z}_{\boldsymbol{k}} = (z_{k_1}, \ldots, z_{k_v})'$ on the v-dimensional quadrature grid. This non-adaptive Gaussian quadrature approximation is available in NLMIXED using the NOAD option.

Following Liu and Pierce (1994), Pinheiro and Bates (1995), Pinheiro and Chao (2006) and Molenberghs and Verbeke (§4.5, 2005), one can improve on standard Gaussian quadrature by expressing the marginal density of $\boldsymbol{y}_i$ using Gauss-Hermite quadrature with q quadrature points in $\boldsymbol{b}_i$ centered about the posterior mode $\tilde{\boldsymbol{b}}_i(\boldsymbol{\tau})$ and scaled to $(-p_iL_i'')^{-1/2}$. Specifically, we know that to a second-order approximation, the integrand of (5.61) is proportional to a $N(\tilde{\boldsymbol{b}}_i(\boldsymbol{\tau}), \boldsymbol{V}_i(\boldsymbol{\tau}))$ distribution where $\boldsymbol{V}_i(\boldsymbol{\tau}) = [-p_iL_i''(\boldsymbol{\tau}, \tilde{\boldsymbol{b}}_i(\boldsymbol{\tau}))]^{-1}$. Based on this, we can center and scale $\boldsymbol{b}_i$ so that the integrand $\pi(\boldsymbol{y}_i;\boldsymbol{\beta},\boldsymbol{\theta}_w|\boldsymbol{b}_i)\pi(\boldsymbol{b}_i;\boldsymbol{\theta}_b) = (2\pi)^{-v/2}\exp\{p_iL_i(\boldsymbol{\tau},\boldsymbol{b}_i)\}$ will be sampled in an appropriate region (see, for example, Molenberghs and Verbeke, §4.5, 2005). This is accomplished by first making a change in variable, $\boldsymbol{z} = \boldsymbol{V}_i(\boldsymbol{\tau})^{-1/2}(\boldsymbol{b}_i - \tilde{\boldsymbol{b}}_i(\boldsymbol{\tau}))$, and noting the Jacobian of the inverse transformation $\boldsymbol{b}_i = \tilde{\boldsymbol{b}}_i(\boldsymbol{\tau}) + \boldsymbol{V}_i(\boldsymbol{\tau})^{1/2}\boldsymbol{z}$ is $|\boldsymbol{V}_i(\boldsymbol{\tau})^{1/2}| = |-p_iL_i''(\boldsymbol{\tau},\tilde{\boldsymbol{b}}_i(\boldsymbol{\tau}))|^{-1/2}$. An adaptive Gaussian quadrature based on q quadrature points would then be given by

$$\begin{aligned}
\pi(\boldsymbol{y}_i;\boldsymbol{\tau}) &= \int \pi(\boldsymbol{y}_i;\boldsymbol{\beta},\boldsymbol{\theta}_w|\boldsymbol{b}_i)\pi(\boldsymbol{b}_i;\boldsymbol{\theta}_b)d\boldsymbol{b}_i \qquad (5.67)\\
&= \frac{1}{(2\pi)^{v/2}}\int \exp\{p_iL_i(\boldsymbol{\tau},\boldsymbol{b}_i)\}d\boldsymbol{b}_i\\
&= \frac{1}{(2\pi)^{v/2}}\int \Big[\exp\Big\{p_iL_i(\boldsymbol{\tau},\boldsymbol{b}_i)+\frac{1}{2}(\boldsymbol{b}_i-\tilde{\boldsymbol{b}}_i)'\boldsymbol{V}_i^{-1/2}\boldsymbol{V}_i^{-1/2}(\boldsymbol{b}_i-\tilde{\boldsymbol{b}}_i)\Big\}\\
&\qquad\times\exp\Big\{-\frac{1}{2}(\boldsymbol{b}_i-\tilde{\boldsymbol{b}}_i)'\boldsymbol{V}_i^{-1/2}\boldsymbol{V}_i^{-1/2}(\boldsymbol{b}_i-\tilde{\boldsymbol{b}}_i)\Big\}\Big]d\boldsymbol{b}_i\\
&= \frac{|\boldsymbol{V}_i(\boldsymbol{\tau})|^{1/2}}{(2\pi)^{v/2}}\int\Big[\exp\Big\{p_iL_i(\boldsymbol{\tau},\tilde{\boldsymbol{b}}_i+\boldsymbol{V}_i^{1/2}\boldsymbol{z})+\frac{1}{2}\boldsymbol{z}'\boldsymbol{z}\Big\}\exp\Big\{-\frac{1}{2}\boldsymbol{z}'\boldsymbol{z}\Big\}d\boldsymbol{z}\\
&\approx \frac{1}{\left|-p_iL_i''\right|^{1/2}}\sum_{k_1=1}^{q}\cdots\sum_{k_v=1}^{q}\Big[\exp\Big\{p_iL_i(\boldsymbol{\tau},\tilde{\boldsymbol{a}}_{\boldsymbol{k}}(\boldsymbol{\tau}))+\frac{1}{2}\boldsymbol{z}_{\boldsymbol{k}}'\boldsymbol{z}_{\boldsymbol{k}}\Big\}\times\prod_{l=1}^{v}w_{k_l}\Big]\\
&= \frac{1}{\left|-p_iL_i''\right|^{1/2}}\sum_{\boldsymbol{k}}^{q}\exp\{p_iL_i(\boldsymbol{\tau},\tilde{\boldsymbol{a}}_{\boldsymbol{k}}(\boldsymbol{\tau}))\}W_{\boldsymbol{k}}
\end{aligned}$$

where $-p_iL_i''=-p_iL_i''(\boldsymbol{\tau},\tilde{\boldsymbol{b}}_i(\boldsymbol{\tau}))$ is the Hessian matrix (5.64) associated with maximizing the log-posterior (5.62) wrt $\boldsymbol{b}_i$, $\boldsymbol{k}=(k_1,\ldots,k_v)'$ is a subscript index vector such that $\sum_{\boldsymbol{k}}^q=\sum_{k_1=1}^q\cdots\sum_{k_v=1}^q$, $\boldsymbol{z}_{\boldsymbol{k}}=(z_{k_1},\ldots,z_{k_v})'$ is a point on the v-dimensional quadrature grid and

$$\begin{aligned}
\tilde{\boldsymbol{a}}_{\boldsymbol{k}}(\boldsymbol{\tau}) &= \tilde{\boldsymbol{b}}_i(\boldsymbol{\tau})+\boldsymbol{V}_i^{-1/2}(\boldsymbol{\tau})\boldsymbol{z}_{\boldsymbol{k}}\\
&= \tilde{\boldsymbol{b}}_i(\boldsymbol{\tau})+\Big\{-l_i''(\boldsymbol{\beta},\tilde{\boldsymbol{b}}_i(\boldsymbol{\tau}),\boldsymbol{\theta}_w)+\boldsymbol{\Psi}(\boldsymbol{\theta}_b)^{-1}\Big\}^{-1/2}\boldsymbol{z}_{\boldsymbol{k}}\\
W_{\boldsymbol{k}} &= \exp(\frac{1}{2}\boldsymbol{z}_{\boldsymbol{k}}'\boldsymbol{z}_{\boldsymbol{k}})\prod_{l=1}^{v}w_{k_l}=\prod_{l=1}^{v}\exp(\frac{1}{2}z_{k_l}^2)w_{k_l}
\end{aligned}$$

are the centered and scaled abscissas and weights, respectively.

Based on (5.67), ML estimation of $\boldsymbol{\tau}=(\boldsymbol{\beta},\boldsymbol{\theta})$ may be carried out by minimizing minus twice the adaptive Gaussian quadrature (AGQ) approximated log-likelihood function $-2L_{\text{AGQ}}(\boldsymbol{\tau},\tilde{\boldsymbol{b}}(\boldsymbol{\tau});\boldsymbol{y})$ where

$$\begin{aligned}
-2L(\boldsymbol{\beta},\boldsymbol{\theta};\boldsymbol{y}) &= -2L(\boldsymbol{\tau};\boldsymbol{y}) \qquad (5.68)\\
&\approx -2\sum_{i=1}^{n}\Big[\log\Big\{\sum_{\boldsymbol{k}}^{q}\exp\{p_iL_i(\boldsymbol{\tau},\tilde{\boldsymbol{a}}_{\boldsymbol{k}}(\boldsymbol{\tau}))\}W_{\boldsymbol{k}}\Big\}\\
&\qquad -\frac{1}{2}\log\Big(\Big|-p_iL_i''(\boldsymbol{\tau},\tilde{\boldsymbol{b}}_i(\boldsymbol{\tau}))\Big|\Big)\Big]\\
&= -2L_{\text{AGQ}}(\boldsymbol{\tau},\tilde{\boldsymbol{b}}(\boldsymbol{\tau});\boldsymbol{y}).
\end{aligned}$$

When $q=1$, the AGQ log-likelihood (5.68) reduces to the Laplace-approximated log-likelihood (5.65) since, for a single quadrature point, $z_1=0$ and $w_1=w_1^*/\sqrt{\pi}=1$ imply $\boldsymbol{z}_{\boldsymbol{k}}=\boldsymbol{0}$, $W_{\boldsymbol{k}}=1$ and $\tilde{\boldsymbol{a}}_{\boldsymbol{k}}(\boldsymbol{\tau})=\tilde{\boldsymbol{b}}_i(\boldsymbol{\tau})$.

As noted above, adaptive Gaussian quadrature is the default approach in NLMIXED. It utilizes the above formulation for evaluating the marginal pdf of $\boldsymbol{y}_i$ but the number of quadrature points, q, is selected adaptively by evaluating the log-likelihood function at the starting values of the parameters until two successive evaluations have a relative difference less that a set tolerance level as specified by the option QTOL. Alternatively, one can specify a fixed-point Gaussian quadrature by specifying the number of quadratures using the NLMIXED option QPOINTS= (see the NLMIXED syntax).

Methods based on conditional second-order generalized estimating equations (CGEE2)

For parametric GNLME models in which the conditional pdf of $\boldsymbol{y}_i$ is in the quadratic exponential family (5.49), Vonesh et. al. (2002) proposed a doubly iterative estimation scheme for jointly estimating the population parameters, $\boldsymbol{\tau} = (\boldsymbol{\beta}, \boldsymbol{\theta}_b, \boldsymbol{\theta}_w)$, and random effects, $\boldsymbol{b} = (\boldsymbol{b}_1', \ldots, \boldsymbol{b}_n')'$. Estimation involves alternating between jointly estimating $(\boldsymbol{\beta}, \boldsymbol{b}, \boldsymbol{\theta}_w)$ for a current estimate of $\boldsymbol{\theta}_b$ and updating the estimate of $\boldsymbol{\theta}_b$ based on approximations to the posterior mean and variance of $\boldsymbol{b}_i$ $(i = 1, \ldots, n)$. For a given value of $\boldsymbol{\theta}_b$, estimates of $(\boldsymbol{\beta}, \boldsymbol{b}, \boldsymbol{\theta}_w)$ are obtained by solving a set of conditional second-order generalized estimating equations (CGEE2) the solutions of which jointly maximize the log-posterior (5.62) with respect to $(\boldsymbol{\beta}, \boldsymbol{b}, \boldsymbol{\theta}_w)$. Iteratively solving this set of CGEE2 allows one to extend the PQL approach of Breslow and Clayton (1993) to the quadratic exponential family (5.49) by including information about $\boldsymbol{\beta}$, $\boldsymbol{b}_i$ and $\boldsymbol{\theta}_w$ contained in the second-order moments determined by $\boldsymbol{w}_i'\boldsymbol{\gamma}_i$.

Analogous to Breslow and Clayton's PQL approach, the CGEE2 estimation scheme of Vonesh et. al. (2002) can be seen to be a derivation of Green's penalized likelihood estimation (Green, 1987). Specifically, let $\boldsymbol{\varphi} = (\boldsymbol{\beta}, \boldsymbol{\theta}_w)$ denote the parameters of the conditional pdf of $\boldsymbol{y}_i|\boldsymbol{b}_i$ such that the log-posterior (5.62) may be written as

$$p_i L_i(\boldsymbol{\varphi}, \boldsymbol{\theta}_b, \boldsymbol{b}_i) = l_i(\boldsymbol{\varphi}, \boldsymbol{b}_i; \boldsymbol{y}_i) - \frac{1}{2}\log\left|\boldsymbol{\Psi}(\boldsymbol{\theta}_b)\right| - \frac{1}{2}\boldsymbol{b}_i'\boldsymbol{\Psi}(\boldsymbol{\theta}_b)^{-1}\boldsymbol{b}_i \tag{5.69}$$

where $l_i(\boldsymbol{\varphi}, \boldsymbol{b}_i; \boldsymbol{y}_i) = -\log\Delta_i + \boldsymbol{y}_i'\boldsymbol{\lambda}_i + \boldsymbol{w}_i'\boldsymbol{\gamma}_i + c_i(\boldsymbol{y}_i)$ is the log pdf, $\pi(\boldsymbol{y}_i|\boldsymbol{b}_i)$, from (5.49). Based on the second-order Laplace approximation (5.63), the log-likelihood objective function (5.65) may be expressed in terms of $(\boldsymbol{\varphi}, \boldsymbol{\theta}_b)$ as

$$\begin{aligned} L_{\mathrm{LA}}(\boldsymbol{\varphi}, \boldsymbol{\theta}_b; \boldsymbol{y}) &= \sum_{i=1}^{n}\left[p_i L_i(\boldsymbol{\varphi}, \boldsymbol{\theta}_b, \tilde{\boldsymbol{b}}_i) - \frac{1}{2}\log\left(\left|-p_i L_i''(\boldsymbol{\varphi}, \boldsymbol{\theta}_b, \tilde{\boldsymbol{b}}_i)\right|\right)\right] \\ &= \sum_{i=1}^{n}\left[l_i(\boldsymbol{\varphi}, \tilde{\boldsymbol{b}}_i; \boldsymbol{y}_i) - \frac{1}{2}\tilde{\boldsymbol{b}}_i'\boldsymbol{\Psi}(\boldsymbol{\theta}_b)^{-1}\tilde{\boldsymbol{b}}_i \right. \\ &\qquad \left. - \frac{1}{2}\log\left(\left|-l_i''(\boldsymbol{\varphi}, \tilde{\boldsymbol{b}}_i; \boldsymbol{y}_i)\boldsymbol{\Psi}(\boldsymbol{\theta}_b) + \boldsymbol{I}_{p_i}\right|\right)\right] \end{aligned} \tag{5.70}$$

where $\tilde{\boldsymbol{b}}_i = \tilde{\boldsymbol{b}}_i(\boldsymbol{\varphi}, \boldsymbol{\theta}_b)$ is the posterior mode obtained by maximizing (5.69) with respect to $\boldsymbol{b}_i$ and $-p_i L_i''(\boldsymbol{\varphi}, \boldsymbol{\theta}_b, \tilde{\boldsymbol{b}}_i) = -l_i''(\boldsymbol{\varphi}, \tilde{\boldsymbol{b}}_i; \boldsymbol{y}_i) + \boldsymbol{\Psi}(\boldsymbol{\theta}_b)^{-1}$ is the Hessian matrix associated with this maximization. Here we have combined the terms $-\frac{1}{2}\log\left|\boldsymbol{\Psi}(\boldsymbol{\theta}_b)\right|$ and $-\frac{1}{2}\log\left(\left|-p_i L_i''(\boldsymbol{\varphi}, \boldsymbol{\theta}_b, \tilde{\boldsymbol{b}}_i)\right|\right)$ into a single term using the identity $\log|\boldsymbol{A}| + \log|\boldsymbol{B}| = \log|\boldsymbol{AB}|$.

Now rather than maximize (5.70) directly, it may be reasonable to ignore the term $-\frac{1}{2}\log\left(\left|-l_i''(\boldsymbol{\varphi}, \tilde{\boldsymbol{b}}_i; \boldsymbol{y}_i)\boldsymbol{\Psi}(\boldsymbol{\theta}_b) + \boldsymbol{I}_{p_i}\right|\right)$ provided $-l_i''(\boldsymbol{\varphi}, \tilde{\boldsymbol{b}}_i; \boldsymbol{y}_i)$ varies slowly with $\boldsymbol{\tau}$ (e.g., Bates and Watts, 1980; Breslow and Clayton, 1993). If one then holds $\boldsymbol{\theta}_b$ fixed, joint estimates of $\boldsymbol{\varphi} = (\boldsymbol{\beta}, \boldsymbol{\theta}_w)$ may be obtained by maximizing $\sum_{i=1}^{n} l_i(\boldsymbol{\varphi}, \tilde{\boldsymbol{b}}_i; \boldsymbol{y}_i) = \sum_{i=1}^{n} l_i(\boldsymbol{\varphi}, \tilde{\boldsymbol{b}}_i(\boldsymbol{\varphi}, \boldsymbol{\theta}_b); \boldsymbol{y}_i)$ with respect to $\boldsymbol{\varphi}$. Doing so leads to estimates $\widehat{\boldsymbol{\varphi}}(\boldsymbol{\theta}_b)$ and $\widehat{\boldsymbol{b}}_i(\boldsymbol{\theta}_b) = \tilde{\boldsymbol{b}}_i(\widehat{\boldsymbol{\varphi}}(\boldsymbol{\theta}_b))$, $i = 1, \ldots, n$, that jointly maximize Green's penalized likelihood function,

$$l_{\theta_b}(\varphi, \boldsymbol{b}; \boldsymbol{y}) = \sum_{i=1}^{n}\left[l_i(\varphi, \boldsymbol{b}_i; \boldsymbol{y}_i) - \frac{1}{2}\boldsymbol{b}_i'\boldsymbol{\Psi}(\boldsymbol{\theta}_b)^{-1}\boldsymbol{b}_i\right] \tag{5.71}$$

or, equivalently, the log-posterior objective function $\sum_{i=1}^{n}\{p_i L_i(\varphi, \boldsymbol{\theta}_b, \boldsymbol{b}_i)\}$ holding $\boldsymbol{\theta}_b$ fixed. The set of CGEE2 of Vonesh et. al. (2002) are then simply the estimating equations

$$U(\varphi) = \partial\, l_{\theta_b}(\varphi, \boldsymbol{b}; \boldsymbol{y})/\partial\varphi = \boldsymbol{0}$$
$$U(\boldsymbol{b}) = \partial\, l_{\theta_b}(\varphi, \boldsymbol{b}; \boldsymbol{y})/\partial\boldsymbol{b} = \boldsymbol{0}$$

required to jointly maximize (5.71) with respect $(\varphi, \boldsymbol{b})$ holding $\boldsymbol{\theta}_b$ fixed. We can express the CGEE2 in terms of the component vectors $\boldsymbol{\beta}$, $\boldsymbol{\theta}_w$ and $\boldsymbol{b}_i$ as

$$U(\boldsymbol{\beta}, \boldsymbol{\theta}_w) = \begin{pmatrix} \partial\, l_{\theta_b}(\varphi, \boldsymbol{b}; \boldsymbol{y})/\partial\boldsymbol{\beta} \\ \partial\, l_{\theta_b}(\varphi, \boldsymbol{b}; \boldsymbol{y})/\partial\boldsymbol{\theta}_w \end{pmatrix} = \begin{pmatrix} \boldsymbol{0} \\ \boldsymbol{0} \end{pmatrix} \tag{5.72}$$
$$U_1(\boldsymbol{b}_1) = \partial\, l_1(\varphi, \boldsymbol{b}_1; \boldsymbol{y}_1)/\partial\boldsymbol{b}_1 - \boldsymbol{\Psi}(\boldsymbol{\theta}_b)^{-1}\boldsymbol{b}_1 = \boldsymbol{0}$$
$$\vdots$$
$$U_n(\boldsymbol{b}_n) = \partial\, l_n(\varphi, \boldsymbol{b}_n; \boldsymbol{y}_n)/\partial\boldsymbol{b}_n - \boldsymbol{\Psi}(\boldsymbol{\theta}_b)^{-1}\boldsymbol{b}_n = \boldsymbol{0}.$$

To complete one full cycle of the CGEE2 estimation scheme, one would then estimate $\boldsymbol{\theta}_b$ using the Gaussian posterior approximation of Laird and Louis (1982).

This entire estimation scheme is equivalent to linearizing the conditional first- and second-order moments about current estimates of the fixed and random effects and applying standard normal-theory ML or REML estimation to an extended pseudo-LME model to update parameter estimates. As such, this approach mimics the PL/CGEE1 algorithm but with CGEE1 replaced by CGEE2 in order to include the conditional second-order moments into the objective function. When the conditional pdf of $\boldsymbol{y}_i$ is in the linear exponential family, the CGEE2 algorithm of Vonesh et. al. (2002) will yield estimates identical to those obtained with the PL/CGEE1 algorithm of GLIMMIX. This is because under the linear exponential family, the second-order term $\boldsymbol{w}_i'\gamma_i$ in (5.49) vanishes and the CGEE2 approach will be equivalent to the PQL approach of Breslow and Clayton (Vonesh et. al., 2002).

While conceptually appealing, the CGEE2 algorithm of Vonesh et. al. (2002) is extremely difficult to implement since it requires specification of third- and fourth-order conditional moments much like the GEE2 approach requires for marginal models (see §4.4.2). However, as with GEE2, one can overcome this problem by specifying third- and fourth-order conditional moments under the "working" assumption of normality. In doing so, one can apply CGEE2 to obtain moments-based rather than likelihood-based estimates of the model parameters. Vonesh et. al. (2002), for example, utilize their CGEE2 algorithm to estimate the parameters $\boldsymbol{\tau} = (\boldsymbol{\beta}, \boldsymbol{\theta}_b, \boldsymbol{\theta}_w)$ based on jointly minimizing, with respect to φ and $\boldsymbol{b}$, the penalized extended least squares (PELS) objective function,

$$Q_{\text{PELS}}(\boldsymbol{\tau}, \boldsymbol{b}) = \sum_{i=1}^{n}\Big\{(\boldsymbol{y}_i - \boldsymbol{\mu}_i(\boldsymbol{\beta}, \boldsymbol{b}_i))'\boldsymbol{\Lambda}_i(\boldsymbol{\beta}, \boldsymbol{b}_i, \boldsymbol{\theta}_w))^{-1}(\boldsymbol{y}_i - \boldsymbol{\mu}_i(\boldsymbol{\beta}, \boldsymbol{b}_i)) \tag{5.73}$$
$$+ \log\left|\boldsymbol{\Lambda}_i(\boldsymbol{\beta}, \boldsymbol{b}_i, \boldsymbol{\theta}_w)\right| + \boldsymbol{b}_i'\boldsymbol{\Psi}(\boldsymbol{\theta}_b)^{-1}\boldsymbol{b}_i\Big\}$$
$$= \sum_{i=1}^{n}\Big\{q_i(\boldsymbol{\beta}, \boldsymbol{b}_i, \boldsymbol{\theta}_w) + \boldsymbol{b}_i'\boldsymbol{\Psi}(\boldsymbol{\theta}_b)^{-1}\boldsymbol{b}_i\Big\}.$$

where

$$q_i(\boldsymbol{\beta}, \boldsymbol{b}_i, \boldsymbol{\theta}_w) = \Big\{(\boldsymbol{y}_i - \boldsymbol{\mu}_i(\boldsymbol{\beta}, \boldsymbol{b}_i))'\boldsymbol{\Lambda}_i(\boldsymbol{\beta}, \boldsymbol{b}_i, \boldsymbol{\theta}_w))^{-1}(\boldsymbol{y}_i - \boldsymbol{\mu}_i(\boldsymbol{\beta}, \boldsymbol{b}_i)) + \log\left|\boldsymbol{\Lambda}_i(\boldsymbol{\beta}, \boldsymbol{b}_i, \boldsymbol{\theta}_w)\right|\Big\}$$

is the individual ELS objective function for the ith subject and $\boldsymbol{b}_i'\boldsymbol{\Psi}(\boldsymbol{\theta}_b)^{-1}\boldsymbol{b}_i$ is a penalty term reflecting the fact that minimization of $\sum_{i=1}^n q_i(\boldsymbol{\beta}, \boldsymbol{b}_i, \boldsymbol{\theta}_w)$ across subjects without the aggregated penalty term $\sum_{i=1}^n \boldsymbol{b}_i'\boldsymbol{\Psi}(\boldsymbol{\theta}_b)^{-1}\boldsymbol{b}_i$ may be intractable when $N = \sum_{i=1}^n p_i$ is close to nv (see, for example, Demidenko, 2004). The fact that minimizing the PELS objective function (5.73) coincides with maximizing the penalized likelihood function (5.71) under the "working" assumption of Gaussian third- and fourth-order conditional moments allows one to obtain least squares type estimates of the model parameters based solely on specification of the first two conditional moments.

In this book, we utilize a CGEE2 approach to minimize an augmented or modified PELS objective function derived under the same "working" assumption of Gaussian third- and fourth-order conditional moments. However, rather than apply the doubly iterative CGEE2 algorithm of Vonesh et. al. (2002) in which one alternates between jointly minimizing the PELS objective function (5.73) with respect to $\boldsymbol{\varphi}$ and $\boldsymbol{b}$ and then using the Gaussian posterior approximation to estimate $\boldsymbol{\theta}_b$, we simply minimize minus twice the Laplace-based "working" log-likelihood function, $-2L_{\text{LA}}(\boldsymbol{\varphi}, \boldsymbol{\theta}_b; \boldsymbol{y})$, derived under a "working" normality assumption. This approach mimics the GEE2/ELS approach described in §4.4.2 for marginal models in that we utilize a conditional distribution that is in the quadratic exponential family, namely the normal distribution, as a means for jointly modeling the first two conditional moments.

Under a "working" assumption that $\boldsymbol{y}_i|\boldsymbol{b}_i \sim N(\boldsymbol{\mu}_i(\boldsymbol{\beta}, \boldsymbol{b}_i), \boldsymbol{\Lambda}_i(\boldsymbol{\beta}, \boldsymbol{b}_i, \boldsymbol{\theta}_w))$, let $\widehat{\boldsymbol{\tau}} = (\widehat{\boldsymbol{\varphi}}, \widehat{\boldsymbol{\theta}}_b)$ be the estimate of $\boldsymbol{\tau} = (\boldsymbol{\varphi}, \boldsymbol{\theta}_b)$ obtained by minimizing $-2L_{\text{LA}}(\boldsymbol{\varphi}, \boldsymbol{\theta}_b; \boldsymbol{y})$. The estimating equations for minimizing $-2L_{\text{LA}}(\boldsymbol{\varphi}, \boldsymbol{\theta}_b; \boldsymbol{y})$ with respect to $\boldsymbol{\varphi}$ and $\boldsymbol{\theta}_b$ are then given by the modified set of CGEE2

$$U_{\text{CGEE2}}(\boldsymbol{\varphi}, \tilde{\boldsymbol{b}}(\boldsymbol{\varphi}, \boldsymbol{\theta}_b)) = \frac{\partial}{\partial \boldsymbol{\varphi}} \sum_{i=1}^n \Big[-2p_i L_i(\boldsymbol{\varphi}, \boldsymbol{\theta}_b, \tilde{\boldsymbol{b}}_i(\boldsymbol{\varphi}, \boldsymbol{\theta}_b)) + \log\Big|-p_i L_i''(\boldsymbol{\varphi}, \boldsymbol{\theta}_b, \tilde{\boldsymbol{b}}_i(\boldsymbol{\varphi}, \boldsymbol{\theta}_b))\Big|\Big] = \boldsymbol{0} \tag{5.74}$$

$$U_{\text{CGEE2}}(\boldsymbol{\theta}_b, \tilde{\boldsymbol{b}}(\boldsymbol{\varphi}, \boldsymbol{\theta}_b)) = \frac{\partial}{\partial \boldsymbol{\theta}_b} \sum_{i=1}^n \Big[-2p_i L_i(\boldsymbol{\varphi}, \boldsymbol{\theta}_b, \tilde{\boldsymbol{b}}_i(\boldsymbol{\varphi}, \boldsymbol{\theta}_b)) + \log\Big|-p_i L_i''(\boldsymbol{\varphi}, \boldsymbol{\theta}_b, \tilde{\boldsymbol{b}}_i(\boldsymbol{\varphi}, \boldsymbol{\theta}_b))\Big|\Big] = \boldsymbol{0}$$

where, for fixed $\boldsymbol{\varphi}$ and $\boldsymbol{\theta}_b$, the $\tilde{\boldsymbol{b}}_i = \tilde{\boldsymbol{b}}_i(\boldsymbol{\varphi}, \boldsymbol{\theta}_b)$ are the posterior modes that jointly solve the estimating equations

$$U_1(\boldsymbol{b}_1) = \partial\, l_1(\boldsymbol{\varphi}, \boldsymbol{b}_1; \boldsymbol{y}_1)/\partial \boldsymbol{b}_1 - \boldsymbol{\Psi}(\boldsymbol{\theta}_b)^{-1}\boldsymbol{b}_1 = \boldsymbol{0} \tag{5.75}$$

$$\vdots$$

$$U_n(\boldsymbol{b}_n) = \partial\, l_n(\boldsymbol{\varphi}, \boldsymbol{b}_n; \boldsymbol{y}_n)/\partial \boldsymbol{b}_n - \boldsymbol{\Psi}(\boldsymbol{\theta}_b)^{-1}\boldsymbol{b}_n = \boldsymbol{0}.$$

Note that $U_1(\boldsymbol{b}_1), \ldots, U_1(\boldsymbol{b}_1)$ are the same estimating equations for the random effects as used in (5.72) but here these estimating equations are embedded in a suboptimization routine that is required to recalculate $\tilde{\boldsymbol{b}}_i(\boldsymbol{\varphi}, \boldsymbol{\theta}_b)$ during each iteration of the global minimization of $-2L_{\text{LA}}(\boldsymbol{\varphi}, \boldsymbol{\theta}_b; \boldsymbol{y})$.

Since the joint estimating equations (5.74) are derived under the "working" assumption of Gaussian third- and fourth-order conditional moments, this estimation scheme is best viewed as a modified singly iterative CGEE2 algorithm designed to estimate the parameters of the GNLME model (5.47) assuming only that the first two conditional moments have been correctly specified and that the random effects are normally distributed. As such, this scheme, which is easily implemented with GLIMMIX or NLMIXED, provides a natural alternative to the original but computationally more challenging CGEE2/PELS scheme of Vonesh et. al. (2002). One can view this as a modified form of PELS by noting that the "working" objective function, $-2L_{\text{LA}}(\boldsymbol{\varphi}, \boldsymbol{\theta}_b; \boldsymbol{y}) = -2L_{\text{LA}}(\boldsymbol{\tau}; \boldsymbol{y})$, may be expressed in terms of an augmented PELS objective function,

$$
\begin{aligned}
Q^*_{\text{PELS}}(\boldsymbol{\tau}, \tilde{\boldsymbol{b}}) &= \sum_{i=1}^{n}\left[p_i Q_i(\boldsymbol{\tau}, \tilde{\boldsymbol{b}}_i) + \log\left|-p_i Q_i''(\boldsymbol{\tau}, \tilde{\boldsymbol{b}}_i)\boldsymbol{\Psi}(\boldsymbol{\theta}_b) + \boldsymbol{I}_{p_i}\right|\right] \qquad (5.76)\\
&= \sum_{i=1}^{n}\Big[(\boldsymbol{y}_i - \boldsymbol{\mu}_i(\boldsymbol{\beta}, \tilde{\boldsymbol{b}}_i))'\boldsymbol{\Lambda}_i(\boldsymbol{\beta}, \tilde{\boldsymbol{b}}_i, \boldsymbol{\theta}_w)^{-1}(\boldsymbol{y}_i - \boldsymbol{\mu}_i(\boldsymbol{\beta}, \tilde{\boldsymbol{b}}_i))\\
&\quad + \log\left|\boldsymbol{\Lambda}_i(\boldsymbol{\beta}, \tilde{\boldsymbol{b}}_i, \boldsymbol{\theta}_w)\right| + \tilde{\boldsymbol{b}}_i'\boldsymbol{\Psi}(\boldsymbol{\theta}_b)^{-1}\tilde{\boldsymbol{b}}_i\\
&\quad + \log\left|\boldsymbol{\Psi}(\boldsymbol{\theta}_b)\right| + \log\left|-p_i Q_i''(\boldsymbol{\tau}, \tilde{\boldsymbol{b}}_i)\right|\Big]
\end{aligned}
$$

where

$$
\begin{aligned}
p_i Q_i(\boldsymbol{\tau}, \tilde{\boldsymbol{b}}_i) = -2p_i L_i(\boldsymbol{\tau}, \tilde{\boldsymbol{b}}_i) &= (\boldsymbol{y}_i - \boldsymbol{\mu}_i(\boldsymbol{\beta}, \tilde{\boldsymbol{b}}_i))'\boldsymbol{\Lambda}_i(\boldsymbol{\beta}, \tilde{\boldsymbol{b}}_i, \boldsymbol{\theta}_w))^{-1}(\boldsymbol{y}_i - \boldsymbol{\mu}_i(\boldsymbol{\beta}, \tilde{\boldsymbol{b}}_i))\\
&\quad + \log\left|\boldsymbol{\Lambda}_i(\boldsymbol{\beta}, \tilde{\boldsymbol{b}}_i, \boldsymbol{\theta}_w)\right| + \tilde{\boldsymbol{b}}_i'\boldsymbol{\Psi}(\boldsymbol{\theta}_b)^{-1}\tilde{\boldsymbol{b}}_i\\
&= q_i(\boldsymbol{\beta}, \tilde{\boldsymbol{b}}_i, \boldsymbol{\theta}_w) + \tilde{\boldsymbol{b}}_i'\boldsymbol{\Psi}(\boldsymbol{\theta}_b)^{-1}\tilde{\boldsymbol{b}}_i
\end{aligned}
$$

is the ith term in the summand of the PELS objective function (5.73) evaluated at $\tilde{\boldsymbol{b}}_i = \tilde{\boldsymbol{b}}_i(\boldsymbol{\tau})$ and $-p_i Q_i''(\boldsymbol{\tau}, \tilde{\boldsymbol{b}}_i) = -p_i L_i''(\boldsymbol{\tau}, \tilde{\boldsymbol{b}}_i)$ is the Hessian matrix obtained by maximizing the log-posterior (5.62) with respect to $\boldsymbol{b}_i$. Observe that by treating the Hessian terms $-p_i Q_i''(\boldsymbol{\tau}, \tilde{\boldsymbol{b}}_i)$ in (5.76) as constant or slowly changing in $\boldsymbol{\tau}$ (i.e., ignoring the last term in the summand of (5.76)), the set of estimating equations $U_{\text{CGEE2}}(\boldsymbol{\varphi}, \tilde{\boldsymbol{b}}(\boldsymbol{\varphi}, \boldsymbol{\theta}_b))$ in (5.74) reduce to

$$
U_{\text{CGEE2}}(\boldsymbol{\varphi}, \tilde{\boldsymbol{b}}(\boldsymbol{\varphi}, \boldsymbol{\theta}_b)) = \frac{\partial}{\partial \boldsymbol{\varphi}} \sum_{i=1}^{n}\left[q_i(\boldsymbol{\beta}, \tilde{\boldsymbol{b}}_i, \boldsymbol{\theta}_w) + \tilde{\boldsymbol{b}}_i'\boldsymbol{\Psi}(\boldsymbol{\theta}_b)^{-1}\tilde{\boldsymbol{b}}_i\right] \qquad (5.77)
$$

which, when combined with (5.75), are equivalent to jointly solving the original set of CGEE2 (5.72) proposed by Vonesh et. al. (2002) for fixed $\boldsymbol{\theta}_b$.

The CGEE2 approach presented here is within the context of penalized extended least squares (PELS). As such, it is intended to be a supplement to the primary methods of estimation available in SAS, namely the PL/CGEE1 and ML estimation techniques presented earlier. In the following section, we contrast differences between PL/CGEE1 and ML estimation by looking at the strengths and weaknesses of each. We then consider how CGEE2/PELS might fit in with these two estimation schemes.

5.3.5 Comparing different estimators

As with GLME models, the two primary methods of estimation in SAS for GNLME models are those based on pseudo-likelihood (PL) estimation and maximum likelihood (ML) estimation. Pseudo-likelihood estimation is based on approximating the moments using one of two forms of Taylor series linearization, a subject-specific (SS) linearization and a population-averaged (PA) linearization. The former involves expanding the conditional mean about a current estimate of the individual's random effect (i.e., about $\boldsymbol{b}_i = \tilde{\boldsymbol{b}}_i$) and the latter expands the conditional mean about the average random effect (i.e., $\boldsymbol{b}_i = \boldsymbol{0}$). We refer to estimates based on the SS linearization as PL/CGEE1 estimates as they involve iteratively solving a set of CGEE1 while estimates based on the PA linearization we refer to as PL/QELS estimates as they correspond to iteratively solving a set of quasi-extended least squares (QELS) estimating equations suitable for marginal models. Both result in moments-based estimates of the population parameters. On the other hand, ML estimation utilizes numerical integration techniques to approximate an integrated log-likelihood function resulting in likelihood-based estimates. Not surprisingly, how these two methods compare for GLNME models is similar to how they compare for GLME models (§5.1.2). We briefly summarize the advantages and disadvantages of each.

There are several advantages with likelihood-based estimation and inference. First, likelihood-based methods utilize a singly iterative algorithm based on a well-defined objective function and can be easily implemented in either GLIMMIX (for GLME models) or NLMIXED (for GNLME models). Second, under suitable regularity conditions, ML estimation yields consistent, asymptotically normal and asymptotically efficient estimates (e.g., Serfling, 1980). Third, as will be discussed in Chapter 6, valid inference is possible with likelihood-based procedures in the presence of missing data provided such data are ignorable or missing at random (MAR). Consequently, whenever feasible, likelihood-based methods should be the preferred approach one takes to estimation and inference under GLME and GNLME models. With respect to model specification, a chief drawback to likelihood-based estimation as implemented in NLMIXED is the difficulty with which one can accommodate nested random effects though even this limitation may be overcome using ARRAY statements within NLMIXED (see, for example, Littell et. al., 2006, §15.4, pp. 587-589). Perhaps the biggest challenge with a likelihood-based analysis is the difficulty of trying to verify key underlying assumptions required for valid inference. For example, if one correctly specifies the conditional mean of a NLME model but incorrectly specifies the conditional variance as a function of either $\boldsymbol{\beta}$ or $\boldsymbol{b}_i$, the NLMIXED procedure will end up maximizing an incorrect integrated log-likelihood function which in turn could lead to biased estimation and inference.

Moments-based estimation using the pseudo-likelihood (PL) approach is available in GLIMMIX for GLME models and in the macro %NLINMIX for GNLME models. In terms of model specification, the chief advantage with PL estimation is the ease with which one can introduce a variety of covariance structures within a given model. For example, both GLIMMIX and %NLINMIX allow for multiple RANDOM statements making it relatively easy to accommodate nested random effects structures. They also allow one to specify an intra-subject covariance structure using either the RSIDE option of the RANDOM statement from GLIMMIX or the REPEATED statement from PROC MIXED (as embedded within the %NLINMIX macro). The PL approach may also provide some protection against misspecification of the conditional covariance structure since linearization is only about the conditional mean unlike the CGEE2 approach of Vonesh et. al. (2002). However, there are several disadvantages associated with PL estimation. First, PL/QELS estimation generally leads to inconsistent estimates as shown by Breslow and Lin (1995) for GLME models and Demidenko (2004, pp. 456-459) for NLME models. This approach is probably best suited as a means for estimating the parameters of a marginal mean via an intermediary GNLME model as discussed in §5.1.2. The PL/CGEE1 approach in which linearization is about the current random effect can, in certain limited cases, also lead to inconsistent estimates even when the model has been correctly specified (see Appendix B for details). Moreover, in order to ensure valid inference under the PL approach or, for that matter, any non-likelihood approach, one must either have complete data (i.e., data that may be unbalanced with respect to covariates but which do not have any missing values), or incomplete data but with missing values assumed to be missing completely at random (MCAR). Assuming missing values are MCAR is far more restrictive than assuming they are MAR as required for a valid likelihood-based inference. We will address the issue of missing data in more detail in Chapter 6. Finally, both PL and ML estimation are susceptible to bias under random effect misspecification as illustrated in simulations by Vonesh (1992). Moreover, as with any nonlinear regression problem, poor starting values can lead to problems with convergence, etc. which may impact the estimates and their performance. We will address some of these concerns in the following section on computational issues.

In terms of direct comparisons, there are several papers that compare the performance of PL and ML estimates. Using simulation, Pinheiro and Bates (1995) and Wolfinger and Lin (1997) show that PL/CGEE1 estimation produces reasonably unbiased and efficient estimates under various NLME model settings. Wolfinger and Lin (1997) also confirm that PL/CGEE1 estimation based on a SS linearization (expansion around $\boldsymbol{b}_i = \widehat{\boldsymbol{b}}_i$) generally outperforms PL/QELS estimation based on a PA linearization (expansion around $\boldsymbol{b}_i = \mathbf{0}$). Pinheiro and Bates (1995) showed how ML estimation based on the Laplace approximation or adaptive Gaussian quadrature can lead to moderate improvements in the accuracy and precision of estimates compared with PL/CGEE1 estimation although, in many cases, the improvements were quite modest. However their simulations were performed assuming homogeneous intra-subject errors that are completely free of either fixed or random effects parameters. As noted by Wolfinger and Lin (1997), a violation of parameter orthogonality between fixed-effects and variance parameters can lead to increasing bias among estimates of the fixed-effects parameters when using PL/CGEE1. Vonesh et. al. (2002) showed how one can improve on the performance of PL/CGEE1 estimation by incorporating such non-orthogonality, when it exists, directly into the estimation scheme using CGEE2.

Finally, for continuous outcomes and outcomes based on count data, the use of Laplace-based ML estimation will, in many cases, compare quite favorably with more computer intensive ML estimation based on AGQ. Under this same scenario, the modified CGEE2 estimator obtained by minimizing an augmented PELS objective function may perform as well or better than the PL/CGEE1 estimator. Both the PL/CGEE1 estimator and modified CGEE2 estimator are, in general, moments-based estimators. However, when normality assumptions are satisfied, solving the set of modified CGEE2 will yield Laplace-based ML estimates of the model parameters. In this case, or whenever the underlying response variable $\boldsymbol{y}_i$ is continuous and approximately normally distributed, the modified CGEE2/PELS estimator is likely to be more accurate and efficient compared to the PL/CGEE1 estimator (e.g., Vonesh et. al., 2002). It is not clear, however, what improvement if any the modified CGEE2/PELS estimator will have over the CGEE2/PELS estimator of Vonesh et. al. (2002). A major advantage with the modified CGEE2/PELS estimation scheme is that it is easily implemented in GLIMMIX or NLMIXED by simply specifying the conditional mean and variance of $\boldsymbol{y}_i$ and applying Laplace's approximation assuming normality. This has the further advantage of allowing one to model overdispersion in a non-Gaussian mixed-effects model where likelihood-based methods are unavailable. For example, one may wish to include and directly estimate an overdispersion parameter within a Poisson mixed-effects modeling framework (see Example 5.4.5 for an illustration). In such cases, the use of CGEE2/PELS may be an improvement over PL/CGEE1. However, when the conditional pdf of $\boldsymbol{y}_i$ is highly discrete such as for binary outcomes, it is doubtful that CGEE2/PELS will offer any advantage and it should probably be avoided along with PL/CGEE1 and PQL in favor of ML estimation using adaptive Gaussian quadrature. Clearly additional research comparing these various estimation techniques is needed before further recommendations can be made.

5.3.6 Computational issues—starting values

In §4.6, we considered a number of computational issues as they relate to nonlinear optimization techniques used to estimate the parameters of a nonlinear regression model. One area of particular importance that we addressed was the use of good starting values especially with respect to the fixed effects regression parameters $\boldsymbol{\beta}$. While there are a number of options available for choosing good starting values for $\boldsymbol{\beta}$ (see §4.6 and references therein), here we face the additional challenge of providing good starting values for the variance-covariance parameters $\boldsymbol{\theta}_b$ of the random effects. As we shall demonstrate in several examples, having a systematic approach for

1) determining exactly what random effects structure to start with, and
2) determining good starting values for $\boldsymbol{\theta}_b$ given that structure

is crucial if we are to avoid getting entangled in a seemingly never-ending "trial and error" approach to determining good starting values. Item 1) above is vital if we are to avoid misspecification of the random effects as such misspecification could lead to biased inference (Vonesh, 1992). Item 2) is needed to help speed up the convergence process and help protect the algorithm from converging to a wrong set of values. To help overcome these challenges, a SAS macro %COVPARMS was developed to help decide which random effects one should include in a given model and what would be reasonable starting values for the variance-covariance parameters of those random effects. While extremely flexible, the macro has only been tested and evaluated in GNLME models for which the predicted value of $\boldsymbol{y}_i|\boldsymbol{b}_i$ is determined exclusively by some specified function of $\boldsymbol{\beta}$ and $\boldsymbol{b}_i$, usually the conditional mean $\boldsymbol{\mu}_i(\boldsymbol{\beta}, \boldsymbol{b}_i) = f(\boldsymbol{x}_{ij}', \boldsymbol{\beta}, \boldsymbol{b}_i)$. This of course includes the class of NLME models (5.51) and the class of semiparametric GNLME models (5.47) where the latter are fit using augmented CGEE2/PELS. Below is a brief description of how the macro works.

Let $\widehat{y}_{ij} = \widehat{y}_{ij}(\widehat{\boldsymbol{\beta}}_{\text{OLS}}, \boldsymbol{0}) = f(\boldsymbol{x}_{ij}', \widehat{\boldsymbol{\beta}}_{\text{OLS}}, \boldsymbol{b}_i)\Big|_{\boldsymbol{b}_i=\boldsymbol{0}}$ be the predicted value of the response variable y_{ij} from the ith subject taken on the jth occasion evaluated at the OLS estimate $\widehat{\boldsymbol{\beta}}_{\text{OLS}}$ and $\boldsymbol{b}_i = \boldsymbol{0}$ and let $\tilde{\boldsymbol{z}}_{ij}(\widehat{\boldsymbol{\beta}}_{\text{OLS}}, \boldsymbol{0}) = \partial f(\boldsymbol{x}_{ij}', \widehat{\boldsymbol{\beta}}_{\text{OLS}}, \boldsymbol{b}_i)/\partial \boldsymbol{b}_i\Big|_{\boldsymbol{b}_i=\boldsymbol{0}}$ be the corresponding $v \times 1$ derivative vector of this predicted value. Letting $r_{ij} = y_{ij} - \widehat{y}_{ij}$ denote the individual residuals, the %COVPARMS macro uses the GLIMMIX procedure to fit the linear random-effects model

$$r_{ij} = \tilde{\boldsymbol{z}}_{ij}(\boldsymbol{\beta}, \boldsymbol{0})'\boldsymbol{b}_i + e_{ij}, \quad i = 1, \ldots, n; \; j = 1, \ldots, p_i \qquad (5.78)$$

from which initial estimates of $\boldsymbol{\theta}_b$ of $Var(\boldsymbol{b}_i) = \boldsymbol{\Psi}(\boldsymbol{\theta}_b)$ are obtained. The macro then creates a SAS dataset containing values $(\widehat{\boldsymbol{\beta}}_{\text{OLS}}, \widehat{\boldsymbol{\theta}}_w, \widehat{\boldsymbol{\theta}}_b)$ where $\widehat{\boldsymbol{\theta}}_w$ is also the OLS estimate of $\boldsymbol{\theta}_w$.

In order to obtain the initial OLS estimates of $\boldsymbol{\beta}$ and $\boldsymbol{\theta}_w$, one will need to make an initial call to NLMIXED using the target GNLME model of interest including specification of the random effects to be included in the model. In doing so, one would supply starting values for $\boldsymbol{\beta}$ and $\boldsymbol{\theta}_w$ using, for example, some of the guidelines discussed in §4.6. However, rather than supply starting values for the variance-covariance parameters, $\boldsymbol{\theta}_b$, of the random effects within a PARMS statement, one would set all of the elements of $\boldsymbol{\theta}_b$ to 0 when specifying the parameters of the assumed normal distribution in the RANDOM statement. Using a PREDICT statement in combination with an ID statement, one would then output a SAS dataset containing the residuals, r_{ij}, as well as the derivatives $\tilde{\boldsymbol{z}}_{ij}(\widehat{\boldsymbol{\beta}}_{\text{OLS}}, \boldsymbol{0})$, from which one can then fit the linear random-effects model (5.78) using PROC GLIMMIX. For example, suppose one wished to fit a simple exponential decay mixed model

$$y_{ij} = (\beta_1 + b_{i1}) \exp\{-(\beta_2 + b_{i2})t_{ij}\} + \epsilon_{ij}$$

where it is assumed initially that $\begin{pmatrix} b_{i1} \\ b_{i2} \end{pmatrix} \sim \text{iid } N\left[\begin{pmatrix} 0 \\ 0 \end{pmatrix} \begin{pmatrix} \psi_{11} & \psi_{12} \\ \psi_{21} & \psi_{22} \end{pmatrix}\right]$ and $\epsilon_{ij} \sim \text{iid } N(0, \sigma_w^2)$. An initial call to NLMIXED might look like the following:

Program 5.9

```
proc nlmixed data=a;
 parms beta1 = 1 beta2 = 1 sigsq_w = 1;
 intercept = beta1+bi1;
 rate = beta2+bi2;
 predmean = intercept*exp(-rate);
 resid=y-predmean;
 model y~normal(predmean, sigsq_w);
 random bi1 bi2 ~ normal([0,0], [0,0,0]) subject=id;
 id resid;
 predict predmean out=predout der;
run;
```

Assuming the starting values in the PARMS statement are reasonable, this call to NLMIXED will yield overall OLS estimates of $\boldsymbol{\beta}' = (\beta_1, \beta_2)$ and σ_w^2 as well as the residuals (using the ID statement) and derivatives (using the DER option of the PREDICT statement) necessary to run the linear random-effects model (5.78).

The macro %COVPARMS takes the output from the PREDICT statement and fits the linear random-effects model to obtain estimates of the variance-covariance parameters $\boldsymbol{\theta}_b = (\psi_{11}, \psi_{21}, \psi_{22})'$ and then amends these back into a final SAS dataset which one would use in a second call to NLMIXED where one would replace the above syntax with the following:

Program 5.10

```
proc nlmixed data=a;
 parms /data=initial;
 intercept = beta1+bi1;
 rate = beta2+bi2;
 predmean = intercept*exp(-rate);
 resid=y-predmean;
 model y~normal(predmean, sigsq_w);
 random bi1 bi2 ~ normal([0,0], [psi11,psi21,psi22]) subject=id;
run;
```

where `data=initial` is the user-supplied name of a SAS dataset created in the macro %COVPARMS containing the OLS estimates of $\boldsymbol{\beta}' = (\beta_1, \beta_2)$ and σ_w^2 as well as initial estimates of $\boldsymbol{\theta}_b = (\psi_{11}, \psi_{21}, \psi_{22})'$ using a naming convention `psi11, psi21, psi22` which is also specified by the user. For a more complete description of macro %COVPARMS, the reader is referred to the author's Web page.

By fitting the linear random effects model (5.78) using an unstructured variance-covariance matrix for the random effects, one can examine the output generated from %COVPARMS to identify whether some of the random effects can be ignored and also help identify a possible covariance structure for the final set of random effects to be included in the model. Examples illustrating the use of this macro are presented in §5.4 and Chapter 7.

5.3.7 Inference and test statistics

For the GNLME model, inference in the form of test statistics, confidence intervals, etc., is based on the same large-sample theory and arguments as presented in §5.1.3 for the GLME model. However, there are some essential differences worth noting. For example, under the GLME model inference is restricted to estimable functions $\boldsymbol{l}'\boldsymbol{\beta}$ of the fixed effects or predictable functions $\boldsymbol{l}'\boldsymbol{\beta}+\boldsymbol{m}'\boldsymbol{b}$ of both fixed and random effects. Given the one-to-one relationship between the conditional linear predictor, $\boldsymbol{\eta}_i = \boldsymbol{X}_i\boldsymbol{\beta}+\boldsymbol{Z}_i\boldsymbol{b}_i$, and conditional mean, $\boldsymbol{\mu}_i(\boldsymbol{\beta},\boldsymbol{b}_i) = \boldsymbol{g}^{-1}(\boldsymbol{\eta}_i) = \boldsymbol{g}^{-1}(\boldsymbol{X}_i\boldsymbol{\beta}+\boldsymbol{Z}_i\boldsymbol{b}_i)$, such inference may be carried out either on the linear predictor scale (i.e., link scale) or the mean scale (i.e., inverse link scale). Conversely, inference under the GNLME model is not restricted to linear combinations of fixed and/or random effects. For example, inference with respect to fixed effects may be in the form of a joint test of hypothesis on one or more estimable nonlinear functions of $\boldsymbol{\beta}$, (i.e., $H_0 : \boldsymbol{l}(\boldsymbol{\beta}) = \boldsymbol{0}$ versus $H_1 : \boldsymbol{l}(\boldsymbol{\beta}) \neq \boldsymbol{0}$ where $\boldsymbol{l}(\boldsymbol{\beta})$ is any once continuously differentiable vector-valued function of $\boldsymbol{\beta}$ of order $r \times 1$ ($r \leq s$). Occasionally, inference may focus on a simple linear combination $\boldsymbol{l}(\boldsymbol{\beta}) = \boldsymbol{l}'\boldsymbol{\beta}$ of $\boldsymbol{\beta}$ but more often than not it will focus on some nonlinear function $\boldsymbol{l}(\boldsymbol{\beta})$ of $\boldsymbol{\beta}$ such as a scalar function representing an area under a curve or a peak concentration value as occurs in pharmacokinetic applications. GNLME models may also focus on general nonlinear predictors that represent any valid SAS programming expression, linear or nonlinear, across all of the observations involving the input data set variables, parameters, and random effects. The ability to conduct inference on nonlinear functions of the model parameters is only possible when one uses the NLMIXED procedure while inference is restricted to linear combinations of $\boldsymbol{\beta}$ and/or $\boldsymbol{b}$ when one chooses to use PL estimation in combination with the %NLINMIX macro. We briefly consider inference in each case.

Case 1. Inference based on pseudo-likelihood estimation: When one uses PL estimation as implemented in %NLINMIX to estimate the parameters of a GNLME model, inference is restricted to estimable linear combinations, $\boldsymbol{l}'\boldsymbol{\beta}$, of the fixed effects or to predictable linear combinations, $\boldsymbol{l}'\boldsymbol{\beta}+\boldsymbol{m}'\boldsymbol{b}$, of fixed and random effects. This is because PL estimation is based on a linearization of the model with respect to $\boldsymbol{\beta}$ and $\boldsymbol{b}_i$. Mimicking what was done for the GLME model, a one-step Gauss-Newton solution to the set of PL/CGEE1 estimating equations required to jointly estimate $\boldsymbol{\beta}$ and $\boldsymbol{b}$ for fixed $\boldsymbol{\theta}$ may be cast within a LME framework as the solution to the pseudo-linear mixed-model equations

$$\begin{pmatrix} \tilde{\boldsymbol{X}}'\tilde{\boldsymbol{\Lambda}}_D^{-1}\tilde{\boldsymbol{X}} & \tilde{\boldsymbol{X}}'\tilde{\boldsymbol{\Lambda}}_D^{-1}\tilde{\boldsymbol{Z}} \\ \tilde{\boldsymbol{Z}}'\tilde{\boldsymbol{\Lambda}}_D^{-1}\tilde{\boldsymbol{X}} & \tilde{\boldsymbol{Z}}'\tilde{\boldsymbol{\Lambda}}_D^{-1}\tilde{\boldsymbol{Z}}+\boldsymbol{\Psi}_D^{-1} \end{pmatrix} \begin{pmatrix} \widehat{\boldsymbol{\beta}}(\boldsymbol{\theta}) \\ \widehat{\boldsymbol{b}}(\boldsymbol{\theta}) \end{pmatrix} = \begin{pmatrix} \tilde{\boldsymbol{X}}'\tilde{\boldsymbol{\Lambda}}_D^{-1}\tilde{\boldsymbol{y}} \\ \tilde{\boldsymbol{Z}}'\tilde{\boldsymbol{\Lambda}}_D^{-1}\tilde{\boldsymbol{y}} \end{pmatrix} \tag{5.79}$$

where

$\widehat{\boldsymbol{\beta}}(\boldsymbol{\theta}) = (\tilde{\boldsymbol{X}}'\tilde{\boldsymbol{\Sigma}}_D^{-1}\tilde{\boldsymbol{X}})^{-1}\tilde{\boldsymbol{X}}'\tilde{\boldsymbol{\Sigma}}_D^{-1}\tilde{\boldsymbol{y}}$ is the GLS estimate (BLUE) of the fixed-effects parameters $\boldsymbol{\beta}$ under (5.56),
$\widehat{\boldsymbol{b}}(\boldsymbol{\theta}) = \boldsymbol{\Psi}_D\tilde{\boldsymbol{Z}}'\tilde{\boldsymbol{\Sigma}}_D^{-1}(\tilde{\boldsymbol{y}} - \tilde{\boldsymbol{X}}\widehat{\boldsymbol{\beta}}(\boldsymbol{\theta}))$ is the best linear unbiased predictor (BLUP) of the random-effects $\boldsymbol{b}' = \begin{bmatrix}\boldsymbol{b}_1' \cdots \boldsymbol{b}_n'\end{bmatrix}$ under (5.56),
$\tilde{\boldsymbol{X}}' = \begin{bmatrix}\tilde{\boldsymbol{X}}_1' \cdots \tilde{\boldsymbol{X}}_n'\end{bmatrix}$, $\tilde{\boldsymbol{Z}}' = Diag\{\tilde{\boldsymbol{Z}}_1', \ldots, \tilde{\boldsymbol{Z}}_n'\}$, $\tilde{\boldsymbol{y}}' = \begin{bmatrix}\tilde{\boldsymbol{y}}_1' \cdots \tilde{\boldsymbol{y}}_n'\end{bmatrix}$,
$\tilde{\boldsymbol{\Lambda}}_D = Diag\{\tilde{\boldsymbol{W}}_1^{-1/2}\boldsymbol{\Lambda}_1(\boldsymbol{\theta}_w)\tilde{\boldsymbol{W}}_1^{-1/2}, \ldots, \tilde{\boldsymbol{W}}_n^{-1/2}\boldsymbol{\Lambda}_n(\boldsymbol{\theta}_w)\tilde{\boldsymbol{W}}_n^{-1/2}\}$
$\boldsymbol{\Psi}_D = Diag\{\boldsymbol{\Psi}(\boldsymbol{\theta}_b), \ldots, \boldsymbol{\Psi}(\boldsymbol{\theta}_b)\}$, and
$\tilde{\boldsymbol{\Sigma}}_D = \boldsymbol{Z}\boldsymbol{\Psi}_D\boldsymbol{Z}' + \tilde{\boldsymbol{\Lambda}}_D = Diag\{\tilde{\boldsymbol{\Sigma}}_1(\boldsymbol{\theta}), \ldots, \tilde{\boldsymbol{\Sigma}}_n(\boldsymbol{\theta})\}$.

When we apply the PL/CGEE1 algorithm for the GNLME model using either the SS or PA expansion, we can estimate $Var\begin{pmatrix}\widehat{\boldsymbol{\beta}}(\boldsymbol{\theta}) \\ \widehat{\boldsymbol{b}}(\boldsymbol{\theta}) - \boldsymbol{b}\end{pmatrix} = \boldsymbol{\Upsilon}(\widehat{\boldsymbol{\beta}}, \widehat{\boldsymbol{b}})$ at the final solutions $(\widehat{\boldsymbol{\beta}}, \widehat{\boldsymbol{b}}, \widehat{\boldsymbol{\theta}})$

based on inverting the information matrix associated with the pseuodo-LME model equations (5.79). In particular, we have

$$\widehat{\Upsilon}(\widehat{\beta},\widehat{b}) = \begin{pmatrix} \tilde{X}'\widehat{\Lambda}_D^{-1}\tilde{X} & \tilde{X}'\widehat{\Lambda}_D^{-1}\tilde{Z} \\ \tilde{Z}'\widehat{\Lambda}_D^{-1}\tilde{X} & \tilde{Z}'\widehat{\Lambda}_D^{-1}\tilde{Z}+\widehat{\Psi}_D^{-1} \end{pmatrix}^{-1} \tag{5.80}$$

$$= \begin{pmatrix} \widehat{\Omega} & -\widehat{\Omega}\tilde{X}'\widehat{\Sigma}_D^{-1}\tilde{Z}\widehat{\Psi}_D \\ -\widehat{\Psi}_D\tilde{Z}'\widehat{\Sigma}_D^{-1}\tilde{X}\widehat{\Omega} & \widehat{Q}+\widehat{\Psi}_D\tilde{Z}'\widehat{\Sigma}_D^{-1}\tilde{X}\widehat{\Omega}\tilde{X}'\widehat{\Sigma}_D^{-1}\tilde{Z}\widehat{\Psi}_D \end{pmatrix}$$

where $\widehat{\Omega} = \widehat{\Omega}(\widehat{\beta}(\widehat{\theta})) = (\tilde{X}'\widehat{\Sigma}_D^{-1}\tilde{X})^{-1}$ is the estimated variance-covariance matrix of $\widehat{\beta}(\widehat{\theta})$ while $\widehat{Q} = Q(\widehat{\theta}) = (\tilde{Z}'\widehat{\Lambda}_D^{-1}\tilde{Z}+\widehat{\Psi}_D^{-1})^{-1}$ with $\widehat{\Lambda}_D$ and $\widehat{\Psi}_D$ representing estimates of $\tilde{\Lambda}_D$ and Ψ_D at the final estimates $(\widehat{\beta},\widehat{b},\widehat{\theta})$. As with the MIXED procedure, when one specifies the EMPIRICAL option in GLIMMIX, $\widehat{\Omega}$ is replaced by the robust variance-covariance matrix $\widehat{\Omega}_R = \widehat{\Omega}_R(\widehat{\beta}(\widehat{\theta}))$.

At this point, all inference with respect to β and b under the GNLME model would be based on (5.80) with inference based either on the model-based covariance matrix, $\widehat{\Omega}$, or the robust covariance matrix, $\widehat{\Omega}_R$, as computed under a final optimization call to the pseudo-weighted LME model (5.56). As a result, all of the standard inferential techniques described for the LME model in Chapter 3 would apply here. Thus tests of hypotheses and confidence interval coverage involving the fixed-effects parameters β would be the same as described in §2.1.2 (page 29) and §3.1.3 (page 78). Likewise, inference with respect to predictable linear functions of β and b would be based on the same methods described in §3.1.3 (page 78) while inference with respect to the variance-covariance parameters, $\theta = (\theta_b, \theta_w)$, would be based on the methods described for the marginal LM of Chapter 2 (see page 33).

Finally, since PL estimation is based on a doubly iterative algorithm and is not truly likelihood-based, the use of goodness-of-fit criteria and other techniques for assessing model adequacy for a LME model, although available from the final call to PROC MIXED in %NLINMIX, would not be appropriate for assessing adequacy under the GNLME model. Instead, one can utilize the SAS macro %GOF discussed briefly on page 34 of Chapter 2 as a means for assessing model adequacy (see also Appendix D for a description of the macro). An example illustrating the use of this macro was given in the analysis of the LDH enzyme leakage data (page 178).

Case 2: Inference based on likelihood estimation: When the population parameters, $\tau = (\beta, \theta)$, of the GNLME model are estimated via maximum likelihood using the Laplace approximation or numerical quadrature, inference will again mimic that described for the GLME model. Specifically, if we let $L^*(\tau, \tilde{b}(\tau); y)$ be the integrated log-likelihood evaluated using either the Laplace approximation $L_{\text{LA}}(\tau, \tilde{b}(\tau); y)$ in (5.65) or the adaptive Gaussian quadrature $L_{\text{AGQ}}(\tau, \tilde{b}(\tau); y)$ in (5.68), model-based inference with respect to estimable nonlinear functions would be based on the delta method in combination with the asymptotic properties of $\widehat{\tau}$ which, for large sample ML theory, may be summarized as

$$\sqrt{n}(\widehat{\tau}-\tau) \xrightarrow{p} \mathrm{N}(\mathbf{0}, \Omega_0) \tag{5.81}$$

$$Var(\widehat{\tau}) \simeq \widehat{\Omega}(\widehat{\tau}) = \left(\left. \frac{\partial^2 L^*(\tau, \tilde{b}(\tau); y)}{\partial\tau\partial\tau'} \right|_{\tau=\widehat{\tau}} \right)^{-1}$$

$$n\widehat{\Omega}(\widehat{\tau}) \xrightarrow{p} \Omega_0$$

where $\widehat{\boldsymbol{\Omega}}(\widehat{\boldsymbol{\tau}})$ is the model-based estimated asymptotic covariance matrix of $\widehat{\boldsymbol{\tau}}$ obtained as the inverse Hessian matrix associated with the approximate integrated log-likelihood function, $L^*(\boldsymbol{\tau}, \tilde{\boldsymbol{b}}(\boldsymbol{\tau}); \boldsymbol{y})$. Here, as for the GLME model, we have ignored the additional error term $O_p\{\min(p_i)^{-1}\}$ associated with the Laplace ML estimate, $\widehat{\boldsymbol{\tau}}_{\text{LMLE}}$, whereby Vonesh (1996) showed that $\widehat{\boldsymbol{\tau}}_{\text{LMLE}} - \boldsymbol{\tau} = O_p(n^{-1/2}) + O_p\{\min(p_i)^{-1}\}$. Alternatively, we can base inference on the robust sandwich estimator

$$\widehat{\boldsymbol{\Omega}}_R(\widehat{\boldsymbol{\tau}}) = \widehat{\boldsymbol{\Omega}}(\widehat{\boldsymbol{\tau}})\Big(\sum_{i=1}^{n} U_i^*(\widehat{\boldsymbol{\tau}})U_i^*(\widehat{\boldsymbol{\tau}})'\Big)\widehat{\boldsymbol{\Omega}}(\widehat{\boldsymbol{\tau}}) \tag{5.82}$$

where $U_i^*(\widehat{\boldsymbol{\tau}}) = \frac{\partial}{\partial \boldsymbol{\tau}} L_i^*(\boldsymbol{\tau}, \tilde{\boldsymbol{b}}(\boldsymbol{\tau}); \boldsymbol{y}_i)$ is the gradient of the approximate log-likelihood with respect to $\boldsymbol{\tau}$ for the ith experimental unit, subject or cluster.

Suppose we partition $\widehat{\boldsymbol{\Omega}}(\widehat{\boldsymbol{\tau}})$ or $\widehat{\boldsymbol{\Omega}}_R(\widehat{\boldsymbol{\tau}})$ as

$$\widehat{\boldsymbol{\Omega}} = \widehat{\boldsymbol{\Omega}}(\widehat{\boldsymbol{\tau}}) = \begin{pmatrix} \widehat{\boldsymbol{\Omega}}(\widehat{\boldsymbol{\beta}}) & \widehat{\boldsymbol{C}}(\widehat{\boldsymbol{\beta}}, \widehat{\boldsymbol{\theta}}) \\ \widehat{\boldsymbol{C}}(\widehat{\boldsymbol{\beta}}, \widehat{\boldsymbol{\theta}})' & \widehat{\boldsymbol{\Omega}}(\widehat{\boldsymbol{\theta}}) \end{pmatrix} \tag{5.83}$$

$$\widehat{\boldsymbol{\Omega}}_R = \widehat{\boldsymbol{\Omega}}_R(\widehat{\boldsymbol{\tau}}) = \begin{pmatrix} \widehat{\boldsymbol{\Omega}}_R(\widehat{\boldsymbol{\beta}}) & \widehat{\boldsymbol{C}}_R(\widehat{\boldsymbol{\beta}}, \widehat{\boldsymbol{\theta}}) \\ \widehat{\boldsymbol{C}}_R(\widehat{\boldsymbol{\beta}}, \widehat{\boldsymbol{\theta}})' & \widehat{\boldsymbol{\Omega}}_R(\widehat{\boldsymbol{\theta}}) \end{pmatrix},$$

where $\widehat{\boldsymbol{\Omega}}(\widehat{\boldsymbol{\beta}})$ and $\widehat{\boldsymbol{\Omega}}_R(\widehat{\boldsymbol{\beta}})$ are, respectively, the model-based and robust variance-covariance matrices of $\widehat{\boldsymbol{\beta}}$; $\widehat{\boldsymbol{\Omega}}(\widehat{\boldsymbol{\theta}})$ and $\widehat{\boldsymbol{\Omega}}_R(\widehat{\boldsymbol{\theta}})$ are the corresponding estimated covariance matrices of $\widehat{\boldsymbol{\theta}}$; and $\widehat{\boldsymbol{C}}(\widehat{\boldsymbol{\beta}}, \widehat{\boldsymbol{\theta}})$ and $\widehat{\boldsymbol{C}}_R(\widehat{\boldsymbol{\beta}}, \widehat{\boldsymbol{\theta}})$ are the corresponding estimated covariances of the crossed components of $\widehat{\boldsymbol{\beta}}$ and $\widehat{\boldsymbol{\theta}}$. Based on this partitioning, inference with respect to predictable functions of $\boldsymbol{\beta}$ and $\boldsymbol{b}$ may be based on the extended covariance matrix (Booth and Hobert, 1998)

$$\widehat{\boldsymbol{\Upsilon}}(\widehat{\boldsymbol{\tau}}, \widehat{\boldsymbol{b}}) = \begin{pmatrix} \widehat{\boldsymbol{\Omega}}^* & \widehat{\boldsymbol{\Omega}}^*\left(\frac{\partial \tilde{\boldsymbol{b}}(\boldsymbol{\tau})}{\partial \boldsymbol{\tau}}\Big|_{\widehat{\boldsymbol{\tau}}}\right) \\ \left(\frac{\partial \tilde{\boldsymbol{b}}(\boldsymbol{\tau})}{\partial \boldsymbol{\tau}'}\Big|_{\widehat{\boldsymbol{\tau}}}\right)\widehat{\boldsymbol{\Omega}}^* & \widehat{\boldsymbol{\Gamma}}^{-1} + \left(\frac{\partial \tilde{\boldsymbol{b}}(\boldsymbol{\tau})}{\partial \boldsymbol{\tau}'}\Big|_{\widehat{\boldsymbol{\tau}}}\right)\widehat{\boldsymbol{\Omega}}^*\left(\frac{\partial \tilde{\boldsymbol{b}}(\boldsymbol{\tau})}{\partial \boldsymbol{\tau}}\Big|_{\widehat{\boldsymbol{\tau}}}\right) \end{pmatrix} \tag{5.84}$$

where $\widehat{\boldsymbol{\Omega}}^* = \widehat{\boldsymbol{\Omega}}$ or $\widehat{\boldsymbol{\Omega}}^* = \widehat{\boldsymbol{\Omega}}_R$ depending on whether one invokes the EMPIRICAL option of NLMIXED, and $\widehat{\boldsymbol{\Gamma}}^{-1}$ is the $nv \times nv$ block diagonal matrix

$$\widehat{\boldsymbol{\Gamma}}^{-1} = \begin{pmatrix} -p_1 L_1''(\boldsymbol{\tau}, \tilde{\boldsymbol{b}}_1(\boldsymbol{\tau})) & \cdots & \boldsymbol{0} \\ \boldsymbol{0} & \ddots & \boldsymbol{0} \\ \boldsymbol{0} & \cdots & -p_n L_n''(\boldsymbol{\tau}, \tilde{\boldsymbol{b}}_n(\boldsymbol{\tau})) \end{pmatrix}\Bigg|_{\boldsymbol{\tau}=\widehat{\boldsymbol{\tau}}} \tag{5.85}$$

obtained by grouping, along the main diagonal, the second-order derivative matrices $-p_i L_i''(\boldsymbol{\tau}, \tilde{\boldsymbol{b}}_i(\boldsymbol{\tau}))$ associated with maximizing the log-posterior (5.62).

Based on the preceding large-sample theory, tests of hypotheses and confidence interval coverage on estimable nonlinear functions as well as inference with respect to predictable nonlinear functions follows that described for the GNLM in §4.4.3 (see equations 4.47-4.51). Given that the model parameters are estimated by maximizing an integrated log-likelihood, model goodness-of-fit may be based on the usual likelihood information criteria. Finally, if one chooses to estimate the model parameters of a GNLME model using augmented CGEE2/PELS, inference would be based on the above formulation as applied to a "working" Gaussian log-likelihood function in combination with the Laplace approximation. In this case, however, it would be better to evaluate model goodness-of-fit using the methods implemented in the %GOF macro as discussed previously.

5.4 Examples of GNLME models

In this section, several examples are given illustrating how one can fit data to a GNLME model using either the NLMIXED procedure for ML estimation or the %NLINMIX macro for PL estimation. We also illustrate how one can use either NLMIXED or GLIMMIX to estimate parameters via augmented CGEE2/PELS. In our first example, we revisit the orange tree data of Chapter 4 (§4.5.2). Our purpose here is to demonstrate how the macro %COVPARMS can be used to quickly decide on a random-effects structure for this data. We also show how one can incorporate a within-tree autoregressive correlation structure into the model using NLMIXED. In the second example, we fit several NLME models to the soybean growth data described in §1.4 of Chapter 1. The goal is to build a parsimonious NLME model that adequately accounts for any intra-cluster heterogeneity and inter-cluster variation that may be present. In our third example, we fit the high-flux hemodialyzer data of Vonesh and Carter (1992) to a group of candidate nonlinear models, both marginal and mixed, in an attempt to select the most parsimonious model possible. This exercise essentially mimics the analyses conducted by Littell et. al. (§9.5, pp. 374-392, 2006) but here we base the analysis on a nonlinear structural model adapted from the original model proposed by Vonesh and Carter (1992). The fourth example involves fitting pharmacokinetic data to a NLME model. Specifically, using a simple two-stage estimation scheme, we fit plasma concentrations of the cephalosporin antibiotic, cefamandole, to a bi-exponential model allowing for a power of the mean variance structure. We then compare estimates from this two-stage approach to those obtained via the first-order and ML approaches available in NLMIXED. In our final example, several overdispersed Poisson mixed-effects models were fit to the epileptic seizure data described in §1.4. Our goal here is to compare estimates using augmented PELS with those using PL/CGEE1.

5.4.1 Orange tree data—continued

In §4.5 (Example 4.5.2), we fit the orange tree data of Draper and Smith (1981) to a reparameterized version of the nonlinear logistic growth curve model used by Lindstrom and Bates (1990). In analyzing the data, we appealed to arguments by Lindstrom and Bates (1990) and Pinheiro and Bates (2000) to justify fitting the data to a model in which only the asymptote parameter β_1 was assumed to vary from tree to tree. The model, as defined in (4.58), is strictly linear in the random asymptote effect. As a result, exact ML estimation is possible using NLMIXED. We fit the data in NLMIXED using initial estimates close to the final estimates obtained by Pinheiro and Bates (2000). This, of course, is not how one would normally go about estimating the parameters of a model. Indeed, without recourse to some sort of preliminary analysis, one would be hard pressed to first determine exactly what parameters, if any, exhibit some degree of random variation and to then provide reasonable starting values to the variance components of those parameters. Below we illustrate how one might go about analyzing this data without the benefit of knowing the results ahead of time.

In §5.3.6, we discussed how one can use the SAS macro %COVPARMS described in Appendix D as a means for determining good starting values for an assumed set of random effects. To run this macro, one must initially fit the data to the GNLME model of interest using the NLMIXED procedure but with all variance components of the specified random effects $\boldsymbol{b}_i$ set to zero. This will yield OLS estimates $\widehat{\boldsymbol{\beta}}_{\text{OLS}}$ of the regression parameters as well as the necessary residual estimates, $r_{ij} = y_{ij} - \widehat{y}_{ij}(\widehat{\boldsymbol{\beta}}_{\text{OLS}}, \mathbf{0})$, and derivatives of the conditional predicted values with respect to $\boldsymbol{b}_i$, (i.e., $\tilde{\boldsymbol{z}}_{ij}(\widehat{\boldsymbol{\beta}}_{\text{OLS}}, \mathbf{0})$), needed to fit the residuals to the linear random-effects model (5.78). Based on this input, a call to the macro %COVPARMS would then use the GLIMMIX procedure to estimate the variance-covariance parameters $\boldsymbol{\theta}_b$ of the random-effects covariance matrix, $\boldsymbol{\Psi}(\boldsymbol{\theta}_b)$, whose structure is determined by the TYPE= option of the RANDOM statement. In our current example, since we do not know a priori which parameters exhibit random variation, a good strategy would be to initially assume random effects for all three parameters. The following program provides the code necessary to make an initial call to %COVPARMS assuming an unstructured covariance matrix for the random effects.

Program 5.11

```
proc sort data=example4_5_2 out=example5_4_1;
 by Tree Days;
run;
/* Step 1: Run NLMIXED on the proposed model */
ods listing close;
ods output ParameterEstimates=OLSparms;
proc nlmixed data=example5_4_1 qpoints=1;
 parms b1=175 b2=700 b3=300 sigsq_w=10 to 100 by 10;
 num = b1+bi1;
 den = 1 + exp(-(Days-(b2+bi2))/(b3+bi3));
 predmean = num/den;
 resid = y - predmean;
 model y ~ normal(predmean, sigsq_w);
 random bi1 bi2 bi3 ~ normal([0,0,0],
                             [0,
                              0,0,
                              0,0,0])
        subject=Tree;
 predict predmean out=predout der;
 id resid;
run;
ods listing;
/* Step 2: Run the macro COVPARMS based on NLMIXED output */
%covparms(parms=OLSparms, predout=predout, resid=resid,
          method=mspl, random=der_bi1 der_bi2 der_bi3,
          subject=Tree, type=un, covname=psi, output=MLEparms);
proc print data=MLEparms;run;
```

To make the initial call to NLMIXED, we still need starting values for the parameters β_1, β_2, β_3 and σ_w^2 of model (4.58) as represented by the NLMIXED parameters `b1, b2, b3,` and `sigsq_w`. By visual inspection of the data (Figure 1.1), we might expect the average asymptote to be about 175 mm. Based on this, we can then use the relation $\beta_2 - t = \beta_3 \log(175/y - 1)$ to find starting values for β_2 and β_3. Again, by visual inspection of the data, we might guess that at $t = 100$ days, the average value of y is roughly 20 mm, which would yield $\beta_3 \approx (\beta_2 - 100)/2$. Now at $y = 87.5$, we have $\beta_2 - t = 0$ and from Figure 1.1 we might guess this occurs at about 700 days which would yield starting values of 700 for β_2 and 300 for β_3. Finally, we simply use a grid search along feasible values to obtain an initial starting value for σ_w^2. Given these starting values, the NLMIXED procedure yields OLS starting values β_1, β_2, β_3 and σ_w^2 while the call to the %COVPARMS macro yields starting values of the variance-covariance parameters of the random effects. A full description of the macro and the macro variables, PARMS=, PREDOUT=, RESID=, METHOD=, RANDOM=, SUBJECT=, TYPE=, COVNAME=, and OUTPUT= is given in Appendix D. The macro displays the results of fitting the linear random-effects model (5.78) assuming an unstructured covariance matrix as determined by the above macro assignment, `type=un`. It also creates the output dataset MLEparms containing the OLS estimates from initial NLMIXED call combined with estimates of the random-effects covariance matrix using a naming convention of psi11, psi21, psi22, etc. as determined by the macro assignment, `covname=psi`, in combination with the designated covariance structure. The generated output from this initial call to %COVPARMS is shown in Output 5.20 below.

Output 5.20: Initial estimates of the variance-covariance parameters of the random effects assuming an unstructured covariance matrix. Included is the SAS dataset MLEparms containing all of the starting parameters one might consider using if one were to run the full NLME model.

Estimated G Matrix

Effect	Row	Col1	Col2	Col3
Der_bi1	1	1324.17	−1356.78	−2656.77
Der_bi2	2	−1356.78	17.2882	−815.14
Der_bi3	3	−2656.77	−815.14	4.09E−16

Estimated G Correlation Matrix

Effect	Row	Col1	Col2	Col3
Der_bi1	1	1.0000	−8.9673	
Der_bi2	2	−8.9673	1.0000	
Der_bi3	3			

Covariance Parameter Estimates

Cov Parm	Subject	Estimate	Standard Error	Wald 95% Confidence Bounds	
UN(1,1)	Tree	1324.17	1027.62	442.45	15058
UN(2,1)	Tree	−1356.78	1384.03	−4069.43	1355.87
UN(2,2)	Tree	17.2882	4702.67	.	.
UN(3,1)	Tree	−2656.77	1127.30	−4866.23	−447.30
UN(3,2)	Tree	−815.14	2513.66	−5741.81	4111.54
UN(3,3)	Tree	4.09E−16	.	.	.
Residual		55.6490	15.7400	34.2273	106.04

Tests of Covariance Parameters Based on the Likelihood

Label	DF	-2 Log Like	ChiSq	Pr > ChiSq	Note
Diagonal G	3	262.93	12.47	0.0059	--

Obs	Parameter	Estimate	StandardError	Ratio
1	b1	192.69	19.8035	9.72993
2	b2	728.75	104.67	6.96222
3	b3	353.53	79.9191	4.42360
4	sigsq_w	499.43	119.39	4.18333
5	psi11	1324.17	1027.62	1.28859
6	psi21	−1356.78	1384.03	−0.98031
7	psi22	17.2882	4702.67	0.00368
8	psi31	−2656.77	1127.30	−2.35676
9	psi32	−815.14	2513.66	−0.32428
10	psi33	4.09E−16	.	.

A quick glance at the output suggests that the random effects b_{i2} and b_{i3} associated with the intercept parameter β_2 and rate parameter β_3 are not feasible given that the ratio of the estimate to its standard error (`psi22`) or the absolute value of the estimate (`psi33`) is essentially 0.

At this point one could make additional calls to the %COVPARMS macro to evaluate alternative covariance structures based on different subsets of the random effects. For example, the following two calls to %COVAPRMS

Program 5.12

```
%covparms(parms=OLSparms, predout=predout, resid=resid,
          method=mspl, random=der_bi1 der_bi2 der_bi3,
          subject=tree, type=vc, covname=psi, output=MLEparms);
%covparms(parms=OLSparms, predout=predout, resid=resid,
          method=mspl, random=der_bi1 der_bi2,
          subject=tree, type=un, covname=psi, output=MLEparms);
```

will enable one to evaluate whether a variance components structure (type=vc) for the random effects is feasible or whether a reduced subset of random effects is feasible. For this example, both of these calls will only confirm what is evident from the initial call, namely that the only parameter exhibiting any measurable degree of variability is the asymptote parameter β_1. Consequently, it seems reasonable to assume that the only random effect in the model is b_{i1} as suggested by Lindstrom and Bates (1990) and Pinheiro and Bates (2000). We therefore fit the NLME model (4.58) one more time using NLMIXED but with starting values determined from a final call to the macro. The programming code is as follows.

Program 5.13

```
/* Step 3: Final call to macro and final run of NLMIXED */
%covparms(parms=OLSparms, predout=predout, resid=resid,
          method=mspl, random=der_bi1,
          subject=tree, type=un, covname=psi, output=MLEparms);
ods listing close;
ods output parameterestimates=pe;
proc nlmixed data=example5_4_1 qpoints=1 ;
 parms / data=MLEparms;
 num = b1+bi1;
 den = 1 + exp(-(Days-b2)/b3);
 predmean = (num/den);
 model y ~ normal(predmean, sigsq_w);
 random bi1 ~ normal(0, psi11)
        subject=tree;
run;
ods listing;
proc print data=pe noobs;
 var Parameter Estimate StandardError DF tValue Probt;
run;
```

The results from NLMIXED, shown in Output 5.21, are nearly identical with those shown in Output 4.24 based on slightly different starting values.

Output 5.21: Final parameter estimates for the orange tree data based on the NLME model (4.58) and starting values from %COVPARMS.

Obs	Parameter	Estimate	StandardError	DF	tValue	Probt
1	b1	192.05	15.6682	4	12.26	0.0003
2	b2	727.92	35.2508	4	20.65	<.0001
3	b3	348.08	27.0810	4	12.85	0.0002
4	sigsq_w	61.5125	15.8824	4	3.87	0.0179
5	psi11	1003.11	651.53	4	1.54	0.1985

To this point, we have demonstrated how one can use the macro %COVPARMS to systematically determine what a reasonable random-effects structure might be for this data as well as provide reasonable starting values for the model parameters. We now turn our attention to consideration of the assumed intra-subject or within-tree covariance structure. The preceding analysis assumes that, conditional on the random effect b_{i1}, the within-tree errors, ϵ_{ij}, are independent and identically distributed $N(0, \sigma_w^2)$ random variables. However, in our previous analysis of the orange tree data, we found some evidence suggesting the overall covariance structure may be misspecified (see Output 4.26 and related discussion). Lindstrom and Bates (1990), for example, note that because the data are collected over time, there is likely to be some serial correlation among measurements taken on each tree. To take this possibility into account, we ran an analysis assuming an intra-subject autoregressive structure for unequally spaced time points. Specifically, we assume that $\boldsymbol{\epsilon}_i \sim$iid $N(\mathbf{0}, \boldsymbol{\Lambda}_i(\boldsymbol{\theta}_w))$ where $\boldsymbol{\theta}_w = (\sigma_w^2, \rho)$ and the $(j, j')^{\text{th}}$ component of $\boldsymbol{\Lambda}_i(\boldsymbol{\theta}_w)$ is defined by $Cov(\epsilon_{ij}, \epsilon_{ij'}|b_{i1}) = \sigma_w^2 \rho^{|t_{ij}-t_{ij'}|}$. While one can easily fit this model using the PL estimation scheme implemented in the macro %NLINMIX, it does require some additional programming if one wishes to obtain exact ML estimates using the NLMIXED procedure. Below are the additional programming statements required to fit this model in NLMIXED using a set of orthogonal linear random effects as described in §5.3.3.

Program 5.14

```
data example5_4_1;
 set example5_4_1;
 by Tree Days;
 retain index;
 array Ind[7] I1-I7;
 if first.Tree then index=0;
 index+1;
 do i=1 to 7;
  Ind[i]=(index=i);
 end;
 days1= 118; days2= 484; days3= 664; days4=1004;
 days5=1231; days6=1372; days7=1582;
run;
/* Add initial value for the autoregressive correlation */
data rho;
 Parameter='rho';estimate=.9;
run;
data MLEparms_rho;
 set MLEparms rho;
 if Parameter='sigsq_w' then estimate=61.5125;
run;
/* Macro VECH creates a vech representation of the */
/* intra-cluster covariance matrix defined in the  */
/* NLMIXED code for any particular application that*/
/* requires an intra-cluster or within-subject     */
/* covariance structure (see Appendix D of Ch  7   */
/* for a more detailed presentation of the macro)  */
%macro vech(dim=4, cov=cov, vechcov=vechcov, name=c);
    %global n vech;
    %let n=%eval(&dim*(&dim+1)/2);
    array &vechcov.[&n] &name.1-&name.&n;
    %let k=0;
    %do i=1 %to &dim;
     %do j=1 %to &i;
         %let k=%eval(&k+1);
         &vechcov.[&k] = &cov.[&i, &j];
     %end;
    %end;
    %let temp = &name.1;
    %do i=2 %to &n;
     %let temp=&temp,&name.&i;
    %end;
    %let vech = &temp;
%mend vech;
```

```
proc nlmixed data=example5_4_1 df=4 qpoints=1;
 parms / data=MLEparms_rho;
 bounds -1<rho< 1;
 /* The following variable, eij, is a within-tree   */
 /* error term expressed as a linear combination of */
 /* linear random effects that have a first-order   */
 /* autoregressive structure for unequally spaced   */
 /* time points as defined by ARRAY statements      */
 eij = I1*e1+I2*e2+I3*e3+I4*e4+I5*e5+I6*e6+I7*e7;
 num = b1+bi1;
 den = 1 + exp(-(Days-b2)/b3);
 predmean = (num/den) + eij;
 delta=1e-8;
 /* The following array statements define the first */
 /* order autoregressive covariance matrix assuming */
 /* unequally spaced time points within a cluster   */
 array cov[7,7];
 array d[7] days1-days7;
 do i=1 to 7;
  do j=1 to 7;
   cov[i,j] = sigsq_w*(rho**abs(d[i]-d[j]));
  end;
 end;
 /* This call to the macro VECH creates a 7x7 lower */
 /* diagonal autoregressive covariance matrix with  */
 /* successive elements c1, c2, c3,..., c28 being   */
 /* the components of the vech representation of    */
 /* this matrix where c is determined by the macro  */
 /* variable NAME= option. This is then used in the */
 /* specification of a block diagonal covariance    */
 /* structure where the random effect bi1 is assumed*/
 /* orthogonal to (independent of) the linear random*/
 /* effects, e1, e2,...,e7                          */
 %vech(dim=7, cov=cov, name=c);
 model y ~ normal(predmean, delta);
 random bi1 e1 e2 e3 e4 e5 e6 e7 ~
        normal([ 0, 0, 0, 0, 0, 0, 0, 0 ],
               [ psi11,
                   0  , c1 ,
                   0  , c2 , c3 ,
                   0  , c4 , c5 , c6 ,
                   0  , c7 , c8 , c9 , c10,
                   0  , c11, c12, c13, c14, c15,
                   0  , c16, c17, c18, c19, c20, c21,
                   0  , c22, c23, c24, c25, c26, c27, c28 ])
        subject=tree;
run;
```

A number of comments are included in the program to help explain how an intra-subject covariance matrix can be modeled in NLMIXED. We include a BOUNDS statement to ensure the autocorrelation coefficient satisfies the restriction $|\rho| < 1$. Also, because the model uses a set of "dummy" random effects to represent the within-subject errors of an autoregressive structure, the NLMIXED option DF=4 is used to override the default degrees of freedom, $n - v$, that NLMIXED would normally use to conduct inference (with 7 "dummy" random effects, the default DF would be −3 and no tests of hypotheses or confidence intervals would be produced). The results, shown in Output 5.22 below, suggest there is little to gain by including an intra-subject autoregressive correlation in the model. Given the enormous gaps in time between measurements, this result is not surprising since the estimated intra-subject correlation between day 118 and day 484 is essentially 0 even though ρ is estimated to be 0.8421. It would appear, therefore, that the random asymptote effect, b_{i1}, suffices to explain both the observed heterogeneity and correlation in the data.

Output 5.22: Results of fitting the NLME model (4.58) to the orange tree data under the assumption of an intra-subject autoregressive structure.

Specifications

Data Set	WORK.EXAMPLE5_4_1
Dependent Variable	y
Distribution for Dependent Variable	Normal
Random Effects	bi1 e1 e2 e3 e4 e5 e6 e7
Distribution for Random Effects	Normal
Subject Variable	Tree
Optimization Technique	Dual Quasi−Newton
Integration Method	Adaptive Gaussian Quadrature

Dimensions

Observations Used	35
Observations Not Used	0
Total Observations	35
Subjects	5
Max Obs Per Subject	7
Parameters	6
Quadrature Points	1

Iteration History

Iter	Flag1	Calls	NegLogLike	Diff	MaxGrad	Slope
1		4	131.575199	0.03834	0.001623	−0.03616
2		6	131.57302	0.002179	0.001535	−0.00039
3		8	131.572169	0.000852	0.001084	−0.00068
4		9	131.571958	0.00021	0.000473	−0.0003
5		10	131.571929	0.000029	0.000095	−0.00006
6		12	131.571929	4.681E−7	0.000021	−1.42E−6
7		13	131.571927	1.281E−6	0.000126	−1.44E−7

NOTE: GCONV convergence criterion satisfied.

Fit Statistics

−2 Log Likelihood	263.1
AIC (smaller is better)	275.1
AICC (smaller is better)	278.1
BIC (smaller is better)	272.8

Parameter Estimates

Parameter	Estimate	Standard Error	DF	t Value	Pr > |t|
b1	192.05	15.6054	4	12.31	0.0003
b2	727.94	34.1916	4	21.29	<.0001
b3	348.08	26.1946	4	13.29	0.0002
sigsq_w	61.4799	16.6643	4	3.69	0.0210
psi11	1007.38	656.43	4	1.53	0.1997
rho	0.8421	512.45	4	0.00	0.9988

Parameter	Alpha	Lower	Upper	Gradient
b1	0.05	148.72	235.37	−0.00006
b2	0.05	633.01	822.87	0.000039
b3	0.05	275.36	420.81	−0.00004
sigsq_w	0.05	15.2123	107.75	−0.00013
psi11	0.05	−815.15	2829.92	0.000014
rho	0.05	−1421.96	1423.64	2.29E−8

5.4.2 Soybean growth data

In this example, we compare the fit of several nonlinear logistic growth curve models to the soybean growth data described in §1.4 of Chapter 1. Unlike the orange tree data which is strictly longitudinal, the soybean growth data is an example of clustered longitudinal data in that measurements are obtained over time on plants sampled within different plots (clusters) of land. Davidian and Giltinan (1993a) fit a "cell-means" version of the NLME logistic growth curve model (1.2) to the data. They did so under each of two intra-cluster covariance structures, a power of the mean variance structure assuming conditional independence and a power of the mean variance structure assuming autocorrelation. In both cases, an unstructured covariance matrix was assumed for all three random effects, $\boldsymbol{b}_i' = (b_{i1}, b_{i2}, b_{i3})$. The authors fit both models to the data using each of two estimation schemes, a pooled two-stage estimation scheme and an adaptation of a linearization-based scheme proposed by Vonesh and Carter (1992). A description and partial listing of the raw data is available through the author's Web page.

Rather than fit a main-effects logistic growth curve model as defined by model (1.2) of §1.4, here we follow Davidian and Giltinan (1993a) and fit a cell means logistic growth curve model. Using the same notation as in §1.4, such a model may be written as

$$\begin{aligned} y_{ij} &= f(\boldsymbol{x}_{ij}', \boldsymbol{\beta}, \boldsymbol{b}_i) + \epsilon_{ij} \\ &= f(\boldsymbol{x}_{ij}', \boldsymbol{\beta}_i) + \epsilon_{ij} \\ &= \frac{\beta_{i1}}{1+\exp\{\beta_{i3}(t_{ij}-\beta_{i2})\}} + \epsilon_{ij} \end{aligned} \tag{5.86}$$

where y_{ij} is the average leaf weight per plant for the ith plot on the jth occasion as before but where the vector of within- and between-cluster covariates is now defined to be $\boldsymbol{x}_{ij}' = \begin{pmatrix}1 & t_{ij} & x_{1i} & x_{2i} & x_{3i} & x_{4i} & x_{5i} & x_{6i}\end{pmatrix}$ where $x_{1i} = a_{1i}(1-a_{2i}-a_{3i})$, $x_{2i} = (1-a_{1i})(1-a_{2i}-a_{3i})$, $x_{3i} = a_{1i}a_{2i}$, $x_{4i} = (1-a_{1i})a_{2i}$, $x_{5i} = a_{1i}a_{3i}$, $x_{6i} = (1-a_{1i})a_{3i}$ are cell-mean indicator variables that designate which genotype-year combination the response variable y_{ij} corresponds (see §1.4 for a definition of a_{1i}, a_{2i} and a_{3i}). The structural form of the conditional mean, $f(\boldsymbol{x}_{ij}', \boldsymbol{\beta}, \boldsymbol{b}_i) = f(\boldsymbol{x}_{ij}', \boldsymbol{\beta}_i) = \beta_{i1}/\big[1+\exp\{\beta_{i3}(t_{ij}-\beta_{i2})\}\big]$, remains the same but the cluster-specific parameter vector, $\boldsymbol{\beta}_i' = \begin{pmatrix}\beta_{i1} & \beta_{i2} & \beta_{i3}\end{pmatrix}$, is now defined in terms of the cell mean indicator variables as

$$\begin{aligned} \boldsymbol{\beta}_i &= \begin{pmatrix}\beta_{i1}\\ \beta_{i2}\\ \beta_{i3}\end{pmatrix} \\ &= \begin{pmatrix} \beta_{11}x_{1i} + \beta_{12}x_{2i} + \beta_{13}x_{3i} + \beta_{14}x_{4i} + \beta_{15}x_{5i} + \beta_{16}x_{6i} \\ \beta_{21}x_{1i} + \beta_{22}x_{2i} + \beta_{23}x_{3i} + \beta_{24}x_{4i} + \beta_{25}x_{5i} + \beta_{26}x_{6i} \\ \beta_{31}x_{1i} + \beta_{32}x_{2i} + \beta_{33}x_{3i} + \beta_{34}x_{4i} + \beta_{35}x_{5i} + \beta_{36}x_{6i} \end{pmatrix} + \begin{pmatrix}b_{i1}\\ b_{i2}\\ b_{i3}\end{pmatrix}. \end{aligned}$$

The vector $\boldsymbol{\beta}$ of fixed-effects regression parameters is an 18×1 vector with components $\beta_{11}, \beta_{12}, \ldots, \beta_{36}$ while the cluster-specific regression parameters are defined by $\beta_{i1} > 0$ which is a limiting growth value or asymptote for the ith plot, $\beta_{i2} > 0$ which is a soybean "half-life" parameter (i.e., the time at which the soybean reaches half its limiting growth value) and $\beta_{i3} < 0$ which is the corresponding growth rate. As in model (1.2), the first two columns of $\boldsymbol{x}_{ij}'$ represent a vector $\boldsymbol{z}_{ij}' = \begin{pmatrix}1 & t_{ij}\end{pmatrix}$ of the within-cluster intercept and time covariates. The vector $\boldsymbol{b}_i' = \begin{pmatrix}b_{i1} & b_{i2} & b_{i3}\end{pmatrix}$ represents a complete set of possible cluster-specific random effects while the ϵ_{ij} are intra-cluster (within-plot or within-unit) errors. In terms of intra-cluster variation, Davidian and Giltinan (1993a) proposed a power of the mean variance structure as a way of accounting for the increasing variability that is apparent when one examines the set of studentized residuals versus predicted values obtained from individual OLS fits to the data.

Here we attempt to fit the NLME model (5.86) to the soybean growth data assuming a power of the mean variance structure for ϵ_{ij} and an unstructured variance-covariance matrix for the random effects $\boldsymbol{b}_i$. That is, we assume ϵ_{ij} ~ind $N(0, \sigma_w^2 f(\boldsymbol{x}'_{ij}, \boldsymbol{\beta}, \boldsymbol{b}_i)^{2\delta})$ and $\boldsymbol{b}_i$ ~iid $N(\boldsymbol{0}, \boldsymbol{\Psi})$ where $\boldsymbol{\Psi}$ is an arbitrary 3×3 positive-definite covariance matrix. To obtain reasonable starting values, we make an initial call to NLMIXED to get OLS estimates of the regression parameters and we then apply the macro %COVPARMS to the output from NLMIXED to get starting values for the covariance parameters of the random effects. The program for this initial analysis is as follows.

Program 5.15

```
/* Step 1: Arrange Data and Get Starting Values */
data Example5_4_2;
 set SASdata.Soybean_Data;
 Days=D1 ;Weight=W1 ; output;
 Days=D2 ;Weight=W2 ; output;
 Days=D3 ;Weight=W3 ; output;
 Days=D4 ;Weight=W4 ; output;
 Days=D5 ;Weight=W5 ; output;
 Days=D6 ;Weight=W6 ; output;
 Days=D7 ;Weight=W7 ; output;
 Days=D8 ;Weight=W8 ; output;
 Days=D9 ;Weight=W9 ; output;
 Days=D10;Weight=W10; output;
 drop D1-D10 W1-W10;
run;
data Example5_4_2;
 set Example5_4_2;
 Trt=Compress(Trim(Genotype)||'-'||Trim(Year));
 /* Rename response variable as y */
 y=Weight;
 /* Define cell means indicator variables */
 X1=(Trt='F-1988');
 X2=(Trt='P-1988');
 X3=(Trt='F-1989');
 X4=(Trt='P-1989');
 X5=(Trt='F-1990');
 X6=(Trt='P-1990');
 _Days_=Days;
 if y=. then delete;
run;
proc sort data=Example5_4_2;
 by Trt Days subject;
run;
ods listing close;
ods output parameterestimates=SAStemp.OLSparms;
proc nlmixed data=Example5_4_2 method=gauss qpoints=1;
 parms b11=20 b12=20 b13=10 b14=18 b15=15 b16=18
       b21=50 b22=50 b23=50 b24=50 b25=50 b26=50
       b31=-0.1 b32=-0.1 b33=-0.1 b34=-0.1 b35=-0.1 b36=-0.1
       sigsq_w=.05 delta=1;
```

```
beta1 = b11*X1 + b12*X2 + b13*X3 +
        b14*X4 + b15*X5 + b16*X6;
beta2 = b21*X1 + b22*X2 + b23*X3 +
        b24*X4 + b25*X5 + b26*X6;
beta3 = b31*X1 + b32*X2 + b33*X3 +
        b34*X4 + b35*X5 + b36*X6;
beta3 = beta3/100;
pred = (beta1+bi1)/(1+exp((beta3+bi3)*(Days-(beta2+bi2))));
var = sigsq_w*(pred**(2*delta));
resid = y - pred;
model y ~ normal(pred, var);
random bi1 bi2 bi3 ~ normal([0,0,0],
                    [0,
                     0,0,
                     0,0,0])
       subject=Plot;
predict pred out=SAStemp.predout der;
id resid;
run;
ods listing;
%covparms(parms=SAStemp.OLSparms, predout=SAStemp.predout,
         resid=resid, method=mspl,
         random=der_bi1 der_bi2 der_bi3, subject=Plot,
         type=un, covname=psi, output=MLEparms);
```

Output 5.23: Initial estimates of the random effects variance-covariance parameters for the soybean growth data based on the %COVPARMS macro.

Estimated G Matrix

Effect	Row	Col1	Col2	Col3
Der_bi1	1	9.9454	12.7428	0.08735
Der_bi2	2	12.7428	20.8328	0.1658
Der_bi3	3	0.08735	0.1658	0.001276

Estimated G Correlation Matrix

Effect	Row	Col1	Col2	Col3
Der_bi1	1	1.0000	0.8853	0.7752
Der_bi2	2	0.8853	1.0000	1.0169
Der_bi3	3	0.7752	1.0169	1.0000

Covariance Parameter Estimates

Cov Parm	Subject	Estimate	Standard Error	Wald 95% Confidence Bounds	
UN(1,1)	Plot	9.9454	2.5291	6.3893	17.5920
UN(2,1)	Plot	12.7428	3.3420	6.1926	19.2929
UN(2,2)	Plot	20.8328	5.3081	13.3734	36.8977
UN(3,1)	Plot	0.08735	0.02821	0.03206	0.1426
UN(3,2)	Plot	0.1658	0.04537	0.07691	0.2547
UN(3,3)	Plot	0.001276	0.000451	0.000708	0.002958
Residual		1.3268	0.1117	1.1325	1.5760

Tests of Covariance Parameters Based on the Likelihood

Label	DF	-2 Log Like	ChiSq	Pr > ChiSq	Note
Diagonal G	3	1518.65	81.38	<.0001	DF

The results of this initial analysis, displayed in Output 5.23, show the unstructured random-effects covariance matrix is singular due to the high correlation between the random effects, particularly b_{i2} and b_{i3}. Attempts to fit the full model with all three random effects failed and subsequent calls to %COVPARMS suggests either (b_{i1}, b_{i2}) or b_{i1} alone should be included in the model as random effects. In particular, two additional calls to %COVPARMS were made, one with the macro argument `random=der_bi1 der_bi2` and one with the macro argument `random=der_bi1 der_bi3`. When we included b_{i1} and b_{i3} as the only random effects, the variance component for b_{i3} was essentially estimated to be 0 (results not shown) suggesting this is not a feasible structure for the random effects. Using the following call to %COVPARMS, we then examined the covariance structure when b_{i1} and b_{i2} are included as the only random effects.

Program 5.16

```
%covparms(parms=SAStemp.OLSparms, predout=SAStemp.predout,
          resid=resid, method=mspl,
          random=der_bi1 der_bi2, subject=Plot,
          type=un, covname=psi, output=MLEparms);
```

The results, shown in Output 5.24, appear to be reasonable. Specifically, the variance component estimates are bounded away from 0 and there is no indication that the random effects are uncorrelated as indicated by both the estimated correlation matrix and by the likelihood ratio test of whether the off-diagonal elements of the unstructured covariance matrix are 0.

Output 5.24: Initial estimates of the random effects variance-covariance parameters for the soybean growth data assuming b_{i1} and b_{i2} are random.

Estimated G Matrix

Effect	Row	Col1	Col2
Der_bi1	1	4.7461	5.2519
Der_bi2	2	5.2519	9.1854

Estimated G Correlation Matrix

Effect	Row	Col1	Col2
Der_bi1	1	1.0000	0.7954
Der_bi2	2	0.7954	1.0000

Covariance Parameter Estimates

Cov Parm	Subject	Estimate	Standard Error	Wald 95% Confidence Bounds	
UN(1,1)	Plot	4.7461	1.2359	3.0206	8.5280
UN(2,1)	Plot	5.2519	1.5933	2.1290	8.3747
UN(2,2)	Plot	9.1854	2.9470	5.3403	19.3998
Residual		1.6394	0.1334	1.4064	1.9358

Tests of Covariance Parameters Based on the Likelihood

Label	DF	-2 Log Like	ChiSq	Pr > ChiSq	Note
Diagonal G	1	1518.65	26.24	<.0001	DF

We therefore fit the NLME model (5.86) to the data assuming b_{i1} and b_{i2} are the only random effects using the following program.

Program 5.17

```
proc nlmixed data=Example5_4_2 qpoints=1;
 parms /data=MLEparms;
 beta1 = b11*X1 + b12*X2 + b13*X3 +
         b14*X4 + b15*X5 + b16*X6;
 beta2 = b21*X1 + b22*X2 + b23*X3 +
         b24*X4 + b25*X5 + b26*X6;
```

```
beta3 = b31*X1 + b32*X2 + b33*X3 +
        b34*X4 + b35*X5 + b36*X6;
beta3 = beta3/100;
pred = (beta1+bi1)/(1+exp(beta3*(Days-(beta2+bi2))));
pred_avg = beta1/(1+exp(beta3*(Days-beta2)));
var = sigsq_w*(pred**(2*delta));
rho21 = psi21/(sqrt(psi11)*sqrt(psi22));
resid = y - pred;
model y ~ normal(pred, var);
random bi1 bi2 ~ normal([0, 0],
                        [psi11,
                         psi21, psi22])
       subject=Plot;
estimate 'corr(bi1,bi2)=' rho21;
contrast 'Test of Asymptote Effects:'
         b11-b12, b11-b13, b11-b14, b11-b15, b11-b16;
contrast 'Test of Half-Life Effects:'
         b21-b22, b21-b23, b21-b24, b21-b25, b21-b26;
contrast 'Test of Growth Rate Effects:'
         b31-b32, b31-b33, b31-b34, b31-b35, b31-b36;
run;
```

An ESTIMATE statement is included in the program to compute the correlation coefficient between b_{i1} and b_{i2}. In addition, three CONTRAST statements are included to test whether there is any overall effect due to year and/or genotype on the asymptote parameters, $\beta_{11}, \ldots, \beta_{16}$, the half-life parameters, $\beta_{21}, \ldots, \beta_{26}$, or the growth rate parameters, $\beta_{31}, \ldots, \beta_{36}$.

Based on the default options used, the above call to the NLMIXED procedure failed to produce any output and the program instead issued the following error and warning messages:

```
ERROR: QUANEW Optimization cannot be completed.
WARNING: Optimization routine cannot improve the function value.
```

At this point, one might consider further simplifying the model by dropping b_{i2} as a random effect from the model. However, such convergence issues are often related to the default optimization routine and we would strongly encourage the user to run the model again but with the default quasi-Newton optimization technique, TECH=QUANEW, replaced by an alternative optimization technique. We have found that using the Newton-Raphson optimization technique, TECH=NEWRAP, will often succeed when the default technique fails. For example, when we ran the above program again using the NLMIXED options MAXFUNC=1000 and TECH=NEWRAP, the optimization routine converged resulting in the parameter estimates displayed in Output 5.25. Although convergence was achieved, the resulting estimates of the variance-covariance parameters are all deemed nonsignificant. We note, however, that the correlation coefficient between b_{i1} and b_{i2} is estimated to be -0.9974 suggesting a near-singular variance-covariance matrix associated with the random effects may be a problem. Specifically, a near-singular covariance matrix of the random effects may produce an ill-conditioned Hessian matrix which, when inverted, could yield erroneously inflated standard errors. There are a number of options available to users interested in setting singularity criteria for the inversion of the Hessian matrix and the reader is referred to the documentation of NLMIXED for further details.

Output 5.25: Select output when model (5.86) is fit to the soybean growth data assuming b_{i1} and b_{i2} are the only random effects in the model.

Specifications

Data Set	WORK.EXAMPLE5_4_2
Dependent Variable	y
Distribution for Dependent Variable	Normal
Random Effects	bi1 bi2
Distribution for Random Effects	Normal
Subject Variable	Plot
Optimization Technique	Newton−Raphson
Integration Method	Adaptive Gaussian Quadrature

Dimensions

Observations Used	412
Observations Not Used	0
Total Observations	412
Subjects	48
Max Obs Per Subject	10
Parameters	23
Quadrature Points	1

Iteration History

Iter	Flag1	Calls	NegLogLike	Diff	MaxGrad	Slope
1	*	53	313.91404	4.218614	350.0965	−36.9311
2	*	78	309.630377	4.283664	127.6256	−10.3069
3	*	105	301.014512	8.615865	307.2788	−5.7051
4	*	130	298.806749	2.207763	42.81457	−3.08211
5	*	155	294.888082	3.918667	23.07151	−4.17753
6	*	180	294.289558	0.598524	3.189443	−0.66866
7	*	207	290.815033	3.474525	19.50166	−1.94584
8	*	233	289.502196	1.312837	21.4545	−13.7208
9	*	258	288.068096	1.4341	37.8979	−4.54735
10		454	288.034988	0.033108	35.70477	−0.57181
11	*	482	288.027619	0.00737	2.901297	−31.6636
12	*	510	288.027562	0.000057	2.946879	−6.92E−7

NOTE: GCONV convergence criterion satisfied.

Fit Statistics

−2 Log Likelihood	576.1
AIC (smaller is better)	622.1
AICC (smaller is better)	624.9
BIC (smaller is better)	665.1

Parameter Estimates

Parameter	Estimate	Standard Error	DF	t Value	Pr > |t|
b11	18.7104	1.0349	46	18.08	<.0001
b12	21.4629	1.1155	46	19.24	<.0001
b13	10.5045	0.5811	46	18.08	<.0001
b14	18.0661	0.8970	46	20.14	<.0001
b15	15.9210	0.8823	46	18.05	<.0001
b16	17.2811	0.9355	46	18.47	<.0001
b21	54.1948	1.0492	46	51.65	<.0001
b22	53.9983	1.0135	46	53.28	<.0001
b23	52.1099	0.9331	46	55.85	<.0001
b24	51.3977	0.9741	46	52.76	<.0001
b25	49.4072	1.1244	46	43.94	<.0001
b26	48.1897	1.0328	46	46.66	<.0001
b31	−12.5827	0.3477	46	−36.19	<.0001
b32	−12.3505	0.3274	46	−37.73	<.0001
b33	−14.2030	0.3925	46	−36.19	<.0001
b34	−13.9491	0.4264	46	−32.71	<.0001
b35	−13.7472	0.5578	46	−24.64	<.0001
b36	−13.5521	0.4953	46	−27.36	<.0001

sigsq_w	0.04415	0.003987	46	11.07	<.0001
delta	0.9549	0.02921	46	32.70	<.0001
psi11	0.6096	0.9577	46	0.64	0.5276
psi21	−0.2768	0.5480	46	−0.51	0.6159
psi22	0.1263	0.3614	46	0.35	0.7282

Parameter	Alpha	Lower	Upper	Gradient
b11	0.05	16.6272	20.7937	0.042383
b12	0.05	19.2175	23.7083	0.038031
b13	0.05	9.3347	11.6742	0.069212
b14	0.05	16.2606	19.8717	0.001518
b15	0.05	14.1451	17.6970	0.002031
b16	0.05	15.3980	19.1642	0.096086
b21	0.05	52.0829	56.3067	−0.05582
b22	0.05	51.9581	56.0384	−0.03882
b23	0.05	50.2317	53.9880	−0.10839
b24	0.05	49.4369	53.3585	0.017773
b25	0.05	47.1438	51.6705	−0.01614
b26	0.05	46.1108	50.2686	0.026657
b31	0.05	−13.2825	−11.8829	0.130231
b32	0.05	−13.0094	−11.6915	0.113812
b33	0.05	−14.9929	−13.4130	0.240953
b34	0.05	−14.8074	−13.0908	−0.10028
b35	0.05	−14.8701	−12.6243	0.03163
b36	0.05	−14.5492	−12.5551	0.017849
sigsq_w	0.05	0.03613	0.05218	2.025227
delta	0.05	0.8961	1.0137	−2.94688
psi11	0.05	−1.3183	2.5374	0.40221
psi21	0.05	−1.3798	0.8263	0.745997
psi22	0.05	−0.6011	0.8538	−0.53769

Additional Estimates

Label	Estimate	Standard Error	DF	t Value	Pr > |t|
corr(bi1,bi2)=	−0.9974	3.8090	46	−0.26	0.7946

Label	Alpha	Lower	Upper
corr(bi1,bi2)=	0.05	−8.6645	6.6697

Since there were no warnings regarding the Hessian matrix sent to the SAS LOG, we elected to fit the NLME model (5.86) one more time assuming b_{i1} is the sole random effect in the model. We fit this reduced model to the data using the above NLMIXED syntax but with starting values determined by the following call to %COVPARMS

```
%covparms(parms=SAStemp.OLSparms, predout=SAStemp.predout,
          resid=resid, method=mspl,
          random=der_bi1, subject=Plot,
          type=un, covname=psi, output=MLEparms);
```

and with the following two modifications to the above NLMIXED syntax:

```
pred = (beta1+bi1)/(1+exp((beta3)*(days-(beta2))));
random bi1 ~ normal(0, psi11) subject=Plot;
```

The results of this analysis, shown in Output 5.26, indicate there is significant variation in the asymptote parameter as reflected by the random effects variance component, $\widehat{\psi}_{11} = 1.4186$ (95% confidence interval: $[0.3226, 2.5146]$). Moreover, as is evident from the overall contrasts, there is a significant difference between at least two of the six genotype-year combinations for each of the three cluster-specific parameters β_{i1}, β_{i2} and β_{i3}. Further contrasts may be constructed to test whether this is due primarily to differences in soil content, weather, etc. from year to year or whether it is due primarily to differences in genotype.

Output 5.26: Select output when model (5.86) is fit to the soybean growth data assuming b_{i1} is the only random effect in the model.

Specifications

Data Set	WORK.EXAMPLE5_4_2
Dependent Variable	y
Distribution for Dependent Variable	Normal
Random Effects	bi1
Distribution for Random Effects	Normal
Subject Variable	Plot
Optimization Technique	Newton−Raphson
Integration Method	Adaptive Gaussian Quadrature

Dimensions

Observations Used	412
Observations Not Used	0
Total Observations	412
Subjects	48
Max Obs Per Subject	10
Parameters	21
Quadrature Points	1

Iteration History

Iter	Flag1	Calls	NegLogLike	Diff	MaxGrad	Slope
1		46	290.467646	1.294817	275.6121	−5.10296
2		69	289.819911	0.647735	35.48926	−1.15726
3		92	289.787908	0.032003	1.390837	−0.06088
4		115	289.787679	0.000229	0.008651	−0.00046
5		138	289.787679	1.821E−8	5.497E−7	−3.64E−8

NOTE: GCONV convergence criterion satisfied.

Fit Statistics

−2 Log Likelihood	579.6
AIC (smaller is better)	621.6
AICC (smaller is better)	623.9
BIC (smaller is better)	660.9

Parameter Estimates

Parameter	Estimate	Standard Error	DF	t Value	Pr > |t|
b11	18.7042	1.0805	47	17.31	<.0001
b12	21.4178	1.1588	47	18.48	<.0001
b13	10.5102	0.6637	47	15.84	<.0001
b14	18.0692	0.9533	47	18.95	<.0001
b15	15.8742	0.9358	47	16.96	<.0001
b16	17.2285	0.9621	47	17.91	<.0001
b21	54.1632	1.0404	47	52.06	<.0001
b22	53.9440	1.0096	47	53.43	<.0001
b23	52.0672	0.9284	47	56.08	<.0001
b24	51.4267	0.9699	47	53.02	<.0001
b25	49.3478	1.1233	47	43.93	<.0001
b26	48.1015	1.0072	47	47.76	<.0001
b31	−12.6060	0.3493	47	−36.09	<.0001
b32	−12.3722	0.3303	47	−37.45	<.0001
b33	−14.2298	0.3958	47	−35.95	<.0001
b34	−13.8997	0.4261	47	−32.62	<.0001
b35	−13.7424	0.5630	47	−24.41	<.0001
b36	−13.5839	0.4936	47	−27.52	<.0001
sigsq_w	0.04498	0.004019	47	11.19	<.0001
delta	0.9541	0.02865	47	33.30	<.0001
psi11	1.4186	0.5448	47	2.60	0.0123

Parameter	Alpha	Lower	Upper	Gradient
b11	0.05	16.5305	20.8780	5.41E−10
b12	0.05	19.0867	23.7490	2.47E−10
b13	0.05	9.1750	11.8453	6.12E−10
b14	0.05	16.1513	19.9870	−662E−12
b15	0.05	13.9917	17.7568	−1.2E−9
b16	0.05	15.2930	19.1640	−743E−13
b21	0.05	52.0702	56.2562	1.069E−9
b22	0.05	51.9129	55.9751	1.3E−9
b23	0.05	50.1995	53.9350	1.518E−9
b24	0.05	49.4756	53.3778	3.178E−9
b25	0.05	47.0881	51.6076	2.939E−9
b26	0.05	46.0752	50.1278	1.98E−9
b31	0.05	−13.3087	−11.9034	1.894E−9
b32	0.05	−13.0367	−11.7077	6.96E−10
b33	0.05	−15.0262	−13.4335	2.39E−10
b34	0.05	−14.7569	−13.0425	−2.4E−9
b35	0.05	−14.8751	−12.6097	−1.32E−9
b36	0.05	−14.5769	−12.5910	−645E−12
sigsq_w	0.05	0.03689	0.05307	−5.5E−7
delta	0.05	0.8965	1.0118	−3E−8
psi11	0.05	0.3226	2.5146	−2.77E−8

Contrasts

Label	Num DF	Den DF	F Value	Pr > F
Test of Asymptote Effects:	5	47	20.43	<.0001
Test of Half−Life Effects:	5	47	5.97	0.0002
Test of Growth Rate Effects:	5	47	4.13	0.0034

The estimates of the fixed-effects regression parameters are similar to those obtained by Davidian and Giltinan (1993a) despite assuming a much reduced random-effects structure. As suggested by Davidian and Giltinan (1993a), there is evidence of model misspecification with respect to the inclusion of b_{i3} as a random effect owing to the high correlation observed between b_{i2} and b_{i3}. However, this analysis suggests further model misspecification may be present if one also includes b_{i2} in the model. Indeed, a comparison of the information criteria between the two models suggests the reduced model with b_{i1} as the sole random effect provides a more parsimonious fit to the data. In terms of the covariance parameters, $\boldsymbol{\theta} = (\psi_{11}, \sigma_w^2, \delta)$, the reduced model yields a substantially lower estimate of $\psi_{11} = Var(b_{i1})$ compared with the non-likelihood based estimates obtained by Davidian and Giltinan (1993a). In contrast, the MLE of the power of the mean variance parameter, δ, is considerably higher than that of Davidian and Giltinan (1993a). Indeed, based on an estimated value of $\widehat{\delta} = 0.9541$ and a 95% confidence interval of $(0.8965, 1.0118)$, there is sufficient evidence to think the model can be reduced further by assuming a constant coefficient of variation (CV) within plots. To confirm this, we fit the model assuming $\delta = 1.00$ by removing the parameter `'delta'` from the dataset MLEparms and redefining the intra-cluster variance via the programming statement

```
var = sigsq_w*(pred**2);
```

The likelihood-based fit statistics and parameter estimates from this model, as displayed in Output 5.27, are similar to those obtained assuming a power of the mean variance structure (Output 5.26). Indeed, a likelihood ratio test comparing the two model yields a chi-square test statistic of 2.5 on 1 degree of freedom (p=0.11385).

Output 5.27: Select output when model (5.86) is fit to the soybean growth data assuming b_{i1} is the only random effect and assuming a constant coefficient of variation structure for the within-plot errors.

Fit Statistics

−2 Log Likelihood	582.1
AIC (smaller is better)	622.1
AICC (smaller is better)	624.3
BIC (smaller is better)	659.5

Parameter Estimates

Parameter	Estimate	Standard Error	DF	t Value	Pr > |t|
b11	18.3366	1.0742	47	17.07	<.0001
b12	21.0744	1.1768	47	17.91	<.0001
b13	10.3921	0.6703	47	15.50	<.0001
b14	17.7748	0.9634	47	18.45	<.0001
b15	15.4528	0.8994	47	17.18	<.0001
b16	16.8883	0.9599	47	17.59	<.0001
b21	53.7340	1.0090	47	53.26	<.0001
b22	53.6372	1.0119	47	53.00	<.0001
b23	51.7884	0.8964	47	57.78	<.0001
b24	50.9833	0.9200	47	55.41	<.0001
b25	48.5351	0.9908	47	48.98	<.0001
b26	47.5908	0.9629	47	49.43	<.0001
b31	−12.7332	0.3335	47	−38.19	<.0001
b32	−12.4522	0.3232	47	−38.53	<.0001
b33	−14.3169	0.3729	47	−38.40	<.0001
b34	−14.0766	0.3999	47	−35.20	<.0001
b35	−14.1562	0.5068	47	−27.93	<.0001
b36	−13.8232	0.4710	47	−29.35	<.0001
sigsq_w	0.04199	0.003209	47	13.08	<.0001
psi11	1.4117	0.5347	47	2.64	0.0112

Based on these analyses, it would appear that a NLME logistic growth curve model with a random asymptote parameter and a constant coefficient of variation (CV) structure within plots may provide the best fit to the data. However, all of the models considered assume the within-plot errors are conditionally independent given the random effects. To account for possible serial correlation over time on observations within the same plot, Davidian and Giltinan (1993a) fit a NLME model assuming an intra-cluster autocorrelation structure. We know from our previous example with the orange tree data that, in certain cases, it is possible to model intra-cluster autocorrelation in NLMIXED using "dummy" random effects. In this example, owing to the unbalanced and unequally spaced time intervals, such an endeavor is highly problematic. However, one can accommodate such a model using the macro %NLINMIX by simply including a REPEATED statement when specifying the PROC MIXED statements that are to be evaluated at each iteration of the PL/CGEE1 algorithm. To illustrate, we use %NLINMIX to fit two NLME logistic growth curve models to the soybean growth data. The first model (Model 1) is the same model we just fit using NLMIXED; namely the logistic growth curve model with a random asymptote effect (b_{i1}) and a constant coefficient of variation (CV) structure within plots. In this case, the within-plot errors are assumed to be conditionally independent. The second model (Model 2) assumes the same random asymptote effect as well as a constant CV structure within plots but it assumes the within-plot errors are autocorrelated. Below is the %NLINMIX macro code required to fit Model 2 using PL/CGEE1. Because the time intervals are unequally spaced, autocorrelation was modeled assuming a spatial power law covariance structure defined by $Cov(\epsilon_{ij}, \epsilon_{ij'}|b_{i1}) = \sigma_w^2 \rho^{|t_{ij}-t_{ij'}|}$. One can request this structure within the %NLINMIX code by specifying the TYPE=SP(POW) option in the REPEATED statement. If one chooses to assume an AR(1) structure as done by

Davidian and Giltinan (1993a), one would specify TYPE=AR(1) instead. For Model 1 in which the errors are assumed conditionally independent, we simply repeat the following %NLINMIX code but with the REPEATED statement excluded.

Program 5.18

```
%nlinmix(data=example5_4_2,
  procopt=method=ml covtest cl,
  parms=%str( b11=18 b12=21 b13=10 b14=18 b15=15 b16=17
              b21=54 b22=54 b23=52 b24=51 b25=49 b26=48
              b31=-13 b32=-12 b33=-14 b34=-14 b35=-14 b36=-14),
  model=%str(
   beta1 = b11*X1 + b12*X2 + b13*X3 +
           b14*X4 + b15*X5 + b16*X6;
   beta2 = b21*X1 + b22*X2 + b23*X3 +
           b24*X4 + b25*X5 + b26*X6;
   beta3 = b31*X1 + b32*X2 + b33*X3 +
           b34*X4 + b35*X5 + b36*X6;
   beta3=beta3/100;
   predv = (beta1+bi1)/(1+exp(beta3*(Days-beta2)));
   pred_avg = beta1/(1+exp(beta3*(Days-beta2)));
  ),
  weight=%str(
   _weight_= 1/predv;
  ),
  stmts=%str(
   class Plot _Days_ ;
   model pseudo_y = d_b11 d_b12 d_b13 d_b14 d_b15 d_b16
                    d_b21 d_b22 d_b23 d_b24 d_b25 d_b26
                    d_b31 d_b32 d_b33 d_b34 d_b35 d_b36 /
                    noint notest s cl;
   random d_bi1 / subject=Plot type=un s;
   repeated _Days_ / subject=Plot type=SP(POW) (Days);
   weight _weight_;
   ods output SolutionF=SAStemp.pe_Model4_PL;
   ods output CovParms=SAStemp.cov_Model4_PL;
  ),
  expand=eblup
);
run;
```

Using ODS OUTPUT statements, we collected the parameter estimates from the most recent call to NLMIXED (corresponding to Model 1 as shown in Output 5.27) as well as the two calls to %NLINMIX, one assuming a conditionally independent error structure with constant CV (Model 1) and one assuming an autoregressive error structure with constant CV (Model 2). The parameter estimates from each analysis are summarized in Output 5.28 using PROC REPORT (SAS code not shown).

Output 5.28: A comparison of Laplace ML estimates from NLMIXED versus PL/CGEE1 estimates from %NLINMIX for a NLME logistic growth curve model assuming a random asymptote and either a constant CV structure with conditionally independent errors (Model 1) or a constant CV structure with autocorrelated errors (Model 2).

Parameter Estimates

	Method					
	Laplace MLE (Model 1)		PL/CGEE1 (Model 1)		PL/CGEE1 (Model 2)	
Parameter	Estimate	StdErr	Estimate	StdErr	Estimate	StdErr
b11	18.3366	1.0742	20.8330	0.9754	20.8327	0.9756
b12	21.0744	1.1768	23.0561	0.9597	23.0569	0.9600
b13	10.3921	0.6703	11.0438	0.6467	11.0438	0.6467
b14	17.7748	0.9634	19.5287	0.7820	19.5289	0.7821
b15	15.4528	0.8994	18.5162	0.8884	18.5164	0.8885
b16	16.8883	0.9599	19.3984	0.8441	19.3990	0.8441
b21	53.7340	1.0090	56.8411	1.0143	56.8409	1.0146
b22	53.6372	1.0119	55.9999	0.9265	56.0007	0.9268
b23	51.7884	0.8964	54.0507	1.1316	54.0510	1.1317
b24	50.9833	0.9200	54.6215	0.9279	54.6220	0.9281
b25	48.5351	0.9908	54.5503	1.0948	54.5511	1.0950
b26	47.5908	0.9629	52.1638	0.9743	52.1654	0.9746
b31	−12.7332	0.3335	−11.4929	0.4733	−11.4930	0.4735
b32	−12.4522	0.3232	−11.3221	0.4299	−11.3218	0.4301
b33	−14.3169	0.3729	−13.0657	0.7161	−13.0655	0.7164
b34	−14.0766	0.3999	−12.2124	0.5072	−12.2120	0.5074
b35	−14.1562	0.5068	−11.1921	0.5162	−11.1921	0.5163
b36	−13.8232	0.4710	−11.4053	0.5069	−11.4051	0.5070
psi11	1.4117	0.5347	1.7746	0.5793	1.7744	0.5794
rho	.	.	.	.	0.3404	1.7551
sigsq_w	0.04199	0.003209	0.1603	0.01188	0.1603	0.01188

In comparing results, there are several things worth noting. First, in comparing the PL/CGEE1 estimates between Models 1 and 2, we find that accounting for possible intra-cluster autocorrelation had little or no impact on the remaining parameter estimates. Indeed, there is little evidence of any significant autocorrelation present given that ρ is estimated to be 0.3404 ± 1.7551. The fact that we achieve a positive autocorrelation coefficient compared with a negative value as reported by Davidian and Giltinan (1993a) can be attributed primarily to the use of a spatial power law covariance structure as opposed to an AR(1) structure. The next thing we observe is that there is a considerable discrepancy between parameter estimates obtained via ML estimation using Laplace's approximation versus PL/CGEE1 estimation using a SS linearization approach. The ML estimates of the asymptote parameters, half-life parameters and growth rate parameters are all consistently lower than the corresponding PL/CGEE1 estimates. The same holds true for the variance components, ψ and σ^2_w. To be sure this is not due to any measurable error associated with Laplace's single quadrature point approximation, we repeated the likelihood-based analysis for Model 1 using adaptive Gaussian quadrature requiring 5 quadrature points. While not shown here, the results were virtually unchanged (users are encouraged to verify this for themselves) suggesting a very real difference exists between ML and PL estimation in this example.

The reader may wonder why such a discrepancy in estimates occurs here when no such discrepancy was observed when we fit essentially the same NLME logistic growth curve model to the orange tree data (compare Output 4.24 to Output 4.27 in §4.5.2). In both examples, we employ a NLME logistic growth curve model assuming a random asymptote effect. The key difference between the two examples is that while the random effect is strictly linear with respect to the conditional mean in both cases, it is not strictly linear in the overall model in both cases. In our current example, we have a within-plot multiplicative error structure which we can write explicitly as

$$y_{ij} = f(\boldsymbol{x}'_{ij}, \boldsymbol{\beta}, b_{i1})(1 + \epsilon_{ij})$$

where both the random effect b_{i1} and the vector of fixed effects $\boldsymbol{\beta}$ appear in the conditional variance structure. As a result the model is not strictly linear in the random effect whereas for the orange tree data, we fit a model with an additive error structure expressed here as

$$y_{ij} = f(\boldsymbol{x}'_{ij}, \boldsymbol{\beta}, b_{i1}) + \epsilon_{ij}.$$

In this case, the model is strictly linear in b_{i1} and the parameters of the conditional variance of y_{ij} will be strictly orthogonal to the parameters of the conditional mean. Since the linearization-based PL/CGEE1 estimation scheme implemented in %NLINMIX essentially treats the parameters of the conditional variance as being orthogonal to the parameters of the conditional mean, the degree to which this conditional orthogonality "assumption" is violated can lead to an increase in bias and inefficiency as was demonstrated in limited simulations by Vonesh et. al. (2002). For GNLME models that are strictly linear in the random effects, we can usually express the model in terms of a marginal GNLM (such as was done for the orange tree data in §4.5.2). In such cases estimation methods such as PL/QELS will yield consistent estimates of the regression parameters provided the conditional mean has not been misspecified.

In summary, we conclude from the analyses presented here that the NLME logistic growth curve model (5.86) with random asymptote and constant coefficient of variation provides a good balance between having a parsimonious model and a model that adequately fits the soybean growth data. Moreover, given the multiplicative error structure of this model along with the relatively large number of observations per plot ($p_i = 8$ to 10), we recommend that estimation be based on maximizing the integrated log-likelihood using Laplace's approximation.

5.4.3 High-flux hemodialyzer data

Vonesh and Carter (1992) analyzed *in vitro* data on high-flux hemodialyzers in which the ultrafiltration performance of 20 dialyzers was assessed. High-flux dialyzers are used in hemodialysis to treat patients with end-stage renal disease. A key component in that treatment is the removal of excess water. The water transport kinetics of high-flux dialyzers are characterized by a functional relationship between the ultrafiltration rate (UFR ml/hr) at which water is removed and the transmembrane pressure (TMP mmHg) that is exerted on the dialyzer membrane at a fixed blood flow rate (Qb ml/min). The 20 dialyzers were evaluated *in vitro* using a single source of bovine blood at blood flow rates of 200 and 300 ml/min. The essential features of the study are summarized as follows (the SAS variable names are shown in parentheses):

- The experimental unit is a high-flux hemodialyzer
- Dialyzer ultrafiltration rates were measured at 7 different transmembrane pressures with 10 dialyzers evaluated at a 200 ml/min blood flow rate and 10 dialyzers evaluated at a 300 ml/min blood flow rate
- Response variable:
  - y_{ij} = ultrafiltration rate in L/hour (UFR) measured on the ith dialyzer at the jth transmembrane pressure in mmHg ($i = 1, \ldots, 20$ dialyzers; 10 dialyzers per each of two blood flow rates; $j = 1, \ldots, 7$ measurements per dialyzer).
- One within-unit covariate:
  - x_{ij} = transmembrane pressure (TMP) measured and recorded for the ith dialyzer on the jth occasion with target values of TMP set at 25, 50, 100, 150, 200, 250, 300 mmHg.

Figure 5.9 Nonparamteric loess curves depicting the regression surface of UFR (L/hour) versus TMP (mmHg) by blood flow rate (Qb ml/min)

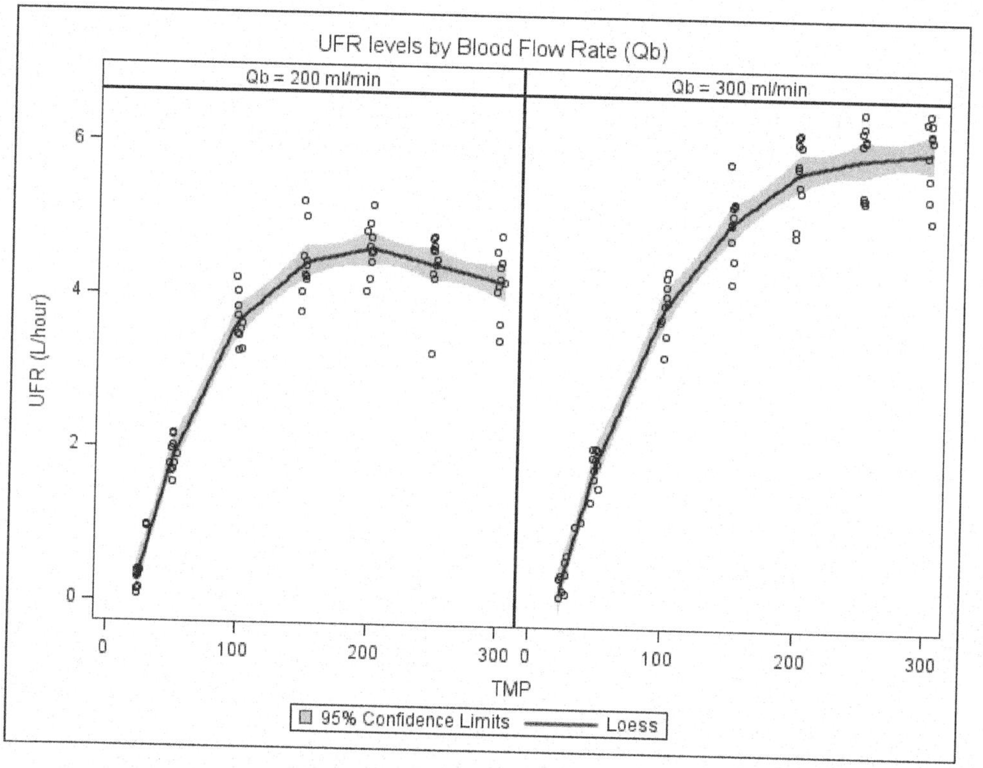

- One between-unit covariate:
  - Blood flow rate (Qb) fixed at 200 ml/min or 300 ml/min and denoted by
  $$a_i = \begin{cases} 0, & \text{if Qb} = 200 \text{ ml/min} \\ 1, & \text{if Qb} = 300 \text{ ml/min} \end{cases} \quad \text{(this indicator is labeled QB\_300)}$$
- Goal: structurally characterize the ultrafiltration profile of the typical dialyzer as a function of transmembrane pressure and blood flow rate and select a suitable covariance structure.

A description and partial listing of the raw data is available through the author's Web page. To provide greater stability in the estimation algorithms, we rescale UFR to be in liters/hour by dividing the original values by 1000. Shown in Figure 5.9 are nonparametric loess response curves depicting a nonlinear relationship between UFR and TMP at each blood flow rate. Vonesh and Carter (1992) note that such a nonlinear relationship can be explained, in part, by a phenomenon known as protein polarization (Henderson, 1996). At high levels of TMP, protein polarization results in ultrafiltration rates that are nearly constant suggesting the possibility that a nonlinear asymptotic regression model may provide a good fit to the data. Moreover, it is clear from Figure 5.9 that variation in UFR increases with increasing TMP. Taking these two features into account, Vonesh and Carter (1992) proposed the use of a mixed-effects asymptotic exponential growth curve model as a means for accounting for the observed heterogeneity in the data as well as providing a structural interpretation to the model parameters. They found that among those versions of the model where one or more of the parameters are assumed random, the following asymptotic exponential growth curve model provided the best fit to the data.

Model 1: Asymptotic exponential growth model

$$y_{ij} = \beta_{i1}(1 - \exp\{-\beta_{i2}(x_{ij} - \beta_{i3})\}) + \epsilon_{ij}, \ (i = 1, \ldots, 20; j = 1, \ldots, 7) \quad (5.87)$$
$$\beta_{i1} = \beta_{11} + \beta_{12}a_i + b_{i1}$$
$$\beta_{i2} = \beta_{21} + \beta_{22}a_i + b_{i2}$$
$$\beta_{i3} = \beta_{31} + \beta_{32}a_i$$

where $\epsilon_{ij} \sim N(0, \sigma_w^2)$ and $\begin{pmatrix} b_{i1} \\ b_{i2} \end{pmatrix} \sim N\left[\begin{pmatrix} 0 \\ 0 \end{pmatrix}, \begin{pmatrix} \psi_{11} & \psi_{12} \\ \psi_{21} & \psi_{22} \end{pmatrix}\right]$. Structurally, the parameter β_{i1} is a dialyzer-specific asymptote parameter representing the maximum UFR achievable due to protein polarization at extreme values of TMP, the parameter β_{i2} is a dialyzer-specific hydraulic permeability transport rate, and the parameter β_{i3} is an "intercept" parameter representing the TMP required to offset patient oncotic pressure. The fact that there was no evidence of a random "intercept" effect was expected since, by design, a pooled source of bovine blood was used for the entire study and this pooled blood source was corrected to have a constant oncotic pressure of about 20-25 mmHg. Vonesh and Carter (1992) fit this model to the data using an estimated generalized least squares (EGLS) procedure based on a first-order PA Taylor series approximation to the model.

While model (5.87) provides a biophysical interpretation to the parameters describing the UFR profile, the nonparametric loess regression curves depicted in Figure 5.9 suggest that UFR values may actually decrease after some threshold value of TMP is reached. This is particularly evident at the lower blood flow rate of 200 ml/min. Littell et. al. (2006), for example, found that a quartic polynomial growth curve model provides a reasonably good fit to the data and one which better reflects the observed decrease in UFR at very high values of TMP. They also determined that among a number of candidate models for the covariance structure, a model with a heterogeneous AR(1) structure provided the most parsimonious fit to the data. Although fitting a high-degree polynomial curve may provide an excellent fit to the data, the lack of any biophysical interpretation to the parameters makes it difficult to characterize the ultrafiltration performance of dialyzers on the basis of any mechanistic forces at work. Consequently, with a goal to better characterize the ultrafiltration performance both mechanistically and stochastically, we consider fitting a modified version of model (5.87) by allowing for a linear asymptote effect as follows.

Model 2: Linear asymptotic exponential growth model

$$y_{ij} = \beta_{i1}(1 - \exp\{-\beta_{i2}(x_{ij} - \beta_{i3})\}) + \epsilon_{ij}, \ (i = 1, \ldots, 20; j = 1, \ldots, 7) \quad (5.88)$$
$$\beta_{i1} = (\beta_{11} + \beta_{12}a_i) + (\beta_{13} + \beta_{14}a_i)x_{ij} + b_{i1}$$
$$\beta_{i2} = \beta_{21} + \beta_{22}a_i + b_{i2}$$
$$\beta_{i3} = \beta_{31} + \beta_{32}a_i.$$

Here the dialyzer-specific asymptote parameter, β_{i1}, is modeled as a linear function of both blood flow rate (a_i) and transmembrane pressure (x_{ij}). If the asymptote "slope" parameters β_{13} and/or β_{14} are significantly less than 0, we could conceivably interpret this "linear" component of the asymptote as representing some sort of phenomenological effect in which a plateau in the ultrafiltration rate occurs at some threshold TMP value beyond which a gradual linear decline in the ultrafiltration rate begins to occur. Such a phenomenological effect, however, is understood to occur only within the range of the observed data.

Using the macro %COVPARMS, we confirmed that the "intercept" parameter may be treated as fixed in keeping with the design of the study (results not shown). We therefore fit the linear asymptotic exponential growth curve model (5.88) to the data and compared its fit to that of the original model (5.87) of Vonesh and Carter. We did so by applying the Laplace approximation to compute ML estimates. The programming code for fitting model (5.88) to the data is as follows.

Program 5.19

```
data Example5_4_3;
 set SASdata.Hemodialyzer_Data;
 /* Re-express UFR in liters/hr*/
 UFR = UFR/1000;
 /* Target TMP mmHg values      */
 TMP_TARG = round(TMP,25);
 /* The correct target TMP for */
 /* lowest measured TMP for     */
 /* dialyzer K80798 was 25 mmHg*/
 if Dialyzer='K80798' and TMP=40.0 then TMP_TARG=25;
 I1=(TMP_TARG=25 );
 I2=(TMP_TARG=50 );
 I3=(TMP_TARG=100);
 I4=(TMP_TARG=150);
 I5=(TMP_TARG=200);
 I6=(TMP_TARG=250);
 I7=(TMP_TARG=300);
 /* Define indicator variable  */
 /* for the two Qb values      */
 Qb_300=(Qb=300);
run;
proc sort data=example5_4_3;
 by Dialyzer TMP;
run;
ods output parameterestimates=pe_model2;
proc nlmixed data=example5_4_3 qpoints=1
             tech=newrap maxfunc=1000;
 /* Rescale TMP to units of dmHg */
 TMPd = TMP/100;
 /* Model description */
 Model='Model (5.88)';
 beta1 = b11 + b12*Qb_300 + b13*TMPd + b14*TMPd*Qb_300;
 beta2 = b21 + b22*Qb_300;
 beta3 = b31 + b32*Qb_300;
 pred = (beta1+bi1)*(1 - exp(-(beta2+bi2)*(TMPd-beta3)));
 pred_avg = beta1*(1 - exp(-beta2*(TMPd-beta3)));
 parms b11=5 b12=0 b13=0 b14=0 b21=1 b22=0 b31=.20 b32=0
       psi11=.1 to 1 by .1 psi21=-0.1 -0.01 0 0.01 0.1
       psi22=.1 to 1 by .1 sigsq_w=.060 / best=1;
 model UFR~normal(pred, sigsq_w);
 random bi1 bi2 ~ normal([0,0], [psi11,
                                 psi21, psi22])
        subject=Dialyzer;
 estimate 'Asymptote Intercept, Qb=200:' b11;
 estimate 'Asymptote Intercept, Qb=300:' b11+b12;
 estimate 'Asymptote Slope, Qb=200:' b13;
 estimate 'Asymptote Slope, Qb=300:' b13+b14;
 estimate 'Hydraulic Rate, Qb=200' b21;
```

```
  estimate 'Hydraulic Rate, Qb=300' b21+b22;
  estimate 'Intercept TMP, Qb=200' b31;
  estimate 'Intercept TMP, Qb=300' b31+b32;
  predict pred_avg out=SAStemp.model2;
  id TMPd Model;
run;
```

The created variable TMP_TARG in the above code represents the machine setting at which transmembrane pressures were targeted during each dialysis run while the measured transmembrane pressures were those recorded as part of the raw data. To provide greater stability in the optimization procedure, we used a rescaled transmembrane pressure, TMPd, which is simply the measured transmembrane pressure divided by 100. Starting values of the model parameters were based on rounded values from results published by Vonesh and Carter (1992) but rescaled here to reflect the use of rescaled TMP values. Included in the NLMIXED syntax are ESTIMATE statements providing the parameter estimates of the typical or average dialyzer for each of the two blood flow rates. Although not shown here, the NLMIXED syntax for fitting model (5.87) is similar to the above except that the asymptote parameter, beta1, would no longer involve the "slopes" associated with TMP.

Output 5.29: Select output from NLMIXED when we fit model (5.88) to the high flux ultrafiltration data.

Specifications

Data Set	WORK.EXAMPLE5_4_3
Dependent Variable	UFR
Distribution for Dependent Variable	Normal
Random Effects	bi1 bi2
Distribution for Random Effects	Normal
Subject Variable	Dialyzer
Optimization Technique	Newton−Raphson
Integration Method	Adaptive Gaussian Quadrature

Dimensions

Observations Used	140
Observations Not Used	0
Total Observations	140
Subjects	20
Max Obs Per Subject	7
Parameters	12
Quadrature Points	1

Iteration History

Iter	Flag1	Calls	NegLogLike	Diff	MaxGrad	Slope
1	*	30	53.1251778	28.95115	283.4274	−121.51
2	*	46	39.8889661	13.23621	447.1929	−58.0745
3	*	60	18.7729911	21.11597	185.8794	−30.1029
4	*	74	15.677592	3.095399	64.66282	−3.89965
5	*	88	13.3344769	2.343115	139.4538	−6.11809
6	*	102	10.2399962	3.094481	47.39925	−4.099
7	*	116	6.96377056	3.276226	28.02409	−4.29755
8	*	130	4.90305244	2.060718	27.30298	−3.03859
9	*	144	4.27533397	0.627718	8.75198	−0.96357
10	*	158	4.14818969	0.127144	2.293162	−0.20424
11	*	172	4.13375957	0.01443	0.273905	−0.02396
12	*	186	4.13286842	0.000891	0.017807	−0.00155
13	*	200	4.13284623	0.000022	0.000489	−0.00004
14	*	214	4.13284606	1.688E−7	0.000021	−3.23E−7
15	*	228	4.13284606	3.52E−10	4.905E−7	−688E−12

NOTE: GCONV convergence criterion satisfied.

Fit Statistics

-2 Log Likelihood	8.3
AIC (smaller is better)	32.3
AICC (smaller is better)	34.7
BIC (smaller is better)	44.2

Parameter Estimates

Parameter	Estimate	Standard Error	DF	t Value	Pr > |t|
b11	7.7930	0.6301	18	12.37	<.0001
b12	0.2077	1.0132	18	0.20	0.8399
b13	−1.1082	0.1837	18	−6.03	<.0001
b14	0.6010	0.2783	18	2.16	0.0445
b21	1.0118	0.1236	18	8.19	<.0001
b22	−0.06603	0.1734	18	−0.38	0.7078
b31	0.2060	0.008881	18	23.19	<.0001
b32	0.009280	0.01232	18	0.75	0.4609
psi11	0.2196	0.08844	18	2.48	0.0231
psi21	−0.04367	0.02448	18	−1.78	0.0913
psi22	0.02488	0.01109	18	2.24	0.0377
sigsq_w	0.03358	0.004771	18	7.04	<.0001

Parameter	Alpha	Lower	Upper	Gradient
b11	0.05	6.4693	9.1167	−1.97E−7
b12	0.05	−1.9210	2.3364	4.905E−7
b13	0.05	−1.4941	−0.7223	5.339E−8
b14	0.05	0.01637	1.1857	−1.33E−7
b21	0.05	0.7523	1.2714	3.089E−8
b22	0.05	−0.4303	0.2983	−8.2E−8
b31	0.05	0.1873	0.2246	5.571E−9
b32	0.05	−0.01659	0.03515	−3.83E−9
psi11	0.05	0.03384	0.4055	1.916E−9
psi21	0.05	−0.09511	0.007767	2.012E−8
psi22	0.05	0.001584	0.04817	9.771E−9
sigsq_w	0.05	0.02356	0.04361	−1.17E−8

Additional Estimates

Label	Estimate	Standard Error	DF	t Value	Pr > |t|
Asymptote Intercept, Qb=200:	7.7930	0.6301	18	12.37	<.0001
Asymptote Intercept, Qb=300:	8.0007	0.9234	18	8.66	<.0001
Asymptote Slope, Qb=200:	−1.1082	0.1837	18	−6.03	<.0001
Asymptote Slope, Qb=300:	−0.5072	0.2483	18	−2.04	0.0561
Hydraulic Rate, Qb=200	1.0118	0.1236	18	8.19	<.0001
Hydraulic Rate, Qb=300	0.9458	0.1461	18	6.47	<.0001
Intercept TMP, Qb=200	0.2060	0.008881	18	23.19	<.0001
Intercept TMP, Qb=300	0.2152	0.009182	18	23.44	<.0001

Label	Alpha	Lower	Upper
Asymptote Intercept, Qb=200:	0.05	6.4693	9.1167
Asymptote Intercept, Qb=300:	0.05	6.0607	9.9407
Asymptote Slope, Qb=200:	0.05	−1.4941	−0.7223
Asymptote Slope, Qb=300:	0.05	−1.0289	0.01458
Hydraulic Rate, Qb=200	0.05	0.7523	1.2714
Hydraulic Rate, Qb=300	0.05	0.6388	1.2528
Intercept TMP, Qb=200	0.05	0.1873	0.2246
Intercept TMP, Qb=300	0.05	0.1959	0.2345

The results of fitting model (5.88) to the data are shown in Output 5.29. There was a significant linear downward trend associated with the asymptotic behavior of ultrafiltration rates with increasing TMP. This trend was most pronounced at the lower blood flow rate of 200 ml/min where it is estimated that the average dialyzer's UFR will decrease at a rate of -1.108 L/hour per 100 mmHg increase in TMP ($\equiv 1$ dmHg increase in TMP) after having reached its peak value. This trend was also present for dialyzers evaluated at a 300 ml/min blood flow rate but the effect was about half that estimated for the 200 ml/min blood flow rate (a rate decrease of -0.5072 L/hour per 100 mmHg increase in TMP). In Figure 5.10, we compare the fit of this linear asymptotic regression model to that of the asymptotic

Figure 5.10 Predicted fit and confidence bands for the average dialyzer of Model (5.87) versus Model (5.88)

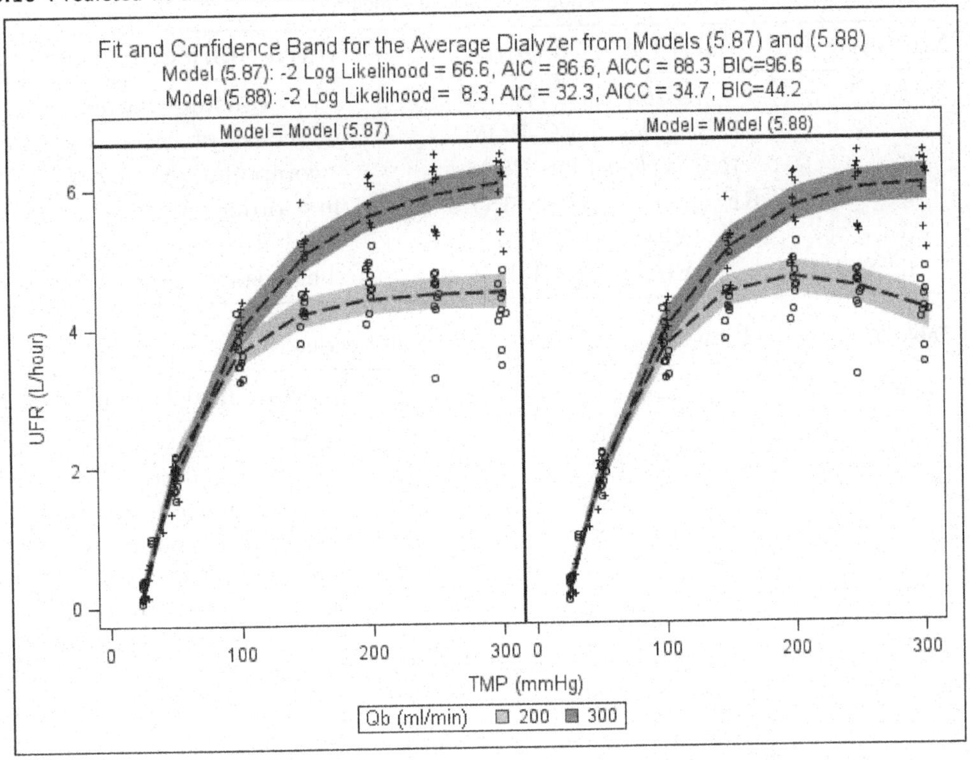

regression model (5.87). As can be seen, the inclusion of a linear asymptote effect within an asymptotic exponential growth curve model provides a visually better fit to observed data. This is further confirmed when we compare the marginal fits of the two models using the various likelihood information criteria available from NLMIXED (Figure 5.10).

From the preceding analysis, it is clear that for this data, a linear asymptotic growth curve model like (5.88) will improve the overall fit of the mean compared to a standard asymptotic growth curve model like (5.87). What is not clear is whether the observed heterogeneity in the data can be adequately explained through specification of the random effects b_{i1} and b_{i2} alone or whether a better fit to the data is possible under some alternative covariance structure. To that end, we evaluate several covariance structures including some of the marginal covariance structures evaluated by Littell et. al. (2006) in their analysis of the high-flux dialyzer data. However, rather than fit a quartic polynomial growth curve model assuming different covariance structures as was done by Littell et. al. (2006), we fit both mixed-effects and marginal versions of the linear asymptotic growth curve model assuming different covariance structures.

We fit six different mixed-effects models starting with 1) the NLME linear asymptotic growth curve model (5.88) with a constant intra-subject variance components structure, VC. We then attempt to fit three multiplicative versions of the same model by replacing the constant within-subject variance structure with 2) a constant variance—power of the mean variance structure, VC-POM, 3) a compound symmetric—power of the mean variance structure, CS-POM, and 4) a first-order autoregressive—power of the mean variance structure, AR(1)-POM. In all three case, the inverse weight matrix that defines $\boldsymbol{\Lambda}_i(\boldsymbol{\beta}, \boldsymbol{b}_i, \boldsymbol{\theta}_w) = \boldsymbol{W}_i(\boldsymbol{X}_i, \boldsymbol{\beta}, \boldsymbol{b}_i)^{-1/2}\boldsymbol{\Lambda}_i(\boldsymbol{\theta}_w)\boldsymbol{W}_i(\boldsymbol{X}_i, \boldsymbol{\beta}, \boldsymbol{b}_i)^{-1/2}$ is given by $\boldsymbol{W}_i(\boldsymbol{X}_i, \boldsymbol{\beta}, \boldsymbol{b}_i)^{-1} = Diag\{f(\boldsymbol{x}'_{ij}, \boldsymbol{\beta}, \boldsymbol{b}_i)^{2\delta}\}$ where $f(\boldsymbol{x}'_{ij}, \boldsymbol{\beta}, \boldsymbol{b}_i)$ is the conditional mean and δ is a power of the mean parameter. The compound symmetric and first-order autoregressive structures that define $\boldsymbol{\Lambda}_i(\boldsymbol{\theta}_w)$ are those shown in Table 3.1. While the first-order autoregressive structure is typically used with equally spaced time series or longitudinal

Table 5.2 Random effects linear asymptotic growth curve models evaluated. An unstructured variance-covariance matrix of the random effects is assumed.

Model	RE $\mid \boldsymbol{\Lambda}_i(\boldsymbol{\beta}, \boldsymbol{b}_i, \boldsymbol{\theta}_w)$	Intra-Subject Covariance Structure
1	RE$(b_{i1}, b_{i2}) \mid$ VC	constant variance component
2	RE$(b_{i1}, b_{i2}) \mid$ VC-POM	constant variance, power of mean
3	RE$(b_{i1}, b_{i2}) \mid$ CS-POM	compound symmetric, power of mean
4	RE$(b_{i1}, b_{i2}) \mid$ AR(1)-POM	first-order autoregressive, power of mean
5	RE$(b_{i1}) \mid$ CSH	heterogeneous compound symmetric
6	RE$(b_{i1}) \mid$ ARH(1)	heterogeneous first-order autoregressive

Table 5.3 Marginal linear asymptotic growth curve models evaluated.

Model	$\boldsymbol{\Sigma}_i(\boldsymbol{\beta}, \boldsymbol{\theta})$	Marginal Covariance Structure
1	UN	unstructured variance-covariance matrix
2	CS	compound symmetric
3	CSH	heterogeneous compound symmetric
4	CS-POM	compound symmetric, power of mean
5	AR(1)	first-order autoregressive
6	ARH(1)	heterogeneous first-order autoregressive
7	AR(1)-POM	first-order autoregressive, power of mean

data, here we follow Littell et. al. (2006) and use the AR(1) structure to account for possible intra-subject correlation associated with serially measured TMP values that are nearly equally spaced. We also evaluate two reduced versions of model (5.88) in which the asymptote parameter is the only random effect (i.e., we drop b_{i2} as a random effect from the NLME model). Under this reduced NLME model, intra-subject heterogeneity and correlation are introduced through specification of 5) a heterogeneous compound symmetric covariance structure, CHS, and 6) a heterogeneous first-order autoregressive structure, ARH(1). A more complete description of these NLME models is presented in Table 5.2.

We also fit seven marginal versions of model (5.88) obtained by dropping the random effects b_{i1} and b_{i2} from the mean structure and modeling the observed heterogeneity directly through specification of a marginal covariance matrix. The modeled covariance structures of these GNLM's include 1) an unstructured covariance matrix (UN), 2) a compound symmetric structure (CS), 3) a heterogeneous compound symmetric structure (CSH), 4) a compound symmetric—power of the mean structure (CS-POM), 5) a first-order autoregressive structure (AR(1)), 6) a heterogeneous first-order autoregressive structure (ARH(1)), and 7) a first-order autoregressive—power of the mean structure (AR(1)-POM). In terms of rationale for some of these covariance structures, it should be noted that heterogeneity can be explained either directly via heterogeneous variances (the CHS and ARH(1) structures) or indirectly via power of the mean (POM) variances. Table 5.3 provides a more complete description of these GNLM's.

For both the NLME models of Table 5.2 and GNLM's of Table 5.3, the mean structure is of that of the linear asymptotic growth curve model, $\mu_{ij}(\boldsymbol{\beta}, \boldsymbol{b}_i) = \beta_{i1}(1 - \exp\{-\beta_{i2}(x_{ij} - \beta_{i3})\}$, with or without random effects. All models were fit via ML estimation in NLMIXED using either the Laplace approximation to the integrated log-likelihood for the NLME models or by exact ML estimation for the GNLM's. In the latter case, the covariance structures are defined through the use of linear random effects. Alternatively, one could use the macro %NLINMIX in conjunction with the REPEATED statement to obtain MLE's of all GNLM's except those that involve a power of the mean variance (recall from §4.4.2 that PL/QELS≡MLE whenever the marginal covariance matrix of $\boldsymbol{y}_i$ does not depend on $\boldsymbol{\beta}$). All of the programs used to fit these models are available at the author's Web page (**http://support.sas.com/publishing/authors/vonesh.html**). Here we present the programs used to fit two of the models, the NLME model 6 (Table 5.2) and the GNLM model 3 (Table 5.3).

The following is the NLMIXED syntax for fitting model 6 of Table 5.2. Under this model, the observed heterogeneity is modeled as a function of a between-dialyzer random asymptote effect b_{i1} and a within-dialyzer covariance structure with heterogeneous variances and a first-order autocorrelation coefficient.

Program 5.20

```
data runmodel;
 length Model $7 Structure $25;
 set example5_4_3;
 /* Model description */
 Model='NLME 6';
 /* Intra-subject covariance structure description */
 Structure='RE(bi1)|ARH(1)';
run;
ods output parameterestimates=SAStemp.pe_NLME_6;
ods output FitStatistics=SAStemp.fit_NLME_6;
proc nlmixed data=runmodel qpoints=1 tech=newrap maxfunc=1000;
 by Model Structure;
 bounds -1<rho< 1;
 /* Rescale TMP to units of dmHg */
 TMPd = TMP/100;
 eij = I1*e1+I2*e2+I3*e3+I4*e4+I5*e5+I6*e6+I7*e7;
 beta1 = b11 + b12*Qb_300 + b13*TMPd + b14*TMPd*Qb_300;
 beta2 = b21 + b22*Qb_300;
 beta3 = b31 + b32*Qb_300;
 pred = (beta1+bi1)*(1 - exp(-beta2*(TMPd-beta3)));
 pred_avg = beta1*(1 - exp(-beta2*(TMPd-beta3)));
 array cov[7,7];
 array sigsq_w[7] sigsq_w1-sigsq_w7;
 do i=1 to 7;
  do j=1 to 7;
   cov[i,j] = sqrt(sigsq_w[i]*sigsq_w[j])*(rho**abs(i-j));
  end;
 end;
 %vech(dim=7, cov=cov, name=c);
 _delta_=1e-8;
 mu = pred + eij;
 parms b11=5 b12=0 b13=0 b14=0 b21=1 b22=0 b31=.20 b32=0
       psi11=.1 to 1 by .1
       sigsq_w1=.1 sigsq_w2=.1 sigsq_w3=.1 sigsq_w4=.1
       sigsq_w5=.1 sigsq_w6=.1 sigsq_w7=.1
       rho=.01 .1 .5 / best=1;
 model UFR~normal(mu, _delta_);
 random bi1 e1 e2 e3 e4 e5 e6 e7 ~
        normal([ 0, 0, 0, 0, 0, 0, 0, 0 ],
               [ psi11,
                  0  , c1 ,
                  0  , c2 , c3 ,
                  0  , c4 , c5 , c6 ,
                  0  , c7 , c8 , c9 , c10,
```

```
                    0  , c11, c12, c13, c14, c15,
                    0  , c16, c17, c18, c19, c20, c21,
                    0  , c22, c23, c24, c25, c26, c27, c28 ])
         subject=Dialyzer;
 predict pred_avg out=SAStemp.pred_NLME_6;
run;
```

We chose to display the program for this model as the code for the other NLME models will be similar in most cases. For all of the models, we specify two ODS OUTPUT statements (see above code) in order to save information on final parameter estimates and likelihood information criteria which we can use later to summarize the various fits. As was done with the orange tree data, we use ARRAY statements together with the macro %VECH as a means for specifying the variance-covariance structure of the model in the RANDOM statement. Note that the within-dialyzer heterogeneous AR(1) structure, $Cov(y_{ij}, y_{ij'}) = \sigma_{w,j}\sigma_{w,j'}\rho^{|j-j'|}$, requires specifying seven heterogeneous variances (denoted above by `sigsq_w1, sigsq_w2,...,sigsq_w7`) and a single correlation coefficient (`rho`). The linear random effects, `e1 e2 e3 e4 e5 e6 e7`, are chosen to be orthogonal to the model random effect b_{i1} (denoted above by `bi1`) as reflected in the specification of the covariance matrix of the RANDOM statement.

Next is the program used to fit the GNLM model 3 (Table 5.3). Under this marginal model, observed heterogeneity in the data is accounted for through specification of a heterogeneous compound symmetric covariance structure in which $Cov(y_{ij}, y_{ij'}) = \sigma_{w,j}\sigma_{w,j'}\rho^{d_{jj'}}$ where $d_{jj'} = \begin{cases} 0 & \text{if } j = j' \\ 1 & \text{if } j \neq j' \end{cases}$. Since ML estimation can be achieved using either %NLINMIX or NLMIXED, the programming syntax for each follows.

Program 5.21

```
%nlinmix(data=example5_4_3,
  procopt=method=ml covtest cl,
  parms=%str(b11=5 b12=0 b13=0 b14=0
             b21=1 b22=0 b31=.20 b32=0),
  model=%str(
   TMPd = TMP/100;
   beta1 = b11 + b12*Qb_300 + b13*TMPd + b14*TMPd*Qb_300;
   beta2 = b21 + b22*Qb_300;
   beta3 = b31 + b32*Qb_300;
   predv = beta1*(1 - exp(-beta2*(TMPd-beta3)));
  ),
  stmts=%str(
   class Dialyzer TMP_TARG;
   model pseudo_UFR = d_b11 d_b12 d_b13 d_b14
                      d_b21 d_b22 d_b31 d_b32 /
                      noint notest s cl;
   repeated TMP_TARG / subject=Dialyzer type=CSH;
   ods output SolutionF=SAStemp.pe_GNLM_3_initial;
   ods output CovParms=SAStemp.cov_GNLM_3_initial;
  ),
  expand=eblup,
  options=
);
run;
```

```
data runmodel;
 length Model $7 Structure $25;
 set example5_4_3;
 /* Model description */
 Model='GNLM 3';
 /* Covariance structure description */
 Structure='CSH';
run;
ods output parameterestimates=SAStemp.pe_GNLM_3;
ods output FitStatistics=SAStemp.fit_GNLM_3;
proc nlmixed data=runmodel qpoints=1 tech=newrap maxfunc=1000;
 by Model Structure;
 bounds -1<rho< 1;
 /* Rescale TMP to units of dmHg */
 TMPd = TMP/100;
 eij = I1*e1+I2*e2+I3*e3+I4*e4+I5*e5+I6*e6+I7*e7;
 beta1 = b11 + b12*Qb_300 + b13*TMPd + b14*TMPd*Qb_300;
 beta2 = b21 + b22*Qb_300;
 beta3 = b31 + b32*Qb_300;
 pred = beta1*(1 - exp(-beta2*(TMPd-beta3)));
 pred_avg = pred;
 array cov[7,7];
 array sigsq_w[7] sigsq_w1-sigsq_w7;
 do i=1 to 7;
  do j=1 to 7;
   dij = 1-(i=j);
   cov[i,j] = sqrt(sigsq_w[i]*sigsq_w[j])*(rho**dij);
  end;
 end;
 %vech(dim=7, cov=cov, name=c);
 _delta_=1e-8;
 mu = pred + eij;
 parms b11=5 b12=0 b13=0 b14=0 b21=1 b22=0 b31=.20 b32=0
       sigsq_w1=.1 sigsq_w2=.1 sigsq_w3=.1 sigsq_w4=.1
       sigsq_w5=.1 sigsq_w6=.1 sigsq_w7=.1
       rho=.01 .1 .5 / best=1;
 model UFR~normal(mu, _delta_);
 random e1 e2 e3 e4 e5 e6 e7 ~
        normal([ 0, 0, 0, 0, 0, 0, 0 ],
               [ c1 ,
                 c2 , c3 ,
                 c4 , c5 , c6 ,
                 c7 , c8 , c9 , c10,
                 c11, c12, c13, c14, c15,
                 c16, c17, c18, c19, c20, c21,
                 c22, c23, c24, c25, c26, c27, c28 ])
        subject=Dialyzer;
run;
```

We note that while the ML estimates of the model parameters as well as the information criteria based on the −2 log-likelihood value will be the same using %NLINMIX and NLMIXED, the standard errors for the regression parameter estimates $\widehat{\boldsymbol{\beta}}$ will differ. This is because the PL/QELS estimation scheme implemented in the above call to %NLINMIX uses a Gauss-Newton or, equivalently, Fisher scoring algorithm for estimating the regression parameters $\boldsymbol{\beta}$. Consequently, the standard errors of $\widehat{\boldsymbol{\beta}}$ are based on Fisher's expected information matrix. In contrast, estimation under NLMIXED is based on a Newton-Raphson algorithm and the standard errors will all be based on the observed second-order Hessian matrix. While ML estimation of any GNLM like model 3 of Table 5.3 is easily implemented using the %NLINMIX macro, the NLMIXED procedure offers users the additional flexibility of testing general nonlinear hypotheses, a feature that is not available using %NLINMIX. Moreover, %NLINMIX cannot be used to obtain MLE's of model parameters when the marginal covariance matrix depends on $\boldsymbol{\beta}$.

One last observation is worth noting. The ease with which one can compute MLE's using %NLINMIX whenever $Var(\boldsymbol{y}_i) = \boldsymbol{\Sigma}_i(\boldsymbol{\theta})$ makes it extremely useful when dealing with complex covariance structures. For example, with an unstructured covariance matrix such as for model 1 of Table 5.3, estimating the 8 regression parameters and 28 covariance parameters using PROC NLMIXED can require prohibitively excessive CPU time if one uses crude starting values. If one wishes to use some of the additional features of NLMIXED not available with the macro, then we suggest first computing the ML estimates using %NLINMIX, then store those estimates in a dataset and finally use those values as starting values in PROC NLMIXED via the DATA= option of the PARMS statement. This is precisely what was done when we attempted to fit the GNLM model 1 with unstructured covariance matrix and convergence was not achieved following an overnight run of NLMIXED. In this case, when we used the MLE's from %NLINMIX as starting values, the NLMIXED program "converged" in a single iteration.

Using similar programming statements as shown above, we ran all of the models in Tables 5.2 and 5.3. Several attempts to fit the NLME Model 4 with nonlinear random effects b_{i1} and b_{i2} and an intra-subject AR(1)-POM structure failed and so this model is not summarized here. Convergence was achieved with all other models and the goodness-of-fit statistics based on likelihood information criterion from the ODS OUTPUT FitStatistics= statements were combined into a single dataset. The information criteria, which includes Akaike's information criterion (AIC), a finite-sample corrected version of AIC (AICC) and Schwarz's Bayesian information criterion (BIC), were each rank ordered from smallest to largest across all of the models and the results summarized in Output 5.30. The results are very much in agreement with the rank ordering of similar models considered by Littell et. al. (2006). Specifically, based on all three information criteria considered, the marginal GNLM Model 6 with a heterogeneous first-order autoregressive structure, ARH(1), provided the best fit to the data followed closely by either the NLME Model 6 with random asymptote b_{i1} and intra-subject ARH(1) structure and the GNLM Model 7 with a first-order autoregressive structure coupled with a power of the mean variance (AR(1)-POM). The worse fitting models were those GNLM's in which repeated measurements from the same dialyzer are assumed to be equicorrelated. These include the CS, CSH and CS-POM structures (Models 2-4 of Table 5.3). Interestingly, the NLME Model 1 with random effects b_{i1} and b_{i2} and constant variance ranks midway among the 12 models evaluated.

Output 5.30: Rank order of model goodness-of-fit according to three different likelihood information criterion, AIC, AICC and BIC, where smaller values are indicative of a better fit.

			Information Criterion					
			AIC		AICC		BIC	
	Model	Covariance Structure	Value	Rank	Value	Rank	Value	Rank
1	NLME 1	RE(bi1,bi2)|VC	32.27	7	34.72	6	44.21	6
	NLME 2	RE(bi1,bi2)|VC-POM	24.45	5	27.34	4	37.39	4
	NLME 3	RE(bi1,bi2)|CS-POM	24.70	6	28.06	5	38.64	5
	NLME 5	RE(bi1)|CSH	42.47	9	47.49	9	59.40	9
	NLME 6	RE(bi1)|ARH(1)	6.38	2	11.40	2	23.31	3

2	GNLM 1	UN	15.35	4	41.22	7	51.20	7
	GNLM 2	CS	77.58	12	79.28	12	87.54	12
	GNLM 3	CSH	54.13	11	58.55	11	70.06	11
	GNLM 4	CS−POM	52.57	10	54.63	10	63.52	10
	GNLM 5	AR(1)	41.37	8	43.07	8	51.33	8
	GNLM 6	ARH(1)	4.91	1	9.34	1	20.84	1
	GNLM 7	AR(1)−POM	11.09	3	13.15	3	22.04	2

As noted by Chi and Reinsel (1989) and Jones (1990), the use of a smaller number of random effects combined with some form of autoregressive correlation often provides a parsimonious fit to longitudinal data. However, both authors point to the difficulties associated with the identifiability of parameters under such models. Here, for example, we find that a NLME model with a single random asymptote parameter along with an intra-subject ARH(1) structure (model 6 of Table 5.2) provides a better fit to the high-flux dialyzer data than a marginal model with heterogeneous power of the mean variances and AR(1) correlation (model 7 of Table 5.3). However, when compared against the marginal GNLM with ARH(1) structure (model 6 of Table 5.3), there is little evidence to suggest the inclusion of a random asymptote effect will help explain any additional heterogeneity that may be present in the data. Both the NLME model 6 and GNLM model 6 provide a good fit to the data with the NLME model having a lower value of −2 log likelihood (−27.6 versus −27.1, respectively).

When choosing between a NLME model and a marginal GNLM each having a comparable fit to the data, careful consideration should be given to the interpretation of the regression parameters and the underlying objectives of the study. In this example, we have two such competing models, the NLME model 6 of Table 5.2 and the GNLM model 6 of Table 5.3. Because the NLME model 6 is strictly linear in the random asymptote effect, b_{i1}, the regression parameters of this model will have both a PA and SS interpretation. We can therefore visually compare the two models in terms of how well their predicted marginal means and associated confidence bands fit the data. Under the NLME model 6, the PA mean response will be equivalent to the average dialyzer mean response. The latter is obtained by setting $b_{i1} = 0$ and evaluating the predicted responses of each dialyzer according to each dialyzer's fixed blood flow rate. These predicted responses are represented by the variable, `pred_avg`, which is defined above in the NLMIXED programming statements for NLME Model 6. Values of this variable are then placed in a dataset using a PREDICT statement as shown in the above code for this model. A similar PREDICT statement is used to output the PA predicted responses under the GNLM model 6 of Table 5.3. The two output data sets are combined and the PA mean response and 95% confidence band plotted for each model using PROC SGPANEL. Based on the mean response profiles shown in Figure 5.11, both models adequately characterize the average UFR performance of high-flux dialyzers at both blood flow rates. However, since the PA mean response for the NLME model 6 is also the average dialyzer's mean response, we prefer this model over the marginal GNLM model 6 since it allows us to draw inference based on one of the stated goals of the study, namely to structurally characterize the ultrafiltration profile of the typical dialyzer.

5.4.4 Cefamandole pharmacokinetic data

Davidian and Giltinan (1993b) present the results of two different NLME estimation methods used in an analysis of pharmacokinetic data investigating the plasma concentration profile of the cephalosporin antibiotic, cefamandole. A description and partial listing of the data are available through the author's Web page. The primary features of the study are as follows.

Figure 5.11 Predicted fit and confidence band for the PA mean response of NLME model 6 having a random asymptote effect and an intra-subject ARH(1) covariance structure and for GNLM model 6 with a marginal ARH(1) covariance structure.

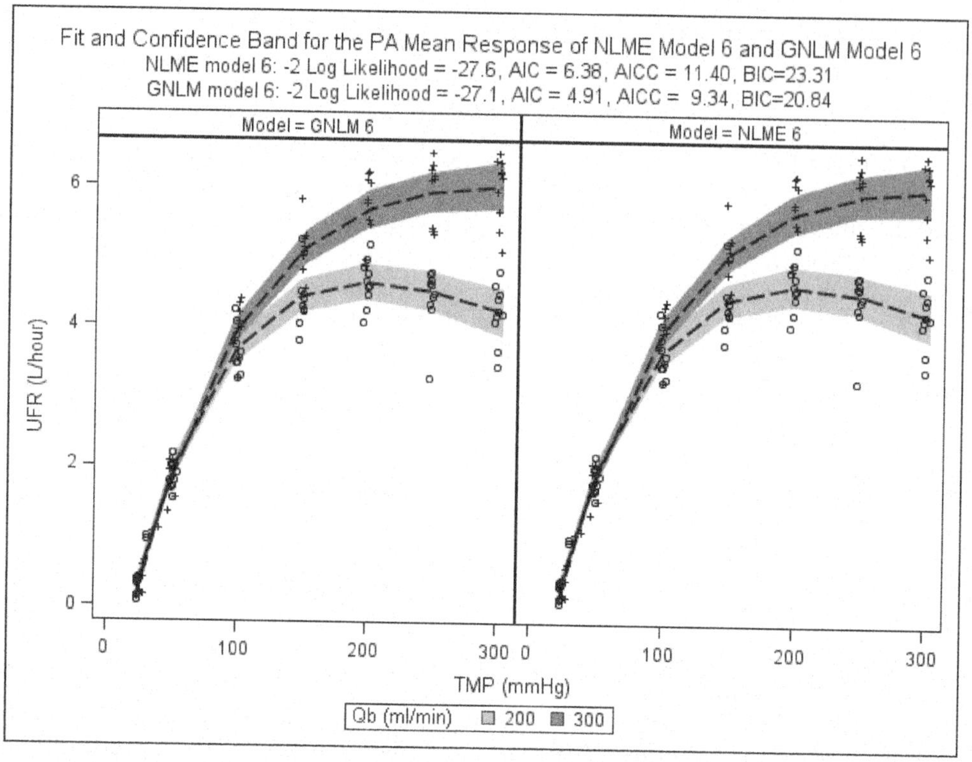

- Study of kinetics of the cephalosporin antibiotic, cefamandole
- Experimental unit or cluster is an individual subject
  - Six healthy male volunteers
- Response variable:
  - y_{ij} = drug concentration in plasma (μg/mL) measured on the ith subject on the jth occasion ($i = 1, \ldots, 6$; $j = 1, \ldots, 14$ time points)
  - measured via high-performance liquid chromatography, HPLC
- One within-unit covariate:
  - Time: t_{ij} =time following a 10-minute intravenous infusion (14 time points over a 6 hour period including time 10 minutes; t_{ij} = 10, 15, 20, 30, 45, 60, 75, 90, 120, 150, 180, 240, 300, 360 minutes)
- Structural model: $y = \beta_1 e^{-\beta_2 t} + \beta_3 e^{-\beta_4 t}$
- Goal: Estimate the average elimination rate parameters β_2 and β_4 of the structural bi-exponential model

The authors fit the cefamandole plasma concentration profiles shown in Figure 5.12 to a NLME bi-exponential model as a means for describing the plasma elimination of the drug following a 10-minute intravenous infusion of cefamandole at a dose level of 15 mg/kg body weight. To ensure positive parameter estimates and account for intra-subject heterogeneity, they used the following parameterized version of a bi-exponential model with power of the mean variance structure.

$$\begin{aligned} y_{ij} &= f(t_{ij}, \boldsymbol{\beta}, \boldsymbol{b}_i) + \epsilon_{ij} \qquad (5.89) \\ &= e^{\beta_{i1}} \exp(-e^{\beta_{i2}} t_{ij}) + e^{\beta_{i3}} \exp(-e^{\beta_{i4}} t_{ij}) + \epsilon_{ij} \end{aligned}$$

Figure 5.12 Cefamandole concentration profiles. Left panel is concentration and right panel is log(concentration)

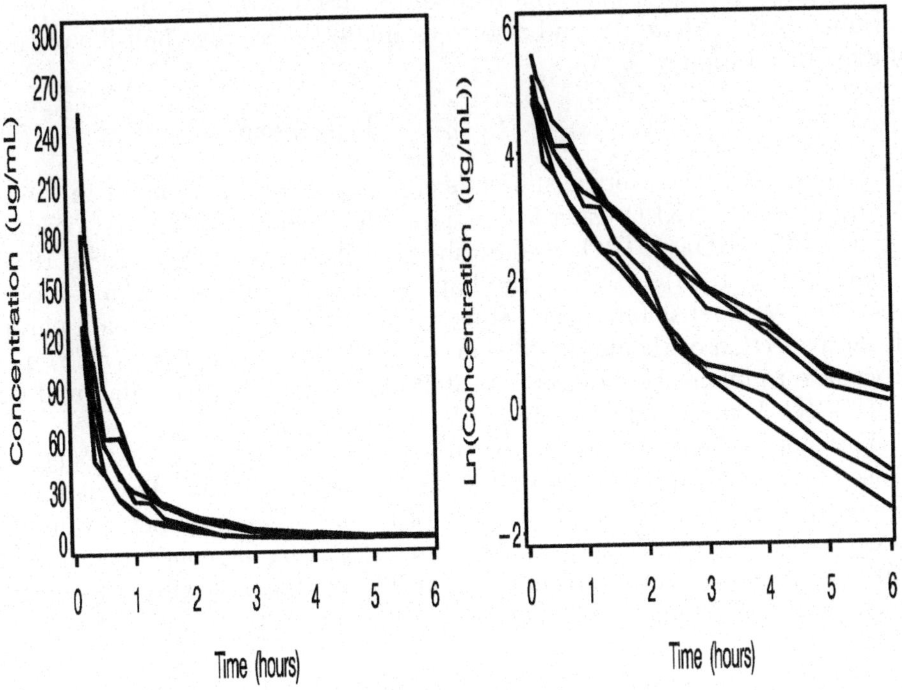

where

$\exp(\beta_{i1}) = \exp(\beta_1 + b_{i1})$ is the ith subject's initial phase average starting concentration
$\exp(\beta_{i2}) = \exp(\beta_2 + b_{i2})$ is the ith subject's rate of elimination during the initial phase
$\exp(\beta_{i3}) = \exp(\beta_3 + b_{i3})$ is the ith subject's terminal phase average starting concentration
$\exp(\beta_{i4}) = \exp(\beta_4 + b_{i4})$ is the ith subject's rate of elimination during the terminal phase
$\epsilon_{ij} \sim\text{ind}\, N\Big(0, \sigma_w^2 f(t_{ij}, \boldsymbol{\beta}, \boldsymbol{b}_i)^{2\delta}\Big)$
$\boldsymbol{b}_i \sim\text{iid}\, N(\mathbf{0}, \boldsymbol{\Psi})$ with $\boldsymbol{\Psi}$ allowed to be an unstructured 4×4 covariance matrix of the random effects vector $\boldsymbol{b}_i' = (b_{i1}, b_{i2}, b_{i3}, b_{i4})$.

Davidian and Giltinan (1993b) estimated the population parameters using both a pooled two-stage (PST) estimation algorithm and a linear mixed-effects algorithm. The latter is similar to the PL/CGEE1 algorithm implemented in the macro %NLINMIX but with an EGLS estimation scheme adapted from Vonesh and Carter (1992) used in place of the pseudo-likelihood approach of Wolfinger (1993) and Wolfinger and Lin (1997).

Here, we fit the same model to the data but rather than assume an unstructured variance-covariance matrix for the random effects, we assume a diagonal variance-components structure

$$\boldsymbol{\Psi}(\boldsymbol{\theta}_b) = \begin{pmatrix} \psi_{11} & 0 & 0 & 0 \\ 0 & \psi_{22} & 0 & 0 \\ 0 & 0 & \psi_{33} & 0 \\ 0 & 0 & 0 & \psi_{44} \end{pmatrix}$$

where $\boldsymbol{\theta}_b = (\psi_{11}, \psi_{22}, \psi_{33}, \psi_{44})'$ is the vector of variance components. Since there are only six subjects, we know from standard multivariate analysis of variance that we can not fit a model with an unstructured covariance matrix since it would require a minimum of 11 subjects to estimate the 10 variance-covariance parameters of $\boldsymbol{\Psi}$ assuming $\boldsymbol{\Psi}$ is arbitrary positive definite.

Our goals are to estimate the regression parameters $\boldsymbol{\beta} = (\beta_1, \beta_2, \beta_3, \beta_4)'$ and variance parameters $\boldsymbol{\theta} = (\boldsymbol{\theta}_b, \sigma_w^2, \delta)$ and also to estimate the average half-life of the terminal phase of drug disposition,

$$t_{1/2,term} = \log(2)/\exp(\beta_4) \tag{5.90}$$

where $\exp(\beta_4)$ is the average subject's elimination rate parameter in the terminal phase. We do so in NLMIXED using ML estimation based on both a first-order approximation method (METHOD=FIRO) as described by Sheiner and Beal (1980, 1981) and Beal and Sheiner (1982, 1988) as well as the Laplace approximation (see §5.3.4 for details of both methods). We first obtain initial starting estimates of the regression parameters $\boldsymbol{\beta}$ via visual inspection of the data and then use a standard two-stage (STS) estimation procedure to refine these estimates (Steimer et. al., 1984; Davidian and Giltinan, 1995, pp.137-138; Vonesh and Chinchilli, 1997, pp. 331-332).

The plot of the concentration profiles (left panel of Figure 5.12) suggests the terminal phase of elimination begins around 30 minutes post infusion. By examining a plot of $\log_e$ concentrations versus time (right panel of Figure 5.12), parameter values for the initial phase may be "guesstimated" by assuming a starting log concentration of 5.5 at 10 minutes and an ending log concentration of 4.0 at 30 minutes. Likewise, a log concentration of 0 is reached around 5 hours (300 minutes). From these estimates we have

$$\beta_1 \approx 5.5,\ \beta_2 \approx \log\{(5.5-4)/[(30\text{-}10)/60]\} = \log(4.5) = 1.5,$$
$$\beta_3 \approx 4.0,\ \beta_4 \approx \log\{(4.0-0)/[(300-30)/60]\} = \log(0.89) = -0.12$$

Using these as initial values, we fit the data to each individual using NLMIXED from which OLS estimates of $\boldsymbol{\beta} = (\beta_1, \beta_2, \beta_3, \beta_4)'$ and σ_w^2 and δ were obtained. The individual OLS estimates of the model parameters and programming statements used to generate these estimates are shown below.

Program 5.22

```
data example5_4_4;
 set cefamandole_data;
 log_conc=log(Conc);
 hour=time/60;
run;
proc sort data=example5_4_4 ;
 by Subject time;
run;
ods listing close;
ods output parameterestimates=OLSpe;
proc nlmixed data=example5_4_4 ;
 by Subject;
 parms beta1=5.5 beta2=1.5 beta3=4 beta4=-0.12
       sigsq_w=1 delta=1;
 func = exp(beta1)*exp(-exp(beta2)*hour) +
        exp(beta3)*exp(-exp(beta4)*hour);
 var_conc = sigsq_w*(func**(2*delta));
 model conc ~ normal(func, var_conc);
run;
```

```
ods listing;
data OLSpe1; set OLSpe;
 by Subject;
 if Parameter = 'beta4' then do;
    Parameter = 't_half'; Estimate=log(2)/exp(Estimate);
 end;
 if Parameter = 't_half';
run;
data OLSpe2; set OLSpe OLSpe1;
 by Subject;
run;
proc transpose data=OLSpe2 out=OLS;
 by subject;
 var estimate;
 id Parameter;
run;
proc print data=OLS noobs;
 var Subject beta1-beta4 sigsq_w delta t_half;
run;
```

Output 5.31: Individual OLS estimates of β_{i1}, β_{i1}, β_{i1}, β_{i1}, σ_w^2 and δ as well as the half-life of the terminal phase of drug disposition, $t_{1/2,term}$

Subject	beta1	beta2	beta3	beta4	sigsq_w	delta	t_half
1	4.8698	1.0003	2.8066	−0.4414	0.02131	1.0902	1.0778
2	5.2608	1.5394	3.6404	−0.00510	0.03395	0.4294	0.6967
3	5.5673	1.7992	4.0675	−0.3115	0.1300	0.5978	0.9465
4	6.2043	2.2605	4.6176	0.2795	0.5334	0.2820	0.5241
5	5.7346	1.0927	3.8646	−0.4491	0.02630	0.8553	1.0861
6	4.9526	0.5649	3.3475	−0.5563	0.01369	0.9614	1.2090

Output 5.32: Standard two-stage (STS) estimates of the model parameters.

Analysis Variable : Estimate

Parameter	N	Mean	Median	Variance	Std Error
beta1	6	5.4316	5.4141	0.2563	0.2067
beta2	6	1.3762	1.3161	0.3735	0.2495
beta3	6	3.7240	3.7525	0.3853	0.2534
beta4	6	−0.2473	−0.3765	0.1028	0.1309
delta	6	0.7027	0.7265	0.1006	0.1295
sigsq_w	6	0.1264	0.0301	0.0416	0.0833
t_half	6	0.9233	1.0121	0.0686	0.1070

Using PROC MEANS, we computed the mean, median, variance and standard errors of the individual OLS estimates to obtain standard two-stage (STS) estimates of the model parameters shown in Output 5.32. These STS estimates were then used as starting values for the more advanced first-order approximation and Laplace approximation methods implemented in NLMIXED. Initial values of the random effect variances were taken to be the estimated variances of the individual OLS estimates. The programming statements used to generate the STS estimates as well as the NLMIXED syntax used to compute the Laplace approximated ML estimates are as follows.

Program 5.23

```
proc means data=OLSpe2 nonobs n mean median var
                         stderr nway maxdec=4;
 class Parameter;
 var Estimate;
 output out=STSresults mean=Estimate Median=Median
                         var=Variance stderr=SE;
run;
data STSparms; set STSresults;
 if Parameter='beta1' then output;
 if Parameter='beta2' then output;
 if Parameter='beta3' then output;
 if Parameter='beta4' then output;
 if Parameter='beta1' then do;
    Parameter='psi11'; Estimate=Variance;output;
 end;
 if Parameter='beta2' then do;
    Parameter='psi22'; Estimate=Variance;output;
 end;
 if Parameter='beta3' then do;
    Parameter='psi33'; Estimate=Variance;output;
 end;
 if Parameter='beta4' then do;
    Parameter='psi44'; Estimate=Variance;output;
 end;
 if Parameter='delta' then do;
    Parameter='delta'; Estimate=Median;output;
 end;
 if Parameter='sigsq_w' then do;
    Estimate=Median;output;
 end;
 if Parameter='t_half' then delete;
run;
proc nlmixed data=example5_4_4 qpoints=1;
 parms /data=STSparms;
 beta1i = beta1 + b1i; beta2i = beta2 + b2i;
 beta3i = beta3 + b3i; beta4i = beta4 + b4i;
 t_half = log(2)/exp(beta4);
 func = exp(beta1i)*exp(-exp(beta2i)*hour) +
        exp(beta3i)*exp(-exp(beta4i)*hour);
 var_conc = sigsq_w*(func**(2*delta));
 model conc ~ normal(func, var_conc);
 random b1i b2i b3i b4i ~ normal([0,0,0,0],
                                 [psi11,
                                    0  , psi22,
                                    0  ,   0  , psi33,
                                    0  ,   0  ,   0  , psi44])
        subject=Subject;
 estimate 't(1/2) - terminal' t_half;
 predict func out=predplot;id func;
run;
```

Table 5.4 Summary of population parameter estimates using STS, First-Order and Laplace approximated ML estimation. † corresponds to NLMIXED option METHOD=FIRO, †† corresponds to NLMIXED option QPOINTS=1.

	Method					
	STS		First-Order†		Laplace††	
Parameter	Estimate	SE	Estimate	SE	Estimate	SE
β_1	5.4316	0.2067	3.8804	0.4994	5.1207	0.1434
β_2	1.3762	0.2495	0.9978	0.0755	0.9497	0.1866
β_3	3.7240	0.2534	3.0004	0.2122	3.2686	0.3537
β_4	-0.2473	0.1309	-0.4695	0.0477	-0.4171	0.1001
$t_{1/2,term}$	0.9233	0.1070	1.1085	0.0529	1.0519	0.1053
ψ_{11}	0.2563	-	0.0060	0.0073	0.0571	0.0418
ψ_{22}	0.3735	-	-0.0127	0.0054	0.0135	0.0250
ψ_{33}	0.3853	-	0.0974	0.0742	0.2190	0.1912
ψ_{44}	0.1028	-	0.0484	0.0306	0.0085	0.0093
σ_w^2	0.1264	0.0833	0.0499	0.0226	0.0325	0.0171
δ	0.7027	0.1295	1.1278	0.1130	0.9686	0.0990

The ML estimates based on the first-order approximation method utilize the same NLMIXED syntax as shown above but with METHOD=FIRO specified as the integration option in the call to NLMIXED.

The parameter estimates from the three methods of estimation; STS, first-order and Laplace, are summarized in Table 5.4. In comparing methods, the STS approach results in much higher estimates of the random effects variance components, $\boldsymbol{\theta}_b = (\psi_{11}, \psi_{22}, \psi_{33}, \psi_{44})'$, compared to either the first-order or Laplace approximation techniques. This is not unexpected since STS estimation ignores any uncertainty in the individual OLS estimates of $\boldsymbol{\beta}_i$ that are used to estimate $\boldsymbol{\Psi}(\boldsymbol{\theta}_b)$. This results in upwardly biased estimates of the variance components as shown by Davidian and Giltinan (1995, pp. 138). In contrast, both the first-order and Laplace approximations result in very imprecise ML estimates of ψ_{11}, ψ_{22}, ψ_{33} and ψ_{44} as judged by how large the standard errors of these estimates are. This is not too surprising since there are only six (6) subjects available with which to estimate between-subject variability. There are some differences between the three methods with respect to the regression parameter estimates although the first-order and Laplace methods appear to be fairly consistent with each other. All three methods yield an estimate of the elimination phase half-life of cefamandole to be about one hour and they each demonstrate a significant power of the mean variance structure that is suggestive of a constant coefficient of variation ($\widehat{\delta} \approx 1.00$ for the first-order and Laplace methods). Shown in Figure 5.13 are the individual fits to the data based on estimates from the Laplace approximation. The predicted curves are computed by combining the Laplace ML estimates of the population parameters with the subject-specific posterior modes of the random effects. As these plots indicate, there is excellent agreement between the observed and fitted values based on a bi-exponential model with intra-subject power of the mean variances.

In summary, the STS estimates provide reasonable starting values for more "advanced" methods such as ML estimation using either a first-order approximation or a Laplace approximation. For small random-effect variance components, the first-order approximation will produce estimates of the population PK parameters that will be comparable to ML estimates (e.g., Solomon and Cox, 1992). This appears to be the case in this example. The fact that the first-order approximation yields a negative variance estimate for b_{i2} while the other estimates are very near the boundary value of 0 may simply reflect the uncertainty there is when estimating these variance components with only six subjects. One can prevent negative variance component estimates by using a BOUNDS statement with

Figure 5.13 Observed versus predicted cefamandole concentrations by subject. Predicted concentrations are based on Laplace approximated ML estimates of the population parameters and posterior modes of the random effects.

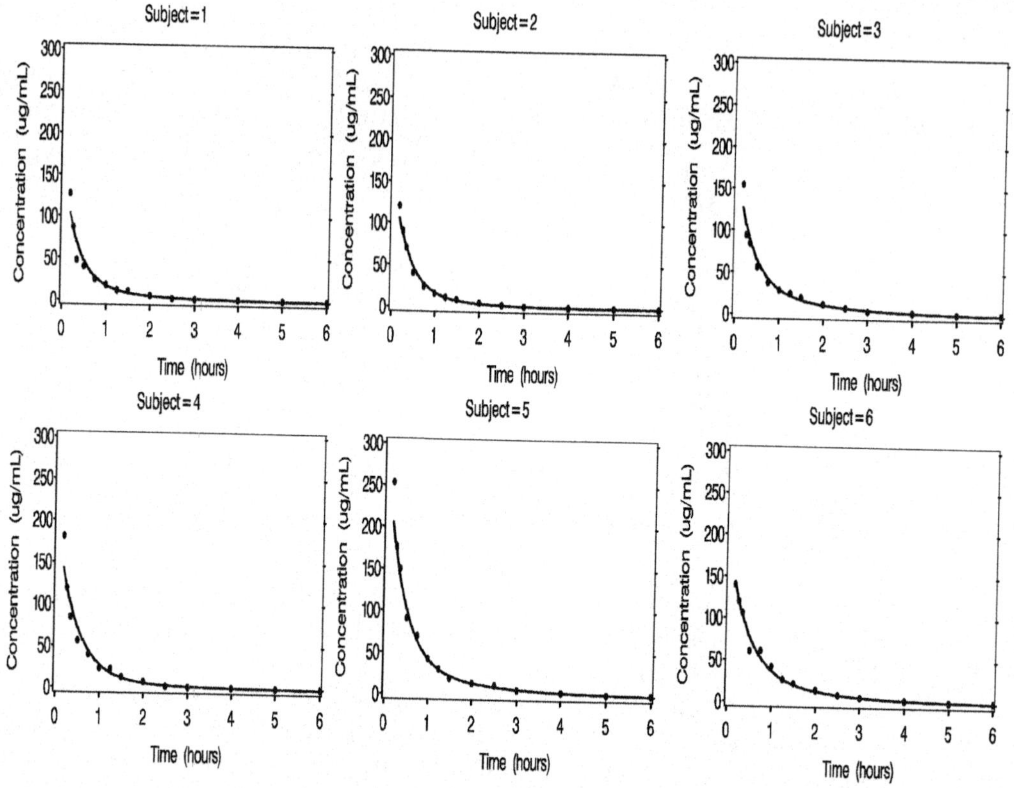

NLMIXED. When we did so in this example, we ended up with an estimate of $\widehat{\psi}_{22}$ which was essentially 0 (results not shown). Of the three estimation methods considered, ML estimation based on Laplace's approximation will generally outperform both the STS and first-order methods, particularly when the number of observations per subject is large as is the case here. Thus for small phase 1 studies like this where one has an adequate sample size per subject but just a few subjects, the Laplace approximation would be the preferred approach to estimation over the STS and first-order methods.

5.4.5 Epileptic seizure data—continued

We continue here with analyses of the epileptic seizure data described in §1.4 (page 15) by fitting an overdispersed version of model IV of Brelsow and Clayton (1993) to the data. We do so under each of two random effects structures: 1) assuming a random intercept effect only and 2) assuming a random intercept effect and a random slope effect. Allowing for intra-subject overdispersion, model IV of Breslow and Clayton (1993) may be written in terms of the GLME model

$$\begin{aligned} E(y_{ij}|\boldsymbol{b}_i) &= \mu_{ij}(\boldsymbol{\beta}, \boldsymbol{b}_i) = g^{-1}(\boldsymbol{x}'_{ij}\boldsymbol{\beta} + \boldsymbol{b}_i) = \exp(\boldsymbol{x}'_{ij}\boldsymbol{\beta} + \boldsymbol{b}_i) \\ Var(y_{ij}|\boldsymbol{b}_i) &= \phi\mu_{ij}(\boldsymbol{\beta}, \boldsymbol{b}_i) \end{aligned} \tag{5.91}$$

where $\boldsymbol{x}'_{ij} = \begin{pmatrix}1 & a_{1i} & \log(a_{2i}) & \log(a_{3i}) & a_{1i}\log(a_{3i}) & v^*_{ij}/10\end{pmatrix}$ is a vector of within- and between-subject covariates as defined in §1.4 (page 15) except that v_{ij} is replaced here by v^*_{ij} where $v^*_{ij} = -3, -1, 1,$ or 3 corresponding to visits 1, 2, 3, or 4, respectively. For the random intercept structure (designated Model IV-a), the conditional linear predictor is given by $\boldsymbol{x}'_{ij}\boldsymbol{\beta} + \boldsymbol{b}_i = (\beta_0 + b_{i1}) + \beta_1 a_{1i} + \beta_2\log(a_{2i}) + \beta_3\log(a_{3i}) + \beta_4 a_{1i}\log(a_{3i}) + \beta_5 v^*_{ij}/10$ with $b_{i1} \sim$iid $N(0, \psi_{11})$ while for the random intercept and random slope structure (Model

IV-b), the conditional linear predictor is $\boldsymbol{x}_{ij}'\boldsymbol{\beta} + \boldsymbol{b}_i = (\beta_1 + b_{i1}) + \beta_2 a_{1i} + \beta_3 \log(a_{2i}) + \beta_4 \log(a_{3i}) + \beta_5 a_{1i} \log(a_{3i}) + (\beta_6 + b_{i2}) v_{ij}^*/10$ with $\boldsymbol{b}_i = (b_{i1}, b_{i2})' \sim$iid $N(0, \boldsymbol{\Psi}(\boldsymbol{\theta}_b))$. Based on the results of Breslow and Clayton (1993), the covariance matrix $\boldsymbol{\Psi}(\boldsymbol{\theta}_b)$ is assumed to be a 2×2 diagonal matrix with variance components ψ_{11} and ψ_{22} (i.e., $Cov(b_{i1}, b_{i2}) = \psi_{12} = 0$). As in §4.3.3, the covariates a_{1i}, a_{2i}, and $\log(a_{3i})$ represent the treatment group indicator (`Trt`), age (`Age`), log normalized baseline count (`y0`).

Since inclusion of the intra-subject overdispersion parameter ϕ precludes the use of ML estimation, these two GNLME models were fit to the data using both the PQL estimation scheme of Breslow and Clayton (1993) and the modified PELS estimation scheme (§5.3.4). Both methods of estimation can be implemented within the GLIMMIX procedure (either PQL or modified PELS) or within the %NLINMIX macro (PQL) and NLMIXED procedure (modified PELS). For this particular application, we take advantage of the fact that the conditional mean is a monotonic function of a linear predictor and use the GLIMMIX procedure to estimate the model parameters. The SAS program used to fit these models is as follows.

Program 5.24

```
data example5_4_5;
 set example4_3_3;
 by ID Visit;
 if Visit=1 then Vtime=-3; ** As per Breslow and Clayton **;
 if Visit=2 then Vtime=-1;
 if Visit=3 then Vtime= 1;
 if Visit=4 then Vtime= 3;
 Vtime=Vtime/10;
run;
data pe;
data cov;
%macro FitModels(method=rspl, random=intercept,
                 model=Model 1, parms=pe1, covparms=cov1);
 %let method=%qupcase(&method);
 ods listing close;
 ods output parameterestimates=&parms;
 ods output covparms=&covparms;
 proc glimmix data=example5_4_5 method=&method empirical;
  class id;
  model y=Trt LogAge y0 y0*Trt Vtime / link=log dist=normal s;
  _variance_=_phi_*_mu_;
  nloptions technique=newrap maxiter=500;
  random &random / subject=id type=vc g;
 run;
 data &parms;
  set &parms;
  Model=''&model'';
  estimation=''&method'';
  if estimation='RSPL'    then Method='         PQL ';
  if estimation='LAPLACE' then Method='Modified PELS';
 run;
 data &covparms;
  set &covparms;
  Model=''&model'';
  estimation=''&method'';
```

```
   if estimation='RSPL'    then Method='         PQL ';
   if estimation='LAPLACE' then Method='Modified PELS';
  run;
  data pe;
   set pe &parms;
   Parameter='Fixed ';
   if estimate=. then delete;
  run;
  data cov;
   set cov &covparms;
   Parameter='Random';
   if estimate=. then delete;
   if CovParm='Intercept' then CovParm='psi11';
   if CovParm='Vtime    ' then CovParm='psi22';
   if CovParm='Scale    ' then CovParm='Scale';
  run;
  ods listing;
%mend;
%FitModels(method=rspl, random=Intercept,
           model=Model IV-a, parms=pe1, covparms=cov1);
%FitModels(method=laplace, random=Intercept,
           model=Model IV-a, parms=pe2, covparms=cov2);
%FitModels(method=rspl, random=Intercept Vtime,
           model=Model IV-b, parms=pe3, covparms=cov3);
%FitModels(method=laplace, random=Intercept Vtime,
           model=Model IV-b, parms=pe4, covparms=cov4);
data all;
 set pe cov(rename=(CovParm=Effect));
run;
proc report data=all HEADLINE HEADSKIP SPLIT='|' nowindows;
 column Model Parameter Effect Method, (Estimate StdErr);
 define Model / Group 'Model' width=10 center;
 define Parameter / Group 'Effect' width=9 center;
 define Effect / GROUP ``Variable'' center width=9 ;
 define Method / Across ``Method of Estimation'' width=25 center;
 define Estimate / MEAN FORMAT=7.4 'Estimate'
                   width=9 NOZERO spacing=1;
 define StdErr   / MEAN FORMAT=7.4 'Robust SE'
                   width=9 NOZERO center spacing=1;
run;
quit;
```

We first create the dataset, `example5_4_5`, with the revised variable, `Vtime`, representing the within-subject covariate $v_{ij}^*/10$ used in Model IV of Breslow and Clayton (1993). We then fit the count data to the two overdispersed mixed-effects models with the help of the macro %FitModels. This macro allows one to specify, via macro variables, the method of estimation to be used (METHOD=), the random effects to be modeled (RANDOM=), a model identifier (MODEL=) and two datasets containing the estimated regression parameters (PARMS=) and estimated variance components (COVPARMS=). The rest of the macro involves the GLIMMIX syntax and programming statements needed to summarize the results using PROC REPORT. The MODEL statement of GLIMMIX always specifies the log link function (LINK=LOG) and the normal distribution (DIST=NORMAL) when fitting each of the models to the data. Also, the overdispersion parameter and intra-subject variance structure is specified directly via the variance assignment statement,

`_variance_=_phi_*_mu_;`

This is done so that when one specifies METHOD=RSPL, the GLIMMIX procedure will ignore the DIST=NORMAL option of the model statement and fit the data via the PL/CGEE1 algorithm described in §5.3.4. This results in the PQL estimation procedure of Breslow and Clayton (1993) with REPL estimates of the variance components. However, when one specifies METHOD=LAPLACE, the GLIMMIX procedure will fit the data via ML estimation under the "working" assumption of normality. A log link function is used to link the mean to the linear predictor and the intra-subject variance is defined directly as a function of the corresponding inverse link function (i.e., `_phi_*_mu_`). This results in the modified PELS estimator as described in §5.3.4. In both cases, inference is based on the use of robust standard errors as specified by the EMPIRICAL option of GLIMMIX. The results from fitting these two models to the data using PQL and modified PELS is summarized in Output 5.33.

Output 5.33: PQL and Modified PELS estimates of two overdispersed mixed-effects Poisson models applied to the epileptic seizure data.

			Method of Estimation			
			PQL		Modified PELS	
Model	Effect	Variable	Estimate	Robust SE	Estimate	Robust SE
Model IV−a	Fixed	Intercept	−1.4878	0.9625	−1.5515	1.0739
		LogAge	0.5297	0.2843	0.4745	0.3159
		Trt	−0.9118	0.3981	−0.9511	0.4285
		Trt*y0	0.3392	0.1986	0.3348	0.2049
		Vtime	−0.2936	0.1750	−0.3333	0.1720
		y0	0.8812	0.1096	0.9764	0.1338
	Random	Scale	1.9932	0.2091	1.9567	0.4116
		psi11	0.2234	0.0608	0.2707	0.0756
Model IV−b	Fixed	Intercept	−1.4492	0.9621	−1.4695	1.0535
		LogAge	0.5184	0.2845	0.4553	0.3111
		Trt	−0.9081	0.3952	−0.9339	0.4208
		Trt*y0	0.3366	0.1976	0.3312	0.2029
		Vtime	−0.2694	0.1622	−0.2876	0.1635
		y0	0.8791	0.1091	0.9644	0.1291
	Random	Scale	1.8497	0.2092	1.7940	0.3739
		psi11	0.2283	0.0611	0.2670	0.0737
		psi22	0.2363	0.2056	0.3234	0.1883

In comparing the two estimation techniques, we find the PQL and modified PELS estimates are remarkably similar under both models. Moreover, there is evidence of significant within-subject variation as estimated by an intra-subject overdispersion parameter, $\widehat{\phi}$, that is nearly two-fold higher than what would be expected assuming a conditional Poisson distribution for the number of epileptic seizures. There is also evidence of significant correlation in the form of random between-subject variability though such variability appears limited primarily to constant subject to subject variation rather than variation between subjects that changes with time (i.e., we observe a significant random intercept effect but an insignificant random slope effect).

The similarity between PQL and modified PELS estimates may be due to how well a normal distribution approximates the underlying conditional distribution of individual counts over time. We re-ran the random intercept model (Model IV-a) using PQL and requested a plot of the conditional residuals using the GLIMMIX option PLOTS=RESIDUALPANEL(BLUP ILINK). The results, shown in Figure 5.14, suggest that a "working" assumption of normality for the underlying conditional distribution is not unreasonable for this particular application which helps explain why the PQL and modified PELS estimates are so similar.

Figure 5.14 Panel of conditional residual plots to assess a working normality assumption

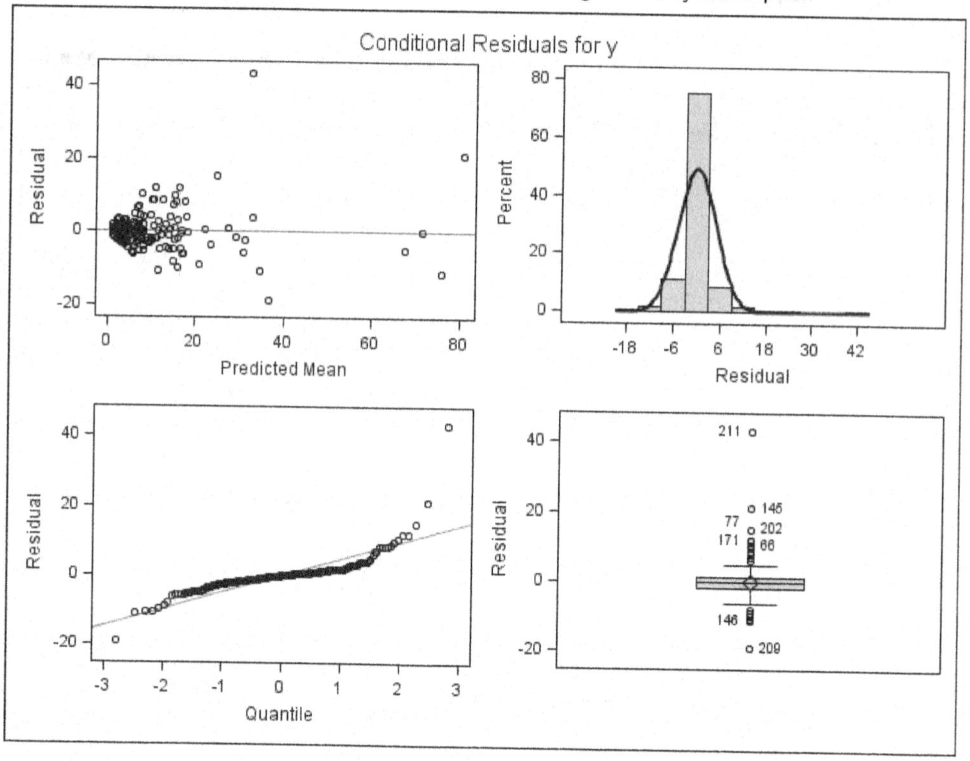

5.5 Summary

In this chapter we have presented two classes of mixed-effects models for correlated response data: the generalized linear mixed-effects (GLME) model and the generalized nonlinear mixed-effects (GNLME) model. Unlike linear mixed-effects models, these models are characterized by conditional means (and possibly conditional variances) that are nonlinear in both the fixed effects and random effects parameters. Consequently, inference is strictly subject-specific (SS) in scope which, at times, can be quite challenging to explain. However, there are instances where one can provide both a subject-specific and a marginal or population-averaged (PA) interpretation of results from fitting a nonlinear mixed model. This was illustrated, for example, in our analysis of the respiratory disorder data (§5.2.1) and epileptic seizure data (§5.2.2).

The GLME model is widely used for applications involving discrete and/or continuous outcomes for which the assumption of normality is unlikely to hold. It encompasses distributions from the regular exponential family and accounts for correlated responses through specification of one or two random effects (usually a simple random intercept effect). As with GLIM's, a potential drawback to a GLME model is its restriction to a conditional mean that is a monotonic invertible function of a linear predictor. The GNLME model extends the GLME model by 1) allowing for a broader class of distributions including the quadratic exponential family and 2) allowing for a more general nonlinear structure that does not require the conditional mean to be a function of a linear predictor. The family of GNLME models includes, as a special case, the important class of normal-theory nonlinear mixed effects (NLME) models.

Several methods of estimation for GLME and GNLME models were presented including moments-based linearization methods using PL/CGEE1 or PL/QELS, and likelihood-based methods using Laplace's approximation or Gauss-Hermite quadrature (see Table 5.5). For GLME models, these methods are available as options within the GLIMMIX procedure. For GNLME models, the moments-based linearization methods may be implemented using

Table 5.5 SAS procedures/macros for GLME and GNLME models, their key statements and/or options and select remarks

PROC (Model)	Key Statements/Options	Remarks
GLIMMIX (GLME)	/METHOD=RSPL(or MSPL)	$(\boldsymbol{\beta}, \boldsymbol{\theta})$ estimated via PL/CGEE1
	/METHOD=RMPL(or MMPL)	$(\boldsymbol{\beta}, \boldsymbol{\theta})$ estimated via PL/QELS
	RANDOM/RSIDE TYPE=	Required to estimate $\boldsymbol{\theta}_w$
	RANDOM "effects" / options	Required to estimate $\boldsymbol{b}_i$ and $\boldsymbol{\theta}_b$
	/METHOD=Laplace(or QUAD)	$(\boldsymbol{\beta}, \boldsymbol{\theta})$ estimated via ML
	RANDOM "effects" / options	Required to estimate $\boldsymbol{b}_i$ and $\boldsymbol{\theta}_b$
	/METHOD=Laplace LINK=	$(\boldsymbol{\beta}, \boldsymbol{\theta})$ estimated via PELS
	MODEL/DIST=NORMAL	Assumes "working" normality
	variance= user-specified	If $(y_{ij}\|\boldsymbol{b}_i)$ depends on $(\boldsymbol{\beta}, \boldsymbol{b}_i)$
%NLINMIX (GNLME)	MIXED/METHOD=ML(or REML)	$\boldsymbol{\theta}$ estimated via PL (or REPL)
	macro EXPAND=EBLUP	$\boldsymbol{\beta}$ estimated via PL/CGEE1
	macro EXPAND=ZERO	$\boldsymbol{\beta}$ estimated via PL/QELS
	REPEATED "effects" / TYPE=	Required to estimate $\boldsymbol{\theta}_w$
	RANDOM "effects" / options	Required to estimate $\boldsymbol{b}_i$ and $\boldsymbol{\theta}_b$
NLMIXED (GNLME)	/METHOD=GAUSS (default)	$(\boldsymbol{\beta}, \boldsymbol{\theta})$ estimated via ML
	/ QPOINTS=1	Required for Laplace MLE
	MODEL y ~ general(ll) or defaults	Specifies conditional pdf
	RANDOM "specfications" options	Required to estimate $\boldsymbol{b}_i$ and $\boldsymbol{\theta}_b$
	/QPOINTS=1	$(\boldsymbol{\beta}, \boldsymbol{\theta})$ estimated via PELS (or
	MODEL y ~ normal(mu, var)	Laplace MLE for NLME models)
	RANDOM "specfications" options	Required to estimate $\boldsymbol{b}_i$ and $\boldsymbol{\theta}_b$

the %NLINMIX macro while the likelihood-based methods are available within the NLMIXED procedure. Although NLMIXED is designed to handle nonlinear mixed models assuming observations within subjects/clusters are independent, we show how one can model intra-subject correlation (i.e., specify an R-side covariance structure) in NLMIXED through specification of linear random effects that are orthogonal to the nonlinear random effects of the model (see the orange tree and high flux dialyzer examples).

In applications involving moderate to large cluster sizes (i.e., within-subject sample sizes), we have found that ML estimation using the Laplace approximation will often produce parameter estimates nearly identical with those obtained using Gauss-Hermite quadrature. However, for highly discrete data and/or sparse within-subject data, one may need to use numerical quadrature to adequately approximate the integrated log-likelihood. Whenever possible, we recommend using one of the likelihood-based estimation methods available in GLIMMIX or NLMIXED versus one of the pseudo-likelihood based methods available in GLIMMIX or %NLINMIX. This, of course, is provided one can specify a conditional probability density function for the model. Shown in Table 5.5 is a summary of the different estimation techniques available in SAS including the procedure or macro, the class of models handled by that procedure or macro, some of the key statements and/or options one would use for a given estimation technique along with some general remarks.

Finally, a number of unique and, in some cases, challenging examples have been provided illustrating the various tools available in SAS for analyzing nonlinear mixed models. In several of those examples, we demonstrate how one can use the macro %COVPARMS (see Appendix D) to obtain good starting values for the model parameters as well as greatly reduce the amount of time spent trying to decide on a reasonable random effects structure for a given application. In the following chapters, we consider a number of case studies illustrating how one can apply the tools described in Chapters 2-5 to more complex problems including, among other things, the problem of missing data (Chapter 6) and the problem of non-Gaussian random effects (Chapter 7).

Part III

Further Topics

6

Missing Data in Longitudinal Clinical Trials

In Chapters 2-5, a number of examples from different disciplines are presented illustrating the types of models and methods of analysis that are available in SAS for analyzing correlated response data. Particular emphasis is placed on applications requiring the use of generalized linear and nonlinear models, both marginal and mixed. Methods of estimation and inference for these models may be classified as either semiparametric or parametric in nature. The semiparametric methods are moments-based and entail solving a set of first- or second-order generalized estimating equations while the parametric methods are likelihood-based and require specification of a well-defined probability density function. Both approaches require certain assumptions be satisfied for valid inference to take place. One implicit but very important assumption made is that any missing data that may be present can in some sense be ignored. In this chapter, we explore this assumption and its implications more closely. We do so by considering the different mechanisms that relate to why missing values occur and by considering the various analytic methods that are used when confronted with missing values.

6.1 Background

When there are no missing values present, all of the statistical methods described thus far allow valid inference to be made when analyzing correlated response data provided the complete data modeling assumptions are met. By complete data modeling assumptions, we mean those assumptions required for valid inference when one has complete but possibly irregularly spaced data (i.e., there are no missing values among those planned for). For example, moments-based methods like GEE1 or GEE2 require that the mean (GEE1) or both the mean and variance (GEE2) be correctly specified while likelihood-based methods require the correct specification of the probability density function including the mean and variance. With complete data, we can test these assumptions using various goodness-of-fit procedures, model diagnostics, etc. However, when we have incomplete data due to missing values, we run into the problem of never being able to verify these assumptions for the unobserved data. Of course even with complete data there are limitations with respect to the degree to which underlying assumptions can be verified. For example, in a simple linear regression setting we may, by design, measure a continuous response variable y at pre-determined time points $t_1 < t_2 < \ldots < t_p$. In the absence of any missing values, we might fit the linear regression model

$$y_i = \beta_1 + \beta_2 t_i + \epsilon_i;\ i = 1, 2, \ldots, p$$

to the observed data assuming $\epsilon_i \sim$iid $N(0, \sigma^2)$. With respect to the observed values, we can test the assumption of linearity using the usual F-test for linearity provided $\epsilon_i \sim$iid $N(0, \sigma^2)$ holds true. We can also use various model diagnostics, residual plots, and goodness-of-fit tests on the residuals to check our assumptions regarding ϵ_i. However, we also make the implicit assumption that the underlying linear relationship between y and t is continuous over the range $[t_1,\ t_p]$ such that for any intermediate time point $t_i < t^* < t_{i+1}$,

$$y(t^*) = \beta_1 + \beta_2 t^* + \epsilon(t^*)$$

with $E(y(t^*)) = \beta_1 + \beta_2 t^*$, $Var(\epsilon(t^*)) = \sigma^2$ and $\epsilon(t^*) \perp \epsilon_i$ for all i. In this case, we can never fully verify the underlying model assumptions as we never "observe" all possible values of $y(t)$. Nevertheless, this implicit assumption makes complete sense provided all other assumptions appear satisfied. This type of implicit assumption readily extends to the longitudinal data setting when we view the longitudinal data as a sequence of independent time series across individuals.

Valid inference may also be possible in the presence of missing values depending on the underlying reason for missing data and the underlying assumptions one is willing to make. In the following sections we briefly review the various mechanisms leading to missing data and we briefly describe several strategies for analyzing data in the presence of missing values. We do so strictly within the framework of missing data in longitudinal clinical trials. Our goal is to simply provide an overview of some of the techniques for handling missing data and demonstrate how one can apply some of these techniques using SAS. It is not our goal to provide an exhaustive treatment of missing data methodology. For a more comprehensive treatment of methods for missing data, the reader is referred to the books by Little and Rubin (2002), Fitzmaurice, Laird and Ware (2004), Molenberghs and Verbeke (2005), and Daniels and Hogan (2008) as well as the tutorial paper by Hogan et. al. (2004).

6.2 Missing data mechanisms

Missing data is frequently encountered in longitudinal studies involving serial measurements over time. This occurs most often in prospective longitudinal clinical trials where observations are planned at specified times during the course of follow-up. In this setting, missing values may occur intermittently when individuals miss one or more planned visits or they may occur when individuals withdraw from the study prior to its conclusion. The former leads to non-monotonic patterns of missing data while missing values due to dropout alone result in monotonic patterns of missing data provided one has a balanced design with respect to planned visits. Table 6.1 illustrates what these two patterns of missing data might look like in a typical clinical trial.

Several different methods of estimation exist for handling missing data depending on the pattern of missing values and the assumption one makes regarding the underlying

Table 6.1 Example patterns of missing data from a longitudinal study involving four planned visits per subject. Observed values are marked with an x. The intermittent or non-monotonic patterns reflect both intermittent missing values and missing values due to dropout while the monotonic patterns reflect missing values due to dropout only.

	Patterns of Missing Data							
	Intermittent				Monotonic			
	Planned Visit				Planned Visit			
Subject	1	2	3	4	1	2	3	4
1	x	x	x	x	x	x	x	x
2	x		x		x	x	x	
3	x	x		x	x	x		
4	x	x			x			

mechanism for why missing values occur. Following Little and Rubin (2002) and others as well as recent guidelines on the prevention and treatment of missing data in clinical trials (National Research Council, 2010), missing data mechanisms may be classified as 1) missing completely at random (MCAR), 2) missing at random (MAR) and 3) missing not at random (MNAR). We describe each of these mechanisms below, but first we introduce some additional notation and assumptions that we will need.

Missing data notation and assumptions:

In order to characterize the different mechanisms that relate to why values are missing, we extend the basic notation and assumptions we have been using for the analysis of correlated response data to the longitudinal data setting involving incomplete data. We do so assuming the complete data are from a balanced design with p intended observations per subject/cluster. Note, however, that one can easily extend this notation along with the ensuing definitions and assumptions to those cases involving an unbalanced study design (e.g., Molenberghs and Verbeke, 2005, Chapter 26).

- Let $\boldsymbol{y}_i = (y_{i1}, \ldots, y_{ip})'$ be the $p \times 1$ "complete-data" vector of outcomes for the ith subject/cluster ($i = 1, \ldots, n$). Let $\boldsymbol{y}_i^o = (y_{i1}, \ldots, y_{ip_i})'$ be the observed data assuming the first observation is always observed (i.e., $1 \le p_i \le p$) and let $\boldsymbol{y}_i^m$ be the unobserved or missing data.
- Let $\boldsymbol{r}_i = (r_{i1}, \ldots, r_{ip})'$ be a $p \times 1$ vector of random variables that summarizes the response/non-response or missing data behavior for the ith subject such that $r_{ij} = \begin{cases} 1 & \text{if } y_{ij} \text{ is observed} \\ 0 & \text{if } y_{ij} \text{ is missing} \end{cases}$. Note that $\boldsymbol{r}_i$ is completely observed and that $r_{i1} = 1$ for all i. The complete data is given by $(\boldsymbol{y}_i, \boldsymbol{r}_i)$ while the observed data is given by $(\boldsymbol{y}_i^o, \boldsymbol{r}_i)$. In many cases, a random variable or vector other than $\boldsymbol{r}_i$ may be used to describe the missing data behavior. Examples might include
  - T_i = time to dropout or censoring (for monotone missing data)
  - M_i = the number of missed visits (for intermittent missing data)
- Let $\boldsymbol{X}_i = \begin{pmatrix} \boldsymbol{x}_{i1}' \\ \vdots \\ \boldsymbol{x}_{ip}' \end{pmatrix}$ be a $p \times u$ design matrix of fully observed explanatory variables, covariates or design variables associated with the ith subject that one intends to condition on or adjust for in the final analysis. These include known time-independent baseline covariates (e.g., baseline treatment group, participating center, baseline characteristics that serve as effect modifiers such as age, gender or race, etc.) and possibly known time-dependent covariates (e.g., treatment indicator variables

representing planned treatment assignments over time, planned visits or follow-up times, planned dosing levels over time, etc.).

- Let $\boldsymbol{b}_i$ be a vector of unobserved random-effects (if applicable).
- Let $\boldsymbol{Z}_i$ be a $p \times v$ design matrix of fully known covariates linked to a set of v random effects (if applicable). Typically $\boldsymbol{Z}_i$ is a subset of $\boldsymbol{X}_i$.
- Let the unconditional probability of response be described by $\pi(\boldsymbol{r}_i)$, the probability density function (pdf) of $\boldsymbol{r}_i$, and let $\pi(\boldsymbol{r}_i|\boldsymbol{X}_i, \boldsymbol{Z}_i, \boldsymbol{y}_i, \boldsymbol{b}_i)$ be the conditional pdf of $(\boldsymbol{r}_i|\boldsymbol{X}_i, \boldsymbol{Z}_i, \boldsymbol{y}_i, \boldsymbol{b}_i)$ based on the joint distribution of $(\boldsymbol{y}_i, \boldsymbol{r}_i, \boldsymbol{b}_i|\boldsymbol{X}_i, \boldsymbol{Z}_i)$ assuming random effects are applicable, i.e.,

$$\pi(\boldsymbol{y}_i, \boldsymbol{r}_i, \boldsymbol{b}_i|\boldsymbol{X}_i, \boldsymbol{Z}_i) = \pi(\boldsymbol{y}_i|\boldsymbol{X}_i, \boldsymbol{Z}_i, \boldsymbol{b}_i)\pi(\boldsymbol{b}_i|\boldsymbol{Z}_i)\pi(\boldsymbol{r}_i|\boldsymbol{X}_i, \boldsymbol{Z}_i, \boldsymbol{y}_i, \boldsymbol{b}_i).$$

  When there are no random effects, we simply remove $\boldsymbol{Z}_i$ and $\boldsymbol{b}_i$ from the notation and consider the conditional distribution of $\boldsymbol{r}_i|\boldsymbol{y}_i$ alone.
- Let $\boldsymbol{\tau} = (\boldsymbol{\beta}, \boldsymbol{\theta})$ index the set of parameters for $(\boldsymbol{y}_i|\boldsymbol{X}_i)$. Under a GNLME model with random effects, $\boldsymbol{b}_i$, we have $\boldsymbol{\tau} = (\boldsymbol{\beta}, \boldsymbol{\theta}_b, \boldsymbol{\theta}_w)$ where $(\boldsymbol{\beta}, \boldsymbol{\theta}_w)$ are the parameters of $\pi(\boldsymbol{y}_i|\boldsymbol{b}_i)$ and $\boldsymbol{\theta}_b$ are the parameters of $Var(\boldsymbol{b}_i) = \boldsymbol{\Psi}(\boldsymbol{\theta}_b)$. In the case of a GNLM, $\boldsymbol{\tau} = (\boldsymbol{\beta}, \boldsymbol{\theta})$.
- Let $\boldsymbol{\eta}$ index the set of parameters for $(\boldsymbol{r}_i|\boldsymbol{X}_i, \boldsymbol{Z}_i, \boldsymbol{y}_i, \boldsymbol{b}_i)$
- Assume $\boldsymbol{\tau}$ and $\boldsymbol{\eta}$ are separable or distinct in that one is not a function of the other and vice versa (Rubin, 1976; Little and Rubin, 2002).

One may also wish to include a matrix of auxiliary variables, $\boldsymbol{U}_i$ representing time-independent and/or time-dependent covariates that are distinct from $\boldsymbol{X}_i$. These variables would be tracked together with $\boldsymbol{y}_i$ and used to help draw inference from incomplete data. In some instances, $\boldsymbol{U}_i$ may consist of time-dependent covariates that may be subject to missing values (e.g., due to dropout) in which case we assume that $\boldsymbol{U}_i = (\boldsymbol{U}_i^o, \boldsymbol{U}_i^m)$ and $\boldsymbol{y}_i = (\boldsymbol{y}_i^o, \boldsymbol{y}_i^m)$ have the same missing data pattern though this assumption can be relaxed if required. Let $\boldsymbol{V}_i = (\boldsymbol{X}_i, \boldsymbol{Z}_i, \boldsymbol{U}_i)$ denote the complete-data matrix of covariates for the ith subject and let the intended complete-data vector of outcomes across subjects be represented by $\boldsymbol{y} = (\boldsymbol{y}_1', \ldots, \boldsymbol{y}_n')'$ with the observed and missing data components represented by $\boldsymbol{y}_{\text{obs}} = (\boldsymbol{y}_1^{o\prime}, \ldots, \boldsymbol{y}_n^{o\prime})'$ and $\boldsymbol{y}_{\text{miss}} = (\boldsymbol{y}_1^{m\prime}, \ldots, \boldsymbol{y}_n^{m\prime})'$, respectively. Similarly, let $\boldsymbol{r} = (\boldsymbol{r}_1', \ldots, \boldsymbol{r}_n')'$ be the complete-data vector of response/non-response variables and $\boldsymbol{V} = (\boldsymbol{V}_1, \ldots, \boldsymbol{V}_n)$ be the complete matrix of covariates. In the following sections, we describe the different missing data mechanisms based on the distribution of $(\boldsymbol{r}_i|\boldsymbol{V}_i, \boldsymbol{y}_i, \boldsymbol{b}_i)$.

6.2.1 Missing Completely at Random (MCAR)

Missing data are said to be *missing completely at random*, or MCAR, when missingness is independent of both observed and unobserved data. More formally, we define the missing data mechanism as MCAR if the probability of response/non-response is independent of the complete data, i.e.

$$\pi(\boldsymbol{r}_i|\boldsymbol{V}_i, \boldsymbol{y}_i, \boldsymbol{b}_i) = \pi(\boldsymbol{r}_i). \tag{6.1}$$

If, instead, we have

$$\pi(\boldsymbol{r}_i|\boldsymbol{V}_i, \boldsymbol{y}_i, \boldsymbol{b}_i) = \pi(\boldsymbol{r}_i|\boldsymbol{X}_i, \boldsymbol{Z}_i) \tag{6.2}$$

then the missing data mechanism is conditional MCAR and missing values are said to be *covariate-dependent missing values* which one can treat as MCAR provided one has conditioned on the appropriate set of covariates $(\boldsymbol{X}_i, \boldsymbol{Z}_i)$ (Little, 1995). When missingness is the result of dropout, Little refers to the conditional MCAR mechanism as *covariate-dependent dropout* (Little, 1995).

Examples of when values might be missing completely at random include 1) administrative censoring such as would occur when a study is terminated on a scheduled date resulting in incomplete data for subjects who entered the study late due to staggered entry, and 2) dropout for external reasons completely unrelated to study participation. In the latter case, it is essential that the cause of dropout be fully documented. The assumption that missing values are MCAR is extremely strong and seldom likely to hold for most of the missing values that occur in a clinical trial. One can effectively test against a null hypothesis that missing values are MCAR against an alternative that they are at least missing at random (MAR) using methods described by Little (1988), Diggle (1989), Ridout (1991), Park and Davis (1993), and Park and Lee (1997). Such tests are important if one wishes to at least partially justify ignoring missing values in one's analysis regardless of the method used to analyze the data. This is because only when the MCAR assumption holds can one be assured of drawing valid inference in the presence of incomplete data using any of the semiparametric or parametric methods described in Chapters 2-5. Other methods valid under MCAR assumptions include complete case analysis, available case analysis, and some relatively simple single imputation methods though these methods tend to be less efficient and in some cases require even stronger assumptions (e.g., Little and Rubin, 2002; Molenberghs and Verbeke, 2005).

6.2.2 Missing at Random (MAR)

Missing data are said to be *missing at random*, or MAR, when missingness is independent of missing responses, $\boldsymbol{y}_i^m$, conditionally on observed responses, $\boldsymbol{y}_i^o$. More formally, we define the missing data mechanism as MAR if the probability of response/non-response depends only on observed values, i.e.,

$$\pi(\boldsymbol{r}_i|\boldsymbol{V}_i,\boldsymbol{y}_i,\boldsymbol{b}_i) = \pi(\boldsymbol{r}_i|\boldsymbol{V}_i,\boldsymbol{y}_i^o,\boldsymbol{y}_i^m,\boldsymbol{b}_i) = \pi(\boldsymbol{r}_i|\boldsymbol{V}_i^o,\boldsymbol{y}_i^o) \tag{6.3}$$

where $\boldsymbol{V}_i^o = (\boldsymbol{X}_i, \boldsymbol{Z}_i, \boldsymbol{U}_i^o)$.

When missingness is the result of dropout, the MAR mechanism is often referred to as *missing at random dropout* or simply *random dropout* (Little, 1995). In this context, it is useful to distinguish between MAR and *sequential* MAR, or S-MAR, as is done in Robins et. al. (1995) and Hogan et. al. (2004). Here we briefly describe the S-MAR mechanism as defined by Hogan et. al. (2004) assuming a balanced longitudinal study design in which planned observations $\boldsymbol{y}_i = (y_{i1}, y_{i2}, \ldots, y_{ip})'$ are to be taken at fixed time points $t_1 < t_2 < \ldots < t_p$. Assuming a monotone pattern of missingness such that $r_{i1} = 1$ and $r_{ik} = 1$ implies $r_{i(k-1)} = 1$ for all $k = 2, \ldots, p$, Hogan et. al. (2004) define the MAR mechanism as *sequential* MAR, or S-MAR, if the hazard of dropout at time t_k $(1 < k \leq p)$ does not depend on current or future responses, $y_{ik}, y_{i(k+1)}, \ldots, y_{ip}$, i.e.,

$$\Pr(r_{ik} = 0|r_{i(k-1)} = 1, \boldsymbol{V}_{ik}, \boldsymbol{y}_i, \boldsymbol{b}_i) = \Pr(r_{ik} = 0|r_{i(k-1)} = 1, \boldsymbol{V}_{ik}, \boldsymbol{y}_{i(k-1)}) \tag{6.4}$$

where $\boldsymbol{V}_{ik}$ denotes the sub-matrix of $\boldsymbol{V}_i$ consisting of row vectors, $\boldsymbol{v}_{ij}' = (\boldsymbol{x}_{ij}', \boldsymbol{z}_{ij}', \boldsymbol{u}_{ij}')$, taken from the design matrices $\boldsymbol{X}_i$ and $\boldsymbol{Z}_i$, and the auxiliary matrix $\boldsymbol{U}_i$ up through and including time t_k (i.e., from $j = 1, \ldots, k$). The vector $\boldsymbol{y}_{i(k-1)} = (y_{i1}, y_{i2}, \ldots, y_{i(k-1)})'$ is the vector of observed outcomes up to and including time t_{k-1}. As defined by Hogan et. al. (2004), S-MAR differs from MAR in that dropout at time t_k is only allowed to depend on the elements of $\boldsymbol{y}_i$ up through time t_{k-1}, and on $\boldsymbol{V}_i = (\boldsymbol{X}_i, \boldsymbol{Z}_i, \boldsymbol{U}_i)$ up through time t_k. This is analogous to the assumption of *non-informative censoring* as defined and used in survival analysis in that, given $r_{i(k-1)} = 1$, the future but potentially incomplete responses $(y_{ik}, y_{i(k+1)}, \ldots, y_{ip})$ are independent of r_{ik} given the past (Hogan et. al., 2004). We defer any further discussion of S-MAR to §6.3 wherein we examine, at length, different dropout mechanisms that give rise to monotone patterns of missing data.

If MAR as defined in (6.3) holds, then valid likelihood inference for the parameter $\boldsymbol{\tau}$ may be based on $\pi(\boldsymbol{y}_i^o|\boldsymbol{X}_i, \boldsymbol{Z}_i; \boldsymbol{\tau})$ since, upon integrating over $\boldsymbol{y}_i^m$ and $\boldsymbol{b}_i$, the contribution of the ith subject to the joint likelihood of the observed data may be factored as

$$\begin{aligned}\pi(\boldsymbol{y}_i^o, \boldsymbol{r}_i|\boldsymbol{V}_i; \boldsymbol{\tau}, \boldsymbol{\eta}) &= \int\!\!\int \pi(\boldsymbol{y}_i^o,\, \boldsymbol{y}_i^m,\, \boldsymbol{b}_i|\boldsymbol{X}_i,\, \boldsymbol{Z}_i)\; \pi(\boldsymbol{r}_i|\boldsymbol{V}_i, \boldsymbol{y}_i^o,\, \boldsymbol{y}_i^m,\, \boldsymbol{b}_i) d\boldsymbol{y}_i^m d\boldsymbol{b}_i \\ &= \pi(\boldsymbol{r}_i|\boldsymbol{V}_i^o, \boldsymbol{y}_i^o) \int\!\!\int \pi(\boldsymbol{y}_i^o,\, \boldsymbol{y}_i^m,\, \boldsymbol{b}_i|\boldsymbol{X}_i,\, \boldsymbol{Z}_i)\; d\boldsymbol{y}_i^m d\boldsymbol{b}_i \\ &= \pi(\boldsymbol{r}_i|\boldsymbol{V}_i^o, \boldsymbol{y}_i^o; \boldsymbol{\eta})\pi(\boldsymbol{y}_i^o|\boldsymbol{X}_i, \boldsymbol{Z}_i; \boldsymbol{\tau}).\end{aligned} \tag{6.5}$$

assuming $\boldsymbol{\tau}$ and $\boldsymbol{\eta}$ are separable (sometimes referred to as the *separability* condition). It is only when both MAR and separability between $\boldsymbol{\tau}$ and $\boldsymbol{\eta}$ hold that one may classify the missing data mechanism as *ignorable missing data.* This is because valid likelihood-based inference for $\boldsymbol{\tau}$ based on the joint log-likelihood function $L(\boldsymbol{\tau}, \boldsymbol{\eta}; \boldsymbol{y}_{\text{obs}}, \boldsymbol{r}) = \sum_{i=1}^n \log\{\pi(\boldsymbol{y}_i^o, \boldsymbol{r}_i|\boldsymbol{V}_i; \boldsymbol{\tau}, \boldsymbol{\eta})\}$ will be equivalent to likelihood-based inference for $\boldsymbol{\tau}$ based solely on the log-likelihood function $L(\boldsymbol{\tau}; \boldsymbol{y}_{\text{obs}}) = \sum_{i=1}^n \log\{\pi(\boldsymbol{y}_i^o|\boldsymbol{X}_i, \boldsymbol{Z}_i; \boldsymbol{\tau})\}$ obtained by *ignoring* the response/non-response model $\pi(\boldsymbol{r}_i|\boldsymbol{V}_i^o, \boldsymbol{y}_i^o; \boldsymbol{\eta})$ under MAR (Rubin, 1976; Little and Rubin, 2002; Hogan et. al., 2004). In this case, an analysis of incomplete data may be based on any of the likelihood methods described in Chapters 2-5 provided one has correctly specified the complete data likelihood and inference is based on the observed rather than the expected information matrix (Little and Rubin, 2002; Kenward and Molenberghs, 1998). Other methods leading to valid inference are also possible under MAR and some of these will be discussed briefly in §6.4.

In most cases, it is impossible to discern whether missing data are MAR or whether missingness truly depends on unobserved values. However, there are settings in which missing values are planned for by design. For example, in a longitudinal clinical trial with repeated outcome measurements, the protocol might specify that when certain criteria based on the outcome variable are met at pre-determined times (e.g., some predefined threshold value is reached on any planned visit), the subject is to be withdrawn from the study prior to completion and placed in an open follow-up phase. In this case, missingness may be classified as MAR since subject withdrawal resulting in incomplete data is a direct consequence of a recorded intermediate outcome value and is therefore a function of an observed value. An example of one such trial involving two anti-hypertensive agents in patients with mild to moderate essential hypertension was described by Murray and Findlay (1988).

6.2.3 Missing Not at Random (MNAR)

Missing data are said to be *missing not at random,* or MNAR, if the MAR assumption is violated and missingness depends on unobserved values. More formally, we define the missing data mechanism as MNAR if the probability of response/non-response depends either on unobserved outcome data, i.e.,

$$\begin{aligned}\pi(\boldsymbol{r}_i|\boldsymbol{V}_i, \boldsymbol{y}_i, \boldsymbol{b}_i) &= \pi(\boldsymbol{r}_i|\boldsymbol{V}_i, \boldsymbol{y}_i) \\ &= \pi(\boldsymbol{r}_i|\boldsymbol{V}_i, \boldsymbol{y}_i^o, \boldsymbol{y}_i^m)\end{aligned} \tag{6.6}$$

or on unobserved random effects, i.e.,

$$\pi(\boldsymbol{r}_i|\boldsymbol{V}_i, \boldsymbol{y}_i, \boldsymbol{b}_i) = \pi(\boldsymbol{r}_i|\boldsymbol{V}_i, \boldsymbol{b}_i). \tag{6.7}$$

Data that are MNAR are frequently referred to as *non-ignorable missing* (NIM) data or as *non-ignorable dropout* or *informative dropout* when missingness is due to dropout. This is because under a MNAR mechanism, likelihood-based inference that ignores missing values will be subject to bias. Following the nomenclature of Little (1995), we find it convenient to refer to MNAR data under (6.6) as *non-ignorable outcome-dependent missing data* or as *non-ignorable outcome-dependent dropout* so as to emphasize that missingness depends on the components of the complete data $\boldsymbol{y}_i$ including at least one or more of the missing data components of $\boldsymbol{y}_i^m$ (i.e., missingness depends on $\boldsymbol{y}_i^m$ but is independent of the unobserved random effects). Similarly, under (6.7) we refer to MNAR data as *non-ignorable random-effects based missing data* or as *non-ignorable random-effects dropout* so as emphasize the dependence of missingness is on the unobserved random effects only.

When missing values are MNAR, valid inference is only possible by correctly modeling the joint distribution of the complete data $(\boldsymbol{y}_i, \boldsymbol{r}_i)$ under unverifiable assumptions about the missing data. To illustrate, suppose we have a marginal model for the response variable and there are no auxiliary covariates (i.e., $\boldsymbol{V}_i = \boldsymbol{X}_i$). Then under the non-ignorable outcome-dependent missing data mechanism (6.6), the complete-data model for $(\boldsymbol{y}_i, \boldsymbol{r}_i)$ may be factored as

$$\begin{aligned}\pi(\boldsymbol{y}_i, \boldsymbol{r}_i|\boldsymbol{X}_i) &= \pi(\boldsymbol{y}_i^o, \boldsymbol{y}_i^m, \boldsymbol{r}_i|\boldsymbol{X}_i)\\ &= \pi(\boldsymbol{y}_i^o, \boldsymbol{r}_i|\boldsymbol{X}_i)\pi(\boldsymbol{y}_i^m|\boldsymbol{X}_i, \boldsymbol{y}_i^o, \boldsymbol{r}_i).\end{aligned}$$

Based on this factorization, we see which components of the complete-data model can be inferred from observed data and which cannot. For instance, we can maximize a joint log-likelihood function of the observed data

$$L(\boldsymbol{\tau}, \boldsymbol{\eta}; \boldsymbol{y}_{\text{obs}}, \boldsymbol{r}) = \sum_{i=1}^{n} \log\{\pi(\boldsymbol{y}_i^o, \boldsymbol{r}_i|\boldsymbol{X}_i; \boldsymbol{\tau}, \boldsymbol{\eta})\}$$

under parametric modeling assumptions about $\pi(\boldsymbol{y}_i^o, \boldsymbol{r}_i|\boldsymbol{X}_i)$ that can be evaluated using goodness-of-fit methods. However, we have no way to assess the validity of the complete-data model since we have no way to verify assumptions we might make regarding the conditional predictive distribution, $\pi(\boldsymbol{y}_i^m|\boldsymbol{X}_i, \boldsymbol{y}_i^o, \boldsymbol{r}_i)$, that describes how missing data can be extrapolated from the observed data.

Here, we briefly describe three modeling strategies that accomplish this goal albeit under unverifiable assumptions. The three strategies are broadly classified under what are known as selection models, pattern-mixture models and shared parameter models. We do so assuming either that $\boldsymbol{V}_i = (\boldsymbol{X}_i, \boldsymbol{Z}_i)$ or that any auxiliary covariates that make up $\boldsymbol{U}_i$ are completely observed and independent of $\boldsymbol{y}_i$ although one could certainly relax either assumption if need be (for example, by including $\boldsymbol{U}_i$ in the definition of the conditional pdf of $\boldsymbol{y}_i$).

The missing data dependency under (6.6) will coincide with a class of models known as marginal *selection models* whenever the joint distribution of $(\boldsymbol{y}_i, \boldsymbol{r}_i)$ is factored as

$$\begin{aligned}\pi(\boldsymbol{y}_i, \boldsymbol{r}_i|\boldsymbol{V}_i) &= \pi(\boldsymbol{r}_i|\boldsymbol{V}_i, \boldsymbol{y}_i)\int \pi(\boldsymbol{y}_i|\boldsymbol{X}_i, \boldsymbol{Z}_i, \boldsymbol{b}_i)\pi(\boldsymbol{b}_i|\boldsymbol{Z}_i)d\boldsymbol{b}_i \qquad (6.8)\\ &= \pi(\boldsymbol{r}_i|\boldsymbol{V}_i, \boldsymbol{y}_i)\pi(\boldsymbol{y}_i|\boldsymbol{X}_i, \boldsymbol{Z}_i).\end{aligned}$$

Since $\boldsymbol{r}_i$ is independent of $\boldsymbol{b}_i$ under (6.6), it may be more convenient to work with the alternative factorization

$$\begin{aligned}\pi(\boldsymbol{y}_i, \boldsymbol{r}_i|\boldsymbol{V}_i) &= \pi(\boldsymbol{y}_i|\boldsymbol{X}_i, \boldsymbol{Z}_i, \boldsymbol{r}_i)\pi(\boldsymbol{r}_i|\boldsymbol{V}_i)\\ &= \pi(\boldsymbol{y}_i^m|\boldsymbol{X}_i, \boldsymbol{Z}_i, \boldsymbol{y}_i^o, \boldsymbol{r}_i)\pi(\boldsymbol{y}_i^o|\boldsymbol{X}_i, \boldsymbol{Z}_i, \boldsymbol{r}_i)\pi(\boldsymbol{r}_i|\boldsymbol{V}_i)\end{aligned}$$

which corresponds to a class of marginal models known as *pattern-mixture models*. In either case, we run into the problem that inference based on the observed data $(\boldsymbol{y}_i^o, \boldsymbol{r}_i)$ cannot be extrapolated to the complete data $(\boldsymbol{y}_i, \boldsymbol{r}_i)$ without making certain unverifiable assumptions. For example, under the marginal selection model (6.8), we can postulate a "selection model," $\pi(\boldsymbol{r}_i|\boldsymbol{V}_i, \boldsymbol{y}_i) = \pi(\boldsymbol{r}_i|\boldsymbol{V}_i, \boldsymbol{y}_i^o, \boldsymbol{y}_i^m)$, that describes the probability of missingness as a function of covariates and the response variable and we can postulate a "response model," $\pi(\boldsymbol{y}_i|\boldsymbol{X}_i, \boldsymbol{Z}_i) = \pi(\boldsymbol{y}_i^o, \boldsymbol{y}_i^m|\boldsymbol{X}_i, \boldsymbol{Z}_i)$, that describes the complete-data response model of interest. However, we cannot verify the assumptions under either model because we never fully observe $\boldsymbol{y}_i$ for all subjects. Similarly, under the pattern-mixture model, we can postulate and assess the goodness-of-fit of the observed data by specifying parametric models for $\pi(\boldsymbol{y}_i^o|\boldsymbol{X}_i, \boldsymbol{Z}_i, \boldsymbol{r}_i)$ and $\pi(\boldsymbol{r}_i|\boldsymbol{V}_i)$. However, as noted above, we have no way to verify assumptions we might make regarding the conditional predictive distribution, $\pi(\boldsymbol{y}_i^m|\boldsymbol{X}_i, \boldsymbol{Z}_i, \boldsymbol{y}_i^o, \boldsymbol{r}_i)$, and so inference based on the complete data can only be made on the basis of unverifiable assumptions.

Alternatively, the unobserved random-effects based dependency under (6.7) coincides with a class of models known as *shared parameter models* in which $\boldsymbol{y}_i$ and $\boldsymbol{r}_i$ are assumed to be conditionally independent given the random effects, i.e.,

$$\begin{aligned}\pi(\boldsymbol{y}_i, \boldsymbol{r}_i|\boldsymbol{V}_i, \boldsymbol{b}_i) &= \pi(\boldsymbol{y}_i|\boldsymbol{X}_i, \boldsymbol{Z}_i, \boldsymbol{b}_i)\pi(\boldsymbol{r}_i|\boldsymbol{V}_i, \boldsymbol{y}_i, \boldsymbol{b}_i) \\ &= \pi(\boldsymbol{y}_i|\boldsymbol{X}_i, \boldsymbol{Z}_i, \boldsymbol{b}_i)\pi(\boldsymbol{r}_i|\boldsymbol{V}_i, \boldsymbol{b}_i).\end{aligned} \tag{6.9}$$

Note that under (6.7) or, equivalently, the shared parameter model (6.9), missing data are non-ignorable because missingness will depend on $\boldsymbol{y}_i^m$ after integrating the distribution $\pi(\boldsymbol{r}_i|\boldsymbol{V}_i, \boldsymbol{y}_i, \boldsymbol{b}_i)$ over the unknown random effects $\boldsymbol{b}_i$. This is perhaps best demonstrated by showing that the posterior distribution of the random effects will depend on $\boldsymbol{y}_i^m$. Specifically, ignoring the notational dependence on $\boldsymbol{V}_i$ and integrating over the random effects, we have

$$\begin{aligned}\pi(\boldsymbol{r}_i|\boldsymbol{y}_i) = \pi(\boldsymbol{r}_i|\boldsymbol{y}_i^o, \boldsymbol{y}_i^m) &= \frac{\int_{\boldsymbol{b}} \pi(\boldsymbol{r}_i, \boldsymbol{y}_i^o, \boldsymbol{y}_i^m|\boldsymbol{b}_i)\pi(\boldsymbol{b}_i)d\boldsymbol{b}_i}{\int_{\boldsymbol{b}} \pi(\boldsymbol{y}_i^o, \boldsymbol{y}_i^m|\boldsymbol{b}_i)\pi(\boldsymbol{b}_i)d\boldsymbol{b}_i} \\ &= \frac{\int_{\boldsymbol{b}} \pi(\boldsymbol{r}_i|\boldsymbol{b}_i)\pi(\boldsymbol{y}_i^o, \boldsymbol{y}_i^m|\boldsymbol{b}_i)\pi(\boldsymbol{b}_i)d\boldsymbol{b}_i}{\int_{\boldsymbol{b}} \pi(\boldsymbol{y}_i^o, \boldsymbol{y}_i^m|\boldsymbol{b}_i)\pi(\boldsymbol{b}_i)d\boldsymbol{b}_i} \\ &= \int_{\boldsymbol{b}} \pi(\boldsymbol{r}_i|\boldsymbol{b}_i)\pi(\boldsymbol{b}_i|\boldsymbol{y}_i^o, \boldsymbol{y}_i^m)d\boldsymbol{b}_i\end{aligned} \tag{6.10}$$

so that, in general, the conditional distribution of $\boldsymbol{r}_i|\boldsymbol{y}_i$ depends on $\boldsymbol{y}_i^m$ through the posterior distribution of $\boldsymbol{b}_i$. Thus inference under the shared parameter model (6.9) does not require missing data to be MCAR or MAR. However, we are still confronted with the problem of dealing with unverifiable assumptions. For example, the assumption is usually made that the components of the complete data vector $\boldsymbol{y}_i = (\boldsymbol{y}_i^o, \boldsymbol{y}_i^m)$ are conditionally independent given the random effects $\boldsymbol{b}_i$. This assumption allows us to integrate over the missing data components and write the joint likelihood for the observed data for the ith subject as

$$\begin{aligned}\pi(\boldsymbol{y}_i^o, \boldsymbol{r}_i|\boldsymbol{V}_i) &= \int_{\boldsymbol{y}^m}\int_{\boldsymbol{b}} \pi(\boldsymbol{y}_i^o, \boldsymbol{y}_i^m, \boldsymbol{r}_i|\boldsymbol{b}_i)\pi(\boldsymbol{b}_i)d\boldsymbol{b}_i d\boldsymbol{y}_i^m \\ &= \int_{\boldsymbol{y}^m}\left\{\int_{\boldsymbol{b}} \pi(\boldsymbol{y}_i^o|\boldsymbol{b}_i)\pi(\boldsymbol{r}_i|\boldsymbol{b}_i)\pi(\boldsymbol{b}_i)d\boldsymbol{b}_i\right\}\pi(\boldsymbol{y}_i^m|\boldsymbol{b}_i)d\boldsymbol{y}_i^m \\ &= \int_{\boldsymbol{b}} \pi(\boldsymbol{y}_i^o|\boldsymbol{b}_i)\pi(\boldsymbol{r}_i|\boldsymbol{b}_i)\pi(\boldsymbol{b}_i)d\boldsymbol{b}_i\left\{\int_{\boldsymbol{y}^m} \frac{\pi(\boldsymbol{y}_i^m, \boldsymbol{b}_i)}{\pi(\boldsymbol{b}_i)}d\boldsymbol{y}_i^m\right\} \\ &= \int_{\boldsymbol{b}} \pi(\boldsymbol{y}_i^o|\boldsymbol{X}_i, \boldsymbol{Z}_i, \boldsymbol{b}_i)\pi(\boldsymbol{r}_i|\boldsymbol{V}_i, \boldsymbol{b}_i)\pi(\boldsymbol{b}_i)d\boldsymbol{b}_i.\end{aligned}$$

While this assumption can only be verified for the observed data $(\boldsymbol{y}_i^o, \boldsymbol{r}_i)$, it does seem improbable that it would not also hold true for the missing data if verified for the observed data.

6.3 Dropout mechanisms

In most clinical trials, subject withdrawal or dropout is the primary source of missing values and much of the literature on missing data is devoted to methods for handling missingness due to dropout (see, for example, the tutorial paper by Hogan et. al., 2004 and references therein). Minimizing dropout is important because the interpretation of study results will be manifestly more difficult when subjects fail to complete a study. For example, randomized controlled trials can be subject to post-randomization selection bias resulting from dropout and this can seriously undermine the study results and conclusions. Therefore every effort should be made to avoid dropout all together. This can be done in the planning stages of a clinical trial using various incentives and strategies to retain subjects (National Research Council, 2010). When dropout does occur, every effort should be made to continue measuring the outcome of interest in accordance with intent-to-treat principles (National Research Council, 2010). Despite these efforts, missing values from dropout do occur and methods are needed for handling such missing data. In this section, we consider the problem of ignorable versus non-ignorable dropout mechanisms as they relate to longitudinal studies that involve fitting a structured regression model to repeated outcome measurements. By ignorable dropout, we are referring to missing at random dropout wherein valid likelihood-based inference is possible under a MAR mechanism assuming $\boldsymbol{\tau}$ and $\boldsymbol{\eta}$ are separable (§6.2.2). Non-ignorable dropout refers to any mechanism that is not MAR. Hence, in addition to the more general MAR and MNAR mechanisms defined in §6.2, we introduce a *sequential at random dropout*, or S-RD, mechanism and a *sequential not at random dropout*, or S-NARD, mechanism both of which are defined in terms of a hazard function associated with the time-to-dropout. Of course dropout that is completely at random is also ignorable and in fact this is the only dropout mechanism under which standard non-likelihood methods like GEE remain valid.

In the case of non-ignorable dropout, it is difficult to think of a model where dropout depends on unobserved values of the outcome variable that occur after dropout. For example, dropout due to death cannot be related to or depend on "potential" future observations whether planned or not as these values are no longer possible. Depending on the underlying cause of dropout and the inferential objectives of a study, a better conceptual framework for non-ignorable dropout might be one that is cast within the context of a time series as was done for the S-MAR mechanism (e.g., Henderson et. al., 2000; Lin and Ying, 2001; Tsiatis and Davidian, 2004). To that end, it is useful to first distinguish between dropout events and non-informative censoring events. The former consists of events that may be related to the outcome measurements and/or design variables while the latter are those events that are independent of both the outcome variable and design variables. Dropout events may be further classified according to whether an event is a *terminal dropout event* or a *non-terminal dropout event.* We consider each separately.

We define a *terminal dropout event* to be any event beyond which the outcome variable is either no longer potentially observable (for example, due to death), or the subject is withdrawn from the study and the outcome is no longer measured because the "terminal" event is known to effectively alter the measured response beyond the framework of the scientific inquiry. Hence, either by design or necessity, one would stop follow-up on individuals who experience a terminal event. Even when measurements remain possible, it makes no sense to continue measuring the response following a terminal event since the event is known *a priori* to alter the hypothetical mean response profile in a manner that is outside the scope of the investigation. An example illustrating this is the MDRD study presented in §6.6.2. Often times terminal events may be of interest in and of themselves as secondary endpoints. The study might then focus on comparing whether treatments jointly effect both the terminal event time and any serial trends in the primary response variable leading up to the terminal event. This occurs, for example, in clinical trials that investigate whether a serial marker can effectively serve as a surrogate for a well-defined terminal event (e.g., Tsiatis and Davidian, 2004).

Conversely, a *non-terminal dropout event* is any event that is not terminal in nature but which does result in the failure to measure the outcome of interest following the event. Reasons for non-terminal dropout include recovery, lack of efficacy, side-effects, and subject "burnout" from the demands of the study (e.g., too many visits, medical procedures, self-administered questionnaires, etc.). These can all be classified as non-terminal events in that measuring the outcome is still possible following dropout and is in fact still within the scope of the study. It should be noted that in some cases, a subject may experience an adverse event or some other event that results in withdrawal from the assigned treatment. However, within the context of missing data, this would only be classified as a non-terminal dropout event if, in addition to withdrawal from treatment, the subject withdraws from the study resulting in a failure to measure the outcome variable of interest.

Collectively, terminal and non-terminal dropouts are referred to as *analysis dropouts* by the National Research Council (2010) since both lead to a failure to measure the outcome of interest. However, following Dufouil et. al. (2004), Kurkland and Heagerty (2005), and Diggle et. al. (2007), we prefer to distinguish between terminal and non-terminal dropouts as a means for clarifying the scientific objectives of a study. For example, with a terminal dropout event as defined here, future observations are either impossible or irrelevant to the study objectives. In this case it seems reasonable that one would target inference over the time frame that a subject remains event-free (e.g., Kurkland and Heagerty, 2005). Moreover, the usual mechanistic framework involving "missing data due to dropout" might be better replaced with a mechanistic framework involving "missing or unobserved data leading to dropout." With non-terminal dropout events, one may well be interested in the hypothetical response individuals would have produced had they not dropped out. In this case, we can think of a mechanistic framework that involves both missing or unobserved data leading to dropout and missing data resulting from dropout. Finally, when both terminal and non-terminal dropout events occur in a study, one may choose to target inference on the hypothetical response one would expect in the absence of a non-terminal dropout event but conditional on a subject not having experienced a terminal event (Diggle et. al., 2007). Various methods have been proposed for these different objectives and we refer the reader to the papers by Dufouil et. al. (2004), Kurkland and Heagerty (2005), and Diggle et. al. (2007) for a more complete discussion regarding underlying objectives and assumptions when analyzing longitudinal data with terminal and non-terminal dropouts events. In this section, we formulate ignorable versus non-ignorable dropout mechanisms in terms of a hazard function associated with time-to-dropout conditional on observed and unobserved measurements from an underlying stochastic process. We do so more from a mechanistic framework that involves "missing or unobserved data leading to dropout" rather than "missing data due to dropout." To facilitate this approach, we adopt a framework involving continuous time processes such as described by Tsiatis and Davidian (2004) for the joint modeling of longitudinal data and event time data.

6.3.1 Ignorable versus non-ignorable dropout

Since there are no guarantees against dropout, analytic methods are needed if one is to address concerns over the possibility of dropout from terminal and/or non-terminal events. In this section, longitudinal data in the form of repeated outcome measurements together with time-independent and time-dependent covariates are represented in terms of a structured stochastic process involving a regression model for the outcomes given a set of known explanatory variables and, if applicable, a vector of random effects. Dropout is cast within the framework of time-to-event outcomes as determined by terminal and non-terminal dropout events and non-informative censoring events.

Event time framework for dropout:

Assume each of $i = 1, 2, \ldots, n$ individuals are followed over an interval $[0, \mathcal{T})$ where $\mathcal{T}$ is a study truncation time. To better characterize the different dropout mechanisms, we introduce the following notation and assumptions for time to dropout events.

- Let T_{1i} be a latent event time random variable that defines the time to a terminal dropout event. Examples might include time to some cause-specific death or time to onset of some disease as each might relate to some corresponding outcome variable of interest. In the example given in §6.6.2, T_{1i} represents the time to death, transplantation or kidney dialysis, whichever occurs first, among chronic kidney disease patients for whom serial measurements of the glomerular filtration rate of the kidney are being modeled over time.
- Let T_{2i} be a latent event time random variable that defines time to a non-terminal dropout event resulting in missing values. Examples might include time to hospitalization for some adverse event requiring that a subject be removed from a study or it may be time to study withdrawal for reasons that are not that well documented (e.g., a subject's unwillingness to continue in a study).
- Let C_i denote a censoring time associated with non-informative or independent censoring events. An example might be censoring for death assuming death is completely unrelated to the outcome or treatment under investigation (e.g., death due to an auto accident). In this case, we have a non-informative but terminal censoring event. A more common example would be administrative censoring due to staggered entries into a study or censoring as a result of successfully completing the study at time $\mathcal{T}$ without experiencing a terminal or non-terminal event.
- Let $T_i = \min\{T_{1i}, T_{2i}\}$ be the time to the first dropout event and let $T_i^o = \min\{T_i, C_i\}$ and $\delta_i = I(T_i \leq C_i)$ be the observation time and dropout event indicator variable, respectively, for the ith subject. It is assumed that once a dropout event or censoring event occurs, all follow-up on the individual ceases.

Regression modeling framework for longitudinal responses:

To accommodate either a continuous or discrete outcome measure for the longitudinal response of interest and allow for either a marginal or mixed-effects regression model, we formulate a regression model framework for the repeated outcome measurements using a structured stochastic process as follows.

- Let $y_{i1}, y_{i2}, \ldots$ be a sequence of realized outcomes taken at times $t_{i1}, t_{i2}, \ldots$ from a stochastic process $\{Y_i(t); t \in [0, \mathcal{T})\}$ satisfying

  $$Y_i(t) = \mu_i(t) + \epsilon_i(t)$$

  where $\mu_i(t)$ is the mean response and $\epsilon_i(t)$ represents possible measurement error from a random measurement error process (e.g., from a stationary Gaussian process with mean 0 and constant variance, σ^2). It is assumed the mean response $\mu_i(t)$ can be described by a possibly generalized nonlinear mixed-effects model

  $$\mu_i(t) = \mu_i(\boldsymbol{x}_i(t), \boldsymbol{\beta}, \boldsymbol{z}_i(t), \boldsymbol{b}_i)$$

  where $\boldsymbol{x}_i(t)$ is a covariate vector of known possibly time-varying explanatory variables linked to a vector of fixed-effects parameters $\boldsymbol{\beta}$ and, when applicable, $\boldsymbol{z}_i(t)$ is a vector of known possibly time-varying covariates linked to a random effects parameter vector $\boldsymbol{b}_i$.
- The covariate process $\{\boldsymbol{x}_i(t); t \in [0, \mathcal{T})\}$ consists of fully observable explanatory variables, covariates, or design variables that one intends to condition on or adjust for in

the final analysis. These include known baseline covariates that do not change with time (e.g., baseline treatment group, participating center, baseline characteristics that serve as effect modifiers such as age, gender, etc.) as well as known or defined time-dependent covariates (e.g., treatment indicator variables representing planned treatment assignments at scheduled times, planned visits or scheduled times of follow-up, known dosing levels at scheduled times, a subject's age at scheduled times, etc.). When applicable, the covariate process $\{\boldsymbol{z}_i(t); t \in [0, \mathcal{T})\}$ also consists of fully known covariates linked to a vector $\boldsymbol{b}_i$ of random effects. Within the context of survival analysis, the components of both $\boldsymbol{x}_i(t)$ and $\boldsymbol{z}_i(t)$ may be classified as external time-independent or external time-dependent covariates (e.g., Kalbfleisch and Prentice, 2002, §6.3; see also pp. 349).

- In addition, let $\boldsymbol{u}_{i1}, \boldsymbol{u}_{i2}, \ldots$ be a sequence of vector-valued covariates taken at times $t_{i1}, t_{i2}, \ldots$from an auxiliary covariate process $\{\boldsymbol{u}_i(t); t \in [0, \mathcal{T})\}$ consisting of auxiliary time-independent and time-dependent variables not included in the regression model for the primary outcome variable of interest but which are measured and recorded along with the observed outcomes as a means for gathering additional information on the possible dropout mechanism. With respect to time to dropout, the covariates that comprise $\boldsymbol{u}_i(t)$ can include both external and internal time-dependent covariates (e.g., Kalbfleisch and Prentice, 2002, §6.3; see also pp. 349).

Suppose, as before, we have a balanced longitudinal study design such that a set of p measurements $\{y_i(t_j), \boldsymbol{x}_i(t_j), \boldsymbol{z}_i(t_j), \boldsymbol{u}_i(t_j); j = 1, \ldots, p\}$ are *intended* to be taken at fixed scheduled times $t_1, t_2, \ldots, t_p$ within the interval $[0, \mathcal{T})$. The study truncation time $\mathcal{T}$ will most often extend for a period of time beyond t_p so as to capture any additional adverse events, etc., associated with the treatment following the last response measurement. This is particularly true in clinical trials of a regulatory nature. We assume the scheduled times are selected independently of $Y_i(t)$ and T_i and that the initial or baseline visit occurs at $t_1 = 0$ with y_{i1} observed for all $i = 1, \ldots, n$ subjects. Let $\boldsymbol{t}_p = (t_1, t_2, \ldots, t_p)'$ denote the complete vector of fixed times for each subject and let $\boldsymbol{t}_k = (t_1, t_2, \ldots t_k)$ denote the vector of time points corresponding to the first k measurements $(1 \le k \le p)$. For ease of notation, let

$$\boldsymbol{y}_{ik} = \boldsymbol{y}_i(\boldsymbol{t}_k) = (y_{i1}, y_{i2}, \ldots, y_{ik})'$$

denote the vector of realized outcome measurements, $y_{ij} = y_i(t_j)$, at times $t = t_j$ $(j = 1, \ldots, k)$ and likewise let $\boldsymbol{V}_{ik} = (\boldsymbol{X}_{ik}, \boldsymbol{Z}_{ik}, \boldsymbol{U}_{ik})$ where

$$\boldsymbol{X}_{ik} = \boldsymbol{X}_i(\boldsymbol{t}_k) = \begin{pmatrix} \boldsymbol{x}'_{i1} \\ \vdots \\ \boldsymbol{x}'_{ik} \end{pmatrix} = \begin{pmatrix} \boldsymbol{x}'_i(t_1) \\ \vdots \\ \boldsymbol{x}'_i(t_k) \end{pmatrix}$$

$$\boldsymbol{Z}_{ik} = \boldsymbol{Z}_i(\boldsymbol{t}_k) = \begin{pmatrix} \boldsymbol{z}'_{i1} \\ \vdots \\ \boldsymbol{z}'_{ik} \end{pmatrix} = \begin{pmatrix} \boldsymbol{z}'_i(t_1) \\ \vdots \\ \boldsymbol{z}'_i(t_k) \end{pmatrix}$$

and

$$\boldsymbol{U}_{ik} = \boldsymbol{U}_i(\boldsymbol{t}_k) = \begin{pmatrix} \boldsymbol{u}'_{i1} \\ \vdots \\ \boldsymbol{u}'_{ik} \end{pmatrix} = \begin{pmatrix} \boldsymbol{u}'_i(t_1) \\ \vdots \\ \boldsymbol{u}'_i(t_k) \end{pmatrix}$$

are the associated design matrices and auxiliary matrix, respectively. The *intended* complete data regression setup with respect to $\{Y_i(t)\}$ may then be written as

$(\boldsymbol{y}_i, \boldsymbol{X}_i, \boldsymbol{Z}_i) = (\boldsymbol{y}_i(\boldsymbol{t}_p), \boldsymbol{X}_i(\boldsymbol{t}_p), \boldsymbol{Z}_i(\boldsymbol{t}_p))$ while the observed data regression setup may be written as $(\boldsymbol{y}_i^o, \boldsymbol{X}_{ip_i}, \boldsymbol{Z}_{ip_i}) = (\boldsymbol{y}_i(\boldsymbol{t}_{p_i}), \boldsymbol{X}_i(\boldsymbol{t}_{p_i}), \boldsymbol{Z}_i(\boldsymbol{t}_{p_i}))$ where

$$\boldsymbol{y}_i^o = \boldsymbol{y}_i(\boldsymbol{t}_{p_i}) = (y_{i1}, y_{i2}, \ldots, y_{ip_i})'$$

is the vector of observed responses $y_i(t_j)$ taken at times $\boldsymbol{t}_{p_i} = (t_1, t_2, \ldots, t_{p_i})'$ and $t_{p_i} = \max\{t_j : 0 \leq t_j < T_i^o; j = 1, \ldots, p\}$ is the time of the last observed outcome.

Dropout mechanisms based on survival analysis principles

With this combined repeated measures regression and survival time framework, one can formulate ignorable and non-ignorable dropout mechanisms guided by principles from survival analysis with time-dependent covariates. The first principle is that a terminal event (or absorbing state) like death cannot depend causally or stochastically on future "potential" observations since such observations are no longer possible or relevant. The second principle from survival analysis is that one cannot *predict* an individual's current dropout status *from future* as yet unrealized values of a time-dependent covariate. This is based on the fundamental principle that causal mechanisms flow forward through time. When modeling the event time random variable T_i as a function of time-dependent covariates including the repeated outcome measurements, the hazard function associated with dropout at time t can only causally depend on information from the past, relative to time t, and not on the future. This is not to say that future responses cannot be associated with *non-terminal* dropout times. Indeed future values may be affected by a non-terminal event or they may be associated with dropout on the basis of how strongly they are correlated with responses leading up to dropout. Rather, it simply means that within the context of missing data *due to* dropout, a dropout event that occurs at time t^* cannot have current or future values of $\{Y_i(t), t \geq t^*\}$ on the *causal pathway* leading to the event.

A third principle deals with the use of both external and internal time-dependent covariates as defined by Kalbfleisch and Prentice (2002, §6.3). An *external time-dependent covariate* has its values generated over time in a manner that does not require the individual to remain free of the event (i.e., "survive") nor does the occurrence of the event affect future values of the covariate. In that regard, it is *possible* to treat the repeated outcome measurements in our setting as values from an external time-dependent covariate to be used in an event-time analysis of *non-terminal dropout* events. This is because the outcome or response variable of interest does not require the individual to be "event free" in order to be observed. An *internal time-dependent covariate*, on the other hand, would be any time-dependent covariate that does not satisfy the conditions for being an external covariate (e.g., see equations 6.12 or 6.13). Typically an internal time-dependent covariate has its values generated over time by the individual and the existence of those values requires the individual to remain free of the event (i.e., "survive"). For example, in a traditional survival analysis involving the comparison of mortality rates, any time-dependent covariate like blood pressure, serum cholesterol level, etc., that is generated by the subject over time would be classified as an internal time-dependent covariate. In our current setting, the repeated outcome measurements would be values of an internal time-dependent covariate relative to an event-time analysis of *terminal dropout* events. This is because the individual must remain free of the event in order for a value of the covariate to exist or to have any kind of "meaningful" existence with respect to scientific inquiry (see the MDRD example in §6.6.2). As we will see, this distinction between treating the response variable of interest as an external versus internal time-dependent covariate for the overall dropout process will be necessary when we formalize definitions for MAR and MNAR dropout mechanisms. A related issue is the fact that inclusion of internal time-dependent covariates requires one to specify an event-time model on the basis of the hazard function as opposed to a probability density function and survivor function (e.g., Kalbfleisch and Prentice, 2002, §6.3). Finally, it is worth noting that although the response process $\{Y_i(t)\}$ is defined over the interval $[0, \mathcal{T})$, actual measurements are confined by the occurrence of a terminal event to be within the interval $[0, \mathcal{T}_i)$ where $\mathcal{T}_i = \min(T_{1i}, \mathcal{T})$.

Guided by these principles, it is typical and indeed necessary to formulate a dropout mechanism relative to observed and unobserved outcome measurements on the basis of the hazard function associated with dropout. To that end, let $\boldsymbol{v}_i(t)' = (\boldsymbol{x}_i(t)', \boldsymbol{z}_i(t)', \boldsymbol{u}_i(t)')$ be the combined vector of all explanatory and auxiliary variables and let $\mathcal{Z}_i(t) = \{\boldsymbol{v}_i(s), y_i(s); s \in [0, t)\}$ be the combined covariate and response history up to time t which we partition as

$$\mathcal{Z}_i(t) = \begin{pmatrix} \mathcal{Z}_{\boldsymbol{v}_i}(t) \\ \mathcal{Z}_{y_i}(t) \end{pmatrix}$$

where $\mathcal{Z}_{\boldsymbol{v}_i}(t)$ represents the covariate history and $\mathcal{Z}_{y_i}(t)$ the response history of both observed and unobserved responses. Also, let $\mathcal{Z}_{y_i}(\mathcal{T}) = \{y_i(u); u \in [0, \mathcal{T})\}$ be the total potential response history defined up to the study termination time $\mathcal{T}$. This represents all potential observed and unobserved responses including any unobserved responses that occur following a non-terminal dropout event (i.e., when $T_i^o = T_{2i} < \mathcal{T}$). Finally, we will let $\mathbb{Z}_i(t)$ denote a vector of derived covariates whose elements consist of functions of $\mathcal{Z}_i(t)$ and t (Kalbfleisch and Prentice, 2002, §6.3-6.4). For example, the derived covariates that make up $\mathbb{Z}_i(t)$ may include lagged values and/or time averaged values taken from the response history $\mathcal{Z}_{y_i}(t)$ as a means for testing whether dropout is completely at random or not.

Based on our preceding discussion, it is clear that terminal and non-terminal events present themselves as competing risks for overall dropout. It is against that background that we have defined two latent event-time random variables, T_{1i} and T_{2i}, as a means for helping differentiate possible dropout mechanisms. As noted by Kalbfleisch and Prentice (2002, §8.2.4), formulating a strategy to a competing risks problem on the basis of latent event-time random variables introduces a number of challenges. These include issues of identifiability with respect to the marginal distributions of the unobserved random variables T_{1i} and T_{2i} and the inability to distinguish between an independent competing risk model and an infinitude of dependent competing risk models all of which give rise to the same set of cause-specific hazard functions. In this chapter, we avoid such issues by choosing to model the hazard function associated with overall dropout as measured by the observable quantities, (T_i^o, δ_i). We nevertheless consider how and when the cause-specific hazard functions associated with terminal and non-terminal events might be used as a means for assessing whether dropout is completely at random. We also show how the competing risks problem formulated on the basis of the latent event-times, T_{1i} and T_{2i}, is useful when explaining why MAR and MNAR dropout mechanisms require unverifiable assumptions.

Conditional on the combined covariate and response history as well as any latent random effects $\boldsymbol{b}_i$ that may apply, the hazard function associated with overall dropout is defined quite generally by

$$\lambda_i(t|\mathcal{Z}_i(t), \boldsymbol{b}_i)dt = \Pr[t \le T_i < t + dt|\mathcal{Z}_i(t), \boldsymbol{b}_i, T_i \ge t] \tag{6.11}$$

where the censoring mechanism associated with the censoring time, C_i, is assumed to be conditionally independent of the dropout mechanism, i.e.,

$$\begin{aligned} &\Pr[t \le T_i < t + dt|\mathcal{Z}_i(t), \boldsymbol{b}_i, T_i \ge t, C_i \ge t] \\ &= \Pr[t \le T_i < t + dt|\mathcal{Z}_i(t), \boldsymbol{b}_i, T_i \ge t] \end{aligned}$$

(e.g., Kalbfleisch and Prentice, 2002, §1.3). Of course we also assume that the censoring mechanism is completely independent of the response process $\{Y_i(t)\}$. The fact that the overall hazard function (6.11) is the sum of the estimable cause-specific hazard functions,

$$\begin{aligned} \lambda_{1i}(t|\mathcal{Z}_i(t), \boldsymbol{b}_i)dt &= \Pr[t \le T_{1i} < t + dt|\mathcal{Z}_i(t), \boldsymbol{b}_i, T_i \ge t] \\ \lambda_{2i}(t|\mathcal{Z}_i(t), \boldsymbol{b}_i)dt &= \Pr[t \le T_{2i} < t + dt|\mathcal{Z}_i(t), \boldsymbol{b}_i, T_i \ge t], \end{aligned}$$

allows one to partially investigate whether dropouts due to terminal or non-terminal events are *completely at random* (in the sense of MCAR). For example, since a terminal event rules out potential future observations from our missing data framework, a test based on the hazard rate $\lambda_{1i}(t|\mathcal{Z}_i(t), \boldsymbol{b}_i)$ for terminal events can be used to rule out whether terminal dropouts are completely at random. However, in the absence of any random effects, this is only partially possible with non-terminal dropout events since responses following dropout are not made available. In particular, suppose dropout is independent of any latent random effects $\boldsymbol{b}_i$ or that there are no random effects in the specification of the regression model $\mu_i(t)$ for $Y_i(t)$. A test based on the hazard rate $\lambda_{2i}(t|\mathcal{Z}_i(t))$ for non-terminal events can be used to rule out whether outcome measurements *leading up to* dropout are independent of dropout given the covariate history to time t. In order to rule out whether unobserved outcome measurements *following* non-terminal dropout are also independent of the dropout mechanism, one would need to verify that the hazard function associated with the non-terminal event time, T_{2i}, satisfies the condition

$$\begin{aligned}\lambda^*_{2i}(t|\mathcal{Z}_i(t))dt &= \lambda^*_{2i}(t|\mathcal{Z}_{\boldsymbol{v}_i}(t), \mathcal{Z}_{y_i}(t))dt \\ &= \Pr[t \le T_{2i} < t + dt|\mathcal{Z}_{\boldsymbol{v}_i}(t), \mathcal{Z}_{y_i}(t), T_{2i} \ge t] \\ &= \Pr[t \le T_{2i} < t + dt|\mathcal{Z}_{\boldsymbol{v}_i}(t), \mathcal{Z}_{y_i}(u), T_{2i} \ge t]\end{aligned} \tag{6.12}$$

for all t, u such that $0 < t \le u$. But (6.12) is precisely the condition required in order for covariates based on $\mathcal{Z}_{y_i}(t) = \{y_i(s); 0 \le s < t\}$ to be regarded as external time-dependent covariates (Kalbfleisch and Prentice, 2002, §6.3). A condition equivalent to (6.12), also given by Kalbfleisch and Prentice (2002, §6.3), is that

$$\Pr[Y_i(u)|Y_i(t),\, T_{2i} \ge t] = \Pr[Y_i(u)|Y_i(t),\, T_{2i} = t], \quad 0 < t \le u. \tag{6.13}$$

These equivalent conditions reflect the very plausible but, in our setting, unverifiable assumption that the response variable may exert an influence on the rate of non-terminal dropout events over time but the response profiles $\{Y_i(u); u \in [t, T)\}$ following such events remain unaffected. In fact it is precisely the unverifiable condition (6.12) (or 6.13) that one must assume in order to extend the S-MAR mechanism (6.4) to the current continuous event-time setting involving terminal and non-terminal dropout events.

Given the above event-time and longitudinal data framework, one can formulate ignorable and non-ignorable dropout mechanisms on the basis of (6.11) and (6.12) (or 6.11 and 6.13) as follows. If the hazard function (6.11) depends only on the observed history, say $\mathcal{Z}^o_i(t)$, up to time t, then we have a *sequential partially random dropout* (S-PRD) mechanism. If, in addition, covariates based on $\mathcal{Z}_{y_i}(t)$ satisfy the requirements of being external time-dependent covariates for the latent event-time T_{2i}, i.e., if $\mathcal{Z}_{y_i}(t)$ satisfies condition (6.12), then we have a *sequential random dropout* (S-RD) mechanism. If either of these two conditions fail, we then have a *sequential not at random dropout* (S-NARD) mechanism. A more formal definition follows.

Sequential partially random dropout (S-PRD):

Assuming the overall hazard function in (6.11) does not depend on any latent random effects, $\boldsymbol{b}_i$, we designate the missing data mechanism *leading up to* dropout as *sequential partially random dropout* (S-PRD) if the hazard function satisfies

$$\begin{aligned}\lambda_i(t|\mathcal{Z}_i(t))dt &= \Pr[t \le T_i < t + dt|\mathcal{Z}_i(t), T_i \ge t] \\ &= \Pr[t \le T_i < t + dt|\mathcal{Z}^o_i(t), T_i \ge t] \\ &= \lambda_i(t|\mathcal{Z}^o_i(t))dt \\ &= \lambda_i(t|\mathcal{Z}^o_{\boldsymbol{v}_i}(t), \mathcal{Z}^o_{y_i}(t))dt\end{aligned} \tag{6.14}$$

where $\mathcal{Z}_i^o(t) = \{\boldsymbol{v}_i(t_j), y_i(t_j); t_j \in [0,t)\} = (\mathcal{Z}_{\boldsymbol{v}_i}^o(t), \mathcal{Z}_{y_i}^o(t))$ is the observed covariate and response history up to time t. If instead we assume,

$$\begin{aligned}\lambda_i(t|\mathcal{Z}_i(t))dt &= \Pr[t \le T_i < t+dt|\mathcal{Z}_{\boldsymbol{v}_i}^o(t), T_i \ge t] \\ &= \lambda_i(t|\mathcal{Z}_{\boldsymbol{v}_i}^o(t))dt\end{aligned} \quad (6.15)$$

such that the hazard of overall dropout depends only on the observed covariate history, then we may classify the dropout mechanism as *sequential partially complete random dropout*, or S-PCRD, conditional on $\mathcal{Z}_{\boldsymbol{v}_i}^o(t)$.

Sequential random dropout (S-RD):

Under S-PRD, missing or unobserved responses *leading up to* dropout are ignorable. However, this condition alone does not imply complete ignorability with respect to dropout. Suppose that in addition to (6.14), condition (6.12) holds or, equivalently, that

$$\begin{aligned}\lambda_{2i}^*(t|\mathcal{Z}_{\boldsymbol{v}_i}(t), \mathcal{Z}_{y_i}(t), \mathcal{Z}_{y_i}(\mathcal{T}_t))dt &= \lambda_{2i}^*(t|\mathcal{Z}_{\boldsymbol{v}_i}(t), \mathcal{Z}_{y_i}(\mathcal{T}))dt \\ &= \lambda_{2i}^*(t|\mathcal{Z}_{\boldsymbol{v}_i}(t), \mathcal{Z}_{y_i}(t))dt\end{aligned} \quad (6.16)$$

where $\mathcal{Z}_{y_i}(\mathcal{T}_t) = \{y_i(u); u \in [t, \mathcal{T})\}$. In this case, covariates based on $\mathcal{Z}_{y_i}(t)$ will *conceptually* satisfy the conditions required to be both internal time-dependent covariates with respect to the latent terminal event-time T_{1i} and external time-dependent covariates with respect to the latent non-terminal event-time T_{2i}. Assuming (6.14) and (6.16) are both satisfied, we designate the missing data mechanism due to dropout as *sequential random dropout* (S-RD). Informally, (6.16) assumes that conditional on the past history and the full-data response history, non-terminal dropout at time t does not depend on current or future response data including those responses intended to be measured. Suppose instead of (6.14), the overall hazard function satisfies (6.15) and the non-terminal hazard function associated with the latent event-time T_{2i} continues to satisfy (6.16). Then the overall hazard rate for dropout depends only on the observed covariate history and we classify the dropout mechanism as *sequential completely random dropout*, or S-CRD, conditional on $\mathcal{Z}_{\boldsymbol{v}_i}^o(t)$.

Note that S-RD as defined by (6.14) and (6.16) may be viewed as a continuous time-to-dropout analog to the S-MAR mechanism defined in (6.4). In particular, suppose we replace $\boldsymbol{V}_{ik}$ in (6.4) with $\boldsymbol{V}_{i(k-1)}$ and view S-MAR under (6.4) within the context of a discrete time survival model for grouped event-time data. In the absence of non-informative censoring, the conditional probability (hazard) of dropout within the interval $[t_{k-1}, t_k)$ given that the subject has complete data (i.e., "survived") to the beginning of the interval may be expressed in terms of a non-response at t_k given response at t_{k-1} as

$$\Pr(r_{ik} = 0|r_{i(k-1)} = 1, \boldsymbol{V}_{i(k-1)}, \boldsymbol{y}_i, \boldsymbol{b}_i) = \Pr(r_{ik} = 0|r_{i(k-1)} = 1, \boldsymbol{V}_{i(k-1)}, \boldsymbol{y}_{i(k-1)}) \quad (6.17)$$

for $k = 2, \ldots, p$. This is the same definition of S-MAR given in (6.4) except that $\boldsymbol{V}_{ik}$ is replaced by $\boldsymbol{V}_{i(k-1)}$ (i.e., we only condition on covariates up to and including time t_{k-1}). Equivalently, we may express (6.17) in terms of a discrete time hazard function as

$$\begin{aligned}\lambda_i(t_k|\boldsymbol{V}_{i(k-1)}, \boldsymbol{y}_i, \boldsymbol{b}_i) &= \Pr[T_i \in [t_{k-1}, t_k)|T_i \ge t_{k-1}, C_i \ge t_{k-1}, \boldsymbol{V}_{i(k-1)}, \boldsymbol{y}_i, \boldsymbol{b}_i] \\ &= \Pr[T_i \in [t_{k-1}, t_k)|T_i \ge t_{k-1}, \boldsymbol{V}_{i(k-1)}, \boldsymbol{y}_{i(k-1)}] \\ &= \lambda_i(t_k|\boldsymbol{V}_{i(k-1)}, \boldsymbol{y}_{i(k-1)}) \\ &= \lambda_i(t_k|\mathcal{Z}_i^o(t_k))\end{aligned} \quad (6.18)$$

where $\mathcal{Z}_i^o(t_k) = \{\boldsymbol{v}_i(t_j), y_i(t_j); t_j \in [0, t_k), j = 1, \ldots, k-1\} = \{\boldsymbol{V}_{i(k-1)}, \boldsymbol{y}_{i(k-1)}\}$ is the observed history up to but excluding time t_k. We may view (6.18) as satisfying the joint

conditions in (6.14) and (6.16) for grouped event-time data and in fact this is what is implicitly implied in (6.4) for S-MAR. Following Robins et. al. (1995), we prefer to condition on $\boldsymbol{V}_{i(k-1)}$ in (6.17) or (6.18) as opposed to $\boldsymbol{V}_{ik}$ as done in (6.4) when defining S-MAR as this avoids any problems associated with possible auxiliary time-dependent covariates in $\boldsymbol{u}_i(t)$ generated by the individual. Under this general framework, the observed data setup $\{\boldsymbol{y}_i^o, \boldsymbol{X}_{ip_i}, \boldsymbol{Z}_{ip_i}, \boldsymbol{U}_{ip_i}\}$ at $T_i^o = t_i^o$ may be expressed as
$\mathcal{Z}_i^o(t_i^o) = \mathcal{Z}_i^o(t_{p_i+1}) = \{\boldsymbol{x}_i(t_j), \boldsymbol{z}_i(t_j), \boldsymbol{u}_i(t_j), y_i(t_j); t_j \in [0, t_{p_i+1}), j = 1,\ldots,p_i\}$.

Sequential not at random dropout (S-NARD)

If the dropout mechanism is not S-RD, then it is S-NARD. However, it is useful to distinguish between S-NARD associated with unobserved responses, $y_i(t)$, versus S-NARD associated with unobserved random effects, $\boldsymbol{b}_i$. Specifically, one can define the dropout mechanism as *sequential not at random dropout* (S-NARD) if either

$$\begin{aligned}\lambda_i(t|\mathcal{Z}_i(t), \boldsymbol{b}_i)dt &= \Pr[t \le T_i < t + dt|\mathcal{Z}_{\boldsymbol{v}_i}(t), \mathcal{Z}_{y_i}(t), \boldsymbol{b}_i, T_i \ge t] \\ &= \Pr[t \le T_i < t + dt|\mathcal{Z}_{\boldsymbol{v}_i}(t), \mathcal{Z}_{y_i}(t), T_i \ge t] \\ &= \lambda_i(t|\mathcal{Z}_{\boldsymbol{v}_i}(t), \mathcal{Z}_{y_i}(t))dt \\ \lambda_{2i}^*(t|\mathcal{Z}_{\boldsymbol{v}_i}(t), \mathcal{Z}_{y_i}(T))dt &\ne \lambda_{2i}^*(t|\mathcal{Z}_{\boldsymbol{v}_i}(t), \mathcal{Z}_{y_i}(t))dt\end{aligned} \tag{6.19}$$

or

$$\begin{aligned}\lambda_i(t|\mathcal{Z}_i(t), \boldsymbol{b}_i)dt &= \Pr[t \le T_i < t + dt|\mathcal{Z}_{\boldsymbol{v}_i}(t), \mathcal{Z}_{y_i}(t), \boldsymbol{b}_i, T_i \ge t] \\ &= \Pr[t \le T_i < t + dt|\mathcal{Z}_{\boldsymbol{v}_i}(t), \boldsymbol{b}_i, T_i \ge t] \\ &= \lambda_i(t|\mathcal{Z}_{\boldsymbol{v}_i}(t), \boldsymbol{b}_i)dt.\end{aligned} \tag{6.20}$$

We refer to the dropout mechanism defined by (6.19) as non-ignorable outcome dependent dropout and the dropout mechanism defined by (6.20) as non-ignorable random effects dropout in a manner analogous to MNAR dropout mechanisms (§6.2.3). The assumption under (6.19) is that the hazard of dropout will depend on unobserved values of the response prior to dropout (i.e., "missing or unobserved values leading to dropout") and/or unobserved values following non-terminal dropouts assuming (6.16) does not hold true. In the case of (6.20), the hazard of dropout will depend on unobserved responses (both those intended to be measured and those that were not) through the posterior distribution of the random effects as shown in (6.10).

6.3.2 Practical issues with missing data and dropout

Under S-RD, dropout at time t is assumed to depend only on the observed measurement history and not on unobserved responses before or, in the case of a non-terminal event, after dropout. Under S-NARD, dropout at time t is assumed to depend on unobserved responses leading up to dropout and possibly on unobserved responses following dropout. This immediately raises a number of issues. The first is how can we know if the underlying mechanism is S-RD or S-NARD. The answer is we cannot. Since we only have observed values available to us, we cannot directly evaluate whether dropout depends on any specific but unobserved value of the response variable leading up to dropout without making additional assumptions such as dependency through some latent random effect. Neither can we verify the assumption in (6.12) or (6.13) that future but potentially incomplete responses following a non-terminal event are conditionally independent of dropout given the past. While it is not necessary to assume (6.12) under the S-NARD mechanism, we still face the challenge of not being able to verify whether unobserved values leading to or following non-terminal dropouts are ignorable. This, of course, is also an issue with the more classical MAR and MNAR mechanisms defined in §6.2. In both settings, one can test for and rule out the possibility that missing data are MCAR or that dropout is completely at random (S-CRD) provided we at least have missing data that are MAR or S-MAR. However, we can never, in general, test for MAR- or MNAR-type mechanisms without making explicit but unverifiable modeling assumptions.

In many respects, the issue of ignorable versus non-ignorable dropout is best understood in light of the traditional MAR/MNAR setting where missing values refer to those outcome measures that one intended to collect but which are no longer available due to dropout. Here we can envision a scenario where the hypothetical response following dropout can no longer be reliably predicted on the basis of the data observed prior to dropout—a consequence of MNAR. However, on the basis of S-RD versus S-NARD as defined in §6.3.1, the idea of "missingness" or "incompleteness" relates primarily to unobserved values of the outcome variable prior to dropout, whether planned or unplanned. This brings us to another issue, namely what constitutes non-ignorable missing data prior to dropout? In the case of non-ignorable outcome-dependent dropout as defined by (6.19), missing values can refer to realized values of the response variable that were not planned for within the study and were not available to the study investigator but which were available to individuals outside the study and used to possibly justify subject withdrawal from the study. An example of this kind of situation will be presented in §6.6.2 when we analyze data from the Modification of Diet in Renal Disease (MDRD) study. Alternatively, missingness may refer to some unmeasured threshold value determined by a subject's "true" response trajectory that somehow leads to dropout (Tsiatis and Davidian, 2004). There are other scenarios as well. It is possible, for example, that a S-PRD mechanism is at work prior to dropout but that the effect of a particular non-terminal dropout event may alter the response profile following dropout resulting in a violation of (6.12) or (6.13) and hence yielding a S-NARD mechanism.

In the end, we face the same challenges whether we have unplanned and unobserved responses leading to dropout or planned but missing responses resulting from dropout. These lead to other issues one must address when jointly analyzing longitudinal data and dropout. For example, what assumptions should one make with regards to the hazard function? Does one use a parametric, semiparametric or nonparametric hazard model? How does one ultimately treat the use of the longitudinal responses as time-dependent covariates in the dropout process when those responses may be viewed conceptually as either internal or external time-dependent covariates depending on whether dropout is terminal or non-terminal? This is a particularly vexing problem in the competing risk literature for survival analysis (e.g., Latouche et. al., 2005) and one might expect this will also carry over when discussing methods for handling dropout due to competing terminal versus non-terminal events (e.g., Dupuy and Mesbah, 2002). As we shall discuss in §6.4, this could have implications with respect to the method of analysis one chooses under MAR assumptions. In general, great care is needed when modeling dropout as a function of repeated outcome measurements when treated as time-dependent covariates. In terms of the longitudinal response model, we are faced with the challenge of specifying a marginal or random-effects regression model for $\{Y_i(t)\}$. Such a choice should always be guided by the objectives of the study and one will need to decide if the primary goal involves population-average (PA) or subject-specific (SS) inferential claims. Other issues include deciding on what assumptions one should make when performing a joint analysis of longitudinal responses and dropout and how sensitive the results are to violations of such assumptions?

6.3.3 Developing an analysis plan for missing data

In general, one will need to develop an analysis plan that deals with how missing data and dropout will be handled. Typically, this would start with a set of primary analyses in which the analyst would focus on valid inference under MAR assumptions. The advantages with this strategy are that 1) a primary analysis assuming MAR can be stipulated in advance of the data, 2) an MAR analysis can usually be carried out using standard analytical methods (e.g., likelihood-based methods), 3) any modeling assumptions one makes under MAR may be verified based on the observed data, and 4) an MAR analysis does not depend on any additional and unverifiable assumptions about the missing data that one would otherwise require for valid inference under MNAR. This last point is crucial particularly for clinical trials of a regulatory nature where there is a strong emphasis on minimizing the assumptions required for valid inference.

Following on this, the analyst will then want to include a number of secondary analyses that focus on how sensitive the primary analyses are to departures from the underlying MAR assumption. Since we have no way of testing whether missing values are MAR or MNAR, such secondary analyses will help when evaluating the strength of evidence from the primary analysis. For example, one might postulate one or more models under different MNAR mechanisms. By then fitting these models to the observed data, one can determine to what degree the underlying MNAR assumptions yield results that agree or disagree with MAR results. We will not present in-depth methods for conducting various types of sensitivity analyses but rather refer the reader to literature on this important topic (e.g., Scharfstein et. al., 1999; Daniels and Hogan, 2008; National Research Council, 2010 and references therein). Our goal here is to simply illustrate how one might go about analyzing data in the presence of missing values particularly as they pertain to dropout. To that end, a very brief overview of some of the methods advocated for the analysis of longitudinal data with missing values is provided in the following two sections. Section 6.4 deals with methods of analysis under MAR which, in most cases, would constitute the primary analysis. In section 6.5, we consider methods for performing sensitivity analyses under MNAR. These sections are by no means intended to be comprehensive in scope but rather are intended to provide the basic concepts behind each of the methods. The reader who is already familiar with these methods may wish to skip directly to the case studies where we illustrate how some of these techniques can be implemented in SAS.

6.4 Methods of analysis under MAR

Methods for handling missing data depend heavily on the underlying study objectives, scientific inquiry, study design, the scope and nature of the missing data, and ultimately on the underlying assumptions one is willing to make. In this section, we review some of the more common methods of analysis used when missing data are assumed to be MAR. These include fully parametric likelihood methods, methods based on single and multiple imputation and methods based on inverse probability of weighting. These methods are valid under both MAR and MCAR but other methods valid under MCAR will not be valid under MAR (examples include available-case and complete-case analysis using standard GEE). Even when MCAR holds true, methods like complete-case (CC) analysis in which one restricts the analysis to those subjects with complete data can be terribly inefficient. We will therefore focus only on those methods valid under the MAR assumption with the understanding that standard moment-based methods like GEE or CGEE presented in Chapters 3-5 will only be valid when there are no missing data or when missing data are MCAR. However, we do show how one can adapt standard GEE methods to incomplete MAR data through the use of multiple imputation and inverse probability weighting.

6.4.1 Likelihood-based methods

As was shown in §6.2.2 (equation 6.5), when missing values are ignorable, i.e., data are MAR and the parameters $\boldsymbol{\tau}$ and $\boldsymbol{\eta}$ are separable, valid inference based on all observed data is possible using likelihood-based methods provided inference is conducted using the observed rather than the expected information matrix (see Little and Rubin, 2002;

Kenward and Molenberghs, 1998). This includes maximum likelihood (ML) methods for linear, generalized linear and generalized nonlinear models for correlated response data, both marginal and mixed, as described in Chapters 2-5. It also includes methods based on Bayesian posterior inference which we do not address here (e.g., Carlin and Louis, 2000; Gelman et. al., 2004; National Research Council, 2010). Here, we discuss some of the advantages and disadvantages of likelihood-based methods.

Perhaps the biggest advantage with likelihood-based methods is the ease with which one can implement an analysis using the various procedures available in SAS. A second advantage is that one can usually test various modeling assumptions under MAR based on standard goodness-of-fit techniques. A third big advantage is that under mild regularity conditions, ML estimators are consistent asymptotically normal (CAN) and asymptotically efficient (e.g., Serfling, 1980 pp. 144-149). However, these advantages do come with a cost. First, likelihood-based methods are fully parametric and will therefore require the correct specification of the probability density function (pdf) for valid inference under MAR. Second, as with almost all of the methods described in Chapters 4-5, likelihood-based inference is almost always based on large sample theory meaning that one should have a reasonably large sample size to begin with. This of course raises the usual question of how large is large? Third, when dealing with mixed-effects models, linear or nonlinear, we need to consider whether the primary focus of the study is on a population-averaged (PA) estimand or a subject-specific (SS) estimand. While this is typically not an issue with LME models, it can be crucial when dealing with GLME or GNLME models. For example, if the focus is on a PA estimand like say the difference in marginal means between a treatment and control group at some given time, we must reconcile ourselves to the fact that such an estimand may be very difficult to evaluate under a nonlinear mixed-effects model without recourse to some numerical integration procedure. Indeed, estimands based on marginal moments are usually formulated on the basis of a marginal likelihood function for which we generally have no closed form expression when nonlinear random effects are present. This makes it extremely difficult to formulate the PA estimand let alone conduct inference on that estimand. Another issue with mixed-effects models is the fact that one need correctly specify the conditional pdf of $\boldsymbol{y}_i|\boldsymbol{b}_i$ as well as the pdf of $\boldsymbol{b}_i$, both of which require parametric assumptions. In the case of LME models, one will need to correctly model the variance-covariance structure which means correctly choosing the random effects and their covariance structure as well as specifying the correct intra-subject covariance structure. In short, likelihood-based inferences will be far more sensitive to model misspecifications. These and other issues need to be thought through very carefully when planning for and conducting a likelihood-based analysis under MAR assumptions.

One need also consider the problem of incorporating auxiliary variables into a likelihood-based analysis. In §6.2.2, if one assumes that $\boldsymbol{V}_i = (\boldsymbol{X}_i, \boldsymbol{Z}_i)$ or that the matrix of auxiliary covariates $\boldsymbol{U}_i$ consists of external covariates that are independent of $\boldsymbol{y}_i$, then one can conduct valid likelihood-based inference under MAR ignoring $\boldsymbol{U}_i$. However, if one wished to relax the latter assumption about $\boldsymbol{U}_i$ and incorporate time-dependent auxiliary variables into an analysis, one would generally be required to specify a joint distribution with respect to $(\boldsymbol{y}_i, \boldsymbol{r}_i, \boldsymbol{U}_i)$ and then integrate over those auxiliary variables within $\boldsymbol{U}_i$ to achieve valid inference (e.g., Hogan et. al., 2004; National Research Council, 2010). As discussed below, methods based on multiple imputation and inverse probability of weighting allow one to more easily accommodate such auxiliary information into an analysis under MAR.

6.4.2 Imputation-based methods

Analyses based on methods that impute missing values are frequently used when the focus of a study is on marginal or population-averaged inference. This is particularly true when a non-likelihood based analysis like GEE is the preferred method of analysis over a fully parametric likelihood-based analysis. Imputation-based analyses rely on filling in missing values using some form of imputation followed by an analysis of the complete data using an appropriate analytical procedure. There are a number of imputation methods one can use including relatively simple single imputation techniques as well as more sophisticated multiple imputation (MI) techniques. Not all of these techniques are guaranteed to work under a MAR mechanism. In particular, intent-to-treat analyses in clinical trials with dropout commonly use single imputation techniques like last observation carried forward (LOCF), baseline observation carried forward (BOCF), and worst observation carried forward (WOCF). However, it is known that these techniques can lead to biased results under MAR and even MCAR assumptions unless additional unverifiable assumptions are satisfied. For example, the MCAR assumption, while necessary, is not sufficient to guarantee unbiased inference under the very popular LOCF analysis (Molenberghs et. al., 2004). In this case, one also needs the rather strong assumption that the response does not change following dropout. Even when the necessary assumptions appear reasonable, one should still take into account the possibility that the statistical uncertainty associated with the estimand of interest may be underestimated (Little and Rubin, 2002). Given their less than satisfactory performance compared to likelihood- and MI-based methods, we forgo any further discussion of these simple single imputation techniques and instead refer the reader to the aforementioned literature on missing data in §6.1 as well as papers by Shao and Zhong (2003), Molenberghs et. al. (2004) and Siddiqui et. al. (2009) for a further critique of these methods. The focus here will be on methods that rely on multiple imputation as developed originally by Rubin (1978, 1987).

Multiple imputation (MI) in combination with non-likelihood based methods like GEE provide a broad spectrum of alternative approaches for handling missing data under MAR. The MI-based approach can be particularly appealing when dealing with non-Gaussian responses such as repeated binary outcomes and repeated count data where GEE and other methods are often preferred to likelihood-based methods. The basic idea behind MI is to "fill in" or replace missing values with $m > 1$ simulated values resulting in m simulated complete datasets. One then analyzes each of these datasets using standard methods (e.g., GEE, GEE2, WLS for repeated categorical data, etc.) from which results are combined to produce final estimates and standard errors for conducting valid inference under MAR. This approach is easily implemented in SAS by applying the procedures MI and MIANALYZE using the following three steps:

1. Missing values are imputed ("filled in") m times resulting in m complete datasets using PROC MI.
2. Perform an analysis on each of the m completed datasets using a BY statement in conjunction with an appropriate analytic procedure (MIXED, GENMOD, etc.) and make sure to output parameter estimates, standard errors, etc. to a dataset using, for example, an ODS OUTPUT statement.
3. Combine the results from the m complete datasets to produce valid inferential results under MAR assumptions using PROC MIANALYZE.

One might think this three-step approach would be extremely computer intensive for large datasets with considerable missing data. However, the number of imputed values, m, that one specifies will typically be somewhere between 5 and 20 with m increasing to say 20 as the percentage of missing data increases (Rubin, 1987, 1996). In fact, Rubin (1987) shows that the asymptotic efficiency of the repeated-imputation estimator based on m finite imputations relative to the estimator based on $m = \infty$ imputations is approximately $[1 + (f/m)]^{-1/2}$ in units of standard deviations where f is the fraction of missing data. This translates to a value of 0.95 when 50% of the data are missing and one carries out just 5 imputations.

A principled approach to multiple imputation that is grounded in Bayesian principles is the repeated imputation approach described by Rubin (1987, 1996). To frame this approach, suppose the complete data $(\boldsymbol{y}_i, \boldsymbol{X}_i)$ within our longitudinal clinical trial setting (see §6.2) may be described by a parametric model, $\pi(\boldsymbol{y}_i|\boldsymbol{X}_i;\boldsymbol{\tau})$, which we will call the *analysis model*. Here we assume there are no missing values for the covariates (i.e., $\boldsymbol{X}_i$ is completely observed) and we assume $\pi(\boldsymbol{y}_i|\boldsymbol{X}_i;\boldsymbol{\tau})$ represents the underlying marginal model obtained by integrating over the distribution of any random effects that may be present. We also assume missing data are ignorable in that missing values are MAR and the model parameters $\boldsymbol{\tau}$ are distinct from the parameters $\boldsymbol{\eta}$ of the probability model for missingness, $\pi(\boldsymbol{r}_i|\boldsymbol{V}_i, \boldsymbol{y}_i^o;\boldsymbol{\eta})$,as defined in (6.5) of §6.2.2. Under this framework, repeated imputations are draws from the posterior predictive distribution of the missing values $\boldsymbol{y}_i^m$ conditional on the observed values $\boldsymbol{y}_i^o$, i.e., $\pi(\boldsymbol{y}_i^m|\boldsymbol{y}_i^o, \boldsymbol{X}_i)$. The posterior predictive distribution is given by

$$\pi(\boldsymbol{y}_i^m|\boldsymbol{y}_i^o, \boldsymbol{X}_i) = \int \pi(\boldsymbol{y}_i^m|\boldsymbol{y}_i^o, \boldsymbol{X}_i;\boldsymbol{\tau})\pi(\boldsymbol{\tau}|\boldsymbol{y}_i^o, \boldsymbol{X}_i)d\boldsymbol{\tau} \tag{6.21}$$

where $\pi(\boldsymbol{\tau}|\boldsymbol{y}_i^o, \boldsymbol{X}_i)$ is the observed-data posterior on which Bayesian inference for $\boldsymbol{\tau}$ is based assuming an appropriate prior $\pi(\boldsymbol{\tau})$.

By noting that the observed-data posterior is, in turn, related to the posterior predictive distribution by

$$\begin{aligned}\pi(\boldsymbol{\tau}|\boldsymbol{y}_i^o, \boldsymbol{X}_i) &= \int \pi(\boldsymbol{\tau}|\boldsymbol{y}_i^o, \boldsymbol{y}_i^m, \boldsymbol{X}_i)\pi(\boldsymbol{y}_i^m|\boldsymbol{y}_i^o, \boldsymbol{X}_i)d\boldsymbol{y}_i^m \\ &= E_{y^m|y^o}\{\pi(\boldsymbol{\tau}|\boldsymbol{y}_i^o, \boldsymbol{y}_i^m, \boldsymbol{X}_i)\},\end{aligned}$$

Tanner and Wong (1987) describe a data augmentation algorithm which can be used to calculate the observed-data posterior distribution. In turn, this can be used create multiple imputations through the generation of predicted values of $\boldsymbol{y}_i^m$ from the posterior predictive distribution (6.21) based on Markov chain Monte Carlo simulation. Below is the basic data augmentation algorithm as set forth by Schafer (1997, §3.4.2).

A data augmentation algorithm for imputing missing values

Step 1(Imputation I-step): Given a current estimate $\widehat{\boldsymbol{\tau}}^{(k)}$ of the parameters, first simulate a draw from the conditional predictive distribution of $\boldsymbol{y}_i^m$ given $\boldsymbol{y}_i^o$ and $\widehat{\boldsymbol{\tau}}^{(k)}$,

$$\boldsymbol{y}_i^{m^{(k+1)}} \sim \pi(\boldsymbol{y}_i^m|\boldsymbol{y}_i^o, \boldsymbol{X}_i;\ \widehat{\boldsymbol{\tau}}^{(k)}).$$

Step 2(Posterior P-step): Given a complete sample $(\boldsymbol{y}_i^o, \boldsymbol{y}_i^{m^{(k+1)}})$, take a random draw from the complete-data posterior

$$\widehat{\boldsymbol{\tau}}^{(k+1)} \sim \pi(\boldsymbol{\tau}|\boldsymbol{y}_i^o, \boldsymbol{y}_i^{m^{(k+1)}}, \boldsymbol{X}_i).$$

By repeating these two steps from a starting value $\widehat{\boldsymbol{\tau}}^{(0)}$, one will have created a Markov chain, $\{\widehat{\boldsymbol{\tau}}^{(k)}, \boldsymbol{y}_i^{m^{(k)}}; k = 1, 2, \ldots\}$ whose stationary distribution is $\pi(\boldsymbol{\tau}, \boldsymbol{y}_i^m|\boldsymbol{y}_i^o, \boldsymbol{X}_i)$ and whose subsequences, $\{\widehat{\boldsymbol{\tau}}^{(k)}; k = 1, 2, \ldots\}$ and $\{\boldsymbol{y}_i^{m^{(k)}}; k = 1, 2, \ldots\}$ have $\pi(\boldsymbol{\tau}|\boldsymbol{y}_i^o, \boldsymbol{X}_i)$ and $\pi(\boldsymbol{y}_i^m|\boldsymbol{y}_i^o, \boldsymbol{X}_i)$ as their stationary distributions (Schafer, 1997). This is essentially the Markov chain Monte Carlo (MCMC) method used in the SAS procedure PROC MI for performing multiple imputation for arbitrary missing data assuming $\boldsymbol{y}_i = (\boldsymbol{y}_i^o, \boldsymbol{y}_i^m)$ has a multivariate normal distribution.

Alternative approaches to multiple imputation have been suggested. In many cases, these require one to specify both an *analysis model*, $\pi(\boldsymbol{y}_i|\boldsymbol{X}_i;\boldsymbol{\tau})$, for the intended complete data and an *imputation model* that is used to impute missing values. The imputation model, written here as $\pi(\boldsymbol{y}_i|\boldsymbol{X}_i, \boldsymbol{U}_i, \boldsymbol{\phi})$, may and probably should include auxiliary information from the matrix $\boldsymbol{U}_i$ defined in §6.2 particularly when that information is thought to be related to the imputed response variable and/or to the missingness of the imputed response variable (Schafer 1997, pp. 143; see also Molenberghs and Kenward, 2007; National Research Council, 2010 and references therein). However, we must then consider whether the analysis model and imputation model are compatible. Ideally, we would choose an imputation model such that when the auxiliary variables in $\boldsymbol{U}_i$ are integrated out of the imputation model, we end up with the analysis model. Other discrepancies may exist between the analysis model and imputation model that may or may not impact inference by multiple imputation (e.g., Schafer, 1997, §4.5.4). For example, if the imputation model is based on a broader set of assumptions compared to the analysis model, inference under the analysis model may still be valid provided its narrower set of assumptions are met.

Once a set of m complete datasets or imputations have been created using PROC MI, we proceed to fit the analysis model to each imputed dataset using a BY statement together with whatever SAS procedure is required to fit the analysis model of interest. We then use PROC MIANALYZE to combine results across the m imputations in order to base inference using Bayesian principles as outlined by Rubin (1987, 1996).

To illustrate, let $\xi = \xi(\boldsymbol{\tau})$ be a scalar estimand of interest which is a function of the parameters of the complete-data analysis model $\pi(\boldsymbol{y}_i|\boldsymbol{X}_i;\boldsymbol{\tau})$. Let $(\boldsymbol{Y}^{(k)}, \boldsymbol{X}) = (\boldsymbol{Y}_{\text{obs}}, \boldsymbol{Y}^{(k)}_{\text{miss}}, \boldsymbol{X})$ be the kth imputed dataset with respect to the intended response vectors $\boldsymbol{y}_i = (\boldsymbol{y}_i^o, \boldsymbol{y}_i^m)$ and covariates $\boldsymbol{X}_i$ and let

$$\widehat{\xi}^{(k)} = \xi(\widehat{\boldsymbol{\tau}}^{(k)}) \text{ and } \widehat{\omega}^{(k)} = Var(\xi(\widehat{\boldsymbol{\tau}}^{(k)}))$$

be the complete-data point estimate of ξ and its associated variance estimate. Following Rubin (1987, 1996) and Schafer (1997), the repeated-imputation estimate of ξ is simply the average of the m complete-data estimates

$$\bar{\xi} = \frac{1}{m}\sum_{k=1}^{m}\widehat{\xi}^{(k)}.$$

If we let $\bar{\omega} = \frac{1}{m}\sum_{k=1}^{m}\widehat{\omega}^{(k)}$ be within-imputation variability based on the complete-data variance estimates $\widehat{\omega}^{(k)}$ and let $\omega_b = \frac{1}{m-1}\sum_{k=1}^{m}(\widehat{\xi}^{(k)} - \bar{\xi})^2$ be between-imputation variability of the complete-data point estimates, then the total variance associated with the repeated-imputation estimate $\bar{\xi}$ is

$$\Omega = \bar{\omega} + (1 + m^{-1})\omega_b.$$

Inference about ξ can then be based on the approximation

$$\Omega^{-1/2}(\xi - \bar{\xi}) \sim t(v)$$

where the Student-t degrees of freedom v is given by

$$v = (m-1)\Big[1 + \frac{\bar{\omega}}{(1+m^{-1})\omega_b}\Big]^2$$

(Rubin, 1987; Schafer, 1997). As noted by Rubin (1987, 1996), the derivation of these expressions follows from a Bayesian perspective in which one treats ξ and $\widehat{\xi}$ as unobserved random variables with normal conditional distributions given the observed values $\{\widehat{\xi}^{(1)}, \ldots, \widehat{\xi}^{(m)}\}$ and $\{\widehat{\omega}^{(1)}, \ldots, \widehat{\omega}^{(m)}\}$.

In the end, the goal is to carry out valid repeated-imputation inference about the unknown parameter vector $\boldsymbol{\tau}$ or some estimand of interest, $\xi = \xi(\boldsymbol{\tau})$, that is a based on a correctly specified complete-data model $\pi(\boldsymbol{y}_i|\boldsymbol{X}_i;\boldsymbol{\tau})$. A comprehensive treatment of

Table 6.2 Multiple Imputation Methods in PROC MI

Missing Data Pattern	Type of Variable To Be Imputed	Recommended Method of Multiple Imputation (MI)
Monotone	Continuous	Regression Predicted Mean Matching Propensity Score
Monotone	Categorical (Ordinal)	Logistic Regression
Monotone	Categorical (Nominal)	Logistic Regression (Binary) Discriminant Function Method
Arbitrary	Continuous	MCMC Full-data MI MCMC Monotone MI

multiple imputation methods including the theory of repeated-imputation inference may be found in Rubin (1987, 1996). Other key sources of information on multiple imputation can be found in Schafer (1997, 1999), Molenberghs and Verbeke (2005), Molenberghs and Kenward (2007) and references therein. Key salient points of MI are also covered in the SAS documentation of MI and MIANALYZE.

Since choosing an appropriate imputation technique is probably the most important of the three steps for a valid MI-based analysis, we briefly describe some of options PROC MI offers for imputing missing values. Our focus here is on imputing missing values of the response variable assuming all covariates are fully observed (i.e., there are no missing covariate values) though one can certainly use MI to impute missing covariates as well. The MI procedure in SAS offers several imputation techniques depending on the pattern of missing data (monotone versus non-monotone). For monotone missing data, one can impute missing values using a parametric regression method or a closely related predictive mean matching method for continuous variables, a nonparametric propensity score method for continuous variables, a logistic method for binary or ordinal variables and a discriminant function method for nominal categorical variables. For arbitrary non-monotone patterns of missing data, PROC MI offers a Monte Carlo Markov Chain (MCMC) method for imputing missing values for continuous variables assuming a multivariate normal distribution. One can also use MCMC to impute only intermittent missing values leaving one with a dataset having strictly monotone missing data patterns. This is useful when analyzing data with both intermittent missing values and missing values due to dropout. A summary of the various methods available in PROC MI is shown in Table 6.2 (taken from the SAS documentation for PROC MI).

Given the scope of this book deals with correlated response data, i.e., response data that are essentially multivariate in nature, the use of the MCMC method for imputing missing response values is probably the method of choice. This is particularly true for repeated measurements and clustered data applications involving a continuous response variable. Even when the response variable is not continuous, one might be able to use the MCMC method in combination with the TRANSFORM statement of PROC MI to achieve approximate normality and thereby create multiple imputations. The reader is referred to Schafer (1997, §5.1) for further discussion on the impact that deviations from normality may have on multiple imputation. A fairly simple and straightforward example illustrating the use of multiple imputation based on the MCMC method is presented in §6.4.4 while in §6.6 we apply multiple imputation to the analysis of the bone mineral density data of Chapter 2.

6.4.3 Inverse probability of weighting (IPW)

In the case of monotone missing data due to dropout, Robins et. al. (1995) show how one can conduct valid semiparametric inference using weighted GEE under the S-MAR mechanism (6.18). Given the data structure as outlined in §6.2 and assuming the marginal mean of the complete vector of repeated outcome measurements follows the semiparametric marginal GLIM (4.8) defined in §4.2, the GEE estimator $\widehat{\boldsymbol{\beta}}$ of $\boldsymbol{\beta}$ for the complete data is that solution to the GEE

$$U_{\boldsymbol{\beta}}(\boldsymbol{\beta},\widehat{\boldsymbol{\theta}}(\boldsymbol{\beta})) = \sum_{i=1}^{n} U_i(\boldsymbol{\beta},\widehat{\boldsymbol{\theta}}(\boldsymbol{\beta})) = \sum_{i=1}^{n} \boldsymbol{D}_i'\boldsymbol{\Sigma}_i(\boldsymbol{\beta},\widehat{\boldsymbol{\theta}}(\boldsymbol{\beta}))^{-1}(\boldsymbol{y}_i - \boldsymbol{\mu}_i(\boldsymbol{\beta})) = \mathbf{0} \tag{6.22}$$

where $E(\boldsymbol{y}_i|\boldsymbol{X}_i) = \boldsymbol{\mu}_i(\boldsymbol{\beta}) = \boldsymbol{g}^{-1}(\boldsymbol{X}_i\boldsymbol{\beta}), \boldsymbol{D}_i = \boldsymbol{D}_i(\boldsymbol{\beta}) = \frac{\partial \boldsymbol{\mu}_i(\boldsymbol{\beta})}{\partial \boldsymbol{\beta}'}$, $\boldsymbol{\Sigma}_i(\boldsymbol{\beta},\widehat{\boldsymbol{\theta}}(\boldsymbol{\beta})) = \boldsymbol{V}_i(\boldsymbol{\mu}_i)^{1/2}\boldsymbol{\Sigma}_i^*(\widehat{\boldsymbol{\theta}}(\boldsymbol{\beta}))\boldsymbol{V}_i(\boldsymbol{\mu}_i)^{1/2}$, $\boldsymbol{V}_i(\boldsymbol{\mu}_i)$ is the diagonal variance function matrix and $\widehat{\boldsymbol{\theta}}(\boldsymbol{\beta})$ is any consistent estimator of the parameter $\boldsymbol{\theta}$ assumed in the working covariance matrix $\boldsymbol{\Sigma}_i(\boldsymbol{\theta})$. Based on (6.22), the generalized estimating equation for the observed data may be written as

$$\begin{aligned} U^o_{\boldsymbol{\beta}}(\boldsymbol{\beta},\widehat{\boldsymbol{\theta}}(\boldsymbol{\beta})) &= \sum_{i=1}^{n} U_i^o(\boldsymbol{\beta},\widehat{\boldsymbol{\theta}}(\boldsymbol{\beta})) \\ &= \sum_{i=1}^{n} \boldsymbol{D}_i'\boldsymbol{\Sigma}_i(\boldsymbol{\beta},\widehat{\boldsymbol{\theta}}(\boldsymbol{\beta}))^{-1}\boldsymbol{M}_i(\boldsymbol{y}_i - \boldsymbol{\mu}_i(\boldsymbol{\beta})) \\ &= \mathbf{0} \end{aligned} \tag{6.23}$$

where $\boldsymbol{M}_i = Diag\{r_{i1}, r_{i2}, \ldots, r_{ip}\}$ is the $p \times p$ diagonal matrix of response/non-response indicator variables indicating the monotone pattern of missing data for the ith subject (i.e., $\boldsymbol{M}_i\boldsymbol{y}_i = \boldsymbol{y}_i^o$, etc.). When the data are MCAR such that $\boldsymbol{M}_i \perp \boldsymbol{y}_i|\boldsymbol{X}_i$, the GEE defined by (6.23) will be unbiased estimating equations for $\boldsymbol{\beta}$ in that $E\{\boldsymbol{D}_i'\boldsymbol{\Sigma}_i(\boldsymbol{\beta},\widehat{\boldsymbol{\theta}}(\boldsymbol{\beta}))^{-1}\boldsymbol{M}_i(\boldsymbol{y}_i - \boldsymbol{\mu}_i(\boldsymbol{\beta}))\} = \mathbf{0}$ and their solution $\widehat{\boldsymbol{\beta}}$ will be $\sqrt{n}$–consistent for $\boldsymbol{\beta}$. However, this is not true when data are MAR since $\boldsymbol{M}_i$ is no longer independent of $\boldsymbol{y}_i$ given $\boldsymbol{X}_i$.

Under S-MAR, Robins et. al. (1995) showed that when the discrete hazard function $\lambda_i(t_k|\boldsymbol{V}_{i(k-1)}, \boldsymbol{y}_{i(k-1)})$ in (6.18) is bounded away from 1 and can be formulated in terms of a known hazard function $\lambda(t_k|\boldsymbol{V}_{i(k-1)}, \boldsymbol{y}_{i(k-1)}; \boldsymbol{\eta})$ of $\boldsymbol{V}_{i(k-1)}$, $\boldsymbol{y}_{i(k-1)}$ and parameter $\boldsymbol{\eta}$, such that

$$\lambda_i(t_k|\boldsymbol{V}_{i(k-1)}, \boldsymbol{y}_{i(k-1)}) = \lambda(t_k|\boldsymbol{V}_{i(k-1)}, \boldsymbol{y}_{i(k-1)}; \boldsymbol{\eta}), \tag{6.24}$$

then a consistent estimate of $\boldsymbol{\beta}$ can be obtained by solving the weighted GEE

$$\begin{aligned} U^w_{\boldsymbol{\beta}}(\boldsymbol{\beta},\widehat{\boldsymbol{\theta}}(\boldsymbol{\beta}),\widehat{\boldsymbol{\eta}}) &= \sum_{i=1}^{n} U_i^w(\boldsymbol{\beta},\widehat{\boldsymbol{\theta}}(\boldsymbol{\beta}),\widehat{\boldsymbol{\eta}}) \\ &= \sum_{i=1}^{n} \boldsymbol{D}_i'\boldsymbol{\Sigma}_i(\boldsymbol{\beta},\widehat{\boldsymbol{\theta}}(\boldsymbol{\beta}))^{-1}\boldsymbol{M}_i(\widehat{\boldsymbol{\eta}})(\boldsymbol{y}_i - \boldsymbol{\mu}_i(\boldsymbol{\beta})) \\ &= \mathbf{0} \end{aligned} \tag{6.25}$$

where $\widehat{\boldsymbol{\eta}}$ is a consistent estimator of $\boldsymbol{\eta}$ under a correctly specified hazard function in (6.24), and where

$$\begin{aligned} \boldsymbol{M}_i(\boldsymbol{\eta}) &= Diag\{r_{i1}/\pi_{i1}(\boldsymbol{\eta}), r_{i2}/\pi_{i2}(\boldsymbol{\eta}), \ldots, r_{ip}/\pi_{ip}(\boldsymbol{\eta})\} \\ &= Diag\{1, r_{i2}/\pi_{i2}(\boldsymbol{\eta}), \ldots, r_{ip}/\pi_{ip}(\boldsymbol{\eta})\} \end{aligned}$$

is an inverse probability of weighting (IPW) matrix formulated from the probability of remaining uncensored (i.e., free from dropout) at time t_k given the prior history, $\boldsymbol{V}_{i(k-1)}$ and $\boldsymbol{y}_{i(k-1)}$, i.e.,

$$\pi_{ik}(\boldsymbol{\eta}) = \Pr[r_{ik} = 1|\boldsymbol{V}_{i(k-1)}, \boldsymbol{y}_{i(k-1)}; \boldsymbol{\eta}] = \prod_{j=1}^{k}\{1 - \lambda(t_j|\boldsymbol{V}_{i(j-1)}, \boldsymbol{y}_{i(j-1)}; \boldsymbol{\eta})\}. \tag{6.26}$$

By convention we set $\lambda(t_1|\boldsymbol{V}_{i0}, \boldsymbol{y}_{i0}; \boldsymbol{\eta}) \equiv 0$ so that $\pi_{i1}(\boldsymbol{\eta}) = 1$ is consistent with our assumption that y_{i1} is observed for all i. Let $\widehat{\boldsymbol{\beta}}$ be a solution to the weighted GEE (6.25). Robins et. al. proved that $\sqrt{n}(\widehat{\boldsymbol{\beta}} - \boldsymbol{\beta})$ will be asymptotically normally distributed as $N(\boldsymbol{0}, \boldsymbol{\Gamma}^{-1}\boldsymbol{\Omega}\boldsymbol{\Gamma}^{-1'})$ where

$$\begin{aligned}
\boldsymbol{\Gamma} &= E\{\partial U_i^w(\boldsymbol{\beta}, \widehat{\boldsymbol{\theta}}(\boldsymbol{\beta}), \boldsymbol{\eta})/\partial\boldsymbol{\beta}'\}, \\
\boldsymbol{\Omega} &= \boldsymbol{\Omega}_R - \boldsymbol{B}\boldsymbol{\Omega}_\eta\boldsymbol{B}' \\
\boldsymbol{\Omega}_R &= E\{U_i^w(\boldsymbol{\beta}, \widehat{\boldsymbol{\theta}}(\boldsymbol{\beta}), \boldsymbol{\eta})U_i^w(\boldsymbol{\beta}, \widehat{\boldsymbol{\theta}}(\boldsymbol{\beta}), \boldsymbol{\eta})'\}, \\
\boldsymbol{B} &= E\{\partial U_i^w(\boldsymbol{\beta}, \widehat{\boldsymbol{\theta}}(\boldsymbol{\beta}), \boldsymbol{\eta})/\partial\boldsymbol{\eta}'\}
\end{aligned}$$

and $\boldsymbol{\Omega}_\eta$ is the asymptotic variance of $\sqrt{n}(\widehat{\boldsymbol{\eta}} - \boldsymbol{\eta})$ (Robins et. al., 1995, Theorem 1). Inference can be carried out using standard error estimates computed analytically using the above formulas or through the use of bootstrap techniques.

Note that the above *interpretation* of $\pi_{ik}(\boldsymbol{\eta})$ in (6.26) as a discrete survivor function, i.e., as the probability of remaining uncensored at time t_k given the prior history $\boldsymbol{V}_{i(k-1)}$ and $\boldsymbol{y}_{i(k-1)}$, assumes there are no internal time-dependent covariates in the conditioning arguments to the hazard function (6.17) or (6.18). This assumption is only possible if, under (6.17) or (6.18), the matrix of auxiliary variables, $\boldsymbol{U}_i$, does not contain any internal time-dependent covariates. This occurs when one replaces the S-MAR condition (6.18) with the stronger condition

$$\begin{aligned}
\lambda_i(t_k|\boldsymbol{\mathcal{Z}}_i(T)) = \lambda_i(t_k|\boldsymbol{V}_i, \boldsymbol{y}_i) &= \Pr[T_i \in [t_{k-1}, t_k)|T_i \geq t_{k-1}, C_i \geq t_{k-1}, \boldsymbol{V}_i, \boldsymbol{y}_i] \\
&= \Pr[T_i \in [t_{k-1}, t_k)|T_i \geq t_{k-1}, \boldsymbol{V}_{i(k-1)}, \boldsymbol{y}_{i(k-1)}] \\
&= \lambda_i(t_k|\boldsymbol{V}_{i(k-1)}, \boldsymbol{y}_{i(k-1)}) \\
&= \lambda_i(t_k|\boldsymbol{\mathcal{Z}}_i^o(t_k))
\end{aligned} \tag{6.27}$$

where the hazard at time t_k conditional on the entire intended history $\boldsymbol{\mathcal{Z}}_i(T) = \{\boldsymbol{V}_i, \boldsymbol{y}_i\}$ is independent of present and future values in both $\boldsymbol{V}_i$ and $\boldsymbol{y}_i$. The stronger condition given by (6.27) is equivalent to the condition specified by equation (3) of Robins et. al. (1995). It is important to note, however, that (6.27) need not hold in order for one to define and use $\pi_{ik}(\widehat{\boldsymbol{\eta}})$ in (6.25) since the solution $\widehat{\boldsymbol{\beta}}$ to the IPW GEE (6.25) will still produce a consistent and asymptotically normal estimator of the estimand $\boldsymbol{\beta}$ provided (6.18), or equivalently (2a) of Robins et. al. (1995), holds (see the discussion immediately preceding Theorem 1 of Robins et. al., 1995).

To implement the IPW approach of Robins et. al. (1995) in SAS, one would first fit an appropriate discrete time failure model for dropout based on some semiparametric specification of the discrete hazard function as indicated in (6.24). Possible choices include modeling the hazard function using a binary regression model with either a complementary log-log link function or a logit link function. The former corresponds to a discrete time proportional hazards model (or grouped relative risk model) and the latter to the discrete time logistic hazards model (or proportional odds model) due to Cox (e.g., Kalbfleisch and Prentice, 2002, §4.8). For example, let $\mathbb{Z}_i(t_k)$ be a vector of covariates formed from the measurement history $\mathcal{Z}_i^o(t_{k+1}) = \{\boldsymbol{v}_i(t_j), y_i(t_j); t_j \in [0, t_{k+1}), j = 1, \ldots, k\} = \{\boldsymbol{V}_{ik}, \boldsymbol{y}_{ik}\}$ up to an including time t_k. Under the discrete time proportional hazards model with baseline hazard rate

$$\lambda_{0k} = \lambda_0(t_k) = \Pr[T_i \in [t_{k-1}, t_k)|T_i \geq t_{k-1}],$$

the hazard function would be modeled as

$$\begin{aligned}\lambda_i(t_k|\mathbb{Z}_i(t_{k-1})) &= 1 - (1 - \lambda_{0k})^{\exp\{\mathbb{Z}_i(t_{k-1})'\boldsymbol{\eta}_1\}} \\ &= 1 - \exp\{-\exp(\eta_{0k} + \mathbb{Z}_i(t_{k-1})'\boldsymbol{\eta}_1)\}\end{aligned}$$

where $\eta_{0k} = \log[-\log(1 - \lambda_{0k})]$. Conversely, under the discrete time logistic hazards model, the hazard function satisfies

$$\frac{\lambda_i(t_k|\mathbb{Z}_i(t_{k-1}))}{1 - \lambda_i(t_k|\mathbb{Z}_i(t_{k-1}))} = \frac{\lambda_0(t_k)}{1 - \lambda_0(t_k)} \exp\{\mathbb{Z}_i(t_{k-1})'\boldsymbol{\eta}_1)$$

where $\lambda_i(t_k|\mathbb{Z}_i(t_{k-1}))$ and $\lambda_0(t_k)$ are individual and baseline hazards at t_k with $\text{logit}(\lambda_i(t_k|\mathbb{Z}_i(t_{k-1}))) = \eta_{0k} + \mathbb{Z}_i(t_{k-1})'\boldsymbol{\eta}_1$ and η_{0k} =logit(λ_{0k}) is the logit of the baseline hazard $\lambda_0(t_k) = \lambda_{0k}$. One can fit a discrete time failure model under either of the two discrete hazard functions using GENMOD or GLIMMIX. In either case, one would partition the parameter vector as $\boldsymbol{\eta} = (\boldsymbol{\eta}_0, \boldsymbol{\eta}_1)$ where $\boldsymbol{\eta}_0$ is the vector of baseline parameters, η_{0k}.

Having fit a discrete time failure model to estimate $\boldsymbol{\eta}$, one would then compute the diagonal elements of the IPW matrix $\boldsymbol{M}_i(\boldsymbol{\eta})$ using the estimated parameter $\widehat{\boldsymbol{\eta}}$ to form the necessary scalar weights for the weighted GEE. One can then carry out an IPW analysis using GENMOD, GLIMMIX or NLMIXED depending on the modeling assumptions one wishes to make (i.e., one could use either GENMOD or GLIMMIX for GEE-based analyses and either GLIMMIX or NLMIXED for GEE2-based analyses). To complete the analysis, one would compute standard error estimates using the above analytical formulas or by using bootstrap techniques.

In summary, valid inference based on a semiparametric GEE approach using IPW is possible whenever the following key assumptions are met:

1) the complete-data mean $E(\boldsymbol{y}_i|\boldsymbol{X}_i) = \boldsymbol{\mu}_i(\boldsymbol{\beta}) = \boldsymbol{g}^{-1}(\boldsymbol{X}_i\boldsymbol{\beta})$ is correctly specified
2) missing data due to dropout are S-MAR in accordance with (6.18) and the hazard function $\lambda_i(t_k|\boldsymbol{V}_{i(k-1)}, \boldsymbol{y}_{i(k-1)})$ in (6.18) is bounded away from 1
3) the known hazard function $\lambda(t_k|\boldsymbol{V}_{i(k-1)}, \boldsymbol{y}_{i(k-1)}; \boldsymbol{\eta})$ in (6.24) is correctly specified, and
4) consistent estimates of $\widehat{\boldsymbol{\theta}}(\boldsymbol{\beta})$ and $\widehat{\boldsymbol{\eta}}$ are available.

Of course one need not restrict the IPW analysis to a marginal GLIM. One could just as easily apply the estimated weights to a GNLM using PL/QELS or GEE2 although valid inference for the latter would require correct specification of both the marginal mean and variance-covariance matrix.

The advantage of the IPW approach to the analysis of longitudinal data is that one need not specify a fully parametric likelihood function in order to guarantee valid inference under S-MAR. As such, IPW allows one to more easily accommodate auxiliary information into one's analysis including the use of internal time-dependent auxiliary variables in calculating the weights although, as noted above, $\pi_{ik}(\boldsymbol{\eta})$ in (6.26) can no longer be interpreted as the probability that subject i remains free from dropout through time t_k (see the discussion immediately preceding Theorem 1 of Robins et. al., 1995). In contrast, when incorporating auxiliary variables into a likelihood-based analysis, one would have to specify a joint distribution with respect to $(\boldsymbol{y}_i, \boldsymbol{r}_i, \boldsymbol{U}_i)$ and then integrate over the auxiliary variables within $\boldsymbol{U}_i$ to achieve valid inference (e.g., Hogan et. al., 2004; National Research Council, 2010). While feasible theoretically, such an approach would be unrealistic given both the enormous programming challenges one would have to overcome as well the additional parametric assumptions one would have to make. Even in the absence of auxiliary information, a semiparametric regression model based on a marginal mean structure makes it possible to carry out valid inference under S-MAR when exact likelihood functions for discrete or continuous outcomes may be difficult if not impossible to specify.

There are, of course, disadvantages with IPW-based analyses. One does need to correctly specify the hazard model to ensure valid inference using weighted GEE although augmented IPW estimation with its double robustness property does offer some potential protection (e.g., van der Laan and Robins, 2003). There is also the potential for numerical instability with weighted GEE as a result of using unstable weights caused by values of $\pi_{ik}(\boldsymbol{\eta})$ near 0. Another disadvantage is that IPW GEE seems best suited to monotone missing data due to dropout. Although extensions to arbitrary missing data patterns are discussed by Robins et. al. (1995), these may not be as straightforward to implement in practice. Moreover, even with monotone missing data, the use of IPW GEE is not particularly well suited to a continuous time-to-dropout setting such as described in §6.3.1 particularly when the longitudinal response data are irregularly spaced over time.

In this and the previous two sections, we have provided a brief overview of methods commonly used to analyze longitudinal data under MAR. We next illustrate how one can apply each of these methods to a simple repeated measures analysis of covariance using the available tools within SAS.

6.4.4 Example: A repeated measures ANCOVA

Consider the simple repeated measurements data shown in Table 6.3. This balanced complete data are from a hypothetical double-blind randomized controlled trial in which a continuous outcome measure Y is to be compared between an active drug and a placebo control at the conclusion of a four week follow-up period. Measurements at baseline (week 0) and end of study (week 4) were randomly generated assuming $\boldsymbol{y}_{ik} = \begin{pmatrix} y_{ik0} \\ y_{ik1} \end{pmatrix} \sim N_2(\boldsymbol{\mu}_k, \boldsymbol{\Sigma})$, $i = 1, \ldots, 10$; $k = 1, 2$ where y_{ik0} is the baseline response at week 0 for the ith subject in the kth treatment arm, y_{ik1} is that subject's response at week 4, $\boldsymbol{\mu}_1 = \begin{pmatrix}\mu_{10} & \mu_{11}\end{pmatrix}' = \begin{pmatrix}5.0 & 5.0\end{pmatrix}'$ is the true population mean response vector for individuals receiving the placebo control, $\boldsymbol{\mu}_2 = \begin{pmatrix}\mu_{20} & \mu_{21}\end{pmatrix}' = \begin{pmatrix}5.0 & 5.5\end{pmatrix}'$ is the true population mean response vector for those receiving the active drug, and $\boldsymbol{\Sigma}$ is the variance-covariance matrix with a common variance of 0.25 and a correlation of 0.50 between the two repeated measurements.

The goal of the study is to determine if there is a significant increase (improvement) in the mean response among subjects randomized to the active drug relative to the placebo control. This can be accomplished by fitting the change from baseline analysis of covariance model

$$y_{ik} = \beta_0 + \beta_d x_{ik} + \beta y_{ik0} + \epsilon_{ik};\ i = 1, \ldots, 10;\ k = 1, 2$$

to the data where $y_{ik} = y_{ik1} - y_{ik0}$ is the change from baseline, y_{ik0} is the baseline covariate, x_{ik} is a treatment group indicator with $x_{i1} = 0$ for subjects with placebo control and $x_{i2} = 1$ for subjects with active drug. The estimand of interest is β_d which is the treatment effect at week 4. It measures the difference in the mean change from baseline between active drug and placebo.

The data in Table 6.3 reflect the complete data one would observe if everyone completed the study. However, in the last two columns of Table 6.3, we have included two scenarios in

Table 6.3 A simulated dataset representing repeated measurements of an outcome measure, Y, measured at week 0 (Y0) and week 4 (Y1) for subjects randomized to receive a placebo control (Placebo) or active drug (Drug).

Complete Data				Missing Data Cases	
Subject	Trt	Y0 (Week 0)	Y1 (Week 4)	Y1-MAR	Y1-MNAR
1	Placebo	5.36	4.64		
2	Placebo	5.10	4.70		
3	Placebo	5.49	5.42		
4	Placebo	6.04	5.45	x	
5	Placebo	5.36	5.65		
6	Placebo	5.07	4.55		
7	Placebo	4.99	4.52		
8	Placebo	4.61	5.53		
9	Placebo	5.67	5.82	x	
10	Placebo	5.45	5.36		
1	Drug	5.35	5.62		
2	Drug	5.39	5.07		
3	Drug	4.28	5.23		
4	Drug	5.48	5.09		
5	Drug	5.51	6.83		xx
6	Drug	4.23	5.37		xx
7	Drug	5.33	5.72		
8	Drug	4.50	4.67		
9	Drug	5.90	6.24	x	
10	Drug	5.37	6.21		

which missing values in the response occur as a result of subjects failing to complete the study. In the first scenario, we consider missing values at week 4 (marked by a single x) under a strictly MAR mechanism in which dropout occurs if a subject's observed baseline response exceeds a value of 5.6 (i.e., whenever Y0$>$ 5.6). This could reflect a situation where there is a baseline threshold value in the response beyond which subjects feel "well enough" that they no longer wish to continue in the study. In the second scenario, we consider missing values at week 4 (marked with an xx) under a MNAR mechanism in which dropout at week 4 occurs whenever the response at week 4 exceeds the baseline response by 1.00 (i.e., whenever Y1$-$Y0$>$ 1.00). In this case, the MNAR mechanism coincides with a greater than 2 standard deviation increase over the individual's baseline response. This reflects an alternative threshold effect wherein a favorable but unobserved post-baseline response to treatment explains why some subjects fail to show up for their final visit.

In this section, we use PROC MIXED to fit the simple repeated measures ANCOVA model to the complete data as well as to the incomplete data generated under the MAR mechanism described above. Our goal is to illustrate how one can use SAS to fit a repeated measures ANCOVA model to incomplete data under the assumption of MAR when the estimand of interest, β_d, is estimated using

1) likelihood-based methods based on the incomplete data alone (§6.4.1),
2) methods based on multiple imputation (MI, §6.4.2),
3) methods based on inverse probability of weighting (IPW, §6.4.3).

The following programming statements fit the simple repeated measures ANCOVA model to the complete data from Table 6.3.

Program 6.1

```
data RM_ANCOVA;
input Subject Trt $ Y0 Y1 @@;
cards;
1  Placebo 5.36 4.64 2  Placebo 5.10 4.70 3  Placebo 5.49 5.42
4  Placebo 6.04 5.45 5  Placebo 5.36 5.65 6  Placebo 5.07 4.55
7  Placebo 4.99 4.52 8  Placebo 4.61 5.53 9  Placebo 5.67 5.82
10 Placebo 5.45 5.36 1  Drug    5.35 5.62 2  Drug    5.39 5.07
3  Drug    4.28 5.23 4  Drug    5.48 5.09 5  Drug    5.51 6.83
6  Drug    4.23 5.37 7  Drug    5.33 5.72 8  Drug    4.50 4.67
9  Drug    5.90 6.24 10 Drug    5.37 6.21
;
/* Center the baseline covariate */
%macro X0(data=_last_);
 proc means data=&data nway;
  where Yd>.;
  var Y0;
  output out=basemean mean=MeanY0;
 run;
 data &data;
  set &data;
  if _n_=1 then set basemean;
 run;
 data &data;
  set &data;
  X0 = Y0 - MeanY0; ** Centered covariate;
 run;
%mend X0;
/* Part 1: ANCOVA based on complete data */
data complete;
 length Type $8;
 set RM_ANCOVA;
 Type='Complete'; ** Type of data;
 Yd = Y1-Y0;      ** Change from baseline;
run;
%X0(data=complete);
proc sort data=complete;
 by Trt Subject;
run;
ods output Estimates=est_complete_REML;
ods select Estimates;
proc mixed data=complete;
 class Trt Subject;
 model Yd = Trt X0 /solution;
 estimate 'Trt Effect' Trt 1 -1;
run;
```

In the above code, we use the macro %X0 to center the baseline covariate Y0 around the observed mean resulting in the use of covariate X0 within the call to PROC MIXED. In this way, one can use ESTIMATE statements rather than LSMEANS statements to directly compute least squares means of interest if one chooses to fit a more complex model (e.g., a model that includes a Trt*Y0 interaction). We then store the estimated value of β_d (i.e., the `'Trt Effect'` estimate) and its standard error within the dataset, `est_complete`, created from the above ODS OUTPUT statement in conjunction with the ESTIMATE statement. We do this so that we can later summarize and compare the results from the complete-data analysis with analyses from the incomplete data generated under the MAR mechanism.

Next, we define an incomplete dataset based on whether the observed baseline response exceeds 5.6. In doing so, we define the response/non-response indicator variable r1 where r1=1 when Y1 is observed and r1=0 if Y1 and hence Yd is missing. We then fit the ANCOVA model to the incomplete data using the default "available-case" likelihood-based analysis of PROC MIXED. The programming statements are as follows.

Program 6.2

```
/* Part 2: ANCOVA based on incomplete MAR data */
data MAR;
 length Type $8;
 set RM_ANCOVA;
 Type='MAR     ';
 if Y0 > 5.6 then Y1=.;
 Yd = Y1-Y0;
 r1=1;  ** Response indicator variable at week 4;
 if Yd=. then r1=0;
run;
%X0(data=MAR);
proc sort data=MAR;
 by Trt Subject;
run;
ods output Estimates=est_MAR_REML;
ods select Estimates;
proc mixed data=MAR;
 class Trt Subject;
 model Yd = Trt X0 /solution;
 estimate 'Trt Effect' Trt 1 -1;
run;
```

Here, we store the likelihood-based REML estimate of β_d and its standard error within the dataset, `est_MAR_REML`.

With the following code, we next perform a repeated-imputation analysis to this incomplete data using PROC MI and PROC MIANALYZE. We do so using the default MCMC method within PROC MI to impute missing values assuming (correctly in this case) the repeated measurements, (y_{ik0}, y_{ik1}), are distributed as bivariate normal. We chose 15 imputations and stored the 15 imputed datasets into a master dataset, `MI_out`. We then fit the ANCOVA model to each of the 15 imputed complete-data sets using PROC MIXED together with a BY statement.

Program 6.3

```
/* Part 3: MI ANCOVA based on incomplete MAR data     */
/* Create datasets with imputed missing data          */
proc mi data=MAR seed=8957565 nimpute=15
        out=MI_out;
 by Trt;
 mcmc chain=multiple displayinit initial=em(itprint);
 var Y0 Y1;
run;
```

```
data MI_out;
 set MI_out;
 Yd = Y1-Y0;
 drop X0 MeanY0;
run;
%X0(data=MI_out);
proc sort data=MI_out;
 by _Imputation_ Trt Subject;
run;
ods output Estimates=est_MI;
ods select Estimates;
proc mixed data=MI_out;
 by _Imputation_;
 class Trt Subject;
 model Yd = Trt X0 /solution;
 estimate 'Trt Effect' Trt 1 -1;
run;
ods output ParameterEstimates=est_MAR_MI;
ods select ParameterEstimates;
proc mianalyze data=est_MI;
 modeleffects Estimate;
 stderr StdErr;
run;
```

We store the individual complete-data point estimates of β_d for the 15 imputed datasets in the dataset, `est_MI`, and apply PROC MIANALYZE to this dataset to obtain the multiple imputation point estimate of β_d and its associated variance estimate as described in §6.4.2. This is then saved in the dataset, `est_MAR_MI`.

Finally, we use inverse probability of weighting (IPW) to analyze the incomplete data. In the code shown below, we first use PROC LOGISTIC to predict the probability of completing the trial given a subject's treatment group (i.e., Pr(r1=1|Trt)). Ideally we would include both Trt and Y0 as covariates within the logistic regression model. However, with this small sample size and only three missing values, we get the following warning message

```
NOTE: PROC LOGISTIC is modeling the probability that r1=1.
WARNING: There is a complete separation of data points. The maximum
likelihood estimate does not exist.
```

when we attempt to fit the model with both Trt and Y0 included or when we only include Y0. We therefore fit the logistic model using Trt as the only predictor of missingness under MAR. Using the OUTPUT OUT= statement, we save the predicted probability of completing the trial, i.e., $\widehat{\pi}_{i1} = \Pr(r_{ik1} = 1 | r_{ik0} = 1, x_{ik})$, to the dataset, `prob_out`. We then compute the estimated inverse probability of weights, IPW = r1/pi1, as well as the true but "unknown" weights, IPW_true = r1/_pi1_, where _pi1_ is the true but "unknown" probability of completing the trial. Under the MAR mechanism, _pi1_ is given by

$$\Pr(r_{ik1} = 1 | r_{ik0} = 1, x_{ik}, y_{ik0}) = \Pr(y_{ik0} \le 5.6) = \Phi\Big(\frac{5.6 - 5.0}{0.5}\Big) = 0.8849$$

where $\Phi(\cdot)$ is the cumulative distribution function of the standard normal distribution. This yields a true weight of 0 for those who do not complete the trial (r1/_pi1_ = 0/0.8849 = 0) and 1.13 for those who do complete the trial (r1/_pi1_ = 1/0.8849 = 1.13). We then fit the IPW ANCOVA model to the incomplete data in PROC MIXED using both the estimated and true IPW values.

Program 6.4

```
/* Part 4: IPW ANCOVA based on incomplete MAR data     */
/* Compute Prob(r1=0) for IPW approach to missingness */
proc logistic data=MAR;
 class Trt;
 model r1(desc) = Trt / link=logit;
 output out=prob_out pred=pi1;
run;
data prob_out;
 set prob_out;
 IPW = r1/pi1;
 zscore = (5.6-5.0)/0.5;
 _pi1_ = probnorm(zscore); ** True Pr(Y0<=5.6)=Pr(r1=1|Y0);
 IPW_true = r1/_pi1_;
run;
ods output Estimates=est_MAR_IPW;
ods select Estimates;
proc mixed data=prob_out;
 class Trt Subject;
 model Yd = Trt X0 /solution;
 weight IPW;
 estimate 'Trt Effect' Trt 1 -1;
 title 'IPW estimate under MAR';
run;
ods output Estimates=est_MAR_IPW_true;
ods select Estimates;
proc mixed data=prob_out;
 class Trt Subject;
 model Yd = Trt X0 /solution;
 weight IPW_true;
 estimate 'Trt Effect' Trt 1 -1;
 title 'True IPW estimate under MAR';
run;
/* Compile results from the different methods & print */
data MARestimates;
set est_Complete(in=a) est_MAR_REML(in=b)
    est_MAR_MI(in=c)    est_MAR_IPW(in=d)
    est_MAR_IPW_true(in=e);
 if a then Method=''Complete Data ANCOVA          '';
 if b then Method=''Available Case ANCOVA         '';
 if c then Method=''MI-based ANCOVA               '';
 if d then Method=''IPW (Estimated Weights) ANCOVA'';
 if e then Method=''IPW (True Weights) ANCOVA     '';
run;
proc print data=MARestimates noobs split='*';
 var Method Estimate StdErr DF tValue Probt;
run;
```

Lastly, we compile results from each of the methods and print them out. The results are summarized in Table 6.4.

In comparing the different approaches, we find that in this simple setting, there is some loss in efficiency and power with both the available-case ANCOVA and the estimated IPW ANCOVA but less so with the MI-based ANCOVA. This is partially the result of having a

Table 6.4 Summary of different point estimates of treatment effect, β_d, measuring the difference in mean change from baseline between placebo control and active drug. Missing data are truly MAR. [1]The standard error estimate does not reflect the uncertainty associated with using estimated weights from the logistic regression model. An asymptotically valid standard error estimator can be computed analytically or by bootstrap techniques.

ANCOVA Method	β_d Estimate	Standard Error	DF	t Value	p-value
Complete-data	0.5485	0.2368	17	2.32	0.0333
Available-case	0.5551	0.2730	14	2.03	0.0614
MI-based	0.6286	0.2900	502	2.17	0.0307
IPW (estimated)[1]	0.5542	0.2712	14	2.04	0.0603
IPW (true)	0.5551	0.2730	14	2.03	0.0614

reduced number of subjects for analysis with the available-case ANCOVA and IPW ANCOVA although, in the case of the estimated IPW ANCOVA, the standard error estimate does not properly reflect the uncertainty associated with using estimated weights. We note that the IPW ANCOVA based on the true weights is equivalent to the available-case ANCOVA. This is because, in this very simple example, the correct weight among subjects who complete the trial is constant at 1.13 and a constant weight will not affect the parameter estimates or their standard errors. We also note the IPW ANCOVA using estimated weights yields a similar point estimate of β_d as achieved using the true weights. This, again, is because the true probability of completing the trial is constant and therefore independent of treatment group. Consequently, an unbiased estimate of the probability of completing the trial can be achieved by fitting an intercept-only logistic regression model or by fitting a model that includes treatment group as a possible predictor. In the former case, we end up with an estimated probability of $17/20 = 0.85$ while in the latter case we end up with an estimated probability of $8/10 = 0.80$ for subjects within the placebo control group and $9/10 = 0.90$ for subjects randomized to the active drug group. As each of these are unbiased estimates of the true probability, i.e. $\Pr(y_{iko} \leq 5.6) = 0.8849$, the analyst would end up with a consistent point estimator of β_d under the MAR mechanism. Of course the analyst would not be aware of the true underlying MAR mechanism. However, by performing both the available-case ANCOVA and the estimated IPW ANCOVA, the analyst would be reassured of the results under the assumption of MAR.

We next consider three commonly used analytical approaches for conducting sensitivity analyses as well as drawing possible inference from incomplete data when missing values are assumed to be MNAR. The three types of analytic models we consider are selection models, pattern mixture models, and shared parameter models which we briefly touched upon in §6.2.

6.5 Sensitivity analysis under MNAR

When missing data or dropout are ignorable, valid inference based on the observed data is possible using the analytical methods described and illustrated in §6.4. However, in many applications the assumption of MAR may not hold. Since we cannot formally test the assumption of MAR (or of MNAR), we would ideally supplement an MAR analysis with a sensitivity analysis under MNAR. This will allow the analyst to judiciously evaluate how robust results are to departures from MAR assumptions.

Unlike with an MAR-based analysis, an analysis under a MNAR mechanism requires one to model the joint distribution of the repeated outcome measurements, $\boldsymbol{y}_i$, and the response/non-response outcomes, $\boldsymbol{r}_i$. This becomes problematic in that one can only investigate the association between the two based on the observed data, $(\boldsymbol{y}_i^o, \boldsymbol{r}_i)$.

Consequently, specification of the joint distribution of the complete data $(\boldsymbol{y}_i, \boldsymbol{r}_i)$ requires one to make certain unverifiable assumptions in order to draw valid inference about the complete-data model on the basis of the observed data $(\boldsymbol{y}_i^o, \boldsymbol{r}_i)$. In particular, if we assume there are no auxiliary variables present such that $\boldsymbol{V}_i = (\boldsymbol{X}_i, \boldsymbol{Z}_i)$, we can write the complete-data model as

$$\begin{aligned}\pi(\boldsymbol{y}_i, \boldsymbol{r}_i|\boldsymbol{X}_i, \boldsymbol{Z}_i) &= \pi(\boldsymbol{y}_i^o, \boldsymbol{y}_i^m, \boldsymbol{r}_i|\boldsymbol{X}_i, \boldsymbol{Z}_i) \\ &= \pi(\boldsymbol{y}_i^o, \boldsymbol{r}_i|\boldsymbol{X}_i, \boldsymbol{Z}_i)\pi(\boldsymbol{y}_i^m|\boldsymbol{X}_i, \boldsymbol{Z}_i, \boldsymbol{y}_i^o, \boldsymbol{r}_i).\end{aligned} \tag{6.28}$$

From (6.28) it is clear that we can specify and test assumptions about the joint distribution of the observed data, $\pi(\boldsymbol{y}_i^o, \boldsymbol{r}_i|\boldsymbol{X}_i, \boldsymbol{Z}_i)$, using standard goodness-of-fit methods. However, we can never verify any assumptions we might make regarding the predictive distribution, $\pi(\boldsymbol{y}_i^m|\boldsymbol{X}_i, \boldsymbol{Z}_i, \boldsymbol{y}_i^o, \boldsymbol{r}_i)$, of the missing data conditional on the observed data since we never observe $\boldsymbol{y}_i^m$. This makes the methods of analysis under MNAR extremely model-dependent and subject to possible criticism without proper sensitivity analyses. In fact, this is precisely why it is impossible to conduct a definitive analysis under MNAR. We therefore focus our attention in this section on various modeling strategies one can apply when conducting a sensitivity analysis under the assumption that missing values are MNAR.

As briefly discussed in §6.2, there are three basic modeling approaches used when analyzing data under MNAR: 1) selection models, 2) pattern-mixture models and 3) shared parameter models. All three are derived by considering a suitable factorization of the joint distribution of the response data and missing data mechanism. We briefly consider all three approaches. To manage the level of complexity, we do so assuming either that $\boldsymbol{V}_i = (\boldsymbol{X}_i, \boldsymbol{Z}_i)$ or that $\boldsymbol{U}_i$ consists solely of external covariates (Kalbfleisch and Prentice, 2002, §6.3) that are independent of $\boldsymbol{y}_i$.

6.5.1 Selection models

Selection models may be classified as *marginal selection models* and *random-effects selection models* according to whether random effects are present. The marginal selection model entails modeling the joint distribution of $(\boldsymbol{y}_i, \boldsymbol{r}_i)$ while the random-effects selection model entails modeling the joint distribution of $(\boldsymbol{y}_i, \boldsymbol{r}_i, \boldsymbol{b}_i)$. In both cases, selection models are formed by factoring the joint distribution in a manner that separates a "selection model" or "missing data model" that characterizes the probability of missing values as a function of covariates and responses from a "response model" that characterizes the regression model of interest. We thus have the following classification of selection models.

Marginal selection models

$$\pi(\boldsymbol{y}_i, \boldsymbol{r}_i|\boldsymbol{V}_i) = \pi(\boldsymbol{r}_i|\boldsymbol{V}_i, \boldsymbol{y}_i)\pi(\boldsymbol{y}_i|\boldsymbol{X}_i, \boldsymbol{Z}_i) \tag{6.29}$$

Random-effects selection models

$$\begin{aligned}\pi(\boldsymbol{y}_i, \boldsymbol{r}_i, \boldsymbol{b}_i|\boldsymbol{V}_i) &= \pi(\boldsymbol{r}_i|\boldsymbol{V}_i, \boldsymbol{y}_i, \boldsymbol{b}_i)\pi(\boldsymbol{y}_i, \boldsymbol{b}_i|\boldsymbol{X}_i, \boldsymbol{Z}_i) \\ &= \pi(\boldsymbol{r}_i|\boldsymbol{V}_i, \boldsymbol{y}_i, \boldsymbol{b}_i)\pi(\boldsymbol{y}_i|\boldsymbol{X}_i, \boldsymbol{Z}_i, \boldsymbol{b}_i)\pi(\boldsymbol{b}_i|\boldsymbol{Z}_i)\end{aligned} \tag{6.30}$$

which we further categorize as follows:

1) *Non-ignorable outcome-dependent selection models*

$$\pi(\boldsymbol{y}_i, \boldsymbol{r}_i, \boldsymbol{b}_i|\boldsymbol{V}_i) = \pi(\boldsymbol{r}_i|\boldsymbol{V}_i, \boldsymbol{y}_i)\pi(\boldsymbol{y}_i|\boldsymbol{X}_i, \boldsymbol{Z}_i, \boldsymbol{b}_i)\pi(\boldsymbol{b}_i|\boldsymbol{Z}_i) \tag{6.31}$$

2) *Non-ignorable random-effects dependent selection models*

$$\pi(\boldsymbol{y}_i, \boldsymbol{r}_i, \boldsymbol{b}_i|\boldsymbol{V}_i) = \pi(\boldsymbol{r}_i|\boldsymbol{V}_i, \boldsymbol{b}_i)\pi(\boldsymbol{y}_i|\boldsymbol{X}_i, \boldsymbol{Z}_i, \boldsymbol{b}_i)\pi(\boldsymbol{b}_i|\boldsymbol{Z}_i) \tag{6.32}$$

We note that the non-ignorable outcome-dependent selection model (6.31) is equivalent to the marginal selection model (6.29) once we integrate over $\boldsymbol{b}_i$ to obtain the marginal complete data response model, $\pi(\boldsymbol{y}_i|\boldsymbol{X}_i, \boldsymbol{Z}_i) = \int \pi(\boldsymbol{y}_i|\boldsymbol{X}_i, \boldsymbol{Z}_i, \boldsymbol{b}_i)\pi(\boldsymbol{b}_i|\boldsymbol{Z}_i)d\boldsymbol{b}_i$. This is easily accomplished whenever the conditional pdf $\pi(\boldsymbol{y}_i|\boldsymbol{X}_i, \boldsymbol{Z}_i, \boldsymbol{b}_i)$ is that of a Gaussian-based LME model since the marginal response model $\pi(\boldsymbol{y}_i|\boldsymbol{X}_i, \boldsymbol{Z}_i)$ can always be expressed in closed form (see equations 3.3 and 3.4 of §3.1). However, when $\pi(\boldsymbol{y}_i|\boldsymbol{X}_i, \boldsymbol{Z}_i, \boldsymbol{b}_i)$ represents a GLME or GNLME model, the use of outcome-dependent random-effects selection models can pose considerably greater challenges and one will usually be better off by simply postulating a marginal GLIM or GNLM for $(\boldsymbol{y}_i|\boldsymbol{X}_i)$ and working with a marginal selection model directly. The non-ignorable random-effects dependent selection model (6.32) falls under the category of a *shared parameter model* and so we defer discussion of this class of selection models to our treatment of shared parameter models. Here we focus strictly on marginal selection models.

Given the primary interest of longitudinal clinical trials is on estimands formulated from the complete-data response model, the class of marginal selection models (6.29) provides an intuitively appealing approach to the analysis of longitudinal data. Specifically, the factorization in (6.29) allows one to directly specify the complete-data response model of interest, $\pi(\boldsymbol{y}_i|\boldsymbol{X}_i, \boldsymbol{Z}_i)$. However, it also requires one to specify a missing data mechanism through specification of a selection model, $\pi(\boldsymbol{r}_i|\boldsymbol{V}_i, \boldsymbol{y}_i)$. There are two approaches that have been taken to specifying these components of a marginal selection model: 1) parametric and 2) semiparametric. We consider each briefly.

Parametric selection models

Parametric selection models are likelihood-based and require one to specify a parametric complete-data response model, $\pi(\boldsymbol{y}_i|\boldsymbol{X}_i, \boldsymbol{Z}_i; \boldsymbol{\tau})$, and a parametric missing data (i.e., selection) model, $\pi(\boldsymbol{r}_i|\boldsymbol{V}_i, \boldsymbol{y}_i; \boldsymbol{\eta}) = \pi(\boldsymbol{r}_i|\boldsymbol{V}_i, \boldsymbol{y}_i^o, \boldsymbol{y}_i^m; \boldsymbol{\eta})$. Assuming the parameter vectors $\boldsymbol{\tau}$ and $\boldsymbol{\eta}$ are distinct, estimation involves maximizing the log-likelihood function of the observed data,

$$L(\boldsymbol{\tau}, \boldsymbol{\eta}; \boldsymbol{y}_{\text{obs}}, \boldsymbol{r}) = \sum_{i=1}^{n} \log \int_{\boldsymbol{y}_i^m} \pi(\boldsymbol{r}_i|\boldsymbol{V}_i, \boldsymbol{y}_i^o, \boldsymbol{y}_i^m; \boldsymbol{\eta})\pi(\boldsymbol{y}_i|\boldsymbol{X}_i, \boldsymbol{Z}_i; \boldsymbol{\tau})d\boldsymbol{y}_i^m$$

which will often require one to impose additional parametric and structural assumptions and may require one to use specialized software that include numerical integration techniques, an EM algorithm, etc., in order to evaluate the log-likelihood and fit the model to the observed data.

As an example, Diggle and Kenward (1994) [see also Diggle et. al., 1994, Chapter 11] proposed a marginal selection model in which a Gaussian-based multivariate repeated measures regression model is assumed for a continuous response variable and a logistic regression hazards model is assumed for dropout (assuming missing data are due strictly to dropout). In keeping with the basic notation defined in §6.2, their complete-data response model $\pi(\boldsymbol{y}_i|\boldsymbol{X}_i, \boldsymbol{Z}_i)$ may be written as

$$\boldsymbol{y}_i = \boldsymbol{X}_i\boldsymbol{\beta} + \boldsymbol{\epsilon}_i$$

where $\boldsymbol{y}_i = (y_{i1}, \ldots, y_{ip})'$ is the *intended* complete data response vector taken at times $\boldsymbol{t} = (t_1, \ldots, t_p)'$, $\boldsymbol{X}_i$ is the fixed known design matrix, and $\boldsymbol{\epsilon}_i \sim N_p(\boldsymbol{0}, \boldsymbol{\Sigma}_i(\boldsymbol{\theta}))$ for some specified covariance structure $\boldsymbol{\Sigma}_i(\boldsymbol{\theta})$. In the case of a LME model, the marginal model for $\boldsymbol{y}_i$ would assume a marginal covariance structure of the form $\boldsymbol{\Sigma}_i(\boldsymbol{\theta}) = \boldsymbol{Z}_i\boldsymbol{\Psi}(\boldsymbol{\theta}_b)\boldsymbol{Z}_i' + \boldsymbol{\Lambda}_i(\boldsymbol{\theta}_w)$ as described in §3.1 with $\boldsymbol{\theta} = (\boldsymbol{\theta}_b, \boldsymbol{\theta}_w)$. Let $\boldsymbol{y}_i^o = (y_{i1}^o, \ldots, y_{ip}^o)'$ be the vector of observed measurements with missing values coded as zero [i.e., if dropout occurs at time t_{p_i+1}, then $\boldsymbol{y}_i^o = (y_{i1}, y_{i2}, \ldots, y_{ip_i}, 0, \ldots, 0)'$]. Using the hazards structure defined in (6.18), a possible logistic regression hazards model for non-ignorable dropout might be

$$\text{logit}(\lambda_{ik}) = \boldsymbol{x}_{i(k-1)}'\boldsymbol{\eta} + y_{i(k-1)}^o\eta_1^* + y_{ik}^o\eta_2^*$$

where $\lambda_{ik} = \Pr(r_{ik} = 0 | r_{i(k-1)} = 1, \boldsymbol{x}_{i(k-1)}, y^o_{i(k-1)}, y^o_{ik})$ is the hazard for dropout at time t_k conditional on the most recently observed design variables $\boldsymbol{x}_{i(k-1)}$, the most recently observed response, $y^o_{i(k-1)} = y_{i(k-1)}$, and the observed response y^o_{ik} at time t_k. An example illustrating the use of this selection model and how one might go about testing for MAR ($\eta^*_2 = 0$) versus MNAR ($\eta^*_2 \neq 0$) is presented in Diggle and Kenward (1994) and Diggle et. al. (1994, Chapter 11). More recently, Diggle et. al. (2008) consider a marginal selection model under the continuous time-to-dropout setting of §9.3 in which the joint distribution of $(\boldsymbol{y}_i, \log(T_i))$ is assumed to be distributed as multivariate normal with the event time assumed to have a log-normal distribution. As noted by Diggle et. al. (2008), this model is suitable for a balanced clinical trial setting such as described above but with dropout events restricted to non-terminal events.

Molenberghs et. al. (1997) and Molenberghs and Verbeke (2005, Chapter 29) describe a marginal selection model for discrete ordinal outcomes data under MNAR. Their selection model specifies a multivariate Dale model for the longitudinal ordinal data and a logistic regression hazards model for dropout. They apply the EM algorithm in order to maximize the corresponding log-likelihood. Other parametric selection models including selection models for non-monotone missing data patterns for discrete categorical data are discussed in Molenberghs and Verbeke (2005, Chapter 29).

Semiparametric selection models

An alternative approach to fully parametric selection models are the class of semiparametric selection models proposed by Rotnitzky et. al. (1998) and Scharfstein et. al. (1999). These authors assume the complete-data response model for $\boldsymbol{y}_i$ may be adequately characterized by a class of semiparametric marginal models that include the GLIM (4.8) of §4.2 and the GNLM (4.29) of §4.4 as special cases. By specifying either a parametric (Rotnitzky et. al., 1998) or semiparametric (Scharfstein et. al., 1999) selection model $\pi(\boldsymbol{r}_i | \boldsymbol{V}_i, \boldsymbol{y}_i)$ for $\boldsymbol{r}_i$, these authors extend the inverse probability of weighting (IPW) estimation techniques described in §6.4.3 for ignorable missing data to cases involving non-ignorable missing data.

To illustrate the basic idea behind this semiparametric approach, let $\boldsymbol{\tau} = (\boldsymbol{\beta}, \boldsymbol{\theta})$ be the parameters of the marginal mean, $E(\boldsymbol{y}_i | \boldsymbol{X}_i) = \boldsymbol{\mu}_i(\boldsymbol{\beta})$,and variance-covariance, $Var(\boldsymbol{y}_i | \boldsymbol{X}_i) = \boldsymbol{\Sigma}_i(\boldsymbol{\beta}, \boldsymbol{\theta})$, of a semiparametric GLIM and suppose $\boldsymbol{\theta}$ is a nuisance parameter for which we have a consistent estimator $\widehat{\boldsymbol{\theta}}(\boldsymbol{\beta})$. Assuming the response/non-response probabilities $\pi(\boldsymbol{r}_i | \boldsymbol{V}_i, \boldsymbol{y}_i)$ are known functions of $\boldsymbol{V}_i$ and $\boldsymbol{y}_i$ where $\boldsymbol{V}_i = (\boldsymbol{X}_i, \boldsymbol{U}_i)$ includes auxiliary covariates, Rotnitzky et. al., (1998) show how one can apply an inverse probability of weighting (IPW) scheme analogous to that described for the MAR mechanism (§6.4.3) to obtain a $\sqrt{n}$-consistent estimator of $\boldsymbol{\beta}$ using weighted GEE. Specifically, one can perform a "complete-case" analysis by solving the set of IPW GEE

$$\sum_{i=1}^{n} \frac{I(\boldsymbol{r}_i = \boldsymbol{1})}{\pi_i(\boldsymbol{1})} \boldsymbol{D}_i' \boldsymbol{\Sigma}_i(\boldsymbol{\beta}, \widehat{\boldsymbol{\theta}}(\boldsymbol{\beta}))^{-1} (\boldsymbol{y}_i - \boldsymbol{\mu}_i(\boldsymbol{\beta})) = \boldsymbol{0}$$

where $\pi_i(\boldsymbol{1}) = \pi(\boldsymbol{r}_i = \boldsymbol{1} | \boldsymbol{V}_i, \boldsymbol{y}_i) = \Pr(r_{i1} = 1, \ldots, r_{ip} = 1 | \boldsymbol{V}_i, \boldsymbol{y}_i)$ are the "known" response probabilities for completers, $I(\boldsymbol{r}_i = \boldsymbol{1}) = 1$ for those with complete data and $I(\boldsymbol{r}_i = \boldsymbol{1}) = 0$ for those with incomplete data. The solution to this weighted GEE is shown to be a consistent and asymptotically normal estimator of $\boldsymbol{\beta}$. To overcome the inefficiency associated with this "complete-case" analysis in which only data from completers are used to estimate $\boldsymbol{\beta}$, Rotnitzky et. al. (1998) proposed extending this class of IPW estimating equations to a class of augmented IPW estimating equations that account for observed values among individuals with incomplete data.

Of course under the assumption of MNAR, the response probabilities $\pi(\boldsymbol{r}_i | \boldsymbol{V}_i, \boldsymbol{y}_i)$ are never known and so Rotnitzky et. al. (1998) propose solving a set of augmented IPW estimating equations that jointly estimate parameters $(\boldsymbol{\beta}, \boldsymbol{\eta})$ under a semiparametric model for $\boldsymbol{y}_i$ and a fully parametric model, $\pi(\boldsymbol{r}_i|\boldsymbol{V}_i, \boldsymbol{y}_i; \boldsymbol{\eta})$, for $\boldsymbol{r}_i$. However, the authors point out the inherent difficulties with this approach including problems with parameter identifiability and loss of precision when the joint model for $\boldsymbol{y}_i$ and $\boldsymbol{r}_i$ is "richly parameterized." As stressed by the authors, the real utility with using semiparametric selection models in combination with augmented IPW comes with one's ability to perform sensitivity analysis. By holding components of $\boldsymbol{\eta}$ that represent the magnitude of non-ignorable non-response fixed, one can perform a sensitivity analysis to examine how inferences concerning the parameters of interest, $\boldsymbol{\beta}$, vary over a range of plausible values of these components.

In general, selection models pose some serious challenges with regards to parameter identifiability and sensitivity to the underlying model assumptions (Diggle and Kenward, 1994; Little, 1995; Little and Rubin, 2002, Chapter 15; Daniels and Hogan, 2008, Chapter 9). For example, under the selection model of Diggle and Kenward (1998) described above, the parameter η_2^* is identifiable because of the strong but untestable assumption that $\boldsymbol{y}_i \sim N(\boldsymbol{X}_i\boldsymbol{\beta}, \boldsymbol{\Sigma}_i(\boldsymbol{\theta}))$. Here parameter identifiability and an untestable assumption are intertwined suggesting that one would best be served by including a sensitivity analysis that targets the underlying assumptions (e.g., Kenward, 1998). This is most evident with likelihood-based selection models where one must rely on strong and unverifiable parametric assumptions. Semiparametric selection models do not assume a fully parametric model for $\boldsymbol{y}_i$ and so are less sensitive to these types of assumptions. Moreover, through the use of augmented IPW, they lend themselves more naturally to various forms of sensitivity analysis. In most cases, both parametric and semiparametric selection models will require specialized software making implementation less than routine. Nevertheless, their appeal lies in the ability to factor the joint distribution of $(\boldsymbol{y}_i, \boldsymbol{r}_i)$ into two components one of which is the complete-data response model $\pi(\boldsymbol{y}_i|\boldsymbol{X}_i, \boldsymbol{Z}_i)$ that is of interest. In the following section we consider an alternative factorization of the complete-data model for $(\boldsymbol{y}_i, \boldsymbol{r}_i)$ that, in many cases, makes sensitivity analysis easier to implement using existing software.

6.5.2 Pattern mixture models

Pattern-mixture models for the repeated measures setting were proposed by Little (1993, 1994) and are based on an alternative factorization of the joint distribution of $(\boldsymbol{y}_i, \boldsymbol{r}_i, \boldsymbol{b}_i)$. As with selection models, one can classify pattern-mixture models into *marginal pattern-mixture models* and *random-effects pattern-mixture models* as follows:

Marginal pattern-mixture models

$$\pi(\boldsymbol{y}_i, \boldsymbol{r}_i|\boldsymbol{V}_i) = \pi(\boldsymbol{y}_i|\boldsymbol{X}_i, \boldsymbol{Z}_i, \boldsymbol{r}_i)\pi(\boldsymbol{r}_i|\boldsymbol{V}_i) \tag{6.33}$$

Random-effects pattern-mixture models

$$\pi(\boldsymbol{y}_i, \boldsymbol{r}_i, \boldsymbol{b}_i|\boldsymbol{V}_i) = \pi(\boldsymbol{y}_i|\boldsymbol{X}_i, \boldsymbol{Z}_i, \boldsymbol{r}_i, \boldsymbol{b}_i)\pi(\boldsymbol{b}_i|\boldsymbol{Z}_i, \boldsymbol{r}_i)\pi(\boldsymbol{r}_i|\boldsymbol{V}_i) \tag{6.34}$$

which, following Little (1995), we further categorize as follows:

1) *Non-ignorable outcome-dependent pattern-mixture models*

$$\pi(\boldsymbol{y}_i, \boldsymbol{r}_i, \boldsymbol{b}_i|\boldsymbol{V}_i) = \pi(\boldsymbol{y}_i|\boldsymbol{X}_i, \boldsymbol{Z}_i, \boldsymbol{r}_i, \boldsymbol{b}_i)\pi(\boldsymbol{b}_i|\boldsymbol{Z}_i)\pi(\boldsymbol{r}_i|\boldsymbol{V}_i) \tag{6.35}$$

2) *Non-ignorable random-effects dependent pattern-mixture models*

$$\pi(\boldsymbol{y}_i, \boldsymbol{r}_i, \boldsymbol{b}_i|\boldsymbol{V}_i) = \pi(\boldsymbol{y}_i|\boldsymbol{X}_i, \boldsymbol{Z}_i, \boldsymbol{b}_i)\pi(\boldsymbol{b}_i|\boldsymbol{Z}_i, \boldsymbol{r}_i)\pi(\boldsymbol{r}_i|\boldsymbol{V}_i). \tag{6.36}$$

Under (6.35), the distribution of $\boldsymbol{b}_i$ is independent of dropout whereas under (6.36) the distribution of $\boldsymbol{b}_i$ may depend on dropout patterns.

As the name suggests, pattern-mixture models describe the full data distribution as a mixture over missing data patterns described by $\pi(\boldsymbol{r}_i|\boldsymbol{V}_i)$. This provides the analyst an alternative and sometimes advantageous framework from which to view the "effects" of missing data. For example, one can easily stratify individuals into sub-populations based on their missing data patterns and then check if there are major differences in the stratum-specific means. Also, one can factor the marginal pattern-mixture model (6.33) as

$$\pi(\boldsymbol{y}_i, \boldsymbol{r}_i|\boldsymbol{V}_i) = \pi(\boldsymbol{y}_i^m|\boldsymbol{X}_i, \boldsymbol{Z}_i, \boldsymbol{y}_i^o, \boldsymbol{r}_i)\pi(\boldsymbol{y}_i^o|\boldsymbol{X}_i, \boldsymbol{Z}_i, \boldsymbol{r}_i)\pi(\boldsymbol{r}_i|\boldsymbol{V}_i). \tag{6.37}$$

In doing so, one may view $\pi(\boldsymbol{y}_i^m|\boldsymbol{X}_i, \boldsymbol{Z}_i, \boldsymbol{y}_i^o, \boldsymbol{r}_i)$ as a predictive distribution of the missing data conditional on all of the observed data including the missing data patterns determined by $\boldsymbol{r}_i$. This provides an imputation-based perspective in that one can impute missing values from a specified predictive distribution. This, in turn, allows one to more easily conduct sensitivity analyses using multiple imputation strategies since (6.37) allows one to explicitly separate the observed data distribution from the predictive distribution of missing data given the observed data (e.g., Little and Wang, 1996; Thijs et. al. 2002; Molenberghs and Verbeke, 2005).

There are two major issues one must contend with when applying pattern-mixture models. The first deals with the fact that one must integrate (average) over the missing data patterns or strata in order to obtain an estimate of the marginal treatment effect or estimand of interest. In some cases this can present computational difficulties particularly when the pattern mixture model involves fitting regression models within each strata (National Research Council, 2010). In other cases, the marginal treatment effect or estimand can be computed by simply weight averaging stratum-specific estimates using the observed proportions for each missing data stratum as the weights. Even in this case, one must still compute an appropriate standard error of the weight average estimand taking into account the uncertainty in the estimated weights. This is most often accomplished using the delta method (e.g., Hogan et. al., 2004).

The second major issue deals with the fact that pattern mixture models are generally overspecified or underidentified with respect to model parameters (e.g., Little, 1993, 1994; Little and Wang, 1996; Thijs et. al., 2002; Little and Rubin, 2002; Molenberghs and Verbeke, 2005). For example, consider a simple longitudinal study with two observations per subject; a baseline value y_{i1} which is observed for all subjects and a follow-up value y_{i2} which is missing in a select percentage of the subjects. Let $\boldsymbol{y}_i = (y_{i1}, y_{i2})'$ be the intended bivariate response vector for the ith subject and let r_i be the binary response indicator for the follow-up observation such that $r_i = 1$ if y_{i2} is observed and $r_i = 0$ if y_{i2} is missing. Little (1994) defines a bivariate normal pattern-mixture model for this setting as

$$(m_i|\pi) \sim \text{iid Bernoulli}(\pi),\ \pi = \Pr(m_i = 1)$$

$$(\boldsymbol{y}_i|m_i = k, \boldsymbol{\tau}^k) \sim N_2(\boldsymbol{\mu}^k, \boldsymbol{\Sigma}^k);\ k = 0, 1$$

where $m_i = 1 - r_i$ is the missing data indicator, and

$$\boldsymbol{\mu}^k = \begin{pmatrix} \mu_1^k \\ \mu_2^k \end{pmatrix},\ \boldsymbol{\Sigma}^k = \begin{pmatrix} \sigma_{11}^k & \sigma_{12}^k \\ \sigma_{12}^k & \sigma_{22}^k \end{pmatrix};\ k = 0, 1$$

are the mean and variance-covariance structure under the missing data stratum defined by subjects with $m_i = k = 0$ and $m_i = k = 1$, respectively. Under this model, the parameters $\boldsymbol{\tau}^k = (\boldsymbol{\mu}^k, \boldsymbol{\Sigma}^k)$ are underidentified by virtue of the fact that one cannot estimate the parameters $\mu_2^1, \sigma_{12}^1, \sigma_{22}^1$ for the missing data stratum ($m_i = k = 1$) based on the observed data (Little, 1994).

There are two basic strategies that have been advocated for handling the problem of parameter identifiability in pattern-mixture models. The first is the use of parameter constraints or restrictions for identifying all relevant parameters (Little, 1993, 1994). To illustrate this approach, Little (1994) introduced into the above example the assumption that missingness depends on the function $y_{i1} + \lambda y_{i2}$ for known λ. He showed how one can use this restriction to identify and estimate the parameters for known λ and pointed out that although one cannot estimate λ from the data, one can perform a sensitivity analysis by simply varying the value of λ to assess how sensitive inference is to differing missing data mechanisms (i.e., λ serves as an index of non-ignorability with $\lambda = 0$ corresponding to the ignorable MAR mechanism and $\lambda > 0$ corresponding to differing MNAR conditions). This approach has been described in detail by Little (1993, 1994), Little and Wang (1996), Thijs et. al. (2002) and in the texts by Little and Rubin (2002) and Molenberghs and Verbeke (2005). It will not be discussed here other than to mention that it is most useful when conducting sensitivity analyses as indicated in the above example.

The second strategy is to use a regression modeling approach in which the pattern-mixture means, for example, may be predicted (i.e., extrapolated) from a specified regression model in which time to dropout and other covariates are used to predict the mean or estimand of interest albeit under a set of unverifiable assumptions. This latter approach includes the work of Wu and Bailey (1988, 1989), Mori et. al. (1992), Fitzmaurice and Laird (2000) and Fitzmaurice et. al. (2001) among others. It can be particularly useful when performing a sensitivity analysis that compares pattern-mixture models with selection models. In this chapter we restrict our attention to this second approach as it involves straightforward extensions of the models and SAS procedures presented in previous chapters.

To illustrate the basic idea behind the regression modeling approach to pattern-mixture models, consider the following simple adaptation of the conditional linear model of Wu and Bailey (1989) which is a special case of a random-effects pattern-mixture model.

A conditional linear model

$$\begin{aligned}
&\text{Stage 1: } y_{ij}|\beta_{i1} = \beta_0 + \beta_{i1} t_{ij} + e_{ij};\ (i = 1\ldots, n,\ j = 1, \ldots, p) \\
&\text{Stage 2: } \beta_{i1}|(T_i^o, b_i) = \beta_{11} + \beta_{12} T_i^o + b_i \\
&\text{Stage 3: } T_i^o = \min(T_i, C_i) = \text{observed dropout time with } T_i \sim \text{iid } F_T(\cdot)
\end{aligned} \quad (6.38)$$

where

Stage 1 assumes a simple linear regression model relating the response y_{ij} to time t_{ij} over a fixed interval $[0, \mathcal{T})$ and assuming $e_{ij} \sim$iid $N(0, \sigma^2)$,
Stage 2 assumes a simple linear regression model relating the ith subject's slope β_{i1} as a linear function of the observed dropout time T_i^o with random error $b_i \sim$iid $N(0, \psi)$ distributed independently of e_{ij},
Stage 3 assumes the time-to-dropout T_i is a continuous event time random variable from a specified distribution F_T with pdf $\pi_T(t)$ and survivor function $S_T(t)$, and C_i is an independent censoring variable. The observed dropout time is $T_i^o = \min(T_i, C_i)$ with $\delta_i = I(T_i < C_i)$ denoting the event indicator variable. If censoring is due to end of follow-up $[0, \mathcal{T})$, it might be helpful to consider averaging over the distribution of T_i using the restricted mean life time defined as $E(\min(T_i, \mathcal{T})) = \mu_T = \int_0^{\mathcal{T}} S_T(t)dt$ (e.g., Karrison, 1987).

Several points are worth noting regarding this conditional linear model. The first point is that model (6.38) corresponds to the non-ignorable random-effects dependent pattern-mixture model (6.36) with $\boldsymbol{r}_i$ replaced by T_i. In particular, Stage 1 of (6.38) defines the conditional model $\pi(\boldsymbol{y}_i|\boldsymbol{X}_i, \boldsymbol{Z}_i, \boldsymbol{b}_i)$ with $\boldsymbol{b}_i = \beta_{i1}$ while Stage 2 defines the conditional model $\pi(\boldsymbol{b}_i|\boldsymbol{Z}_i, T_i)$ and Stage 3 defines the unconditional model $\pi(T_i|\boldsymbol{V}_i)$ defined by $\pi_T(T_i)$

or more generally by $\pi_T(T_i^o)^{\delta_i} S_T(T_i^o)^{1-\delta_i}$ when censoring is present. The second point is that model (6.38) may also be expressed in terms of (6.35) by combining the first two stages into a linear model for y_{ij} conditional on T_i^o and b_i as

$$\begin{aligned} y_{ij} &= \beta_0 + (\beta_{11} + \beta_{12}T_i^o + b_i)t_{ij} + e_{ij} \\ &= (\beta_0 + \beta_{11}t_{ij}) + \beta_{12}T_i^o t_{ij} + b_i t_{ij} + e_{ij} \end{aligned} \tag{6.39}$$

and then specifying the distribution $\pi(b_i)$ and $\pi_T(T_i)$ of the random effect b_i and dropout time T_i, respectively. By further integrating out b_i, we can express the joint model for $(\boldsymbol{y}_i, T_i)$ in terms of the marginal pattern-mixture model

$$\pi(\boldsymbol{y}_i, T_i|\boldsymbol{X}_i, \boldsymbol{Z}_i) = \pi(\boldsymbol{y}_i|\boldsymbol{X}_i, \boldsymbol{Z}_i, T_i)\pi_T(T_i)$$

where $\pi(\boldsymbol{y}_i|\boldsymbol{X}_i, \boldsymbol{Z}_i, T_i)$ is the pdf corresponding to the multivariate normal linear mixture model

$$\boldsymbol{y}_i = \boldsymbol{X}_i\boldsymbol{\beta} + \boldsymbol{\epsilon}_i;\ i = 1, \ldots, n$$

with $\boldsymbol{\epsilon}_i \sim N_p(\boldsymbol{0}, \boldsymbol{\Sigma}_i(\boldsymbol{\theta}))$ where $\boldsymbol{\Sigma}_i(\boldsymbol{\theta}) = \boldsymbol{Z}_i\psi\boldsymbol{Z}_i' + \sigma^2\boldsymbol{I}_p$ and

$$\boldsymbol{X}_i = \begin{pmatrix} 1 & t_{i1} & T_i^o t_{i1} \\ 1 & t_{i2} & T_i^o t_{i2} \\ \vdots & \vdots & \vdots \\ 1 & t_{ip} & T_i^o t_{ip} \end{pmatrix}, \boldsymbol{Z}_i = \begin{pmatrix} t_{i1} \\ t_{i2} \\ \vdots \\ t_{ip} \end{pmatrix}, \boldsymbol{\beta} = \begin{pmatrix} \beta_0 \\ \beta_{11} \\ \beta_{12} \end{pmatrix}.$$

A third point is that the underlying but unverifiable assumptions of this model imply that, conditional on T_i, the unobserved post-dropout observations follow the same distribution as the observed pre-droput observations for non-terminal dropout events. This means that missing values in $\boldsymbol{y}_i$ are MAR conditional on T_i (e.g., Hogan and Laird, 1997a, 1997b). In fact, because the complete-data response vector $\boldsymbol{y}_i$ follows a multivariate normal linear model conditional on T_i, we can integrate over the missing data components of $\boldsymbol{y}_i$ and write the joint log-likelihood for the observed data as

$$\begin{aligned} \sum_{i=1}^n \log\{\pi(\boldsymbol{y}_i^o, T_i)\} &= \sum_{i=1}^n \left[\log \int_{\boldsymbol{y}_i^m} \pi(\boldsymbol{y}_i^o, \boldsymbol{y}_i^m|T_i)d\boldsymbol{y}_i^m + \log\{\pi_T(T_i)\}\right] \\ &= \sum_{i=1}^n \log\{\pi(\boldsymbol{y}_i^o|T_i)\} + \sum_{i=1}^n \log\{\pi_T(T_i)\} \end{aligned}$$

where $\sum \log\{\pi(\boldsymbol{y}_i^o|T_i)\}$ is the conditional log-likelihood component of the pattern mixture model based on the multivariate normal pdf of $(\boldsymbol{y}_i^o|T_i)$ and $\sum \log\{\pi_T(T_i)\}$ is the marginal log-likelihood component of the model based on the pdf of T_i (or, more generally, $\sum\{\delta_i \log \pi_T(T_i^o) + (1-\delta_i)\log(S_T(T_i^o)\}$ in the presence of censoring). We can maximize the conditional log-likelihood component $\sum \log\{\pi(\boldsymbol{y}_i^o|T_i)\}$ by fitting the conditional LME model (6.39) to the observed data using PROC MIXED.

Finally, given that the estimand of interest is the rate of change as measured by the slope β_{i1} from (6.38), we need to distinguish between the conditional and unconditional slope over the interval $[0, \mathcal{T})$,

$$\text{Conditional slope: } E(\beta_{i1}|T_i^o) = \beta_{11} + \beta_{12}T_i^o$$
$$\text{Unconditional slope: } E(\beta_{i1}) = E_T(E(\beta_{i1}|T_i^o)) = \beta_{11} + \beta_{12}\mu_T.$$

In most clinical applications, unconditional inference about a treatment effect or a population trend, etc. will be of primary interest. Here, for example, the unconditional slope $\beta_{11} + \beta_{12}\mu_T$ might be the primary estimand of interest. If so, then we must be able to average over the missing data patterns which, in this case, are uniquely determined by the observed dropout times, T_i^o. This points to the first issue we raised regarding the challenges of fitting pattern-mixture models to data, namely that we must average over the patterns of missing data to estimate the marginal treatment effect of interest.

The simple conditional LME model (6.39) above is a special case of a general linear mixture model for discrete dropout times described by Hogan and Laird (1997a) and later expanded on by Fitzmaurice et. al. (2001). Here we let T_i be a discrete or continuous dropout time. Under the assumptions described by Hogan and Laird (1997a) and Fitzmaurice et. al. (2001), we can write a general linear mixture model in terms of the observed data as

$$\boldsymbol{y}_i^o = \boldsymbol{X}_i\boldsymbol{\beta} + \boldsymbol{D}_i\boldsymbol{\beta}_d + \boldsymbol{\epsilon}_i, \ (i = 1, \ldots, n) \tag{6.40}$$

where

$\boldsymbol{y}_i^o = (y_{i1}, \ldots, y_{ip_i})$ is the observed response vector for the ith subject
$\boldsymbol{X}_i$ is a $p_i \times s$ design matrix of known fully observed between-subject and within-subject design variables and covariates
$\boldsymbol{\beta}$ is the parameter vector of interest associated with $\boldsymbol{X}_i$
$\boldsymbol{D}_i = \boldsymbol{D}_i(\boldsymbol{d}_i)$ is a $p_i \times r$ design matrix that depends on an $m \times 1$ vector of covariates, $\boldsymbol{d}_i = (d_{1i}, \ldots, d_{mi})' = (d_{1i}(T_i^o, \delta_i), \ldots, d_{mi}(T_i^o, \delta_i))'$, whose elements are functions of (T_i^o, δ_i). We allow $\boldsymbol{D}_i$ to also depend on select interactions between elements of $\boldsymbol{d}_i$ and $\boldsymbol{X}_i$ resulting in a general $p_i \times r$ dimension for $\boldsymbol{D}_i$. For example, $\boldsymbol{D}_i$ may consist of a single covariate $d_{1i}(T_i^o, \delta_i) = T_i^o$ and/or its interactions with select components of $\boldsymbol{X}_i$ as in the simple conditional linear model (6.39).
$\boldsymbol{\beta}_d$ is an $r \times 1$ parameter vector associated with $\boldsymbol{D}_i$
$T_i^o = \min(T_i, C_i)$ is the observed dropout time with T_i representing a continuous or discrete event time random variable and C_i is an independent censoring variable. In addition, let $\delta_i = I(T_i < C_i)$ denote the event indicator variable. We assume T_i can be modeled parametrically or semiparametrically as a function of select components of $\boldsymbol{X}_i$ (e.g., baseline covariates - see Fitzmaurice et. al., 2001).
$\boldsymbol{\epsilon}_i \sim \text{ind } N(\boldsymbol{0}, \boldsymbol{\Sigma}_i(\boldsymbol{\theta}))$.

Note that by setting $\boldsymbol{\epsilon}_i = \boldsymbol{Z}_i\boldsymbol{b}_i + \boldsymbol{e}_i$ where $\boldsymbol{b}_i \sim \text{iid } N(\boldsymbol{0}, \boldsymbol{\Psi}(\boldsymbol{\theta}_b))$ is a vector of random effects and $\boldsymbol{e}_i \sim \text{ind } N(\boldsymbol{0}, \boldsymbol{\Lambda}_i(\boldsymbol{\theta}_w))$ is a vector of within-subject errors, the general linear mixture model (6.40) will also encompass the random-effects general linear mixture model

$$\boldsymbol{y}_i^o = \boldsymbol{X}_i\boldsymbol{\beta} + \boldsymbol{D}_i\boldsymbol{\beta}_d + \boldsymbol{Z}_i\boldsymbol{b}_i + \boldsymbol{e}_i$$

where $\boldsymbol{\Sigma}_i(\boldsymbol{\theta}) = \boldsymbol{Z}_i\boldsymbol{\Psi}(\boldsymbol{\theta}_b)\boldsymbol{Z}_i' + \boldsymbol{\Lambda}_i(\boldsymbol{\theta}_w)$.

As noted above with the simple conditional LME model (6.39), the primary challenge with the general linear mixture model (6.40) is that inference is restricted to estimable functions of the conditional mean

$$E(\boldsymbol{y}_i^o|\boldsymbol{X}_i, \boldsymbol{D}_i) = \boldsymbol{X}_i\boldsymbol{\beta} + \boldsymbol{D}_i\boldsymbol{\beta}_d \tag{6.41}$$

rather than to the unconditional or marginal mean

$$E(\boldsymbol{y}_i^o|\boldsymbol{X}_i) = E_D(E(\boldsymbol{y}_i^o|\boldsymbol{X}_i, \boldsymbol{D}_i)) = \boldsymbol{X}_i\boldsymbol{\beta} + E_D(\boldsymbol{D}_i)\boldsymbol{\beta}_d. \tag{6.42}$$

To overcome this problem, Fitzmaurice et. al. (2001) suggest replacing the functional covariates $d_{ki} = d_{ki}(T_i^o, \delta_i)$ that define $\boldsymbol{D}_i$ with the residual covariates $d_{ki}^* = d_{ki} - E(d_{ki})$. For example, if $\boldsymbol{d}_i$ consists of a single covariate, $d_{1i}(T_i^o, \delta_i) = T_i^o$, then one would replace $d_{1i} = T_i^o$ with the residual dropout time, $d_{1i}^* = T_i^* = T_i^o - E(T_i^o)$, or alternatively with the restricted residual dropout time, $T_i^* = T_i^o - \mu_T$ where $\mu_T = \int_0^T S_T(t)dt$ is the restricted mean over some defined period $[0, T)$ (e.g., Irwin, 1949; Karrison, 1987; Chen and Tsiatis, 2001; Meier et. al., 2004). In either case, one would then fit the observed data to the modified general linear mixture model

$$\boldsymbol{y}_i^o = \boldsymbol{X}_i\boldsymbol{\beta} + \boldsymbol{D}_i^*\boldsymbol{\beta}_d + \boldsymbol{\epsilon}_i,\ (i = 1, \ldots, n) \tag{6.43}$$

where $\boldsymbol{D}_i^* = \boldsymbol{D}_i(\boldsymbol{d}_i^*)$ is the design matrix $\boldsymbol{D}_i$ expressed as a function of the residual covariates $\boldsymbol{d}_i^* = (d_{1i}^*, d_{2i}^*, \ldots, d_{mi}^*)'$ where $d_{ki}^* = d_{ki} - E(d_{ki})$. By doing so, one can take advantage of how $\boldsymbol{D}_i^*$ is constructed together with the fact that $E(\boldsymbol{d}_i^*) = \boldsymbol{0}$ to show that $E_D\{\boldsymbol{D}_i^*\boldsymbol{\beta}_d\} = \boldsymbol{0}$ and hence that $E(\boldsymbol{y}_i^o|\boldsymbol{X}_i) = \boldsymbol{X}_i\boldsymbol{\beta}$. The obvious advantage with this approach is that inference can then be expressed in terms of estimable linear functions of $\boldsymbol{X}_i\boldsymbol{\beta}$ which will be the estimands of interest in most clinical trial settings. Another advantage is the ease with which one can conduct various sensitivity analyses based on parametric or semiparametric models for T_i (Fitzmaurice et. al., 2001). There are of course some technical challenges one must overcome such as specifying an appropriate set of covariates d_{ki} and their means when forming $d_{ki}^* = d_{ki} - E(d_{ki})$. In §6.6, we illustrate how one can overcome some of these challenges using NLMIXED.

We have attempted here to briefly describe some of the basic strategies and issues associated with pattern-mixture models. The literature is rich in applications illustrating the use of pattern-mixture models including material in the texts by Little and Rubin (2002), Molenberghs and Verbeke (2005), Molenberghs and Kenward (2007), and Daniels and Hogan (2008) as well as papers by Hedeker and Gibbons (1997), Hogan and Laird (1997a, 1997b) and Hogan et. al. (2004). In §6.6, we consider one such application based on an analysis of the MDRD data as presented by Li and Schluchter (2004).

6.5.3 Shared parameter (SP) models

In our discussion of selection models, we noted that non-ignorable random-effects dependent selection models (6.32) are commonly classified as *shared parameter models* (e.g., Follmann and Wu, 1995; Vonesh et. al., 2006). The term "shared parameter" stems from the fact that an unobserved latent random effects vector $\boldsymbol{b}_i$ is "shared" between the complete-data response model, $\pi(\boldsymbol{y}_i|\boldsymbol{X}_i, \boldsymbol{Z}_i, \boldsymbol{b}_i)$, and the "missing data" or "selection" model, $\pi(\boldsymbol{r}_i|\boldsymbol{V}_i, \boldsymbol{b}_i)$. This sharing of $\boldsymbol{b}_i$ between the two models induces a marginal association or correlation between the response/non-response indicator vector $\boldsymbol{r}_i$ (or a time-to-dropout variable T_i) and the complete data response vector $\boldsymbol{y}_i = (\boldsymbol{y}_i^o, \boldsymbol{y}_i^m)$ through the posterior distribution of the random effects (see equation 6.10). Thus shared parameter (SP) models do not require one to assume missing data are MCAR or MAR. However, like the other approaches already discussed, the SP model does make somewhat strong and unverifiable assumptions that can only be checked based on observed data.

While SP models can be used in a variety of settings, we will focus in this chapter on a class of shared parameter models that are particularly useful when analyzing complex longitudinal clinical trials with rather long follow-up times and which are subject to missing or incomplete data due to dropout. This class of SP models, sometimes referred to as shared frailty models, have also been used strictly within the context of survival analysis where the focus is on jointly modeling repeated measurements and an event time in an effort to characterize the relationship between a longitudinal response profile and the time-to-event (e.g., DeGruttola and Tu, 1994; Faucett and Thomas, 1996; Tsiatis and Davidian, 2004). An example illustrating this type of application of the SP model will be given in Chapter 7 where we jointly model patient survival and the rate of decline in glomerular filtration rates among incident patients with end-stage renal disease.

Here, specific attention will be given to the application of SP models to the longitudinal clinical trial setting where the primary goal is to estimate and compare serial trends over time while adjusting for possible informative censoring due to patient dropout. Ignoring notational dependence on covariates and using the basic setup and notation of §6.3.1, the class of SP models considered here can be characterized by factoring the joint distribution of the response variable and time-to-dropout as

$$\pi(\boldsymbol{y}_i, T_i) = \int_{\boldsymbol{b}} \pi(\boldsymbol{y}_i, T_i|\boldsymbol{b}_i)\pi(\boldsymbol{b}_i)d\boldsymbol{b}_i = \int_{\boldsymbol{b}} \pi(\boldsymbol{y}_i|\boldsymbol{b}_i)\pi(T_i|\boldsymbol{b}_i)\pi(\boldsymbol{b})d\boldsymbol{b}_i \tag{6.44}$$

where $\boldsymbol{y}_i = \boldsymbol{y}_i(\boldsymbol{t}_{P_i}) = (y_{i1}, \ldots, y_{iP_i})'$ is the *complete* but possibly unbalanced vector of repeated measurements on the ith subject taken at times $\boldsymbol{t}_{P_i} = (t_1, \ldots, t_{P_i})'$, $T_i = \min(T_{1i}, T_{2i})$ is the time to dropout based on a terminal or non-terminal event, and $\boldsymbol{b}_i$ is a shared random effects vector. Here we adapt the notation of §6.3.1 to a complete but unbalanced longitudinal data setup by defining P_i to be the number of observations intended for the ith subject and by letting $\boldsymbol{y}_i = \boldsymbol{y}_i(\boldsymbol{t}_{P_i})$ denote the complete data setup for the ith subject. Possible correlation (association) between observed and unobserved values of the repeated measurements and event times is induced through the shared random effects. Observe that under this SP model, dropout may be classified as ignorable provided $\pi(T_i|\boldsymbol{b}_i) = \pi(T_i)$ since $\boldsymbol{y}_i$ and T_i will then be independently distributed.

Assuming the y_{ij} are conditionally independent given $\boldsymbol{b}_i$, $j = 1, \ldots, P_i$, we can integrate over the missing or incomplete responses $\boldsymbol{y}_i^m$ and write the SP model for the observed data including covariates as

$$\begin{aligned}\pi(\boldsymbol{y}_i^o, T_i|\boldsymbol{X}_i, \boldsymbol{Z}_i) &= \int_{\boldsymbol{b}}\int_{\boldsymbol{y}_i^m} \pi(\boldsymbol{y}_i^o, \boldsymbol{y}_i^m, T_i|\boldsymbol{X}_i, \boldsymbol{Z}_i, \boldsymbol{b}_i)\pi(\boldsymbol{b}_i|\boldsymbol{Z}_i)d\boldsymbol{b}_i d\boldsymbol{y}_i^m \\ &= \int_{\boldsymbol{b}} \pi(\boldsymbol{y}_i^o|\boldsymbol{X}_i, \boldsymbol{Z}_i, \boldsymbol{b}_i)\pi(T_i|\boldsymbol{X}_i, \boldsymbol{Z}_i, \boldsymbol{b}_i)\pi(\boldsymbol{b}_i|\boldsymbol{Z}_i)d\boldsymbol{b}_i\end{aligned} \tag{6.45}$$

where $\boldsymbol{y}_i^o = \boldsymbol{y}_i(\boldsymbol{t}_{p_i}) = (y_{i1}, \ldots, y_{ip_i})'$ is the *observed* vector of repeated measurements on the ith subject taken at times $\boldsymbol{t}_{p_i} = (t_1, \ldots, t_{p_i})'$. In accordance with the setup defined in §6.3.1, the elements of the ith subject's observed vector $\boldsymbol{y}_i^o$ are realized values from the stochastic process $\{Y_i(t); t \in [0, \mathcal{T})\}$ where $\mathcal{T}$ is a fixed study termination time point while $\boldsymbol{X}_i$ and $\boldsymbol{Z}_i$ are fixed known design matrices from covariate processes $\{\boldsymbol{x}_i(t); t \in [0, \mathcal{T})\}$ and $\{\boldsymbol{z}_i(t); t \in [0, \mathcal{T})\}$, respectively. The vector $\boldsymbol{b}_i$ serves as a set of random effects for the response model $\pi(\boldsymbol{y}_i^o|\boldsymbol{X}_i, \boldsymbol{Z}_i, \boldsymbol{b}_i)$ and as a set of time-independent "covariates" for the event-time model $\pi(T_i|\boldsymbol{X}_i, \boldsymbol{Z}_i, \boldsymbol{b}_i)$.

Wu and Carroll (1988) first utilized this class of SP models as a means for comparing rates of change in forced expiratory volume (FEV) among individuals with PiZ emphysema. They used a linear random effects model for the response and a probit discrete time survival model for dropout. Follmann and Wu (1995) describe a general SP model setup for missing data in which they specify a GLME model for $\pi(\boldsymbol{y}_i|\boldsymbol{b}_i)$ and a GLIM for $\pi(T_i|\boldsymbol{b}_i)$ in which T_i can be a discrete or continuous event time random variable. Ten Have et. al. (1998) fit a SP model to a randomized controlled trial investigating the effect on pain relief of two pain relievers and a placebo assuming a mixed-effects logistic regression model for $(\boldsymbol{y}_i|\boldsymbol{b}_i)$ and a discrete time hazards model for $(T_i|\boldsymbol{b}_i)$. Vonesh et. al. (2006) extended the SP model of Follmann and Wu by allowing the response model $\pi(\boldsymbol{y}_i|\boldsymbol{b}_i)$ to be a GNLME model and the event-time model $\pi(T_i|\boldsymbol{b}_i)$ to be any parametric or semiparametric accelerated failure-time (AFT) model. An example illustrating this approach is given in §6.6. Other studies that have used SP models or variations thereof include Wu and Follmann (1999), Schluchter et. al. (2001), and Rademaker et. al. (2003).

While SP models offer considerable flexibility in terms of their ability to model the response data without assuming MCAR or MAR, they nevertheless require strong parametric or semiparametric assumptions on both the response model and event-time model as well as other unverifiable assumptions. For example, the SP model (6.45) requires the assumption of conditional independence between measured responses given the random effects, an assumption that can only be verified for the observed data. As another example, one may fit a SP model and find the specified random effects are not associated with the dropout time T_i. One may take this as evidence of an ignorable dropout mechanism but this may not necessarily follow since one may have misspecified the distribution of the random effects (e.g., assumed $\boldsymbol{b}_i$ is Gaussian), misspecified the random effects structure (e.g., $\boldsymbol{Z}_i$) or it may be that missingness is related to unobserved values of $\boldsymbol{y}_i$ rather than to unobserved random effects (i.e., the non-ignorable outcomes dependent selection model (6.31) describes the MNAR mechanism).

Shared parameter models also require some form of numerical integration in order to maximize the joint log-likelihood, $\sum \log\{\pi(\boldsymbol{y}_i^o, T_i|\boldsymbol{X}_i, \boldsymbol{Z}_i)\}$. Until recently, this has been a major hurdle for those wishing to use SP models but with the release of the NLMIXED procedure in SAS, this is generally no longer an issue (e.g., Rademaker et. al., 2003; Guo and Karlin, 2004; Vonesh et. al., 2006; Littell et. al., 2006). Since the response model can come from the family of LME, GLME or GNLME models, one need also determine if the primary objective of the study is to draw inference about a well-defined subject-specific (SS) estimand or a well-defined population-averaged (PA) estimand. If a PA estimand is the primary focus and the model is not a LME model, then there will certainly be additional computational hurdles to overcome (e.g., one may require numerical integration techniques to compute the marginal means and perhaps bootstrap methods to compute the corresponding standard errors).

Finally, SP models are more apt to be used in large complex and often times lengthy clinical trials where the primary goal is to investigate serial trends over time in the response variable of interest taking into account the possibility of non-ignorable dropout. Such studies are likely to focus on within-subject patterns of change in the response over time. Consequently, the parameters or estimands of interest can be quite complex and very often are subject-specific in scope. Typical of this situation would be studies that investigate individual rates of change over time as estimated by subject-specific slopes from a random-effects linear regression model.

In contrast, SP models are less likely to be used in a clinical regulatory setting where the focus is usually on a summary measure of efficacy in the target population at some pre-specified period of follow-up. Here there is a premium on minimizing the number of assumptions required to demonstrate such efficacy. As an example, one might conduct a three month double-blind randomized controlled trial comparing the mean change from baseline at each month between a treatment and control group. The final summary measure of efficacy is likely to be based on the difference in the 3-month mean change from baseline between the two groups. In that case, one might fit a repeated measures ANCOVA with baseline response as a covariate. Assuming an unstructured covariance matrix, one could estimate differences in mean change from baseline at each month but summarize overall efficacy as the difference in mean change from baseline at 3 months (e.g., Suddiqui et. al., 2009). It is clear that in this type of regulatory setting, the use of a SP model should probably be avoided unless there are compelling reasons why one might consider such a model in a sensitivity analysis.

6.5.4 A repeated measures ANCOVA—continued

To illustrate how one might conduct a sensitivity analysis when missing values are thought to be MNAR, we revisit the repeated measures ANCOVA example described in §6.4.4. Here we use PROC MIXED to fit the ANCOVA model to the incomplete data generated under the MNAR mechanism whereby missing values occur whenever the change from baseline value exceeds 1.00 (i.e., whenever Y1−Y0> 1.00)—see Table 6.3. We do so based on

1) a likelihood-based analysis assuming the missing values are MAR,
2) an analysis using last observation carried forward (LOCF) to impute missing values,
3) a sensitivity analysis based on a pattern mixture model (§6.5.3).

Given that there are only two missing values, we elect not to fit a marginal selection model or perform sensitivity analysis using IPW methods.

The following code creates an incomplete dataset (`data MNAR`) under the MNAR mechanism where dropout at week 4 occurs if Y1−Y0> 1.00, and an imputed dataset (`data LOCF`) in which we carry forward the baseline value Y0 to week 4 to define change from baseline response to be 0. A LOCF analysis or baseline observation carried forward (BOCF) analysis is often used as a type of "worst-case" sensitivity analysis although this makes certain unverifiable assumptions. The SAS code also includes the PROC MIXED statements required to fit the repeated measures ANCOVA model to each dataset with the estimand of interest, $\widehat{\beta}_d$, saved to an output dataset.

Program 6.5

```
/* Part 5: ANCOVA based on incomplete MNAR data */
data MNAR;
 length Type $8;
 set RM_ANCOVA;
 Type='MNAR    ';
 Yd = Y1-Y0;
 if Yd > 1.0 then do;
  Y1=.;Yd=.;
 end;
 r1=1;  ** Response indicator variable at week 4;
 if Yd=. then r1=0;
run;
%X0(data=MNAR);
proc sort data=MNAR;
 by Trt Subject;
run;
data LOCF;
 set MNAR;
 by Trt Subject;
 if Yd = . then Yd=0;
 drop X0 MeanY0;
run;
%X0(data=LOCF);
/* Fit available-case ANCOVA and LOCF ANCOVA */
ods output Estimates=est_MNAR_REML;
ods select Estimates;
proc mixed data=MNAR;
 class Trt Subject;
 model Yd = Trt X0 /solution;
 lsmeans Trt / diff;
 estimate 'Trt Effect' Trt 1 -1;
run;
ods output Estimates=est_LOCF_REML;
ods select Estimates;
proc mixed data=LOCF;
 class Trt Subject;
 model Yd = Trt X0 /solution;
 lsmeans Trt / diff;
 estimate 'Trt Effect' Trt 1 -1;
run;
```

Letting $r_{ik} = \begin{cases} 1 & \text{if } y_{ik} \text{ is observed} \\ 0 & \text{if } y_{ik} \text{ is missing} \end{cases}$ be the response/non-response indicator variable for the change from baseline response variable, $y_{ik} = y_{ik1} - y_{ik0}$, we also perform a sensitivity analysis based on the pattern mixture model

$$\pi(y_{ik}, r_{ik}|x_{ik}, x_{ik0}) = \pi(y_{ik}|x_{ik}, x_{ik0}, r_{ik})\pi(r_{ik})$$

where $x_{ik0} = (y_{ik0} - \bar{y}_0)$ is the centered baseline covariate centered about the mean among subjects with complete data

$$\bar{y}_0 = \sum_{i=1}^{10}\sum_{k=1}^{2} r_{ik}y_{ik0} / \sum_{i=1}^{10}\sum_{k=1}^{2} r_{ik}$$

and $\pi(y_{ik}|x_{ik}, x_{ik0}, r_{ik})$ is the pdf of the ANCOVA mixture model

$$y_{ik} = \beta_0 + (\beta_d + (1 - r_{ik})\Delta)x_{ik} + \beta x_{ik0} + \epsilon_{ik};\ i = 1, \ldots, 10;\ k = 1, 2.$$

Under this model, Δ is a sensitivity parameter specific to the active drug group since there are no missing values in the placebo control group (i.e., $r_{i1} = 1$ for $i = 1, \ldots, 10$). It measures the difference in the treatment effect between responders and non-responders. Specifically, the marginal treatment effect under this pattern mixture model is

$$\begin{aligned} E(y_{i2} - y_{i1}) &= \pi E(y_{i2}|r_{i2} = 1) + (1 - \pi)E(y_{i2}|r_{i2} = 0) - E(y_{i1}) \\ &= \pi\beta_d + (1 - \pi)(\beta_d + \Delta) = \beta_d + (1 - \pi)\Delta \end{aligned}$$

where $\pi = \Pr(r_{i2} = 1)$. It is clear that the parameters of this ANCOVA mixture model are underidentified in that Δ cannot be estimated from the observed data. Nevertheless, by holding Δ fixed and varying its value from -1.00 to 1.00 in 0.25 increments, we can assess what impact departures from MAR ($\Delta = 0$) have on the treatment effect. Negative values of Δ reflect a lowering of the estimated treatment effect while positive values reflect an increase in the estimated treatment effect. The following code allows us to perform this sensitivity analysis taking into account the uncertainty we have when estimating the mixing proportion, π.

Program 6.6

```
/* Part 6: PMM ANCOVA based on incomplete MNAR data    */
/* Pattern Mixture Model(PMM) for sensitivity analysis*/
data MNAR_sensitivity;
 set MNAR;
 do Delta = -1 to 1 by .25;
 output;
 end;
run;
data MNAR_sensitivity;
 set MNAR_sensitivity;
 ind=1; response=r1; output;
 ind=0; response=Yd; output;
run;
data MNAR_sensitivity;
 set MNAR_sensitivity;
 if ind=1 and Trt='Placebo' then delete;
 if ind=1 and Yd=. then Yd=0;
run;
```

```
proc sort data=MNAR_sensitivity;
 by Delta ind Trt Subject;
run;
ods output additionalestimates=ae;
ods select additionalestimates;
proc nlmixed data=MNAR_sensitivity df=15;
 by Delta;
 parms Beta0=2.0131 Beta_d=0.3848 Beta=-0.4071
       eta0=1;
 PredMean = Beta0 + Beta_d*(Trt=''Drug'') + Beta*X0;
 SigmaSq=0.2194;
 linpred = eta0;
 p1 = 1/(1+exp(-linpred));
 ll_r = r1*log(p1) + (1-r1)*log(1-p1);
 ll_Y = (- 0.5*log(2*CONSTANT('PI'))
         - 0.5*((Yd - PredMean)**2)/SigmaSq
         - 0.5*log(SigmaSq));
 ll = ll_r*ind + ll_Y*(1-ind);
 Trt_Effect = p1*Beta_d + (1-p1)*(Beta_d+Delta);
 Trt_Effect_Unadj = 0.80*Beta_d + 0.20*(Beta_d+Delta);
 model response ~ General(ll);
 estimate ''Trt Effect (Adjusted SE)''   Trt_Effect;
 estimate ''Trt Effect (Unadjusted SE)'' Trt_Effect_Unadj;
run;
proc print data=ae;
 title ''PMM with adjusted and unadjusted standard errors'';
run;
```

In order to compute standard error estimates of the estimated treatment effect adjusted for the estimation of $\pi = \Pr(r_{i2} = 1)$, we fit the joint pattern mixture model by estimating $\Pr(r_{i2} = 1)$ using an intercept-only logistic regression model fit to the active drug group only. This is done by defining an extended dataset, `data MNAR_sensitivity`, which outputs an indicator variable, `ind`, that takes a value of 1 when modeling the response/non-response indicator, `r1`, and a value of 0 when modeling the response variable, `Yd`. Finally, in order to compare results against the REML output under PROC MIXED, we held the variance estimate fixed at 0.2194 and the degrees of freedom for inference was held fixed at DF=15 (using the DF=option of NLMIXED).

The results of these different analyses, including the results one would have obtained if one had observed the complete data, are shown in Table 6.5. Of course, the analyst never sees the complete-data results and, instead, must decide on the basis of the incomplete data what set of results makes the most sense for the data at hand. Given the plausible range of values for Δ and assuming the model fit to the data is otherwise valid (an unverifiable assumption), it would appear reasonable to assume that the unobserved responses from the two subjects who failed to complete the study would not have led to a statistically significant negative treatment effect if in fact their responses had been observed. In fact, given a Δ of -1.00 still results in a positive treatment effect of 0.1848 suggests some benefit to treatment. However, without additional information related to cause of dropout, it would be impossible to discern whether the true treatment effect is closer to 0.50 than it is to 0.3848 as estimated under the MAR assumption. For example, information related to reasons for withdrawal from the study as well as external information from other related studies might be used to better judge the efficacy of treatment. However, in the absence of such information, the sensitivity analysis can be used to at least "point" the investigator in the right direction for future studies—studies

Table 6.5 Summary of different point estimates of treatment effect based on the difference in mean change from baseline between placebo control and active drug. Missing data are truly MNAR. [1]A pattern mixture model (PMM) was used to perform sensitivity analysis to departures from MAR ($\Delta = 0$).

ANCOVA Method	Δ	$E(y_{i2} - y_{i1})$ Estimate	Standard Error	DF	t Value	p-value
Complete-data	–	0.5485	0.2368	17	2.32	0.0333
REML - MAR	–	0.3848	0.2240	15	1.72	0.1064
REML - LOCF	–	0.3243	0.2105	17	1.54	0.1419
PMM[1]	-1.00	0.1848	0.2573	15	0.72	0.4835
	-0.75	0.2348	0.2433	15	0.97	0.3497
	-0.50	0.2848	0.2328	15	1.22	0.2400
	-0.25	0.3348	0.2263	15	1.48	0.1596
	0	0.3848	0.2240	15	1.72	0.1064
	0.25	0.4348	0.2263	15	1.92	0.0738
	0.50	0.4848	0.2328	15	2.08	0.0548
	0.75	0.5348	0.2433	15	2.20	0.0440
	1.00	0.5848	0.2573	15	2.27	0.0382

guided by possibly increasing the sample size, formulating incentives to minimize dropout, and capturing auxiliary data that may be useful in modeling the dropout mechanism.

In the following section, we turn our attention to several more complex case studies involving missing data due to dropout. Two of the case studies will utilize a SP model approach assuming parametric and semiparametric survival models for time-to-dropout. In one case the response model is that of a LME model (the MDRD study) and in the second case the response model is a GLME model (the psychiatric data). In both cases, the joint model (6.45) falls under the category of a GNLME model since $\pi(T_i|\boldsymbol{b}_i)$ will be inherently nonlinear in the random effects regardless of what form the response model $\pi(\boldsymbol{y}_i|\boldsymbol{b}_i)$ takes.

6.6 Missing data—case studies

In this section, we consider three case studies involving missing data related to dropout. These studies all tend to be larger-scale studies and all involve fitting some form of regression model to the outcome variable of interest. In all three cases dropout is substantial with at least 18% of the subjects failing to compete the trial. The first study involves re-analyzing the bone mineral density data of §2.2.2 using multiple imputation to take into account missing data due to dropout. The second study compares the rate of change in glomerular filtration rates among chronic kidney disease patients randomized to one of four interventions in the Modification of Diet in Renal Disease (MDRD) study. Here we apply a random-effects pattern-mixture model and a shared parameter (SP) model to compare the rate of change in the presence of non-ignorable dropout. The third study also uses a SP model in a re-analysis of data from the NIMH Schizophrenia study as presented in §5.2.3. Specifically, we apply a SP model with the goal of comparing the rate of change in disease severity among patients randomized to receive an active antipsychotic drug versus placebo control taking into account a differential dropout between active treatment versus placebo control.

6.6.1 Bone mineral density data—continued

The first case study we consider is the bone mineral density study presented and analyzed in (§2.2.2). Recall the data consist of total body bone mineral density, or TBBMD (g/cm^2), measured roughly every 6 months over a two-year period in a group of 112 healthy adolescent women. The women were randomized to receive a daily calcium supplement (500 mg calcium citrate malate) or placebo over the course of the two-year study and the goal was to determine if the administration of the daily calcium supplement would improve TBBMD over time. Of the 112 women who started the study, only 91 completed the trial resulting in incomplete data in 18.75% of the participating women. In §2.2.2 we performed both a complete-case and available-case analysis of the data by fitting a GMANOVA model assuming a linear growth curve over time. The complete-case analysis is restricted to the 91 women with complete data and is valid under MCAR while the available-case analysis is on all 112 women and is valid under MAR.

Under the balanced data setup described in §2.2.2, we define $y_{ij} = y_i(t_j)$ to be the observed value of TBBMD for the ith woman measured at time t_j where $t_j = 0, 0.5, 1.0, 1.5, 2.0$ are the observation times rounded to the nearest half-year. To investigate possible mechanisms resulting in incomplete data, we first examine the different patterns of missing data within each treatment group (C = daily calcium supplement treatment group, P = placebo control group). This is easily accomplished using the MI procedure of SAS as follows.

Program 6.7

```
proc sort data=TBBMD out=example6_6_1;
 by group subject;
run;
ods listing close;
ods output MissPattern = MissPattern;
proc mi data=example6_6_1 nimpute=0;
 by group;
 var tbbmd1-tbbmd5;
run;
ods listing;
proc print data=MissPattern noobs split='|';
 by group; id group;
 var group2 tbbmd1_miss tbbmd2_miss tbbmd3_miss tbbmd4_miss
     tbbmd5_miss Freq Percent;
 label group='Group' group2='Pattern';
 title 'Missing data patterns by treatment group';
proc print data=MissPattern noobs split='|';
 by group; id group;
 var group2 tbbmd1 tbbmd2 tbbmd3 tbbmd4 tbbmd5;
 label group='Group' group2='Pattern';
 format tbbmd1-tbbmd5 8.6;
 title 'Mean TBBMD by treatment and missing data patterns';
run;
```

The pattern of missing data shown in Output 6.1 is strictly monotonic indicating missing data is the result of subject withdrawal from the study.

Output 6.1: Results from PROC MI showing the monotonic patterns of missing data by treatment group and what the mean TBBMD values are according to treatment and pattern of missing data.

Missing data patterns by treatment group

Group	Pattern	tbbmd1	tbbmd2	tbbmd3	tbbmd4	tbbmd5	Freq	Percent
C	1	X	X	X	X	X	44	80.00
	2	X	X	X	X	.	2	3.64
	3	X	X	X	.	.	2	3.64
	4	X	X	.	.	.	4	7.27
	5	X	.	.	.	.	3	5.45
P	1	X	X	X	X	X	47	82.46
	2	X	X	X	X	.	1	1.75
	3	X	X	X	.	.	3	5.26
	4	X	X	.	.	.	2	3.51
	5	X	.	.	.	.	4	7.02

Mean TBBMD by treatment and missing data patterns

Group	Pattern	tbbmd1	tbbmd2	tbbmd3	tbbmd4	tbbmd5
C	1	0.881250	0.908909	0.938045	0.965227	0.988159
	2	0.853500	0.893500	0.934000	0.948500	.
	3	0.888500	0.921000	0.947000	.	.
	4	0.829500	0.837250	.	.	.
	5	0.949333	.	.	.	.
P	1	0.870340	0.890681	0.915574	0.941787	0.958234
	2	0.840000	0.861000	0.904000	0.935000	.
	3	0.876667	0.890000	0.895333	.	.
	4	0.857500	0.880000	.	.	.
	5	0.875750	.	.	.	.

It is not clear from looking at the mean TBBMD profiles across the different patterns in Output 6.1 whether there is any indication that missing data may be MCAR. One might therefore consider applying Ridout's logistic regression to test whether the assumption that missing data are MCAR can be ruled out (Ridout, 1991). To perform Ridout's test for MCAR, let

$$r_{ij} = \begin{cases} 1 & \text{if } y_i(t_j) \text{ is measured at } t_j \\ 0 & \text{if } y_i(t_j) \text{ is missing at } t_j \end{cases}, \; j = 1, 2, 3, 4, 5. \tag{6.46}$$

be the binary indicator of response/non-response at each visit as defined in §6.2. Following Ridout (1991) [see also Diggle et. al., 1994, Chapter 11, pp. 211-216 and Curran et. al., 1998], one can test against the assumption of MCAR by formulating the logistic hazard rate

$$\text{logit}\{\Pr[r_{ij} = 0 | r_{i(j-1)} = 1, \boldsymbol{X}_i, h(\boldsymbol{y}_i(\boldsymbol{t}_{j-1}))]\} \tag{6.47}$$

where $\boldsymbol{t}_{j-1} = (t_1, t_2, \ldots, t_{j-1})'$ is the vector of observation times through time t_{j-1}, $\boldsymbol{y}_i(\boldsymbol{t}_{j-1}) = (y_{i1}, y_{i2}, \ldots, y_{i(j-1)})'$ is the vector of observed responses up to and including time t_{j-1} and $h(\boldsymbol{y}_i(\boldsymbol{t}_{j-1})) = \mathbb{Z}_i(t_{j-1})$ is any covariate function of the observed responses prior to time t_j. Setting $\boldsymbol{X}_i$ to be the matrix of baseline covariates defined by treatment group and visit, we fit model (6.47) to the following model specifications:

Model 1: $h(\boldsymbol{y}_i(\boldsymbol{t}_{j-1})) = \sum_{k=1}^{j-1} y_{ik}/(j-1)$ is the average TBBMD up through time t_{j-1} (labeled as `tbbmd_avg`),
Model 2: $h(\boldsymbol{y}_i(\boldsymbol{t}_{j-1})) = y_{i(j-1)}$ is the last observed TBBMD prior to visit j (labeled as `tbbmd_last`),
Model 3: $h(\boldsymbol{y}_i(\boldsymbol{t}_{j-1})) = (y_{i(j-1)} - y_{i1})$ is the change from baseline TBBMD at time t_{j-1} (labeled as `tbbmd_change`).

The three models were fit using logistic regression as implemented in the GENMOD procedure and summarized using the REPORT procedure. The programming statements required to define the TBBMD-based covariates as well as arrange the data for GENMOD and perform Ridout's test are as follows.

Program 6.8

```
data example6_6_1_down;
 set example6_6_1;
 by group subject;
 array t tbbmd1-tbbmd5;
 array tsum tbbmd_sum1-tbbmd_sum5;
 lastvisit=1;
 if t[5]>. then lastvisit=5;
 else if t[4]>. and t[5]=. then lastvisit=4;
 else if t[3]>. and t[4]=. then lastvisit=3;
 else if t[2]>. and t[3]=. then lastvisit=2;
 tsum[1]=t[1]; tsum[2]=t[1];
 do i=3 to 5 by 1;
    tsum[i]=(tsum[i-2]+t[i-1]);
 end;
 tbbmd_last=t[1]; tbbmd_change=t[1];
 do visit=1 to 5 by 1;
    tbbmd=t[visit];
    tbbmd_avg=tsum[visit]/max(1,(visit-1));
    response=(t[visit]>.);
    years=0.5*(visit-1);
    if visit> 1 then do;
       tbbmd_last=t[visit-1];
       tbbmd_change=(t[visit-1]-t[1]);
    end;
    keep subject group data visit lastvisit years tbbmd
         tbbmd_last tbbmd_avg tbbmd_change response;
    output;
 end;
run;
/*--- A macro to perform the Ridout test for MCAR ---*/
/*--- MODEL:      defines a model ID number        ---*/
/*--- VISIT1:    defines the first visit to be     ---*/
/*---             included in the model            ---*/
/*--- COVARIATE: defines a TBBMD-based covariate  ---*/
%macro Ridout(model=1, visit1=2, covariate=tbbmd_avg);
 ods listing close;
 ods output parameterestimates=peout;
 proc genmod data=example6_6_1_down;
  where &visit1 <= visit <= lastvisit+1;
  class group visit subject;
  model response = group visit &covariate / dist=bin;
 run;
 data peout;
  length covariate $12;
  set peout;
  Model=&model;
  Covariate=''&covariate'';
 run;
```

```
data pe;
 set pe peout;
 if Parameter='Scale' then delete;
run;
ods listing;
%mend Ridout;
data pe;
%Ridout(model=3, visit1=3, covariate=tbbmd_change);
%Ridout(model=2, visit1=2, covariate=tbbmd_last);
%Ridout(model=1, visit1=2, covariate=tbbmd_avg);
proc sort data=pe;
 by Model;
run;
proc report data=pe headskip split='*' nowindows ;
 column Model Parameter Level1 DF Estimate StdErr
        ChiSq ProbChiSq;
 define Model  / group  'Model';
 define Parameter / 'Parameter' width=12 display;
 define Level1 / 'Value' width=5 display;
 define DF / display;
 define Estimate / display;
 define StdErr / 'Std Error' width=9 display;
 define ChiSq / 'ChiSq' display;
 define ProbChiSq / 'p-value' width=7 display;
run;
quit;
```

Using ARRAY statements, we first define the functional covariate values of the TBBMD response variable, i.e., `tbbmd_avg`, `tbbmd_last`, `tbbmd_change`. We then output these values so the data are arranged vertically as required for GENMOD. The macro %Ridout then applies logistic regression to fit the logistic hazards model for dropout (6.47) using each of the TBBMD covariate functions. Note that for model 3, we specify `visit1=3` in the macro call to %Ridout so as to restrict the logistic regression to historical values of TBBMD between visit 3 and the last observed visit. This is because the first change from baseline value starts with visit 2 and is therefore undefined for those who dropout at visit 2.

The results from the three logistic regression models are summarized in Output 6.2. Although not significant at the 0.05 level, there is some indication that missing values may not be MCAR based on model 3. Under model 3, the observed changes from baseline appear to be marginally associated

Output 6.2: Ridout's test of MCAR for select covariate functions of TBBMD.

Model	Parameter	Value	DF	Estimate	Std Error	ChiSq	p-value
1	Intercept		1	−2.7880	2.7847	1.00	0.3167
	group	C	1	0.1526	0.4518	0.11	0.7356
	group	P	0	0.0000	0.0000	.	.
	visit	2	1	0.9027	1.0466	0.74	0.3884
	visit	3	1	0.8165	1.0859	0.57	0.4521
	visit	4	1	0.3956	0.8130	0.24	0.6265
	visit	5	0	0.0000	0.0000	.	.
	tbbmd_avg		1	−1.0303	4.0261	0.07	0.7980
2	Intercept		1	−2.0288	3.2969	0.38	0.5383
	group	C	1	0.1679	0.4539	0.14	0.7114
	group	P	0	0.0000	0.0000	.	.
	visit	2	1	0.5847	0.7527	0.60	0.4372

	visit	3	1	0.5210	0.7475	0.49	0.4858
	visit	4	1	0.4379	0.7507	0.34	0.5596
	visit	5	0	0.0000	0.0000	.	.
	tbbmd_last		1	−1.5460	3.4537	0.20	0.6544
3	Intercept		1	−2.4574	1.0451	5.53	0.0187
	group	C	1	0.5190	0.5724	0.82	0.3646
	group	P	0	0.0000	0.0000	.	.
	visit	3	1	−0.2614	0.9576	0.07	0.7848
	visit	4	1	0.0522	0.8089	0.00	0.9486
	visit	5	0	0.0000	0.0000	.	.
	tbbmd_change		1	−16.9284	13.1017	1.67	0.1963

with dropout although a large standard error, due possibly to the reduced number of observations, does accurately reflect the degree of uncertainty associated with this particular covariate effect (-16.9284 ± 13.1017, $p = 0.1963$). Of course one could formulate any number of alternative logistic hazard models designed to test the MCAR assumption using different TBBMD-based covariates. For example, one could include models that examine interactions between the TBBMD covariate and other covariates in the model (e.g., treatment group or visit).

Based on the results from the main-effects model 3 (Output 6.2), there is at least a hint of evidence to suggest that missing data may not be MCAR. To gain further insight into the missing data mechanism, we re-analyzed the data under the GMANOVA linear growth curve model of §2.2.2 by applying a complete-case (CC) analysis, an available-case (AC) analysis and an analysis based on multiple imputation (MI). In each case, we applied MLE to estimate the model parameters. The estimand of interest is the difference in the population slopes between active treatment (calcium supplement) and placebo control groups. The CC-based MLE of this estimand will be unbiased and asymptotically normally distributed under MCAR while the AC- and MI-based MLE's will both be unbiased and asymptotically normal under either MCAR or MAR. The SAS programming statements required to compare the three estimates of the slope differences are shown below. Multiple imputation was carried out using the default settings of the MCMC method but with 10 imputations (PROC MI option `nimpute=10`). The MCMC method assumes the complete data model follows a multivariate normal distribution with an unstructured covariance matrix.

Program 6.9

```
** Create datasets with imputed missing data **;
proc mi data=example6_6_1 seed=9385672 nimpute=10
        out=example6_6_1_MI;
 by group;
 mcmc chain=multiple displayinit initial=em(itprint);
 var tbbmd1-tbbmd5;
run;
proc sort data=example6_6_1_MI;
 by _imputation_ group subject;
run;
** Convert horizontal data to vertical format **;
data example6_6_1_MI_down;
 set example6_6_1_MI;
 by _imputation_ group subject;
 array d date1-date5;
 array t tbbmd1-tbbmd5;
```

```
 do visit=1 to 5 by 1;
    date=d[visit];
    tbbmd=t[visit];
    years=0.5*(visit-1);
    format date date7.;
    keep _imputation_ group subject date
         visit years tbbmd;
    output;
 end;
run;
** Perform a complete-case analysis and output    **;
** the difference in slopes (the target estimand) **;
ods listing close;
ods output Estimates=CCestimates;
proc mixed data=example6_6_1_down method=ml;
 where data='Complete';
 class subject group visit;
 model tbbmd=group group*years /noint solution;
 repeated visit / subject=subject type=unr;
 estimate 'slope diff' group*years 1 -1;
run;
** Perform an available-case analysis and output  **;
** the difference in slopes (the target estimand) **;
ods output Estimates=ACestimates;
proc mixed data=example6_6_1_down method=ml;
 where tbbmd>.;
 class subject group visit;
 model tbbmd=group group*years /noint solution;
 repeated visit / subject=subject type=unr;
 estimate 'slope diff' group*years 1 -1;
run;
** Perform MI analyses by imputed datasets and output **;
** the differences in slopes (the target estimand)    **;
ods output Estimates=MIestimates;
proc mixed data=example6_6_1_MI_down method=ml;
 by _imputation_;
 class subject group visit;
 model tbbmd=group group*years /noint solution;
 repeated visit / subject=subject type=unr;
 estimate 'slope diff' group*years 1 -1;
run;
ods listing;
** Summarize the MI-based estimands using MIANALYZE **;
ods output parameterestimates=MIestimates_final;
proc mianalyze data=MIestimates;
 modeleffects Estimate;
 stderr StdErr;
run;
** Compare CC, AC and MI based ML estimates of    **;
** the difference in slopes (the target estimand) **;
data estimates;
 set CCestimates(in=a) ACestimates(in=b)
     MIestimates(in=c) MIestimates_final(in=d);
```

```
  Nimpute=_imputation_;
  if a then Method=''CC-MLE              '';
  if b then Method=''AC-MLE              '';
  if c then Method=''MI-individual MLE's'';
  if d then Method=''MI-summary MLE      '';
run;
proc print data=estimates noobs;
 var Method Nimpute Estimate StdErr DF tValue Probt;
run;
```

It is difficult to determine based on the results shown in Output 6.3 whether or not the MCAR assumption is viable for this particular study. There appear to be very minor differences between the complete-case (CC) and available-case (AC) results versus the results obtained by multiple imputation (MI). Given how close the three methods agree, it may well be that the dropout mechanism is a mixture of MCAR for some women and MAR for others but without further information about the cause of dropout, etc., it would be virtually impossible to determine what that mixture might be.

Output 6.3: Summary results comparing complete-case (CC), available-case (AC) and multiple imputation (MI) based MLE's of the difference in population slopes for the bone mineral density data

Method	Nimpute	Estimate	StdErr	DF	tValue	Probt
CC−MLE	.	0.008597	0.003090	89	2.78	0.0066
AC−MLE	.	0.008704	0.003012	110	2.89	0.0047
MI−individual MLE's	1	0.006525	0.002843	110	2.30	0.0236
MI−individual MLE's	2	0.007908	0.002823	110	2.80	0.0060
MI−individual MLE's	3	0.007470	0.002912	110	2.57	0.0116
MI−individual MLE's	4	0.01004	0.002782	110	3.61	0.0005
MI−individual MLE's	5	0.009591	0.002892	110	3.32	0.0012
MI−individual MLE's	6	0.007568	0.002875	110	2.63	0.0097
MI−individual MLE's	7	0.008609	0.002732	110	3.15	0.0021
MI−individual MLE's	8	0.007919	0.002946	110	2.69	0.0083
MI−individual MLE's	9	0.01097	0.002663	110	4.12	<.0001
MI−individual MLE's	10	0.008295	0.002933	110	2.83	0.0056
MI−summary MLE	.	0.008490	0.003172	230	2.68	0.0080

In this example, we have presented three different methods of analysis of the bone mineral density data; a complete-case likelihood analysis, an available-case likelihood analysis and a likelihood analysis based on multiple imputation. The first method provides for valid inference under the MCAR mechanism while the latter two methods provide for valid inference under a MAR mechanism. Since one can only use Ridout's test to rule out the possibility that missing data are MCAR, the above analyses do not prove dropout is completely at random. However, there is very little evidence to suggest anything to the contrary. Given the relatively low frequency of dropout and the lack of any auxiliary information that may help in modeling the dropout mechanism, we avoided using inverse probability of weighting as a means for analyzing the data though one could certainly apply this approach as well.

6.6.2 MDRD study—GFR data

Beck et. al. (1991) describe the design and statistical analysis plan for a multi-center randomized controlled trial investigating the effects of diet and blood pressure control on the progression of renal disease, the so-called Modification of Diet in Renal Disease (MDRD) study. The study was designed to investigate whether a modification of dietary protein and phosphorous intake and/or reduction of blood pressure will significantly reduce the rate of progression of chronic kidney disease (CKD) as measured by a subject's glomerular filtration rate (GFR). The overall trial was divided into two separate studies, Study A was restricted to CKD patients with a baseline GFR of 25-55 ml/min/1.73 m^2body surface area and Study B was restricted to CKD patients with a baseline GFR of 13-24 ml/min/1.73 m^2 reflecting a more advanced stage of CKD. Final results of the trial, as summarized by Klahr et. al. (1994), indicate that among patients with moderate renal insufficiency (Study A), there was a slower decline in GFR among those randomized to a low-protein diet. There was no evidence of any effect due to dietary or blood pressure control on the rate of decline in GFR among patients with more severe renal insufficiency (Study B).

In the planning stages of the trial, Beck et. al. (1991) recognized the need for an analysis that takes into account the possibility that non-ignorable dropout due to a terminal event (dialysis, transplantation, a serious medical condition or death) may impact the results from Study B since this study was restricted to patients with a more advanced stage of CKD. This possibility is made even stronger given that both clinical center personnel and study subjects were masked to the results of GFR during the course of follow-up (Beck et. al., 1991). To avoid possible bias due to non-ignorable dropout, Klahr et. al. (1994) used a single-slope informative censoring model described by Schluchter (1992) to compare the rate of decline in GFR among patients.

In this section, we examine several strategies for comparing the mean rate of decline in GFR among patients in MDRD Study B. The primary objective of Study B was to estimate and compare the rate of decline in GFR among more severely impaired CKD patients randomized to one of four interventions. The four intervention groups correspond to a 2×2 factorial design with one intervention consisting of patients receiving a low-protein diet (Diet L) or a very-low-protein diet (Diet K) and the other intervention consisting of patients receiving either normal blood pressure control (Normal BP) or low blood pressure control (Low BP). Normal BP control had a target mean arterial blood pressure (MAP) of 107 mmHg and Low BP control had a target MAP of 92 mmHg. The decline in GFR was modeled as a linear random effects model with random slopes representing patient-specific rates of decline in GFR (Schluchter et. al., 2001). A simple profile plot of GFR versus time (Figure 6.1) confirms this choice of model. The following describe some of the essential characteristics of the MDRD Study B dataset as presented and analyzed by Schluchter et. al. (2001) including a description of the causes of dropout (see also the author's Web page). A more thorough description of the study design, etc. can be found in Beck et. al. (1991).

- Experimental unit or cluster is an individual patient with CKD
  - 255 patients with more severe renal impairment were enrolled
- Response variable:
  - y_{ij} = GFR (ml/min/1.73m^2) measured on the ith subject on the jth occasion ($i = 1, \ldots, 255$ subjects; j ranging from 1 to possibly 13 time points). GFR measured via urinary clearance of [^{125}I]iothalamate.
- One within-unit covariate:
  - t_{ij} =planned follow-up times with times varying by subject due to staggered entry into the study and possibly on intermittently missed visits ($t_{ij} = 0, 2, 4, 8, 12$ months and every 4 months thereafter until the study end; planned follow-up was from 1.5 to 4 years)
- Five between-subject covariates:
  - G_{1i} =Indicator variable for Group 1 = Diet K, Low BP control
  - G_{2i} =Indicator variable for Group 2 = Diet K, Normal BP control
  - G_{3i} =Indicator variable for Group 3 = Diet L, Low BP control
  - G_{4i} =Indicator variable for Group 4 = Diet L, Normal BP control
  - u_i =log baseline urine protein concentration
- Structural model: $y = \beta_1 + \beta_2 t$ (see Figure 6.1)

Figure 6.1 Patient-specific profile plots depicting GFR trends over time

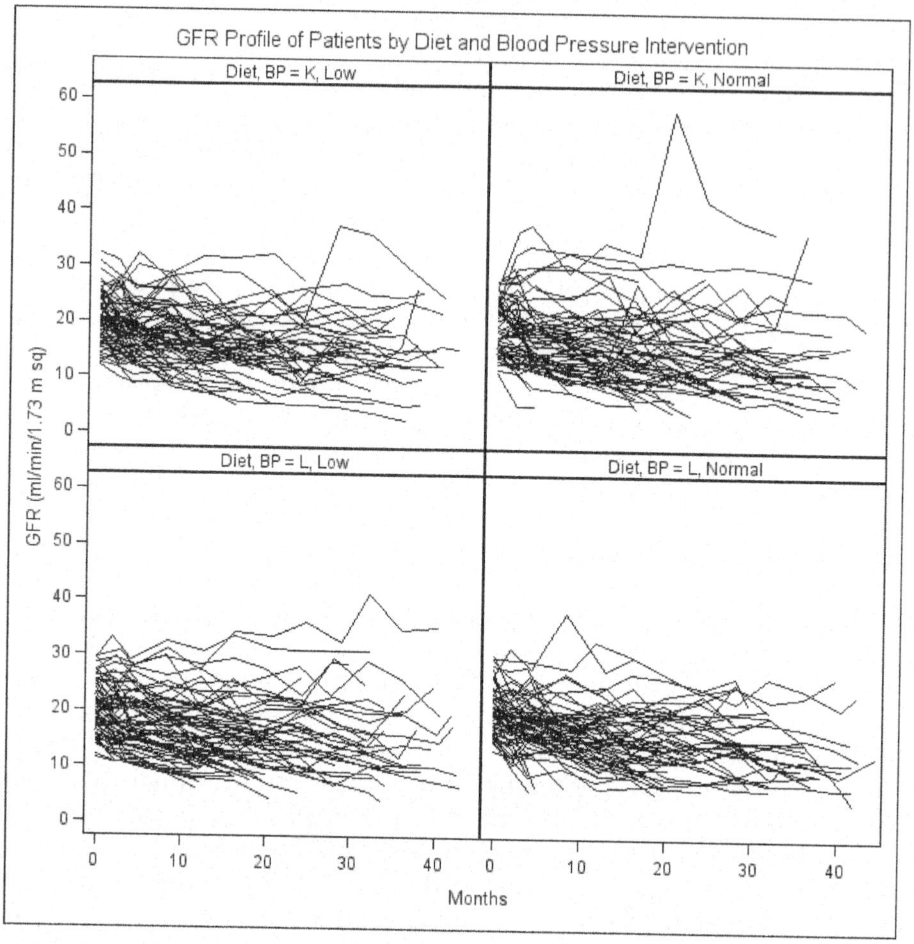

- Goal: Estimate and compare the mean rate of decline in GFR (β_2) among the four intervention groups
- Primary Issue: Patient dropout was significant with nearly 40% of the patients failing to complete the study for reasons indicated in the following table

Cause of dropout	N	%
Dialysis	81	32%
Kidney Transplant	11	4%
Death	5	2%
Other Medical	4	2%
Total	101	40%

With the possible exception of other medical complications, the above dropout events can all be considered as terminal events. In particular, dialysis, transplantation and death are all events that either make measuring GFR impossible (death and transplantation) or are known to alter kidney function in a manner that is no longer consistent with the underlying objective of the study. For example, a successful transplantation results in the removal of the patient's diseased kidneys making any subsequent measure of GFR on those kidneys impossible. Likewise, patients that begin dialysis do so because they have made a transition from chronic kidney disease to end-stage renal disease in which the kidney can no longer effectively filter toxins and excess water. In this case, the initiation of dialysis results in a sharp exponential rate of decline in kidney function (e.g., Lysaght et. al., 1991). This nonlinear decay rate no longer reflects the mere progression of chronic kidney disease but rather it reflects the combined effects of dialysis (e.g., nephrotoxic, ischemic and hydrodynamic effects of hemodialysis) on the decline of whatever kidney function remains among patients with end-stage renal disease (Lysaght et. al., 1991). In fact, many hemodialysis patients become functionally anuric within one year following initiation of dialysis meaning the kidney is no longer functional (e.g., Moist et. al., 2000).

In studies like the MDRD Study B where the estimand of interest is a rate of change in the response variable, there is a real possibility that dropout will be non-ignorable. For example, among CKD patients, dialysis is usually started when a patient's GFR reaches a certain threshold value, typically between 5-10 ml/min/1.73m^2. Such a threshold value is predicated on the patient's intrinsic rate of change in GFR over time, i.e., the patient-specific slope. Consequently, with 40% (101 of 255) of the patients unable to complete their planned follow-up due to renal-related terminal events like dialysis, transplantation or death, the likelihood that dropout is non-ignorable is extremely high. To address this concern, we apply a standard LME model to the analysis of the MDRD data assuming an ignorable dropout mechanism (either S-RD or S-CRD) and compare the results of this analysis to analyses assuming a non-ignorable dropout mechanism (S-NARD). Analyses under non-ignorable dropout are carried out using both a random-effects pattern-mixture modeling approach and a random-effects selection modeling approach. The former is based on a modification of the conditional linear model of Wu and Bailey (1988, 1989) while the latter is based on a shared parameter model such as described by Vonesh et. al. (2006).

A partial listing of the data provided for this analysis is shown in Output 6.4 (excluding log protein concentrations). The 101 dropout events were lumped into a single event category for analysis as was done by Schluchter et. al. (2001) and Vonesh et. al. (2006). As such, the observed follow-up time for each patient is defined in terms of an event time random variable $T_i = \min(T_{1i}, T_{2i})$ which is the time from enrollment into the study to early stopping due either to dialysis, transplantation, or death (T_{1i}) or to other medical complications (T_{2i}). In addition, let C_i be a non-informative censoring time for patients who do not dropout due to one of the defined events. In particular, C_i is the time from enrollment to administrative termination of patient follow-up or until follow-up is terminated for reasons felt by the investigators to be unrelated to the outcome of interest (Schluchter et. al., 2001). The observed event times are therefore defined by $T_i^o = \min(T_i, C_i)$ and δ_i where $\delta_i = 1$ if $T_i^o = T_i$ and $\delta_i = 0$ if $T_i^o = C_i$. Shown in Output 6.4 is a printout of data for two patients (ptid=1 and ptid=7).

Output 6.4: Partial listing of data for two patients. The treatment indicator variables are G1 (diet K, low BP; N=65 or 25.5%), G2 (diet K, normal BP; N=61 or 23.9%), G3 (diet L, low BP; N=67 or 26.3%) and G4 (diet L, normal BP; N=62 or 24.3%). The SAS variable `T` is the observed follow-up time and `Dropout` is a dropout indicator. The data are provided courtesy of G. Beck from the Cleveland Clinic Foundation on behalf of the MDRD Study Group.

ptid	G1	G2	G3	G4	Months	GFR	T	Dropout
1	0	0	0	1	0	24.33	38.37	0
1	0	0	0	1	1.74	23.75	38.37	0
1	0	0	0	1	4.5	21.84	38.37	0
1	0	0	0	1	7.95	18.72	38.37	0
1	0	0	0	1	11.89	15.51	38.37	0
1	0	0	0	1	16.26	14.97	38.37	0
1	0	0	0	1	20.14	12.67	38.37	0

1	0	0	0	1	24.34	12.08	38.37	0
1	0	0	0	1	28.42	14.15	38.37	0
1	0	0	0	1	32.16	10.57	38.37	0
1	0	0	0	1	35.84	10.27	38.37	0
7	0	0	0	1	0	16.45	10.41	1
7	0	0	0	1	2.07	15.31	10.41	1
7	0	0	0	1	4.7	12.54	10.41	1
7	0	0	0	1	8.11	6.61	10.41	1

Assuming intermittent missing values are MAR and in the absence of dropout or under the assumption that dropout is ignorable, the rate of change in GFR over time can be modeled via the LME model

$$\begin{aligned}\text{Stage 1: } & y_{ij} = \beta_{1i} + \beta_{2i}t_{ij} + \epsilon_{ij},\ \epsilon_{ij} \sim \text{iid } N(0,\sigma^2) \\ \text{Stage 2: } & \beta_{1i} = \beta_{11}G_{1i} + \beta_{12}G_{2i} + \beta_{13}G_{3i} + \beta_{14}G_{4i} + b_{1i} \\ & \beta_{2i} = \beta_{21}G_{1i} + \beta_{22}G_{2i} + \beta_{23}G_{3i} + \beta_{24}G_{4i} + b_{2i}\end{aligned} \tag{6.48}$$

where G_{1i}, G_{2i}, G_{3i} and G_{4i} are group indicator variables defining which diet and blood pressure group the ith patient belongs $(i = 1,\ldots,n)$. For notational convenience, we will let $\boldsymbol{y}_i = (y_{i1}, y_{i2},\ldots,y_{ip_i})'$ denote the vector of observed GFR values for the ith patient at times t_{ij}, $(j = 1,\ldots,p_i)$ rather than $\boldsymbol{y}_i^o$ as defined and used in §6.2-§6.5. Throughout this example, we will assume intermittent missing values are MAR and that the random effects are assumed to be multivariate normal with mean $\mathbf{0}$ and unstructured covariance matrix $\boldsymbol{\Psi}$, i.e.,

$$\begin{pmatrix} b_{1i} \\ b_{2i} \end{pmatrix} \sim \text{iid } N_2\left[\begin{pmatrix} 0 \\ 0 \end{pmatrix}, \begin{pmatrix} \psi_{11} & \psi_{12} \\ \psi_{21} & \psi_{22} \end{pmatrix}\right].$$

In terms of the primary goal of MDRD Study B, this LME model may be regarded as the primary response model of interest and the primary estimands of interest are the treatment group slopes $\beta_{21}, \beta_{22}, \beta_{23}, \beta_{24}$ and their differences. One may include the log baseline urine protein concentration as a covariate in the model but we chose not to for this example.

While the parameters of the LME model (6.48) are of primary interest, we need consider modeling the joint distribution of $(\boldsymbol{y}_i, T_i^o, \delta_i)$ if we are to account for the substantial dropout that occurred over the course of the study. Since dropout is specified in terms of a continuous event time, we can model the joint distribution of the longitudinal response process and dropout process through an appropriate choice of the joint log-likelihood function, $\sum_{i=1}^n L_i(\boldsymbol{\tau},\boldsymbol{\eta};\boldsymbol{y}_i,T_i^o,\delta_i)$. In this example, we modeled the joint log-likelihood of the data $(\boldsymbol{y}_i, T_i^o, \delta_i)$ assuming both ignorable and non-ignorable dropout.

Under the assumption of ignorable dropout (either S-RD or S-CRD), we can factor the ith individual's contribution to the joint log-likelihood as

$$\begin{aligned} L_i(\boldsymbol{\tau},\boldsymbol{\eta};\boldsymbol{y}_i,T_i^o,\delta_i) &= \log \int \pi(\boldsymbol{y}_i|\boldsymbol{b}_i)\pi(\boldsymbol{b}_i)\pi_T(T_i^o|\boldsymbol{y}_i)^{\delta_i} S_T(T_i^o|\boldsymbol{y}_i)^{1-\delta_i}\, d\boldsymbol{b}_i \\ &= \log\left[\pi(\boldsymbol{y}_i;\boldsymbol{\tau})\pi_T(T_i^o|\boldsymbol{y}_i;\boldsymbol{\eta})^{\delta_i} S_T(T_i^o|\boldsymbol{y}_i;\boldsymbol{\eta})^{1-\delta_i}\right] \quad \text{(S-RD)} \\ &= \log\left[\pi(\boldsymbol{y}_i;\boldsymbol{\tau})\pi_T(T_i^o|\boldsymbol{\eta})^{\delta_i} S_T(T_i^o|\boldsymbol{\eta})^{1-\delta_i}\right] \quad \text{(S-CRD)} \end{aligned} \tag{6.49}$$

where $\boldsymbol{\tau} = (\boldsymbol{\beta}, \psi_{11}, \psi_{12}, \psi_{22}, \sigma^2) = (\boldsymbol{\beta}, \boldsymbol{\theta})$ are the parameters for the LME model (6.48) and $\boldsymbol{\eta}$ is a vector of parameters for the underlying distribution of the event time T_i having pdf $\pi_T(\cdot)$ and survivor function $S_T(\cdot)$. Given our primary goal is to estimate and compare the mean rate of change in GFR between treatment groups, we fit the data based on the log-likelihood function (6.49) assuming a S-CRD mechanism. We do so assuming T_i can be adequately characterized by a piecewise exponential survival model although, under S-CRD, any parametric or semiparametric event time model could be used without affecting our estimate of $\boldsymbol{\tau}$. The semiparametric piecewise exponential model has been described by a number of authors including Holford (1980), Karrison (1987) and, in the context of the MDRD study, Vonesh et. al. (2006). In the absence of covariates, it yields a smoothed version of the actuarial life-table estimator of survival but it can also incorporate covariates into an actuarial life-table approach to survival analysis in much the same way that the Cox model incorporates covariates into a Kaplan-Meier approach to survival analysis. It can be formulated by first partitioning the time scale into k disjoint intervals, say $(t_0, t_1]$, $(t_1, t_2]$, $(t_2, t_3]$, $\ldots$, $(t_{k-1}, t_k]$. The piecewise exponential model is obtained by assuming the underlying hazard rate, $\lambda_0(t)$, varies from interval to interval but is constant within each interval according to

$$\lambda_0(t) = \sum_{h=1}^{k} \lambda_{0h} I(t \in (t_{h-1}, t_h]) \tag{6.50}$$

where $I(t \in (t_{h-1}, t_h])$ is the indicator function taking value 1 when $t \in (t_{h-1}, t_h]$ and 0 otherwise. By convention, one would set $t_0 = 0$ and $t_k = \infty$ although in practical settings like the MDRD study one would set $t_k = \mathcal{T}$ corresponding to the study termination point. To fit a proportional hazards piecewise exponential model for a given set of external covariates $\boldsymbol{v}_i$, one would then simply specify the hazard function $\lambda_i(t; \boldsymbol{v}_i) = \lambda_0(t)\exp\{\boldsymbol{v}_i'\boldsymbol{\eta}\}$.

Assuming S-CRD conditional on the treatment groups, we modeled the piecewise exponential hazard function for the MDRD data as

$$\begin{aligned} \lambda_i(t; \boldsymbol{x}_i) &= \lambda_0(t)\exp\{\boldsymbol{x}_i'\boldsymbol{\eta}_G\} \qquad (6.51) \\ &= \exp\Big\{\sum_{h=1}^{7} \eta_h I(t \in (t_{h-1}, t_h]) + \boldsymbol{x}_i'\boldsymbol{\eta}_G\Big\}. \end{aligned}$$

where the η_h are the baseline log-hazard rates within the intervals $(t_{h-1}, t_h]$ for patients randomized to diet L-normal BP ($G_{4i} = 1$). The η_h are assumed constant within the follow-up intervals $(t_{h-1}, t_h]$, $h = 1, \ldots 7$; with $t_0 = 0$, $t_1 = 6$, $t_2 = 12$, $t_3 = 18$, $t_4 = 24$, $t_5 = 30$, $t_6 = 36$ and $t_7 = 45$ where 45 marks the study truncation time $\mathcal{T}$ (the longest follow-up time was 44.48 months). The vector $\boldsymbol{x}_i' = (G_{1i}, G_{2i}, G_{3i})$ is the vector of treatment group indicator variables with G_{4i} serving as the reference or control group, while $\boldsymbol{\eta}_G' = (\eta_{G1}, \eta_{G2}, \eta_{G3})$ is the vector of treatment group log hazard ratios (relative to treatment group G_{4i}). Likelihood-based methods for estimating the parameters of this model using either a piecewise exponential representation or an interval Poisson representation are described by Allison (1995).

Since under S-CRD, the log-likelihood (6.49) is the sum of two independent log-likelihood functions, i.e., $L_i(\boldsymbol{\tau}, \boldsymbol{\eta}; \boldsymbol{y}_i, T_i^o, \delta_i) = L_i(\boldsymbol{\tau}; \boldsymbol{y}_i) + L_i(\boldsymbol{\eta}; T_i^o, \delta_i)$, we can fit the two log-likelihood functions separately to the data. Specifically, we can fit the LME model (6.48) to the GFR data using PROC MIXED and the piecewise exponential model to the dropout data using PROC GENMOD. In the latter case, we use the fact that the kernel of the likelihood function of a piecewise exponential model will be identical to that of an interval Poisson model and so we fit the event time data to the piecewise exponential model using PROC GENMOD (see Allison, 1995, pp. 104-109). The SAS programming code for analyzing the joint distribution of GFR and dropout under S-CRD is as follows.

Program 6.10

```
data example6_6_2(rename=(DietK_LowBP=G1
                          DietK_NormBP=G2
                          DietL_LowBP=G3
                          DietL_NormBP=G4
                          FollowupMonths=T));
 set SASdata.MDRD_data;
 by ptid Months;
run;
proc print data=example6_6_2 noobs;
 where ptid in (1 7);
 var ptid G1 G2 G3 G4 Months GFR T Dropout;
run;
/* Compute an estimated mean follow-up time */
/* to be used in defining an empirically    */
/* derived residual dropout time covariate  */
proc means data=example6_6_2 noprint; where Months=0;
 var T;
 output out=meanT mean=meanT;
run;
data example6_6_2; set example6_6_2;
 if _n_=1 then set meanT;
 If G1=1 then DietBP='K, Low   ';
 If G2=1 then DietBP='K, Normal';
 If G3=1 then DietBP='L, Low   ';
 If G4=1 then DietBP='L, Normal';
 If G1=1 or G2=1 then Diet='K';
 if G3=1 or G4=1 then Diet='L';
 DietK=(Diet='K'); DietL=(Diet='L');
 /* Define Td to be an empirically derived   */
 /* residual dropout time having a mean of 0 */
 meanT=round(meanT,0.0001);
 Td = T-meanT;
 /* The following variables need to be defined */
 /* in order to run a SP model in NLMIXED      */
 Indicator=0; Response=GFR; t1=0; t2=0; Risktime=0; Event=0;
run;
proc sort data=example6_6_2;
 by ptid months;
run;
/*============================================================
 * The following programming statements create intervals
 * for use with piecewise exponential survival analysis in
 * combination with MDRD nonignorable missing data analyses
 * Key    t1 = Defines starting time point for the different
                intervals being created
          t2 = Defines ending time point for the different
                intervals being created
       Event = Defines a 0-1 indicator which is 0 in each
                interval until the interval in which the
                actual dropout time occurs and then the value
                of Event equals the value of Dropout
    Risktime = Defines the time at risk within each interval
============================================================*/
```

```
data dropout;
 set example6_6_2;
 by ptid Months;
 if last.ptid;
run;
data dropout1;
 set dropout;
 do t1=0 to 36 by 6;
    if t1< 36 then t2 = t1+6;
    if t1=36 then t2 = 45;
    Event=0; Risktime=t2-t1;
    if t1<T<=t2 then do;
       Event=Dropout; Risktime=T-t1;
    end;
    output;
 end;
run;
data dropout1;
 set dropout1;
 Interval=t1;
 if t1>T then delete;
 Log_risktime=log(Risktime);
 Response=Event;
 Indicator=1;
run;
proc sort data=dropout1;
 by ptid t1 t2;
run;
/* Dataset used for fitting all SP models */
data example6_6_2_SP;
 set example6_6_2 dropout1;
run;
proc sort data=example6_6_2_SP;
 by ptid Indicator Months;
run;
/* These MIXED and GENMOD calls are used */
/* to obtain joint estimates of the model*/
/* parameters for GFR and T under S-CRD  */
/* The results from these two models are */
/* then used to define starting values   */
/* for subsequent analyses using NLMIXED */
ods output SolutionR=re;
proc mixed data=example6_6_2 method=ML noclprint=5;
 class DietBP ptid;
 model GFR = DietBP DietBP*Months / noint s;
 random intercept Months / subject=ptid type=un s;
 contrast 'No Overall Diet & BP Intercept Effects'
          DietBP 1 -1  0  0,
          DietBP 1  0 -1  0,
          DietBP 1  0  0 -1;
 contrast 'No Overall Diet & BP Slope Effects'
          DietBP*Months 1 -1  0  0,
          DietBP*Months 1  0 -1  0,
          DietBP*Months 1  0  0 -1;
```

```
 estimate 'Diet K Slope:'
          DietBP*Months .516 .484    0    0 ;
 estimate 'Diet L Slope:'
          DietBP*Months    0    0 .519 .481 ;
 estimate 'Difference (K-L):'
          DietBP*Months .516 .484 -.519 -.481;
run;
ods select ParameterEstimates Type3;
proc genmod data=example6_6_2_SP;
 where Indicator=1;
 class t1 DietBP;
 model Event = t1 DietBP / noint dist=Poisson
                         offset=log_risktime
                         type3;
run;
```

A number of comments are included in the SAS program to help explain some of the logic behind the code. The dataset, `example6_6_2_SP`, contains both the GFR and dropout data. The dropout data are structured to allow one to fit dropout times to a piecewise exponential model assuming piecewise constant hazard rates within consecutive 6 month intervals: $(0,6]$, $(6,12]$, $(12,18]$, $(18,24]$, $(24,30]$, $(30,36]$. The last interval for follow-up is defined to be $(36,45]$ where 45 marks the study truncation time $\mathcal{T}$ (the largest follow-up time was 44.48 months). An indicator variable (`Indicator`) defines whether the response variable is GFR (`Indicator=0`) or the interval-based dropout times and event indicators from the piecewise exponential model (`Indicator=1`).

The results of fitting the LME model (6.48) to the GFR data are displayed in Output 6.5. In addition to testing whether there are any differences in the intercepts and slopes between the four treatment groups, we also included estimates of the average slope of patients according to which diet they received. Averages were based on the marginal frequency of patients within each of the four treatment groups (see Output 6.4). For example, among the 126 patients randomized to receive diet K, 65 patients were randomized to the low BP intervention (51.6% of diet K patients) and 61 were randomized to the normal BP intervention (48.4%) yielding a marginal mean slope for diet K of $0.516 \times -0.2511 + 0.484 \times -0.2516 = -0.2513$ ml/min/1.73m^2 per month. A similar calculation among the 129 patients on diet L yields a marginal mean slope of -0.3086 ml/min/1.73m^2 per month with a mean difference between diet K versus diet L of 0.05726 ($p = 0.0819$, Output 6.5). Hence, while there were no overall differences in the rate of decline in GFR between the four treatment groups ($p = 0.3156$, Output 6.5), when we combined patients according to the diet they received, there was evidence of a marginal reduction in the rate of decline in GFR among patients randomized to the very-low-protein diet K though this did not reach significance at the 5% level.

Output 6.5: Select output from PROC MIXED. Shown are the results we obtain when we fit the LME model (6.48) to the GFR data.

```
                 Model Information
Data Set                      WORK.EXAMPLE6_6_2
Dependent Variable            GFR
Covariance Structure          Unstructured
Subject Effect                ptid
Estimation Method             ML
Residual Variance Method      Profile
Fixed Effects SE Method       Model-Based
Degrees of Freedom Method     Containment
```

```
            Class Level Information
Class     Levels    Values
DietBP         4    K, Low K, Normal L, Low L, Normal
ptid         255    not printed
          Dimensions
Covariance Parameters          4
Columns in X                   8
Columns in Z Per Subject       2
Subjects                     255
Max Obs Per Subject           13
        Number of Observations
Number of Observations Read           1988
Number of Observations Used           1988
Number of Observations Not Used          0
              Iteration History
Iteration   Evaluations       -2 Log Like       Criterion
        0             1    12701.22934389
        1             2    10321.05645934      0.01116417
        2             1    10272.92213274      0.00512681
        3             1    10251.57296401      0.00161133
        4             1    10245.22221011      0.00023536
        5             1    10244.36913156      0.00000682
        6             1    10244.34618478      0.00000001
Convergence criteria met.
Covariance Parameter Estimates
Cov Parm    Subject      Estimate
UN(1,1)     ptid          19.9335
UN(2,1)     ptid          0.06126
UN(2,2)     ptid          0.05007
Residual                   5.2287
          Fit Statistics
-2 Log Likelihood              10244.3
AIC (smaller is better)        10268.3
AICC (smaller is better)       10268.5
BIC (smaller is better)        10310.8
Null Model Likelihood Ratio Test
DF    Chi-Square          Pr > ChiSq
 3       2456.88              <.0001
                        Solution for Fixed Effects
                                             Standard
Effect            DietBP         Estimate       Error     DF   t Value    Pr > |t|
DietBP            K, Low          19.3841      0.5794   1481     33.45      <.0001
DietBP            K, Normal       18.9982      0.5988   1481     31.73      <.0001
DietBP            L, Low          19.5835      0.5714   1481     34.27      <.0001
DietBP            L, Normal       19.7172      0.5927   1481     33.27      <.0001
Months*DietBP     K, Low          -0.2511     0.03290   1481     -7.63      <.0001
Months*DietBP     K, Normal       -0.2516     0.03347   1481     -7.52      <.0001
Months*DietBP     L, Low          -0.2922     0.03171   1481     -9.21      <.0001
Months*DietBP     L, Normal       -0.3263     0.03355   1481     -9.72      <.0001
                                   Estimates
Label                      Estimate   Standard Error     DF   t Value   Pr > |t|
Diet K Slope:               -0.2513          0.02346   1481    -10.71     <.0001
Diet L Slope:               -0.3086          0.02305   1481    -13.39     <.0001
Difference (K-L):           0.05726          0.03289   1481      1.74     0.0819
                                   Contrasts
Label                                      Num DF   Den DF   F Value   Pr > F
No Overall Diet & BP Intercept Effects          3     1481      0.28   0.8415
No Overall Diet & BP Slope Effects              3     1481      1.18   0.3156
```

Output 6.6 displays the results of fitting a proportional hazards piecewise exponential model to the dropout data. The log hazard ratios (HR) comparing dropout rates for the diet K-low BP group (log HR: $-0.0986, p = 0.7249$), the diet K-normal BP group (log HR: $-0.1464, p = 0.6054$) and the diet L-low BP group (log HR: $-0.1489, p = 0.5916$) to the diet L-normal BP group were not different from 0 suggesting there is no difference in dropout rates between the four groups. This is further confirmed when we apply an overall chi-square test comparing dropout rates between all four groups (Chi-Square of 0.37 on 3 DF, $p = 0.9471$ - Output 6.6).

Output 6.6: Select output from PROC GENMOD. Shown are the results we obtain when we fit the piecewise exponential model to the dropout data.

Analysis Of Maximum Likelihood Parameter Estimates

Parameter		DF	Estimate	Standard Error	Wald Chi-Square	Pr > ChiSq
Intercept		0	0.0000	0.0000	.	.
t1	0	1	−6.1357	0.6010	104.24	<.0001
t1	6	1	−4.3494	0.2959	216.02	<.0001
t1	12	1	−3.6158	0.2455	216.99	<.0001
t1	18	1	−3.5801	0.2620	186.75	<.0001
t1	24	1	−4.1813	0.3471	145.16	<.0001
t1	30	1	−3.8956	0.3592	117.62	<.0001
t1	36	1	−4.2103	0.5951	50.06	<.0001
DietBP	K, Low	1	−0.0986	0.2804	0.12	0.7249
DietBP	K, Normal	1	−0.1464	0.2833	0.27	0.6054
DietBP	L, Low	1	−0.1489	0.2775	0.29	0.5916
DietBP	L, Normal	0	0.0000	0.0000	.	.
Scale		0	1.0000	0.0000	_	_

LR Statistics For Type 3 Analysis

Source	DF	Chi-Square	Pr > ChiSq
t1	6	41.49	<.0001
DietBP	3	0.37	0.9471

Based on these analyses, one might conclude there is no effect due to either diet or blood pressure control on the rate of decline in GFR. However, as noted previously, patient dropout to dialysis or transplantation is likely to depend on a patient's intrinsic starting GFR (intercept) and/or rate of decline in GFR (slope) thus violating the assumption that dropout is completely at random. As an informal check on the assumption of S-CRD, we plotted the empirical Bayes estimates (or empirical best linear unbiased predictors, EBLUP's) of the random slope errors b_{2i} versus follow-up time T^o together with nonparametric loess curves according to whether patients experienced a dropout event ($\delta_i = 1$) or not ($\delta_i = 0$). The results, shown in Figure 6.2, suggest that dropout is not completely at random as otherwise one would expect the residual slopes to be distributed randomly about a mean of 0 with no discernible difference between dropouts and non-dropouts. This analysis is therefore likely to be biased with respect to inference about dropout. It is also possible that inference with respect to the rate of decline in GFR could be biased depending on whether the assumption of S-RD is viable or not (i.e., whether dropout depends only on observed data or not).

In an attempt to address this concern, we fit a conditional LME model to the GFR data in which we estimate and compare the slopes conditional on the observed follow-up time T_i^o. Such a model is similar to the conditional linear model of Wu and Bailey (1989). However, here we fit a conditional LME model using the general linear mixture model (6.43) setup described in §6.5. Specifically, let T_i^o designate the overall retention time in the study, i.e., the time to study termination or "dropout" including "dropout" due to censoring. If we then replace T_i^o with the residual "dropout" time, $T_i^* = T_i^o - E(T_i^o)$, we can extend the LME model (6.48) to the general linear mixture model setup (6.43) by simply including T_i^* as a covariate within model (6.48). Doing so results in fitting the

Figure 6.2 EBLUPs of random slope errors b_{2i} vs. time of follow-up

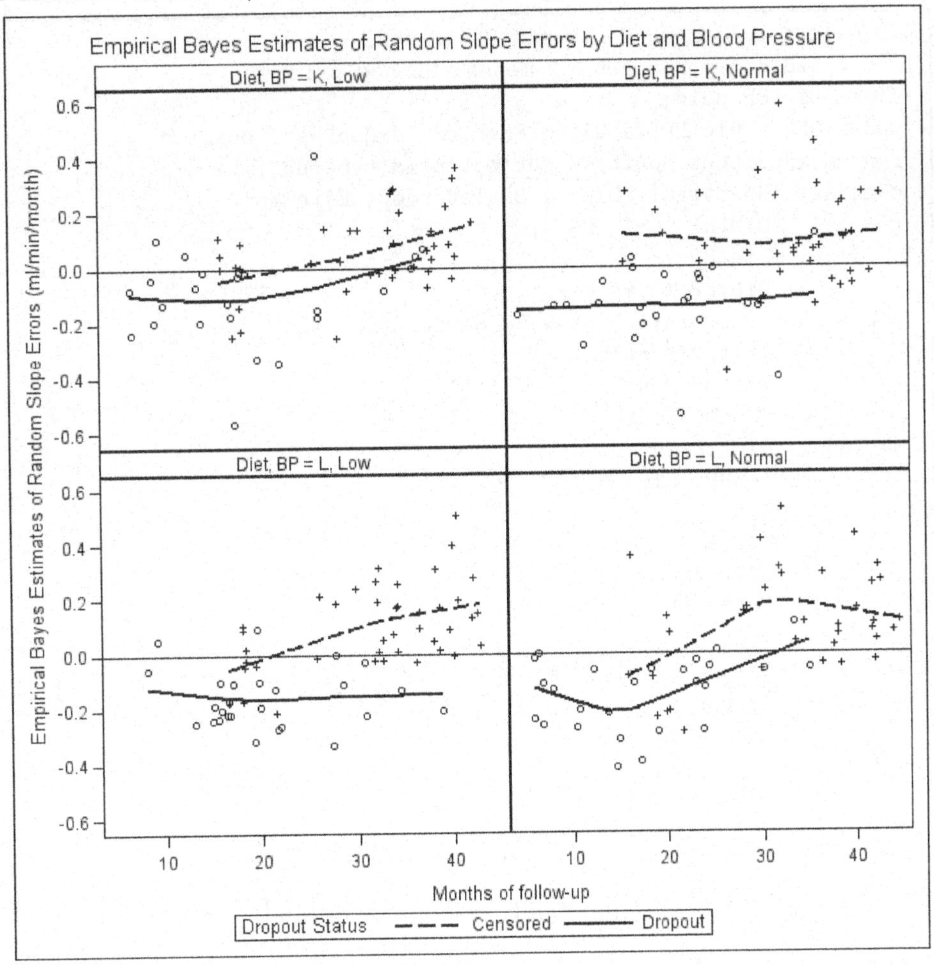

conditional LME model (conditional on both $\boldsymbol{b}_i$ and T_i^*),

$$\text{Stage 1: } y_{ij} = \beta_{1i} + \beta_{2i} t_{ij} + \epsilon_{ij},\ \epsilon_{ij} \sim \text{iid}\, N(0, \sigma^2) \tag{6.52}$$

$$\text{Stage 2: } \beta_{1i} = \beta_{11}G_{1i} + \beta_{12}G_{2i} + \beta_{13}G_{3i} + \beta_{14}G_{4i} + \beta_{15}T_i^* + b_{1i}$$

$$\beta_{2i} = \beta_{21}G_{1i} + \beta_{22}G_{2i} + \beta_{23}G_{3i} + \beta_{24}G_{4i} + \beta_{25}T_i^* + b_{2i}$$

to the GFR data. There are several ways one can fit this conditional LME model. One is to empirically define the residual "dropout" or retention time as $T_i^* = T_i^o - \widehat{\mu}_{T^o}$ where $\widehat{\mu}_{T^o} = \sum_{i=1}^n T_i^o/n$ is the simple average of the observed follow-up times including both dropout times and censored times. One can then simply fit model (6.52) to the GFR data using PROC MIXED. The code is essentially identical to that shown when we fit model (6.48) assuming S-CRD except that we now include the residual covariate $T_i^* = T_i^o - \widehat{\mu}_{T^o}$ in the model. The PROC MIXED syntax required to fit this conditional LME model is as follows:

Program 6.11

```
ods select SolutionF Contrasts Estimates;
proc mixed data=example6_6_2 method=ML noclprint=5;
 class DietBP ptid;
 model GFR = DietBP Td DietBP*Months Td*Months / noint s;
 random intercept Months / subject=ptid type=un;
 contrast 'No Overall Diet & BP Intercept Effects'
           DietBP 1 -1  0  0,
           DietBP 1  0 -1  0,
           DietBP 1  0  0 -1;
 contrast 'No Overall Diet & BP Slope Effects'
           DietBP*Months 1 -1  0  0,
           DietBP*Months 1  0 -1  0,
           DietBP*Months 1  0  0 -1;
 estimate 'Diet K Intercept:'
          DietBP .516 .484    0     0 ;
 estimate 'Diet L Intercept:'
          DietBP     0     0 .519 .481 ;
 estimate 'Difference (K-L):'
          DietBP .516 .484 -.519 -.481;
 estimate 'Diet K Slope:'
          DietBP*Months .516 .484     0     0 ;
 estimate 'Diet L Slope:'
          DietBP*Months     0     0 .519 .481 ;
 estimate 'Difference (K-L):'
          DietBP*Months .516 .484 -.519 -.481;
run;
```

The covariate, Td, is the empirically estimated residual "dropout" time defined in the data programming statements preceding the analysis shown in Output 6.5. We included in this analysis three additional ESTIMATE statements allowing us to estimate and compare the average intercepts for patients randomized to diet K versus diet L. The results of fitting this conditional LME model to the data are shown in Output 6.7.

This analysis mimics the basic pattern-mixture modeling approach used by Li and Schluchter (2004) to analyze the MDRD data but modified here to incorporate the use of a residual covariate T_i^* as suggested by Fitzmaurice et. al. (2001). The estimated slopes from this conditional LME model are greater, in absolute value, than those obtained under S-CRD (Output 6.6) by a factor of about 0.04 to 0.05 ml/min/1.73m^2/month. This together with the

Output 6.7: Select output from fitting the conditional LME model (6.52) to the GFR data using PROC MIXED. Note that the standard errors of the regression parameters do not take into account the fact that the residual covariate T_i^*, defined by the SAS variable, Td, is based on an estimated mean time of follow-up.

Solution for Fixed Effects

Effect	DietBP	Estimate	Standard Error	DF	t Value	Pr > |t|
DietBP	K, Low	19.6004	0.5630	1481	34.81	<.0001
DietBP	K, Normal	19.0525	0.5823	1481	32.72	<.0001
DietBP	L, Low	19.6596	0.5555	1481	35.39	<.0001
DietBP	L, Normal	19.9622	0.5761	1481	34.65	<.0001
Td		0.08889	0.02764	1481	3.22	0.0013

Months*DietBP	K, Low	−0.2926	0.02858	1481	−10.24	<.0001
Months*DietBP	K, Normal	−0.3018	0.02916	1481	−10.35	<.0001
Months*DietBP	L, Low	−0.3280	0.02753	1481	−11.92	<.0001
Months*DietBP	L, Normal	−0.3690	0.02929	1481	−12.60	<.0001
Td*Months		0.01601	0.001604	1481	9.98	<.0001

Estimates

Label	Estimate	Standard Error	DF	t Value	Pr > |t|
Diet K Intercept:	19.3352	0.4048	1481	47.76	<.0001
Diet L Intercept:	19.8052	0.4000	1481	49.52	<.0001
Difference (K-L):	−0.4700	0.5683	1481	−0.83	0.4084
Diet K Slope:	−0.2971	0.02081	1481	−14.28	<.0001
Diet L Slope:	−0.3477	0.02042	1481	−17.03	<.0001
Difference (K-L):	0.05061	0.02808	1481	1.80	0.0716

Contrasts

Label	Num DF	Den DF	F Value	Pr > F
No Overall Diet & BP Intercept Effects	3	1481	0.43	0.7333
No Overall Diet & BP Slope Effects	3	1481	1.44	0.2287

fact the residual "dropout" time T_i^* was found to be significantly associated with both the GFR intercepts (`Td` estimated effect: 0.08889 ± 0.02764, $p = 0.0013$) and slopes (`Td*Months` estimated effect: 0.01601 ± 0.001604, $p < 0.0001$) proves that dropout is not completely at random. Indeed, the results suggest that dropout may in fact be non-ignorable although this is impossible to test based on the observed data. Finally when we group patients according to whether they were randomized to diet K or diet L, there remains some evidence to suggest a slower progression of CKD among those randomized to the very-low-protein diet K (estimated difference in mean slopes (K-L): 0.05061 ± 0.02808, $p = 0.0716$).

As pointed out by both Fitzmaurice et. al. (2001) and Li and Schluchter (2004), the problem with the above analysis is that the estimated standard errors from PROC MIXED will underestimate the true standard errors of the regression parameter estimates. This is because PROC MIXED will treat $T_i^* = T_i^o - \widehat{\mu}_{T^o}$ as known rather than estimated (i.e., it fails to account for the additional variability associated with estimating μ_{T^o}). Therefore, while the estimated intercepts and slopes for the four treatment groups are unbiased estimates of the population intercepts and slopes obtained by averaging over the distribution of T_i^o, bias inference in the form of inflated type I errors is possible as a result of underestimating the standard errors of the parameter estimates. To overcome this, one can apply the delta method to compute corrected standard errors as suggested by Fitzmaurice et. al. (2001) and Li and Schluchter (2004). An example of this approach including SAS IML code is given by Hogan et. al. (2004).

An alternative approach to estimating the parameters of interest is to jointly fit the conditional LME model (6.52) together with a specified marginal model for T_i^o using PROC NLMIXED. For example, if we assume $T_i^o \sim \text{iid}\, N(\mu_{T^o}, \sigma^2_{T^o})$, then the MLE of $E(T_i^o) = \mu_{T^o}$ will be the sample mean $\widehat{\mu}_{T^o} = \sum_{i=1}^n T_i^o/n$ and the variance of $\widehat{\mu}_{T^o}$ will be the uncorrected sample variance divided by n, i.e., $Var(\widehat{\mu}_{T^o}) = \widehat{\sigma}^2_{T^o}/n = n^{-1}\Big(\sum_{i=1}^n (T_i^o - \widehat{\mu}_{T^o})^2/n \Big)$. By jointly fitting the conditional LME model (6.52) together with a Gaussian-based marginal model for T_i^o using NLMIXED, we not only obtain unbiased estimates $\widehat{\beta}_{21}$, $\widehat{\beta}_{22}$, $\widehat{\beta}_{23}$ and $\widehat{\beta}_{24}$ of the parameters of interest, namely the population-averaged slopes, but also consistent estimates of their respective standard errors. This is because NLMIXED will automatically compute the variance-covariance matrix of $\widehat{\boldsymbol{\beta}}$ using the delta method based on the joint log-likelihood of $(\boldsymbol{y}_i, T_i^o)$. In fact, this is precisely the approach described by Fitzmaurice et. al. (2001) for obtaining asymptotically valid ML estimates of the parameters of the general linear mixture model (6.43). We applied this alternative approach using the following NLMIXED code.

Program 6.12

```
proc nlmixed data=example6_6_2 start
             technique=newrap
             qpoints=1;
 ************************************************
 The following define the model parameters:
   -(beta11, beta12,...) define the LME regression
    parameters as rounded from PROC MIXED output
   -(psi11, psi12, psi22) define the random
    effects covariance parameters (also rounded)
   -Sigma_Sq defines the within-subject variance,
   -MuT and VarT define the sample mean and variance
    of the follow-up times T assuming normality
   -(b1i, b2i) define the random intercept and slope
 ************************************************;
parms beta11=19.4 beta12=19.0
      beta13=19.6 beta14=19.7 beta15=0
      beta21=-.25 beta22=-.25
      beta23=-.29 beta24=-.33 beta25=0
      psi11=19.9335 psi12=0.0613 psi22=0.05007
      Sigma_Sq=5.2287
      MuT=30 VarT=1;
** Define SS intercepts and slopes **;
beta1i=beta11*G1 + beta12*G2 + beta13*G3 +
       beta14*G4 + beta15*(T-MuT) + b1i;
beta2i=beta21*G1 + beta22*G2 + beta23*G3 +
       beta24*G4 + beta25*(T-MuT) + b2i;
MeanGFR = beta1i + beta2i*Months;
VarGFR  = Sigma_Sq;
** ll_y is the conditional log-likelihood of GFR **;
ll_y = (- 0.5*log(2*CONSTANT('PI'))
        - 0.5*((GFR - MeanGFR)**2)/(VarGFR)
        - 0.5*log(VarGFR));
*************************************************
ll_T is the marginal log-likelihood of T assuming
normality with mean MuT and variance VarT. T-MuT then
serves as a single residual covariate for the general
linear mixture model (conditional LME model) for GFR
*************************************************;
ll_T = (Months=0)*
       (- 0.5*log(2*CONSTANT('PI'))
        - 0.5*((T - MuT)**2)/(VarT)
        - 0.5*log(VarT));
** ll is the joint log-likelihood of (GFR,T) **;
ll = ll_y + ll_T;
** response is a dummy response variable **;
model response ~ general(ll);
random b1i b2i ~ normal([0,0],[psi11,
                               psi12, psi22])
       subject=ptid;
** Frequency of patients in each Diet and BP group **;
f1=65;f2=61;f3=67;f4=62;
** Diet K and Diet L average intercepts and slopes **;
```

```
estimate 'Diet K Intercept:'
         (f1*beta11+f2*beta12)/(f1+f2);
estimate 'Diet L Intercept:'
         (f3*beta13+f4*beta14)/(f3+f4);
estimate 'Difference (K-L):'
         (f1*beta11+f2*beta12)/(f1+f2) -
         (f3*beta13+f4*beta14)/(f3+f4);
estimate 'Diet K Slope:'
         (f1*beta21+f2*beta22)/(f1+f2);
estimate 'Diet L Slope:'
         (f3*beta23+f4*beta24)/(f3+f4);
estimate 'Difference (K-L):'
         (f1*beta21+f2*beta22)/(f1+f2) -
         (f3*beta23+f4*beta24)/(f3+f4);
contrast 'No Overall Intercept Effect:'
          beta11-beta12, beta11-beta13, beta11-beta14;
contrast 'No Overall Slope Effect:'
          beta21-beta22, beta21-beta23, beta21-beta24;
run;
```

We included a number of comments within the NLMIXED code to help explain how one can fit random-effects pattern mixture models using the general linear mixture model framework given in (6.43). In this particular example, we define both the conditional log-likelihood of the response variable `GFR` given the random effects and the residual covariate, `T-MuT`, and we define the marginal log-likelihood of the "dropout" or retention time `T`. The former, denoted by `ll_y`, is based on the conditional LME model (6.52) while the latter, denoted by `ll_T`, is required in order to estimate $E(T_i^o) = \mu_{T^o}$ (denoted by `MuT`). By specifying the joint conditional log-likelihood, denoted by `ll = ll_y + ll_T`, we are effectively using NLMIXED to maximize the marginal log-likelihood of $(\boldsymbol{y}_i, T_i^o)$. Specifically, if we let $\boldsymbol{\tau} = (\boldsymbol{\beta}, \boldsymbol{\theta})$ and $\boldsymbol{\eta} = (\mu_{T^o}, \sigma^2_{T^o})'$, our objective is to maximize the joint log-likelihood function

$$\sum_{i=1}^{n} L_i(\boldsymbol{\tau}, \boldsymbol{\eta}; \boldsymbol{y}_i, T_i^o) = \sum_{i=1}^{n} L_i(\boldsymbol{\tau}; \boldsymbol{y}_i | T_i^*) + \sum_{i=1}^{n} L_i(\boldsymbol{\eta}; T_i^o)$$

where $L_i(\boldsymbol{\eta}; T_i^o)$ is the ith patient's marginal log-likelihood function for T_i^0 assuming $T_i^o \sim \text{iid}\, N(\mu_{T^o}, \sigma^2_{T^o})$ and

$$L_i(\boldsymbol{\tau}; \boldsymbol{y}_i | T_i^*) = \log \int_b \exp\Big\{\log(\pi(\boldsymbol{y}_i | \boldsymbol{b}_i, T_i^*)) + \log(\pi(\boldsymbol{b}_i))\Big\} d\boldsymbol{b}_i$$

is the conditional log-likelihood function of $(\boldsymbol{y}_i | T_i^*)$ based on the conditional LME model (6.52) and the assumption that $\boldsymbol{b}_i \perp T_i^o$. In terms of the NLMIXED code, `ll_y` is the conditional log-likelihood, $\log(\pi(\boldsymbol{y}_i | \boldsymbol{b}_i, T_i^*))$, under model (6.52) with $T_i^* = (T_i^o - \mu_{T^o})$ held fixed and `ll_T` is the marginal log-likelihood function $L_i(\boldsymbol{\eta}; T_i^o)$.

Because the conditional log-likelihood function of $(\boldsymbol{y}_i | T_i^*)$ depends on μ_{T^o} through T_i^*, this approach will be equivalent to the approach taken by Li and Schluchter (2004) in that it will indirectly apply their delta method for obtaining corrected standard errors of the marginal intercept and slope estimates (see, also, Fitzmaurice et. al., 2001). In fact, one can fit the conditional LME model in NLMIXED using `T` rather than `T-MuT` as the covariate and then apply the delta method of Li and Schluchter directly by using ESTIMATE statements. One feature worth noting is that in order to replicate the parameter estimates obtained using PROC MIXED, one must specify the NLMIXED option TECHNIQUE=NEWRAP as this is the default method used in PROC MIXED (using the default QUANEW technique in NLMIXED will result in slightly different estimates). Shown in Output 6.8 are the results when we use this joint modeling approach.

Output 6.8: Select output from NLMIXED when we jointly fit the conditional LME model (6.52) for y_i (GFR) and a Gaussian-based marginal model for T_i^o (T). Standard errors are correctly adjusted for the fact that μ_{T^o} (MuT) is estimated.

Dimensions

Observations Used	1988
Observations Not Used	0
Total Observations	1988
Subjects	255
Max Obs Per Subject	13
Parameters	16
Quadrature Points	1

Fit Statistics

−2 Log Likelihood	12051
AIC (smaller is better)	12083
AICC (smaller is better)	12083
BIC (smaller is better)	12139

Parameter Estimates

Parameter	Estimate	Standard Error	DF	t Value	Pr > |t|
beta11	19.6004	0.5659	253	34.63	<.0001
beta12	19.0525	0.5851	253	32.56	<.0001
beta13	19.6596	0.5585	253	35.20	<.0001
beta14	19.9622	0.5790	253	34.48	<.0001
beta15	0.08889	0.02764	253	3.22	0.0015
beta21	−0.2926	0.03041	253	−9.62	<.0001
beta22	−0.3018	0.03104	253	−9.72	<.0001
beta23	−0.3280	0.02947	253	−11.13	<.0001
beta24	−0.3690	0.03119	253	−11.83	<.0001
beta25	0.01601	0.001621	253	9.88	<.0001
psi11	18.7282	1.8148	253	10.32	<.0001
psi12	−0.01802	0.06150	253	−0.29	0.7697
psi22	0.03399	0.004308	253	7.89	<.0001
Sigma_Sq	5.1591	0.1885	253	27.37	<.0001
MuT	26.8291	0.6474	253	41.44	<.0001
VarT	106.88	9.4659	253	11.29	<.0001

Parameter	Alpha	Lower	Upper	Gradient
beta11	0.05	18.4858	20.7149	1.62E−13
beta12	0.05	17.9001	20.2049	−146E−15
beta13	0.05	18.5597	20.7595	−303E−16
beta14	0.05	18.8219	21.1025	−296E−16
beta15	0.05	0.03445	0.1433	−152E−14
beta21	0.05	−0.3525	−0.2327	2E−13
beta22	0.05	−0.3629	−0.2407	−61E−15
beta23	0.05	−0.3860	−0.2699	2.12E−12
beta24	0.05	−0.4304	−0.3075	−161E−14
beta25	0.05	0.01282	0.01921	−119E−12
psi11	0.05	15.1543	22.3022	−227E−17
psi12	0.05	−0.1391	0.1031	1.93E−13
psi22	0.05	0.02550	0.04247	−941E−14
Sigma_Sq	0.05	4.7878	5.5303	−236E−15
MuT	0.05	25.5541	28.1041	6.51E−10
VarT	0.05	88.2426	125.53	−1.62E−8

Contrasts

Label	Num DF	Den DF	F Value	Pr > F
No Overall Diet & BP Intercept Effects:	3	253	0.43	0.7334
No Overall Diet & BP Slope Effects:	3	253	1.44	0.2321

Additional Estimates

Label	Estimate	Standard Error	DF	t Value	Pr > |t|
Diet K Intercept:	19.3351	0.4089	253	47.28	<.0001
Diet L Intercept:	19.8050	0.4041	253	49.01	<.0001
Difference (K-L):	-0.4699	0.5683	253	-0.83	0.4091
Diet K Slope:	-0.2971	0.02329	253	-12.76	<.0001
Diet L Slope:	-0.3477	0.02299	253	-15.12	<.0001
Difference (K-L):	0.05060	0.02809	253	1.80	0.0729

Label	Alpha	Lower	Upper
Diet K Intercept:	0.05	18.5298	20.1404
Diet L Intercept:	0.05	19.0091	20.6010
Difference (K-L):	0.05	-1.5892	0.6494
Diet K Slope:	0.05	-0.3429	-0.2512
Diet L Slope:	0.05	-0.3929	-0.3024
Difference (K-L):	0.05	-0.00472	0.1059

By applying this approach in NLMIXED, the standard errors of the regression parameters will be corrected for the fact the residual covariate T_i^* is estimated and not a fixed known covariate. Indeed, when we compare the NLMIXED results above with those from PROC MIXED (Output 6.7), we see that while the individual regression parameter estimates are the same in both cases, the standard errors based on NLMIXED are slightly greater due to its correction for the sampling error associated with estimating μ_{T^o}. Moreover, the slopes under the conditional LME model (6.52) assuming non-ignorable dropout are 12-20% greater in absolute value compared to those estimated under the LME model (6.48) assuming ignorable dropout (Output 6.5). This suggests that the true rate of decline in GFR will be underestimated (closer to 0) if one ignores dropout in the analysis. We note, however, that there was no overall significant difference in slopes between the four treatment arms under any of the preceding analyses. This reflects the fact that the estimated bias due to dropout is in the same direction for all four treatment groups. Finally, the results in Output 6.8 comparing dietary intervention continue to suggest that patients in the diet L group have a more rapid decline in GFR compared to those in the diet K group although this again did not reach statistical significance at the 0.05 level. It is interesting to note that the results of fitting the conditional LME model using MIXED with its naive standard errors and larger default degrees of freedom are very nearly identical to those of NLMIXED with its corrected standard errors and lower default degrees of freedom. This is particularly true with respect to the comparative effect of dietary intervention on the progression of CKD.

The preceding analysis utilizes a random-effects pattern mixture modeling approach to analyze the MDRD data under non-ignorable dropout. The results suggest that dropout may be non-ignorable assuming we have correctly conditioned on the correct residual covariate, namely $T_i^* = T_i^o - \mu_{T^o}$. While the results seem plausible, a sensitivity analysis based on fitting the data under alternative modeling assumptions is recommended as a means for checking the consistency of results across different sets of assumptions. For example, under the conditional LME model (6.52), we treat the observed follow-up time T_i^o as an overall retention time, i.e., the time to study termination or "dropout." In doing so, we are effectively treating censored ($\delta_i = 0$) and uncensored ($\delta_i = 1$) values of T_i^o the same resulting in an estimated mean $\widehat{\mu}_{T^o} = 26.8291$ that is likely to underestimate the true mean time to dropout when dropout is defined by the occurrence of a terminal or non-terminal event.

As an alternative to the conditional LME model, we fit a shared parameter (SP) model to the data assuming the joint distribution of GFR and time to dropout, $\pi(\boldsymbol{y}_i, T_i)$, may be characterized by a hazard rate for dropout that depends on the patient-specific intercepts and slopes of the LME model (6.48). Specifically, following the approach taken by Vonesh et. al. (2006), we fit a SP model assuming the joint log-likelihood function for the ith

patient may be expressed as

$$L_i(\boldsymbol{\tau},\boldsymbol{\eta};\boldsymbol{y}_i,T_i^o,\delta_i) = \log\int \exp\Big\{L_i(\boldsymbol{\beta},\boldsymbol{\theta}_w;\boldsymbol{y}_i|\boldsymbol{b}_i) + L_i(\boldsymbol{\eta};T_i^o,\delta_i|\boldsymbol{b}_i) + L_i(\boldsymbol{\theta}_b;\boldsymbol{b}_i)\Big\}d\boldsymbol{b}_i \qquad (6.53)$$

where

$$L_i(\boldsymbol{\beta},\boldsymbol{\theta}_w;\boldsymbol{y}_i|\boldsymbol{b}_i) = \log\Big\{\pi(\boldsymbol{y}_i|\boldsymbol{b}_i,\boldsymbol{\beta},\boldsymbol{\theta}_w)\Big\}$$
$$L_i(\boldsymbol{\eta};T_i^o,\delta_i|\boldsymbol{b}_i) = \log\Big\{\pi_T(T_i^o|\boldsymbol{b}_i,\boldsymbol{\eta})^{\delta_i}S_T(T_i^o|\boldsymbol{b}_i,\boldsymbol{\eta})^{1-\delta_i})\Big\}$$
$$L_i(\boldsymbol{\theta}_b;\boldsymbol{b}_i) = \log\Big\{\pi(\boldsymbol{b}_i;\boldsymbol{\theta}_b)\Big\}$$

are, respectively, the conditional log-likelihood function with respect to the GFR response variable $\boldsymbol{y}_i$ given $\boldsymbol{b}_i$, the conditional log-likelihood function with respect to the dropout time variable T_i given $\boldsymbol{b}_i$, and the marginal log-likelihood function with respect to the random effect $\boldsymbol{b}_i$. Here $L_i(\boldsymbol{\beta},\boldsymbol{\theta}_w;\boldsymbol{y}_i|\boldsymbol{b}_i)$ is based on the LME model (6.48) while $L_i(\boldsymbol{\eta};T_i^o,\delta_i|\boldsymbol{b}_i)$ is based on a specified pdf, $\pi_T(T_i^o|\boldsymbol{b}_i)$, and survivor function, $S_T(T_i^o|\boldsymbol{b}_i)$, of the time-to-dropout random variable T_i. Vonesh et. al. (2006), for example, fit a SP model to the MDRD data assuming both a Weibull and piecewise exponential model for $T_i|\boldsymbol{b}_i$.

Here we fit the SP model (6.53) assuming a piecewise exponential model for $T_i|\boldsymbol{b}_i$ that is similar to the one used by Vonesh et. al. (2006). Specifically, we fit the SP model assuming a piecewise exponential model having the same hazard function as defined in (6.51) for the S-CRD model but extended to include the residual random intercepts b_{1i} and slopes b_{2i} as additional time-independent covariates. Thus the piecewise exponential hazard function, written in exponential form, is

$$\begin{aligned}\lambda_i(t;\boldsymbol{x}_i,\boldsymbol{b}_i) &= \lambda_0(t)\exp\{\boldsymbol{x}_i'\boldsymbol{\eta}_G + \boldsymbol{b}_i'\boldsymbol{\eta}_b\} \qquad (6.54)\\ &= \exp\Big\{\sum_{h=1}^{7}\eta_h I(t\in(t_{h-1},t_h]) + \boldsymbol{x}_i'\boldsymbol{\eta}_G + \boldsymbol{b}_i'\boldsymbol{\eta}_b\Big\}\end{aligned}$$

where $\boldsymbol{x}_i'\boldsymbol{\eta}_G + \boldsymbol{b}_i'\boldsymbol{\eta}_b = \eta_{G_1}G_{1i} + \eta_{G_2}G_{2i} + \eta_{G_3}G_{3i} + \eta_{b_1}b_{1i} + \eta_{b_2}b_{2i}$. The parameter vector $\boldsymbol{\eta} = (\boldsymbol{\eta}_0',\boldsymbol{\eta}_G',\boldsymbol{\eta}_b')'$ consists of the baseline hazard rate parameters, $\boldsymbol{\eta}_0 = (\eta_1,\eta_2,\eta_3,\eta_4,\eta_5,\eta_6,\eta_7)'$, the treatment group parameters, $\boldsymbol{\eta}_G = (\eta_{G_1},\eta_{G_2},\eta_{G_3})'$, and random-effect parameters, $\boldsymbol{\eta}_b = (\eta_{b_1},\eta_{b_2})'$. This SP model differs from the one used by Vonesh et. al. (2006) in that we explicitly account for treatment group effects within the hazard function and we replace the random intercepts and slopes, β_{1i} and β_{2i}, as covariates with the residual random intercepts and slopes, b_{1i} and b_{2i}. The NLMIXED code required to fit this SP model is as follows.

Program 6.13

```
proc nlmixed data=example6_6_2_SP start
             technique=newrap
             qpoints=1;
 **************************************************
 The following defines the model parameters:
   -(beta11, beta12,...) define the LME regression
    parameters including random effects,
   -(psi11, psi12, psi22) define the random
    effects covariance parameters,
```

```
  -Sigma_Sq defines the within-subject variance,
  -(eta0 eta1...eta_G1 eta_G2 eta_G3 eta_b1i eta_b2i)
   define the PE regression parameters under
   the assumption of non-ignorable dropout,
  -(b1i, b2i) define the random intercept and slope
**************************************************;
parms beta11=19.4 beta12=19.0 beta13=19.6 beta14=19.7
      beta21=-.25 beta22=-.25 beta23=-.29 beta24=-.33
      psi11=19.9335 psi12=0.0613 psi22=0.05007
      Sigma_Sq=5.2287
      eta1=-6.14 eta2=-4.35 eta3=-3.62 eta4=-3.58
      eta5=-4.18 eta6=-3.90 eta7=-4.21
      eta_G1=-.1 eta_G2=-.1 eta_G3=-.1
      eta_b1i=0 eta_b2i=0;
** Define SS intercepts and slopes **;
beta1i = beta11*G1 + beta12*G2 + beta13*G3 + beta14*G4 + b1i;
beta2i = beta21*G1 + beta22*G2 + beta23*G3 + beta24*G4 + b2i;
MeanGFR = (beta1i + beta2i*Months);
VarGFR  = Sigma_Sq;
** Define the log-hazard rate for a PE model **;
eta_i = eta1*(t1=0) + eta2*(t1=6) + eta3*(t1=12) +
        eta4*(t1=18) +  eta5*(t1=24) + eta6*(t1=30) +
        eta7*(t1=36) +
        eta_G1*G1 + eta_G2*G2 + eta_G3*G3 +
        eta_b1i*b1i + eta_b2i*b2i;
** Lambda_i defines the PE hazard rate per unit time **;
Lambda_i = exp(eta_i);
** ll_y is the conditional log-likelihood of y=GFR **;
ll_y = (1-Indicator)*
       (- 0.5*log(2*CONSTANT('PI'))
        - 0.5*((GFR - MeanGFR)**2)/(VarGFR)
        - 0.5*log(VarGFR));
************************************************
** ll_T is the conditional log-likelihood of T
   Here T is distributed across intervals according
   to the variable risktime which is defined as the
   amount of time at risk within each interval (see
   above code defining the dataset example6_6_2_SP)
   Note that b1i and b2i act as covariates for T
**************************************************;
ll_T = indicator*( Event*(eta_i) - Lambda_i*risktime );
** ll is the joint log-likelihood of (y,T) **;
ll = ll_y + ll_T;
** response is a dummy response variable **;
model response ~ general(ll);
random b1i b2i ~ normal([0,0],[psi11,
                                psi12, psi22])
       subject=ptid;
** Frequency of patients in each Diet and BP group **;
f1=65;f2=61;f3=67;f4=62;
** Diet K and Diet L average intercepts and slopes **;
estimate 'Diet K Intercept:'
       (f1*beta11+f2*beta12)/(f1+f2);
estimate 'Diet L Intercept:'
       (f3*beta13+f4*beta14)/(f3+f4);
```

```
  estimate 'Difference (K-L):'
           (f1*beta11+f2*beta12)/(f1+f2) -
           (f3*beta13+f4*beta14)/(f3+f4);
  estimate 'Diet K Slope:'
           (f1*beta21+f2*beta22)/(f1+f2);
  estimate 'Diet L Slope:'
           (f3*beta23+f4*beta24)/(f3+f4);
  estimate 'Difference (K-L):'
           (f1*beta21+f2*beta22)/(f1+f2) -
           (f3*beta23+f4*beta24)/(f3+f4);
  contrast 'No Overall Diet & BP Intercept Effects:'
           beta11-beta12, beta11-beta13, beta11-beta14;
  contrast 'No Overall Diet & BP Slope Effects:'
           beta21-beta22, beta21-beta23, beta21-beta24;
  contrast 'No Treatment Effect on Dropout:'
           eta_G1, eta_G2, eta_G3;
run;
```

We again include a number of comments within the SAS code to help explain how one can fit shared parameter models using NLMIXED. The programming required to fit a SP model is similar to that used to fit a general linear mixture model. Both require one to maximize a joint log-likelihood function. Where they differ is in how one factors the joint distribution function of $(\boldsymbol{y}_i, T_i)$. Under SP models with no censoring, we write $\pi(\boldsymbol{y}_i, T_i)$ based on the integrated likelihood $\pi(\boldsymbol{y}_i, T_i) = \int \pi(\boldsymbol{y}_i|\boldsymbol{b}_i)\pi(T_i|\boldsymbol{b}_i)\pi(\boldsymbol{b}_i)d\boldsymbol{b}_i$ where as under the general linear mixture model we use the factorization $\pi(\boldsymbol{y}_i, T_i) = \pi(\boldsymbol{y}_i|T_i)\pi(T_i)$. To fit our SP model to the MDRD data, we set `ll_y` above to be the conditional log-likelihood of $(\boldsymbol{y}_i|\boldsymbol{b}_i)$ based on the LME model (6.48) and we set `ll_T` to be the conditional log-likelihood of $(T_i|\boldsymbol{b}_i)$ based on the piecewise exponential model with hazard function (6.54). In the latter case, the conditional log-likelihood function $L_i(\boldsymbol{\eta}; T_i^o, \delta_i|\boldsymbol{b}_i)$ may be expressed in terms of the hazard function and cumulative hazard function as

$$
\begin{aligned}
L_i(\boldsymbol{\eta}; T_i^o, \delta_i|\boldsymbol{b}_i) &= \log\Big\{\lambda_i(T_i^o; \boldsymbol{x}_i, \boldsymbol{b}_i)^{\delta_i} \exp(-\Lambda_i(T_i^o; \boldsymbol{x}_i, \boldsymbol{b}_i))\Big\} \qquad (6.55)\\
&= \delta_i\Big\{\sum_{h=1}^{7} \eta_h I(T_i^o \in (t_{h-1}, t_h]) + \boldsymbol{x}_i'\boldsymbol{\eta}_G + \boldsymbol{b}_i'\boldsymbol{\eta}_b\Big\}\\
&\quad - \Lambda_i(T_i^o; \boldsymbol{x}_i, \boldsymbol{b}_i)
\end{aligned}
$$

where $\Lambda_i(T_i^o; \boldsymbol{x}_i, \boldsymbol{b}_i)$ is the piecewise exponential cumulative hazard function

$$
\begin{aligned}
\Lambda_i(t; \boldsymbol{x}_i, \boldsymbol{b}_i) &= \int_0^t \lambda_i(s; \boldsymbol{x}_i, \boldsymbol{b}_i)ds \qquad (6.56)\\
&= \sum_{h=1}^{7}\Big\{I(t > t_h)\lambda_i(t_h; \boldsymbol{x}_i, \boldsymbol{b}_i)(t_h - t_{h-1}) +\\
&\quad I(t \in (t_{h-1}, t_h])\lambda_i(t; \boldsymbol{x}_i, \boldsymbol{b}_i)(t - t_{h-1})\Big\}\\
&= \sum_{h=1}^{7}\Big\{I(t > t_h)\exp\Big\{\eta_h + \boldsymbol{x}_i'\boldsymbol{\eta}_G + \boldsymbol{b}_i'\boldsymbol{\eta}_b\Big\}(t_h - t_{h-1}) +\\
&\quad I(t \in (t_{h-1}, t_h])\exp\Big\{\eta_h + \boldsymbol{x}_i'\boldsymbol{\eta}_G + \boldsymbol{b}_i'\boldsymbol{\eta}_b\Big\}(t - t_{h-1})\Big\}
\end{aligned}
$$

evaluated at $t = T_i^o$. We then use the Laplace approximation (QPOINTS=1) to approximate the integrated log-likelihood function (6.53) taking advantage of the SP model assumptions that 1) the y_{ij} are conditionally independent given $\boldsymbol{b}_i$ and 2) the response vector $\boldsymbol{y}_i$ and event time T_i are conditionally independent given $\boldsymbol{b}_i$. Finally, we note that unlike Vonesh et. al. (2006), we have included the constant $-\frac{1}{2}\log(2\pi)$ when evaluating the individual's contribution to the conditional log-likelihood for $(y_{ij}|\boldsymbol{b}_i)$ which explains why the information criteria reported here differs from that given by Vonesh et. al. (2006).

The results of fitting this SP model are shown in Output 6.9. The estimated intercepts and slopes for the four treatment groups are very similar to those obtained under the conditional LME model (6.52) shown in Output 6.8. Likewise, we again find no difference in either the intercepts or slopes between the four groups. However the standard errors of the estimated intercepts and slopes are all greater than what was estimated under the pattern-mixture model (6.52). Comparing the estimated variance-covariance parameters of $\boldsymbol{b}_i$ between the two models, the increase in the estimated standard errors under the SP model can largely be explained by the increase in the estimated variances of both the random intercepts b_{1i} and slopes b_{2i}. Such an increase in the random effects variances may also explain why there is a noticeably less pronounced effect of dietary intervention on the rate of decline in GFR (Dietary slope difference (K-L): 0.04521 ± 0.03492, $p = 0.1965$).

Output 6.9: Select output from NLMIXED based on the SP model (6.53) defined by the LME model (6.48) for the pdf of $(\boldsymbol{y}_i|\boldsymbol{b}_i)$ and the piecewise exponential model with hazard function (6.54) for the pdf and survivor function of $T_i|\boldsymbol{b}_i$.

Dimensions

Observations Used	3247
Observations Not Used	0
Total Observations	3247
Subjects	255
Max Obs Per Subject	20
Parameters	24
Quadrature Points	1

Fit Statistics

-2 Log Likelihood	11059
AIC (smaller is better)	11107
AICC (smaller is better)	11108
BIC (smaller is better)	11192

Parameter Estimates

Parameter	Estimate	Standard Error	DF	t Value	Pr > |t|
beta11	19.5439	0.5727	253	34.13	<.0001
beta12	19.0942	0.5918	253	32.27	<.0001
beta13	19.6792	0.5647	253	34.85	<.0001
beta14	19.8545	0.5859	253	33.89	<.0001
beta21	-0.3104	0.03520	253	-8.82	<.0001
beta22	-0.2914	0.03573	253	-8.16	<.0001
beta23	-0.3151	0.03416	253	-9.22	<.0001
beta24	-0.3803	0.03623	253	-10.50	<.0001
psi11	19.4760	1.8810	253	10.35	<.0001
psi12	0.1465	0.07924	253	1.85	0.0656
psi22	0.06333	0.007847	253	8.07	<.0001
Sigma_Sq	5.1312	0.1859	253	27.60	<.0001
eta1	-9.3856	0.9913	253	-9.47	<.0001
eta2	-6.3736	0.6141	253	-10.38	<.0001
eta3	-4.7721	0.5111	253	-9.34	<.0001
eta4	-3.6888	0.4768	253	-7.74	<.0001
eta5	-3.4281	0.5210	253	-6.58	<.0001
eta6	-2.9137	0.5321	253	-5.48	<.0001
eta7	-2.8374	0.7159	253	-3.96	<.0001

eta_G1	−0.4573	0.5981	253	−0.76	0.4452
eta_G2	−0.8396	0.6075	253	−1.38	0.1682
eta_G3	−0.7249	0.5942	253	−1.22	0.2236
eta_b1i	−0.2420	0.03959	253	−6.11	<.0001
eta_b2i	−9.7398	1.1424	253	−8.53	<.0001

Parameter	Alpha	Lower	Upper	Gradient
beta11	0.05	18.4161	20.6717	−577E−13
beta12	0.05	17.9288	20.2596	−83E−12
beta13	0.05	18.5670	20.7913	−713E−13
beta14	0.05	18.7007	21.0082	−755E−13
beta21	0.05	−0.3798	−0.2411	−8.95E−9
beta22	0.05	−0.3618	−0.2210	−7.95E−9
beta23	0.05	−0.3824	−0.2478	−9.05E−9
beta24	0.05	−0.4517	−0.3090	−1.02E−8
psi11	0.05	15.7716	23.1804	8.37E−13
psi12	0.05	−0.00953	0.3026	1.514E−9
psi22	0.05	0.04788	0.07878	−2.05E−7
Sigma_Sq	0.05	4.7651	5.4973	−633E−12
eta1	0.05	−11.3379	−7.4333	2E−11
eta2	0.05	−7.5830	−5.1641	−331E−12
eta3	0.05	−5.7787	−3.7656	−804E−12
eta4	0.05	−4.6278	−2.7497	−356E−12
eta5	0.05	−4.4542	−2.4020	−504E−13
eta6	0.05	−3.9615	−1.8658	1.57E−11
eta7	0.05	−4.2473	−1.4274	1.27E−11
eta_G1	0.05	−1.6351	0.7205	−348E−12
eta_G2	0.05	−2.0361	0.3569	−393E−12
eta_G3	0.05	−1.8952	0.4454	−379E−12
eta_b1i	0.05	−0.3200	−0.1641	2.06E−10
eta_b2i	0.05	−11.9896	−7.4900	5.95E−10

Contrasts

Label	Num DF	Den DF	F Value	Pr > F
No Overall Diet & BP Intercept Effects:	3	253	0.31	0.8214
No Overall Diet & BP Slope Effects:	3	253	1.19	0.3143
No Treatment Effect on Dropout:	3	253	0.76	0.5167

Additional Estimates

Label	Estimate	Standard Error	DF	t Value	Pr > |t|
Diet K Intercept:	19.3262	0.4117	253	46.95	<.0001
Diet L Intercept:	19.7634	0.4068	253	48.58	<.0001
Difference (K−L):	−0.4372	0.5782	253	−0.76	0.4502
Diet K Slope:	−0.3012	0.02530	253	−11.90	<.0001
Diet L Slope:	−0.3464	0.02510	253	−13.80	<.0001
Difference (K−L):	0.04521	0.03492	253	1.29	0.1965

Label	Alpha	Lower	Upper
Diet K Intercept:	0.05	18.5154	20.1369
Diet L Intercept:	0.05	18.9622	20.5646
Difference (K−L):	0.05	−1.5759	0.7015
Diet K Slope:	0.05	−0.3511	−0.2514
Diet L Slope:	0.05	−0.3959	−0.2970
Difference (K−L):	0.05	−0.02355	0.1140

Given the somewhat discrepant results between the conditional LME model (6.52) and shared parameter model (6.53), particularly with respect to the estimated variation in random effects, we performed an additional set of analyses aimed at examining the sensitivity of using an alternative pattern-mixture model versus a SP model. As noted previously, one of the problems with fitting the conditional LME model (6.52) based on the overall study termination time T_i^o is that it lumps both censored and uncensored dropout times into the single residual covariate $T_i^* = T_i^o - \mu_{T^o}$. As suggested by the plots in Figure 6.2 and pointed out by Li and Schluchter (2004), this effectively ignores the possible separate effects that administratively censored follow-up times C_i and actual dialysis-related dropout times T_i may have on the estimated slopes. Li and Schluchter (2004) report on several alternative linear mixture models each of which examines different

Table 6.6 A description of the pattern-mixture LME model of Li and Schluchter (2004). This is the pattern-mixture model 4 of Tables I and II of the authors' paper.

Pattern	**Dropout Status** (δ_i)	**Follow-up Time** (T_i^o)
1	Dropout ($\delta_i = 1$)	$T_i^o \geq 12$ months
2	Dropout ($\delta_i = 1$)	$12 < T_i^o \leq 18$ months
3	Dropout ($\delta_i = 1$)	$18 < T_i^o \leq 24$ months
4	Dropout ($\delta_i = 1$)	$T_i^o > 24$ months
5	Censored ($\delta_i = 0$)	$T_i^o \leq 30$ months
6	Censored ($\delta_i = 0$)	$T_i^o > 30$ months

effects associated with censored times versus dropout times. Here we follow their approach and focus on comparing the effects of dietary intervention on the rate of decline in GFR. We do so by fitting Li and Schluchter's pattern-mixture LME model (model 4 in their paper) to the MDRD data but within the framework of the general linear mixture model (6.43). Li and Schluchter's pattern-mixture LME model is formulated by grouping patients into one of six unique patterns of dropout as defined by length of follow-up T_i^o and whether study termination was due to censoring ($\delta_i = 0$) or dropout ($\delta_i = 1$). The six patterns are described in Table 6.6.

Their pattern-mixture LME model can be cast within the framework of the general linear mixture model (6.43) as follows. Let $a_i = \begin{cases} 1 & \text{if Diet K} \\ 0 & \text{if Diet L} \end{cases}$ be an indicator variable reflecting which dietary intervention the ith patient was randomized to and let r_i denote a nominal dropout response variable indicating which pattern the ith patient falls (i.e., $r_i = k$ for one of $k = 1, \ldots, 6$ as defined in Table 6.6). In terms of programming within NLMIXED, it will be convenient to work with the set of indicator variables $d_{ki} = \begin{cases} 1 & \text{if } r_i = k \\ 0 & \text{otherwise} \end{cases}$ reflecting which dropout pattern the ith patient falls. Let $\boldsymbol{d}_i = (d_{1i}, d_{2i}, d_{3i}, d_{4i}, d_{5i}, d_{6i})'$ be the vector of dropout pattern indicator variables and let $\boldsymbol{x}_i' = (1, a_i)$. If we let $\pi_{ki}(\boldsymbol{x}_i) = \Pr(d_{ki} = 1|\boldsymbol{x}_i) = \Pr(r_i = k|\boldsymbol{x}_i)$ denote the marginal probability of the ith patient's dropout pattern given a_i, we can model the marginal distribution of $\boldsymbol{d}_i$ subject to the usual linear constraint $\sum_{k=1}^{6} \pi_{ki}(\boldsymbol{x}_i) = 1$ based on the generalized logit model

$$\pi_{6i}(\boldsymbol{x}_i) = \frac{1}{1 + \sum_{k=1}^{5} \exp(\boldsymbol{\phi}_k' \boldsymbol{x}_i)} \tag{6.57}$$

$$\pi_{ki}(\boldsymbol{x}_i) = \pi_{6i}(\boldsymbol{x}_i) \exp(\boldsymbol{\phi}_k' \boldsymbol{x}_i), \quad k = 1, \ldots, 5$$

where $\boldsymbol{\phi}_k' = (\phi_{0k}, \phi_k)'$ and $\boldsymbol{\phi}_k' \boldsymbol{x}_i = (\phi_{0k} + \phi_k a_i)$, $k = 1, \ldots, 5$. Under model (6.57), we set dropout pattern 6 as the reference category in which case

$$\log\left[\frac{\Pr(r_i = k|\boldsymbol{x}_i)}{\Pr(r_i = 6|\boldsymbol{x}_i)}\right] = \log\left[\frac{\Pr(d_{ki} = 1|\boldsymbol{x}_i)}{\Pr(d_{6i} = 1|\boldsymbol{x}_i)}\right] = \log\left[\frac{\pi_{ki}(\boldsymbol{x}_i)}{\pi_{6i}(\boldsymbol{x}_i)}\right] = \boldsymbol{\phi}_k' \boldsymbol{x}_i;\ 1 \leq k < 6.$$

is the kth logit (e.g., Agresti, 1990, §9.2).

Under the generalized logit model (6.57), the marginal mean of $\boldsymbol{d}_i$ would be

$$E(\boldsymbol{d}_i|\boldsymbol{x}_i) = \boldsymbol{\pi}_i(\boldsymbol{\phi}; \boldsymbol{x}_i) = \begin{pmatrix} \pi_1(\boldsymbol{\phi}; \boldsymbol{x}_i) \\ \pi_2(\boldsymbol{\phi}; \boldsymbol{x}_i) \\ \vdots \\ \pi_6(\boldsymbol{\phi}; \boldsymbol{x}_i) \end{pmatrix} = \begin{pmatrix} \pi_{6i}(\boldsymbol{x}_i) \exp(\boldsymbol{\phi}_1' \boldsymbol{x}_i) \\ \pi_{6i}(\boldsymbol{x}_i) \exp(\boldsymbol{\phi}_2' \boldsymbol{x}_i) \\ \vdots \\ \left\{1 + \sum_{k=1}^{5} \exp(\boldsymbol{\phi}_k' \boldsymbol{x}_i)\right\}^{-1} \end{pmatrix}$$

where $\boldsymbol{\phi} = (\boldsymbol{\phi}_1', \boldsymbol{\phi}_2', \boldsymbol{\phi}_3', \boldsymbol{\phi}_4', \boldsymbol{\phi}_5')'$ is the vector of logit parameters. If we now let $\boldsymbol{d}_i^* = \boldsymbol{d}_i - E(\boldsymbol{d}_i|\boldsymbol{x}_i) = \boldsymbol{d}_i - \boldsymbol{\pi}_i(\boldsymbol{\phi};\boldsymbol{x}_i)$, then we can cast the pattern-mixture model of Li and Schluchter (2004) within the framework of the general linear mixture model (6.43) as follows:

$$\text{Stage 1: } y_{ij} = \beta_{1i} + \beta_{2i}t_{ij} + \epsilon_{ij},\ \epsilon_{ij} \sim \text{iid}\, N(0,\sigma^2) \tag{6.58}$$

$$\text{Stage 2: } \beta_{1i} = \left[\beta_{1L} + \sum_{k=1}^{5}\beta_{1Lk}d_{ki}^*\right](1-a_i) + \left[\beta_{1K} + \sum_{k=1}^{5}\beta_{1Kk}d_{ki}^*\right]a_i + b_{1i}$$

$$\beta_{2i} = \left[\beta_{2L} + \sum_{k=1}^{5}\beta_{2Lk}d_{ki}^*\right](1-a_i) + \left[\beta_{2K} + \sum_{k=1}^{5}\beta_{2Kk}d_{ki}^*\right]a_i + b_{2i}$$

where (β_{1L}, β_{1K}) are the population intercepts for diet L and diet K, and similarly (β_{2L}, β_{2K}) are the population slopes for diet L and diet K, respectively. By combining and rearranging terms, one can write model (6.58) in the form of the general linear mixture model (6.43). In particular, if we let

$$\boldsymbol{Z}_i = \begin{pmatrix} 1 & t_{i1} \\ \vdots & \vdots \\ 1 & t_{ip_i} \end{pmatrix} \text{ and } \boldsymbol{A}_i = \begin{pmatrix} 1-a_i & a_i & 0 & 0 \\ 0 & 0 & 1-a_i & a_i \end{pmatrix},$$

then we can write model (6.58) as

$$\boldsymbol{y}_i = \boldsymbol{X}_i\boldsymbol{\beta} + \boldsymbol{D}_i^*\boldsymbol{\beta}_d + (\boldsymbol{Z}_i\boldsymbol{b}_i + \boldsymbol{\epsilon}_i)$$

where $\boldsymbol{X}_i = \boldsymbol{Z}_i\boldsymbol{A}_i$, $\boldsymbol{\beta} = (\beta_{1L}, \beta_{1K}, \beta_{2L}, \beta_{2K})'$ and all other fixed-effects terms can be grouped collectively into a term $\boldsymbol{D}_i^*\boldsymbol{\beta}_d$.

In order to fit the pattern-mixture LME model defined by (6.58), we need to jointly estimate the parameters of the generalized logit model (6.57) together with the parameters of the pattern-mixture LME model (6.58) using NLMIXED. The approach for doing this is similar to that used to fit the conditional LME model (6.52) to the MDRD data. We first use PROC MIXED and PROC GLIMMIX to obtain initial starting values for the parameters $\boldsymbol{\tau} = (\boldsymbol{\beta}, \boldsymbol{\theta})$ and $\boldsymbol{\phi}$, respectively, based on the following SAS code.

Program 6.14

```
data example6_6_2_PM;
 set example6_6_2;
 /* Define Missing Data Patterns  */
 d1=0;d2=0;d3=0;d4=0;d5=0;d6=0;
 if dropout=1 then do;
  if T<=12 then d1=1;
  if 12<T<=18 then d2=1;
  if 18<T<=24 then d3=1;
  if T> 24 then d4=1;
 end;
 if dropout=0 then do;
  if T<=30 then d5=1;
  if T> 30 then d6=1;
 end;
```

```
 if d1=1 then Pattern=1;
 if d2=1 then Pattern=2;
 if d3=1 then Pattern=3;
 if d4=1 then Pattern=4;
 if d5=1 then Pattern=5;
 if d6=1 then Pattern=6;
run;
proc sort data=example6_6_2_PM;
 by Diet ptid months;
run;
proc mixed data=example6_6_2_PM method=ml;
 class Diet ptid;
 model GFR=Diet Diet*Months / noint s;
 random intercept Months / subject=ptid type=un;
 contrast 'No Diet effect on slopes' Diet*Months 1 -1;
 title 'Model 5 of Li and Schluchter (2004, Table I)';
run;
proc mixed data=example6_6_2_PM method=ml;
 class Diet Pattern ptid;
 model GFR = Diet*Pattern Diet*Pattern*Months / noint s;
 random intercept Months / subject=ptid type=un;
 title 'Model 4 of Li and Schluchter (2004, Table I)';
run;
proc glimmix data=example6_6_2_PM;
 where months=0;
 model Pattern = DietK / dist=multinomial link=glogit s;
run;
```

The first call to PROC MIXED is used to get starting values for the overall population intercepts and slopes for diet L and diet K under the pattern-mixture LME model (6.58) and also to provide estimates of the intercepts and slopes assuming ignorable dropout. The call to GLIMMIX is used to get starting values for the parameters of the generalized logit model (6.57). Finally, the second call to PROC MIXED fits the pattern-mixture LME model of Li and Schluchter (2004) to the MDRD data and yields estimates of the pattern-specific intercepts and slopes for the two dietary intervention groups identical to those shown in Table I of Li and Schluchter (2004). However, this analysis does not provide one with the parameter estimates and correct standard errors of interest, namely the marginal intercepts and slopes and their standard errors for the two dietary intervention groups.

Using the starting values from PROC MIXED and PROC GLIMMIX, we use NLMIXED to fit the pattern-mixture LME model (6.58) to the MDRD data by maximizing the joint log-likelihood function

$$\sum_{i=1}^{n} L_i(\boldsymbol{\tau}, \boldsymbol{\phi}; \boldsymbol{y}_i, \boldsymbol{d}_i) = \sum_{i=1}^{n} L_i(\boldsymbol{\tau}; \boldsymbol{y}_i | \boldsymbol{d}_i^*) + \sum_{i=1}^{n} L_i(\boldsymbol{\phi}; \boldsymbol{d}_i)$$

where $L_i(\boldsymbol{\tau}; \boldsymbol{y}_i|\boldsymbol{d}_i^*) = \log[\pi(\boldsymbol{y}_i|\boldsymbol{d}_i^*)]$ is the marginal log-likelihood of $(\boldsymbol{y}_i|\boldsymbol{d}_i^*)$ and $L_i(\boldsymbol{\phi}; \boldsymbol{d}_i) = \log[\pi(\boldsymbol{d}_i)]$ is the marginal log-likelihood of $\boldsymbol{d}_i$. The log-likelihood functions of $(\boldsymbol{y}_i|\boldsymbol{d}_i^*)$ and $\boldsymbol{d}_i$ are defined by

$$L_i(\boldsymbol{\tau}; \boldsymbol{y}_i|\boldsymbol{d}_i^*) = \log \int_b \exp\Big\{\log(\pi(\boldsymbol{y}_i|\boldsymbol{b}_i, \boldsymbol{d}_i^*)) + \log(\pi(\boldsymbol{b}_i))\Big\} d\boldsymbol{b}_i$$

$$L_i(\boldsymbol{\phi}; \boldsymbol{d}_i) = \sum_{k=1}^{6} d_{ki} \log(\pi_{ki}(\boldsymbol{x}_i))$$

where $\log(\pi(\boldsymbol{y}_i|\boldsymbol{b}_i, \boldsymbol{d}_i^*))$ is the conditional log-likelihood of the pattern-mixture LME model (6.58) and $\log(\pi(\boldsymbol{b}_i))$ is the log-likelihood of the Gaussian random-effects. The programming for performing these tasks within NLMIXED is not shown here owing to its length. However, a complete listing of the SAS programming code including numerous comments is available at the author's Web page.

Based on ESTIMATE statements from NLMIXED, Output 6.10 displays select results from fitting the pattern-mixture LME model (6.58) to the MDRD data. Shown are the pattern-specific intercepts and slopes according to each of the six patterns of dropout listed in Table 6.6 as well as the primary parameters of interest, namely the population intercepts (β_{1L}, β_{1K}) for diet L and diet K, and the corresponding population slopes (β_{2L}, β_{2K}). The estimated pattern-specific intercepts and slopes are identical to those obtained by Li and Schluchter (2004, Table I) and in fact are exactly the same estimates one obtains when making the second call to PROC MIXED shown above. However, the results for the marginal intercepts and slopes for diet L and diet K differ slightly from those obtained by Li and Schluchter (2004, Table II). We elected to average over the distribution of $\boldsymbol{d}_i$ based on modeled proportions from the generalized logit model (6.57) whereas Li and Schluchter used the sample proportion within each pattern stratified by which diet group the patients belong. Nevertheless, the results agree quite closely with those of Li and Schluchter.

Output 6.10: Select output from PROC NLMIXED. Shown are the results obtained when we fit the pattern-mixture model (6.58) to the GFR data using the general linear mixture model setup of Fitzmaurice et. al. (2001).

```
                              Specifications

  Data Set                                 WORK.EXAMPLE6_6_2_PM
  Dependent Variable                       Response
  Distribution for Dependent Variable      General
  Random Effects                           b1i b2i
  Distribution for Random Effects          Normal
  Subject Variable                         ptid
  Optimization Technique                   Newton-Raphson
  Integration Method                       Adaptive Gaussian Quadrature
              Dimensions

  Observations Used          1988
  Observations Not Used         0
  Total Observations         1988
  Subjects                    255
  Max Obs Per Subject          13
  Parameters                   38
  Quadrature Points             1
                             Iteration History
  Iter   Flag1   Calls   NegLogLike        Diff     MaxGrad       Slope
     1              83   5495.14211    35.54859    211.6217    -1328.92
     2   *         123   5476.07512    19.06698    227.7773     -181.52
     3   *         163   5440.93266    35.14246    466.5937    -106.932
     4   *         203    5431.3608    9.571867    153.0304      -18.72
     5             243   5431.20091    0.159882    13.02604     -0.3048
     6             283   5431.19981    0.001107    0.117519     -0.0022
     7             323   5431.19981    8.679E-8    6.257E-6    -1.74E-7

  NOTE: GCONV convergence criterion satisfied.
             Fit Statistics

  -2 Log Likelihood             10862
  AIC (smaller is better)       10938
  AICC (smaller is better)      10940
  BIC (smaller is better)       11073
```

Additional Estimates

Label	Estimate	Standard Error	DF	t Value	Pr > |t|
Intercept L-Pattern 1:	18.0350	1.4575	253	12.37	<.0001
Intercept L-Pattern 2:	18.8094	1.1267	253	16.69	<.0001
Intercept L-Pattern 3:	18.3536	1.1872	253	15.46	<.0001
Intercept L-Pattern 4:	20.2010	1.2711	253	15.89	<.0001
Intercept L-Pattern 5:	21.0733	0.8679	253	24.28	<.0001
Intercept L-Pattern 6:	20.4023	0.6138	253	33.24	<.0001
Intercept K-Pattern 1:	17.8743	1.4554	253	12.28	<.0001
Intercept K-Pattern 2:	16.4316	1.1258	253	14.59	<.0001
Intercept K-Pattern 3:	17.8268	1.3453	253	13.25	<.0001
Intercept K-Pattern 4:	19.9357	1.2827	253	15.54	<.0001
Intercept K-Pattern 5:	20.6055	0.8849	253	23.29	<.0001
Intercept K-Pattern 6:	20.2005	0.6064	253	33.31	<.0001
Intercept L:	19.9154	0.4007	253	49.70	<.0001
Intercept K:	19.3852	0.4155	253	46.66	<.0001
Difference (L-K):	0.5302	0.5772	253	0.92	0.3592
Slope L-Pattern 1:	-1.2066	0.1961	253	-6.15	<.0001
Slope L-Pattern 2:	-0.7319	0.07034	253	-10.40	<.0001
Slope L-Pattern 3:	-0.5246	0.05848	253	-8.97	<.0001
Slope L-Pattern 4:	-0.4122	0.05216	253	-7.90	<.0001
Slope L-Pattern 5:	-0.2995	0.03885	253	-7.71	<.0001
Slope L-Pattern 6:	-0.1428	0.02431	253	-5.88	<.0001
Slope K-Pattern 1:	-0.9902	0.1735	253	-5.71	<.0001
Slope K-Pattern 2:	-0.4836	0.07103	253	-6.81	<.0001
Slope K-Pattern 3:	-0.5271	0.06764	253	-7.79	<.0001
Slope K-Pattern 4:	-0.3126	0.05705	253	-5.48	<.0001
Slope K-Pattern 5:	-0.2333	0.03958	253	-5.89	<.0001
Slope K-Pattern 6:	-0.1430	0.02384	253	-6.00	<.0001
Slope L:	-0.3964	0.03540	253	-11.20	<.0001
Slope K:	-0.3211	0.03079	253	-10.43	<.0001
Difference (L-K):	-0.07529	0.04691	253	-1.61	0.1097

We also fit a SP model to the MDRD data assuming the LME model

$$\begin{aligned} \text{Stage 1: } & y_{ij} = \beta_{1i} + \beta_{2i}t_{ij} + \epsilon_{ij},\ \epsilon_{ij} \sim \text{iid}\, N(0,\sigma^2) \\ \text{Stage 2: } & \beta_{1i} = \beta_{11}(1-a_i) + \beta_{12}a_i + b_{1i} \\ & \beta_{2i} = \beta_{21}(1-a_i) + \beta_{22}a_i + b_{2i} \end{aligned} \tag{6.59}$$

for the pdf of $(\boldsymbol{y}_i|\boldsymbol{b}_i)$ where a_i is the diet K indicator variable defined above and assuming the piecewise exponential model (6.54) for $(T_i|\boldsymbol{b}_i)$ but with a modified hazards function, $\lambda_0(t)\exp(a_i\eta_K + \boldsymbol{b}_i'\boldsymbol{\eta}_b)$. The SAS code, which is available on the author's Web page, is very similar to that used to fit the SP model corresponding to Output 6.9.

Selected results from fitting this SP model to the MDRD data are shown in Output 6.11. The estimated slope or rate of change in GFR for patients in the diet L group was -0.3460 ± 0.02514 ml/min/1.73m^2/month and -0.3012 ± 0.02535 ml/min/1.73m^2/month for those patients in the diet K group. In both cases, the slope was greater in absolute value (i.e., a higher

Output 6.11: Select output from NLMIXED for a SP model defined by the LME model (6.59) for $(\boldsymbol{y}_i|\boldsymbol{b}_i)$ and a piecewise exponential model for $(T_i|\boldsymbol{b}_i)$ with proportional hazards function $\lambda_0(t)\exp(a_i\eta_K + \boldsymbol{b}_i'\boldsymbol{\eta}_b)$.

Fit Statistics

-2 Log Likelihood	11063
AIC (smaller is better)	11099
AICC (smaller is better)	11099
BIC (smaller is better)	11163

Parameter Estimates

Parameter	Estimate	Standard Error	DF	t Value	Pr > |t|
beta11	19.7407	0.4100	253	48.15	<.0001
beta12	19.3257	0.4149	253	46.58	<.0001
beta21	−0.3460	0.02514	253	−13.76	<.0001
beta22	−0.3012	0.02535	253	−11.88	<.0001
psi11	19.8084	1.9385	253	10.22	<.0001
psi12	0.1441	0.08059	253	1.79	0.0750
psi22	0.06357	0.007884	253	8.06	<.0001
Sigma_Sq	5.1306	0.1859	253	27.60	<.0001
eta1	−9.6330	0.9645	253	−9.99	<.0001
eta2	−6.7071	0.5613	253	−11.95	<.0001
eta3	−5.1367	0.4294	253	−11.96	<.0001
eta4	−4.0661	0.3811	253	−10.67	<.0001
eta5	−3.8091	0.4200	253	−9.07	<.0001
eta6	−3.2871	0.4355	253	−7.55	<.0001
eta7	−3.1784	0.6403	253	−4.96	<.0001
eta_DietK	−0.2764	0.4205	253	−0.66	0.5116
eta_b1i	−0.2414	0.03953	253	−6.11	<.0001
eta_b2i	−9.6793	1.1419	253	−8.48	<.0001

Parameter	Alpha	Lower	Upper	Gradient
beta11	0.05	18.9332	20.5482	−0.14754
beta12	0.05	18.5086	20.1427	0.00622
beta21	0.05	−0.3955	−0.2965	0.197243
beta22	0.05	−0.3511	−0.2513	−0.19162
psi11	0.05	15.9909	23.6260	0.083938
psi12	0.05	−0.01461	0.3028	0.06907
psi22	0.05	0.04804	0.07909	−0.2487
Sigma_Sq	0.05	4.7645	5.4966	−0.00989
eta1	0.05	−11.5325	−7.7336	0.0602
eta2	0.05	−7.8126	−5.6016	−0.03344
eta3	0.05	−5.9823	−4.2911	−0.02261
eta4	0.05	−4.8166	−3.3156	0.009736
eta5	0.05	−4.6363	−2.9819	−0.02115
eta6	0.05	−4.1448	−2.4294	−0.09091
eta7	0.05	−4.4395	−1.9173	0.096477
eta_DietK	0.05	−1.1044	0.5517	0.031561
eta_b1i	0.05	−0.3192	−0.1636	−0.19298
eta_b2i	0.05	−11.9281	−7.4306	−0.02099

Additional Estimates

Label	Estimate	Standard Error	DF	t Value	Pr > |t|
Intercept L:	19.7407	0.4100	253	48.15	<.0001
Intercept K:	19.3257	0.4149	253	46.58	<.0001
Difference (L−K):	0.4150	0.5827	253	0.71	0.4770
Slope L:	−0.3460	0.02514	253	−13.76	<.0001
Slope K:	−0.3012	0.02535	253	−11.88	<.0001
Difference (L−K):	−0.04476	0.03498	253	−1.28	0.2018

Label	Alpha	Lower	Upper
Intercept L:	0.05	18.9332	20.5482
Intercept K:	0.05	18.5086	20.1427
Difference (L−K):	0.05	−0.7325	1.5626
Slope L:	0.05	−0.3955	−0.2965
Slope K:	0.05	−0.3511	−0.2513
Difference (L−K):	0.05	−0.1137	0.02412

rate of decline) compared to that obtained under the standard LME modelassuming ignorable dropout which, based on the first call to MIXED in the preceding SAS code, was -0.3082 ± 0.02303 ml/min/1.73m^2/month for diet L and -0.2512 ± 0.02344 ml/min/1.73m^2/month for diet K (see also model 5 results in Table II of Li and Schluchter, 2004). These results are consistent with our previous results based on comparing all four treatment groups (i.e., Output 6.5 assuming ignorable dropout under a LME model and Output 6.9 assuming non-ignorable dropout under a SP model).

We can gain further insight into why the SP model yields estimated slopes suggestive of a more rapid decline in GFR compared to the standard LME model by also examining the hazard ratios associated with the residual intercepts and slopes under the piecewise exponential hazards model. While the estimated log hazard ratio, $\widehat{\eta}_K$, for diet K (denoted `eta_DietK` in Output 6.11) was not significant ($-0.2764 \pm 0.4205, p = 0.5116$), the log-hazard ratios for both the patient-specific residual intercept b_{1i} and slope b_{2i} were significantly associated with an increased risk of dropout. In terms of the residual slope, the log hazard ratio was estimated to be -9.6793 ± 1.1419 with 95% confidence interval $(-11.9281, -7.4306)$. This corresponds to an estimated hazard ratio of 2.63 for every 0.1 ml/min/1.73m^2/month *further decrease* in the rate of decline in GFR. For example, the average rate of decline is -0.3012 ml/min/1.73m^2/month for patients in the diet K group so that patients with a 0.1 ml/min/1.73m^2/month *further decrease* have a rate of decline of -0.4012 ml/min/1.73m^2/month and this corresponds to a $2.63 = \exp(-9.6793 \times -0.1)$ higher rate of dropout. This result, while not providing definitive proof, does provide reasonable evidence to suggest that dropout is non-ignorable and that one need take this possibility into account when estimating the rate at which GFR declines over time.

In summary, several models were fit to the MDRD data each aimed at estimating the rate of decline in GFR assuming non-ignorable dropout. In each case, the estimated slopes decreased by an additional -0.03 to -0.05 ml/min/1.73m^2/month compared to the slopes estimated assuming dropout is ignorable. These models all provide sufficient evidence to reject the hypothesis that dropout is completely at random. However, since certain assumptions for a given pattern-mixture or shared parameter model can only be verified for observed values of the response variable y, in this case GFR, we are somewhat restricted in what we can claim about the underlying dropout mechanism. For example, under the shared parameter models, we can only verify the assumption that $(\boldsymbol{y}_i|\boldsymbol{b}_i)$ and $(T_i|\boldsymbol{b}_i)$ are conditionally independent based on observed values of the response variable y. Consequently, the strength of evidence supporting the assumption of non-ignorability will ultimately rest on 1) the ability to verify key modeling assumptions one makes based on observed values and 2) the ability to assess how sensitive one's results are to various model misspecifications. In this example, we have seen how two different pattern-mixture models and a shared parameter model all point to a conclusion that dropout is non-ignorable. While numerically the results from each model differ slightly, qualitatively they are all in the same direction. Examples of additional sensitivity analyses for this data can be found in Vonesh et. al. (2006) for SP models and Li and Schluchter (2004) for pattern-mixture models. Each of these analyses also lend credence to a non-ignorable dropout mechanism.

6.6.3 Schizophrenia data—continued

In example 5.2.3 of §5.2, we fit a mixed-effects cumulative logit model with random intercepts and slopes to the four-point ordinal response data on overall severity of illness from the NIMH Schizophrenia collaborative study (see example 4.3.4 of §4.3 for a complete description of the data, study design, key variables, etc.). The analysis was done assuming missing data, particularly missing data due to dropout, are MAR. However, as Hedeker and Gibbons (1997) note, there were a substantial number of patients that failed to complete the six-week study with the rate of dropout being greater in those patients randomized to the placebo control group. In an attempt to assess what impact dropout might have on the study results, Hedeker and Gibbons (1997) fit a pattern-mixture LME model to the data treating the original seven-point severity of illness score from Item 79 of the Inpatient Multidimensional Psychiatric Scale (IMPS) as continuous response. They found substantial evidence indicating that dropout is significantly related to IMPS79 scores although when averaged over the patterns of dropout, the comparative results were qualitatively similar to those from a standard LME model assuming ignorable dropout.

In this section, we examine what possible impact non-ignorable dropout might have on the mixed-effects ordinal regression analysis of example 5.2.3. We do so by fitting a shared parameter (SP) model to the four-point ordinal response data assuming a mixed-effects cumulative logit model with random intercepts and slopes for the ordinal response and a discrete-time proportional hazards model for dropout with the random intercepts and slopes serving as covariates. To perform this analysis, we must first create a dataset with the necessary structure to jointly model both the ordinal response and dropout using NLMIXED. The SAS programming code to create the necessary dataset along with a partial listing of the data follows.

Program 6.15

```
/* Create primary response variable dataset */
data example6_6_3;
 set example4_3_4;
 /* Ind is a 0-1 indicator variable for a SP model */
 /* Ind=0 implies we are modeling IMPS79o response */
 Ind=0;
run;
/* Calculate the maximum value of WEEK for each subject */
/* This is the last week at which IMPS79o was measured  */
/* prior to dropout or upon completion of the study     */
proc means data=example6_6_3 noprint nway;
 class ID Trt;
 var Week;
 output out=temp1 max=LastWeek;
run;
/* Create discrete time survival dataset */
data temp2;
 set temp1;
 /* Ind is a 0-1 indicator variable for a SP model */
 /* Ind=1 implies we are modeling Dropout response */
 Ind=1;
 Keep ID Trt LastWeek Ind;
run;
data dropout;
 set temp2;
 /* Time represents the interval dropout time    */
 Time = LastWeek+1;
 do Week=2 to Time;
  if (Week< 7) then do;
     Dropout=(Week=Time); t1=Week-1; t2=Week;
     if Week=2 then t1=0;
     Interval=compress('['||left(t1)||','||left(t2)||')');
     output;
  end;
 end;
run;
/* Dataset required to fit a shared parameter model */
data example6_6_3;
 set example6_6_3 dropout;
 if Ind=0 then do;
  Response=IMPS79o;
 end;
```

```
 if Ind=1 then do;
  Response=Dropout;
  SWeek=0;
 end;
run;
proc sort data=example6_6_3;
 by ID Ind Week;
run;
proc print data=example6_6_3 noobs;
 where ID in (1103 1105 1118 2118);
 var ID Trt Ind LastWeek Week SWeek IMPS79o
     Interval Dropout Response;
run;
```

Unlike with the MDRD data, the schizophrenia data does not have an actual date of dropout. Rather it contains the last week at which the primary response variable of interest, the IMPS79 ordinal score, was measured. The above program computes the last week at which the IMPS79 ordinal score was measured and this information is then used to construct a dataset which one can use to fit a discrete time proportional hazards model to the incidence of dropout. The original dataset containing the IMPS79 ordinal scores and the dataset containing the dropout data are then merged to form a dataset suitable for fitting a shared parameter model. Shown in Output 6.12 is a partial listing of this combined dataset.

Output 6.12: Partial listing of the combined dataset required to fit a shared parameter model to the four-point ordinal response data on overall severity of illness from the NIMH Schizophrenia collaborative study. The variable Ind is the 0-1 indicator variable that determines whether the outcome variable (labeled `Response`) is the ordinal response variable `IMPS79o` or the dropout response (defined by `LastWeek` and `Dropout`).

ID	Trt	Ind	Last Week	Week	SWeek	IMPS79o	Interval	Dropout	Response
1103	1	0	.	0	0.0000	4		.	4
1103	1	0	.	1	1.0000	2		.	2
1103	1	0	.	3	1.7321	2		.	2
1103	1	0	.	6	2.4495	2		.	2
1103	1	1	6	2	0.0000	.	[0,2)	0	0
1103	1	1	6	3	0.0000	.	[2,3)	0	0
1103	1	1	6	4	0.0000	.	[3,4)	0	0
1103	1	1	6	5	0.0000	.	[4,5)	0	0
1103	1	1	6	6	0.0000	.	[5,6)	0	0
1105	1	0	.	0	0.0000	2		.	2
1105	1	0	.	1	1.0000	2		.	2
1105	1	0	.	3	1.7321	1		.	1
1105	1	1	3	2	0.0000	.	[0,2)	0	0
1105	1	1	3	3	0.0000	.	[2,3)	0	0
1105	1	1	3	4	0.0000	.	[3,4)	1	1
1118	1	0	.	0	0.0000	4		.	4
1118	1	0	.	1	1.0000	4		.	4
1118	1	1	1	2	0.0000	.	[0,2)	1	1
2118	0	0	.	0	0.0000	4		.	4
2118	0	0	.	1	1.0000	4		.	4
2118	0	0	.	3	1.7321	4		.	4
2118	0	0	.	5	2.2361	4		.	4
2118	0	1	5	2	0.0000	.	[0,2)	0	0
2118	0	1	5	3	0.0000	.	[2,3)	0	0
2118	0	1	5	4	0.0000	.	[3,4)	0	0
2118	0	1	5	5	0.0000	.	[4,5)	0	0
2118	0	1	5	6	0.0000	.	[5,6)	1	1

Table 6.7 Frequency of longitudinal measurements of IMPS79 ordinal responses per week and frequency of dropout/completion. The number of dropouts at a given week are the number of patients who started the week and had their IMPS79 ordinal score measured but who subsequently did not return (if week< 6) or who completed the trial (week=6).

		Week						
Treatment	**Frequency**	0	1	2	3	4	5	6
Placebo ($n = 108$)	IMPS79	107	105	5	87	2	2	70
	Dropouts	0	13	5	16	2	2	70
Drug ($n = 329$)	IMPS79	327	321	9	287	9	7	265
	Dropouts	0	24	5	26	3	6	265

Table 6.7 summarizes the ordinal response sample size per week as well as the frequency of patients according to their last week of follow-up prior to dropout or study completion. From this table we see that repeated measurements of the IMPS79 ordinal score are irregularly spaced over the six week study with the bulk of measurements occurring at weeks 0, 1, 3 and 6. The imbalance in measurements could be due to missed visits or because of scheduling conflicts. For example, three subjects (1 in the placebo control group and two in the drug group) had their first IMPS79 ordinal score measured at week 1 rather than week 0. In subsequent analyses, we treat any imbalances in measurements over time as due either to a lagged effect in scheduled visits or possibly as intermittent missing values which we assume to be MAR.

To examine whether dropout is non-ignorable we first fit the data assuming dropout is completely at random. We do so by jointly fitting the mixed-effects cumulative logit model for the ordinal response IMPS79 (see Model 2, page 246) together with a discrete time proportional hazards (PH) model with weekly baseline hazard rates

$$\lambda_{0k} = \lambda_0(t_k) = \Pr[T_i \in [t_{k-1}, t_k) | T_i \geq t_{k-1}],\ k = 1, \ldots, 5. \tag{6.60}$$

Based on the pattern of dropout from Table 6.7, we set $t_0 = 0$, $t_1 = 2$, $t_2 = 3$, $t_3 = 4$, $t_4 = 5$, $t_5 = 6$. We combined the two weekly intervals, $[0, 1)$ and $[1, 2)$, into one interval, $[0, 2)$, since there were no dropouts between baseline (week=0) and the start of week 1. Hence the intervals within which dropouts are known to occur are $[0, 2)$, $[2, 3)$, $[3, 4)$, $[4, 5)$, $[5, 6)$. For patients who completed the study, we would define an interval of follow-up to be $[6, 7)$ but since we do not know whether these individuals would have "survived" the interval $[6, 7)$ or not, we must censor them at the end of the previous interval, namely $[5, 6)$.

Under this setting, we can think of a latent continuous dropout time, T_i, that is grouped according to the last week on which observations are taken on an individual. For a subject (e.g., ID 1105, Output 6.12) whose last week of follow-up is week 3, dropout is known to have occurred after the start of week 3 but before the start of week 4. This subject would then contribute three discrete time intervals to the dropout model: $[0, 2)$, $[2, 3)$ and $[3, 4)$. Based on the discrete intervals defined above, we fit a discrete time PH model to the data assuming the overall hazard function is

$$\begin{aligned} \lambda_i(t_k|a_{1i}) &= 1 - (1 - \lambda_{0k})^{\exp\{a_{1i}\eta_{\text{Trt}}\}} \\ &= 1 - \exp\{-\exp(\eta_{0k} + a_{1i}\eta_{\text{Trt}})\} \end{aligned} \tag{6.61}$$

where $\eta_{0k} = \log[-\log(1 - \lambda_{0k})]$ are nuisance parameters describing the baseline hazard rates (6.60) over the intervals $[t_{k-1}, t_k)$, a_{1i} is the treatment group indicator variable ($a_{1i} = 1$ if drug, 0 if placebo control as defined in example 4.3.4 of §4.3) and η_{Trt} is the log-hazard ratio for a treatment effect from the continuous-time relative risk (proportional hazards) model for T_i (e.g., Kalbfleisch and Prentice, 2002, §2.4 and §4.8).

Assuming dropout is completely at random, we jointly fit the GLME cumulative logit model with random intercepts and slopes to the repeated IMPS79 ordinal response data and the discrete time PH model (6.61) to the dropout data using NLMIXED. Starting values for the parameters of the GLME logit model are based on our previous analysis (Output 5.11, example 5.2.3) while starting values for the parameters of the discrete PH model (6.61) are based on a preliminary analysis of dropout using PROC LOGISTIC. The SAS code used to jointly model IMPS79 and dropout under S-CRD are shown below.

Program 6.16

```
proc format;
 value Trt 0 = 'Placebo' 1 = 'Drug';
run;
/* Discrete time proportional hazards model for dropout  */
/* This model is used to get starting values for NLMIXED */
proc logistic data=dropout;
 class Interval /param=glm;
 model Dropout(event='1')= Interval Trt /
       noint link=cloglog technique=newton;
run;
/* Joint model for IMPS79o and Dropout under S-CRD */
ods output parameterestimates=SAStemp.peCRD;
proc nlmixed data=example6_6_3 qpoints=5 tech=newrap;
 parms beta11=-7.3 beta12=-3.4 beta13=-0.8
       beta_Trt = -0.06 beta_SWeek = 0.88
       beta_Trt_x_SWeek = 1.68
       psi11=6.85 psi12=-1.44 psi22=1.95
       eta1=-1.9 eta2=-3.2 eta3=-1.7
       eta4=-3.7 eta5=-3.2 eta_Trt=-0.69;
 /* Define SS intercepts and slopes */
 bi = b1i + b2i*SWeek;
 /* Define log-likelihood for the GLME logit model */
 if Ind=0 then do;
 /* Define linear predictor as defined in GLIMMIX  */
 /* In the comments below the variable y = IMPS79o */
 /* is the ordinal response variable of interest   */
 linpred = beta_Trt*Trt + beta_SWeek*SWeek +
           beta_Trt_x_SWeek*Trt*SWeek + bi;
 Z1 = (beta11 + linpred);
 Z2 = (beta12 + linpred);
 Z3 = (beta13 + linpred);
 /* Define the cumulative response probabilities   */
 P1 = exp(Z1)/(1+exp(Z1)); ** = Pr[y<=1]=Pr[y=1];
 P2 = exp(Z2)/(1+exp(Z2)); ** = Pr[y<=2];
 P3 = exp(Z3)/(1+exp(Z3)); ** = Pr[y<=3];
 P4 = 1;                   ** = Pr[y<=4];
 /* Define the individual response probabilities   */
 pi1 = P1;                 ** = Pr[y=1];
 pi2 = P2-P1;              ** = Pr[y=2];
 pi3 = P3-P2;              ** = Pr[y=3];
 pi4 = 1-P3;               ** = Pr[y=4]=1-Pr[y<=3];
 P = (IMPS79o=1)*pi1 + (IMPS79o=2)*pi2 +
     (IMPS79o=3)*pi3 + (IMPS79o=4)*pi4;
 LogL = log(P);
 end;
```

```
 /* Define log-likelihood for discrete time PH model */
 if Ind=1 then do;
 Lambda_i = eta1*(Week=2) + eta2*(Week=3) + eta3*(Week=4) +
            eta4*(Week=5) + eta5*(Week=6) + eta_Trt*Trt;
 ** Complementary log-log link = Hazard function;
 Hazard_i = 1 - exp(-exp(Lambda_i));
 ** Binary(1,H_i) log likelihood;
 LogL = response*log(Hazard_i/(1-Hazard_i)) + log(1-Hazard_i);
 end;
 model response ~ general(LogL);
 random b1i b2i ~ normal([0,0],[psi11,
                                psi12, psi22]) subject=ID;
 estimate 'CumLogOR(Week=0)'
  beta_Trt;
 estimate 'CumLogOR(Week=1)'
  beta_Trt + beta_Trt_x_SWeek*1;
 estimate 'CumLogOR(Week=3)'
  beta_Trt + beta_Trt_x_SWeek*1.7321;
 estimate 'CumLogOR(Week=6)'
  beta_Trt + beta_Trt_x_SWeek*2.4495;
 estimate 'CumOR(Week=0)'
  exp(beta_Trt);
 estimate 'CumOR(Week=1)'
  exp(beta_Trt + beta_Trt_x_SWeek*1);
 estimate 'CumOR(Week=3)'
  exp(beta_Trt + beta_Trt_x_SWeek*1.7321);
 estimate 'CumOR(Week=6)'
  exp(beta_Trt + beta_Trt_x_SWeek*2.4495);
run;
```

In order to replicate the results shown in Output 5.11, we specified the use of 5 quadrature points (QPOINTS=5) in NLMIXED as this is the number of quadrature points GLIMMIX used when evaluating the integrated log-likelihood of example 5.2.3 (Note: when we ran the default adaptive Gaussian quadrature in NLMIXED, it utilized 7 quadrature points which produced slightly different estimates from those shown here). Select results from this joint analysis assuming S-CRD are shown in Output 6.13. The parameter estimates for the ordinal GLME logit model are exactly the same as obtained in example 5.2.3 (Output 5.11). However, the standard errors of the parameter estimates shown here are the model-based likelihood estimates which differ slightly from the empirical robust standard errors used when we fit the model using GLIMMIX (see Output 5.11). Both the parameter estimates and their standard errors obtained from fitting the discrete time PH model (6.61) are the same as those obtained using the LOGISTIC procedure (results of the latter are not shown here). Using ODS OUTPUT (`ods output parameterestimates=SAStemp.peCRD;`), we stored the parameter estimates from this analysis into a temporary SAS dataset, `SAStemp.peCRD`, which we can use to define starting values in subsequent calls to NLMIXED. This can be an extremely useful programming tip when one anticipates fitting a series of increasingly more complex GNLME models.

Output 6.13: Select output from NLMIXED when we jointly analyze the repeated IMPS79 ordinal response data and time to dropout assuming S-CRD.

Specifications

Data Set	WORK.EXAMPLE6_6_3
Dependent Variable	Response
Distribution for Dependent Variable	General
Random Effects	b1i b2i
Distribution for Random Effects	Normal
Subject Variable	ID
Optimization Technique	Newton–Raphson
Integration Method	Adaptive Gaussian Quadrature

Dimensions

Observations Used	3521
Observations Not Used	0
Total Observations	3521
Subjects	437
Max Obs Per Subject	10
Parameters	15
Quadrature Points	5

Iteration History

Iter	Flag1	Calls	NegLogLike	Diff	MaxGrad	Slope
1		34	2028.83014	0.077248	0.046935	−0.15276
2		51	2028.8301	0.000036	0.00003	−0.00007
3		68	2028.8301	1.57E−11	3.24E−11	−264E−13

NOTE: GCONV convergence criterion satisfied.

Fit Statistics

−2 Log Likelihood	4057.7
AIC (smaller is better)	4087.7
AICC (smaller is better)	4087.8
BIC (smaller is better)	4148.9

Parameter Estimates

Parameter	Estimate	Standard Error	DF	t Value	Pr > |t|
beta11	−7.2832	0.4673	435	−15.59	<.0001
beta12	−3.3996	0.3825	435	−8.89	<.0001
beta13	−0.8057	0.3487	435	−2.31	0.0213
beta_Trt	−0.05630	0.3881	435	−0.15	0.8847
beta_SWeek	0.8785	0.2160	435	4.07	<.0001
beta_Trt_x_SWeek	1.6840	0.2499	435	6.74	<.0001
psi11	6.8471	1.2818	435	5.34	<.0001
psi12	−1.4470	0.5152	435	−2.81	0.0052
psi22	1.9488	0.4036	435	4.83	<.0001
eta1	−1.9487	0.2066	435	−9.43	<.0001
eta2	−3.1933	0.3409	435	−9.37	<.0001
eta3	−1.6789	0.2023	435	−8.30	<.0001
eta4	−3.7339	0.4667	435	−8.00	<.0001
eta5	−3.2434	0.3781	435	−8.58	<.0001
eta_Trt	−0.6934	0.2050	435	−3.38	0.0008

Parameter	Alpha	Lower	Upper	Gradient
beta11	0.05	−8.2016	−6.3648	1.04E−12
beta12	0.05	−4.1514	−2.6478	2.2E−13
beta13	0.05	−1.4909	−0.1204	−274E−15
beta_Trt	0.05	−0.8192	0.7066	5.81E−14
beta_SWeek	0.05	0.4539	1.3031	4.7E−13
beta_Trt_x_SWeek	0.05	1.1929	2.1751	4.77E−13
psi11	0.05	4.3278	9.3664	1.61E−13
psi12	0.05	−2.4596	−0.4344	5.78E−13
psi22	0.05	1.1556	2.7420	7.67E−13
eta1	0.05	−2.3549	−1.5426	3.24E−11

```
eta2      0.05   -3.8633   -2.5234    -69E-15
eta3      0.05   -2.0766   -1.2813    3.5E-13
eta4      0.05   -4.6512   -2.8166   1.25E-12
eta5      0.05   -3.9865   -2.5002   5.25E-12
eta_Trt   0.05   -1.0964   -0.2904    2.1E-11
                      Additional Estimates
Label               Estimate   Standard Error   DF    t Value    Pr > |t|
CumLogOR(Week=0)    -0.05630           0.3881   435     -0.15      0.8847
CumLogOR(Week=1)      1.6277           0.3210   435      5.07      <.0001
CumLogOR(Week=3)      2.8605           0.3838   435      7.45      <.0001
CumLogOR(Week=6)      4.0686           0.5056   435      8.05      <.0001
CumOR(Week=0)         0.9453           0.3669   435      2.58      0.0103
CumOR(Week=1)         5.0921           1.6344   435      3.12      0.0020
CumOR(Week=3)        17.4707           6.7056   435      2.61      0.0095
CumOR(Week=6)        58.4757          29.5663   435      1.98      0.0486
Label               Alpha      Lower      Upper
CumLogOR(Week=0)     0.05    -0.8192     0.7066
CumLogOR(Week=1)     0.05     0.9968     2.2585
CumLogOR(Week=3)     0.05     2.1062     3.6149
CumLogOR(Week=6)     0.05     3.0749     5.0624
CumOR(Week=0)        0.05     0.2241     1.6664
CumOR(Week=1)        0.05     1.8798     8.3043
CumOR(Week=3)        0.05     4.2913    30.6501
CumOR(Week=6)        0.05     0.3651     116.59
```

Based on the estimated log hazard ratio for treatment of $\widehat{\eta}_{\text{Trt}} = -0.6934 \pm 0.2050$, $p = 0.0008$ ($\widehat{\eta}_{\text{Trt}}$ is labeled as **eta_Trt** in Output 6.13), the hazard ratio for dropout is estimated to be 0.50 indicating that dropout among patients in the active drug group is half that seen in the placebo control patients. To investigate whether this differential dropout rate is truly non-ignorable, we fit a shared parameter model to the data by including as covariates within the hazard function (6.61), the residual patient-specific intercepts b_{1i} and slopes b_{2i}. The resulting hazard function is therefore

$$\lambda_i(t_k|a_{1i}, \boldsymbol{b}_i) = 1 - \exp\{-\exp(\eta_{0k} + a_{1i}\eta_{\text{Trt}} + b_{1i}\eta_{b1} + b_{2i}\eta_{b2})\}. \tag{6.62}$$

We can fit this SP model in NLMIXED using the same code as above but with the modified hazard function (6.62) defined by the statements

```
Lambda_i = eta1*(Week=2) + eta2*(Week=3) + eta3*(Week=4) +
           eta4*(Week=5) + eta5*(Week=6) + eta_Trt*Trt +
           eta_b1i*b1i + eta_b2i*b2i;
** Complementary log-log link = Hazard function **;
Hazard_i = 1 - exp(-exp(Lambda_i));
```

and by specifying starting values using the PARMS statement,

```
parms / data=initial;
```

where the dataset initial is defined as follows:

```
data initial;
 Parameter='eta_b1i';Estimate=0; output;
 Parameter='eta_b2i';Estimate=0; output;
run;
data initial;
 set SAStemp.peCRD initial;
run;
```

Here, we set the starting values for the log hazard ratios, η_{b1} and η_{b1}, (labeled as eta_b1i and eta_b2i) to zero. By making these changes and re-running NLMIXED, we get the following results shown in Output 6.14. We find that there is no effect associated with either the patient-specific residual intercepts b_{1i} or slopes b_{2i} on the rate of dropout.

Output 6.14: Results from a shared parameter model (SP Model 1) obtained by adding the random intercepts b_{1i} and slopes b_{2i} as covariates to the hazard rate resulting in a discrete time PH model with hazard rate (6.62).

Specifications

Data Set	WORK.EXAMPLE6_6_3
Dependent Variable	Response
Distribution for Dependent Variable	General
Random Effects	b1i b2i
Distribution for Random Effects	Normal
Subject Variable	ID
Optimization Technique	Newton-Raphson
Integration Method	Adaptive Gaussian Quadrature

Dimensions

Observations Used	3521
Observations Not Used	0
Total Observations	3521
Subjects	437
Max Obs Per Subject	10
Parameters	17
Quadrature Points	5

Iteration History

Iter	Flag1	Calls	NegLogLike	Diff	MaxGrad	Slope
1		38	2027.774	1.0561	4.012316	−2.75734
2		57	2027.57674	0.197265	0.41865	−0.38727
3		76	2027.5753	0.001441	0.001889	−0.00289
4		95	2027.5753	9.535E−8	2.263E−7	−1.91E−7

NOTE: GCONV convergence criterion satisfied.

Fit Statistics

−2 Log Likelihood	4055.2
AIC (smaller is better)	4089.2
AICC (smaller is better)	4089.3
BIC (smaller is better)	4158.5

Parameter Estimates

Parameter	Estimate	Standard Error	DF	t Value	Pr > |t|
beta11	−7.3633	0.4759	435	−15.47	<.0001
beta12	−3.4754	0.3915	435	−8.88	<.0001
beta13	−0.8743	0.3570	435	−2.45	0.0147
beta_Trt	−0.01649	0.3942	435	−0.04	0.9666
beta_SWeek	0.9746	0.2275	435	4.28	<.0001
beta_Trt_x_SWeek	1.6345	0.2537	435	6.44	<.0001
psi11	7.0485	1.3236	435	5.33	<.0001
psi12	−1.5498	0.5384	435	−2.88	0.0042
psi22	2.0065	0.4161	435	4.82	<.0001
eta1	−1.9875	0.2140	435	−9.29	<.0001
eta2	−3.2257	0.3446	435	−9.36	<.0001
eta3	−1.7069	0.2072	435	−8.24	<.0001
eta4	−3.7565	0.4684	435	−8.02	<.0001
eta5	−3.2635	0.3800	435	−8.59	<.0001
eta_Trt	−0.6958	0.2087	435	−3.33	0.0009
eta_b1i	−0.00165	0.05078	435	−0.03	0.9741
eta_b2i	0.2123	0.1399	435	1.52	0.1298

Additional Estimates

Label	Estimate	Standard Error	DF	t Value	Pr > |t|
CumLogOR(Week=0)	−0.01649	0.3942	435	−0.04	0.9666
CumLogOR(Week=1)	1.6180	0.3224	435	5.02	<.0001
CumLogOR(Week=3)	2.8145	0.3847	435	7.32	<.0001
CumLogOR(Week=6)	3.9871	0.5080	435	7.85	<.0001
CumOR(Week=0)	0.9836	0.3877	435	2.54	0.0115
CumOR(Week=1)	5.0428	1.6260	435	3.10	0.0021
CumOR(Week=3)	16.6856	6.4186	435	2.60	0.0097
CumOR(Week=6)	53.8987	27.3801	435	1.97	0.0496

Since neither the random intercepts or slopes are predictive of dropout, one might conclude there is no evidence to indicate dropout is non-ignorable. Such a conclusion would be valid provided the specified SP model is in fact the true model. However, the discrete hazard function (6.62) assumes any effects due to the patient-specific intercepts b_{1i} and/or slopes b_{2i} are the same for patients in the active drug group versus those in placebo control group. Given the differential dropout rate between the two groups and the fact that there is a significant difference in the treatment slopes of the average patient in each group (i.e., a significant treatment by time interaction as determined by the `beta_Trt_x_SWeek` parameter in Output 6.14), it is more likely that any effect the patient-specific intercepts and slopes might have on dropout will differ according to which treatment group the patient is in. We therefore added the interaction terms, $a_{1i}b_{1i}\eta_{11} + a_{1i}b_{2i}\eta_{12}$, to the hazard function (6.62) by specifying the following NLMIXED programming statements

```
Lambda_i = eta1*(Week=2) + eta2*(Week=3) + eta3*(Week=4) +
           eta4*(Week=5) + eta5*(Week=6) + eta_Trt*Trt +
           eta_b1i*b1i + eta_b2i*b2i +
           eta_Trt_b1i*Trt*b1i + eta_Trt_b2i*Trt*b2i;
** Complementary log-log link = Hazard function **;
Hazard_i = 1 - exp(-exp(Lambda_i));
```

where η_{11} is denoted by `eta_Trt_b1i` and η_{12} by `eta_Trt_b2i`. We then re-ran NLMIXED based on the PARMS statement,

```
parms / data=initial2;
```

where the dataset initial2 is defined as follows:

```
data initial1;
 Parameter='eta_Trt_b1i';Estimate=0; output;
 Parameter='eta_Trt_b2i';Estimate=0; output;
run;
data initial2;
 set initial initial1;
run;
```

The results of fitting this second SP model (SP Model 2) are shown in Output 6.15. While there are minor differences between the joint model assuming S-CRD (Output 6.13, AIC= 4087.7) and the shared parameter model with hazard function (6.62) assuming non-ignorable dropout (Output 6.14, AIC= 4089.2), inclusion of the interaction effects, $a_{1i}b_{1i}\eta_{11} + a_{1i}b_{2i}\eta_{12}$, into the hazard function (6.62) results in a significantly better fit to the data (Output 6.15, AIC= 4071.4). The likelihood ratio test comparing the two SP models is highly significant (likelihood ratio chi-square= 21.8, DF= 2, $p < 0.0001$) suggesting that dropout may indeed be non-ignorable.

Output 6.15: Select results from a shared parameter model (SP Model 2) obtained by adding the interaction terms, $a_{1i}b_{1i}\eta_{11} + a_{1i}b_{2i}\eta_{12}$, to the hazard function (6.62) of SP Model 1.

Iteration History

Iter	Flag1	Calls	NegLogLike	Diff	MaxGrad	Slope
1		43	2019.96627	8.863835	15.48576	−32.756
2		64	2016.87244	3.093834	1.855836	−5.61311
3		85	2016.70649	0.165947	0.103776	−0.31653
4		106	2016.7052	0.001292	0.000966	−0.00256
5		127	2016.7052	1.951E−7	1.93E−7	−3.9E−7

NOTE: GCONV convergence criterion satisfied.

Fit Statistics

−2 Log Likelihood	4033.4
AIC (smaller is better)	4071.4
AICC (smaller is better)	4071.6
BIC (smaller is better)	4148.9

Parameter Estimates

Parameter	Estimate	Standard Error	DF	t Value	Pr > |t|
beta11	−7.2665	0.4703	435	−15.45	<.0001
beta12	−3.3649	0.3863	435	−8.71	<.0001
beta13	−0.7551	0.3527	435	−2.14	0.0328
beta_Trt	−0.1596	0.3932	435	−0.41	0.6850
beta_SWeek	0.7357	0.2277	435	3.23	0.0013
beta_Trt_x_SWeek	1.9289	0.2645	435	7.29	<.0001
psi11	7.0324	1.3158	435	5.34	<.0001
psi12	−1.5102	0.5317	435	−2.84	0.0047
psi22	2.0633	0.4183	435	4.93	<.0001
eta1	−2.1672	0.2637	435	−8.22	<.0001
eta2	−3.3680	0.3716	435	−9.06	<.0001
eta3	−1.7965	0.2427	435	−7.40	<.0001
eta4	−3.8053	0.4839	435	−7.86	<.0001
eta5	−3.3042	0.3990	435	−8.28	<.0001
eta_Trt	−0.7232	0.2743	435	−2.64	0.0087
eta_b1i	−0.2340	0.09071	435	−2.58	0.0102
eta_b2i	−0.5093	0.2780	435	−1.83	0.0676
eta_Trt_b1i	0.3835	0.1148	435	3.34	0.0009
eta_Trt_b2i	1.0297	0.3300	435	3.12	0.0019

Additional Estimates

Label	Estimate	Standard Error	DF	t Value	Pr > |t|
CumLogOR(Week=0)	−0.1596	0.3932	435	−0.41	0.6850
CumLogOR(Week=1)	1.7693	0.3276	435	5.40	<.0001
CumLogOR(Week=3)	3.1814	0.4018	435	7.92	<.0001
CumLogOR(Week=6)	4.5651	0.5359	435	8.52	<.0001
CumOR(Week=0)	0.8525	0.3352	435	2.54	0.0113
CumOR(Week=1)	5.8665	1.9220	435	3.05	0.0024
CumOR(Week=3)	24.0798	9.6749	435	2.49	0.0132
CumOR(Week=6)	96.0763	51.4870	435	1.87	0.0627

Both the SS intercepts and slopes are significantly associated with dropout but in a manner that differs according to which treatment group an individual is assigned. Specifically, patients in the active drug group with a higher slope (better outcome) are more likely to dropout than those in the placebo control group. For example, from the estimates in Output 6.15, we can estimate the probability that the typical or average placebo control patient will complete the six-week trial based on the baseline "survival" estimate

$$\Pr[T_i \geq 6 | a_{1i} = 0, \boldsymbol{b}_i = \boldsymbol{0}] = \widehat{S}_0(6) = \prod_{k=1}^{5} \exp\left[-\exp(\widehat{\eta}_{0k})\right]. \quad (6.63)$$

Substituting the baseline hazard parameters $\widehat{\eta}_{0k}$ from Output 6.15 into (6.63), we estimate the average patient will complete the trial with probability of 0.688 which is reasonably close to the observed proportion of $70/108 = 0.648$ patients who completed the study. We can then estimate the probability that a select individual with specified covariates $\boldsymbol{x}'_k = (a_{1k}, b_{1k}, b_{2k}, a_{1k}b_{1k}, a_{1k}b_{2k})$ will complete the trial by evaluating the relative risk function

$$\exp\{\boldsymbol{x}'_k\widehat{\boldsymbol{\eta}}\} = \exp\{a_{1k}\widehat{\eta}_{\text{Trt}} + b_{1k}\widehat{\eta}_{b1} + b_{2k}\widehat{\eta}_{b2} + a_{1k}b_{1k}\widehat{\eta}_{11} + a_{1k}b_{2k}\widehat{\eta}_{12}\}$$

and computing $\widehat{S}_k(6|\boldsymbol{x}_k) = \widehat{S}_0(6)^{\exp\{\boldsymbol{x}'_k\widehat{\boldsymbol{\eta}}\}}$. To illustrate, the typical or average patient in the active drug group has covariates $\boldsymbol{x}'_k = (1, 0, 0, 0, 0)$ such that this patient would have an estimated probability of completing the trial of $(0.688)^{\exp(-0.7232)} = 0.834$ which is higher than the marginal proportion of $265/329 = 0.806$ patients. Similar calculations will show that a patient in the active drug group with an average intercept of $b_{1k} = 0$ and a slope that is one standard deviation above the mean (i.e., $b_{2k} = 0 + \sqrt{\widehat{\psi}_{22}} = 1.436$) has an estimated probability of $(0.688)^{\exp(-0.7232-0.5093\times1.436+1.0297\times1.436)} = 0.682$ while a patient with the same intercept and slope in the placebo control group has an estimated probability of $(0.688)^{\exp(-0.5093\times1.436)} = 0.835$.

The impact of the interaction between treatment group and the random intercepts and slopes on dropout also affects the estimated IMPS79 ordinal response profile as evidenced by the higher cumulative odds ratios in Output (6.15). To gain a better perspective of this effect, a side-by-side comparison of the two SP models is provided in Output 6.16. When we compare the regression parameter estimates from the two SP models, we find that the inclusion of the interaction effects, $a_{1i}b_{1i}\eta_{11} + a_{1i}b_{2i}\eta_{12}$, to the hazard function (6.62) results in a greater subject-specific (SS) effect associated with the treatment by time interaction (an effect of 1.6345±0.2537 versus 1.9289±0.2645). This further increases the SS cumulative odds ratio at weeks 1,3 and 6. At week 6, for example, the estimated odds that a typical patient in the active drug group will have a lower IMPS79 score compared to the typical patient in the placebo control group increases from 53.8987 to 96.0763. Qualitatively, the two SP models along with the joint model assuming S-CRD all demonstrate a significant reduction in severity of illness over time among patients treated with one of the three active drugs.

Output 6.16: Side-by-side comparison of the two SP models.

Comparison of Shared Parameter Model Results

	Model					
	SP Model 1 (AIC=4089.2)			SP Model 2 (AIC=4071.4)		
Parameter	Estimate	StdErr	Pr>|t|	Estimate	StdErr	Pr>|t|
beta11	−7.3633	0.4759	<.0001	−7.2665	0.4703	<.0001
beta12	−3.4754	0.3915	<.0001	−3.3649	0.3863	<.0001
beta13	−0.8743	0.3570	0.0147	−0.7551	0.3527	0.0328
beta_SWeek	0.9746	0.2275	<.0001	0.7357	0.2277	0.0013
beta_Trt	−0.0165	0.3942	0.9666	−0.1596	0.3932	0.6850
beta_Trt_x_SWeek	1.6345	0.2537	<.0001	1.9289	0.2645	<.0001
eta1	−1.9875	0.2140	<.0001	−2.1672	0.2637	<.0001
eta2	−3.2257	0.3446	<.0001	−3.3680	0.3716	<.0001
eta3	−1.7069	0.2072	<.0001	−1.7965	0.2427	<.0001
eta4	−3.7565	0.4684	<.0001	−3.8053	0.4839	<.0001
eta5	−3.2635	0.3800	<.0001	−3.3042	0.3990	<.0001
eta_Trt	−0.6958	0.2087	0.0009	−0.7232	0.2743	0.0087
eta_Trt_b1i	.	.	.	0.3835	0.1148	0.0009
eta_Trt_b2i	.	.	.	1.0297	0.3300	0.0019
eta_b1i	−0.0017	0.05078	0.9741	−0.2340	0.09071	0.0102
eta_b2i	0.2123	0.1399	0.1298	−0.5093	0.2780	0.0676
psi11	7.0485	1.3236	<.0001	7.0324	1.3158	<.0001
psi12	−1.5498	0.5384	0.0042	−1.5102	0.5317	0.0047
psi22	2.0065	0.4161	<.0001	2.0633	0.4183	<.0001

As noted in our discussion of shared parameter models (§6.5) and illustrated with the schizophrenia data in §5.2.3, one of the issues we face is the challenge of interpreting results given the presence of random effects. Since SP models involve random effects, the regression parameters of interest require a subject-specific (SS) interpretation rather than a marginal or population-averaged (PA) interpretation. While this is not an issue when the response model of interest is linear in the random effects (as with the MDRD data), it can become an issue when the response model is nonlinear in the random effects (as with this example). While one can certainly employ graphical and numerical techniques such as presented in §5.2.3 to compare the average patient's response to the PA response obtained by averaging over individuals, these comparisons do not reflect uncertainty in the PA estimates. Other means are required if one wishes to formally compare treatment groups on the basis of some well-defined marginal or PA estimand. For example, if one chooses to fit a marginal regression model to the IMPS79 ordinal responses, then a marginal selection model or marginal pattern-mixture model would be required when assuming non-ignorable dropout.

6.7 Summary

In this chapter, we address various issues related to missing data in longitudinal clinical trials. We chose to focus on longitudinal clinical trials as this is one area where missing data is not only common but also because it has received considerable attention in the literature. Another area that routinely involves missing data and which has also received considerable attention is the area of sample surveys. For that reason, we have spent a considerable portion of this chapter describing the various mechanisms leading to missing data. These include a description of the more general MCAR, MAR and MNAR mechanisms (§6.2) as well mechanisms specific to dropout in clinical trials, namely CRD, S-RD and S-NARD (§6.3). The latter is an area with which this author is most familiar and hence considerable attention has been paid to the problem of ignorable versus non-ignorable dropout. This includes a rather detailed and somewhat theoretical discussion of the underlying assumptions and issues one need consider when distinguishing between ignorable and non-ignorable dropout. Among them is the issue of determining whether an event leading to dropout is a terminal or non-terminal event. Such a distinction is necessary when considering whether one has missing data leading to dropout or missing data resulting from dropout.

Following the rather lengthy discussion involving various missing data mechanisms, we summarize some of the common approaches used to analyze data when missing values are MAR (§6.4) or MNAR (§6.5). These include the use of likelihood-based methods as well as semiparametric GEE-based methods for MAR data; and either selection models, pattern-mixture models or shared parameter models for MNAR data. While a comprehensive treatment of these different approaches is beyond the scope of this book, we have tried to emphasize some of the more practical issues one must confront with each approach. When missingness is ignorable (MAR), parametric likelihood-based analyses and semiparametric GEE analyses using multiple imputation or inverse probability of weighting all yield valid inference. As such, each of these methods provides a definitive approach for how to handle missing data provided such data are MAR. Determining which approach one should choose rests with how comfortable one is with the underlying assumptions each involves. However, when missingness is non-ignorable, there is no definitive approach one can take to the analysis of longitudinal data. Selection models, pattern-mixture models and shared parameter models all require one to make some additional unverifiable assumptions about the response variable of interest, $\boldsymbol{y}_i$, and its relation to the response/non-response variable, $\boldsymbol{r}_i$, that describes missingness. More often than not, one will need to perform several analyses using possibly different approaches as a means for assessing how sensitive one's results are to different assumptions.

Finally, we present three different case studies involving missing data. In each case, we illustrate how one can use SAS to analyze longitudinal data when missing values are present. Our goals in the bone mineral density example were to demonstrate how one can use GENMOD to test whether missing values are MCAR using Ridout's test as well as how one can use the MI and MIANALYZE procedures to analyze the data under MAR using multiple imputation techniques. Our goals with the MDRD example were two-fold: 1) to illustrate how one can use NLMIXED to fit a random effects pattern-mixture model to data in such a way as to provide unbiased estimates of both the marginal estimand of interest and its standard error assuming one has the correct model, and 2) to illustrate how one can use NLMIXED to fit a shared parameter model to data that results in valid inference provided the underlying model assumptions are correct. In the last example, we again illustrate how one can fit a shared parameter model to the schizophrenia data but in this case, both the response model of interest and the model for dropout are GLME models that involve sharing a set of nonlinear random effects between the two conditional models. In each example, we performed some degree of sensitivity analysis but by no means was this analysis meant to be comprehensive.

In the end, there are numerous and complex issues one must address when faced with missing data, particularly when missing data are due to dropout. We have attempted to describe some of these issues and at least demonstrate how one might go about analyzing such data using SAS. Nevertheless, the biggest challenge we face with missing data is that we can never fully verify the assumptions that must be made in order to analyze the data. Inference in the presence of missing data must be made on the basis of the evidence at hand, i.e., on the basis of what the observed data tell us, and also on a firm grasp of the science involved in the particular application. Hopefully, the discussion and examples serve to illustrate how one might go about handling missing data in longitudinal clinical trials.

7

Additional Topics and Applications

In this chapter, we consider four additional applications in which we apply some of the techniques described in previous chapters as well as a technique described by Nelson et. al. (2006) for fitting mixed-effects models with non-Gaussian random effects. We start in §7.1 by considering how one can use NLMIXED to fit mixed-effects models to correlated data assuming non-Gaussian random effects. This occurs, for example, in applications where one wishes to estimate the incidence rate of some event over time assuming the rate varies from subject to subject according to a gamma distribution. An example of this was already presented in Chapter 5 when we fit a multi-level random effects model to the ADEMEX hospitalization data (example 5.2.4, §5.2). In that example, we fit a random-effects negative binomial model to the data in an attempt to account for unexplained heterogeneity by assuming two sources of variation: a random center effect (assuming normality) and a random subject effect (assuming a gamma-distributed random rate per subject). At the patient level, the number of hospitalizations are assumed to be conditionally distributed as Poisson with a conditional mean and variance determined by a patient-specific random hospitalization rate. After integrating over the random rate, we end up with a negative binomial model for the counts conditional on a center-specific random effect. In §7.1.1 we illustrate how one can fit a non-Gaussian random effects model in NLMIXED. Specifically, we jointly model the ADEMEX peritonitis and hospitalization data of §4.3.1 and §5.2.4, respectively, using a shared parameter model in which a gamma distributed random effect serves as the shared parameter.

In §7.2, we consider two applications involving pharmacokinetic (PK) data. The first application involves fitting a NLME model to theophylline PK data. Our goals with this example are to demonstrate how to obtain good starting values for NLMIXED and also how one can use the ESTIMATE statement to estimate key pharmacokinetic summary parameters like area under the curve (AUC), apparent volume of distribution (V) and the peak concentration time (T_{MAX}). In the second example, we compare and contrast the fit of different NLME models to phenobarbital PK data. This application illustrates some of the challenges of fitting PK data when one has rather sparse intra-subject data as well as intermittent dosing. Our goal here is to estimate the population PK parameters as possible functions of covariates while accounting for between-subject and within-subject errors on either an additive or multiplicative scale. Finally, in §7.3 we conclude with an example from the ADEMEX study in which we jointly model longitudinal trends in glomerular filtration rates (GFR) and survival data using a shared parameter (SP) model. Unlike Chapter 6 where SP models are used to address issues of non-ignorable dropout, here the primary focus is on estimating the rate of decline in GFR and characterizing its relationship with patient survival.

7.1 Mixed models with non-Gaussian random effects

In Chapters 3 and 5, we present methods of estimation and inference for a class of linear (LME) and nonlinear (GLME, GNLME) mixed-effects models all of which assume random effects are normally distributed. In this section, we consider applications where random effects are likely to be non-Gaussian in nature. We already mentioned example 5.2.4 where, at the patient level, the rate of hospitalization is assumed to be randomly distributed according a gamma distribution. Another situation would be frailty survival models where a subject-specific or cluster-specific random effect is introduced within the hazard function as a means for accounting for unobserved heterogeneity and/or correlation. Heterogeneity, for example, may be due to the exclusion of unmeasured risk factors or it may simply reflect sampling from a heterogenous population of individuals each with his/her own hazard. Correlation, on the other hand, may occur among individuals within clusters such as patients within a center all of whom receive a similar standard of care. Under a proportional hazards or relative risk model, the random-effect or frailty term is assumed to have a multiplicative effect on the baseline hazard function and in many cases a gamma distribution is often assumed for the frailty term. Liu and Huang (2008), for example, illustrate how one can fit frailty models assuming either normal or gamma random effects using NLMIXED.

One can use NLMIXED to fit mixed-effects models assuming non-Gaussian random effects based on work by Nelson et. al. (2006). Specifically, consider the class of GNLME models (Chapter 5, §5.3.1) in which the conditional pdf of $\boldsymbol{y}_i|\boldsymbol{b}_i$ is given by

$$\pi(\boldsymbol{y}_i;\boldsymbol{\beta},\boldsymbol{\theta}_w|\boldsymbol{b}_i) = \prod_{j=1}^{p_i} \pi(y_{ij};\boldsymbol{\beta},\boldsymbol{\theta}_w|\boldsymbol{b}_i) \tag{7.1}$$

where $\pi(y_{ij};\boldsymbol{\beta},\boldsymbol{\theta}_w|\boldsymbol{b}_i)$ is any general parametric density function with location and scale parameters $\boldsymbol{\beta}$ and $\boldsymbol{\theta}_w$, and $\boldsymbol{b}_i$ is a vector of random effects. As in §5.3.1, we assume the y_{ij} are conditionally independent given $\boldsymbol{b}_i$ but we now consider cases where the pdf of $\boldsymbol{b}_i$, say $\pi(\boldsymbol{b}_i;\boldsymbol{\theta}_b)$, is non-Gaussian. Following Nelson et. al. (2006), we assume the components of $\boldsymbol{b}_i = (b_{i1},\ldots,b_{iv})'$ are independently distributed continuous random variables with pdf's $\pi(b_{ik};\boldsymbol{\theta}_{b_k})$, $k=1,\ldots,v$ such that

$$\pi(\boldsymbol{b}_i;\boldsymbol{\theta}_b) = \prod_{k=1}^{v} \pi(b_{ik};\boldsymbol{\theta}_{b_k}). \tag{7.2}$$

The assumption that the components of $\boldsymbol{b}_i$ are independent allows one to apply probability integral transformations when evaluating the integrated log-likelihood function

$$\begin{aligned} L(\boldsymbol{\beta},\boldsymbol{\theta};\boldsymbol{y}) &= \sum_{i=1}^{n} \log \int_b \exp\Big\{\log\Big[\pi(\boldsymbol{y}_i;\boldsymbol{\beta},\boldsymbol{\theta}_w|\boldsymbol{b}_i)\pi(\boldsymbol{b}_i;\boldsymbol{\theta}_b)\Big]\Big\} d\boldsymbol{b}_i \qquad (7.3) \\ &= \sum_{i=1}^{n} \log \int_b \exp\left[\sum_{j=1}^{p_i} l_{ij}(\boldsymbol{\beta},\boldsymbol{b}_i,\boldsymbol{\theta}_w;y_{ij}) + \sum_{k=1}^{v} l^*_{ik}(\boldsymbol{\theta}_{b_k};b_{ik})\right] d\boldsymbol{b}_i \end{aligned}$$

where $l_{ij}(\boldsymbol{\beta}, \boldsymbol{b}_i, \boldsymbol{\theta}_w; y_{ij}) = \log[\pi(y_{ij}; \boldsymbol{\beta}, \boldsymbol{\theta}_w|\boldsymbol{b}_i)]$ is the conditional log-likelihood function of $y_{ij}|\boldsymbol{b}_i$ and $l^*_{ik}(\boldsymbol{\theta}_{b_k}; b_{ik}) = \log[\pi(b_{ik}; \boldsymbol{\theta}_{b_k})]$ is the log-likelihood function of b_{ik}.

The probability integral transformation states that for a continuous random variable Y with cumulative distribution function (CDF) $F(\cdot)$ and pdf $\pi_Y(y)$, the random variable $X = F(Y) \sim U(0,1)$ has a uniform distribution over the interval $(0,1)$ (e.g., Hoel, Port and Stone, 1971, §5.4). Likewise, if $X \sim U(0,1)$ then $Y = F^{-1}(X)$ is a random variable with pdf $\pi_Y(y)$. Therefore, by successive application of the probability integral transformation, a single non-Gaussian random effect, $b_i \sim \pi(b_i; \boldsymbol{\theta}_b)$, can be written as

$$b_i = F^{-1}_{\boldsymbol{\theta}_b}(\Phi(z_i)) = F^{-1}_{\boldsymbol{\theta}_b}(u_i) \tag{7.4}$$

where $F_{\boldsymbol{\theta}_b}(\cdot)$ is the cumulative distribution function (CDF) of b_i, and $u_i = \Phi(z_i) \sim U(0,1)$ is a uniform$(0,1)$ random variable obtained as the probability integral transform $\Phi(z_i)$ where $z_i \sim N(0,1)$ and $\Phi(\cdot)$ is the standard normal CDF. Hence for a single random effect b_i, the integrated log-likelihood (7.3) may be written as

$$\begin{aligned} L(\boldsymbol{\beta}, \boldsymbol{\theta}; \boldsymbol{y}) &= \sum_{i=1}^{n} \log \int_b \exp\Big\{\log\Big[\pi(\boldsymbol{y}_i; \boldsymbol{\beta}, \boldsymbol{\theta}_w|b_i)\pi(b_i; \boldsymbol{\theta}_b)\Big]\Big\} db_i \qquad (7.5)\\ &= \sum_{i=1}^{n} \log \int_b \exp\left[\sum_{j=1}^{p_i} l_{ij}(\boldsymbol{\beta}, b_i, \boldsymbol{\theta}_w; y_{ij}) + l^*_i(\boldsymbol{\theta}_b; b_i)\right] db_i \\ &= \sum_{i=1}^{n} \log \int_z \exp\left[\sum_{j=1}^{p_i} l_{ij}(\boldsymbol{\beta}, F^{-1}_{\boldsymbol{\theta}_b}(\Phi(z_i)), \boldsymbol{\theta}_w; y_{ij}) + \log(\pi(z_i))\right] dz_i \end{aligned}$$

where $\pi(z_i)$ is the pdf of a $N(0,1)$ random variable. One can then maximize (7.5) by applying Gaussian quadrature to evaluate the integral.

One can certainly extend (7.5) to include more than one random effect provided the random effects are all independently distributed according to specified pdf's $\pi(b_{ik}; \boldsymbol{\theta}_{b_k})$, $k = 1, \ldots, v$ as indicated in (7.2) above. However when correlation is present among the components of $\boldsymbol{b}_i$, one will need to use a more sophisticated multivariate probability integral transformation approach as noted by Nelson et. al. (2006). Since in many applications, a single non-Gaussian random effect is all that is required, we will restrict our discussion to those cases where one has a single random effect or possibly two or three independently distributed random effects. One can provide some protection against erroneously assuming the random effects are mutually independent by carrying out inference using robust standard errors computed with the EMPIRICAL option of NLMIXED. In the following section, we illustrate the use of the probability integral transformation by fitting a shared parameter model to the ADEMEX peritonitis and hospitalization data of §4.3.1 and §5.2.4 assuming a gamma distributed random effect.

7.1.1 ADEMEX peritonitis and hospitalization data

In example 5.2.4 of §5.2, we fit several models to the ADEMEX hospitalization data in an attempt to explain possible sources of overdispersion observed when we fit a simple Poisson model to the data. The models considered included both a single and two multi-level random effects models. In this section, we examine whether the overdispersion seen in the hospitalization data can be explained by considering an additional patient-specific risk factor not accounted for in the previous analyses—namely patient-specific rates of peritonitis. Specifically, since an episode of peritonitis can lead to hospitalization, we examine whether or not patients who have an intrinsically higher rate of peritonitis might also have a higher rate of hospitalization. We can accomplish this by jointly modeling the number of hospitalizations (our "primary" response variable of interest) and the number of episodes of peritonitis. We do so by assuming the rate of peritonitis varies from patient-to-patient according to a gamma distribution. We then fit a shared parameter model to the joint data by assuming a gamma-Poisson random effects model for the peritonitis data that is identical to the negative binomial model of §4.3.1, and a Poisson model for the hospitalization data that includes all of the risk factors used in the initial Poisson model (see §5.2.4, Output 5.13) but which also includes the random patient-specific peritonitis rates as an additional covariate.

We start by first jointly fitting a gamma-Poisson random-effects model to the peritonitis data and a fixed-effects Poisson model to the hospitalization data using NLMIXED. Specifically, suppose that for the ith subject, the distribution of peritonitis counts, y_{i1}, conditional on a log-gamma random effect b_i, is Poisson with conditional mean and variance

$$\begin{aligned} E(y_{i1}|b_i) &= \mu_{i1}(\boldsymbol{\eta})\exp(b_i) = \exp\{\boldsymbol{x}'_i\boldsymbol{\eta} + \log(t_i)\}\exp(b_i) \\ &= \exp\{\boldsymbol{x}'_i\boldsymbol{\eta} + b_i + \log(t_i)\} \\ &= Var(y_{i1}|b_i) \end{aligned} \tag{7.6}$$

where $\mu_{i1}(\boldsymbol{\eta}) = \exp\{\boldsymbol{x}'_i\boldsymbol{\eta} + \log(t_i)\}$ and

$$\boldsymbol{x}'_i\boldsymbol{\eta} = \eta_0 + \eta_1 x_{1i} + \eta_2 x_{2i} + \eta_3 x_{3i} + \eta_4 x_{4i} + \eta_5 x_{5i} + \eta_6 x_{6i}$$

is the linear predictor (excluding the offset $\log(t_i)$) based on the patient-specific covariates $x_{1i}, x_{2i}, \ldots, x_{6i}$ defined in §4.3.1. Let $\lambda_i = \exp(b_i)$ follow a gamma distribution $G(\gamma, 1/\alpha)$ as defined in (5.44) but with $\gamma = 1/\alpha$ for identifiability purposes (Nelson et. al., 2006). The pdf of λ_i is then

$$\pi(\lambda_i;\alpha) = \frac{1}{\alpha^{\frac{1}{\alpha}}\Gamma(\frac{1}{\alpha})}\lambda_i^{(\frac{1}{\alpha}-1)}\exp\{-\frac{1}{\alpha}\lambda_i\} \tag{7.7}$$

with $E(\lambda_i) = 1$ and $Var(\lambda_i) = \alpha$. Upon integrating over λ_i, the marginal distribution of y_{1i} can be shown to be negative binomial with mean $\mu_{1i}(\boldsymbol{\eta})$ and variance $\mu_{1i}(\boldsymbol{\eta})(1+\alpha\mu_{1i}(\boldsymbol{\eta}))$.

Next, we assume that the marginal distribution of the hospitalization counts, y_{i2}, is Poisson with marginal mean and variance

$$\begin{aligned} E(y_{i2}) &= \mu_{i2}(\boldsymbol{\beta}) = \exp\{\boldsymbol{x}'_i\boldsymbol{\beta} + \log(t_i)\} \\ &= Var(y_{2i}). \end{aligned} \tag{7.8}$$

where

$$\boldsymbol{x}'_i\boldsymbol{\beta} = \beta_0 + \beta_1 x_{1i} + \boldsymbol{\beta}_2 x_{2i} + \beta_3 x_{3i} + \beta_4 x_{4i} + \beta_5 x_{5i} + \beta_6 x_{6i}$$

is the linear predictor (excluding the offset $\log(t_i)$) associated with hospitalizations. Note that for both peritonitis and hospitalization, we model the individual event rates using the same set of risk factors. However, we allow for overdispersion in the form of a gamma random effect when modeling the peritonitis rates.

The SAS code required to jointly fit the peritonitis and hospitalization data assuming y_{i1} and y_{i2} are independently distributed is shown below. We elected to fit the gamma-Poisson random effects model to the peritonitis data directly rather than fit the corresponding marginal negative binomial model. We do so to illustrate how one can use NLMIXED to fit non-Gaussian random effects models in SAS. To help speed up convergence, we start by running PROC GENMOD for the two models and using the parameter estimates as starting values in NLMIXED.

Program 7.1

```
data example7_1_1;
 set SASdata.ADEMEX_Peritonitis_Data;
 log_time=log(MonthsAtRisk); ** Offset;
 Center=scan(ptid, 1);
run;
proc sort data=example7_1_1;
 by ptid;
run;
ods listing close;
ods output parameterestimates=pe1;
proc genmod data=example7_1_1;
  model Episodes=Trt Sex Age Diabetic Albumin PriorMonths
       / dist=NB link=log offset=log_time;
run;
ods output parameterestimates=pe2;
proc genmod data=example7_1_1;
  model Hosp=Trt Sex Age Diabetic Albumin PriorMonths
       / dist=Poisson link=log offset=log_time;
run; ods listing;
data pe1;
 length Parameter $16;
 set pe1;
 Parameter=compress('eta_'||left(Parameter));
 if Parameter='eta_Dispersion' then Parameter='alpha';
run;
data pe2;
 length Parameter $16;
 set pe2;
 Parameter=compress('beta_'||left(Parameter));
 if Parameter='beta_Scale' then delete;
run;
data peINITIAL;
 set pe1 pe2;
run;
/*==========================================================
  Fit a joint model to peritonitis (gamma-Poisson model)
  and hospitalization (Poisson model) data assuming the
  two sets of count data are independently distributed.
==========================================================*/
data example7_1_1_SP;
 set example7_1_1;
 response=hosp;    ind=0;output;
 response=episodes;ind=1;output;
run;
proc sort data=example7_1_1_SP;
 by ptid ind;
run;
ods output ParameterEstimates=SAStemp.peJOINT;
proc nlmixed data=example7_1_1_SP method=Gauss
             tech=newrap noad qpoints=20;
  parms /data=peINITIAL;
  ui = CDF('NORMAL',zi);
  if ui> 0.999999 then ui=0.999999;
```

```
 /*=========================================================
  -alpha is the negative binomial dispersion parameter
   in GENMOD (see equations 5.43-5.46 for a derivation)
  -gi~G(1/alpha, 1/alpha) has the gamma distribution as
   defined by the pdf (7.7) with E(gi)=1 Var(gi)=alpha
  -bi is log-gamma distributed (Nelson et. al., 2006)
  =========================================================*/
 gi1 = quantile('GAMMA', ui, 1/alpha);
 gi = alpha*gi1;
 bi = log(gi);
 /* Define conditional mean count for peritonitis data    */
 x_eta = eta_Intercept + eta_Trt*Trt + eta_Sex*Sex +
         eta_Age*Age + eta_Diabetic*Diabetic +
         eta_Albumin*Albumin + eta_PriorMonths*PriorMonths +
         log_time;
 mu1_i = exp(x_eta + bi);
 /* Define the marginal mean for hospitalization data     */
 x_beta = beta_Intercept + beta_Trt*Trt + beta_Sex*Sex +
          beta_Age*Age + beta_Diabetic*Diabetic +
          beta_Albumin*Albumin + beta_PriorMonths*PriorMonths +
          log_time;
 mu2_i = exp(x_beta);
 /*=========================================================
  Define the numerator of the Poisson overdispersion
  parameter phi as defined in equation (4.6) of Ch. 4
  for the hospitalization data
  =========================================================*/
 phi_num = (((response-mu2_i)**2)/mu2_i)*(ind=0);
 /*=========================================================
  Define the joint independent means for y1i and y2i
  where y1i is the number of episodes of peritonitis
  (response=Episodes) and y2i is the number of hospital
  admissions (response=Hosp) from the ADEMEX dataset.
  =========================================================*/
 mu = mu1_i*(ind=1) + mu2_i*(ind=0);
 model response ~ Poisson(mu);
 random zi ~ N(0,1) subject=ptid;
 id mu1_i mu2_i phi ;
 predict mu2_i out=pred;
run;
/*=========================================================
 Calculate the Poisson overdispersion parameter and
 associated scale (square root) parameter as a means
 for assessing model goodness-of-fit with respect to
 the hospitalization data
 =========================================================*/
proc means data=pred sum noprint;
 where ind=0;
 var phi_num;
 output out=phi n=n sum=phi_num;
run;
data phi; set phi;
 phi = phi_num/(n-7);
 Scale = sqrt(phi);
run;
```

```
proc print data=phi noobs split='|';
 var phi Scale;
 label phi='Poisson overdispersion estimate (phi)'
       Scale='Poisson scale parameter';
run;
```

Some comments regarding the above program are in order. First, rather than use the NLMIXED `general(ll)` specification for the model statement, we take advantage of the fact that the gamma-Poisson model for peritonitis is in fact a conditional random-effects Poisson model that is independent of the Poisson model for hospitalization. Hence we can use the indicator variable, `ind`, defined in the data programming step,

```
data example7_1_1_SP;
 set example7_1_1;
 response=hosp;     ind=0;output;
 response=episodes;ind=1;output;
run;
```

to help specify that the conditional joint distribution of y_{i1} and y_{i2} will itself be Poisson. In this case, the two outcome variables y_{i1} and y_{i2} must be properly defined using a single variable which we define above by the SAS variable `response`. This differs from when one specifies the joint likelihood using the NLMIXED model specification:

```
model response~general(loglike);
```

in that the outcome variable, `response`, is treated simply as a dummy variable by NLMIXED and is never actually used in the optimization unless it is also used to define the general log-likelihood function, `loglike`, above. In fact, one can simply define a dummy outcome variable, `dummy=1`, within the dataset and use the specification,

```
model dummy~general(loglike);
```

to achieve the correct ML estimation provided one has correctly specified the log-likelihood function, `loglike`.

A second thing to note is the use of the NOAD option when running NLMIXED. The NOAD option requests that the Gaussian quadrature be non-adaptive (see §5.3.4); that is, the quadrature points are centered at zero for each of the random effects and the current random-effects variance matrix (in this case, 1) is used as the scale matrix. The reason we request non-adaptive Gaussian quadrature is that we "know" the normal random effect z_i is distributed as a standard normal and the NOAD makes use of that knowledge to compute the correct quadrature points required to define the non-Gaussian random effect. In fact, if one were to re-run the above call to NLMIXED without the NOAD and QPOINTS=20 options, one would get the following error message in the SAS log:

```
ERROR: Quadrature accuracy of 0.000100 could not be achieved with 31 points.
The achieved accuracy was 1.000000.
```

Finally, we have found that when modeling a single random effect for subjects with a single observation such as with the gamma-Poisson random effects model, one need specify a large number of quadrature points to accurately reproduce the results of fitting the same model in GENMOD or GLIMMIX based on fitting the marginal negative binomial model. In this example, we used QPOINTS=20 which takes additional CPU time to run but which gives a much better approximation than what one obtains using 5 or even 10 quadrature points.

The results of running the above program are shown in Output 7.1. Included is the estimated dispersion parameter $\widehat{\phi}$ and corresponding scale parameter, $\sqrt{\widehat{\phi}}$, associated with the fixed-effects Poisson model for the hospitalization data. As noted in example 5.2.4, there is evidence of significant overdispersion ($\widehat{\phi} = 3.029 > 1$) relative to the assumed Poisson model. In comparing the results in Output 7.1 with those of Output 4.3 of example 4.3.1 (§4.3) and Output 5.13 of example 5.2.4 (§5.2), we find that fitting the gamma-Poisson and Poisson models to the peritonitis and hospitalization data jointly gives nearly identical results to those obtained when we separately fit a marginal negative binomial model to the peritonitis data (Output 4.3) and an overdispersed Poisson model to the hospitalization data (Output 5.13). The joint fit to the data is summarized by the various fit statistics in Output 7.1 (e.g., $-2\,$log-likelihood$=5820.5$ and AIC$=5850.5$).

Output 7.1: Results from jointly modeling peritonitis rates and hospitalization rates using a gamma-Poisson random effects model for peritonitis rates and a fixed-effects Poisson model for hospitalization rates.

Specifications

Data Set	WORK.EXAMPLE7_1_1_SP
Dependent Variable	response
Distribution for Dependent Variable	Poisson
Random Effects	zi
Distribution for Random Effects	Normal
Subject Variable	ptid
Optimization Technique	Newton−Raphson
Integration Method	Gaussian Quadrature

Dimensions

Observations Used	1754
Observations Not Used	176
Total Observations	1930
Subjects	1754
Max Obs Per Subject	1
Parameters	15
Quadrature Points	20

Iteration History

Iter	Flag1	Calls	NegLogLike	Diff	MaxGrad	Slope
1		34	2910.25167	1.18E−10	2.946E−8	−236E−12

NOTE: GCONV convergence criterion satisfied.

Fit Statistics

−2 Log Likelihood	5820.5
AIC (smaller is better)	5850.5
AICC (smaller is better)	5850.8
BIC (smaller is better)	5932.5

Parameter Estimates

Parameter	Estimate	Standard Error	DF	t Value	Pr > |t|
eta_Intercept	−2.3006	0.3646	1753	−6.31	<.0001
eta_Trt	0.04967	0.1034	1753	0.48	0.6310
eta_Sex	0.04655	0.1064	1753	0.44	0.6618
eta_Age	−0.00249	0.004393	1753	−0.57	0.5711
eta_Diabetic	0.1900	0.1233	1753	1.54	0.1236
eta_Albumin	−0.3043	0.08812	1753	−3.45	0.0006
eta_PriorMonths	0.000262	0.002144	1753	0.12	0.9025
alpha	0.7779	0.1229	1753	6.33	<.0001
beta_Intercept	−1.3326	0.1785	1753	−7.47	<.0001
beta_Trt	0.1220	0.05091	1753	2.40	0.0167

beta_Sex	0.1369	0.05186	1753	2.64	0.0084
beta_Age	−0.00820	0.002272	1753	−3.61	0.0003
beta_Diabetic	0.4893	0.06258	1753	7.82	<.0001
beta_Albumin	−0.3553	0.04293	1753	−8.28	<.0001
beta_PriorMonths	0.002177	0.000988	1753	2.20	0.0277

Parameter	Alpha	Lower	Upper	Gradient
eta_Intercept	0.05	−3.0157	−1.5854	−421E−12
eta_Trt	0.05	−0.1531	0.2525	−216E−12
eta_Sex	0.05	−0.1621	0.2553	−164E−12
eta_Age	0.05	−0.01111	0.006128	1.792E−9
eta_Diabetic	0.05	−0.05187	0.4318	−209E−12
eta_Albumin	0.05	−0.4771	−0.1314	−1.29E−9
eta_PriorMonths	0.05	−0.00394	0.004467	1.736E−9
alpha	0.05	0.5370	1.0189	2.946E−8
beta_Intercept	0.05	−1.6827	−0.9826	1.35E−10
beta_Trt	0.05	0.02210	0.2218	6.73E−11
beta_Sex	0.05	0.03522	0.2387	6.72E−11
beta_Age	0.05	−0.01266	−0.00374	8.257E−9
beta_Diabetic	0.05	0.3665	0.6120	4.6E−11
beta_Albumin	0.05	−0.4395	−0.2711	4.02E−10
beta_PriorMonths	0.05	0.000239	0.004114	1.547E−8

Poisson overdispersion estimate (phi)	Poisson scale parameter
3.02934	1.74050

Next we fit a shared parameter model to the peritonitis and hospitalization data by simply adding the log-gamma random effect b_i from the gamma-Poisson random effects model as an additional covariate or risk factor to the Poisson model for hospitalization. In this case, the gamma-Poisson random effects model for peritonitis remains exactly the same but the Poisson model for hospitalization is now modified by setting the linear predictor to be

$$\boldsymbol{x}_i'\boldsymbol{\beta} = \beta_0 + \beta_1 x_{1i} + \boldsymbol{\beta}_2 x_{2i} + \beta_3 x_{3i} + \beta_4 x_{4i} + \beta_5 x_{5i} + \beta_6 x_{6i} + \beta_7 b_i$$

where the log-gamma random effect from the gamma-Poisson model, b_i, is included as a possible risk factor for increased hospitalization. Using the parameter estimates from the previous call to NLMIXED stored in the dataset `SAStemp.peJOINT` as created by the ODS statement:

```
ods output ParameterEstimates=SAStemp.peJOINT;
```

we re-ran the above code by simply replacing the PARMS statement

```
 parms /data=peINITIAL;
```

with

```
 parms /data=peJOINT1;
```

where the dataset `peJOINT1` is created with the following statements

```
data pe_bi;
 Parameter='beta_bi';estimate=0.5;
run;
data peJOINT1;
 set SAStemp.peJOINT pe_bi;
run;
```

We then simply modified the Poisson model for hospitalization by setting the mean to

```
mu2_i = exp(x_beta + beta_bi*bi);
```

and re-ran the preceding NLMIXED program with these changes. As can be seen from the results in Output 7.2, we achieved a significantly better fit with the shared parameter model by virtue of a lower log-likelihood and AIC (Output 7.2: -2 log-likelihood= 5093.7, AIC= 5125.7) compared with the joint model under independence (Output 7.1: -2 log-likelihood= 5820.5, AIC= 5850.5). The patient-specific log-gamma random effect for peritonitis is significantly associated with an increased risk for hospitalization (beta_bi: $1.0843 \pm 0.1052, p < 0.0001$) indicating that incidence of peritonitis and hospitalization are highly correlated.

Output 7.2: Results from fitting a joint shared parameter model to the peritonitis and hospitalization data assuming a gamma-Poisson random effects model for the peritonitis rates and a "fixed" Poisson model for the hospitalization rates in which the patient-specific log-gamma random effect from the gamma-Poisson model serves as a covariate or risk factor for hospitalization. See author's web page at http://support.sas.com/publishing/authors/vonesh.html for corrections to Output 7.2.

Specifications

Data Set	WORK.EXAMPLE7_1_1_SP
Dependent Variable	response
Distribution for Dependent Variable	Poisson
Random Effects	zi
Distribution for Random Effects	Normal
Subject Variable	ptid
Optimization Technique	Newton−Raphson
Integration Method	Gaussian Quadrature

Dimensions

Observations Used	1754
Observations Not Used	176
Total Observations	1930
Subjects	1754
Max Obs Per Subject	1
Parameters	16
Quadrature Points	20

Iteration History

Iter	Flag1	Calls	NegLogLike	Diff	MaxGrad	Slope
1		37	2661.1274	101.2361	1289.231	−377.866
2		56	2658.97577	2.151628	1154.987	−20.28
3		74	2656.05164	2.92414	673.5001	−9.25268
4		92	2655.70676	0.344871	13.34389	−0.70793
5		110	2655.70603	0.000731	0.012859	−0.00146
6		128	2655.70603	1.889E−8	8.519E−6	−3.8E−8

NOTE: GCONV convergence criterion satisfied.

Fit Statistics

−2 Log Likelihood	5311.4
AIC (smaller is better)	5343.4
AICC (smaller is better)	5343.7
BIC (smaller is better)	5430.9

Parameter Estimates

Parameter	Estimate	Standard Error	DF	t Value	Pr > |t|
eta_Intercept	−2.2987	0.3680	1753	−6.25	<.0001
eta_Trt	0.04923	0.1044	1753	0.47	0.6371
eta_Sex	0.04737	0.1074	1753	0.44	0.6592
eta_Age	−0.00248	0.004432	1753	−0.56	0.5755

```
eta_Diabetic         0.1910     0.1244    1753    1.54   0.1249
eta_Albumin         -0.3046    0.08894    1753   -3.42   0.0006
eta_PriorMonths    0.000282   0.002164    1753    0.13   0.8964
alpha                0.8132     0.1215    1753    6.69   <.0001
beta_Intercept      -1.0983     0.3113    1753   -3.53   0.0004
beta_Trt             0.1087    0.08906    1753    1.22   0.2224
beta_Sex             0.1259    0.09182    1753    1.37   0.1705
beta_Age           -0.00726   0.003872    1753   -1.87   0.0610
beta_Diabetic        0.4979     0.1078    1753    4.62   <.0001
beta_Albumin        -0.4071    0.07538    1753   -5.40   <.0001
beta_PriorMonths   0.002046   0.001826    1753    1.12   0.2626
beta_bi              1.0843     0.1052    1753   10.31   <.0001
Parameter          Alpha      Lower      Upper     Gradient
eta_Intercept       0.05    -3.0204    -1.5770    5.921E-9
eta_Trt             0.05    -0.1554     0.2539    2.444E-9
eta_Sex             0.05    -0.1632     0.2580    2.353E-9
eta_Age             0.05   -0.01117   0.006210    1.457E-6
eta_Diabetic        0.05   -0.05299     0.4350    2.914E-9
eta_Albumin         0.05    -0.4790    -0.1302    1.694E-8
eta_PriorMonths     0.05   -0.00396   0.004527    5.103E-7
alpha               0.05     0.5749     1.0514    8.519E-6
beta_Intercept      0.05    -1.7089    -0.4878    -1.58E-8
beta_Trt            0.05   -0.06597     0.2834    -8.32E-9
beta_Sex            0.05   -0.05419     0.3060    -6.23E-9
beta_Age            0.05   -0.01485   0.000336    -3.96E-7
beta_Diabetic       0.05     0.2866     0.7092    -5.19E-9
beta_Albumin        0.05    -0.5550    -0.2593    -4.98E-8
beta_PriorMonths    0.05   -0.00153   0.005627    -2.87E-7
beta_bi             0.05     0.8779     1.2906    -8.69E-8
Poisson overdispersion estimate (phi)   Poisson scale parameter
                              3.47986                   1.86544
```

Based on the corrected results to Output 7.2 (see author's web page for corrections), this SP model provides a better fit compared to the joint model under independence; however, there is still no evidence of a treatment difference in hospitalization rates (rate ratio: exp(0.09595) = 1.10, p = 0.2894). Moreover, under this SP model, we estimate the "overdispersion" parameter φ for the hospitalization data to be φ = 3 .49703, suggesting that the inclusion of the log-gamma peritonitis rate as a covariate does not help explain the overdispersion seen in the number of hospital admissions. If we combine the log-likelihood and AIC values from the gamma-Poisson model for peritonitis (Output 4.3) and the random effects negative binomial model for hospitalization (Output 5.18), we find this alternative joint independent model provides a fit between the two models just considered although the SP model does provide the best overall fit (see Table 7.1).

Table 7.1 Likelihood-based goodness-of-fit measures comparing the fit of two joint independent models and a shared parameter model to the peritonitis and hospitalization data. [1]The gamma-Poisson model is a random effects Poisson model with a gamma-distributed random effect for the rate of peritonitis as defined in (7.6) and which is equivalent to the negative binomial model of example 4.3.1 (Output 4.3); [2]The Poisson model used here is a fixed-effects Poisson model with linear predictor defined in (7.8); [3]The random-effects negative binomial (RE Neg. Bin.) model is a two-level random effects model defined in (5.43) of example 5.2.4; [4]The Poisson model used here is a fixed-effects Poisson model (7.8) but with the linear predictor modified to include the log-gamma rate parameter, b_i, from the gamma-Poisson model for peritonitis as a covariate.

Joint models for peritonitis and hospitalization			**Goodness-of-fit**	
Model	Peritonitis	Hospitalization	$-2\,L$	AIC
Independent	gamma-Poisson[1]	Poisson[2]	5820.5	5850.5
Independent	gamma-Poisson[1]	RE Neg. Bin.[3]	5303.2	5337.2
Correlated (SP)	gamma-Poisson[1]	Poisson[4]	5093.7	5125.7

Of course, one can try to fit an even more complex SP model. For example, one might include the log-gamma peritonitis rate parameter b_i from the gamma-Poisson model as a covariate within the random-effects negative binomial model (5.43) of example 5.2.4. However, one would need to define a nested random-effects structure with a normally distributed random effect at the center level for the hospitalization data and a gamma-distributed peritonitis rate at the patient level for the peritonitis data. This could be quite challenging and would likely require the use of array statements as illustrated by Littell et. al. (2006, §15.4, pp. 587-589). We leave the possibility of fitting this alternative SP model and/or other SP models to those readers interested in exploring the full capabilities of NLMIXED.

7.2 Pharmacokinetic applications

In this section, we consider two additional applications that involve fitting NLME models to pharmacokinetic (PK) data. The first application involves the pharmacokinetics of the anti-asthmatic drug theophylline as analyzed by Boeckmann et. al. (1992) and subsequently by Pinheiro and Bates (1995) and Davidian and Giltinan (1995). It entails fitting a one-compartment open model with first-order absorption and elimination to a group of 12 individuals sampled at ten time points over a 25-hour period. The second application involves the pharmacokinetics of the seizure prevention drug phenobarbital as first analyzed by Grasela and Donn (1985) and later by Davidian and Giltinan (1995). It entails fitting a one-compartment open model with intravenous administration and first-order elimination to a group of 59 preterm (neonatal) infants. Both examples have their own unique set of challenges such as the selection of good starting values in the first example and the selection of an appropriate random-effects structure in the second example.

7.2.1 Theophylline data

Boeckmann et. al. (1992) and subsequently Pinheiro and Bates (1995) and Davidian and Giltinan (1995) present methods for analyzing the kinetics of the anti-asthmatic drug theophylline. The drug was administered orally to 12 individuals and serum concentrations were measured at ten time points per subject over a 25-hour period. The concentration-time profiles of all 12 subjects are shown in Figure 7.1. Based on these profiles and the fact that the route of administration was oral, a one-compartment open model with first-order absorption and elimination seems appropriate for these data. A brief description summarizing key aspects of the data and the PK structural model is shown below with the SAS variables shown in parentheses.

- Study of kinetics of the anti-asthmatic drug theophylline
- Experimental unit or cluster is an individual ($n = 12$)
- Response variable:
  - y_{ij} = serum concentration (mg/L) - (conc).
- One within-unit covariate:
  - t_{ij} =time following oral administration (11 time points over a 25 hour period including time 0) - (time)
- Two between-unit covariates:
  - Dose of drug: D_i = dose per kg body weight (mg/kg) - (dose)
  - Weight: w_i =weight (kg) - (wt)
- Structural model:
  At time t, the serum concentration of the drug following oral administration may be described by the following one-compartment open model with first-order absorption and

Figure 7.1 Individual concentration versus time profiles for the theophylline PK data

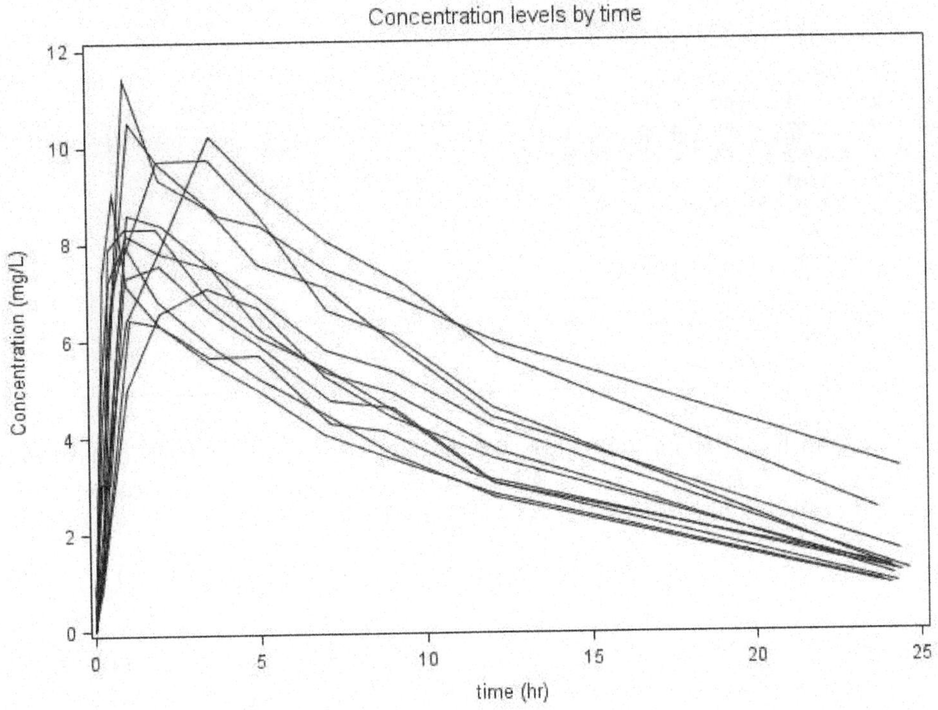

elimination:

$$y(t) = \frac{F \times Dose \times k_a}{V \times (k_a - k_e)} \times \left\{ e^{-k_e t} - e^{-k_a t} \right\}$$

where

F = fraction of drug available ($F \equiv 1$, complete absorption),
k_a = a first-order absorption constant,
$k_e = Cl/V$ is a first-order elimination constant,
Cl = drug clearance rate,
V = Apparent volume of distribution.

- Goal: Estimate the population PK parameters.

Output 7.3 provides a partial listing of the data for the first two subjects (see also the author's Web page). Serum concentrations were measured at time zero and at ten time points following oral administration of the drug. Three of the subjects had a zero concentration at time zero (e.g., subject 2) which can greatly complicate any analysis that models the within-subject mean-variance relation. Subsequently, as was done by Davidian and Giltinan (1995), we exclude the zero time point from the analyses reported here.

Output 7.3: Partial listing of the theophylline PK data.

subject	time	conc	dose	wt
1	0.00	0.74	4.02	79.6
1	0.25	2.84	4.02	79.6
1	0.57	6.57	4.02	79.6
1	1.12	10.50	4.02	79.6
1	2.02	9.66	4.02	79.6
1	3.82	8.58	4.02	79.6

1	5.10	8.36	4.02	79.6
1	7.03	7.47	4.02	79.6
1	9.05	6.89	4.02	79.6
1	12.12	5.94	4.02	79.6
1	24.37	3.28	4.02	79.6
2	0.00	0.00	4.40	72.4
2	0.27	1.72	4.40	72.4
2	0.52	7.91	4.40	72.4
2	1.00	8.31	4.40	72.4
2	1.92	8.33	4.40	72.4
2	3.50	6.85	4.40	72.4
2	5.02	6.08	4.40	72.4
2	7.03	5.40	4.40	72.4
2	9.00	4.55	4.40	72.4
2	12.00	3.01	4.40	72.4
2	24.30	0.90	4.40	72.4

One can fit a PK model to the theophylline data by setting the structural PK parameters Cl, k_a and V to be subject-specific random effect parameters, say Cl_i, k_{ai} and V_i, respectively. This yields the NLME model

$$y_{ij} = \frac{FD_i k_{ai}}{V_i(k_{ai} - k_{ei})}\Big\{\exp(-k_{ei}t_{ij}) - \exp(-k_{ai}t_{ij})\Big\} + \epsilon_{ij} \tag{7.9}$$

where $k_{ei} = Cl_i/V_i$ and the ϵ_{ij} are intra-subject errors with some specified intra-subject covariance structure ($i = 1, \ldots, 12; j = 1, \ldots, 10$). One might consider setting $Cl_i = Cl + b_{i1}$, $k_{ai} = k_a + b_{i2}$ and $V_i = V + b_{i3}$ where Cl, k_a and V are the population PK parameters and $\boldsymbol{b}_i = (b_{i1}, b_{i2}, b_{i3})'$ are normally distributed random effects but this does not guarantee that the SS parameters Cl_i, k_{ai} and V_i will all be positive as required structurally.

To ensure the SS parameters are positive, it is best to reparameterize the model by defining the conditional mean in terms of population parameters $\boldsymbol{\beta} = (\beta_{Cl}, \beta_{ka}, \beta_V)'$ and random effects $\boldsymbol{b}_i = (b_{i1}, b_{i2}, b_{i3})'$ as:

$$E(y_{ij}|\boldsymbol{b}_i) = \mu_{ij}(\boldsymbol{\beta}, \boldsymbol{b}_i) = \frac{FD_i k_{ai}}{V_i(k_{ai} - k_{ei})}\Big\{\exp(-k_{ei}t_{ij}) - \exp(-k_{ai}t_{ij})\Big\} \tag{7.10}$$

where $k_{ei} = Cl_i/V_i$ is the SS elimination rate expressed as a function of the SS clearance and volume parameters and where Cl_i, k_{ai} and V_i are parameterized in terms of the nonlinear functions

$$\begin{aligned} Cl_i &= \exp(\beta_{Cl} + b_{i1}) \\ k_{ai} &= \exp(\beta_{ka} + b_{i2}) \\ V_i &= \exp(\beta_V + b_{i3}) \end{aligned} \tag{7.11}$$

as was done by Davidian and Giltinan (1995). To complete the specification of the NLME model (7.9), we assume that, conditional on $\boldsymbol{b}_i$, the within-subject errors ϵ_{ij} are independently distributed $N(0, \sigma_{ij}^2)$ random variables with a power-of-the-mean variance structure,

$$\sigma_{ij}^2 = \sigma^2 \mu_{ij}(\boldsymbol{\beta}, \boldsymbol{b}_i)^{2\delta} \tag{7.12}$$

as suggested by Davidian and Giltinan (1995).

To obtain initial starting values, we fit the structural model defined by (7.10)-(7.12) to each individual separately and computed the naive standard two-stage (STS) estimators of the model parameters by simply averaging the individual estimates. The data and SAS code for this preliminary fit as well as the results displayed in Figure 7.1 and Output 7.3 are shown below.

Program 7.2

```
data theoph;
    input subject time conc dose wt;
    datalines;
 1  0.00  0.74 4.02 79.6
 1  0.25  2.84 4.02 79.6
 1  0.57  6.57 4.02 79.6
 :
title1 ''Concentration levels by time'';
proc sgplot data=theoph noautolegend;
 series x=time y=conc / group=subject
                        lineattrs=(color=black pattern=1);
 label conc='Concentration (mg/L)' time='time (hr)';
run;
proc print data=theoph(obs=22) noobs;
 title 'Partial listing of the theophylline PK data';
run;
proc means data=theoph noprint;
 where time=0;
 var dose;
 output out=MeanDose(keep=MeanDose) mean=MeanDose;
run;
proc sort data=theoph;
 by subject time;
run;
/* Exclude time zero and merge in mean dose from above */
data example7_2_1;
 set theoph;
 by subject time;
 if _n_=1 then set MeanDose;
 if time> 0;
run;
/*===============================================================
  Step 1: Compute unweighted standard two-stage estimates of
  the population PK parameters as starting values
===============================================================*/
ods listing close; ods output parameterestimates=OLSpe;
proc nlmixed data=example7_2_1 qpoints=1;
 by subject;
 parms beta_Cl=-10 to -1 by 1
       beta_ka=0.1 to 1 by 0.1
       beta_V=-5 to -1 by 1
       sigsq=.1 to 1 by .1 delta=1;
 Cl = exp(beta_Cl);
 ka = exp(beta_ka);
 V  = exp(beta_V );
 ke = Cl/V;
 expfunc = exp(-ke*time) - exp(-ka*time);
 predmean = (dose*ka/(V*(ka-ke))) * expfunc;
 predvar = sigsq*(predmean**(2*delta));
 model conc ~ normal(predmean,predvar);
run; ods listing;
```

```
proc means data=OLSpe mean median var nway;
 class Parameter;
 var Estimate;
 title 'STS estimates of theophylline PK parameters';
run;
```

Shown in Output 7.4 are the STS estimates of the model parameters. To fit the NLME model (7.9) under the assumptions in (7.10)-(7.12) and assuming $\boldsymbol{b}_i \sim N_3(\boldsymbol{0}, \boldsymbol{\Psi})$ where $\boldsymbol{\Psi}$ is an unstructured covariance matrix, we set the variances of b_{i1}, b_{i2}, and b_{i3} to be the sample variances of the SS parameters (i.e., beta_Cl, beta_ka, beta_V) in Output 7.4.

Output 7.4: STS estimates of theophylline PK parameters.

Analysis Variable : Estimate				
Parameter	N Obs	Mean	Median	Variance
beta_Cl	12	−3.2144272	−3.1465755	0.0822079
beta_V	12	−0.7635894	−0.7734059	0.0255956
beta_ka	12	0.4748828	0.4080376	0.4547722
delta	12	1.9951368	0.5853918	9.7814747
sigsq	12	2.8651274	0.0846392	86.5371664

Next we tried to fit the full NLME model assuming $\boldsymbol{\Psi}$ is unstructured using the following NLMIXED statements.

```
ods output parameterestimates=MLEparms;
proc nlmixed data=example7_2_1 qpoints=1;
 parms beta_Cl=-3.214 beta_ka=0.475 beta_V=-0.764
       sigsq=2.865 delta=2.00
       psi11=0.082 psi22=0.455 psi33=0.026
       psi21=0 psi31=0 psi32=0;
 Cl = exp(beta_Cl + bi1);
 ka = exp(beta_ka + bi2);
 V  = exp(beta_V  + bi3);
 ke=Cl/V;
 expfunc = exp(-ke*time) - exp(-ka*time);
 predmean = (dose*ka/(V*(ka-ke))) * expfunc;
 predvar = sigsq*(predmean**(2*delta));
 model conc ~ normal(predmean,predvar);
 random bi1 bi2 bi3 ~ normal([0,0,0],[psi11,
                                  psi21, psi22,
                                  psi31, psi32, psi33])
        subject=subject;
run;
```

However, the program failed to converge and the following error and warning were issued in the SAS log:

```
ERROR: QUANEW Optimization cannot be completed.
WARNING: Optimization routine cannot improve the function value.
```

To make sure this failure is not the result of using the default quasi-Newton optimization algorithm, we re-ran the code using the Newton-Raphson algorithm with a maximum of 1,000 function calls (via the NLMIXED options: `tech=newrap` maxfunc=1000), but the program again met with failure. While it is never clear what might trigger such a failure, one possibility is that the covariance structure may be misspecified. To check this, we ran the %COVPARMS macro described in §5.3.6 of Chapter 5 to determine if one or more of the random effects could be the problem. To call the macro, we first re-ran the above NLMIXED program but with the covariance parameters all set equal to zero. We include the statements

```
resid = conc - predmean;
predict predmean out=predout der;
id resid;
```

just before the RUN statement so as to generate the required dataset as input into the %COVPARMS macro. We then call the macro with the following syntax:

```
%covparms(parms=MLEparms, predout=predout,
          resid=resid, method=mspl,
          random=der_bi1 der_bi2 der_bi3,
          subject=subject, type=un, covname=psi,
          output=MLEparms1);
```

The macro generates the following results (Output 7.5).

Output 7.5: Initial estimates of the random effects variance-covariance parameters for the theophylline PK data assuming b_{i1}, b_{i2} and b_{i3} are all random.

Estimated G Matrix

Effect	Row	Col1	Col2	Col3
Der_bi1	1	0.09931	−0.09007	0.03554
Der_bi2	2	−0.09007	0.5311	−0.00127
Der_bi3	3	0.03554	−0.00127	0.01456

Estimated G Correlation Matrix

Effect	Row	Col1	Col2	Col3
Der_bi1	1	1.0000	−0.3922	0.9346
Der_bi2	2	−0.3922	1.0000	−0.01450
Der_bi3	3	0.9346	−0.01450	1.0000

Covariance Parameter Estimates

Cov Parm	Subject	Estimate	Standard Error	Wald 95% Confidence Bounds	
UN(1,1)	subject	0.09931	0.05958	0.04016	0.5251
UN(2,1)	subject	−0.09007	0.09487	−0.2760	0.09588
UN(2,2)	subject	0.5311	0.2500	0.2515	1.7660
UN(3,1)	subject	0.03554	0.01842	−0.00056	0.07164
UN(3,2)	subject	−0.00127	0.03582	−0.07149	0.06894
UN(3,3)	subject	0.01456	0.009424	0.005580	0.09303
Residual		0.5684	0.08823	0.4285	0.7903

Tests of Covariance Parameters Based on the Likelihood

Label	DF	-2 Log Like	ChiSq	Pr > ChiSq	Note
Diagonal G	3	354.15	10.71	0.0134	DF

As suggested by the results in Output 7.5, the estimated variance component of the volume parameter random effect b_{i3} is a) close to zero and b) highly correlated with the clearance parameter random effect b_{i1} ($\widehat{\psi}_{33} = 0.01456 \pm 0.009424$, $Corr(b_{i1}, b_{i3}) = 0.9346$). These are possible indications of an ill-conditioned covariance matrix due possibly to overspecification (see also Pinheiro and Bates, 1995, pp. 363-365 and Davidian and Giltinan, 1995, §5.5). When we attempt to fit the NLME model using the parameter estimates from Output 7.5 as starting values for an unstructured covariance matrix, NLMIXED again failed to converge with an error message sent to the SAS log stating that no valid parameter points were found. We therefore dropped the volume parameter random effect b_{i3} from the model and fit the reduced NLME model to the theophylline data using the following SAS code.

Program 7.3

```
data MLEparms2;
 set MLEparms1;
 if Parameter in ('psi31' 'psi32' 'psi33') then delete;
run;
ods output parameterestimates=peLMLE;
ods output additionalestimates=aeLMLE;
proc nlmixed data=example7_2_1 qpoints=1 tech=newrap;
 parms /data=MLEparms2;
 Cl = exp(beta_Cl + bi1);
 ka = exp(beta_ka + bi2);
 V  = exp(beta_V);
 ke=Cl/V;
 Cl_mean = exp(beta_Cl+.5*psi11);
 ka_mean = exp(beta_ka+.5*psi22);
 ke_mean = exp(beta_Cl+.5*psi11-beta_V);
 Tmax = log(ka_mean/ke_mean)/(ka_mean-ke_mean);
 AUC_mean = MeanDose/(V*ke_mean);
 expfunc = exp(-ke*time) - exp(-ka*time);
 predmean = (dose*ka/(V*(ka-ke))) * expfunc;
 predvar = sigsq*(predmean**(2*delta));
 resid = conc - predmean;
 model conc ~ normal(predmean,predvar);
 random bi1 bi2 ~ normal([0,0],[psi11,
                                psi21, psi22])
        subject=subject;
 estimate 'Cl' Cl_mean;
 estimate 'V' V;
 estimate 'ka' ka_mean;
 estimate 'ke' ke_mean;
 estimate 'Tmax' Tmax;
 estimate 'AUC' AUC_mean;
run;
```

We initially attempted to fit this reduced model with the above call to NLMIXED using the default quasi-Newton optimization algorithm but the program failed and the error message

```
ERROR: QUANEW Optimization cannot be completed.
```

was written out to the SAS log. We then ran the above version of NLMIXED using the TECH=NEWRAP option and achieved convergence in 7 iterations as indicated in Output 7.6. We included within the above NLMIXED code, additional programming statements combined with ESTIMATE statements so that we could summarize estimates of the population PK clearance rate, $Cl = E_b(Cl_i) = E_b\{\exp(\beta_{Cl} + b_{i1})\} = \exp(\beta_{Cl} + \frac{1}{2}\psi_{11})$, the apparent volume of distribution, $V = E_b(V_i) = \exp(\beta_V)$, and population PK absorption and elimination rates, $k_a = E_b(k_{ai}) = E_b\{\exp(\beta_{ka} + b_{i2})\} = \exp(\beta_{ka} + \frac{1}{2}\psi_{22})$ and $k_e = E_b(k_{ei}) = E_b(Cl_i/V_i) = E_b\{\exp(\beta_{Cl} + b_{i1} - \beta_V)\} = \exp(\beta_{Cl} + \frac{1}{2}\psi_{11} - \beta_V)$. In addition, we use ESTIMATE statements to also compute the population PK area under the curve

(AUC) and peak concentration time (T_{MAX}) as defined by:

$$AUC = F\bar{D}/Cl = \bar{D}/(Vk_e)$$

$$T_{MAX} = \frac{\log(k_a) - \log(k_e)}{(k_a - k_e)}$$

where $\bar{D}$ is the mean dose across the 12 individuals and F is the fraction of the administered dose available which we have assumed here to be 1 (Gibaldi and Perrier, 1982).

Output 7.6: Select output from NLMIXED when we fit the NLME model defined by (7.9-7.12) to the theophylline PK data but restricted to only b_{i1} and b_{i2} as the random effects. Results are based on the Laplace approximation as specified by the option QPOINTS=1.

Specifications

Data Set	WORK.EXAMPLE7_2_1
Dependent Variable	conc
Distribution for Dependent Variable	Normal
Random Effects	bi1 bi2
Distribution for Random Effects	Normal
Subject Variable	subject
Optimization Technique	Newton-Raphson
Integration Method	Adaptive Gaussian Quadrature

Dimensions

Observations Used	120
Observations Not Used	0
Total Observations	120
Subjects	12
Max Obs Per Subject	10
Parameters	8
Quadrature Points	1

Iteration History

Iter	Flag1	Calls	NegLogLike	Diff	MaxGrad	Slope
1	*	28	183.590803	22.63655	127.299	-4265.84
2	*	38	177.577973	6.01283	24.97026	-10.9921
3	*	48	176.695688	0.882285	7.82645	-1.47114
4	*	58	176.537956	0.157732	3.422878	-0.26642
5	*	68	176.521193	0.016763	0.3606	-0.03055
6	*	78	176.520834	0.000359	0.01558	-0.0007
7	*	88	176.520833	4.461E-7	0.000053	-8.85E-7

NOTE: GCONV convergence criterion satisfied.

Fit Statistics

-2 Log Likelihood	353.0
AIC (smaller is better)	369.0
AICC (smaller is better)	370.3
BIC (smaller is better)	372.9

Parameter Estimates

Parameter	Estimate	Standard Error	DF	t Value	Pr > |t|
beta_Cl	-3.2177	0.09225	10	-34.88	<.0001
beta_ka	0.3682	0.2099	10	1.75	0.1099
beta_V	-0.8097	0.02828	10	-28.63	<.0001
sigsq	0.1571	0.08327	10	1.89	0.0885
delta	0.4591	0.1620	10	2.83	0.0178
psi11	0.08974	0.04461	10	2.01	0.0720
psi21	-0.08137	0.07075	10	-1.15	0.2768
psi22	0.4752	0.2232	10	2.13	0.0591

Parameter	Alpha	Lower	Upper	Gradient
beta_Cl	0.05	−3.4232	−3.0121	2.674E−6
beta_ka	0.05	−0.09949	0.8360	−1.73E−6
beta_V	0.05	−0.8727	−0.7467	−1.44E−6
sigsq	0.05	−0.02840	0.3426	−1.37E−6
delta	0.05	0.09804	0.8202	−0.00005
psi11	0.05	−0.00966	0.1891	1.615E−6
psi21	0.05	−0.2390	0.07627	−2.14E−6
psi22	0.05	−0.02202	0.9724	−3.67E−6

Additional Estimates

Label	Estimate	Standard Error	DF	t Value	Pr > |t|
Cl	0.04189	0.003948	10	10.61	<.0001
V	0.4450	0.01258	10	35.36	<.0001
ka	1.8328	0.4467	10	4.10	0.0021
ke	0.09413	0.009569	10	9.84	<.0001
Tmax	1.7076	0.2859	10	5.97	0.0001
AUC	110.44	10.4093	10	10.61	<.0001

Label	Alpha	Lower	Upper
Cl	0.05	0.03309	0.05068
V	0.05	0.4170	0.4730
ka	0.05	0.8374	2.8282
ke	0.05	0.07280	0.1154
Tmax	0.05	1.0705	2.3447
AUC	0.05	87.2464	133.63

The resulting Laplace-based MLE's of the model parameters are similar to results reported by Davidian and Giltinan (1995, §5.5, §6.6) for the same model but with all three random effects included. When reporting the results of a PK analysis such as summarized in Output 7.6, one may choose to emphasize inference based on the parameters Cl, k_a, and V of the underlying PK model rather than the transformed parameters, β_{Cl}, β_{ka} and β_V. However, one must then decide on what basis should estimates $\widehat{Cl}$, $\widehat{k}_a$ and $\widehat{V}$ be reported. We chose to emphasize inference on the basis of the population PK parameters obtained by taking the expected value of the subject-specific clearances, $Cl_i = \exp(\beta_{Cl} + b_{i1})$, the subject-specific absorption rate constants, $k_{ai} = \exp(\beta_{ka} + b_{i2})$, and the subject-specific volumes $V_i = \exp(\beta_V)$, directly. The above NLMIXED programming and ESTIMATE statements implement this approach. The resulting estimates reflect what we expect the average clearance, absorption and volume parameters to be across the population of individuals. An alternative approach would be to base inference on what the average subject's log-clearance, log-absorption and log-volume parameters are and then exponentiate these values to obtain estimates of Cl, k_a, and V for the typical subject. In this case we are estimating $Cl = \exp(E_b(\log(Cl_i)) = \exp(\beta_{Cl})$, $k_a = \exp(E_b(\log(k_{ai})) = \exp(\beta_{ka})$, and $V = \exp(E_b(\log(V_i)) = \exp(\beta_V)$. In comparing the two approaches with respect to say clearance, we see that the population estimate of clearance, $Cl = E_b(Cl_i) = \exp(\beta_{Cl} + \frac{1}{2}\psi_{11})$, takes into account variation in clearances across individuals whereas the subject-specific estimate of clearance, $Cl = \exp(E_b(\log(Cl_i)) = \exp(\beta_{Cl})$, ignores such variation and instead reflects what the clearance would be for the typical (i.e., $b_{i1} = 0$) individual. While both approaches have merit, the population-averaged approach used here seems more consistent with the goal of estimating population PK parameters that are themselves nonlinear functions of $\boldsymbol{b}_i$ such as Cl_i and k_{ai}.

In addition to the Laplace-based MLE's, we fit the NLME model using the default adaptive Gaussian quadrature in which 5 quadrature points were used to evaluate the integrated log-likelihood and we fit the model using the first-order population-averaged approximation obtained by linearizing the model about $\boldsymbol{b}_i = \boldsymbol{0}$. The SAS code is exactly the same as above except that we remove the QPOINTS=1 option in both cases and for the first order approximation, we specify METHOD=FIRO. The model parameter estimates and additional estimates from the ESTIMATE statements were saved to datasets using ODS OUTPUT statements and the results summarized using PROC REPORT.

Output 7.7: A summary comparison of model parameter ML estimates based on the NLMIXED first-order approximation (METHOD=FIRO), the Laplace approximation (QPOINTS=1) and adaptive Gaussian approximation (the default method which, in this example, required 5 quadrature points).

Parameter Estimates

	Method					
	First order		Laplace MLE		MLE (qpoints=5)	
Parameter	Estimate	StdErr	Estimate	StdErr	Estimate	StdErr
beta_Cl	−3.1409	0.09104	−3.2177	0.09225	−3.2177	0.09237
beta_V	−0.7294	0.02610	−0.8097	0.02828	−0.8097	0.02829
beta_ka	1.0054	0.1003	0.3682	0.2099	0.3687	0.2103
delta	0.9903	0.2038	0.4591	0.1620	0.4583	0.1629
psi11	0.1072	0.05569	0.08974	0.04461	0.08997	0.04474
psi21	−0.00209	0.08419	−0.08137	0.07075	−0.08183	0.07103
psi22	0.6461	0.2841	0.4752	0.2232	0.4771	0.2244
sigsq	0.01786	0.01380	0.1571	0.08327	0.1576	0.08393

Parameter Estimates

	Method					
	First order		Laplace MLE		MLE (qpoints=5)	
PK Parameters	Estimate	StdErr	Estimate	StdErr	Estimate	StdErr
AUC	101.39	11.1440	110.44	10.4093	110.43	10.4224
Cl	0.04562	0.005015	0.04189	0.003948	0.04189	0.003954
Tmax	1.0016	0.1447	1.7076	0.2859	1.7058	0.2865
V	0.4822	0.01258	0.4450	0.01258	0.4450	0.01259
ka	3.7753	0.7242	1.8328	0.4467	1.8354	0.4486
ke	0.09461	0.008821	0.09413	0.009569	0.09413	0.009582

The results of all three methods are summarized in Output 7.7. The first thing to note is that there is very little difference between ML estimates based on the Laplace approximation versus adaptive Gaussian quadrature. This is not unexpected given that the Laplace approximation is asymptotically equivalent to ML estimation for large intra-subject sample sizes (Vonesh, 1996). In this case, we have nearly the same number of observations per subject as subjects which is why the two agree so closely with one another. Conversely, the first-order ML estimates are noticeably different from both the Laplace and adaptive Gaussian ML estimates. This is particularly true with respect to the absorption parameter β_{ka} and the intra-subject variance parameters δ and σ^2. As discussed in §5.1.2 and §5.3.5, the first-order method will generally lead to inconsistent estimates regardless of how large the intra-subject sample sizes are. Indeed, Breslow and Lin (1995), Demidenko (2004, pp. 456-459) and Solomon and Cox (1992) all show that linearization-based methods obtained by expanding the conditional moments about zero are biased by an order of magnitude that depends on the variances of the random effects. Only when the random-effect variance components are small can one expect a reasonable degree of agreement between the first-order method and either the Laplace or adaptive Gaussian methods. In this example, the estimated variance component $\widehat{\psi}_{22}$ of the absorption parameter random effect b_{i2} is relatively large compared to the actual parameter estimate, $\widehat{\beta}_{ka}$, and this may explain the rather large discrepancy between $\widehat{\beta}_{ka} = 1.01 \pm 0.10$ for the first-order method versus $\widehat{\beta}_{ka} = 0.37 \pm 0.21$ for the Laplace and adaptive Gaussian methods (rounded here for comparison sake). The apparent bias in the absorption rate parameter using the first-order method leads to an obvious bias in other PK parameters as evidenced by the greater estimated value of k_a and, correspondingly, lower estimated value of T_{MAX} (see Output 7.7). Of course for the first-order method, it might make more sense to estimate k_a by $\exp(\widehat{\beta}_{ka})$ rather than $\exp(\widehat{\beta}_{ka} + \widehat{\psi}_{11})$ since estimation involves expanding about $\boldsymbol{b}_i = \mathbf{0}$. Nevertheless, it would seem that for this particular example, at least, one should avoid using the first-order method.

Figure 7.2 Individual concentration versus time profiles for the phenobarbital PK data

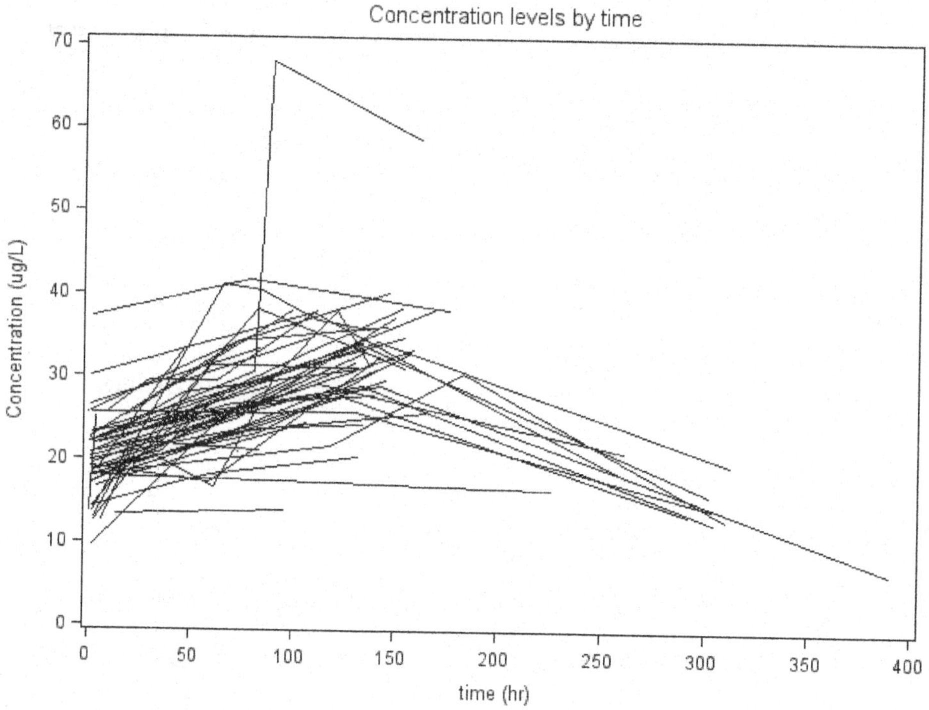

7.2.2 Phenobarbital data

Grasela and Donn (1985) described the neonatal population pharmacokinetics of the seizure prevention drug, phenobarbital. Subsequently, Davidian and Giltinan (1995) applied a number of different NLME models to the phenobarbital PK data in an attempt to best describe the observed heterogeneity in the data through different random effect specifications (e.g., additive versus multiplicative error structures) to account for unexplained random variation, as well different covariate specifications to account for systematic or explained variation. Shown in Figure 7.2 are the individual concentration-time profiles of $n = 59$ preterm infants given phenobarbital for prevention of seizures during the first 16 days after birth. The concentration-time profiles are from routine clinical data in which between 1 to 6 concentration measurements per infant were obtained over a 16 day period after birth. Each infant received an initial dose followed by one or more intermittent sustaining doses by intravenous administration. The birth weight (kg) and a 5 minute Apgar score were also obtained on each infant as possible covariates for inclusion in the PK model. Following Grasela and Donn (1985) and Davidian and Giltinan (1995), the pharmacokinetics of phenobarbital may be described by a one-compartment open model with intravenous administration and first-order elimination. A summary description highlighting key aspects of the data and the PK structural model is shown below. The SAS variable names corresponding to the response variable and covariates are shown in parentheses.

- Study of kinetics of the seizure prevention drug phenobarbital
- Experimental unit or cluster is an individual (59 preterm infants)
- Response variable:
  - y_{ij} = serum concentration (μg/L) - (conc).

- Two within-unit covariates:
  - t_{ij} = time following administration of initial dose - (time)
    > Drug concentrations were obtained at 1 to 6 time points (other than dose times) over a 16 day period after birth
  - Dose of drug: D_{ij} = dose per kg body weight (μg/kg) - (dose)
    > Each individual received an initial dose followed by one or more sustaining doses by intravenous administration
- Two between-unit covariates:
  - Weight: w_i =birth weight (kg) - (weight)
  - Apgar score: $s_i = 5$ minute Apgar score - (apgar)
    > Also analyzed as a dichotomous variable, $\delta_i = I(s_i < 5)$ - (apgarind)
- Up to 155 measurements in 59 preterm infants
  - Sparse intra-subject data
  - Intermittent dosing
- Structural model:
  At time t, the serum concentration may be described by the following one-compartment open model with intravenous administration and first-order elimination

$$y(t) = \frac{D_{t_d}}{V} \times exp\Big\{-\frac{Cl}{V}(t - t_d)\Big\}$$

where

D_{t_d} = a single dose administered at time t_d
Cl = drug clearance rate,
V =apparent volume of distribution.

- Structural model with intermittent dosing:
  The concentration at time t following intermittent administration of sustaining doses may be described by the sum of such terms, i.e.

$$y(t) = \sum_{d:t_d<t} \frac{D_{t_d}}{V} \times exp\Big\{-\frac{Cl}{V}(t - t_d)\Big\}$$

which can also be expressed in recursive form (Grasela and Donn, 1985)

- Goal: Estimate the PK parameters taking into account variation due to subject-specific covariates and between- and within-subject errors.

A partial listing of the data, as made available courtesy of the Resource Facility for Population Kinetics (RFPK) from their website at

http://www.rfpk.washington.edu

is shown below. A description of the data is also available through the author's Web page. We include the data for subject ID 50 as this infant received a second large bolus injection 88 hours following birth which accounts for the unusual spike in concentration seen one hour later in Figure 7.2.

Output 7.8: Partial listing of phenobarbital data for subjects 47 and 50.

ID	time	dose	weight	apgar	conc	eventid
47	0.0	40.0	2.6	9	.	1
47	9.3	6.7	2.6	9	.	1
47	19.3	6.7	2.6	9	.	1
47	33.3	6.7	2.6	9	.	1
47	36.3	6.7	2.6	9	.	1
47	38.3	.	2.6	9	25.1	0

50	0.0	20.0	1.1	6	.	1
50	3.0	.	1.1	6	22.2	0
50	12.5	2.5	1.1	6	.	1
50	24.5	2.5	1.1	6	.	1
50	36.5	2.5	1.1	6	.	1
50	48.0	2.5	1.1	6	.	1
50	60.5	2.5	1.1	6	.	1
50	72.5	2.5	1.1	6	.	1
50	81.0	.	1.1	6	30.5	0
50	84.5	2.5	1.1	6	.	1
50	88.0	30.0	1.1	6	.	1
50	89.0	.	1.1	6	67.9	0
50	96.5	2.5	1.1	6	.	1
50	108.5	2.5	1.1	6	.	1
50	120.5	3.5	1.1	6	.	1
50	132.5	3.5	1.1	6	.	1
50	144.5	3.5	1.1	6	.	1
50	157.0	3.5	1.1	6	.	1
50	162.0	.	1.1	6	58.7	0

As one can see from both Figure 7.2 and Output 7.8, the concentration measurements for individuals are fairly sparse and the times at which a sustaining dose or concentration measurement are taken are irregularly spaced.

Let $\boldsymbol{a}_i = (w_i, s_i, \delta_i)'$ be the vector of between-subject covariates consisting of the birth weight, w_i, the five-minute Apgar score, s_i, and the Apgar score indicator variable, $\delta_i = I(s_i < 5)$. Also, let $\boldsymbol{x}^*_{ij} = (t_{ij}, D_{ij})'$ be the vector of within-subject covariates consisting of the measurement times, t_{ij}, and corresponding dose levels, D_{ij}. Within the program, we shall define D_{ij} to be 0 if, at time t_{ij}, the subject did not receive a sustaining dose. Setting $t_{i1} = 0$ such that $y(t_{i1}) = D_{i1}/V_i$ is the concentration immediately following the initial dose of phenobarbital, it follows from the structural form of the one-compartment open model with intravenous administration and first-order elimination that the ith subject's concentration measured without error may be written in recursive form as

$$\begin{aligned}
y(t_{i1}) &= \frac{D_{i1}}{V_i} \\
y(t_{i2}) &= \frac{D_{i2}}{V_i} + y(t_{i1})\exp\Big\{-\frac{Cl_i}{V_i}(t_{i2}-t_{i1})\Big\} \\
&= \frac{D_{i2}}{V_i} + \frac{D_{i1}}{V_i}\exp\Big\{-\frac{Cl_i}{V_i}(t_{i2}-t_{i1})\Big\} \\
y(t_{i3}) &= \frac{D_{i3}}{V_i} + y(t_{i2})\exp\Big\{-\frac{Cl_i}{V_i}(t_{i3}-t_{i2})\Big\} \\
&= \frac{D_{i3}}{V_i} + \frac{D_{i2}}{V_i}\exp\Big\{-\frac{Cl_i}{V_i}(t_{i3}-t_{i2})\Big\} + \frac{D_{i1}}{V_i}\exp\Big\{-\frac{Cl_i}{V_i}(t_{i3}-t_{i1})\Big\} \\
&\quad\vdots
\end{aligned}$$

and the stochastic model for concentration y_{ij} at time t_{ij} may be written as

$$y_{ij} = y(t_{ij}) + \epsilon_{ij} = \frac{D_{ij}}{V_i} + \sum_{k<j}\frac{D_{ik}}{V_i} \times \exp\Big\{-\frac{Cl_i}{V_i}(t_{ij}-t_{ik})\Big\} + \epsilon_{ij}. \tag{7.13}$$

Following Littell et. al. (2006), it will be convenient from a programming perspective to write this PK model in its recursive form

$$y_{ij} = y(t_{ij}) + \epsilon_{ij} = f(\boldsymbol{x}^{*\prime}_{ij}, \boldsymbol{\beta}_i) + \epsilon_{ij}; \quad (i = 1, \ldots, n;\ j = 1, \ldots, p_i)$$

where $\boldsymbol{\beta}_i = (\beta_{i1}, \beta_{i2})' = (Cl_i, V_i)'$ are the subject-specific clearance and volume parameters for the ith infant, ϵ_{ij} are intra-subject errors, and

$$y(t_{ij}) = f(\boldsymbol{x}^{*\prime}_{ij}, \boldsymbol{\beta}_i) = \frac{D_{ij}}{V_i} + f(\boldsymbol{x}^{*\prime}_{i(j-1)}, \boldsymbol{\beta}_i)\exp\Big\{-\frac{Cl_i}{V_i}(t_{ij} - t_{i(j-1)})\Big\}$$

is the concentration measured without error for the ith infant immediately following the jth "dose" at time t_{ij} (recall we define $D_{ij} = 0$ if no dose was delivered).

To allow for flexible modeling of the SS clearance and volume parameters, $\beta_{i1} = Cl_i$ and $\beta_{i2} = V_i$, as functions of the between subject covariates, $\boldsymbol{a}_i = (w_i, s_i, \delta_i)'$, and random effects, $\boldsymbol{b}_i = (b_{i1}, b_{i2})'$, we modeled the phenobarbital data based on the Gaussian two-stage NLME model

$$\text{Stage 1: } y_{ij} = f(\boldsymbol{x}^{*\prime}_{ij}, \boldsymbol{\beta}_i) + \epsilon_{ij}, \quad \epsilon_{ij} \sim \text{ind}\, N(0, \sigma^2_{ij}) \tag{7.14}$$

$$\text{Stage 2: } \boldsymbol{\beta}_i = \begin{pmatrix} \beta_{i1} \\ \beta_{i2} \end{pmatrix} = \begin{pmatrix} Cl_i \\ V_i \end{pmatrix} = \begin{pmatrix} g_1(\boldsymbol{a}_i, \boldsymbol{\beta}, b_{i1}) \\ g_2(\boldsymbol{a}_i, \boldsymbol{\beta}, b_{i2}) \end{pmatrix}, \quad \boldsymbol{b}_i \sim \text{iid}\, N(\boldsymbol{0}, \boldsymbol{\Psi}(\boldsymbol{\theta}_b)).$$

where $\beta_{i1} = g_1(\boldsymbol{a}_i, \boldsymbol{\beta}, b_{i1})$ and $\beta_{i2} = g_2(\boldsymbol{a}_i, \boldsymbol{\beta}, b_{i2})$ are general nonlinear functions of the covariates and random effects, respectively. Of course, one can write this in terms of the NLME model (5.50) by setting $\boldsymbol{x}'_{ij} = (\boldsymbol{x}^{*\prime}_{ij}, \boldsymbol{a}'_i)$ and defining $\sigma^2_{ij} = \sigma^2 w_{ij}(\boldsymbol{x}'_{ij}, \boldsymbol{\beta}, \boldsymbol{b}_i)$ for some suitable scalar weight function, $w_{ij}(\boldsymbol{x}'_{ij}, \boldsymbol{\beta}, \boldsymbol{b}_i)$.

Davidian and Giltinan (1995, §6.6) fit several different NLME models to the phenobarbital data in an attempt to identify systematic and/or random sources of variation that best explain inter-individual variation. This is accomplished through an appropriate specification of the functions $g_1(\boldsymbol{a}_i, \boldsymbol{\beta}, b_{i1})$ and $g_2(\boldsymbol{a}_i, \boldsymbol{\beta}, b_{i1})$ with systematic variation entering through specification of the covariates $\boldsymbol{a}_i$ and random variation entering through specification of the random effects structure (additive versus multiplicative). They also considered two different specifications of the intra-subject error structure: either a homogeneous additive error structure or a heterogeneous multiplicative error structure. Here we confine our attention to four of the models the authors considered.

Statistical models

The following four models considered by Davidian and Giltinan (1995) were fit to the phenobarbital data using NLMIXED. Under each model, the clearance and volume parameter random effects, b_{i1} and b_{i2}, are assumed to be independent as per Davidian and Giltinan (1995).

Model 1: A model with no covariates and additive error structures

$\boldsymbol{\beta} = (\beta_1, \beta_2)'$,
$\beta_{i1} = g_1(\boldsymbol{a}_i, \boldsymbol{\beta}, \boldsymbol{b}_i) = \beta_1 + b_{i1}$,
$\beta_{i2} = g_2(\boldsymbol{a}_i, \boldsymbol{\beta}, \boldsymbol{b}_i) = \beta_2 + b_{i2}$,
$Var(\epsilon_{ij}) = \sigma^2_{ij} = \sigma^2$

Model 2: A model with no covariates and multiplicative error structures

$\boldsymbol{\beta} = (\beta_1, \beta_2)'$,
$\beta_{i1} = g_1(\boldsymbol{a}_i, \boldsymbol{\beta}, \boldsymbol{b}_i) = \beta_1(1 + b_{i1})$,
$\beta_{i2} = g_2(\boldsymbol{a}_i, \boldsymbol{\beta}, \boldsymbol{b}_i) = \beta_2(1 + b_{i2})$,
$Var(\epsilon_{ij}) = \sigma^2_{ij} = \sigma^2 f(\boldsymbol{x}^{*\prime}_{ij}, \boldsymbol{\beta}_i)^2$

Model 3: A model with birth weight as a covariate for clearance and volume and additive error structures

$\boldsymbol{\beta} = (\beta_{11}, \beta_{12}, \beta_{21}, \beta_{22})'$,
$\beta_{i1} = g_1(\boldsymbol{a}_i, \boldsymbol{\beta}, \boldsymbol{b}_i) = \beta_{11} + \beta_{12} w_i + b_{i1}$,

$\beta_{i2} = g_2(\boldsymbol{a}_i, \boldsymbol{\beta}, \boldsymbol{b}_i) = \beta_{21} + \beta_{22}w_i + b_{i2},$
$Var(\epsilon_{ij}) = \sigma_{ij}^2 = \sigma^2$

Model 4: A model with birth weight as a covariate for clearance and volume and multiplicative error structures

$\boldsymbol{\beta} = (\beta_{11}, \beta_{12}, \beta_{21}, \beta_{22})',$
$\beta_{i1} = g_1(\boldsymbol{a}_i, \boldsymbol{\beta}, \boldsymbol{b}_i) = (\beta_{11} + \beta_{12}w_i)(1 + b_{i1}),$
$\beta_{i2} = g_2(\boldsymbol{a}_i, \boldsymbol{\beta}, \boldsymbol{b}_i) = (\beta_{21} + \beta_{22}w_i)(1 + b_{i2}),$
$Var(\epsilon_{ij}) = \sigma_{ij}^2 = \sigma^2 f(\boldsymbol{x}_{ij}^{*\prime}, \boldsymbol{\beta}_i)^2$

We fit each of these models using NLMIXED. To do so, we first structure the data as in Littell et. al. (2006, Example 15.7, pp. 607-623) so that we can fit the recursive form of the NLME model defined by (7.14) to the data. Specifically, the following programming statements take the dataset downloaded from the RFPK site, http://www.rfpk.washington.edu, and organizes the data in a manner similar to that done by Littell et. al. (2006). Included in the programming statements is a macro %PK which one can use to make repetitive calls to NLMIXED based on the model specifications one chooses.

Program 7.4

```
data example7_2_2;
 set SASdata.phenobarbital_data;
 retain cursub .;
 if cursub ne ID then do;
    newsub = 1;  cursub = ID;
 end;
 else newsub = 0;
 apgarind = (apgar < 5);
 lagtime = lag(time);
 if (newsub=1) then lagtime = 0;
 /* eventid =1 if a sustaining dose is given while */
 /* eventid =0 if a serum concentration is given   */
 if eventid = 0 and dose=. then dose=0;
 drop cursub;
run;
/* PARMS12 is a macro variable for use with macro %PK  */
/* Grid values for PK models 1-2 with no covariates    */
%let parms12=%str(
     parms beta1=0.01 0.5 beta2=.01 .1 1
           psi11=0.000005 0.0005 .05
           psi22=.01 0.1 sigsq=.1 3 /best=1; );
/* PARMS34 is a macro variable for use with macro %PK  */
/* Grid values for PK models 3-4 with covariate=weight */
%let parms34=%str(
     parms beta11=0.01 0.5 beta21=.01 .1 1
           beta12=0 .05 .5 beta22=0 .01 .1
           psi11=0.000005 0.0005 .05
           psi22=.01 0.1 sigsq=.1 3 /best=1; );
/*====================================================
 MACRO PK - A SAS macro that calls NLMIXED to fit a NLME
            model to the phenobarbital data using the
            NLMIXED specifications that one chooses.
 KEY:
```

```
    CLOSE    -defines whether NLMIXED output prints
              CLOSE=CLOSE is default (no printout)
              CLOSE='blank' instructs macro to print
    METHOD   -defines which estimation method to use
              METHOD=GAUSS (with QPOINTS=1) is default
              METHOD=FIRO calls the first-order method
    PARMS    -defines the PARMS statement for starting
              values of the parameters using either
              PARMS=&PARMS12 or PARMS=&PARMS34
    BOUNDS   -defines a bounds statement for NLMIXED
    CL       -defines the clearance parameter as a
              function of parameters, covariates and
              random effects
    V        -defines the volume parameter as a
              function of parameters, covariates and
              random effects
    VARIANCE -defines the intra-subject variance structure
    MODEL    -defines a Model ID number
    FIT      -defines the dataset that saves the likelihood
              fit statistics using ODS OUTPUT FitStatistics=
    OUTPUT   -defines the dataset that saves the parameter
              estimates using ODS OUTPUT ParameterEstimates=
    PREDOUT  -defines a dataset containing predicted SS
              clearances as well as the SS random effects
=======================================================*/
%macro PK(close=close,
          method=Gauss,
          parms=&parms12,
          bounds=%str(bounds beta1>=0, beta2>=0,
                      psi11>=0, psi22>=0, sigsq> 0;),
          Cl= beta1/100 + bi1,
          V = beta2 + bi2,
          variance=sigsq,
          model=Model 1,
          fit=fit,
          output=pe,
          predout=pred);
  %let method=%upcase(&method);
  ods listing &close;
  ods output fitstatistics=&fit;
  ods output parameterestimates=&output;
  proc nlmixed data=example7_2_2 method=&method qpoints=1;
   &parms;
   &bounds;
   /* Clearance and Volume parameters terms expressed */
   /* as functions of parameters and covariates       */
   Cl = &Cl;
   V  = &V;
   func =  exp(-(Cl/V)*(time-lagtime));
   /* PK model expressed as a recursive NLME model    */
   if (newsub = 1) then
      predmean = Dose/V;
   else
      predmean = Dose/V +
                 zlag(predmean)*exp(-(Cl/V)*(time-lagtime));
      predvar  = &variance;
```

```
    model conc ~ normal(predmean,predvar);
    random bi1 bi2 ~ normal([0,0],[psi11,0,psi22]) subject=ID;
    predict Cl out=&predout; id Cl V bi1 bi2;
  run;
  data &pe;
   set &pe;
   Method_=''&method'';
   if Method_=''GAUSS'' then Method='Laplace MLE';
   if Method_=''FIRO''  then Method='First-order';
   Model=''&model'';
   drop Method_;
  run;
  ods listing;
%mend;
```

Based on the macro %PK, we fit the four NLME models to the data using both the first-order approximation method (METHOD=FIRO) and the Laplace approximation (METHOD=GAUSS with a default QPOINTS=1 option in the macro). As was done by Littell et. al. (2006), we re-scaled the clearance parameter β_1 by dividing it by 100 to improve numerical stability. Repetitive calls to the macro %PK were made for each model and the results stored in separate datasets. We used a grid search to obtain starting values for the parameters based on the macro variables &PARMS12 and &PARMS34 defined above. Following recommendations in the SAS on-line documentation for NLMIXED, the recursive function, `predmean`, was defined using the ZLAG function rather than the usual LAG function. A total of eight calls to the %PK macro were made in order to fit the four NLME models to the data using the first-order and Laplace approximation methods. The following call to the macro illustrates how to fit the fourth model listed above using the first-order approximation method.

```
%PK(method=FIRO,
    parms=&parms34,
    bounds=%str(bounds beta11>=0, beta21>=0,
                psi11>=0, psi22>=0, sigsq> 0;),
    Cl = ((beta11 + beta12*weight)/100)*(1 + bi1),
    V  = (beta21 + beta22*weight)*(1 + bi2),
    variance = sigsq*(predmean**2),
    model=Model 4,
    fit=fit41,
    output=pe41
    predout=pred41);
```

While the remaining calls to the macro are not shown here, a complete listing of the SAS program including all eight calls to %PK is available at the author's Web page.

Convergence was achieved in each of the eight calls made to the macro %PK. However, the program did encounter problems when fitting the NLME models with additive error structures (models 1 and 3). For example, in addition to re-scaling the clearance parameter as indicated above, we also found it necessary to use a BOUNDS statement constraining the parameters to be greater than zero whenever we fit the model using the first-order approximation method. Despite these efforts and the fact that convergence was achieved in all cases, problems were still encountered with regards to some of the parameter estimates and their standard errors for models 1 and 3. Specifically, in three of the four calls to %PK involving the two models with additive error structures, SAS issued warnings in the SAS log concerning the Hessian matrix. For example, when we fit model 1 to the data using the

Laplace approximation (METHOD=GAUSS), convergence was achieved but the following warning regarding the Hessian matrix was issued in the SAS log.

```
NOTE: GCONV convergence criterion satisfied.
NOTE: At least one element of the (projected) gradient is greater than 1e-3.
NOTE: Moore-Penrose inverse is used in covariance matrix.
WARNING: The final Hessian matrix is not positive definite, and therefore
the estimated covariance matrix is not full rank and may be unreliable. The
variance of some parameter estimates is zero or some parameters are linearly
related to other parameters.
```

Likewise, when we fit model 3 to the data using the first-order approximation (METHOD=FIRO), the following warning was issued in the SAS log:

```
NOTE: FCONV convergence criterion satisfied.
NOTE: At least one element of the (projected) gradient is greater than 1e-3.
WARNING: The final projected Hessian matrix is full rank but has at least
one negative eigenvalue. Second-order optimality condition violated.
```

Lastly, a warning that the final Hessian matrix was not positive definite was issued when we fit the same model using the Laplace approximation. These computational issues are likely due to unreasonable additive error assumptions under models 1 and 3.

Despite these warnings, ODS OUTPUT datasets containing the parameter estimates and fit statistics from each of the eight calls to %PK were merged and the results summarized via PROC REPORT according to the model and the method of estimation used (code not shown). The results, shown in Output 7.9, reflect the warnings issued by SAS regarding the Hessian matrix, particularly with respect to model 3. For example, the use of the first-order approximation method for model 3 results in the clearance intercept parameter, β_{11}, to be set to its boundary value of 0 with no standard error while no standard errors could be computed for the two random effect variance components, ψ_{11} and ψ_{22}. While the Laplace approximation faired better for model 3, it too resulted in an inestimable standard error for the random variance component, ψ_{11}.

Output 7.9: Summary of model parameter estimates, their standard errors and p-values from the four NLME models fit to the phenobarbital PK data. Also included are likelihood-based fit statistics summarizing the fit of each model according to the method of estimation used.

MLE's for models 1 and 2 (with no covariates)
Method: FIRO=First-order approximation GAUSS=Laplace approximation

		Model					
		Model 1			Model 2		
Method	Parameter	Estimate	StdErr	Pr > t	Estimate	StdErr	Pr > t
FIRO	beta1	0.5468	0.09236	<.0001	0.5406	0.04021	<.0001
	beta2	1.3963	0.07991	<.0001	1.1917	0.04843	<.0001
	psi11	6.591E−6	5.409E−6	0.2281	0.05957	0.02102	0.0064
	psi22	0.2842	0.1061	0.0096	0.03891	0.01039	0.0004
	sigsq	8.0282	1.5258	<.0001	0.01235	0.002245	<.0001
GAUSS	beta1	0.5597	0.04285	<.0001	0.5765	0.03824	<.0001
	beta2	1.5696	0.1666	<.0001	1.5855	0.1127	<.0001
	psi11	1.686E−6	1.261E−6	0.1867	0.08345	0.03981	0.0405
	psi22	1.4992	0.3471	<.0001	0.2773	0.07106	0.0003
	sigsq	23.6502	0.007989	<.0001	0.01859	0.003578	<.0001

MLE's for models 3 and 4 (weight included as a covariate)
Method: FIRO=First-order approximation GAUSS=Laplace approximation

		Model					
		Model 3			Model 4		
Method	Parameter	Estimate	StdErr	Pr > t	Estimate	StdErr	Pr > t
FIRO	beta11	0	.	.	0.04297	0.1025	0.6767
	beta12	0.4768	0.06477	<.0001	0.4357	0.07989	<.0001
	beta21	0.1100	0.04917	0.0292	0.04704	0.09176	0.6102
	beta22	0.9247	0.06928	<.0001	0.9444	0.07373	<.0001
	psi11	1.12E−6	.	.	0.01980	0.009193	0.0355
	psi22	0.07422	.	.	0.01218	0.003621	0.0014
	sigsq	8.9856	1.8341	<.0001	0.01058	0.001904	<.0001
GAUSS	beta11	−0.09939	0.09328	0.2911	−0.03436	0.09430	0.7169
	beta12	0.5450	0.06906	<.0001	0.4948	0.07570	<.0001
	beta21	0.07126	0.1099	0.5193	0.1151	0.08744	0.1932
	beta22	0.9585	0.07531	<.0001	0.9255	0.07127	<.0001
	psi11	1.231E−6	0	<.0001	0.03800	0.01931	0.0540
	psi22	0.05952	0.01867	0.0023	0.02851	0.007843	0.0006
	sigsq	9.5262	0.000978	<.0001	0.01337	0.002319	<.0001

Likelihood-based fit statistics by model and method of estimation
Method: FIRO=First-order approximation GAUSS=Laplace approximation

		Model			
		Model 1	Model 2	Model 3	Model 4
Method	Fit statistic	Value	Value	Value	Value
FIRO	−2 Log Likelihood	1002.1	1029.4	894.1	864.7
	AIC (smaller is better)	1012.1	1039.4	908.1	878.7
	AICC (smaller is better)	1012.5	1039.8	908.9	879.5
	BIC (smaller is better)	1022.5	1049.8	922.7	893.3
GAUSS	−2 Log Likelihood	1146.4	1049.7	896.2	878.7
	AIC (smaller is better)	1156.4	1059.7	910.2	892.7
	AICC (smaller is better)	1156.8	1060.1	910.9	893.4
	BIC (smaller is better)	1166.8	1070.1	924.7	907.2

Judging from the likelihood-based fit statistics in Output 7.9, it is clear that for both methods of estimation, the use of birth weight as a covariate for clearance and volume significantly improves the fit to the data (model 3 versus 1 and model 4 versus 2). Moreover, except for model 1 versus 2 under the first-order method, specifying a multiplicative error structure over an additive error structure also improves the fit to the data. Certainly alternative models can be used to overcome the numerical instabilities encountered under the additive error structure of models 1 and 3. For example, one can reparameterize model 3 as

Model 3a: A model with birth weight as a covariate for log clearance and log volume assuming an additive error structure on the log scale

$\boldsymbol{\beta} = (\beta_{11}, \beta_{12}, \beta_{21}, \beta_{22})'$,
$\beta_{i1} = g_1(\boldsymbol{a}_i, \boldsymbol{\beta}, \boldsymbol{b}_i) = \exp(\beta_{11} + \beta_{12} w_i + b_{i1})$,
$\beta_{i2} = g_2(\boldsymbol{a}_i, \boldsymbol{\beta}, \boldsymbol{b}_i) = \exp(\beta_{21} + \beta_{22} w_i + b_{i2})$,
$Var(\epsilon_{ij}) = \sigma^2_{ij} = \sigma^2$

where $\beta_{11} + \beta_{12} w_i + b_{i1} = \log(Cl_i)$ and $\beta_{21} + \beta_{22} w_i + b_{i2} = \log(V_i)$. Upon making the following call to %PK

```
%PK(method=FIRO,
    parms=&parms34,
    bounds=,
    Cl = exp(beta11 + beta12*weight + bi1),
    V  = exp(beta21 + beta22*weight + bi2),
    variance = sigsq,
    model=Model 3a,
```

```
fit=fit31a,
output=pe31a);
```

and a similar call using METHOD=GAUSS, we were able to fit model 3a to the data with convergence achieved in both cases but without any warnings, etc. in the SAS log regarding the Hessian matrix. The results, summarized in Output 7.10, provide much greater stability to the parameter estimates and their standard errors for both methods. Moreover, the first-order and Laplace approximations yield fairly comparable results. This may be due, in part, to the fact that the estimated random-effect variance components are similar for both estimation methods and the estimates are also relatively small compared to the fixed-effects regression parameters (a low coefficient of variation). As indicated in the previous example, the first-order method will often yield comparable results under these conditions. Another reason could be due to the sparsity of data available for estimation. With sparse data, the posterior modes used to estimate the random effects under the Laplace approximation will effectively be shrunk toward zero resulting in estimates that will be similar to that obtained with the first-order approximation.

Output 7.10: Summary of results when we fit the NLME model 3a to the phenobarbital data.

MLE's for model 3a (weight included as a covariate)
Method: FIRO=First-order approximation GAUSS=Laplace approximation

	Method					
	FIRO			GAUSS		
Parameter	Estimate	StdErr	Pr > t	Estimate	StdErr	Pr > t
beta11	−5.9419	0.1261	<.0001	−5.9903	0.1267	<.0001
beta12	0.5986	0.07594	<.0001	0.6222	0.07637	<.0001
beta21	−0.4857	0.06436	<.0001	−0.4711	0.06723	<.0001
beta22	0.5355	0.03810	<.0001	0.5319	0.04025	<.0001
psi11	0.05792	0.02764	0.0406	0.05169	0.02444	0.0388
psi22	0.02760	0.007908	0.0009	0.02920	0.008293	0.0009
sigsq	7.4616	1.2692	<.0001	7.4120	1.2593	<.0001

Likelihood-based fit statistics for model 3a by method of estimation
Method: FIRO=First-order approximation GAUSS=Laplace approximation

	Method	
	FIRO	GAUSS
Fit statistic	Value	Value
−2 Log Likelihood	876.3	875.2
AIC (smaller is better)	890.3	889.2
AICC (smaller is better)	891.1	890.0
BIC (smaller is better)	904.8	903.7

One can also try alternative fits to the data using the %PK macro. For example, the following call to %PK fits model 6 of Davidian and Giltinan (1995) obtained by including birth weight as a covariate for clearance, and both birth weight and the Apgar indicator variable, δ_i, as covariates for the volume parameter. In addition, rather than assume multiplicative Gaussian random effects, model 6 of Davidian and Giltinan (1995) assumes a log-normal multiplicative random effect for the clearance and volume parameters. The model specifications are as follows.

Model 6: A model with birth weight as a covariate for clearance, birth weight and the Apgar indicator variable for volume and log-normal multiplicative random effects

$$\boldsymbol{\beta} = (\beta_1, \beta_2, \beta_3, \beta_4)',$$
$$\beta_{i1} = g_1(\boldsymbol{a}_i, \boldsymbol{\beta}, \boldsymbol{b}_i) = (\beta_{11} w_i)\exp(b_{i1}),$$
$$\beta_{i2} = g_2(\boldsymbol{a}_i, \boldsymbol{\beta}, \boldsymbol{b}_i) = (\beta_{21} w_i)(1 + \beta_{22}\delta_i)\exp(b_{i2}),$$
$$Var(\epsilon_{ij}) = \sigma_{ij}^2 = \sigma^2 f(\boldsymbol{x}_{ij}^{*\prime}, \boldsymbol{\beta}_i)^2$$

Based on this specification, Laplace-based ML estimates of the model parameters can be computed with the following call to %PK.

```
%PK(close=,
    method=GAUSS,
    parms=%str(
    parms beta11=0.01 0.5
          beta21=.01 .1 1 beta22=0 .01 .1
          psi11=0.000005 0.0005 .05
          psi22=.01 0.1 sigsq=.1 3 /best=1; ),
    bounds=,
    Cl = (beta11/100*weight)*exp(bi1),
    V  = (beta21*weight)*(1 + beta22*apgarind)*exp(bi2),
    variance = sigsq*(predmean**2),
    model=Model 6,
    fit=fit62,
    output=pe62,
    predout=pred62);
```

The results of fitting model 6 to the data are shown in Output 7.11.

Output 7.11: Select output from NLMIXED when we fit Model 6 to the phenobarbital data using the Laplace approximation.

Specifications

Data Set	WORK.EXAMPLE7_2_2
Dependent Variable	conc
Distribution for Dependent Variable	Normal
Random Effects	bi1 bi2
Distribution for Random Effects	Normal
Subject Variable	ID
Optimization Technique	Dual Quasi−Newton
Integration Method	Adaptive Gaussian Quadrature

Dimensions

Observations Used	155
Observations Not Used	589
Total Observations	744
Subjects	59
Max Obs Per Subject	6
Parameters	6
Quadrature Points	1

Iteration History

Iter	Flag1	Calls	NegLogLike	Diff	MaxGrad	Slope
1		58	470.740667	17.29609	459.7428	−2611.85
2		61	466.587665	4.153002	4481.526	−21118.2
3		64	466.297041	0.290625	4247.204	−436.417
4		67	465.001352	1.295689	4018.805	−542.765
5		69	453.010709	11.99064	1804.966	−101.965
6		71	445.658036	7.352673	971.9779	−170.091
7		74	444.084663	1.573373	1269.25	−15.9863
8		76	439.708684	4.375979	726.7145	−9.29371
9		78	437.456454	2.25223	360.4465	−5.71539
10		80	436.216227	1.240227	251.6949	−1.83699
11		82	435.622293	0.593934	91.15802	−1.26652
12		84	435.504471	0.117822	76.2325	−0.18762
13		86	435.475208	0.029263	20.01332	−0.04589
14		88	435.46788	0.007328	21.46141	−0.00479

15	90	435.444436	0.023445	16.82543	−0.00671
16	92	435.432268	0.012168	3.1526	−0.01084
17	94	435.431439	0.000829	1.795839	−0.00115
18	96	435.431409	0.000029	0.074941	−0.00005
19	98	435.431409	2.305E−7	0.022178	−4.2E−7

NOTE: GCONV convergence criterion satisfied.

Fit Statistics

−2 Log Likelihood	870.9
AIC (smaller is better)	882.9
AICC (smaller is better)	883.4
BIC (smaller is better)	895.3

Parameter Estimates

Parameter	Estimate	Standard Error	DF	t Value	Pr > |t|
beta11	0.4616	0.02121	57	21.76	<.0001
beta21	0.9753	0.02775	57	35.14	<.0001
beta22	0.1586	0.07814	57	2.03	0.0471
psi11	0.02937	0.01465	57	2.00	0.0497
psi22	0.02751	0.007130	57	3.86	0.0003
sigsq	0.01328	0.002201	57	6.03	<.0001

The fit of model 6 to the data yields PK estimates similar to that obtained by Littell et. al. (2006) with the difference being the latter allowance for correlation between the random effects b_{i1} and b_{i2}. These results also agree with the first-order estimates from Davidian and Giltinan (1995) which is consistent with our previous remarks regarding the impact that small variance component estimates and sparse data have on first-order versus Laplace approximation methods. The strength of association between birth weight and both clearance and volume is depicted in Figure 7.3 which plots the estimated subject-specific clearances and volumes from dataset **pred62** against the corresponding subject-specific birth weights.

Figure 7.3 Individual PK clearances and volumes plotted against birth weight for the phenobarbital PK data

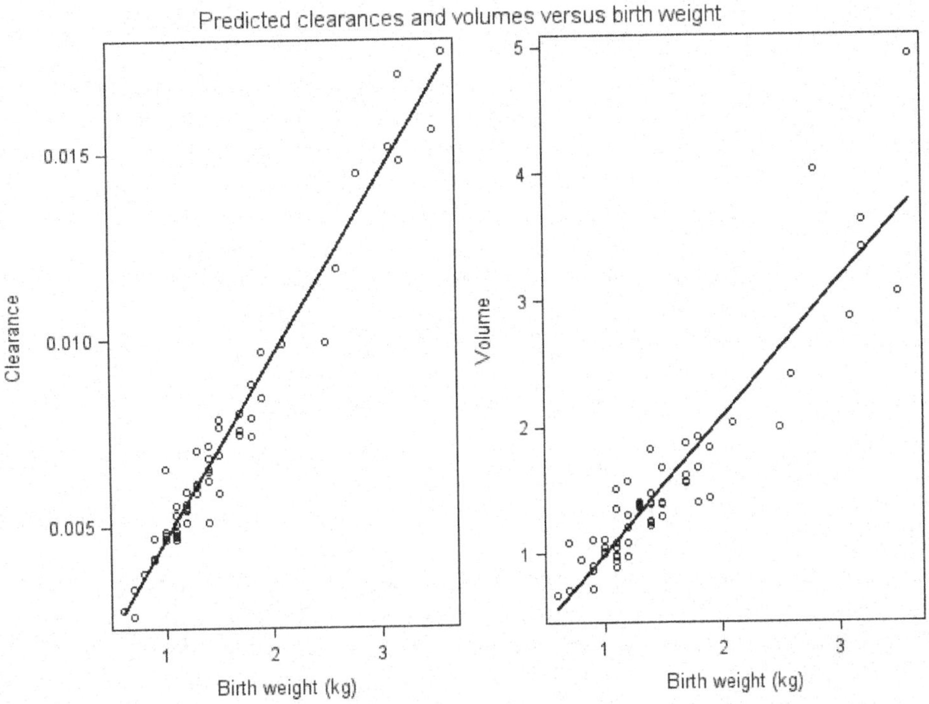

In summary, a number of different NLME models were fit to the phenobarbital data. Based on the results presented here, we can conclude that 1) the use of a strictly additive Gaussian random effects structure should be avoided, and 2) it is essential to include birth weight as a covariate in the PK model if one is to adequately account for the overall variation seen in both the clearance and volume parameters of individuals.

7.3 Joint modeling of longitudinal data and survival data

In this section, we revisit the use of shared parameter models described in Chapter 6 (§6.5) but we do so within the context of applications where the goal is to jointly model longitudinal data and survival data. In many clinical trials, the primary outcome of interest will be patient survival or some other well-defined event time. However, one may also be interested in exploring whether summary trends in the serial measurements of some response variable are related in any way to survival. In a large well-conducted randomized controlled trial of survival, for example, one may wish to investigate whether there is any relationship between trends (e.g., slopes) in some biomarker of disease and patient outcome. An example would be the use of CD4 cells per cubic millimeter of blood as a biomarker for disease progression to AIDS. The study of such relationships can prove useful when evaluating whether a biomarker can effectively serve as a surrogate endpoint (e.g., Pawitan and Self, 1993; Guo and Carlin, 2004). In this section, we consider an application from the ADEMEX study in which we use a shared parameter (SP) model to explore and characterize the relationship between the rate of decline in the glomerular filtration rate (GFR) of incident end-stage renal disease patients and subsequent patient survival. The application of a SP model here therefore differs from the applications considered in Chapter 6 where SP models are used to help one cope with the analysis of longitudinal data in the presence of non-ignorable dropout.

7.3.1 ADEMEX study—GFR data and survival

As introduced in §1.4, the ADEMEX trial was a randomized multi-center trial of 965 Mexican patients designed to compare patient outcomes (e.g., survival, hospitalization, quality of life, etc.) among end-stage renal disease patients randomized to one of two dose levels of continuous ambulatory peritoneal dialysis (CAPD). A detailed description of the study may be found in Paniagua et. al. (2002) as well as in the ADEMEX adequacy example of §2.2.3 (page 55). Patient survival was the primary endpoint in the ADEMEX study and a number of secondary analyses were performed to determine whether survival differed according to select subgroups of patients. One such analysis examined whether there was any difference in patient survival according to either 1) the dose of peritoneal dialysis as measured by peritoneal creatinine clearance or 2) the estimated glomerular filtration rate (GFR) which is a measure of the remaining kidney function of patients with end-stage renal disease (Paniagua et. al., 2002). That analysis, based on a time-dependent Cox proportional hazards model, established that patients with greater time-dependent GFR values had a lower risk of death but peritoneal clearance had no effect on overall patient survival (Paniagua et. al., 2002). In the example presented here, we revisit the ADEMEX study and explore whether there is any relation between a patient's intrinsic rate of decline in GFR and survival. The establishment of such a relationship could prove useful for establishing the use of GFR as a surrogate endpoint in future interventional studies of patient survival.

In order to examine the relationship between the rate of decline in GFR and patient survival, we need to jointly model two outcome variables, namely a patient's GFR over time (denoted by y_{ij}) and the patient's survival time (denoted by T_i). A patient's GFR is a measure of whatever remaining kidney function the patient has following the onset of end-stage renal disease and subsequent start of dialysis. Following standard practice, GFR was measured in the ADEMEX study as the average of urinary creatinine and urea clearances over a 24-hour period. Patient survival times were determined as the elapsed time between when a patient was randomized to either the control or intervention arm and when the patient either died or was censored, whichever occurred first. The primary analysis was intention-to-treat (ITT) with death attributed to the treatment group to which the patient was randomized (control versus treated). Moreover, in keeping with the ITT principle and the desire to avoid informative censoring, every effort was made to assess a patient's survival status up through the study termination date even though some patients dropped out sooner and were no longer monitored for adherence to their prescribed dose of dialysis, etc. Consequently, patient survival times were censored only when patients received a transplant, had a return of kidney function, were truly lost to follow-up or reached the study termination date (Paniagua et. al., 2002).

Since the goal here is to explore the relationship between the rate of decline in GFR and patient survival, the analysis presented here is restricted to incident-only patients who initiated dialysis within three months prior to the date they were randomized and for whom a baseline measurement of GFR> 1.00 ml/min was available prior to randomization. Patients with a baseline GFR≤ 1.00 ml/min were excluded as these patients are generally considered to be functionally anuric (e.g., Paniagua et. al., 2002). Such patients are either emergent dialysis patients or they had already experienced a rapid decline in GFR prior to entry into the study. Indeed, preliminary analysis showed that patients who had been on dialysis 3 months had the lowest baseline GFR (mean baseline GFR of 2.18±1.97 ml/min) compared to those who had been on for 1 month (mean baseline GFR of 4.13±2.44 ml/min) or 2 months (2.91±2.16 ml/min). We also excluded the two patients (one control, one treated) who had a return of renal function resulting in 270 incident patients for this analysis (133 control, 137 treated). Five between-subject covariates were included as possible risk factors for patient survival and a brief description of each is given in §1.4 (see also Appendix C, Dataset C.18, for a partial listing and description of SAS variable names, etc.).

There are any number of possible models one can specify when jointly exploring the relationship between the rate of decline in GFR and patient survival. We elected here to focus attention on 1) whether the rate of decline in GFR differs according to treatment group and 2) whether the latent random effect linked to a patient's rate of decline in GFR is an independent risk factor associated with survival. Consequently, we modeled GFR as a function of time (t_{ij}) and treatment group (a_{1i}) only, while the effects of treatment group (a_{1i}), gender (a_{2i}), age (a_{3i}), diabetes (a_{4i}), baseline albumin (a_{5i}) and baseline nPNA (a_{6i}) were included in the survival model. We first fit a joint model for GFR and survival in which baseline GFR was included as a covariate within the survival model. We then fit a shared parameter model in which we excluded the measured baseline GFR value as a covariate but included a patient's latent random intercept and latent random decay rate from the GFR model as possible covariates in the survival model.

In terms of the survival component of these joint models, we fit a piecewise exponential (PE) model (see, for example, equations 6.50 and 6.51 of example 6.6.2) to the survival times by assuming a piecewise constant mortality rate within each of five non-overlapping intervals of follow-up. The hazard function may be written as

$$\begin{aligned}\lambda_i(t;\boldsymbol{x}_i) &= \lambda_0(t)\exp\{\boldsymbol{x}_i'\boldsymbol{\eta}_x\} \qquad (7.15)\\ &= \exp\left\{\eta_0 + \sum_{h=1}^{4}\eta_h I(t\in(t_h,t_{h+1}]) + \boldsymbol{x}_i'\boldsymbol{\eta}_x\right\}\end{aligned}$$

where η_0 is an overall constant baseline log hazard rate and the η_h are baseline log-hazard rate modifiers within the intervals $(t_h, t_{h+1}]$ for patients randomized to control group ($a_{1i} = 0$). The η_h are assumed constant within the follow-up intervals $(t_h, t_{h+1}]$, $h = 1, \ldots 4$;

with $t_1 = 6$, $t_2 = 12$, $t_3 = 18$, $t_4 = 24$ and $t_5 = 36$ where 36 marks a study truncation time $\mathcal{T}$ that exceeds the longest follow-up time. The term $\boldsymbol{x}_i'\boldsymbol{\eta}_x$ is a linear predictor consisting of a specified vector of covariates, $\boldsymbol{x}_i$, and a vector of corresponding log-hazard ratios, $\boldsymbol{\eta}_x$, depending on what form the joint model takes.

We fit a negative exponential decay model to the GFR data assuming one of two forms. As depicted in Figure 1.6, there is a fairly rapid decline in GFR following initiation of dialysis. To account for this rapid initial decline, we first consider fitting an exponential decay model of the form (1.5) to the GFR data assuming an additive error structure. Since we are only interested in a model that compares the GFR profile of the two treatment groups, we fit the following version of (1.5) to the GFR data

$$\begin{aligned}\textbf{Model 1}.\ \text{Stage 1: } y_{ij} &= f(\boldsymbol{x}_{ij}', \boldsymbol{\beta}, \boldsymbol{b}_i) + \epsilon_{ij} \qquad (7.16)\\ &= f(\boldsymbol{x}_{ij}', \boldsymbol{\beta}_i) + \epsilon_{ij}\\ &= \beta_{i1}\exp(-\beta_{i2}t_{ij}) + \epsilon_{ij}\\ \text{Stage 2: } \boldsymbol{\beta}_i = \begin{pmatrix}\beta_{i1}\\ \beta_{i2}\end{pmatrix} &= \begin{pmatrix}\beta_{11}+\beta_{12}a_{1i}\\ \beta_{21}+\beta_{22}a_{1i}\end{pmatrix} + \begin{pmatrix}b_{i1}\\ b_{i2}\end{pmatrix}.\end{aligned}$$

where $\boldsymbol{x}_{ij}' = (t_{ij}, 1, a_{1i})$. Under this model, we assume an additive error structure with $\epsilon_{ij} \sim \text{iid}\, N(0, \sigma^2)$ independent of the random effects $\boldsymbol{b}_i = (b_{i1}, b_{i2})'$ which we assume to be iid $N(\boldsymbol{0}, \boldsymbol{\Psi}(\boldsymbol{\theta}_b))$ with unstructured variance-covariance matrix, $\boldsymbol{\Psi}(\boldsymbol{\theta}_b) = \begin{pmatrix}\psi_{11} & \psi_{12}\\ \psi_{21} & \psi_{22}\end{pmatrix}$. In addition, we also considered the following two-stage NLME model with multiplicative errors,

$$\begin{aligned}\textbf{Model 2}.\ \text{Stage 1: } y_{ij} &= f(\boldsymbol{x}_{ij}', \boldsymbol{\beta}, \boldsymbol{b}_i) + \epsilon_{ij} f(\boldsymbol{x}_{ij}', \boldsymbol{\beta}, \boldsymbol{b}_i)^{\theta} \qquad (7.17)\\ &= f(\boldsymbol{x}_{ij}', \boldsymbol{\beta}_i) + \epsilon_{ij} f(\boldsymbol{x}_{ij}', \boldsymbol{\beta}_i)^{\theta}\\ &= \beta_{i1}\exp(-\beta_{i2}t_{ij}) + \epsilon_{ij} f(\boldsymbol{x}_{ij}', \boldsymbol{\beta}_i)^{\theta}\\ \text{Stage 2: } \boldsymbol{\beta}_i = \begin{pmatrix}\beta_{i1}\\ \beta_{i2}\end{pmatrix} &= \begin{pmatrix}\exp(\beta_{11}+\beta_{12}a_{1i}+b_{i1})\\ \exp(\beta_{21}+\beta_{22}a_{1i}+b_{i2})\end{pmatrix},\end{aligned}$$

where $\epsilon_{ij} \sim \text{iid}\, N(0, \sigma^2)$ and $\boldsymbol{b}_i \sim \text{iid}\, N(\boldsymbol{0}, \boldsymbol{\Psi}(\boldsymbol{\theta}_b))$ independent of ϵ_{ij} as in (7.16). Under this multiplicative error structure, the conditional variance is given by $Var(y_{ij}|\boldsymbol{b}_i) = \sigma^2 f(\boldsymbol{x}_{ij}', \boldsymbol{\beta}_i)^{2\theta}$. The advantage of (7.17) is that we need not worry about individuals whose predicted GFR profile increases over time since the parameterization forces $\beta_{i1} > 0$ and $\beta_{i2} > 0$.

We first fit a joint model to the GFR and survival data assuming patient survival depends on the covariate vector $\boldsymbol{x}_i' = (a_{1i}, a_{2i}, a_{3i}, a_{4i}, a_{5i}, a_{6i}, y_{i1})$ consisting of the treatment group indicator variable, a_{1i}, as well as the baseline covariates of gender (a_{2i}), age (a_{3i}), diabetes (a_{4i}), albumin (a_{5i}) and nPNA (a_{6i}). We included the baseline starting GFR value y_{i1} as a possible predictor of mortality as well. We fit this joint model assuming both the additive error structure of model (7.16) as well as the multiplicative error structure of model (7.17) for the GFR profiles. The two *joint* models are labeled as Model 1 and 2, respectively. Shown below is the SAS code used to construct the SAS dataset, `example7_3_1_SP`, needed to jointly fit a NLME model for GFR and a piecewise exponential (PE) model for survival. Also included is the code needed to fit a Cox proportional hazards model to the survival data using PHREG. The SAS variable, `ITTtime`, represents the observed survival time $T_i^* = \min(T_i, C_i)$ for the ith patient where T_i is the patient's time to death and C_i is the patient's censoring time, while the SAS variable, `ITTdeath`, represents the censoring indicator variable $\delta_i = \begin{cases}1 & \text{if } T_i \le C_i\\ 0 & \text{if } T_i > C_i\end{cases}$ (see author's Web page for a complete description of all the variables).

Program 7.5

```
proc format;
 value trtfmt
 0='Control' 1='Treated';
run;
proc sort data=SASdata.ADEMEX_GFR_Data
          out=example7_3_1;
 by ptid Months;
run;
data example7_3_1;
 set example7_3_1;
 by ptid Months;
 Indicator=0;
 t1=0;t2=0;
 Risktime=1;
 Response=GFR_ml_min;
run;

/*==========================================================
 * The following programming statements create intervals
 * for use with piecewise exponential survival analysis in
 * combination with analysis of the ADEMEX GFR data
 * Key      t1 = Defines starting time point for the different
                 intervals being created
            t2 = Defines ending time point for the different
                 intervals being created
         Event = Defines a 0-1 indicator which is 0 in each
                 interval until the interval in which the
                 actual survival time occurs and then the value
                 of Event equals the value of ITTdeath
      Risktime = Defines the time at risk within each interval
==========================================================*/
data survival;
 set example7_3_1;
 by ptid Months;
 if last.ptid;
run;
data survival1;
 set survival;
 do t1=0 to 24 by 6;
    if t1< 24 then t2 = t1+6;
    if t1=24 then t2 = 36;
    Event=0;
    Risktime=t2-t1;
    if t1<ITTtime<=t2 then do;
       Event=ITTdeath;
       Risktime=ITTtime - t1;
    end;
    output;
 end;
run;
data survival1;
 set survival1;
 Interval=t1;
 if t1>ITTtime then delete;
```

```
  Response=Event;
  Indicator=1;
run;
proc sort data=survival1;
 by ptid t1 t2;
run;
/* Dataset used for fitting all SP models */
data example7_3_1_SP;
 set example7_3_1 survival1;
run;
proc sort data=example7_3_1_SP;
 by ptid Indicator Months;
run;
proc print data=example7_3_1_SP heading=h rows=page noobs;
 where ptid in ('01 001' '01 028');
 id ptid;
 var Indicator Trt Age Sex Diabetic Albumin0 nPNA0 GFR0
     Months GFR_ml_min ITTdeath ITTtime
     t1 t2 risktime event response;
run;
proc phreg data=example7_3_1_SP;
 where t1=0 and indicator=1;
 model ITTtime*ITTdeath(0) = Trt Age Sex Diabetic
                             Albumin0 nPNA0 GFR0;
 hazardratio Albumin0 / units=0.1;
 hazardratio nPNA0 / units=0.1;
 hazardratio Age / units=10;
run;
```

As we did for the MDRD example (§6.6.2), we included a number of comments to guide one through the dataset structure. Observe that we need to artificially define values of the variables `t1` and `Risktime` within the original dataset `example7_3_1` along with the creation of the variable `Indicator` which is assigned a value of 0 to indicate that the response variable of interest is GFR. The remaining code is similar to that created for the MDRD example. Shown in Output 7.12 is a partial listing of data from the newly created dataset, `example7_3_1_SP`, as well as the results from fitting the Cox proportional hazards (PH) model directly to the survival data. Note that there is a reduction in the number of patients from 270 to 256 as a result of missing baseline values for albumin and/or nPNA.

Output 7.12: Partial listing of the dataset, example7_3_1_SP, required to fit the NLME model to the GFR data and piecewise exponential (PE) model to the survival data jointly. Also shown for comparative purposes are the results of fitting a Cox proportional hazards (PH) model to the patient survival data using the same covariate specifications as used for the piecewise exponential model.

ptid	Indicator	Trt	Age	Sex	Diabetic	Albumin0	nPNA0	GFR0	Months
01 001	0	1	34	1	0	2.96	0.85084	3.3198	0.0000
01 001	1	1	34	1	0	2.96	0.85084	3.3198	0.0000
01 001	1	1	34	1	0	2.96	0.85084	3.3198	0.0000
01 001	1	1	34	1	0	2.96	0.85084	3.3198	0.0000
01 001	1	1	34	1	0	2.96	0.85084	3.3198	0.0000
01 001	1	1	34	1	0	2.96	0.85084	3.3198	0.0000
01 028	0	0	54	0	0	3.48	.	5.1160	0.0000
01 028	0	0	54	0	0	3.48	.	5.1160	5.0000
01 028	0	0	54	0	0	3.48	.	5.1160	8.9145
01 028	0	0	54	0	0	3.48	.	5.1160	13.3224

01 028	0	0	54	0	0	3.48	.	5.1160	16.9737
01 028	0	0	54	0	0	3.48	.	5.1160	20.6908
01 028	0	0	54	0	0	3.48	.	5.1160	25.7895
01 028	1	0	54	0	0	3.48	.	5.1160	25.7895
01 028	1	0	54	0	0	3.48	.	5.1160	25.7895
01 028	1	0	54	0	0	3.48	.	5.1160	25.7895
01 028	1	0	54	0	0	3.48	.	5.1160	25.7895
01 028	1	0	54	0	0	3.48	.	5.1160	25.7895

ptid	GFR_ml_min	ITTdeath	ITTtime	t1	t2	Risktime	Event	Response
01 001	3.3198	0	28.5197	0	0	1.00000	.	3.3198
01 001	3.3198	0	28.5197	0	6	6.00000	0	0.0000
01 001	3.3198	0	28.5197	6	12	6.00000	0	0.0000
01 001	3.3198	0	28.5197	12	18	6.00000	0	0.0000
01 001	3.3198	0	28.5197	18	24	6.00000	0	0.0000
01 001	3.3198	0	28.5197	24	36	4.51974	0	0.0000
01 028	5.1160	0	32.3026	0	0	1.00000	.	5.1160
01 028	10.8301	0	32.3026	0	0	1.00000	.	10.8301
01 028	4.8665	0	32.3026	0	0	1.00000	.	4.8665
01 028	6.3700	0	32.3026	0	0	1.00000	.	6.3700
01 028	2.8615	0	32.3026	0	0	1.00000	.	2.8615
01 028	1.0970	0	32.3026	0	0	1.00000	.	1.0970
01 028	0.8632	0	32.3026	0	0	1.00000	.	0.8632
01 028	0.8632	0	32.3026	0	6	6.00000	0	0.0000
01 028	0.8632	0	32.3026	6	12	6.00000	0	0.0000
01 028	0.8632	0	32.3026	12	18	6.00000	0	0.0000
01 028	0.8632	0	32.3026	18	24	6.00000	0	0.0000
01 028	0.8632	0	32.3026	24	36	8.30263	0	0.0000

Model Information

Data Set	WORK.EXAMPLE7_3_1_SP
Dependent Variable	ITTtime
Censoring Variable	ITTdeath
Censoring Value(s)	0
Ties Handling	BRESLOW

Number of Observations Read 270 Number of Observations Used 256

Summary of the Number of Event and Censored Values

Total	Event	Censored	Percent Censored
256	84	172	67.19

Convergence Status

Convergence criterion (GCONV=1E−8) satisfied.

Model Fit Statistics

Criterion	Without Covariates	With Covariates
-2 LOG L	881.224	814.376
AIC	881.224	828.376
SBC	881.224	845.391

Testing Global Null Hypothesis: BETA=0

Test	Chi-Square	DF	Pr > ChiSq
Likelihood Ratio	66.8480	7	<.0001
Score	58.4048	7	<.0001
Wald	55.6708	7	<.0001

Analysis of Maximum Likelihood Estimates

Parameter	DF	Parameter Estimate	Standard Error	Chi-Square	Pr > ChiSq	Hazard Ratio
Trt	1	−0.37118	0.22651	2.6854	0.1013	0.690
Age	1	0.02596	0.01260	4.2430	0.0394	1.026
Sex	1	0.17490	0.23607	0.5489	0.4588	1.191
Diabetic	1	0.71238	0.29665	5.7667	0.0163	2.039
Albumin0	1	−0.69082	0.17810	15.0446	0.0001	0.501
nPNA0	1	−1.98956	0.57632	11.9177	0.0006	0.137
GFR0	1	0.04395	0.04338	1.0261	0.3111	1.045

Hazard Ratios for Albumin0

Description	Point Estimate	95% Wald Confidence Limits	
Albumin0 Unit=0.1	0.933	0.901	0.966

Hazard Ratios for nPNA0

Description	Point Estimate	95% Wald Confidence Limits	
nPNA0 Unit=0.1	0.820	0.732	0.918

Hazard Ratios for Age

Description	Point Estimate	95% Wald Confidence Limits	
Age Unit=10	1.296	1.013	1.660

We then fit the joint model (Model 1) for GFR and survival assuming the additive error structure (7.16) for GFR. We did so using the Laplace approximation as implemented with the following NLMIXED code.

Program 7.6

```
/* Model 1 - Additive random effects and error */
ods output parameterestimates=SAStemp.pe1_1;
ods output fitstatistics=SAStemp.fit1_1;
proc nlmixed data=example7_3_1_SP start maxiter=500 empirical
             technique=quanew qpoints=1;
 parms beta11=.1 1 3 beta12=0
       beta21=-1 -3 0.1 beta22=0
       eta0=-5 eta1=0 eta2=0 eta3=0 eta4=0
       eta_TRT=0 eta_Diabetic=0 eta_Age=0 eta_Sex=0
       eta_GFR0=0 eta_nPNA0=0 eta_Albumin0=0
       psi11=.001 .01 1 10 100
       psi21=0
       psi22=.001 .01 1 10 100
       sigsq=.001 .01 1 10 100 / best=1;
 Model=1;
 ** Re-scale beta2i to be in units of (1/yr) **;
 ** so that the intercept parameter beta21 & **;
 ** psi22 are on a scale similar to beta11   **;
 beta1i = (beta11 + beta12*Trt + bi1);
 beta2i = (beta21 + beta22*Trt + bi2)/12;
 MeanGFR  = beta1i*exp(-beta2i*months);
 VarGFR = sigsq;
 ** Define the log-hazard rate for a PE model **;
 eta_i = eta0 + eta1*(t1=6) + eta2*(t1=12) +
         eta3*(t1=18) + eta4*(t1=24) + eta_TRT*Trt +
         eta_Diabetic*Diabetic + eta_Age*Age +
         eta_Sex*Sex + eta_nPNA0*nPNA0 +
         eta_Albumin0*Albumin0 + eta_GFR0*GFR0 ;
 ** Lambda_i is the PE hazard rate per unit time **;
 Lambda_i = exp(eta_i);
 ** Hazard_i defines the PE hazard function **;
 Hazard_i = Lambda_i*risktime;
 ** ll_y is the conditional log-likelihood of y=GFR **;
 ll_y = (1-Indicator)*
        (- 0.5*log(2*CONSTANT('PI'))
         - 0.5*((response - MeanGFR)**2)/(VarGFR)
         - 0.5*log(VarGFR));
 **************************************************
 ** ll_T is the conditional log-likelihood of T
 where T is defined as T = Risktime which is the
 amount of time at risk within each interval (see above
 code defining the dataset example7_3_1_SP)
 for the piecewise exponential survival model
 **************************************************;
 ll_T = Indicator*( response*(eta_i) - Lambda_i*risktime );
 ** ll is the joint log-likelihood of (y,T) **;
 ll = ll_y + ll_T;
 model response ~ general(ll);
```

```
random bi1 bi2 ~ normal([0,0],[psi11,
                                psi21, psi22])
                 subject=ptid;
 id Model MeanGFR VarGFR sigsq beta11 beta12 beta21 beta22
    psi11 psi22 beta1i beta2i bi1 bi2
    eta_i Lambda_i Hazard_i ll_y ll_T;
 predict MeanGFR out=SASoutA.nlmix1_1;
run;
```

In fitting the model using NLMIXED, we re-scaled the subject-specific (SS) decay rate parameter β_{i2} (denoted as **beta2i** in the code) by dividing it by 12 so that the SS decay rates are expressed in terms of the inverse unit of time, years^{-1}, rather than the inverse unit of time, months^{-1}. This was done to improve the numerical stability of the algorithm particularly as it pertains to the estimation of the intercept parameter β_{21} and variance component ψ_{22}. Output 7.13 displays the parameter estimates and their robust standard errors (as per the EMPIRICAL option) when we jointly fit the NLME model (7.16) for GFR and the piecewise exponential model (7.15) with $\boldsymbol{x}_i' = (a_{1i}, a_{2i}, a_{3i}, a_{4i}, a_{5i}, a_{6i}, y_{i1})$. Observe that, when compared with the Cox PH model (Output 7.12), the piecewise exponential PH model yields very similar estimates of the log-hazard ratios (see also Output 7.17).

Output 7.13: Select output when we jointly fit the NLME model (7.16) to the GFR data and the piecewise exponential model (7.15) with $\boldsymbol{x}_i' = (a_{1i}, a_{2i}, a_{3i}, a_{4i}, a_{5i}, a_{6i}, y_{i1})$ to the survival data.

Dimensions

Observations Used	2008
Observations Not Used	73
Total Observations	2081
Subjects	256
Max Obs Per Subject	13
Parameters	20
Quadrature Points	1

Fit Statistics

−2 Log Likelihood	4615.0
AIC (smaller is better)	4655.0
AICC (smaller is better)	4655.4
BIC (smaller is better)	4725.9

Parameter Estimates

Parameter	Estimate	Standard Error	DF	t Value	Pr > |t|
beta11	3.5742	0.2213	254	16.15	<.0001
beta12	−0.2048	0.2924	254	−0.70	0.4842
beta21	1.1320	0.1225	254	9.24	<.0001
beta22	0.2279	0.1555	254	1.47	0.1440
eta0	−3.6286	1.0179	254	−3.56	0.0004
eta1	0.8533	0.3590	254	2.38	0.0182
eta2	0.8233	0.3784	254	2.18	0.0305
eta3	1.3694	0.3560	254	3.85	0.0002
eta4	0.5211	0.5015	254	1.04	0.2998
eta_TRT	−0.3718	0.2378	254	−1.56	0.1192
eta_Diabetic	0.7289	0.3131	254	2.33	0.0207
eta_Age	0.02787	0.01406	254	1.98	0.0486
eta_Sex	0.1823	0.2441	254	0.75	0.4560
eta_GFR0	0.03965	0.03775	254	1.05	0.2945
eta_nPNA0	−1.8742	0.5859	254	−3.20	0.0016
eta_Albumin0	−0.6778	0.1871	254	−3.62	0.0004

psi11	4.2187	0.7182	254	5.87	<.0001
psi21	−0.2422	0.1970	254	−1.23	0.2200
psi22	0.4535	0.1143	254	3.97	<.0001
sigsq	1.3787	0.2929	254	4.71	<.0001

Parameter	Alpha	Lower	Upper	Gradient
beta11	0.05	3.1385	4.0100	0.041328
beta12	0.05	−0.7807	0.3710	−0.06758
beta21	0.05	0.8907	1.3733	0.009166
beta22	0.05	−0.07836	0.5342	0.005525
eta0	0.05	−5.6332	−1.6239	−0.21902
eta1	0.05	0.1462	1.5603	0.184343
eta2	0.05	0.07800	1.5685	0.021904
eta3	0.05	0.6682	2.0705	0.207417
eta4	0.05	−0.4666	1.5087	0.051183
eta_TRT	0.05	−0.8402	0.09656	−0.04199
eta_Diabetic	0.05	0.1123	1.3455	−0.11745
eta_Age	0.05	0.000170	0.05556	1.342918
eta_Sex	0.05	−0.2985	0.6630	−0.0865
eta_GFR0	0.05	−0.03468	0.1140	0.092324
eta_nPNA0	0.05	−3.0279	−0.7204	0.298
eta_Albumin0	0.05	−1.0463	−0.3093	0.05794
psi11	0.05	2.8042	5.6332	−0.07741
psi21	0.05	−0.6301	0.1457	0.024892
psi22	0.05	0.2285	0.6786	0.008535
sigsq	0.05	0.8019	1.9555	0.00573

Based on this joint model, there is no evidence of an independent treatment effect on either the rate of decline in GFR nor on patient mortality rates. Note that the mean profiles for the control patients displayed in Figure 1.6 are based on output from this joint model. The PA mean response profile shown in Figure 1.6 is obtained by plotting the mean response $\widehat{y}(t) = n^{-1}\Big(\sum_{i=1}^{n} \widehat{\beta}_{i1} \exp(-\widehat{\beta}_{i2}t)\Big)$ versus time where $\widehat{\beta}_{i1} = \widehat{\beta}_{11} + \widehat{b}_{i1}$ and $\widehat{\beta}_{i2} = \widehat{\beta}_{21} + \widehat{b}_{i2}$. The SS mean response is the plot of $\widehat{y}(t) = \widehat{\beta}_{11} \exp(-\widehat{\beta}_{21}t)$ versus time which is nothing more than the predicted response curve for the typical patient having $\boldsymbol{b}_i = \mathbf{0}$. The code for generating Figure 1.6 is not shown here but is available at the author's Web page.

We next fit the joint model (Model 2) for GFR and survival assuming the multiplicative error structure (7.17) for GFR. The NLMIXED code required to fit this joint model is presented below. Included are additional ESTIMATE statements for estimating and comparing population-averaged (PA) versus typical subject-specific (SS) intercepts and rates of decline in GFR. Specifically, under the NLME model (7.17), the intercept and decay rate parameters, β_{i1} and β_{i2}, are expressed as nonlinear functions of the random effects, b_{i1} and b_{i2}, respectively. Under normality assumptions, we can estimate a PA intercept and PA decay rate for the control and treated patients based on

$$\begin{aligned} E(\beta_{i1}|a_{1i}) &= E_{b_1}\{\exp(\beta_{11} + \beta_{12}a_{1i} + b_{i1})\} \\ &= \exp\Big(\beta_{11} + \beta_{12}a_{1i} + \frac{1}{2}\psi_{11}\Big) \\ E(\beta_{i2}|a_{1i}) &= E_{b_2}\{\exp(\beta_{21} + \beta_{22}a_{1i} + b_{i2})\} \\ &= \exp\Big(\beta_{21} + \beta_{22}a_{1i} + \frac{1}{2}\psi_{22}\Big) \end{aligned} \tag{7.18}$$

which differ from the typical SS intercept and decay rate estimates based on

$$\begin{aligned} (\beta_{i1}|a_{1i}, b_{i1} = 0) &= \exp(\beta_{11} + \beta_{12}a_{1i}) \\ (\beta_{i2}|a_{1i}, b_{i2} = 0) &= \exp(\beta_{21} + \beta_{22}a_{1i}). \end{aligned} \tag{7.19}$$

Of course the estimates based on (7.18) and (7.19) will be similar whenever the offset variance components, $\frac{1}{2}\psi_{11}$ and $\frac{1}{2}\psi_{22}$, are small (i.e., approach 0).

Program 7.7

```
ods output parameterestimates=SAStemp.pe2_1;
ods output AdditionalEstimates=SAStemp.ae2_1;
ods output fitstatistics=SAStemp.fit2_1;
proc nlmixed data=example7_3_1_SP start maxiter=500 empirical
             technique=newrap qpoints=1; ** Requires NEWRAP **;
 parms beta11=.1 1 3 beta12=0
       beta21=-1 -3 0.1 beta22=0
       eta0=-5 eta1=0 eta2=0 eta3=0 eta4=0
       eta_TRT=0 eta_Diabetic=0 eta_Age=0 eta_Sex=0
       eta_GFR0=0 eta_nPNA0=0 eta_Albumin0=0
       theta=.001 .01 1 2
       psi11=.001 .01 1 10 100
       psi21=0
       psi22=.001 .01 1 10 100
       sigsq=.001 .01 1 10 100 / best=1;
 Model=2;
 ** Ensures beta1i> 0 and beta2i> 00 **;
 beta1i = exp(beta11 + beta12*Trt + bi1);
 beta2i = exp(beta21 + beta22*Trt + bi2);
 MeanGFR  = beta1i*exp(-beta2i*months);
 VarGFR = sigsq*(MeanGFR**(2*theta));
 ** Define the log-hazard rate for a PE model **;
 eta_i = eta0 + eta1*(t1=6) + eta2*(t1=12) +
         eta3*(t1=18) + eta4*(t1=24) + eta_TRT*Trt +
         eta_Diabetic*Diabetic + eta_Age*Age +
         eta_Sex*Sex + eta_nPNA0*nPNA0 +
         eta_Albumin0*Albumin0 + eta_GFR0*GFR0 ;
 ** Lambda_i is the PE hazard rate per unit time **;
 Lambda_i = exp(eta_i);
 ** Hazard_i defines the PE hazard function **;
 Hazard_i = Lambda_i*risktime;
 ** ll_y is the conditional log-likelihood of y=GFR **;
 ll_y = (1-Indicator)*
        (- 0.5*log(2*CONSTANT('PI'))
         - 0.5*((response - MeanGFR)**2)/(VarGFR)
         - 0.5*log(VarGFR));
 ** ll_T is the log-likelihood function for T **;
 ll_T = Indicator*( response*(eta_i) - Lambda_i*risktime );
 ** ll is the joint log-likelihood of (y,T) **;
 ll = ll_y + ll_T;
 model response ~ general(ll);
 random bi1 bi2 ~ normal([0,0],[psi11,
                                 psi21, psi22])
                  subject=ptid;
 estimate ''PA Intercept(Control):''
          exp(beta11 + .5*psi11);
 estimate ''PA Intercept(Treated):''
          exp(beta11 + beta12 + .5*psi11);
 estimate ''Diff. in PA Intercepts:''
          exp(beta11 + .5*psi11) -
          exp(beta11 + beta12 + .5*psi11);
```

```
 estimate ''SS Intercept(Control):''
          exp(beta11);
 estimate ''SS Intercept(Treated):''
          exp(beta11 + beta12);
 estimate ''Diff. in SS Intercepts:''
          exp(beta11) - exp(beta11 + beta12);
 estimate ''PA Decay Rate(Control):''
          exp(beta21 + .5*psi22);
 estimate ''PA Decay Rate(Treated):''
          exp(beta21 + beta22 + .5*psi22);
 estimate ''Diff. in PA Decay Rates:''
          exp(beta21 + .5*psi22) -
          exp(beta21 + beta22 + .5*psi22);
 estimate ''SS Decay Rate(Control):''
          exp(beta21);
 estimate ''SS Decay Rate(Treated):''
          exp(beta21 + beta22);
 estimate ''Diff. in SS Decay Rates:''
          exp(beta21) - exp(beta21 + beta22);
 id Model MeanGFR VarGFR sigsq beta11 beta12 beta21 beta22
    theta psi11 psi22 beta1i beta2i bi1 bi2
    eta_i Lambda_i Hazard_i ll_y ll_T;
 predict MeanGFR out=SAStemp.nlmix2_1;
run;
```

To achieve convergence using the Laplace approximation, we had to specify the Newton-Raphson optimization technique (TECHNIQUE = NEWRAP) since the default quasi-Newton optimization technique failed to achieve convergence.

The results from fitting this joint model are shown in Output 7.14. There is again no evidence of a treatment-related effect on the rate of decline in GFR nor on patient survival. Note, however, that because of the difference in optimization techniques used, the parameter estimates associated with the PE survival model do differ slightly from those shown in Output 7.13 despite using the same PE model. Judging from the information criteria for the two models, it appears that an exponential decay model with multiplicative errors provides a better fit to the GFR data than a model with additive errors.

Output 7.14: Select output when we jointly fit the NLME model (7.17) to the GFR data and the piecewise exponential model (7.15) with $x_i' = (a_{1i}, a_{2i}, a_{3i}, a_{4i}, a_{5i}, a_{6i}, y_{i1})$ to the survival data. Included are the PA (7.18) and SS (7.19) estimates of the GFR intercept and decay rate parameters.

```
            Dimensions

Observations Used          2008
Observations Not Used        73
Total Observations         2081
Subjects                    256
Max Obs Per Subject          13
Parameters                   21
Quadrature Points             1
            Fit Statistics

-2 Log Likelihood           3648.9
AIC (smaller is better)     3690.9
AICC (smaller is better)    3691.4
BIC (smaller is better)     3765.4
```

Parameter Estimates

Parameter	Estimate	Standard Error	DF	t Value	Pr > |t|
beta11	1.1021	0.06282	254	17.54	<.0001
beta12	−0.05832	0.08526	254	−0.68	0.4946
beta21	−2.6393	0.1234	254	−21.38	<.0001
beta22	0.1649	0.1506	254	1.10	0.2744
eta0	−3.3088	1.0101	254	−3.28	0.0012
eta1	0.7923	0.3489	254	2.27	0.0240
eta2	0.7721	0.3681	254	2.10	0.0369
eta3	1.3148	0.3458	254	3.80	0.0002
eta4	0.4670	0.4967	254	0.94	0.3480
eta_TRT	−0.3770	0.2376	254	−1.59	0.1139
eta_Diabetic	0.7404	0.3134	254	2.36	0.0189
eta_Age	0.02536	0.01389	254	1.83	0.0690
eta_Sex	0.1610	0.2444	254	0.66	0.5106
eta_GFR0	0.04293	0.03758	254	1.14	0.2544
eta_nPNA0	−2.0149	0.5932	254	−3.40	0.0008
eta_Albumin0	−0.6908	0.1860	254	−3.71	0.0003
theta	0.9285	0.03362	254	27.62	<.0001
psi11	0.3569	0.03478	254	10.26	<.0001
psi21	−0.01895	0.05222	254	−0.36	0.7170
psi22	0.6264	0.07365	254	8.51	<.0001
sigsq	0.1882	0.01171	254	16.08	<.0001

Parameter	Alpha	Lower	Upper	Gradient
beta11	0.05	0.9784	1.2258	4.39E−10
beta12	0.05	−0.2262	0.1096	−218E−12
beta21	0.05	−2.8824	−2.3962	1.59E−10
beta22	0.05	−0.1316	0.4615	−1.33E−9
eta0	0.05	−5.2981	−1.3196	−9.41E−6
eta1	0.05	0.1052	1.4794	9.156E−7
eta2	0.05	0.04721	1.4969	8.668E−7
eta3	0.05	0.6338	1.9957	7.598E−7
eta4	0.05	−0.5112	1.4453	7.826E−7
eta_TRT	0.05	−0.8449	0.09101	2.502E−7
eta_Diabetic	0.05	0.1233	1.3575	2.108E−7
eta_Age	0.05	−0.00199	0.05271	7.65E−6
eta_Sex	0.05	−0.3203	0.6424	9.441E−7
eta_GFR0	0.05	−0.03108	0.1169	4.271E−7
eta_nPNA0	0.05	−3.1831	−0.8467	2.894E−6
eta_Albumin0	0.05	−1.0572	−0.3245	1.031E−6
theta	0.05	0.8623	0.9947	5.88E−10
psi11	0.05	0.2884	0.4254	−1.05E−9
psi21	0.05	−0.1218	0.08388	−1.69E−9
psi22	0.05	0.4813	0.7714	−1.39E−9
sigsq	0.05	0.1652	0.2113	−4.36E−9

Additional Estimates

Label	Estimate	Standard Error	DF	t Value	Pr > |t|
PA Intercept(Control):	3.5985	0.2405	254	14.96	<.0001
PA Intercept(Treated):	3.3946	0.2005	254	16.93	<.0001
PA Intercept(Diff):	0.2039	0.2994	254	0.68	0.4966
SS Intercept(Control):	3.0104	0.1891	254	15.92	<.0001
SS Intercept(Treated):	2.8399	0.1592	254	17.83	<.0001
SS Intercept(Diff):	0.1706	0.2503	254	0.68	0.4962
PA Decay Rate(Control):	0.09768	0.01072	254	9.11	<.0001
PA Decay Rate(Treated):	0.1152	0.01134	254	10.16	<.0001
PA Decay Rate(Diff):	−0.01752	0.01594	254	−1.10	0.2729
SS Decay Rate(Control):	0.07141	0.008815	254	8.10	<.0001
SS Decay Rate(Treated):	0.08422	0.008408	254	10.02	<.0001
SS Decay Rate(Diff):	−0.01281	0.01154	254	−1.11	0.2682

Label	Alpha	Lower	Upper
PA Intercept(Control):	0.05	3.1249	4.0722
PA Intercept(Treated):	0.05	2.9999	3.7894
PA Intercept(Diff):	0.05	−0.3858	0.7936
SS Intercept(Control):	0.05	2.6380	3.3828
SS Intercept(Treated):	0.05	2.5263	3.1534
SS Intercept(Diff):	0.05	−0.3224	0.6635

PA Decay Rate(Control):	0.05	0.07657	0.1188
PA Decay Rate(Treated):	0.05	0.09285	0.1375
PA Decay Rate(Diff):	0.05	−0.04891	0.01388
SS Decay Rate(Control):	0.05	0.05405	0.08877
SS Decay Rate(Treated):	0.05	0.06766	0.1008
SS Decay Rate(Diff):	0.05	−0.03553	0.009922

For comparative purposes, Figure 7.4 plots the patient-specific GFR profiles for the control patients along with their PA mean response curve and two SS mean response curves based on the NLME model (7.17) with multiplicative errors. The SS mean response curve for the typical patient (labeled typ.) corresponds to the predicted GFR profile of an individual with the typical SS intercept and decay rate parameters (7.19) while the SS mean response curve on average (labeled avg.) corresponds to the predicted GFR profile of an individual with the PA intercept and decay rate parameters (7.18). The PA mean response is obtained the same way as described for Figure 1.6 but with $\widehat{y}(t) = n^{-1}\left(\sum_{i=1}^{n} \widehat{\beta}_{i1} \exp(-\widehat{\beta}_{i2}t)\right)$ based on $\widehat{\beta}_{i1} = \exp(\widehat{\beta}_{11} + \widehat{b}_{i1})$ and $\widehat{\beta}_{i2} = \exp(\widehat{\beta}_{21} + \widehat{b}_{i2})$. When compared with Figure 1.6, we see that the PA mean response assuming multiplicative errors continues to trend downward as a result of having individual intercepts and decay rates that are always positive (i.e., $\widehat{\beta}_{i2} > 0$ so that $\exp(-\widehat{\beta}_{i2}t)$ is always decreasing).

Given the degree of variability in the individual profiles, we elected to also plot the average GFR values $\pm$ 1 SD by month of follow-up and superimpose the PA and SS predicted mean response curves on top of the observed means. We did this for both the additive errors model (Model 1) and multiplicative errors model (Model 2) as depicted in Figure 7.5. The SS mean response curve is that of the typical patient (i.e., $\boldsymbol{b}_i = \boldsymbol{0}$). Interestingly, the PA mean response for the NLME model with multiplicative errors

Figure 7.4 SS and PA mean GFR profiles among control patients randomized to the standard dose of dialysis. The mean response profiles are based on the negative exponential decay model (7.17) with multiplicative errors. The SS mean response(avg.) is the mean response based on the PA intercept and decay rate in (7.18) while the SS mean response(typ.) is the mean response based on the typical SS intercept and decay rate in (7.19).

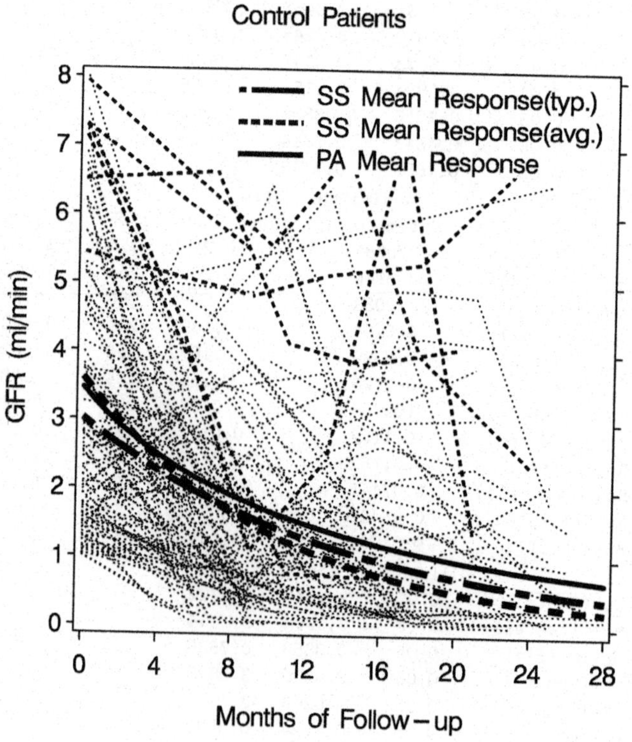

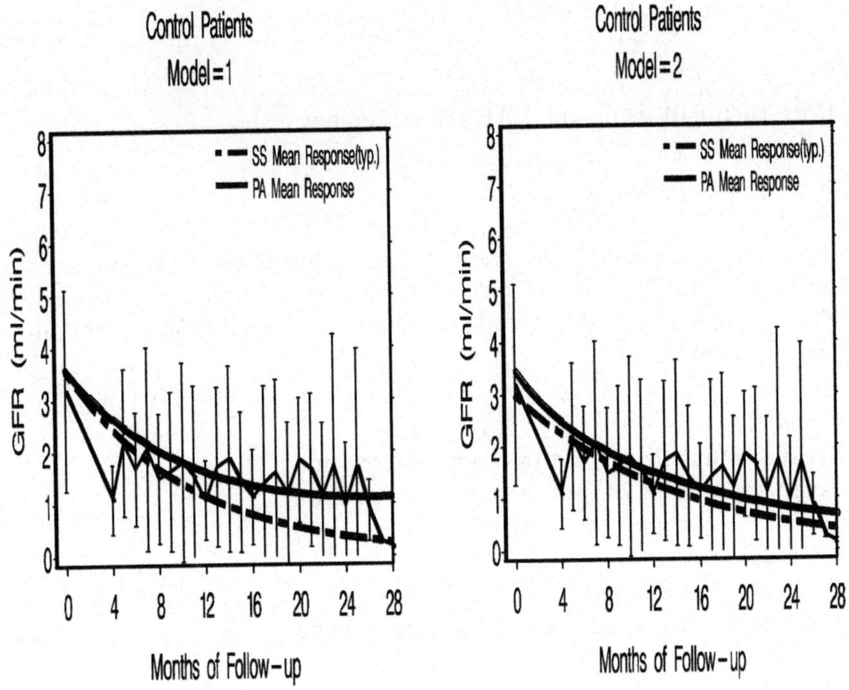

Figure 7.5 Average individual (SS) and marginal (PA) mean response curves under a negative exponential decay model with additive errors (Model 1) versus multiplicative errors (Model 2). The average GFR (+/- 1 SD) by month of follow-up is shown for comparative purposes.

appears to provide a closer fit to the observed means during the first 12 months but a worse fit thereafter compared to the NLME model with additive errors.

Lastly, we fit a joint shared parameter model to the GFR and survival data by including the random effects, b_{i1} and b_{i2}, as covariates under the piecewise exponential hazard function (7.15). We did this for both the exponential decay model with additive errors (model 7.16) which we label as Model 3 and the exponential decay model with multiplicative errors (model 7.17) which we label as Model 4. We can simply modify the previous NLMIXED syntax by including the two random effects in the hazard function via the following assignment statement:

```
** Define the log-hazard rate for a PE model **;
eta_i = eta0 + eta1*(t1=6) + eta2*(t1=12) +
        eta3*(t1=18) + eta4*(t1=24) + eta_TRT*Trt +
        eta_Diabetic*Diabetic + eta_Age*Age +
        eta_Sex*Sex + eta_nPNA0*nPNA0 +
        eta_Albumin0*Albumin0 +
        eta_bi1*bi1 + eta_bi2*bi2;
```

We used the parameter estimates stored in the ODS OUTPUT datasets `SAStemp.pe1_1` and `SAStemp.pe2_1` from the above calls to NLMIXED as starting values but modified to 1) exclude η_{GFR0} as a parameter under the piecewise exponential model and 2) include starting values for η_{b1} and η_{b2}. For example, under the SP model assuming additive errors, we create a dataset, INITIAL, containing starting values using the following code

```
data peNEW;
 Parameter='eta_bi1';estimate=0;output;
 Parameter='eta_bi2';estimate=0;output;
```

```
run;
data initial;
 set SAStemp.pe1_1 peNEW;
 if Parameter in ('eta_GFR0') then delete;
run;
```

which we then call using the PARMS statement

```
parms / data=initial;
```

Similarly, under the SP model assuming multiplicative errors, we set the starting values for η_{b1} (eta_bi1) and η_{b2} (eta_bi2) to be 0.50 as determined by trial and error (the program based on initial estimates of 0 failed and we next tried 0.50 as starting values). The SAS code in this case is:

```
data peNEW;
 Parameter='eta_bi1';estimate=0.50;output;
 Parameter='eta_bi2';estimate=0.50;output;
run;
data initial;
 set SAStemp.pe2_1 peNEW;
 if Parameter in ('eta_GFR0') then delete;
run;
```

which we then call through the `parms /data=initial;` statement from within NLMIXED.

The likelihood-based fit statistics and ML estimates from fitting the two SP models are shown in Output 7.15 and Output 7.16, respectively. There was no treatment-related effect on either the starting GFR or on the rate of decline in GFR for either of the SP models. Under the SP model 3 with additive errors, the patient-specific random decay rate b_{i2} was found to be significantly associated with increased mortality having an estimated hazard ratio (HR) of 1.26 per 0.10 year^{-1} increase in the yearly decay rate (95% CI: 1.15, 1.38). The HR is computed as $\exp(2.3075 \times 0.10) = 1.26$ where 2.3075 is the estimated log HR of b_{i2} (i.e., eta_bi2 in Output 7.15) while the lower and upper 95% confidence limits are computed as $\exp(1.3655 \times 0.10) = 1.15$ and $\exp(3.2495 \times 0.10) = 1.38$ where 1.3655 and 3.2495 are the lower and upper 95% confidence limits of the log HR of b_{i2} (i.e., the values of Lower and Upper for parameter eta_bi2 in Output 7.15). However, there was no association between the patient-specific random intercept, b_{i1}, and survival as determined by an estimated HR of 1.10 (95% CI: 0.97, 1.25) per 0.50 ml/min increase in starting GFR.

Output 7.15: Select output when we fit a joint SP model to the data assuming a NLME model (7.16) with additive errors for the GFR data and a piecewise exponential model (7.15) with $\boldsymbol{x}_i' = (a_{1i}, a_{2i}, a_{3i}, a_{4i}, a_{5i}, a_{6i}, b_{i1}, b_{i2})$ for the survival data.

Fit Statistics

-2 Log Likelihood	4597.3
AIC (smaller is better)	4639.3
AICC (smaller is better)	4639.8
BIC (smaller is better)	4713.7

Parameter Estimates

Parameter	Estimate	Standard Error	DF	t Value	Pr > |t|
beta11	3.5735	0.2213	254	16.15	<.0001
beta12	-0.1894	0.2914	254	-0.65	0.5164
beta21	1.2493	0.1292	254	9.67	<.0001

beta22	0.1948	0.1537	254	1.27	0.2062
eta0	−3.4960	1.4635	254	−2.39	0.0176
eta1	1.3148	0.4211	254	3.12	0.0020
eta2	1.7746	0.5360	254	3.31	0.0011
eta3	2.6162	0.5526	254	4.73	<.0001
eta4	1.9606	0.7401	254	2.65	0.0086
eta_TRT	−0.4412	0.3336	254	−1.32	0.1872
eta_Diabetic	0.4806	0.4465	254	1.08	0.2828
eta_Age	0.03181	0.01674	254	1.90	0.0585
eta_Sex	0.4084	0.3591	254	1.14	0.2565
eta_nPNA0	−2.9006	0.8944	254	−3.24	0.0013
eta_Albumin0	−0.9047	0.3033	254	−2.98	0.0031
psi11	4.1951	0.7254	254	5.78	<.0001
psi21	−0.3085	0.1970	254	−1.57	0.1187
psi22	0.5290	0.1307	254	4.05	<.0001
sigsq	1.3995	0.2857	254	4.90	<.0001
eta_bi1	0.1903	0.1274	254	1.49	0.1363
eta_bi2	2.3075	0.4783	254	4.82	<.0001

Parameter	Alpha	Lower	Upper	Gradient
beta11	0.05	3.1376	4.0094	−0.01429
beta12	0.05	−0.7633	0.3846	0.013772
beta21	0.05	0.9948	1.5037	−0.01157
beta22	0.05	−0.1079	0.4976	−0.022
eta0	0.05	−6.3781	−0.6139	0.03743
eta1	0.05	0.4855	2.1442	0.055711
eta2	0.05	0.7190	2.8302	0.046005
eta3	0.05	1.5279	3.7045	0.065792
eta4	0.05	0.5032	3.4181	−0.03197
eta_TRT	0.05	−1.0983	0.2158	−0.11398
eta_Diabetic	0.05	−0.3988	1.3599	−0.04506
eta_Age	0.05	−0.00115	0.06478	0.178795
eta_Sex	0.05	−0.2988	1.1157	−0.08843
eta_nPNA0	0.05	−4.6620	−1.1391	0.114219
eta_Albumin0	0.05	−1.5019	−0.3075	−0.03562
psi11	0.05	2.7665	5.6238	−0.01807
psi21	0.05	−0.6965	0.07954	0.026742
psi22	0.05	0.2716	0.7863	0.007411
sigsq	0.05	0.8369	1.9622	0.006219
eta_bi1	0.05	−0.06047	0.4411	−0.01346
eta_bi2	0.05	1.3655	3.2495	0.004694

Conversely, when we fit the data to the SP model with a multiplicative error structure for GFR, we found patient survival was not significantly associated with either the log random intercept effect, b_{i1}, or the log random decay rate effect, b_{i2}, although the latter did lean toward significance ($p = 0.0852$, Output 7.16). The HR per 0.10 increase in the log random decay rate is $\exp(0.4153 \times 0.1) = 1.04$ (95% CI: 0.99, 1.09) while the HR per 0.10 increase in the log random intercept effect is 1.01 (95% CI: 0.96, 1.06).

Output 7.16: Select output when we fit a joint SP model to the data assuming a NLME model (7.17) with multiplicative errors for the GFR data and a piecewise exponential model (7.15) with $\boldsymbol{x}_i' = (a_{1i}, a_{2i}, a_{3i}, a_{4i}, a_{5i}, a_{6i}, b_{i1}, b_{i2})$ for the survival data.

Dimensions

Observations Used	2008
Observations Not Used	73
Total Observations	2081
Subjects	256
Max Obs Per Subject	13
Parameters	22
Quadrature Points	1

Fit Statistics

-2 Log Likelihood	3646.0
AIC (smaller is better)	3690.0
AICC (smaller is better)	3690.5
BIC (smaller is better)	3768.0

Parameter Estimates

Parameter	Estimate	Standard Error	DF	t Value	Pr > |t|
beta11	1.1081	0.06358	254	17.43	<.0001
beta12	-0.06033	0.08600	254	-0.70	0.4836
beta21	-2.5761	0.1162	254	-22.16	<.0001
beta22	0.1313	0.1475	254	0.89	0.3742
eta0	-3.4460	1.0497	254	-3.28	0.0012
eta1	0.8058	0.3635	254	2.22	0.0275
eta2	0.8126	0.3888	254	2.09	0.0376
eta3	1.3553	0.3654	254	3.71	0.0003
eta4	0.5084	0.5125	254	0.99	0.3221
eta_TRT	-0.4007	0.2467	254	-1.62	0.1055
eta_Diabetic	0.6891	0.3300	254	2.09	0.0378
eta_Age	0.02566	0.01429	254	1.80	0.0736
eta_Sex	0.2657	0.2632	254	1.01	0.3136
eta_nPNA0	-1.8944	0.5919	254	-3.20	0.0015
eta_Albumin0	-0.6525	0.2074	254	-3.15	0.0019
theta	0.9274	0.03047	254	30.43	<.0001
psi11	0.3548	0.03430	254	10.35	<.0001
psi21	-0.03249	0.04588	254	-0.71	0.4795
psi22	0.6558	0.1199	254	5.47	<.0001
sigsq	0.1861	0.01254	254	14.84	<.0001
eta_bi1	0.1096	0.2375	254	0.46	0.6450
eta_bi2	0.4153	0.2403	254	1.73	0.0852

Parameter	Alpha	Lower	Upper	Gradient
beta11	0.05	0.9829	1.2333	-5.17E-7
beta12	0.05	-0.2297	0.1090	-1.02E-6
beta21	0.05	-2.8050	-2.3472	2.55E-6
beta22	0.05	-0.1591	0.4217	-4.89E-7
eta0	0.05	-5.5133	-1.3788	-2.35E-7
eta1	0.05	0.08989	1.5218	8.867E-8
eta2	0.05	0.04679	1.5783	5.395E-8
eta3	0.05	0.6356	2.0750	6.751E-8
eta4	0.05	-0.5008	1.5176	1.984E-8
eta_TRT	0.05	-0.8864	0.08510	3.178E-7
eta_Diabetic	0.05	0.03922	1.3390	-3.22E-7
eta_Age	0.05	-0.00247	0.05380	-0.00002
eta_Sex	0.05	-0.2526	0.7839	-4.18E-7
eta_nPNA0	0.05	-3.0601	-0.7287	-3.32E-8
eta_Albumin0	0.05	-1.0610	-0.2441	1.714E-7
theta	0.05	0.8674	0.9874	-2.79E-6
psi11	0.05	0.2873	0.4224	-0.00001
psi21	0.05	-0.1229	0.05786	-0.00002
psi22	0.05	0.4197	0.8919	1.185E-7
sigsq	0.05	0.1614	0.2108	-0.00003
eta_bi1	0.05	-0.3582	0.5774	2.796E-7
eta_bi2	0.05	-0.05800	0.8886	3.115E-7

Additional Estimates

Label	Estimate	Standard Error	DF	t Value	Pr > |t|
PA Intercept(Control):	3.6164	0.2422	254	14.93	<.0001
PA Intercept(Treated):	3.4047	0.2008	254	16.96	<.0001
PA Intercept(Diff):	0.2117	0.3033	254	0.70	0.4858
SS Intercept(Control):	3.0285	0.1925	254	15.73	<.0001
SS Intercept(Treated):	2.8512	0.1600	254	17.81	<.0001
SS Intercept(Diff):	0.1773	0.2539	254	0.70	0.4855
PA Decay Rate(Control):	0.1056	0.01408	254	7.50	<.0001
PA Decay Rate(Treated):	0.1204	0.01225	254	9.83	<.0001
PA Decay Rate(Diff):	-0.01481	0.01634	254	-0.91	0.3657
SS Decay Rate(Control):	0.07607	0.008843	254	8.60	<.0001
SS Decay Rate(Treated):	0.08674	0.008632	254	10.05	<.0001
SS Decay Rate(Diff):	-0.01067	0.01190	254	-0.90	0.3709

```
Label                          Alpha      Lower      Upper
PA Intercept(Control):          0.05     3.1394     4.0935
PA Intercept(Treated):          0.05     3.0093     3.8001
PA Intercept(Diff):             0.05    -0.3856     0.8091
SS Intercept(Control):          0.05     2.6493     3.4077
SS Intercept(Treated):          0.05     2.5360     3.1664
SS Intercept(Diff):             0.05    -0.3226     0.6773
PA Decay Rate(Control):         0.05    0.07787     0.1333
PA Decay Rate(Treated):         0.05    0.09628     0.1445
PA Decay Rate(Diff):            0.05   -0.04700    0.01737
SS Decay Rate(Control):         0.05    0.05866    0.09349
SS Decay Rate(Treated):         0.05    0.06974     0.1037
SS Decay Rate(Diff):            0.05   -0.03411    0.01277
```

Of course it is difficult to directly compare estimated hazard ratios associated with the shared random-effect parameters of the two SP models given the difference in parameterization between the two models. However, one can easily perform a side-by-side comparison of the estimated hazard ratios that assess the relative risk of death due to treatment group, age, gender, diabetes, albumin and nPCR. This is most easily accomplished with a call to the macro %HR listed below.

Program 7.8

```
%macro HR(data=, unit_bi1=1, unit_bi2=1);
 ** &unit_bi1 and &unit_bi2 define the unit   **
 ** measure of increase for bi1 and bi2 with **
 ** which to compute the hazard ratio for    **
 ** the latent random effects bi1 and bi2    **;
 data hazards;
  set &data;
  if Parameter IN ('eta_TRT'  'eta_Diabetic'
                   'eta_GFR0' 'eta_Sex')
  then do;
    Units=1.0;
    HR=exp(estimate*Units);
    LCL=exp(lower*Units);
    UCL=exp(upper*Units);
  end;
  if Parameter IN ('eta_Age')
  then do;
    Units=10.0;
    HR=exp(estimate*Units);
    LCL=exp(lower*Units);
    UCL=exp(upper*Units);
  end;
  if Parameter IN ('eta_bi1') then do;
    Units=&unit_bi1;
    HR=exp(estimate*Units);
    LCL=exp(lower*Units);
    UCL=exp(upper*Units);
  end;
  if Parameter IN ('eta_bi2') then do;
    Units=&unit_bi2;
    HR=exp(estimate*Units);
    LCL=exp(lower*Units);
    UCL=exp(upper*Units);
  end;
```

```
  if Parameter IN ('eta_nPNA0' 'eta_Albumin0')
  then do;
    Units=0.1;
    HR=exp(estimate*Units);
    LCL=exp(lower*Units);
    UCL=exp(upper*Units);
  end;
  StdErr=StandardError;
 run;
 ods listing;
 proc print data=hazards noobs;
  where HR ne .;
  var Parameter Estimate StdErr Probt
      Lower Upper Units HR LCL UCL;
  format HR LCL UCL 5.2;
 run;
%mend HR;
```

To call this macro, one will need to specify a value for the macro keyword parameter, `data=` , which defines a dataset containing all relevant parameter estimates from a specific call to NLMIXED (i.e., the dataset as created by the ODS OUTPUT ParameterEstimates= statement). One may also supply values to the macro keyword parameters `unit_bi1=` and `unit_bi2=` each of which defines a unit of measure with which to express the hazard ratios associated with the random effects b_{i1} and b_{i2}, respectively. For example, consider the SP model with additive error structure for GFR and consider an individual whose starting GFR is 0.50 ml/min higher than the average individual and whose yearly decay rate is 0.10 year^{-1} higher than the average individual (both measured without error). Given all other factors are the same, the hazard ratios based on the piecewise exponential proportional hazards model (7.15) would be given by

$$\begin{aligned} \text{HR}(b_{i1}) &= \frac{\lambda_i(t, b_{i1} = 0.50)}{\lambda_i(t, b_{i1} = 0)} \\ &= \exp\{\eta_{b1} \times 0.50\}, \\ \text{HR}(b_{i2}) &= \frac{\lambda_i(t, b_{i2} = 0.10)}{\lambda_i(t, b_{i2} = 0)} \\ &= \exp\{\eta_{b2} \times 0.10\}. \end{aligned} \tag{7.20}$$

One would request these estimates using %HR by specifying `unit_bi1=0.50`, and `unit_bi2=0.10`. If, alternatively, one wished to estimate the hazard ratio for the latent random decay rate on the basis of the monthly decay rate, then one would specify `unit_bi2=0.1/12` so as to define the unit increase in terms of the monthly decay rate (i.e., a 0.1 year^{-1} increase in the yearly decay rate is equivalent to a 0.0083 month^{-1} increase in the monthly decay rate).

To calculate the estimated hazard ratios under the SP model with multiplicative errors, it may be easier to express the hazard ratios for b_{i1} and b_{i2} in terms of relative increases rather than absolute increases. For example, suppose we wish to estimate the hazard ratio for someone whose starting GFR is 20% higher than the typical patient and whose decay rate is 10% higher than the typical patient. In terms of the subject-specific starting GFR, β_{i1}, and decay rate, β_{i2}, this implies

$$\begin{aligned} \beta_{i1} &= 1.20 \times \exp(\beta_{11} + \beta_{12}a_{1i} + E_{b_{i1}}(b_{i1})) \\ &= 1.20 \times \exp(\beta_{11} + \beta_{12}a_{1i}) \\ &= \exp(\beta_{11} + \beta_{12}a_{1i} + \log(1.20)) \end{aligned} \tag{7.21}$$

$$\begin{aligned}\beta_{i2} &= 1.10 \times \exp(\beta_{21} + \beta_{22}a_{1i} + E_{b_{i2}}(b_{i2})) \\ &= 1.10 \times \exp(\beta_{21} + \beta_{22}a_{1i}) \\ &= \exp(\beta_{21} + \beta_{22}a_{1i} + \log(1.10)).\end{aligned} \tag{7.22}$$

These relative increases occur when the subject's latent random intercept is $b_{i1} = \log(1.20)$ higher than the typical patient's value of $b_{i1} = 0$ and when the subject's latent random decay rate is $b_{i2} = \log(1.10)$ higher than the typical patient's value of $b_{i2} = 0$. Thus the hazard ratios corresponding to (7.21) would be given by :

$$\begin{aligned}\text{HR}(b_{i1}) &= \frac{\lambda_i(t, b_{i1} = \log(1.20))}{\lambda_i(t, b_{i1} = 0)} \\ &= \exp\{\eta_{b1} \times \log(1.20)\}. \\ \text{HR}(b_{i2}) &= \frac{\lambda_i(t, b_{i2} = \log(1.10))}{\lambda_i(t, b_{i2} = 0)} \\ &= \exp\{\eta_{b2} \times \log(1.10)\}.\end{aligned} \tag{7.23}$$

One would then request these estimates by calling the macro %HR with the specifications: `unit_bi1=log(1.20)`, and `unit_bi2=log(1.10)`.

The following calls to the macro %HR compare the hazard ratios from the two SP models (Model 3, Output 7.15 and Model 4, Output 7.16) as well as the first joint model for GFR and survival (Model 1, Output 7.13) in which we assume an additive error structure for GFR. For the SP model with additive errors (Model 3, Output 7.15), we estimated the hazard ratios associated with b_{i1} and b_{i2} based on a unit increase of 0.50 ml/min in the latent random intercept b_{i1} and a unit increase of 0.10 year^{-1} = 0.0083 month^{-1} increase in the latent random decay rate b_{i2}. We then translate these into relative increases in β_{i1} and β_{i2} under the SP model with multiplicative errors (Model 4, Output 7.16) as follows. Compared to the typical control patient having a "true" (i.e., measured without error) starting GFR of $\beta_{i1} = \exp(1.1081) = 3.0286$ ml/min (Output 7.16), a 0.50 ml/min increase in β_{i1} is equivalent to a 16.5% increase ($3.5286/3.0286 = 1.165$) in the starting GFR or a log(1.165) increase in b_{i1} over a baseline value of $b_{i1} = 0$. Likewise, compared to the typical control patient having a "true" decay rate of $\beta_{i2} = \exp(-2.5761) = 0.076$ month^{-1} (Output 7.16), a 0.0083 month^{-1} increase in β_{i2} is equivalent to an 11.0% increase ($0.0844/0.076 = 1.11$) in the monthly decay rate or a log(1.11) increase in b_{i2} over a baseline value of $b_{i2} = 0$.

Program 7.9

```
%HR(data=SASoutA.pe1_1);
data HR1;
 set Hazards;
 Model=1;
run;
%HR(data=SASoutA.pe1_2,unit_bi1=0.5, unit_bi2=0.1/12);
data HR3;
 set Hazards;
 Model=3;
 run;
 %HR(data=SASoutA.pe2_2, unit_bi1=log(1.165),
                        unit_bi2=log(1.11));
data HR4;
 set Hazards;
 Model=4
; run;
```

```
data HR;
 set HR1 HR3 HR4;
run;
proc sort data=HR;
 by Model;
run;
proc print data=HR noobs;
 where HR ne .;
 by Model; id Model;
 var Parameter Estimate Lower Upper Units HR LCL UCL;
 format HR LCL UCL 5.2 Units 5.3;
run;
```

Output 7.17: Estimated hazard ratios (HR) and associated 95% confidence limits (LCL, UCL) under the two SP models and the initial joint model with additive errors. Included are the original parameter estimates from all three models together with their lower and upper 95% confidence limits. Model 1 is the initial joint model with additive errors for GFR (Output 7.13), Model 3 is the joint SP model with additive errors (Output 7.15) and Model 4 is the joint SP model with multiplicative errors (Output 7.16).

Model	Parameter	Estimate	Lower	Upper	Units	HR	LCL	UCL
1	eta_TRT	−0.3718	−0.8402	0.09656	1.000	0.69	0.43	1.10
	eta_Diabetic	0.7289	0.1123	1.3455	1.000	2.07	1.12	3.84
	eta_Age	0.02787	0.000170	0.05556	10.00	1.32	1.00	1.74
	eta_Sex	0.1823	−0.2985	0.6630	1.000	1.20	0.74	1.94
	eta_GFR0	0.03965	−0.03468	0.1140	1.000	1.04	0.97	1.12
	eta_nPNA0	−1.8742	−3.0279	−0.7204	0.100	0.83	0.74	0.93
	eta_Albumin0	−0.6778	−1.0463	−0.3093	0.100	0.93	0.90	0.97
3	eta_TRT	−0.4412	−1.0983	0.2158	1.000	0.64	0.33	1.24
	eta_Diabetic	0.4806	−0.3988	1.3599	1.000	1.62	0.67	3.90
	eta_Age	0.03181	−0.00115	0.06478	10.00	1.37	0.99	1.91
	eta_Sex	0.4084	−0.2988	1.1157	1.000	1.50	0.74	3.05
	eta_nPNA0	−2.9006	−4.6620	−1.1391	0.100	0.75	0.63	0.89
	eta_Albumin0	−0.9047	−1.5019	−0.3075	0.100	0.91	0.86	0.97
	eta_bi1	0.1903	−0.06047	0.4411	0.500	1.10	0.97	1.25
	eta_bi2	2.3075	1.3655	3.2495	0.008	1.02	1.01	1.03
4	eta_TRT	−0.4007	−0.8864	0.08510	1.000	0.67	0.41	1.09
	eta_Diabetic	0.6891	0.03922	1.3390	1.000	1.99	1.04	3.82
	eta_Age	0.02566	−0.00247	0.05380	10.00	1.29	0.98	1.71
	eta_Sex	0.2657	−0.2526	0.7839	1.000	1.30	0.78	2.19
	eta_nPNA0	−1.8944	−3.0601	−0.7287	0.100	0.83	0.74	0.93
	eta_Albumin0	−0.6525	−1.0610	−0.2441	0.100	0.94	0.90	0.98
	eta_bi1	0.1096	−0.3582	0.5774	0.153	1.02	0.95	1.09
	eta_bi2	0.4153	−0.05800	0.8886	0.104	1.04	0.99	1.10

The estimated hazard ratios for these three joint models are shown in Output 7.17. We included Model 1, which is the initial joint model with additive errors, so that we can directly compare hazard ratio estimates from this piecewise exponential PH model with that of the Cox PH model. In both these models, patient survival is assumed independent of GFR except through a possible association with baseline GFR values and in both cases, the estimated hazard ratios are all very similar (see Output 7.12 for the hazard ratios under the Cox PH model). The treatment effect on patient survival, although not significant at the 0.05 level of significance, consistently favored the high-dose treatment group for all three models (an estimated HR of between 0.64 and 0.69). Furthermore, elevated values of baseline albumin and nPCR were consistently associated with improved patient survival for all three models. With respect to the impact that the rate of decline in GFR has on patient survival, the two SP models offer differing conclusions. Both models result in estimated hazard ratios that indicate the risk of death is higher among individuals with higher rates of decline in GFR. However, statistical significance was only achieved when we fit the SP model assuming an additive error structure. Based on the superior fit of the SP model with multiplicative errors (Model 4: -2 log-likelihood$=3646.0$, AIC$=3690.0$) compared with the SP assuming additive errors (Model 3: -2 log-likelihood$=4597.3$, AIC$=4639.3$), one might conclude there is no evidence to suggest an association exists between the rate of decline in GFR and subsequent patient survival. Nonetheless, a trend toward significance was seen even under the SP model with multiplicative errors ($p = 0.0852$, Output 7.16). The lack of significance at the 0.05 level may be the result of inadequate power since the trial was never designed to evaluate outcomes in incident-only patients.

While a trend toward reduced mortality was seen among incident patients receiving a higher dose of peritoneal dialysis, one must be wary not to read too much into this trend as well. First, this result is in stark contrast with overall results from the ADEMEX study in which there was absolutely no evidence of a treatment-related effect when all 965 prevalent and incident patients are included in the analysis (HR$=1.00$, 95% CI: 0.80, 1.24, Paniagua et. al., 2002). Second, the findings here are subject to selection bias as a result of restricting the analysis to only those incident patients having a starting GFR of 1.00 ml/min or higher. Of course, one can consider a number of alternative analyses as well. It might prove useful, for example, to repeat some of the analyses presented here using multiple imputation techniques so as to not exclude the 14 patients (9 control, 5 treated) who had missing albumin and/or nPNA values at baseline. One might also consider fitting a shared parameter model with multi-level random effects. For example, one may wish to take into account possible cluster-specific correlation among patients treated within the same center. Such an approach to shared parameter models for the joint analysis of longitudinal and survival data was recently illustrated by Liu et. al. (2008). While the specification of a multi-level random-effects model is fairly straightforward for GLME models as was shown for the ADEMEX hospitalization data (§5.2.4), it is not exactly clear how one would incorporate a two-level random-effects shared parameter model using NLMIXED. Nonetheless, as we hopefully have demonstrated throughout this book, there are many ways one can "trick" SAS into overcoming apparent limitations, particularly when using the very flexible and general NLMIXED procedure.

Part IV

Appendices

Some useful matrix notation and results

Appendix A

For those interested in some of the more technical material in the book, we include in this appendix a short synopsis of key matrix algebra results that may prove useful when working through some of the theory. By no means is this meant to be a comprehensive treatment of matrix algebra. Rather it is hoped that readers with a basic working knowledge of matrix algebra will find this material to be a helpful reminder. For a more thorough treatment of this subject matter, the reader is referred to texts by Rao (1973), Searle, Casella and McCulloch (1992) and Harville (1997) as well as articles by Neudecker (1969), Henderson and Searle (1979) and Magnus and Neudecker (1979).

A.1 Matrix notation and results

Let $\boldsymbol{A} = (a_{ij})_{i=1,\ldots,m;j=1,\ldots,n}$ denote a general $m \times n$ matrix such that

$$\boldsymbol{A} = \begin{pmatrix} a_{11} & a_{12} & \cdots & a_{1n} \\ a_{21} & a_{22} & \cdots & a_{2n} \\ \vdots & \vdots & \ddots & \vdots \\ a_{m1} & a_{m2} & \cdots & a_{mn} \end{pmatrix} = \begin{pmatrix} \boldsymbol{a}_1 & \boldsymbol{a}_2 & \cdots & \boldsymbol{a}_n \end{pmatrix}$$

where $\boldsymbol{a}_j = \begin{pmatrix} a_{1j} \\ a_{2j} \\ \vdots \\ a_{mj} \end{pmatrix}$ is the jth column vector of $\boldsymbol{A}$, $(j = 1, \ldots, n)$.

The transpose of a vector or matrix

We shall denote the transpose of an $m \times 1$ vector $\boldsymbol{a}$ by $\boldsymbol{a}'$ where

$$\boldsymbol{a} = \begin{pmatrix} a_1 \\ a_2 \\ \vdots \\ a_m \end{pmatrix} \text{ and } \boldsymbol{a}' = \begin{pmatrix} a_1 & a_2 & \cdots & a_m \end{pmatrix}.$$

Likewise, the transpose of an $m \times n$ matrix $\boldsymbol{A}$ will be denoted by $\boldsymbol{A}'$ where $\boldsymbol{A}'$ is obtained by simply interchanging the rows and columns of $\boldsymbol{A}$ such that $\boldsymbol{A}' = (a_{i'j'})_{i'=1,\ldots,n;j'=1,\ldots,m}$. It follows from the definition of a transpose that the $m \times n$ matrix $\boldsymbol{A}$ may be written as $\boldsymbol{A} = \begin{pmatrix} \boldsymbol{a}_1 & \boldsymbol{a}_2 & \cdots & \boldsymbol{a}_m \end{pmatrix}'$ where $\boldsymbol{a}_1', \ldots, \boldsymbol{a}_m'$ are the m row vectors of $\boldsymbol{A}$. A useful property of transposes is the fact that $(\boldsymbol{AB})' = \boldsymbol{B}'\boldsymbol{A}'$.

Rank of a Matrix

The rank of an $m \times n$ matrix $\boldsymbol{A}$ is defined as the number of linearly independent column vectors, $\boldsymbol{a}_1, \ldots, \boldsymbol{a}_n$, of $\boldsymbol{A}$. A set of column vectors, $\boldsymbol{a}_1, \ldots, \boldsymbol{a}_n$, is said to be linearly independent if there does not exist a set of scalar coefficients, $c_1, \ldots, c_n$, (not all zero) such that $c_1\boldsymbol{a}_1 + \ldots + c_n\boldsymbol{a}_n = \boldsymbol{0}$. The column space of the matrix $\boldsymbol{A}$ is defined as the vector space spanned by the column vectors of $\boldsymbol{A}$ while the row space of $\boldsymbol{A}$ is defined as the vector space spanned by the row vectors of $\boldsymbol{A}$. The dimension of the column space of $\boldsymbol{A}$ is referred to as the column rank of $\boldsymbol{A}$ or simply as the rank of $\boldsymbol{A}$ while the dimension of the row space of $\boldsymbol{A}$ is referred to as the row rank of $\boldsymbol{A}$. If we let rank($\boldsymbol{A}$) denote the column rank of $\boldsymbol{A}$, then for matrices that conform with respect to matrix multiplication, the following properties hold:

1. rank($\boldsymbol{A}$) =rank($\boldsymbol{A}'$)
2. rank($\boldsymbol{AB}$) $\leq$ min{rank($\boldsymbol{A}$), rank($\boldsymbol{B}$)}
3. For an $m \times n$ matrix $\boldsymbol{A}$ with rank n and an $n \times p$ matrix $\boldsymbol{B}$ with rank p, rank($\boldsymbol{AB}$) $= p$
4. rank($\boldsymbol{AB}$)+rank($\boldsymbol{BC}$) $\leq$rank($\boldsymbol{B}$)+rank($\boldsymbol{ABC}$)

The matrix $\boldsymbol{A}$ is said to be of full column rank if rank($\boldsymbol{A}$) $= n$ and it is said to be of full row rank if the rank($\boldsymbol{A}$) $= m$. If $\boldsymbol{A}$ is an $m \times m$ square matrix with rank($\boldsymbol{A}$) $= m$, then $\boldsymbol{A}$ is said to be of full rank.

Inverse Matrices

Let $\boldsymbol{A}$ be an $m \times m$ square matrix with rank($\boldsymbol{A}$) $= m$. Then $\boldsymbol{A}$ is said to be nonsingular and invertible with the property that there exists a unique inverse matrix, $\boldsymbol{A}^{-1}$, such that $\boldsymbol{A}\boldsymbol{A}^{-1} = \boldsymbol{A}^{-1}\boldsymbol{A} = \boldsymbol{I}_m$ where $\boldsymbol{I}_m$ is the $m \times m$ identity matrix. Some useful results regarding the inverse of matrices are as follows:

1. If $\boldsymbol{A}$ and $\boldsymbol{B}$ are nonsingular, then $(\boldsymbol{AB})^{-1} = \boldsymbol{B}^{-1}\boldsymbol{A}^{-1}$
2. If $\boldsymbol{A}$ is nonsingular and $\boldsymbol{B}$ is any matrix such that the inverse $(\boldsymbol{A}+\boldsymbol{B})^{-1}$ exists, then
$$(\boldsymbol{A}+\boldsymbol{B})^{-1} = \boldsymbol{A}^{-1} - \boldsymbol{A}^{-1}\boldsymbol{B}(\boldsymbol{A}+\boldsymbol{B})^{-1}.$$
If, in addition, $\boldsymbol{B}$ is nonsingular then
$$(\boldsymbol{A}+\boldsymbol{B})^{-1} = \boldsymbol{A}^{-1} - \boldsymbol{A}^{-1}(\boldsymbol{A}^{-1}+\boldsymbol{B}^{-1})^{-1}\boldsymbol{A}^{-1}$$
and
$$(\boldsymbol{I}+\boldsymbol{AB}^{-1})^{-1} = \boldsymbol{B}(\boldsymbol{A}+\boldsymbol{B})^{-1}.$$
3. If $\boldsymbol{A}$ is an $n \times n$ nonsingular matrix and $\boldsymbol{D}$ is an $m \times m$ nonsingular matrix, then for any $m \times n$ matrix $\boldsymbol{B}$ of full column rank n,
$$\boldsymbol{B}'(\boldsymbol{A}+\boldsymbol{BDB}')^{-1}\boldsymbol{B} = [(\boldsymbol{B}'\boldsymbol{A}^{-1}\boldsymbol{B})^{-1}+\boldsymbol{D}]^{-1}.$$

Other useful results can be found in the references at the beginning of this Appendix.

Determinants, Diagonal Matrices, and Traces

The determinant of a square matrix $\boldsymbol{A}$ of order m is denoted by $|\boldsymbol{A}|$ and satisfies the following properties:

1. $\boldsymbol{A}$ is nonsingular if and only if $|\boldsymbol{A}| \neq 0$
2. $|\boldsymbol{AB}| = |\boldsymbol{A}||\boldsymbol{B}|$
3. $|\boldsymbol{A}^{-1}| = 1/|\boldsymbol{A}|$.

The diagonal matrix of an $m \times m$ square matrix $\boldsymbol{A}$ is denoted by $Diag(\boldsymbol{A})$ and is defined as

$$Diag(\boldsymbol{A}) = Diag\begin{pmatrix} a_{11} & a_{12} & \cdots & a_{1m} \\ a_{21} & a_{22} & \cdots & a_{2m} \\ \vdots & \vdots & \ddots & \vdots \\ a_{m1} & a_{m2} & \cdots & a_{mm} \end{pmatrix} = \begin{pmatrix} a_{11} & 0 & \cdots & 0 \\ 0 & a_{22} & \cdots & 0 \\ \vdots & \vdots & \ddots & \vdots \\ 0 & 0 & \cdots & a_{mm} \end{pmatrix}.$$

The trace of an $m \times m$ square matrix $\boldsymbol{A}$, denoted trace($\boldsymbol{A}$), is defined to be the sum of the diagonal elements of $\boldsymbol{A}$, i.e., trace($\boldsymbol{A}$) $= \sum_{i=1}^{m} a_{ii}$, and it satisfies the following properties:

1. trace($\boldsymbol{A} + \boldsymbol{B}$) =trace($\boldsymbol{A}$)+trace($\boldsymbol{B}$)
2. trace($\boldsymbol{AB}$) =trace($\boldsymbol{BA}$)
3. $\boldsymbol{a}'\boldsymbol{A}\boldsymbol{a}$ =trace($\boldsymbol{a}'\boldsymbol{A}\boldsymbol{a}$) =trace($\boldsymbol{A}\boldsymbol{a}\boldsymbol{a}'$) for any $m \times 1$ vector $\boldsymbol{a}$.

Symmetric and Positive-Definite Matrices

An $m \times m$ square matrix $\boldsymbol{A}$ is said to be symmetric if and only if $\boldsymbol{A} = \boldsymbol{A}'$. Let $\boldsymbol{x}$ be any $m \times 1$ vector of real values. A quadratic form in $\boldsymbol{x}$ is a homogeneous quadratic function of the form

$$Q(\boldsymbol{x}) = \boldsymbol{x}'\boldsymbol{A}\boldsymbol{x}$$

for a symmetric matrix $\boldsymbol{A}$ of order m. A symmetric matrix $\boldsymbol{A}$ of order m is said to be positive-definite if for all non-null $\boldsymbol{x}$, $Q(\boldsymbol{x}) = \boldsymbol{x}'\boldsymbol{A}\boldsymbol{x} > 0$, and it is said to be positive-semidefinite if $Q(\boldsymbol{x}) = \boldsymbol{x}'\boldsymbol{A}\boldsymbol{x} \geq 0$ for all non-null $\boldsymbol{x}$.

The $Vec(\cdot)$ and $Vech(\cdot)$ Operators

Let $\boldsymbol{A}$ be any $m \times n$ matrix $(a_{ij})_{i=1,\ldots,m;j=1,\ldots,n}$. The Vec operator creates an $mn \times 1$ column vector from $\boldsymbol{A}$ by stacking the column vectors of $\boldsymbol{A}$ below one another. Hence if $\boldsymbol{A} = \begin{pmatrix} \boldsymbol{a}_1 & \boldsymbol{a}_2 & \cdots & \boldsymbol{a}_n \end{pmatrix}$ then

$$Vec(\boldsymbol{A}) = \begin{pmatrix} \boldsymbol{a}_1 \\ \boldsymbol{a}_2 \\ \vdots \\ \boldsymbol{a}_n \end{pmatrix}.$$

Now let $\boldsymbol{A}$ be any $m \times m$ symmetric matrix $(a_{ij})_{i=1,\ldots,m;j=1,\ldots,m}$. The $Vech$ operator creates a $\frac{1}{2}m(m+1) \times 1$ column vector from $\boldsymbol{A}$ by stacking lower diagonal elements of $\boldsymbol{A}$ below one another. For example, if $\boldsymbol{A} = \begin{pmatrix} a_{11} & a_{12} & a_{13} \\ a_{21} & a_{22} & a_{23} \\ a_{31} & a_{32} & a_{33} \end{pmatrix}$, then

$$Vech(\boldsymbol{A}) = \begin{pmatrix} a_{11} \\ a_{21} \\ a_{31} \\ a_{22} \\ a_{32} \\ a_{33} \end{pmatrix}.$$

For any $m \times m$ symmetric matrix $\boldsymbol{A}$, there exists an $m(m+1)/2 \times m^2$ matrix $\boldsymbol{B}$ of 0's and 1's and an $m^2 \times m(m+1)/2$ matrix $\boldsymbol{C}$ of 0's and 1's such that

1. $Vech(\boldsymbol{A}) = \boldsymbol{B}Vec(\boldsymbol{A})$
2. $Vec(\boldsymbol{A}) = \boldsymbol{C}Vech(\boldsymbol{A})$.

These results are useful when deriving the ML estimating equations or the GEE2 and CGEE2 estimating equations for GNLM's and GNLME models of Chapters 4 and 5.

Direct (Kronecker) Products

We define the direct product (also known as the Kronecker product) of two matrices, $\boldsymbol{A}_{m\times n}$ and $\boldsymbol{B}_{p\times q}$, as follows:

$$\boldsymbol{A} \otimes \boldsymbol{B} = ((a_{ij}\boldsymbol{B}))_{i=1,\ldots,m;j=1,\ldots,n} = \begin{pmatrix} a_{11}\boldsymbol{B} & a_{12}\boldsymbol{B} & \ldots & a_{1n}\boldsymbol{B} \\ a_{21}\boldsymbol{B} & a_{22}\boldsymbol{B} & \ldots & a_{2n}\boldsymbol{B} \\ \vdots & \vdots & \ddots & \vdots \\ a_{m1}\boldsymbol{B} & a_{m2}\boldsymbol{B} & \ldots & a_{mn}\boldsymbol{B} \end{pmatrix}_{mp\times nq}.$$

Some properties related to the use of the direct product are as follows:

1. $(\boldsymbol{A} \otimes \boldsymbol{B}) \otimes \boldsymbol{C} = \boldsymbol{A} \otimes (\boldsymbol{B} \otimes \boldsymbol{C})$
2. $(\boldsymbol{A}_1 + \boldsymbol{A}_2) \otimes \boldsymbol{B} = (\boldsymbol{A}_1 \otimes \boldsymbol{B}) + (\boldsymbol{A}_2 \otimes \boldsymbol{B})$
3. $\boldsymbol{A} \otimes (\boldsymbol{B}_1 + \boldsymbol{B}_2) = (\boldsymbol{A} \otimes \boldsymbol{B}_1) + (\boldsymbol{A} \otimes \boldsymbol{B}_2)$
4. For scalar a, $a \otimes \boldsymbol{A} = \boldsymbol{A} \otimes a = a\boldsymbol{A}$
5. For scalars a and b, $a\boldsymbol{A} \otimes b\boldsymbol{B} = ab\boldsymbol{A} \otimes \boldsymbol{B}$
6. For conforming matrices, $(\boldsymbol{A} \otimes \boldsymbol{B})(\boldsymbol{C} \otimes \boldsymbol{D}) = \boldsymbol{AC} \otimes \boldsymbol{BD}$
7. $(\boldsymbol{A} \otimes \boldsymbol{B})' = (\boldsymbol{A}' \otimes \boldsymbol{B}')$
8. For nonsingular matrices $\boldsymbol{A}$ and $\boldsymbol{B}$, $(\boldsymbol{A} \otimes \boldsymbol{B})^{-1} = \boldsymbol{A}^{-1} \otimes \boldsymbol{B}^{-1}$
9. For $m \times m$ matrix $\boldsymbol{A}$ and $n \times n$ matrix $\boldsymbol{B}$, $|\boldsymbol{A} \otimes \boldsymbol{B}| = |\boldsymbol{A}|^n|\boldsymbol{B}|^m$.

Many of these results are used when deriving some of the estimation schemes described in the book (e.g., ML, GEE2 and CGEE2) or when deriving the asymptotic properties of the resulting estimators.

Additional results on estimation

Appendix B

In this appendix, we present some additional results as they pertain to the large sample behavior of different estimators under the GLME and GNLME models of Chapter 5. The emphasis is on comparing the asymptotic bias associated with the various estimators including those based on approximating the marginal moments using a first-order or second-order Taylor series expansion (i.e., PL/CGEE1, PL/QELS and CGEE2/PELS) and those based on approximating the marginal log-likelihood function using numerical integration such as Laplace-based MLE (LMLE) and MLE based on adaptive Gaussian quadrature (AGQ). We start first with a brief synopsis of each of the estimators.

B.1 The different estimators for mixed-effects models

In Chapter 5, we present two basic approaches to estimation for GLME and GNLME models: 1) a moments-based approach utilizing some form of Taylor-series linearization and 2) a likelihood-based approach using some form of numerical integration. These two approaches yield different sets of estimators which we have classified in Chapter 5 as PL/CGEE1 estimators, PL/QELS estimators, first-order ML estimators, Laplace-based ML estimators, ML estimators based on adaptive Gaussian quadrature (AGQ), and CGEE2/PELS estimators. A synopsis of each of type of estimator is provided below including a brief description of the basic estimation scheme used and the procedure or macro to which each estimator corresponds.

PL/CGEE1

The PL/CGEE1 approach is implemented with GLIMMIX using METHOD = RSPL or METHOD = MSPL, and with %NLINMIX using the macro argument EXPAND = EBLUP together with the MIXED option METHOD = REML or METHOD = ML. This approach involves iteratively taking a first-order Taylor series expansion of the conditional means about an empirical best linear unbiased predictor of the random effects and applying standard LME estimation to obtain iteratively updated parameter estimates. Within the GLME model literature, PL/CGEE1 is frequently referred to as penalized quasi-likelihood (PQL) (e.g., Breslow and Clayton, 1993) and within the NLME model literature as the LME approximation method of Lindstrom and Bates (1990) (e.g., Pinheiro and Bates, 2000).

PL/QELS

The PL/QELS approach is implemented with GLIMMIX using METHOD = RMPL or METHOD = MMPL, and with %NLINMIX using the macro argument EXPAND = ZERO together with the MIXED option METHOD = REML or METHOD = ML. This approach essentially involves iteratively taking a first-order Taylor series expansion of the conditional means about the average of the random effects (i.e., about zero) and applying standard LME estimation to obtain iteratively updated parameter estimates. Within the GLME model literature, PL/QELS is frequently referred to as marginal quasi-likelihood (MQL) (e.g., Breslow and Clayton, 1993).

First-order MLE

The first-order MLE approach is implemented with NLMIXED using METHOD = FIRO. This approach entails taking a one-time first-order Taylor series expansion of the conditional mean and error terms about the average of the random effects (i.e., about zero) and applying standard ML estimation to the ensuing log-likelihood function assuming normality. This approach is commonly referred to simply as the first-order method (e.g., Roe, 1997).

Laplace-based MLE (LMLE)

The Laplace-based MLE approach is implemented with GLIMMIX using METHOD = LAPLACE, and with NLMIXED using QPOINTS = 1 along with METHOD = GAUSS, METHOD = HARDY or METHOD = ISAMP. Under this approach, ML estimation is carried out by minimizing minus twice the log-likelihood function where the log-likelihood is approximated using the second-order Laplace approximation as described in §5.1.1 and §5.3.4.

MLE based on adaptive Gaussian quadrature (AGQ)

ML estimation based on AGQ is implemented in GLIMMIX using METHOD = QUAD, and with NLMIXED using METHOD = GAUSS. This involves numerically integrating the conditional likelihood over the random effects using Gaussian quadrature as described in §5.1.1 and §5.3.4. Alternatively, METHOD = HARDY specifies the use of Hardy quadrature based on an adaptive trapezoidal rule and is available only for one-dimensional integrals (i.e., it is limited to models with a single random effect). The METHOD = ISAMP specifies adaptive importance sampling which closely resembles the adaptive Gaussian quadrature approximation (Pinheiro and Bates, 1995).

CGEE2/PELS

The CGEE2/PELS approach is a moments-based estimation procedure similar in concept to the PL/CGEE1 approach except that it performs an expansion around both the first- and second-order conditional moments. It can be implemented in GLIMMIX and in NLMIXED by specifying a conditional mean and conditional variance and then applying the Laplace approximation under a "working" normality assumption, i.e., by assuming the conditional third- and fourth-order moments of $\boldsymbol{y}_i|\boldsymbol{b}_i$ are the same as that of a Gaussian random vector (see §5.3.4 for further details).

B.2 Comparing large sample properties of the different estimators

In the preceding section, we presented a brief synopsis of each of the estimators available within SAS for GLME and GNLME models. In this section, we compare the large sample behavior of these estimators based on available asymptotic theory from the literature as well as from limited simulations based on discrete binary outcomes. We start by first introducing the little o and big O notation used to describe and compare the large-sample behavior of the different estimators.

Little o and big O notation

The large-sample behavior of an estimator is frequently expressed in terms of its limiting behavior as the sample size goes to ∞. This, in turn, is frequently conveyed using the little

o and big O notation describing the order of magnitude at which parameter estimates either converge in probability to a given value or are bounded in probability by a given value (e.g., Serfling, 1980).

Let $\{Y_n\}, n = 1, 2, \ldots$ be a sequence of random variables. Following Fuller (1976), we say that Y_n converges to a constant c *in probability* (written $Y_n \xrightarrow{p} c$) if for any given $\epsilon > 0$

$$\lim_{n\to\infty} \Pr\Big(|Y_n - c| > \epsilon\Big) = 0.$$

In terms of order of convergence, we write

$$Y_n = c + o_p(a_n) \tag{B.1}$$

whenever $(Y_n - c)/a_n \xrightarrow{p} 0$ for a sequence of positive real numbers, $\{a_n\}$. In terms of estimation, we define an estimate $\widehat{\boldsymbol{\theta}}_n$ of a vector parameter $\boldsymbol{\theta}$ to be consistent (or weakly consistent) if $(\widehat{\boldsymbol{\theta}}_n - \boldsymbol{\theta}) = o_p(1)$.

An alternative concept which is often used to determine whether an estimator is consistent is the concept of a sequence of random variables being bounded in probability. For a sequence of positive real numbers, $\{a_n\}$, and a constant c, we say that $Y_n - c$ is bounded in probability by a_n if, for every $\epsilon > 0$, there exists a positive real number, K_ϵ, such that

$$\Pr\Big(|Y_n - c| \geq a_n K_\epsilon\Big) \leq \epsilon \tag{B.2}$$

for all n. We write this in terms of big O notation as

$$Y_n - c = O_p(a_n). \tag{B.3}$$

An important property that relates the concept of "bounded in probability" to the concept of "converges in probability" is the following result.

Proposition 1 *Suppose there is a sequence of positive real numbers $\{a_n\}$ such that $\lim_{n\to\infty} a_n = 0$. If $\{Y_n\}$ is a sequence of random variables satisfying $Y_n = c + O_p(a_n)$, then $Y_n = c + o_p(1)$.*

This result implies that if there exists a vector estimate $\widehat{\boldsymbol{\theta}}_n$ of a parameter vector $\boldsymbol{\theta}$ such that $\sqrt{n}(\widehat{\boldsymbol{\theta}}_n - \boldsymbol{\theta})$ is bounded in probability, i.e., $\widehat{\boldsymbol{\theta}}_n = \boldsymbol{\theta} + O_p(n^{-1/2})$, then $\widehat{\boldsymbol{\theta}}_n$ is a consistent estimate of $\boldsymbol{\theta}$ with $(\widehat{\boldsymbol{\theta}}_n - \boldsymbol{\theta}) = o_p(1)$. Fuller (1976) provides a number of other useful results pertaining to the order of convergence based on the $o_p(a_n)$ and $O_p(a_n)$ notation many of which are summarized in Vonesh and Chinchilli (1997).

Another useful concept when investigating large sample behavior of a particular estimator is the idea of convergence in distribution. In general, if $\{Y_n\}$ is a sequence of random variables with cumulative distribution functions $\{F_{Y_n}(\cdot)\}$, then Y_n is said to *converge in distribution* to a random variable Y having cumulative distribution function $F_Y(\cdot)$ (written $Y_n \xrightarrow{d} Y$) if $\lim_{n\to\infty} F_{Y_n} \to F_Y$. A second useful result based on convergence in distribution is the following.

Proposition 2 *If $Y_n \xrightarrow{d} Y$ then $Y_n = O_p(1)$*

One can use this result to show that any estimator $\widehat{\boldsymbol{\theta}}_n$ satisfying $\sqrt{n}(\widehat{\boldsymbol{\theta}}_n - \boldsymbol{\theta}) \xrightarrow{d} N(\mathbf{0}, \boldsymbol{\Omega})$ will necessarily be consistent for $\boldsymbol{\theta}$ since $\sqrt{n}(\widehat{\boldsymbol{\theta}}_n - \boldsymbol{\theta}) = O_p(1)$ implies $\widehat{\boldsymbol{\theta}}_n = \boldsymbol{\theta} + O_p(n^{-1/2})$ which implies $\widehat{\boldsymbol{\theta}}_n$ is $\sqrt{n}$ consistent for $\boldsymbol{\theta}$. These and other results are used to derive the asymptotic properties of the estimators discussed in Chapter 2-5 and which are further summarized below.

Asymptotic results

In the ensuing discussion, it is assumed that we are dealing with a GLME or GNLME model having at least one random effect that enters the model nonlinearly as otherwise all of the above estimators can be shown to be consistent and asymptotically normally distributed. With that caveat, it appears that a formal unified theory regarding the asymptotic behavior of the different estimators under GLME and GNLME models is incomplete at best (at least to this author's knowledge). A number of authors have investigated the asymptotic properties of the above estimators but often under rather specialized conditions. One reason for the lack of a unifying theory is that whereas classical asymptotic theory assumes an underlying model that involves random variables that are singly arrayed (i.e., are represented by a single index), the random variables studied in mixed models are, at a minimum, doubly arrayed (i.e., are represented by two or more indices) with at least one array of random variables entering the model nonlinearly. For example, with the single-level random effects models that one encounters with repeated measurements data or singly-clustered data, observations within the same subject or cluster are correlated. This requires a new asymptotic theory; one which examines the properties of estimators for GLME and GNLME models as a function of both the number of subjects/clusters and the number of observations per subject/cluster. Despite these limitations to a unified theory, the asymptotic properties that have been investigated do reveal some important insights into the strengths and weaknesses of each of the estimation techniques available within SAS.

With regards to the PL/CGEE1 class of estimators, the large sample properties of this estimator has been investigated for single-level GLME models by Breslow and Lin (1995) and Bellamy et. al. (2005) and indirectly through single-level GNLME models by Vonesh et. al. (2002). Holding the number of observations per subject or cluster fixed (i.e., holding p_i under the fully parametric GLME model (5.1-5.2) fixed), Breslow and Lin (1995) investigated the asymptotic bias of the PL/CGEE1 estimator (i.e., PQL) assuming a single random effect. Expanding the true log-likelihood and the corresponding PQL approximation to the log-likelihood about $\theta_b = 0$ where $\theta_b = \psi$ is the random effects variance component, they showed that the asymptotic bias of the PL/CGEE1 or PQL estimator of $\boldsymbol{\beta}$ is of order $O_p(\theta_b)$ as $n \to \infty$. Under a fully parametric GNLME model defined by (5.47) and (5.49), Vonesh et. al. (2002) investigated the asymptotic properties of a CGEE2 estimator, $\widehat{\boldsymbol{\beta}}_{\text{CGEE2}}$, that is equivalent to the PL/CGEE1 or PQL estimator under the GLME model. Holding the random-effects variance-covariance components, $\boldsymbol{\theta}_b$, fixed, they showed that under suitable regularity conditions,

$$\widehat{\boldsymbol{\beta}}_{\text{CGEE2}} = \boldsymbol{\beta} + O_p\{\max[n^{-1/2}, \min(p_i)^{-1/2}]\}$$

where n is the number of subjects and p_i the number of observations per subject. This implies that under the appropriate regularity conditions, both the CGEE2 estimator and the PL/CGEE1 or PQL estimator will be consistent, i.e., converge in probability to $\boldsymbol{\beta}$, as n and $\min(p_i) \to \infty$. Similarly, Demindenko (2004) showed that under suitable regularity conditions, the PL/CGEE1 estimator will be a consistent estimator of $\boldsymbol{\beta}$ under the family of normal-theory NLME models provided both n and $\min(p_i) \to \infty$. Finally, Bellamy et. al. (2005) demonstrated that for fixed n and $\boldsymbol{\theta}_b$, the asymptotic bias (i.e., $E(\widehat{\boldsymbol{\beta}} - \boldsymbol{\beta})$) of the PQL estimator is of order $O_p\{\min(p_i)^{-1}\}$.

The PL/QELS or MQL class of estimators as well as the first-order MLE estimator are known to be asymptotically biased as n and/or $\min(p_i) \to \infty$. However, such bias is a function of the random effects variance components and hence will be minimal when $\boldsymbol{\theta}_b \to \mathbf{0}$ (e.g., Solomon and Cox, 1992; Breslow and Lin, 1995; Lin and Breslow, 1996; and Demidenko, 2004). For example, Demidenko (2004, pp. 456-459) demonstrated that for the simple NLME model,

$$y_{ij} = e^{\beta+b_i} + \epsilon_{ij},\ i = 1,\ldots,n;\ j = 1,\ldots,p,$$

the first-order ML estimate is given by

$$\widehat{\beta}_{\mathrm{FO}} = \log\left(\frac{1}{np}\sum_{i=1}^{n}\sum_{j=1}^{p} y_{ij}\right)$$

which he shows has the limiting behavior

$$\lim_{n\to\infty} \widehat{\beta}_{\mathrm{FO}} = \log(e^{\beta+\frac{1}{2}\sigma^2\psi}) = \beta + \frac{1}{2}\sigma^2\psi$$

assuming ϵ_{ij} ~iid $N(0,\sigma^2)$ independent of b_i ~iid $N(0,\sigma^2\psi)$. Here we see that $\widehat{\beta}_{\mathrm{FO}} = \beta + O_p(\sigma^2\psi)$ as both n and $p \to \infty$ but the magnitude of bias disappears as $\psi \to 0$.

The Laplace-based MLE was shown by Vonesh (1996) (see also Demidenko, 2004, §8.8) to be consistent as both n and $p \to \infty$. Specifically, Vonesh showed that under suitable regularity conditions,

$$\widehat{\boldsymbol{\beta}}_{\mathrm{LMLE}} = \boldsymbol{\beta} + O_p\{\max[n^{-1/2}, \min(p_i)^{-1}]\}$$

which is slightly stronger than the results shown for the PL/CGEE1 estimator. Based on the work of Bellamy et. al. (2005) and Demidenko (2004), one might conjecture that the PL/CGEE1 or PQL estimator will achieve the same order of convergence as the Laplace ML estimator although a more formal proof is required. Finally, by taking a sufficient number of quadrature points, one can argue that ML estimates based on Gaussian quadrature will, under the usual regularity conditions for ML estimation, yield consistent and asymptotically efficient estimates of $\boldsymbol{\beta}$ as $n \to \infty$ (e.g., Demindenko, 2004; §8.2 and §8.4).

In comparing the asymptotic behavior of different estimators, Demidenko (2004) showed that for a certain class of normal-theory NLME models, a two-stage estimator will be asymptotically equivalent to the PL/CGEE1 estimator (see Demidenko, 2004; §8.10). Likewise, based on the fact that the set of CGEE1 for $\boldsymbol{\beta}$ are embedded within the Laplace-based ML estimating equations for $\boldsymbol{\beta}$ (see equations (5.20) and (5.21) of §5.1.2), one would expect the performance of the PL/CGEE1 and Laplace ML estimators to be very similar (e.g., Demindenko, 2004, pp. 454). Indeed, limited simulations by Pinheiro and Bates (1995) demonstrate how similar the two can be for several different NLME models. However, these results do not hold in general and there exists subtle but very real differences. For example, the PL/CGEE1 estimator uses a Laplace approximation to approximate a *modified profile likelihood* rather than the true likelihood. This is achieved by applying the Laplace approximation to *both* random- and fixed-effects parameters, $\boldsymbol{b}_i$ and $\boldsymbol{\beta}$, the latter of which is assumed to have a flat prior (Wolfinger, 1993). In contrast, the Laplace-based MLE applies the Laplace approximation to only the random-effects resulting in a direct approximation to the *marginal likelihood* (Vonesh, 1996). As will be shown below, there are NLME models where this difference will result in a biased and inconsistent PL/CGEE1 estimator but an unbiased and consistent Laplace-based ML estimator.

Possible bias with PL/CGEE1 versus Laplace MLE

Piersol (2000) considered several NLME models where the PL/CGEE1 algorithm either completely breaks down or it yields biased and inconsistent estimators of the model parameters. Consider the following very simple NLME model adapted from Piersol (2000):

$$y_{ij} = \beta b_i + \epsilon_{ij},\quad i = 1,\ldots n;\ j = 1,\ldots,p \tag{B.4}$$
$$b_i \sim \text{iid } N(0,1)$$
$$\epsilon_{ij} \sim iid\ N(0,1)$$

where the random effects, b_i, are independent of the within-subject errors ϵ_{ij} and where the variances are both fixed and known. This model is linear in $\beta|b_i$ and linear in $b_i|\beta$ but nonlinear in β and b_i jointly. Moreover, the marginal distribution is given by

$$\boldsymbol{y}_i \sim N_p(\boldsymbol{0}, \beta^2 \boldsymbol{1}_p \boldsymbol{1}_p' + \boldsymbol{I}_p) \tag{B.5}$$

so the only information about β is contained within the second-order moments of $\boldsymbol{y}_i$. As follows from this model, the marginal mean is zero. While we may view this model as being unrealistic, it nevertheless amplifies the technical difficulties in estimation and implied inconsistency as shown below.

Under model (B.4), it can be shown that for any fixed value of β, the posterior mode that maximizes the penalized log-likelihood

$$L_i(\beta, b_i) = -\frac{1}{2}(\boldsymbol{y}_i - \beta b_i \boldsymbol{1}_p)'(\boldsymbol{y}_i - \beta b_i \boldsymbol{1}_p) - \frac{1}{2} b_i^2 \tag{B.6}$$

for the ith subject/cluster is given by

$$\widehat{b}_i = p\bar{y}_i \beta / (1 + p\beta^2). \tag{B.7}$$

Under the PL/CGEE1 approach, we substitute these posterior modes back into the CGEE1 for β and then solve for β. Doing so, we have

$$\begin{aligned}
U_{\text{CGEE1}}(\beta) &= \sum_{i=1}^n \frac{\partial}{\partial \beta}\{L_i(\beta, \widehat{b}_i)\} \qquad\qquad \text{(B.8)}\\
&= \sum_{i=1}^n \left(p\bar{y}_i \widehat{b}_i - p\beta \widehat{b}_i^2\right)\\
&= \sum_{i=1}^n \left(p^2 \bar{y}_i^2 \beta/(1 + p\beta^2) - p\beta\left[p^2 \bar{y}_i^2 \beta^2/(1 + p\beta^2)^2\right]\right)\\
&= \sum_{i=1}^n p^2\, \bar{y}_i^2 \beta / \left[(1 + p\beta^2)^2\right] = 0
\end{aligned}$$

which has solutions $\beta = 0$ or $\beta = \infty$. Substituting either of these solutions to β back into (B.7) yields $\widehat{b}_i = 0$ for all i and we are left with an intractable system of CGEE1's. The PL/CGEE1 estimator of β is therefore both biased and inconsistent.

Under the Laplace-based ML approach, the posterior mode, $\widehat{b}_i$, is again given by (B.7) for any fixed value of β. However, by initially holding β fixed rather than profiling it out as we do under the PL/CGEE1 algorithm, we end up solving the Laplace-based extended estimating equations:

$$\begin{aligned}
U_{\text{LMLE}}(\beta) &= \sum_{i=1}^n \left\{\frac{\partial}{\partial \beta}\{L_i(\beta, \widehat{b}_i)\} + \frac{\partial}{\partial \beta}\log\left(\frac{1}{|-L_i''(\beta, \widehat{b}_i)|}\right)^{1/2}\right\} \qquad \text{(B.9)}\\
&= \sum_{i=1}^n \left\{\left(p\widehat{b}_i \bar{y}_i - p\beta \widehat{b}_i^2\right) + \frac{\partial}{\partial \beta}\log\left(\frac{1}{|1 + p\beta^2|}\right)^{1/2}\right\}\\
&= \sum_{i=1}^n \left\{\left(p\widehat{b}_i \bar{y}_i - p\beta \widehat{b}_i^2\right) - \frac{1}{2}\left(\frac{2p\beta}{1 + p\beta^2}\right)\right\}\\
&= \sum_{i=1}^n \left\{p^2 \bar{y}_i^2 \beta - p\beta(1 + p\beta^2)\right\} = 0
\end{aligned}$$

the solution to which yields the Laplace-based MLE estimate

$$\widehat{\beta}^2_{\text{LMLE}} = \frac{1}{n}\sum_{i=1}^{n} \bar{y}_i^2 - \frac{1}{p}. \tag{B.10}$$

To better understand the root problem, we first note that $E(\bar{y}_i^2) = \frac{1}{p}(1+p\beta^2)$. From this, it is easy to show that the CGEE1 for β is a biased estimating equation with

$$\begin{aligned} E\{U_{\text{CGEE1}}(\beta)\} &= E\Big\{\sum p^2\, \bar{y}_i^2 \beta / \big[(1+p\beta^2)^2\big]\Big\} \\ &= \frac{np\beta}{(1+p\beta^2)} \neq 0 \end{aligned}$$

unless β has the trivial solution $\beta = 0$ or $\beta = \infty$. In contrast, the Laplace-based estimating equation, by virtue of including the Hessian term, is unbiased with

$$\begin{aligned} E\{U_{\text{LMLE}}(\beta)\} &= E\Big\{\sum \big[p^2\bar{y}_i^2\beta - p\beta(1+p\beta^2)\big]\Big\} \\ &= \Big\{\sum p^2\beta E(\bar{y}_i^2)\Big\} - np\beta(1+p\beta^2) = 0. \end{aligned}$$

In this example, the Laplace-based ML approach corresponds to expanding the conditional mean function, $f_{ij}(\beta, b_i) = \beta b_i$, about the posterior mode holding β fixed (Vonesh, 1996). This leads to approximating the marginal density of y_{ij} as:

$$\begin{aligned} y_{ij} &= f_{ij}(\beta, \widehat{b}_i) + \tilde{Z}_{ij}(\beta)(b_i - \widehat{b}_i) + \epsilon_{ij} \\ &= \beta\widehat{b}_i + \beta b_i - \beta\widehat{b}_i + \epsilon_{ij} \\ &= 0 + \beta b_i + \epsilon_{ij} \end{aligned}$$

where $\tilde{Z}_{ij} = \tilde{Z}_{ij}(\beta) = \partial f_{ij}(\beta, b_i)/\partial b_i\Big|_{(\beta=\beta, b_i=\widehat{b}_i)} = \beta$. Thus the joint marginal pdf of $\boldsymbol{y}_i$ may be expressed as

$$\boldsymbol{y}_i \sim N(\boldsymbol{\mu}_i(\beta, \widehat{b}_i),\, \boldsymbol{V}_i(\beta))$$

where $\boldsymbol{\mu}_i(\beta, \widehat{b}_i) = \boldsymbol{f}_i(\beta, \widehat{b}_i) - \tilde{\boldsymbol{Z}}_i(\beta)\widehat{b}_i = \boldsymbol{0}$ and $\boldsymbol{V}_i(\beta) = \tilde{\boldsymbol{Z}}_i(\beta)\tilde{\boldsymbol{Z}}_i(\beta)' + \boldsymbol{I} = \beta^2\boldsymbol{1}\boldsymbol{1}' + \boldsymbol{I}$. As expected for models that are linear in the random effects, this coincides with the exact marginal pdf (B.5). Indeed, it is because the derivative matrix $\tilde{\boldsymbol{Z}}_i$ is evaluated at β that we can recover the marginal pdf and thereby estimate β through the marginal second-order moments.

Conversely, under the PL/CGEE1 algorithm or, equivalently, the NLME algorithm of Lindstrom and Bates (1990), we estimate the unknown parameters iteratively via a LME step in which we approximate a *modified* marginal density. In this example, this is done by profiling β out of the derivative matrices that effectively form the first and second-order moments of the *modified* marginal density. Specifically, under the LME step, we approximate the marginal density of y_{ij} as:

$$\begin{aligned} y_{ij} &= f_{ij}(\beta, \widehat{b}_i) + \tilde{Z}_{ij}(\widehat{\beta})(b_i - \widehat{b}_i) + \epsilon_{ij} \\ &= \beta\widehat{b}_i + \widehat{\beta} b_i - \widehat{\beta}\widehat{b}_i + \epsilon_{ij} \\ &= (\beta - \widehat{\beta})\widehat{b}_i + \widehat{\beta} b_i + \epsilon_{ij} \end{aligned}$$

where $\tilde{Z}_{ij}(\widehat{\beta}) = \partial f_{ij}(\beta, b_i)/\partial b_i\Big|_{(\beta=\widehat{\beta}, b_i=\widehat{b}_i)} = \widehat{\beta}$. In this case, the joint marginal pdf of $\boldsymbol{y}_i$ may again be expressed as $\boldsymbol{y}_i \sim N(\boldsymbol{\mu}_i(\beta, \widehat{b}_i), \boldsymbol{V}_i(\beta))$ but with $\boldsymbol{\mu}_i(\beta, \widehat{b}_i) = (\beta - \widehat{\beta})\widehat{b}_i\boldsymbol{1}_p \neq \boldsymbol{0}$ and $\boldsymbol{V}_i(\beta) = \tilde{\boldsymbol{Z}}_i(\widehat{\beta})\tilde{\boldsymbol{Z}}_i(\widehat{\beta})' + \boldsymbol{I} = \widehat{\beta}^2\boldsymbol{1}\boldsymbol{1}' + \boldsymbol{I}$. Since the derivative matrix, $\tilde{\boldsymbol{Z}}_i(\widehat{\beta})$, is evaluated at $\widehat{\beta}$ and not β, the NLME algorithm leads to an intractable system of CGEE1.

Ostensibly, by profiling out β from the derivative matrix $\tilde{\boldsymbol{Z}}_i$, the PL/CGEE1 algorithm effectively ignores information about β that may be contained in the second-order marginal moments. While this example is somewhat artificial, more realistic examples including an entire class of NLME models are considered by Piersol (2000) in which the PL/CGEE1 algorithm breaks down. These examples further illustrate potential biases that may occur when fitting GNLME models to data using the PL/CGEE1 approach. Despite such problems, PL/CGEE1 estimation remains a viable alternative to likelihood-based methods for most practical applications. Moreover, when specification of a conditional pdf for $\boldsymbol{y}_i|\boldsymbol{b}_i$ is not feasible so that the Laplace approximation or adaptive Gaussian quadrature is not possible, PL/CGEE1 estimation based strictly on specification of first- and second-order conditional moments would seem to provide a reasonable recourse.

Simulation study comparing estimators

We conclude with a brief summary of the results of a simulation study comparing the PL/CGEE1/PQL estimator (PQL), the PL/QELS/MQL estimator (MQL), the Laplace-based ML estimator (LMLE) and the ML estimator using adaptive Gaussian quadrature (MLE). The study was patterned after Breslow and Clayton (1993) and is based on fitting a mixed-effects logistic regression model to simulated data under a variety of scenarios. We chose a mixed-effects logistic regression setting because of the unique challenges it sets for the different estimators (e.g., data that can be both sparse and highly discrete within subjects) and also because mixed-effects logistic regression finds it way into a variety of disciplines that one is likely to encounter in practice.

The simulation study follows the basic study design described by Breslow and Clayton (1993) but restricted to considering only binary outcomes at each visit. The simulations include each of the principal estimation techniques available with GLIMMIX (i.e., PQL, MQL, Laplace and ML via AGQ). We also expand the simulations by varying the size of the random intercept variance component from small to large ($\psi_{11} = 0.25, 0.50, 1.00$). The essential features of this expanded simulation study are as follows.

Simulation study design

- Discrete logistic mixed-effects model (Breslow and Clayton, 1993)
  Case 1:
  $$\log\frac{\pi_{ij}}{1-\pi_{ij}} = (\beta_1 + b_{i1}) + \beta_2 t_{ij} + \beta_3 Trt_i + \beta_4(t_{ij} \times Trt_i)$$
  Case 2:
  $$\log\frac{\pi_{ij}}{1-\pi_{ij}} = (\beta_1 + b_{1i}) + (\beta_2 + b_{i2})t_{ij} + \beta_3 Trt_i + \beta_4(t_{ij} \times Trt_i)$$
  where $\pi_{ij} = E(y_{ij}|b_i)$ is the conditional probability of an event for the ith subject on the jth visit or occasion
- Response variable
  $y_{ij} \sim \text{Binary}(\pi_{ij})$, $\quad (i = 1, 2, \ldots n;\ j = 1, 2, \ldots, 7)$
  $n = 100$ subjects and $p = 7$ visits per subject
- Within-subject covariate t_{ij} with $p = 7$ visits per subject
  $t_{ij} = \{-3, -2, -1, 0, 1, 2, 3\}$

- Between-subject covariate Trt_i
  $$Trt_i = \begin{cases} 0 & i \leq 50 \\ 1 & i > 50 \end{cases}$$
- Regression parameters
  $\beta_1 = -2.50$,
  $\beta_2 = 1.00$,
  $\beta_3 = -1.00$ and
  $\beta_4 = -0.50$
- Random-effects structure
  Case 1:
  $b_{i1} \sim N(0, \psi_{11})$ is a random intercept effect with
  $\psi_{11} = 0.25$ (small random-effect variation)
  $\psi_{11} = 0.50$ (moderate random-effect variation)
  $\psi_{11} = 1.00$ (large random-effect variation)
  Case 2:
  $b_{i1} \sim N(0, \psi_{11})$ is a random intercept effect with $\psi_{11} = 0.50$
  $b_{i2} \sim N(0, \psi_{22})$ is random slope effect independent of b_{i1} with $\psi_{22} = 0.25$
- Estimation technique
  PQL (PL/CGEE1) - METHOD=MSPL in GLIMMIX
  MQL (PL/QELS) - METHOD=MMPL in GLIMMIX
  MLE (using AGQ) - METHOD=QUAD in GLIMMIX
  Laplace (LMLE) - METHOD=Laplace in GLIMMIX

A total of 200 datasets were generated under each of the four scenarios described above (three different scenarios under Case 1 involving the three different values of ψ_{11}, and the one scenario under Case 2). Output B.1 below summarizes the performance of each of the estimators under Case 1.

Output B.1: A comparison of estimates for a mixed-effects binary logistic regression model under Case 1 with $p = 7$ observations per subject.

		Method							
		LMLE		MLE		MQL		PQL	
ψ_{11}	Parameter	Mean	Std	Mean	Std	Mean	Std	Mean	Std
0.25	$\beta_1 = -2.50$	−2.53	0.32	−2.53	0.32	−2.42	0.26	−2.43	0.27
	$\beta_2 = 1.00$	1.01	0.15	1.01	0.15	0.97	0.13	0.98	0.13
	$\beta_3 = -1.00$	−1.04	0.48	−1.04	0.48	−1.03	0.47	−1.02	0.47
	$\beta_4 = -0.50$	−0.51	0.22	−0.51	0.21	−0.47	0.20	−0.48	0.20
	ψ_{11}	0.30	0.35	0.29	0.34	0.20	0.22	0.20	0.21
0.50	$\beta_1 = -2.50$	−2.55	0.33	−2.55	0.32	−2.37	0.26	−2.40	0.27
	$\beta_2 = 1.00$	1.04	0.15	1.03	0.15	0.96	0.12	0.98	0.13
	$\beta_3 = -1.00$	−1.00	0.48	−1.00	0.49	−0.97	0.47	−0.96	0.47
	$\beta_4 = -0.50$	−0.52	0.23	−0.52	0.23	−0.46	0.22	−0.47	0.22
	ψ_{11}	0.49	0.39	0.48	0.39	0.31	0.24	0.31	0.23
1.00	$\beta_1 = -2.50$	−2.52	0.36	−2.52	0.35	−2.21	0.28	−2.28	0.29
	$\beta_2 = 1.00$	1.02	0.15	1.02	0.15	0.90	0.12	0.93	0.12
	$\beta_3 = -1.00$	−1.14	0.55	−1.14	0.56	−1.05	0.50	−1.04	0.50
	$\beta_4 = -0.50$	−0.47	0.23	−0.46	0.23	−0.37	0.21	−0.40	0.21
	ψ_{11}	0.93	0.55	0.93	0.54	0.59	0.29	0.58	0.29

		Method							
		LMLE		MLE		MQL		PQL	
ψ_{11}	Parameter	Bias	MSE	Bias	MSE	Bias	MSE	Bias	MSE
0.25	$\beta_1 = -2.50$	−0.03	0.10	−0.03	0.10	0.08	0.07	0.07	0.08
	$\beta_2 = 1.00$	0.01	0.02	0.01	0.02	−0.03	0.02	−0.02	0.02
	$\beta_3 = -1.00$	−0.04	0.23	−0.04	0.23	−0.03	0.22	−0.02	0.22
	$\beta_4 = -0.50$	−0.01	0.05	−0.01	0.05	0.03	0.04	0.02	0.04
	ψ_{11}	0.05	0.12	0.04	0.12	−0.05	0.05	−0.05	0.05
0.50	$\beta_1 = -2.50$	−0.05	0.11	−0.05	0.11	0.13	0.09	0.10	0.08
	$\beta_2 = 1.00$	0.04	0.02	0.03	0.02	−0.04	0.02	−0.02	0.02
	$\beta_3 = -1.00$	−0.00	0.23	−0.00	0.24	0.03	0.22	0.04	0.22
	$\beta_4 = -0.50$	−0.02	0.05	−0.02	0.05	0.04	0.05	0.03	0.05
	ψ_{11}	−0.01	0.15	−0.02	0.15	−0.19	0.09	−0.19	0.09
1.00	$\beta_1 = -2.50$	−0.02	0.13	−0.02	0.12	0.29	0.16	0.22	0.13
	$\beta_2 = 1.00$	0.02	0.02	0.02	0.02	−0.10	0.02	−0.07	0.02
	$\beta_3 = -1.00$	−0.14	0.32	−0.14	0.33	−0.05	0.25	−0.04	0.25
	$\beta_4 = -0.50$	0.03	0.06	0.04	0.06	0.13	0.06	0.10	0.05
	ψ_{11}	−0.07	0.30	−0.07	0.29	−0.41	0.26	−0.42	0.26

A GLIMMIX convergence status of 1 (i.e., convergence criterion satisfied) was achieved in every case but one involving MLE with $\psi_{11} = 0.50$. The output summarizes both the mean and standard deviation of each of the parameter estimates as well as the mean bias (bias $= \sum(\widehat{\theta}_i - \theta)/200$) and average mean square error (MSE $= \sum(\widehat{\theta}_i - \theta)^2/200$) where θ represents any generic parameter estimate. In comparing the various regression parameter estimates, we find that when $\psi_{11} = 0.25$, the MQL and PQL estimators were comparable to each other and similar to the Laplace (LMLE) and adaptive Gaussian quadrature (MLE) maximum likelihood estimators. This is not unexpected based on the small variance asymptotics (i.e., as $\psi \to 0$) discussed above. However, as the value of the random intercept variance increases (from 0.25 to 1.00), the bias in the regression parameter estimates also increases for both MQL and PQL with the latter being slightly less bias. The LMLE and MLE parameter estimates were similar to each other for all three values of ψ_{11} and generally had lower mean bias than MQL or PQL except when $\psi_{11} = 1.00$ where the LMLE and MLE estimates of the between-subject treatment effect parameter β_3 displayed greater bias compared to MQL and PQL. In terms of average MSE, the MQL and PQL estimators were generally comparable to or had slightly lower MSE's than the LMLE and MLE based procedures.

There was considerable variation in the estimate of the variance component ψ_{11} for all four estimators. In fact, 49 of the 200 replicated datasets (25%) generated with $\psi_{11} = 0.25$ resulted in a non-positive definite estimate of ψ_{11} as underscored by the GLIMMIX message

```
NOTE: Estimated G matrix is not positive definite.
NOTE: The covariance matrix is the zero matrix.
```

issued within the SAS log for all four estimators. This became less of an issue for larger values of ψ_{11} where 23 of 200 replicated datasets (12%) resulted in a non-positive definite estimate of ψ_{11} when $\psi_{11} = 0.50$ and only 2 of 200 (1%) resulted in a non-positive definite estimate of ψ_{11} when $\psi_{11} = 1.00$. Even in those cases where the estimate of ψ_{11} was positive definite, the MQL and PQL methods had a tendency to underestimate the variance component ψ_{11} as shown in Figure B.1.

Shown in Output B.2 are the results from 200 replicated datasets generated under the scenario described above for Case 2 in which we include a random intercept effect and a random slope effect within the mixed-effects logistic regression model. Under Case 2, a GLIMMIX convergence status of 1 (i.e., convergence criterion satisfied) was achieved in every case but one involving ML estimation using adaptive Gaussian quadrature.

Figure B.1 Distribution of the random intercept variance component estimate under Case 1 of the mixed-effects binary logistic regression model

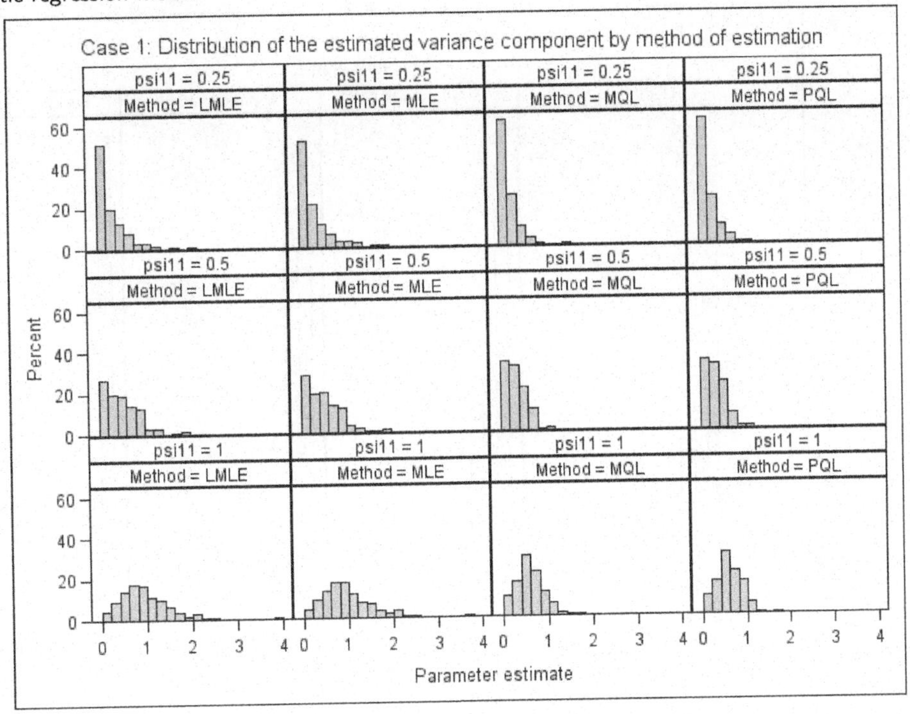

Output B.2: A comparison of estimates for a mixed-effects binary logistic regression model under Case 2 with $p = 7$ observations per subject.

	Method							
	LMLE		MLE		MQL		PQL	
Parameter	Mean	Std	Mean	Std	Mean	Std	Mean	Std
$\beta_1 = -2.50$	−2.58	0.35	−2.54	0.34	−2.21	0.26	−2.25	0.26
$\beta_2 = 1.00$	1.03	0.17	1.01	0.16	0.86	0.12	0.89	0.13
$\beta_3 = -1.00$	−1.03	0.43	−1.01	0.44	−0.83	0.42	−0.84	0.41
$\beta_4 = -0.50$	−0.56	0.23	−0.51	0.22	−0.37	0.20	−0.40	0.21
$\psi_{11} = 0.50$	0.63	0.51	0.51	0.50	0.26	0.27	0.25	0.26
$\psi_{22} = 0.25$	0.28	0.18	0.24	0.16	0.09	0.06	0.11	0.07

	Method							
	LMLE		MLE		MQL		PQL	
Parameter	Bias	MSE	Bias	MSE	Bias	MSE	Bias	MSE
$\beta_1 = -2.50$	−0.08	0.13	−0.04	0.11	0.29	0.15	0.25	0.13
$\beta_2 = 1.00$	0.03	0.03	0.01	0.03	−0.14	0.03	−0.11	0.03
$\beta_3 = -1.00$	−0.03	0.19	−0.01	0.19	0.17	0.20	0.16	0.20
$\beta_4 = -0.50$	−0.06	0.06	−0.01	0.05	0.13	0.06	0.10	0.05
$\psi_{11} = 0.50$	0.13	0.28	0.01	0.25	−0.24	0.13	−0.25	0.13
$\psi_{22} = 0.25$	0.03	0.03	−0.01	0.02	−0.16	0.03	−0.14	0.02

Under this scenario with two independent random effects, the ML estimator based on adaptive Gaussian quadrature clearly outperformed the LMLE, MQL and PQL procedures in terms of mean bias and MSE (except for ψ_{11} where the average MSE was lower for both MQL and PQL). The Laplace approximated ML estimator also outperformed the MQL and PQL estimators in terms of mean bias and MSE (except for ψ_{11} where MQL and PQL had lower MSE). In fact, the Laplace approximated MLE compared favorably with the more computer intensive AGQ-based MLE.

Figure B.2 Distribution of the random intercept variance component estimates under Case 2 of the mixed-effects binary logistic regression model

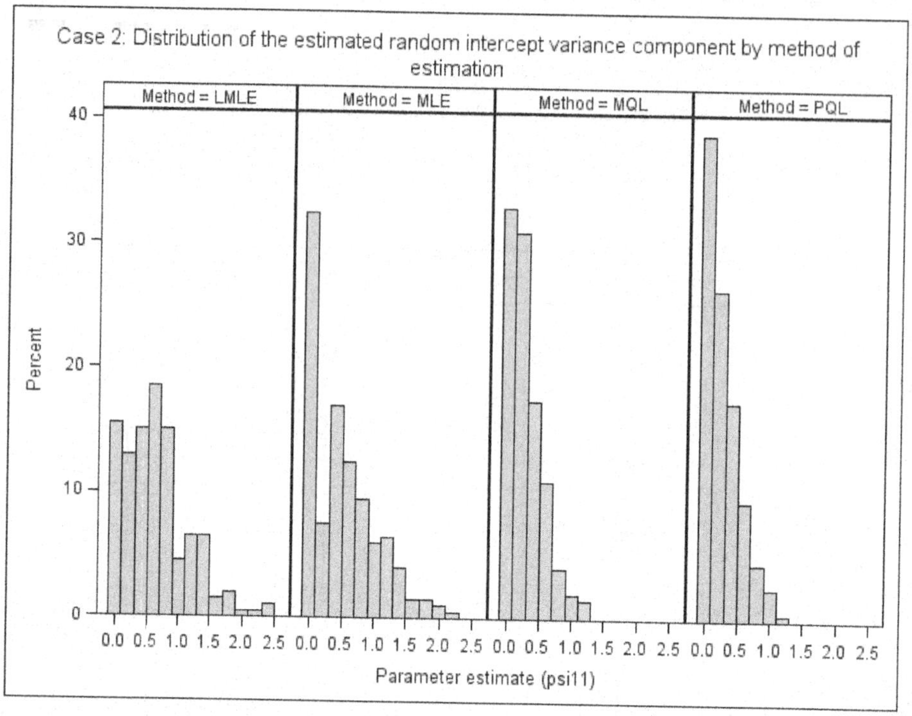

As seen in Case 1 (Output B.1), there was considerable variation in the estimated values of both ψ_{11} and ψ_{22} for all four estimation procedures. Letting $\boldsymbol{\Psi}(\boldsymbol{\theta}_b) = \begin{pmatrix} \psi_{11} & 0 \\ 0 & \psi_{22} \end{pmatrix}$, the GLIMMIX procedure issued a warning that $\boldsymbol{\Psi}(\boldsymbol{\theta}_b)$ was not positive definite with the following frequencies:

Method	Frequency	Percent
LMLE	30/200	15%
MLE	90/200	45%
MQL	68/200	34%
PQL	68/200	34%

such that one or both components of $\boldsymbol{\Psi}(\boldsymbol{\theta}_b)$ were set to zero. The variation in $\widehat{\psi}_{11}$ is displayed graphically in Figure B.2. The MQL and PQL estimation schemes consistently underestimate the random intercept variance ψ_{11} compared with the LMLE and MLE schemes whereas both the LMLE and MLE schemes display a considerably broader range of values which explains why they each exhibit greater MSE's.

Similarly, Figure B.3 displays the variation seen in the random slope variance component estimates, $\widehat{\psi}_{22}$. We again find that the MQL and PQL estimation schemes consistently underestimate the random slope variance ψ_{22} compared with the LMLE and MLE schemes. On the other hand, the LMLE and MLE schemes yield, on average, an unbiased estimate of ψ_{22} although again we find there is considerable variation in the estimates.

Based on this very limited simulation in which we hold n and p fixed but vary the random effect variance components, we find that within a discrete binary logistic regression setting, the ML estimator based on adaptive Gaussian quadrature yields reasonably unbiased estimates of the model parameters. This also holds, albeit to a slightly lesser extent, when ML estimation is based on the Laplace approximation. It is not entirely clear why the ML estimators of the between-subject treatment effect (i.e., β_3) display greater

Figure B.3 Distribution of the random slope variance component estimates under Case 2 of the mixed-effects binary logistic regression model

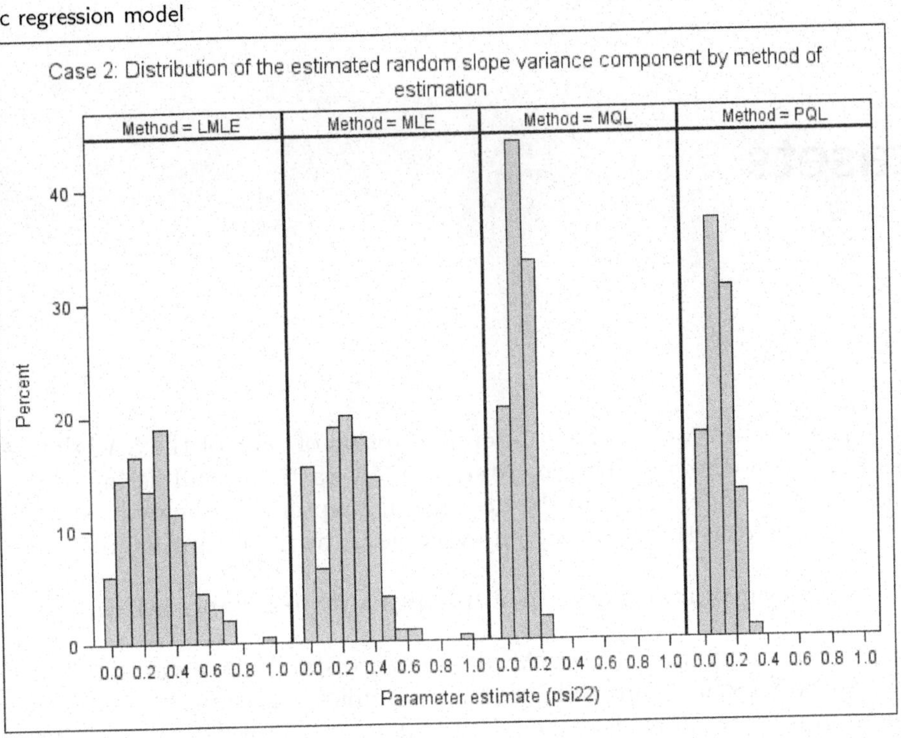

bias compared to either the MQL or PQL estimators when $\psi_{11} = 1.00$ (Output B.1). Some insight into this may be found in the work of Bellamy et. al. (2005) in which the authors show both analytically and through simulation that PQL works well and even outperforms MLE when one has small numbers of subjects/clusters but a large number of observations per subject/cluster. Their simulations are based on a binary logistic regression model with a single random intercept effect and unlike the present study, they vary both the number of clusters (n) and the cluster size (p). However, they do not vary the random intercept variance within their simulations. Further simulations in which (n, p, ψ) are all varied in a systematic fashion may be required to better compare the performance of these different estimators under the discrete binary logistic regression setting and other possibly under other settings involving continuous outcome measurements. Finally, we should keep in mind that simulations are a limited tool for studying the properties of estimators for nonlinear statistical models because they depend heavily on the choice of the parameter values.

Appendix C

Datasets

In this appendix, we provide a brief description of each of the SAS datasets used in the book in the order in which they first occur. A partial or full listing of the data is presented either in this appendix or in the designated chapter as indicated. Also, the datasets and programs used in this book are all available through the author's Web page at

http://support.sas.com/publishing/authors/vonesh.html.

In some of the examples, the SAS datasets are stand alone datasets that are called from within the SAS program while in other examples, the dataset will be provided within the program itself (such as described for the dental growth data below). The output/data analyses for this book were generated using SAS software (Copyright, SAS Institute Inc. SAS and all other SAS Institute Inc. product or service names are registered trademarks or trademarks of SAS Institute Inc., Cary, NC, USA).

C.1 Dental growth data

The dental growth data listed in Output 2.1 of Chapter 2 are from Table 1 of Potthoff and Roy (1964) and are made available here with permission from Oxford University Press.

C.2 Bone mineral density data

The bone mineral density data are from Vonesh and Chinchilli (1997; Table 5.4.1, pp. 228-230) and reproduced here with permission from Taylor & Francis Group, LLC - Books.

C.3 ADEMEX adequacy data

The ADEMEX adequacy data are from Paniagua et. al. (2002) and were made available courtesy of Dr. Ramon Paniagua. The data are available as a SAS dataset, ADEMEX_adequacy_data.

C.4 MCM2 biomarker data

The MCM2 biomarker data are from Helenowski et. al. (2011) and were made available courtesy of Irene Helenowski. The data are available as a SAS dataset, MCM2_data.

C.5 Estrogen hormone data

The estrogen hormone data are from Gann et. al. (2006) and were made available courtesy of Dr. Peter Gann. The data are available as a SAS dataset, Estradiol_data.

C.6 ADEMEX peritonitis and hospitalization data

The ADEMEX peritonitis and hospitalization data (Examples 4.3.1, 5.2.4, and 7.1.1) are from Paniagua et. al. (2002) and were made available courtesy of Dr. Ramon Paniagua. The data are available as a SAS dataset, ADEMEX_Peritonitis_data.

C.7 Respiratory disorder data

The respiratory disorder data are from Stokes et. al. (2000) and is available in the SAS Sample Library under Example 5 for PROC GENMOD. A partial listing of the data as analyzed in the book and which is based on the dataset made available courtesy of SAS Institute Inc., is given in Output 4.5.

C.8 Epileptic seizure data

The epileptic seizure data are from Thall and Vail (1990) and is available in the SAS Sample Library under Example 7 for PROC GENMOD. A partial listing of the data, which were made available courtesy of SAS Institute Inc., is given in Output 4.8.

C.9 Schizophrenia data

The schizophrenia data were made available courtesy of Dr. Donald Hedeker with permission from Dr. Nina Schooler. The data are available as a SAS dataset, Schizophrenia_data. A partial listing of the data is given in Output 4.14.

C.10 LDH enzyme leakage data

The LDH enzyme leakage data are from Gennings, Chinchilli and Carter (1989) and are reproduced here with permission from *The Journal of the American Statistical Association* (Copyright 1989 by the American Statistical Association. All rights reserved). The data are available as a SAS dataset, LDH_Enzyme_data.

C.11 Orange tree data

The orange tree data are from Draper and Smith (1981) and are reproduced here with permission from John Wiley and Sons, Inc. A complete listing of the data is given in Output 4.23 and consists of the tree ID (`Tree`), the days on which trunk circumference was measured (`Days`) and the trunk circumference of the tree measured in mm (`y`).

C.12 Soybean growth data

The soybean growth data are from Davidian and Giltinan (1993a) and are reproduced here with permission from the publisher (Taylor and Francis Group, http://www.informaworld.com). The data are available as a SAS dataset, Soybean_data.

C.13 High flux hemodialyzer data

The high flux hemodialyzer data are from Vonesh and Carter (1992) and the dataset, which is available with the software program MIXNLIN 3.1 (Copyright 1995 by Edward F. Vonesh), was made available here with permission from the author. The data are available as a SAS dataset, Hemodialyzer_data.

C.14 Cefamandole pharmacokinetic data

The cefamandole PK data are from Davidian and Giltinan (1993b) and are reproduced here with permission from Elsevier. The data are available as a SAS dataset, Cefamandole_data.

C.15 MDRD data

The MDRD data are provided courtesy of Dr. Gerald Beck from the Cleveland Clinic Foundation on behalf of the MDRD Study Group. The data are available as a SAS dataset, MDRD_data.

C.16 Theophylline data

The theophylline data are from Boeckmann et. al. (1992) and subsequently Pinheiro and Bates (1995) and is available in the SAS Sample Library under Documentation Example 1 for PROC NLMIXED. A partial listing of the data, as made available courtesy of SAS Institute Inc., is given in Output 7.3.

C.17 Phenobarbital data

The phenobarbital data are from Grasela and Donn (1985) and are made available courtesy of the Resource Facility for Population Kinetics (RFPK) from their website at http://www.rfpk.washington.edu with funding support from NIH/NIBIB grant P41-EB01975. The data were downloaded from the RFPK website and a SAS dataset, `phenobarbital_data`, was created.

C.18 ADEMEX GFR and survival data

The ADEMEX GFR and survival data (Examples 7.3.1) are from Paniagua et. al. (2002) and were made available courtesy of Dr. Ramon Paniagua. The data are available as a SAS dataset, ADEMEX_GFR_data.

Appendix D

Select SAS macros

In this appendix we present select SAS macros that are used in several examples throughout the book. Each macro is available through the author's Web page. The macros are provided as is without any expressed or implied warranty.

D.1 The GOF Macro

The %GOF macro computes summary goodness-of-fit measures based on methods described by Vonesh et. al. (1996) and Vonesh and Chinchilli (1997).

D.2 The GLIMMIX_GOF Macro

The %GLIMMIX_GOF macro computes the same summary goodness-of-fit measures but the %GLIMMIX_GOF macro was written for exclusive use with the GLIMMIX procedure.

D.3 The CCC Macro

The %CCC macro estimates concordance correlations and was used in the analysis of the MCM2 biomarker data (Example 2.2.4).

D.4 The CONCORR Macro

The %CONCORR macro computes pairwise concordance correlations and was also used in the analysis of the MCM2 biomarker data (Example 2.2.4).

D.5 The COVPARMS Macro

The %COVPARMS macro was written exclusively for use with the NLMIXED procedure and is designed to provide users with reasonable starting values for both the fixed-effects parameters and the variance-covariance parameters of a nonlinear or generalized nonlinear mixed-effects model.

D.6 The VECH Macro

The %VECH macro was written exclusively for use with the NLMIXED procedure. It was developed for use in applications where one wishes to fit a GNLME model to a set of data assuming an intrasubject covariance structure other than the default independence structure that NLMIXED assumes (see §5.3.3 for an approach one can take to overcome this limitation of NLMIXED).

References

Aalen, Odd. O., Ornulf Borgan, and Hakon K. Gjessing. 2008. *Survival and Event History Analysis: A Process Point of View*. New York: Springer.

Abramowitz, Milton, and Irene A. Stegun. 1972. *Handbook of Mathematical Functions with Formulas, Graphs, and Mathematical Tables*. New York: Dover Publications.

Agresti, Alan. 1990. *Categorical Data Analysis*. New York: Wiley.

Allison, Paul. 2010. *Survival Analysis Using SAS: A Practical Guide*. Second edition. Cary, North Carolina: SAS Institute Inc.

Anderson, T. W. 1984. *An Introduction to Multivariate Statistical Analysi. Second edition*. New York: Wiley.

Barndorff-Nielsen, Ole E., and D. R. Cox. 1989. *Asymptotic Techniques for Use in Statistics*. New York: Chapman & Hall.

Barnhart, Huiman X., Michael Haber, and Jingli Song. 2002. "Overall Concordance Correlation Coefficient for Evaluating Agreement Among Multiple Observers." *Biometrics* 58:1020–1027.

Barnhart, Huiman X., and John M. Williamson. 2001. "Modeling Concordance Correlation via GEE to Evaluate Reproducibility." *Biometrics* 57:931–940.

Bates, Douglas M., and Donald G. Watts. 1980. "Relative curvature measures of nonlinearity." *Journal of the Royal Statistical Society*, Series B 42:1–25.

_____. 1988. *Nonlinear Regression Analysis and its Applications*. New York: Wiley.

Beal, S. L., and L. B. Sheiner. 1982. "Estimating population kinetics." *Critical Reviews in Biomedical Engineering* 8:195–222.

_____. 1988. "Heteroscedastic nonlinear regression." *Technometrics* 30:327–338.

Beck, G. J., R. L. Berg, and C. H. Coggins et al. 1991. "Design and statistical issues of the Modification of Diet in Renal Disease Trial: The Modification of Diet in Renal Disease Study Group." *Controlled Clinical Trials* 12:566–586.

Boeckmann, Alison J., Lewis B. Sheiner, and Stuart L. Beal. 1994. *NONMEM Users Guide, Part V, Introductory Guide*. San Francisco, California: University of California.

Boos, Dennis D. 1992. "On generalized score tests." *The American Statistician* 46:327–333.

Booth, James G., and James P. Hobert. 1998. "Standard errors of prediction in generalized linear mixed models." *Journal of the American Statistical Association* 93:262–272.

Bradley, Ralph A., and John J. Gart. 1962. "The asymptotic properties of ML estimators when sampling from associated populations." *Biometrika* 49:205–214.

Breslow, N. E., and D. G. Clayton. 1993. "Approximate inference in generalized linear mixed models." *Journal of the American Statistical Association* 88:9–25.

Breslow, Norman E., and Xihong Lin. 1995. "Bias correction in generalized linear mixed models with a single component of dispersion." *Biometrika* 82:81–91.

Carlin, Bradley P., and Thomas A. Louis. 2000. *Bayes and Empirical Bayes Methods for Data Analysis. Second edition.* Boca Raton, Florida: Chapman & Hall.

Carrasco, Josep L., and Lluís Jover. 2003. "Estimating the Generalized Concordance Correlation Coefficient through Variance Components." *Biometrics* 59:849–858.

Carroll, Raymond J., and David Ruppert. 1988. *Transformation and Weighting in Regression.* New York: Chapman & Hall.

Chaganty, N. Rao. 1997. "An alternative approach to the analysis of longitudinal data via generalized estimating equations." *Journal of Statistical Planning and Inference* 63:39–54.

Chen, Pei-Yun, and Anastasios A. Tsiatis. 2001. "Causal inference on the difference of the restricted mean lifetime between two groups." *Biometrics* 57:1030–1038.

Chi, Eric M., and Gregory C. Reinsel. 1989. "Models for longitudinal data with random effects and AR(1) errors." *Journal of the American Statistical Association* 84:452–459.

Cole, J. O., and the NIMH-PSC Collaborative Study Group. 1964. " Phenothiazine treatment in acute schizophrenia: Effectiveness." *Archives of General Psychiatry* 10:246–261.

Cressie, Noel. A. C. 1993. *Statistics for Spatial Data. Revised Edition.* New York: Wiley.

Crowder, Martin. 1995. "On the use of a working correlation matrix in using generalised linear models for repeated measures." *Biometrika* 82:407–410.

Curran, D., M. Bacchi, S. F. Hsu Schmitz, G. Molenberghs, and R. J. Sylvester. 1998. "Identifying the types of missingness in quality of life data from clinical trials." *Statistics in Medicine* 17:739–756.

Daniels, Michael J., and Joseph W. Hogan. 2008. *Missing Data in Longitudinal Studies: Strategies for Bayesian Modeling and Sensitivity Analysis.* Boca Raton, Florida: Chapman & Hall/CRC.

Davidian, M., and A. R. Gallant. 1992. "Smooth nonparametric maximum likelihood estimation for population pharmacokinetics, with application to quinidine." *Journal of Pharmacokinetics and Biopharmaceutics* 20:529–556.

_______. 1993. "The nonlinear mixed effects model with a smooth random effects density." *Biometrika* 80:475–488.

Davidian, Marie, and David M. Giltinan. 1995. *Nonlinear models for repeated measurement data.* London: Chapman & Hall.

_______. 1993a. "Some general estimation methods for nonlinear mixed-effects models." *Journal of Biopharmaceutical Statistics* 3:23–55.

_______. 1993b. "Analysis of repeated measurement data using the nonlinear mixed effects model." *Chemometrics and Intelligent Laboratory Systems* 20:1–24.

Davis, Charles S. 2002. *Statistical Methods for the Analysis of Repeated Measurements.* New York: Springer.

Demidenko, Eugene. 1997. "Asymptotic properties of nonlinear mixed-effects models." In *Modelling Longitudinal and Spatially Correlated Data: Methods, Applications, and Future Directions*, edited by T. G. Gregoire, D. R. Brillinger, P. J. Diggle, E. Russek-Cohen, W. G. Warren, and R. D. Wolfinger, 49–62. New York: Springer.

_______. 2004. *Mixed Models: Theory and Applications.* Hoboken, New Jersey: Wiley.

Diggle, Peter, Daniel Farewell, and Robin Henderson. 2007. "Analysis of longitudinal data with drop-out: objectives, assumptions and a proposal." Applied Statistics 56:499–550.

Diggle, P., and M. G. Kenward. 1994. "Informative drop-out in longitudinal data analysis (with discussion)." *Applied Statistics* 43:49–93.

Diggle, Peter J. 1988. "An approach to the analysis of repeated measurements." *Biometrics* 44:959–971.

_______. 1989. "Testing for random dropouts in repeated measurement data." *Biometrics* 45:1255–1258.

Diggle, Peter J., Kung-Yee Liang, and Scott L. Zeger. 1994. *Analysis of Longitudinal Data*. Oxford: Oxford University Press.

Diggle, Peter J., Ines Sousa, and Amanda G. Chetwynd. 2008. "Joint modelling of repeated measurements and time-to-event outcomes: The fourth Armitage lecture." *Statistics in Medicine* 27:2981–2998.

Draper, Norman R., and H. Smith. 1981. *Applied Regression Analysis*. Second edition. New York: Wiley.

Dupuy, Jean-Francois, and Mounir Mesbah. 2002. "Joint modeling of event time and nonignorable missing longitudinal data." *Lifetime Data Analysis* 8:99–115.

Edwards, Lloyd J., Keith E. Muller, Russell D. Wolfinger, Bahjat F. Qaqish, and Oliver Schabenberger. 2008. "An R2 statistic for effects in the linear mixed model." *Statistics in Medicine* 27:6137–6157.

Fahrmeir, Ludwig, and Gerhard Tutz. 1994. *Multivariate Statistical Modelling Based on Generalized Linear Models*. New York: Springer-Verlag.

Faucett, Cheryl L., and Duncan C. Thomas. 1996. "Simultaneously modeling censored survival data and repeatedly measured covariates: A Gibbs sampling approach." *Statistics in Medicine* 15:1663–1685.

Fearn, T. 1975. "A Bayesian approach to growth curves." *Biometrika* 62:89–100.

Fitzmaurice, Garrett M. 1995. "A caveat concerning independence estimating equations with multivariate binary data." *Biometrics* 51:309–317.

Fitzmaurice, Garrett M., and Nan M. Laird. 2000. "Generalized linear mixture models for handling nonignorable dropouts in longitudinal studies." *Biostatistics* 1:141–156.

Fitzmaurice, Garrett M., Nan M. Laird, and Lucy Shneyer. 2001. "An alternative parameterization of the general linear mixture model for longitudinal data with non-ignorable drop-outs." *Statistics in Medicine* 20:1009–1021.

Fitzmaurice, Garrett M., Nan M. Laird, and James H.Ware. 2004. *Applied Longitudinal Analysis*. Hoboken, New Jersey: Wiley Interscience.

Follmann, Dean, and Margaret Wu. 1995. "An approximate generalized linear model with random effects for informative missing data." *Biometrics* 51:151–168.

Frees, Edward W. 2004. *Longitudinal and Panel Data: Analysis and Applications in the Social Sciences*. New York: Cambridge University Press.

Fuller, Wayne A. 1976. *Introduction to Statistical Time Series*. New York:Wiley.

Gallant, A. Ronald. 1975. "Seemingly unrelated nonlinear regressions." *Journal of Econometrics* 3:35–50.

________. 1987. *Nonlinear Statistical Models*. New York: Wiley.

Geisser, Seymour. 1970. "Bayesian analysis of growth curves." *Sankhya A* 32:53–64.

Gelfand, Alan E., Susan E. Hills, Amy Racine-Poon, and Adrian F. Smith. 1990. "Illustration of Bayesian inference in normal data models using Gibbs sampling." *Journal of the American Statistical Association* 85:972–985.

Gelman, Andrew, J. B. Carlin, Hal S. Stern, and Donald B. Rubin. 2004. *Bayesian Data Analysis*. Second edition. Boca Raton, Florida: Chapman & Hall/CRC.

Gennings, Chris, Vernon M. Chinchilli, and Walter H. Carter. 1989. "Response surface analysis with correlated data: A nonlinear model approach." *Journal of the American Statistical Association* 84:805–809.

Geweke, J. 1989. "Bayesian inference in econometric models using Monte Carlo integration." *Econometrica* 57:1317–1339.

Gibaldi, Milo, and Donald Perrier. 1982. *Pharmacokinetics*. Second edition. New York: Marcel Dekker, Inc.

Goldstein, Harvey. 1991. "Nonlinear multilevel models, with an application to discrete response data." *Biometrika* 78:45–51.

Golub, Gene. H., and John H. Welsch. 1969. "Calculation of Gauss quadrature rules." *Mathematics of Computation* 23:221–230.

Gourieroux, C., A. Monfort, and A. Trognon. 1984. "Pseudo Maximum Likelihood Methods: Theory." *Econometrica* 52:681–700.

Grasela, T. H., and S. M. Donn. 1985. "Neonatal population pharmacokinetics of phenobarbital derived from routine clinical data." *Developmental Pharmacology and therapeutics* 8:374–383.

Graybill, Franklin A. 1976. *Theory and Application of the Linear Model*. North Scituate, Massachusetts: Duxbury Press.

Green, Peter J. 1987. "Penalized likelihood for general semi-parametric regression models." *International Statistical Review* 55:245–259.

Greenhouse, Samuel W., and Seymour Geisser. 1959. "On Methods in the Analysis of Profile Data." *Psychometrika* 32:95–112.

Grizzle, James E., and David M. Allen. 1969. "Analysis of Growth and Dose Response Curves." *Biometrics* 25:357–381.

Gruttola, Victor , and Xin Ming Tu. 1994. "Modelling progression of CD4-lymphocyte count and its relationship to survival time." *Biometrics* 50:1003–1014.

Gumpertz, Marcia L., and Sastry G. Pantula. 1992. "Nonlinear regression with variance components." *Journal of the American Statistical Association* 87:201–209.

Guo, Xu, and Bradley P. Carlin. 2004. "Separate and joint modeling of longitudinal and event time data using standard computer packages." *The American Statistician* 58:16–24.

Hall, Daniel B., and Thomas A. Severini. 1998. "Extended generalized estimating equations for clustered data." *Journal of the American Statistical Association* 93:1365–1375.

Harville, David A. 1976. "Extension of the Gauss-Markov theorem to include the estimation of random effects." *The Annals of Statistics* 4:384–395.

_______. 1977. "Maximum likelihood approaches to variance component estimation and to related problems." *Journal of the American Statistical Association* 72:320–338.

_______. 1997. Matrix Algebra from a Statistician's Perspective. New York: Springer-Verlag.

Hedeker, Donald, and Robert D. Gibbons. 1994. "A random-effects ordinal regression model for multilevel analysis." *Biometrics* 50:933–944.

_______. 1997. "Application of random-effects pattern-mixture models for missing data in longitudinal studies." *Psychological Methods* 2:64–78.

Helenowski, Irene B., Edward F. Vonesh, Hakan Demirtas, Alfred W. Rademaker, Vijayalakshmi Ananthanarayanan, Peter H. Gann, and Borko D. Jovanovic. 2011. "Defining reproducibility statistics as a function of the spatial covariance structures in biomarker studies." *The International Journal of Biostatistics* 7:1–21.

Henderson, C. R. 1953. "Estimation of variance and covariance components." *Biometrics* 9:226–252.

_______. 1963. "Selection index and expected genetic advance." In *Statistical Genetics and Plant Breeding*, edited by W. D. Hanson and H. F. Robinson, 141-163. Washington, D.C.: National Academy of Sciences and National Research Council Publication No 982.

Henderson, Harold V., and S. R. Searle. 1979. "Vec and Vech operators for matrices, with some uses in Jacobians and multivariate statistics." *Canadian Journal of Statistics* 7:65–81.

_______. 1981. "Vec-Permutation matrix, the vec operator and Kronecker products: A review." *Linear and Multilinear* Algebra 9:271–288.

Henderson, L. W. 1996. "Biophysics of ultrafiltration and hemofiltration." In *Replacement of Renal Function by Dialysis*, edited by J. F. Winchester, C. Jacobs, C. M. Kjellstrand, and K. M. Koch, Fourth edition. Boston: Kluwer Academic Publishers.

Henderson, Robin, Peter Diggle, and Angela Dobson. 2000. "Joint modelling of longitudinal measurements and event time data." *Biostatistics* 1:465–480.

Hoel, Paul G., Sidney C. Port, and Charles J. Stone. 1971. *Introduction to Probability Theory*. Boston, Massachusetts: Houghton Mifflin Company.

Hogan, Joseph W., and Nan M. Laird. 1997a. "Mixture models for the joint distribution of repeated measures and event times." *Statistics in Medicine* 16:239–257.

_______. 1997b. "Model-based approaches to analysing incomplete longitudinal and failure time data." *Statistics in Medicine* 16:259–272.

Hogan, Joseph W., Jason Roy, and Christina Korkontzelou. 2004. "Tutorial in biostatistics: Handling drop-out in longitudinal studies." *Statistics in Medicine* 23:1455–1497.

Holford, Theodore R. 1980. "The analysis of rates and of survivorship using log-linear models." *Biometrics* 36:299–305.

Hsiao, Cheng. 2003. *Analysis of Panel Data*. Second edition. New York: Cambridge University Press.

Huynh, Huynh, and Leonard S. Feldt. 1970. "Conditions Under Which Mean Square Ratios in Repeated Measurements Designs Have Exact F-Distributions." *Journal of the American Statistical Association* 65:1582–1589.

_______. 1976. "Estimation of the Box Correction for Degrees of Freedom from Sample Data in Randomized Block and Split-Plot Designs." *Journal of Educational Statistics* 1:69–82.

Irwin, J. O. 1949. "The standard error of an estimate of expectation of life, with special reference to expectation of tumourless life in experiments with mice." *Journal of Hygiene* 47:188–189.

Jennrich, Robert I., and Mark D. Schluchter. 1986. "Unbalanced repeated-measures models with structured covariance matrices." *Biometrics* 42:805–820.

Johnson, Norman L., and Samuel Kotz. 1969. *Discrete Distributions*. New York: Wiley.

Jones, Richard H. 1990. "Serial correlation or random subject effects?" *Communications in Statistics: Simulation and Computation A* 19:1105–1123.

Jones, Richard H., and Francis Boadi-Boateng. 1991. "Unequally spaced longitudinal data with AR(1) serial correlation." *Biometrics* 47:161–175.

Kalbfleisch, John D., and Ross L. Prentice. 2002. *The Statistical Analysis of Failure Time Data*. Second edition. Hoboken, New Jersey: Wiley.

Karrison, Theodore. 1987. "Restricted mean life with adjustment for covariates." *Journal of the American Statistical Association* 82:1169–1176.

Kenward, Michael. G. 1998. "Selection models for repeated measurements with non-random dropout: An illustration of sensitivity." *Statistics in Medicine* 17:2723–2732.

Kenward, M. G., and G. Molenberghs. 1998. "Likelihood based frequentist inference when data are missing at random." *Statistical Science* 13:236–247.

Khatri, C. G. 1966. "A Note on a MANOVA Model Applied to Problems in Growth Curve." *Annals of the Institute of Statistical Mathematics* 18:75–86.

Kim, Kevin, and Neil Timm. 2007. *Univariate and multivariate general linear models: Theory and applications with SAS*. Second edition. Boca Raton, Florida: Chapman & Hall/CRC.

Klahr, S., A. S. Levey, G. J. Beck, A. W. Caggiula, L. Hunsicker, J. W. Kusek, and G. Striker for the Modification of Diet in Renal Disease Study Group. 1994. "The effects of dietary protein restriction and blood-pressure control on the progression of chronic renal disease." *New England Journal of Medicine* 330:877–884.

Koch, Gary G., Gregory J. Carr, Ingrid A. Amara, Maura E. Stokes, and Thomas J. Uryniak. 1990. "Categorical Data Analysis." In *Statistical Methodology in the Pharmaceutical Sciences*, edited by Donald A. Berry, 389–473. New York: Marcel Dekker.

Koch, Gary G., J. Richard Landis, Jean L. Freeman, Daniel H. Freeman, and Robert G. Lehnen. 1977. "A general methodology for the analysis of experiments with repeated measurement of categorical data." *Biometrics* 33:133–158.

Laird, Nan M., Nicholas Lange, and Daniel Stram. 1987. "Maximum likelihood computations with repeated measures: Application of the EM algorithm." *Journal of the American Statistical Association* 82:97–105.

Laird, Nan M., and Thomas A. Louis. 1982. "Approximate posterior distributions for incomplete data problems." *Journal of the Royal Statistical Society* B 44:190–200.

Laird, Nan M., and James H.Ware. 1982. "Random-effects models for longitudinal data." *Biometrics* 38:963–974.

Latouche, A., R. Porcher, and S. Chevret. 2005. "A note on including time-dependent covariate in regression model for competing risks data." *Biometrical Journal* 47:807–814.

Lee, Y., and J. A. Nelder. 1996. "Hierarchical Generalized Linear Models." *Journal of the Royal Statistical Society* B 58:619–678.

________. 2001. "Hierarchical generalised linear models: A synthesis of generalised linear models, random-effect models and structured dispersions." *Biometrika* 88:987–1006.

Leppik, I. E. et al. 1985. "A double-blind crossover evaluation of progabide in partial seizures." Neurology 35:285.

Li, Jingjin, and Mark D. Schluchter. 2004. "Conditional mixed models adjusting for non-ignorable drop-out with administrative censoring in longitudinal studies." *Statistics in Medicine* 23:3489–3503.

Liang, Kung-Yee, and Scott L. Zeger. 1986. "Longitudinal data analysis using generalized linear models." *Biometrika* 73:13–22.

Lin, D. Y., L. J. Wei, and Z. Ying. 2002. "Model-Checking Techniques Based on Cumulative Residuals." *Biometrics* 58:1–12.

Lin, D. Y., and Z. Ying. 2001. "Semiparametric and nonparametric regression analysis of longitudinal data." *Journal of the American Statistical Association* 96:103–126.

Lin, Lawrence I-Kuei. 1989. "A concordance correlation coefficient to evaluate reproducibility." *Biometrics* 45:255–268.

________. 2000. "Correction to 'A concordance correlation coefficient to evaluate reproducibility.'" *Biometrics* 56:324–325.

Lin, Xihong, and Norman E. Breslow. 1996. "Bias correction in generalized linear mixed models with multiple components of dispersion." *Journal of the American Statistical Association* 91:1007–1016.

Lindley, D. V., and A. F. M. Smith. 1972. "Bayes estimates for the linear model (with discussion)." *Journal of the Royal Statistical Society* B 34:1–41.

Lindstrom, Mary J., and Douglas M. Bates. 1988. "Newton-Raphson and EM algorithms for linear mixed-effects models for repeated-measures data." *Journal of the American Statistical Association* 83:1014–1022.

________. 1990. "Nonlinear mixed effects models for repeated measures data." *Biometrics* 46:673–687.

Lipitz, Stuart R., Garrett M. Fitzmaurice, Endel J. Orav, and Nan M. Laird. 1994. "Performance of generalized estimating equations in practical situations." *Biometrics* 50:270–278.

Littell, Ramon C., George A. Milliken, Walter W. Stroup, and Russell D. Wolfinger. 1996. *SAS System for Mixed Models*. Cary, North Carolina: SAS Institute Inc.

Littell, Ramon C., George A. Milliken, Walter W. Stroup, Russell D. Wolfinger, and Oliver. Schabenberger. 2006. *SAS for Mixed Models*. Second edition. Cary, North Carolina: SAS Institute Inc.

Little, Roderick J. A. 1988. "A test of missing completely at random for multivariate data with missing values." *Journal of the American Statistical Association* 83:1198–1202.

_______. 1993. "Pattern-mixture models for multivariate incomplete data." *Journal of the American Statistical Association* 88:125–134.

_______. 1994. "A class of pattern-mixture models for normal incomplete data." *Biometrika* 81:471–483.

_______. 1995. "Modeling the drop-out mechanism in repeated-measures studies." *Journal of the American Statistical Association* 90:1112–1121.

Little, Roderick J. A., and Donald B. Rubin. 2002. *Statistical Analysis with Missing Data*. Second edition. Hoboken, New Jersey: Wiley.

Little, Roderick J. A., and Yongxiao Wang. 1996. "Pattern-mixture models for multivariate incomplete data with covariates." *Biometrics* 52:98–111.

Liu, Lei, and Xuelin Huang. 2008. "The use of Gaussian quadrature for estimation in frailty proportional hazards models." *Statistics in Medicine* 27:2665–2683.

Liu, Lei., Jennie Z. Ma, and John O'Quigley. 2008. "Joint analysis of multi-level repeated measures data and survival: an application to the end stage renal disease (ESRD) data." *Statistics in Medicine* 27:5679–5691.

Liu, Qing, and Donald A. Pierce. 1994. "A note on Gauss-Hermite Quadrature." *Biometrika* 81:624–629.

Lloyd, T., M. B. Andon, N. Rollings, J. K. Martel, J. R. Landis, L. M. Demers, D. F. Eggli, K. Kieselhorst, and H. E. Kulin. 1993. "Calcium supplementation and bone mineral density in adolescent girls." *Journal of the American Medical Association* 270:841–844.

Longford, N. T. 1994. "Logistic regression with random coefficients." *Computational Statistics & Data Analysis* 17:1–15.

Lysaght, M. J., E. F. Vonesh, and F. Gotch et al. 1991. "The influence of dialysis treatment modality on the decline of remaining renal function." *ASAIO Transactions* 37:598–604.

Mátyás, Laszlo, and Patrick Sevestre, eds. 2008. *The Econometrics of Panel Data: Fundamentals and Recent Developments in Theory and Practice*. Third edition. Berlin Heidelberg: Springer.

Magnus, Jan R., and H. Neudecker. 1979. "The commutation matrix: some properties and applications." *The Annals of Statistics* 7:381–394.

Mallet, A. 1986. "A maximum likelihood estimation method for random coefficient regression models." *Biometrika* 73:645–656.

Mallet, A., F. Mentre´, J. L. Steimer, and F. Lokiec. 1988. "Nonparametric maximum likelihood estimation for population pharmacokinetics, with application to Cyclosporine." *Journal of Pharmacokinetics and Biopharmaceutics* 16:311–327.

McCullagh, Peter. 1983. "Quasi-likelihood functions." The Annals of Statistics 11:59–67.

McCullagh, P., and J. A. Nelder. 1989. *Generalized Linear Models*. Second edition. New York: Chapman & Hall.

McCulloch, Charles E. 1997. "Maximum likelihood algorithms for generalized linear mixed models." *Journal of the American Statistical Association* 92:162–170.

Meier, Paul, Theodore Karrison, Rick Chappell, and Hui Xie. 2004. "The price of Kaplan-Meier." *Journal of the American Statistical Association* 99:890–896.

Moist, Louise M., Friedrich K. Port, and Sean M. Orzol et al. 2000. "Predictors of loss of residual renal function among new dialysis patients." *Journal of the American Society of Nephrology* 11:556–564.

Molenberghs, Geert, and Michael G. Kenward. 2007. *Missing Data in Clinical Studies*. Chichester, UK: Wiley.

Molenberghs, G., M. G. Kenward, and E. Lesaffre. 1997. "The analysis of longitudinal ordinal data with nonrandom drop-out." *Biometrika* 84:33–44.

Molenberghs, Geert, and Geert Verbeke. 2005. *Models for Discrete Longitudinal Data.* New York: Springer.

Molenberghs, Geert, Herbert Thijs, Ivy Jansen, Caroline Beunckens, Michael G. Kenward, Craig Mallinckrodt, and Raymond J. Carroll. 2004. "Analyzing incomplete longitudinal clinical trial data." *Biostatistics* 5:445–464.

Mori, Motomi, George G. Woodworth, and Robert F. Woolson. 1992. "Application of empirical Bayes inference to estimation of rate of change in the presence of informative right censoring." *Statistics in Medicine* 11:621–631.

Muller, Keith E., and Paul W. Stewart. 2006. *Linear Model Theory: Univariate, Multivariate, and Mixed Models*. New York: Wiley.

Murray, Gordon D., and Janet G. Findlay. 1988. "Correcting for the bias caused by dropouts in hypertension trials." *Statistics in Medicine* 7:941–946.

National Research Council, 2010. *The Prevention and Treatment of Missing Data in Clinical Trials*. Panel on Handling Missing Data in Clinical Trials. Committee on National Statistics, Division of Behavioral and Social Sciences and Education. Washington, DC: The National Academies Press.

Nelder, J. A., and R. W. M. Wedderburn. 1972. "Generalized linear models." *Journal of the Royal Statistical Society* A 135:370–384.

Nelson, Kerrie P., Stuart R. Lipsitz, Garrett M. Fitzmaurice, Joseph Ibrahim, Michael Parzen, and Robert Strawderman. 2006. "Use of the probability integral transformation to fit nonlinear mixed-effects models with nonnormal random effects." *Journal of Computational and Graphical Statistics* 15:39–57.

Neudecker, H. 1969. "Some theorems on matrix differentiation with special reference to Kronecker matrix products." *Journal of the American Statistical Association* 64:953–963.

Neuhaus, John M., and Mark R. Segal. 1997. "An assessment of approximate maximum likelihood estimators in generalized linear models." In *Modelling Longitudinal and Spatially Correlated Data: Methods, Applications, and Future Directions*, edited by T. G. Gregoire, D. R. Brillinger, P. J. Diggle, E. Russek-Cohen, W. G. Warren, and R. D. Wolfinger, 11–22. New York: Springer.

Orelien, Jean G., and Lloyd J. Edwards. 2008. "Fixed-effect variable selection in linear mixed models using R_2 statistics." *Computational Statistics & Data Analysis* 52:1896–1907.

Pan, Wei. 2001. "Akaike's information criterion in generalized estimating equations." *Biometrics* 57:120–125.

Paniagua, R., D. Amato, E. Vonesh, R. Correa-Rotter, A. Ramos, J. Moran, and S. Mujais. 2002. "Effects of increased peritoneal clearances on mortality rates in peritoneal dialysis: ADEMEX, a prospective, randomized, controlled trial." *Journal of the American Society of Nephrology* 13:1307–1320.

Park, Taesung, and Charles S. Davis. 1993. "A test of the missing data mechanism for repeated categorical data." *Biometrics* 49:631–638.

Park, Taesung, and Seung-Yeoun Lee. 1997. "A test of missing completely at random for longitudinal data with missing observations." *Statistics in Medicine* 16:1859–1871.

Pawitan, Yudi, and Steve Self. 1993. "Modeling disease marker processes in AIDS." *Journal of the American Statistical Association* 88:719–726.

Piersol, Laura Jean. 2000. "Fitting nonlinear mixed effect models by Laplace approximation." Ph.D. diss., University of California.

Pinheiro, José C., and Douglas M. Bates. 1995. "Approximations to the log-likelihood function in the nonlinear mixed-effects model." *Journal of Computational and Graphical Statistics* 4:12–35.

_______. 2000. *Mixed-Effects Models in S and S-PLUS*. New York: Springer-Verlag.

Pinheiro, José C., and Edward C. Chao. 2006. "Efficient Laplacian and adaptive Gaussian quadrature algorithms for multilevel generalized linear mixed models." *Journal of Computational and Graphical Statistics* 15:58–81.

Prentice, Ross L. 1988. "Correlated binary regression with covariates specific to each binary observation." *Biometrics* 44:1033–1048.

Prentice, Ross L., and Lue Ping Zhao. 1991. "Estimating equations for parameters in means and covariances of multivariate discrete and continuous responses." *Biometrics* 47:825–839.

Rademaker, A. W., E. F. Vonesh, and J. A. Logemann et al. 2003. "Eating ability in head and neck cancer patients after treatment with chemoradiation: A 12-month follow-up study accounting for dropout." *Head & Neck* 25:1034–1041.

Rao, C. R. 1965. "The theory of least squares when the parameters are stochastic and its application to the analysis of growth curves." *Biometrika* 52:447–458.

_______. 1972. "Estimation of Variance and Covariance Components in Linear Models." *Journal of the American Statistical Association* 67:112–115.

_______. 1973. *Linear Statistical Inference and Its Applications*. Second edition. New York: Wiley.

Ratkowsky, David A. 1983. *Nonlinear Regression Modeling: A Unified Practical Approach*. New York: Marcel Dekker, Inc.

Raudenbush, Stephen W., Meng-Li Yang, and Matheos Yosef. 2000. "Maximum likelihood for generalized linear models with nested random effects via high-order, multivariate Laplace approximation." *Journal of Computational and Graphical Statistics* 9:141–157.

Ridout, Martin S., and Peter J. Diggle. 1991. "Testing for random dropouts in repeated measurement data." *Biometrics* 47:1617–1621.

Robins, James M., Andrea Rotnitzky, and Lue Ping Zhao. 1995. "Analysis of semiparametric regression models for repeated outcomes in the presence of missing data." Journal of the American Statistical Association 90:106–121.

Rodriguez, Germán, and Noreen Goldman. 1995. "An assessment of estimation procedures for multilevel models with binary responses." *Journal of the Royal Statistical Society, Series A* 158:73–89.

Roe, Denise. J. 1997. "Comparison of population pharmacokinetic modeling methods using simulated data: results from the population modeling workgroup." *Statistics in Medicine* 16:1241–1262.

Rosenberg, Barr. 1973. "Linear regression with randomly dispersed parameters." *Biometrika* 60:65–72.

Rotnitzky, Andrea, and Nicholas P. Jewell. 1990. "Hypothesis testing of regression parameters in semiparametric generalized linear models for cluster Correlated Data." *Biometrika* 77:485.497.

Rotnitzky, Andrea, James M. Robins, and Daniel O. Scharfstein. 1998. "Semiparametric regression for repeated outcomes with nonignorable nonresponse." *Journal of the American Statistical Association* 93:1321–1339.

Rubin, D. B. 1976. "Inference and missing data." *Biometrika* 63:581–592.

_______. 1978. "Multiple imputations in sample surveys—a phenomenological Bayesian approach to nonresponse." In *Imputation and Editing of Faulty or Missing Survey Data*, 1–23. Washington, DC: U.S. Department of Commerce.

_______. 1987. Multiple Imputation for Nonresponse in Surveys. New York: Wiley.

Schafer, Joseph L. 1997. *Analysis of Incomplete Multivariate Data*. New York: Chapman & Hall.

_______. 1999. "Multiple imputation: A Primer." *Statistical Methods in Medical Research* 8:3–15.

Schall, Robert. 1991. "Estimation in generalized linear models with random effects." *Biometrika* 78:719–727.

Scharfstein, Daniel O., Andrea Rotnitzky, and James M. Robins. 1999. "Adjusting for nonignorable drop-out using semiparametric nonresponse models (with discussion)." *Journal of the American Statistical Association* 94:1096–1146.

Schluchter, Mark D. 1992. "Methods for the analysis of informatively censored longitudinal data." *Statistics in Medicine* 11:1861–1870.

Schluchter, Mark D., Tom Greene, and Gerald J. Beck. 2001. "Analysis of change in the presence of informative censoring: application to a longitudinal clinical trial of progressive renal disease." *Statistics in Medicine* 20:989–1007.

Searle, S. R. 1971. *Linear Models*. New York: Wiley.

_______. 1987. *Linear Models for Unbalanced Data*. New York: Wiley.

Searle, Shayle R., George Casella, and Charles E. McCulloch. 1992. *Variance Components*. New York: Wiley.

Serfling, Robert J. 1980. *Approximation Theorems of Mathematical Statistics*. New York: Wiley.

Shao, J., and B. Zhong. 2003. "Last observation carry-forward and last observation analysis." *Statistics in Medicine* 22:2429–2441.

Sheiner, L. B., and S. L. Beal. 1980. "Evaluation of methods for estimating population pharmacokinetic parameters. I. Michaelis-Menten model: routine clinical pharmacokinetic data." *Journal of Pharmacokinetics and Biopharmaceutics* 8:553–571.

_______. 1981. "Evaluation of methods for estimating population pharmacokinetic parameters II. Biexponential model and experimental pharmacokinetic data." *Journal of Pharmacokinetics and Biopharmaceutics* 9:635–651.

Shults, Justine, and N. Rao. Chaganty. 1998. "Analysis of serially correlated data using quasi-least squares." *Biometrics* 54:1622–1630.

Siddiqui, O., H. M. J. Hung, and R. O'Neill. 2009. "MMRM vs. LOCF: A comprehensive comparison based on simulation study and 25 NDA datasets." *Journal of Biopharmaceutical Statistics* 19:227–246.

Solomon, P. J., and D. R. Cox. 1992. "Nonlinear component of variance models." *Biometrika* 79:1–11.

Steimer, J. L., A. Mallet, J. L. Golmard, and J. F. Boisvieux. 1984. "Alternative approaches to estimation of population pharmacokinetic parameters: Comparison with the nonlinear mixed-effect model." *Drug Metabolism Reviews* 15:265–292.

Stiratelli, Robert, Nan M. Laird, and James H. Ware. 1984. "Random-effects models for serial observations with binary response." *Biometrics* 40:961–971.

Stokes, Maura E., Charles S. Davis, and Gary G. Koch. 2000. *Categorical Data Analysis Using the SAS System*. Second edition. Cary, North Carolina: SAS Institute Inc.

Sutradhar, B. C., and K. Das. 1999. "Miscellanea. On the efficiency of regression estimators in generalised linear models for longitudinal data." *Biometrika* 86:459–465.

Swamy, P. A. V. B. 1970. "Efficient inference in a random coefficient regression model." *Econometrica* 38:311–323.

Tanner, Martin A., and Wing Hung Wong. 1987. "The calculation of posterior distributions by data augmentation." *Journal of the American Statistical Association* 82:528–540.

Ten Have, Thomas R., Allen R. Kunselman, Erik P. Pulkstenis, and J. Richard Landis. 1998. "Mixed effects logistic regression models for longitudinal binary response data with informative drop-out." *Biometrics* 54:367–383.

Ten Have, Thomas R., and A. Russell Localio. 1999. "Empirical Bayes estimation of random effects parameters in mixed effects logistic regression models." *Biometrics* 55:1022–1029.

Thall, Peter F., and Stephen C. Vail. 1990. "Some covariance models for longitudinal count data with overdispersion." *Biometrics* 46:657–671.

Tierney, Luke, and Joseph B. Kadane. 1986. "Accurate approximations for posterior moments and marginal densities." *Journal of the American Statistical Association* 81:82–86.

Timm, Neil H. 2002. *Applied Multivariate Analysis*. New York: Springer-Verlag.

Tsiatis, Anastasios A., and Marie Davidian. 2004. "Joint modeling of longitudinal and time-to-event data: An overview." *Statistica Sinica* 14:809–834.

Verbeke, Geert, and Geert Molenberghs. 2009. *Linear Mixed Models for Longitudinal Data*. New York: Springer-Verlag.

Vonesh, Edward F. 1985. "Estimating rates of recurrent peritonitis for patients on CAPD." *Peritoneal Dialysis International* 5:59–65.

_______. 1990. "Modelling peritonitis rates and associated risk factors for individuals on continuous ambulatory peritoneal dialysis." *Statistics in Medicine* 9:263–271.

_______. 1992. "Non-linear models for the analysis of longitudinal data." *Statistics in Medicine* 11:1929–1954.

_______. 1996. "A note on the use of Laplace's approximation for nonlinear mixed-effects models." *Biometrika* 83:447–452.

Vonesh, Edward F., and Randy L. Carter. 1992. "Mixed-effects nonlinear regression for unbalanced repeated measures." *Biometrics* 48:1–17.

Vonesh, Edward F., and Vernon M. Chinchilli. 1997. *Linear and Nonlinear Models for the Analysis of Repeated Measurements*. New York: Marcel Dekker, Inc.

Vonesh, Edward F., Vernon M. Chinchilli, and Kewei Pu. 1996. "Goodness-of-fit in generalized nonlinear mixed-effects models." *Biometrics* 52:572–587.

Vonesh, Edward F., Tom Greene, and Mark D. Schluchter. 2006. "Shared parameter models for the joint analysis of longitudinal data and event times." *Statistics in Medicine* 25:143–163.

Vonesh, Edward F., Hao Wang, and Dibyen Majumdar. 2001. "Generalized least squares, Taylor series linearization, and Fisher's scoring in multivariate nonlinear regression." *Journal of the American Statistical Association* 96:282–291.

Vonesh, Edward F., Hao Wang, Lei Nie, and Dibyen Majumdar. 2002. "Conditional second-order generalized estimating equations for generalized linear and nonlinear mixed-effects models." *Journal of the American Statistical Association* 97:271–283.

Wakefield, J. C., A. F. M. Smith, A. Racine-Poon, and A. E. Gelfand. 1994. "Bayesian analysis of linear and non-linear population models by using the Gibbs sampler." *Applied Statistics* 43:201–221.

Wedderburn, R. W. M. 1974. "Quasi-likelihood functions, generalized linear models, and the Gauss-Newton method." *Biometrika* 61:439–447.

Wolfinger, Russ. 1993. "Laplace's approximation for nonlinear mixed models." *Biometrika* 80:791–795.

Wolfinger, Russ, and Michael O'Connell. 1993. "Generalized linear mixed models: A pseudo-likelihood approach." *Journal of Statistica Computation and Simulation* 48:233–243.

Wolfinger, Russell D., and Xihong. Lin. 1997. "Two Taylor-series approximation methods for nonlinear mixed models." *Computational Statistics & Data Analysis* 25:465–490.

Wolfinger, Russell D., Randy D. Tobias, and John Sall. 1994. "Computing Gaussian Likelihoods and Their Derivatives for General Linear Mixed Models." *SIAM Journal on Scientific Computing* 15:1294–1310.

Wu, Margaret C., and Kent R. Bailey. 1988. "Analysing changes in the presence of informative right censoring caused by death and withdrawal." *Statistics in Medicine* 7:337–346.

_______. 1989. "Estimation and comparison of changes in the presence of informative right censoring: conditional linear model." *Biometrics* 45:939–955.

Wu, Margaret C., and Raymond J. Carroll. 1988. "Estimation and comparison of changes in the presence of informative right censoring by modeling the censoring process." *Biometrics* 44:175–188.

Wu, Margaret C., and Dean A. Follmann. 1999. "Use of summary measures to adjust for informative missingness in repeated measures data with random effects." *Biometrics* 55:75–84.

Yuh, L., S. Beal, M. Davidian, F. Harrison, A. Hester, K. Kowalski, E. Vonesh, and R. Wolfinger. 1994. "Population pharmacokinetic/pharmacodynamic methodology and applications: A bibliography." *Biometrics* 50:566–575.

Zeger, Scott L., and M. Rezaul Karim. 1991. "Generalized linear models with random effects: a Gibbs sampling approach." *Journal of the American Statistical Association* 86:79–86.

Zeger, Scott L., Kung-Yee Liang, and Paul S. Albert. 1988. "Models for longitudinal data: A generalized estimating equation approach." *Biometrics* 44:1049–1060.

Index

D

E

P

W

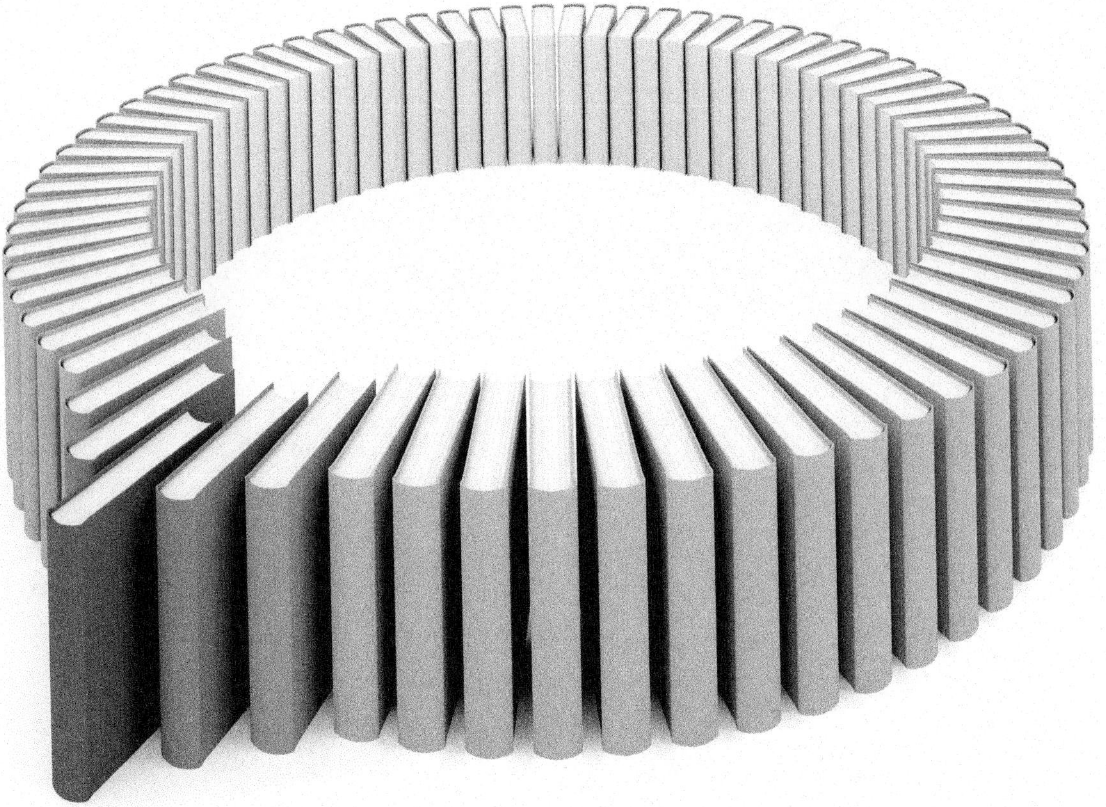

CPSIA information can be obtained
at www.ICGtesting.com
Printed in the USA
BVOW09s2322231216
471671BV00002B/47/P

9 781599 946474